PARTICLES AND NUCLEI

Panic 96

Proceedings of the 14th International Conference on

PARTICLES AND NUCLEI

Panic 96

Williamsburg, Virginia, USA 22 – 28 May 1996

Editors

Carl E. Carlson
College of William & Mary

John J. Domingo
TJNAF, USA

World Scientific
Singapore • New Jersey • London • Hong Kong

Published by

World Scientific Publishing Co. Pte. Ltd.
P O Box 128, Farrer Road, Singapore 912805
USA office: Suite 1B, 1060 Main Street, River Edge, NJ 07661
UK office: 57 Shelton Street, Covent Garden, London WC2H 9HE

British Library Cataloguing-in-Publication Data
A catalogue record for this book is available from the British Library.

PANIC '96
Proceedings of the 14th International Conference on Particles and Nuclei

ISBN 981-02-3003-6

Printed in the United States of America.

PREFACE

These are the proceedings of the XIVth Particles and Nuclei International Conference, PANIC96 for short. The conference was the latest in a series that began at CERN in 1962; the previous meeting was at Perugia in 1993. The conference series was begun and continues with the goal of sharing information between the particle and nuclear physics communities, particularly between members of those communities who work near the interface of their fields. The spectrum of interested physicists is therefore broad, and there were over 600 of us in attendance.

We met in Williamsburg, Virginia from 22 to 28 May 1996, hosted by the College of William and Mary and by the Continuous Electron Beam Accelerator Facility (CEBAF). The name of the latter laboratory changed to the Thomas Jefferson National Accelerator Facility, the "Jefferson Laboratory," during the conference.

These proceedings record the formal talks given at the conference, and thus, we hope, give a reasonably complete idea of the state of the art in studying phenomena involving nucleons and nuclei that are of interest to both nuclear and particle physicists. To all our attendees, speakers, and authors go many thanks. Special thanks are due to the convenors of the parallel sessions, whose help in organizing those sessions was instrumental to the success of the meeting.

Thanks are also due to the people who worked hard—and successfully—to ensure that our stay in Williamsburg was pleasant and productive. Allow us to mention especially the conference services groups at both the Jefferson Laboratory and at William and Mary for the excellence of their work. Persons who worked especially hard on these proceedings include Karen Hokansson, Michelle Martin, and Gladys Smith: many thanks to them.

Finally, we thank the Department of Energy and the National Science Foundation for their support of this conference.

The Editors

PANIC 96:
THE XIVth INTERNATIONAL CONFERENCE ON PARTICLES AND NUCLEI

Carl E. Carlson and John J. Domingo, co-chairs

The International Advisory Committee

J. Appel - FNAL
A. Arima - RIKEN
J. Arvieux - SATURNE
A. Astbury - TRIUMF
P. Barnes - Los Alamos
A. Bromley - Yale
J. Cameron - IUCF
D. Cassel - Cornell
M. Danilov - ITEP
P. De Witt Huberts - NIKHEF
C. Detraz - CNRS
H. Ejiri - Osaka
T. Ericson - CERN
L. Foa - CERN
C. Gelbke - Michigan
J. Haissinski - SACLAY
S. Harar - GANIL
W. Henning - Argonne
E. Iarocci - Frascati
K. Killian - Julich
T. Kirk - Brookhaven
S. Kowalski - MIT
S. Kullander - Uppsala
T. D. Lee - Columbia
D. Leith - SLAC
R. Malfliet - Groningen
E. Moniz - MIT
K. Nakai - KEK
B. Povh - MPI
R. Redwine - MIT
A. Richter - Darmstadt
I. Savin - Dubna
C. Schaerf - Rome
J. Schiffer - Argonne
B. Schoch - Bonn
D. Skopik - Saskatoon
A. Skrinsky - Novosibirsk
H. Specht - GSI
M. Sugawara - Sendai
A. Thomas - Adelaide
E. Vogt - TRIUMF
D. Von Harrach - Mainz
A. Wagner - Desy
H. Walter - PSI
D. Wilkinson - TRIUMF
S. Yamada - INS Tokyo
M. Zeller - Yale
Z. Zhi-Peng - IHEP Beijing
S. Zuxun - IAE Beijing

Program Advisory Committee

E. Beise (Maryland)
D. Bryman (TRIUMPH)
F. Close (Rutherford)
W. Haxton (Washington)
K. Königsman (Freiburg)
F. Lenz (Erlangen)
R. McKeown (Caltech)
E. Moniz (M.I.T.)
K. Nakai (KEK)
J. Stachel (Stony Brook)

Local Advisory Committee

O. K. Baker (Hampton)
M. Eckhause (W & M)
N. Isgur (CEBAF)
S. Kuhn (ODU)
K. McFarlane (NSU)
B. Mecking (CEBAF)
J. D. Walecka (W & M)

Local Organizing Committee

R. Bizot (CEBAF)
M. Eckhause (W & M)
S. Ewing (CEBAF)
C. Garwood (CEBAF)
M. Hightower (CEBAF)
K. Hokansson (CEBAF)
R. Jeffrey (W & M)
J. Leverenz (CEBAF)
J. Prater (CEBAF)
W. Tian (W & M)
J. D. Walecka (W & M)
R. Whitney (CEBAF)

Contents

Session 2: Mesons and Nucleon Interactions with Nucleons and Nuclei

Session 4: The Standard Model

Session 6: Hadron Spectroscopy

Session 7: Neutrino Physics and Weak Interactions

Session 12: New Facilities

Plenary Sessions

NUCLEON SPIN STRUCTURE MEASUREMENTS

B. W. FILIPPONE
W. K. Kellogg Laboratory, California Institute of Technology, Pasadena, CA 91125

Recent measurements of deep inelastic spin structure functions for H, D, and ^{3}He targets are reviewed. These data can be used to measure important matrix elements of the nucleon. The previous data and present state of on-going experiments will be discussed.

1 Introduction

Understanding the underlying composition of the nucleon's spin has proven a significant challenge both theoretically and experimentally. Recent developments in polarization technology have led to important advances in the experimental information that have spurred new insights in the development of the theory. This talk will summarize the experimental status of measurements of the nucleon's spin structure as probed via deep inelastic lepton-nucleon scattering with polarized beams and targets. A detailed discussion of the theory will be presented by Forte[1]. A brief introduction, necessary for understanding the data will be presented below.

Deep inelastic scattering (DIS) with lepton beams has been an important tool over the last 20 years for understanding the electromagnetic sub-structure of the nucleon. With polarized beams and targets the electromagnetic *spin* sub-structure of the nucleon becomes accessible. In a simple constituent model one can decompose the nucleon's spin as

$$J_z^N = S_z^q + L_z^q + S_z^g + L_z^g = \frac{1}{2}. \tag{1}$$

where S_z and L_z are intrinsic and orbital angular momentum for quarks and gluons. A non-relativistic constituent quark model gives directly $S_z^q = \frac{1}{2}$ and the other components $= 0$. QCD may give another answer, although rigorous calculations are not available.

Experimentally the spin structure is probed via inclusive and semi-inclusive scattering of virtual photons (with energy ν and squared 4-momentum transfer Q^2) from quarks. Inclusive polarized scattering, in analogy with the unpolarized case, yields two structure functions: $g_1(x, Q^2)$ and $g_2(x, Q^2)$, with the Bjorken variable $x = Q^2/2M\nu$ specifying the nucleon momentum fraction carried by quarks in the quark-parton model. In this model g_1 is related to the flavor sum of charge-weighted quark spin distributions ($\Delta q_i(\mathrm{x})$), while g_2 probes

$\Delta q_i(x)$ along with higher order contributions (eg. transverse spin). These structure functions are extracted from the measured asymmetry ($A_{||}, A_{\perp}$) of the scattering cross section as the beam or target spin is flipped. These asymmetries are measured with longitudinally polarized beams and longitudinally ($A_{||}$) and transversely ($A_{\perp}$) polarized targets via

$$A_{||} = \frac{N^{\uparrow\downarrow} - N^{\uparrow\uparrow}}{N^{\uparrow\downarrow} + N^{\uparrow\uparrow}}(\frac{1}{fP_bP_T})) + \Delta A_{RC} \tag{2}$$

and a similar expression for $A_{\perp}$. Here the normalized events per spin state ($N^{\uparrow\downarrow}$), target dilution factor (f), beam polarization (P_B), and target polarization (P_T) contribute to the statistical precision of the measurement. ΔA_{RC} is a calculated correction for higher-order electromagnetic radiative effects.

In addition to the x dependence of the spin distributions, the structure function integrals over x can provide important information on basic nucleon matrix elements and sum rules. For example, isospin symmetry and current algebra lead to the Bjorken sum rule: $\int_0^1 (g_1^p(x) - g_1^n)dx = \frac{1}{6}\frac{g_A}{g_V}$ where g_A and g_V are the axial-vector and vector coupling constants from neutron beta decay. At the finite Q^2 of the measurements gluon radiative corrections (from QCD) must be included and result in an additional factor of $(1 - \frac{\alpha_s}{\pi} + O(\alpha_s^2))$ in the sum rule.

In addition, in the quark-parton model the integral over g_1 yields the total charge-weighted contribution of quarks to the nucleon spin. When combined with several assumptions (*uds* flavor symmetry and zero strange quark polarization) and some additional experimental information from hyperon beta-decay, these integrals lead to the Ellis-Jaffe sum rules, with specific predictions for $\Gamma_1^p = \int_0^1 g_1^p(x)dx, \Gamma_1^n, \ldots$ With the above information, these sums provide a measurement of the nucleon matrix element for the isosinglet axial vector current. In the absence of contributions from the axial anomaly in the form of gluon spin (Δg), these measurements give the total contribution of the quarks to the nucleon's spin. Conversely if the quark contribution is found to be small (as suggested from experiment - see below), one interpretation is that a significant amount of the nucleon's spin must result from the gluons.

Semi-inclusive asymmetries (detecting a hadron in coincidence with the scattered lepton) can be used to provide new information on the flavor content of the valence and sea quark contribution to the nucleon spin. With appropriate hadron identification, information on the struck quark can be obtained assuming factorization of the quark fragmentation which has been established in a variety of previous unpolarized tests of the quark-parton model.

Table 1: Summary of DIS Spin Structure Function Measurements

Lab	Exp.	Year	Beam	Target
SLAC	E80	75	16 GeV e^-	H-butanol
	E130	80	23 GeV e^-	H-butanol
	E142	92	25 GeV e^-	^3He
	E143	93	29 GeV e^-	NH_3,ND_3
	E154	95	48 GeV e^-	^3He
	E155	97	48 GeV e^-	NH_3, ND_3, LiD(?)
CERN	EMC	85	100-200 GeV μ^+	NH_3
	SMC	92	100 GeV μ^+	D-butanol
	"	93	190 GeV μ^+	H-butanol
	"	94	190 GeV μ^+	D-butanol
	"	95	190 GeV μ^+	D-butanol
	"	96	190 GeV μ^+	NH_3
DESY	HERMES	95	28 GeV e^+	^3He
		96	28 GeV e^+	H

2 Summary of Experiments

An extensive program of polarized DIS measurements is underway at a variety of laboratories. A summary of the spin structure function measurements is shown in Table 1, where the beams, targets, and energies are listed for each experiment. The experiments complement each other in their kinematic coverage and in their sensitivity to possible systematic errors associated with the measured quantities. For example for the quantities presented in Eq. 2, the experiments span the range $1 - 300 \times 10^6$ DIS events, $0.1 - 1$ for f, $0.4 - 0.8$ for P_B, and $0.3 - 0.9$ for P_T.

The pioneering measurements on the proton were performed at SLAC [2–3] with modest polarizations of beam and target. These first data with limited statistics and kinematic coverage showed agreement with the Ellis-Jaffe sum rule for the proton with some assumptions for the extrapolation to low x. However the later proton data from EMC [4–5] (with extended coverage to low x), while consistent with the SLAC data in the SLAC kinematic range, showed a large deviation from the Ellis-Jaffe sum rule and a corresponding small value for the implied quark contribution to the proton's spin. This result spurred a new generation of experiments at SLAC, CERN, and DESY with significant improvements in precision.

We can summarize the various programs by highlighting the advantages of each: SLAC: high luminosity, fast beam polarization reversal, low - medium Q^2 coverage; CERN: two simultaneous oppositely polarized targets, high Q^2 coverage, semi-inclusive hadrons; DESY: pure atomic targets, semi-inclusive hadrons (with pion identification, kaon identification ~ '98), low - medium Q^2 coverage.

3 Summary of Results

We will focus on the high precision data that have been presented over the last 5 years. To illustrate the different sensitivities of the SMC [6–10] vs. SLAC [11–12] data, we plot both $g_1(x)$ and $xg_1(x)$ for the proton and deuteron. These are shown in figures 1 and 2. The high statistical precision and high x coverage of the SLAC data is evident as is the more extensive coverage of the low x region for SMC.

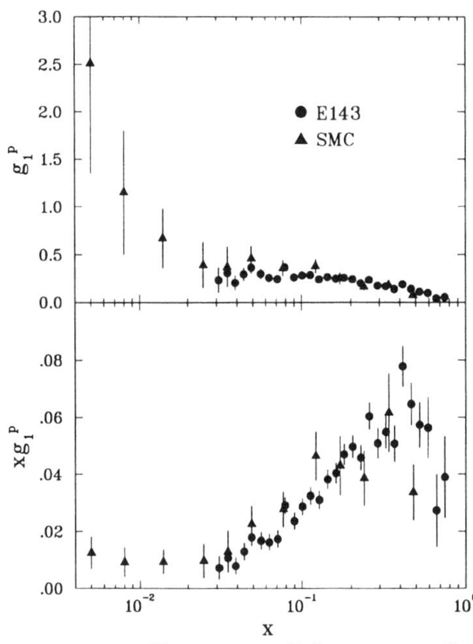

Fig. 1. SLAC E143 [11] and SMC [7–8] data for g_1^p and xg_1^p.

For the neutron, $g_1(x)$ can be determined from a ^{3}He target with some corrections for nuclear effects. Because the wave function for ^{3}He is dominated by the configuration with the protons paired to zero spin, most of the asymmetry from ^{3}He is due to the neutron. This has been confirmed in several detailed nuclear physics calculations where the major nuclear correction is simply dilution due to scattering from the two unpolarized protons. The neutron structure function determined from ^{3}He is shown in Fig. 3 from the SLAC E142[13–14] and HERMES[15] experiments.

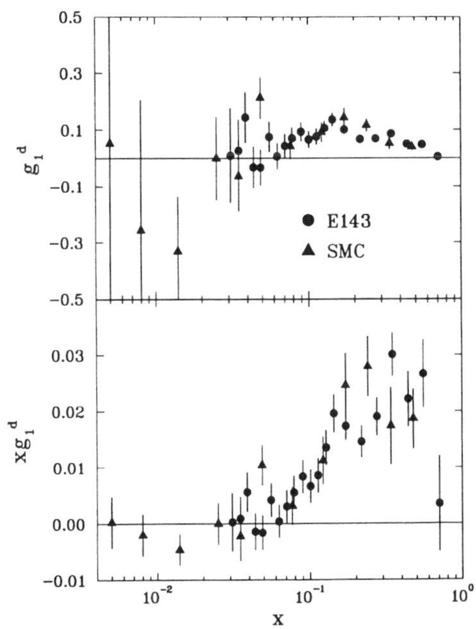

Fig. 2. SLAC E143[12] and SMC[6,9–10] data for g_1^d and xg_1^d.

The x dependence of the data seem to reflect qualitatively what is observed for the unpolarized structure function F_1 - a suggestively large value at low x (gluons?), a peak in the mid x range (dominance of valence quarks), and an approach to zero at large x (vanishing of the quark distributions as $x \to 1$). As in the unpolarized case, the x dependence of the structure functions is a basic property of the nucleon and cannot be compared with any rigorous QCD calculations as it is a result of the non-perturbative confinement. The integral of the structure function however can provide additional insight, as discussed abovc.

Extracting the structure function integrals from the measured data requires two steps. First, because of the nature of the experiments (usually fixed beam energy) the data is not taken at fixed Q^2, while the interpretation of the integrals requires a fixed Q^2. Previously this has involved making assumptions on the Q^2 dependence of the structure function (eg. g_1/F_1 is independent of

Q^2) and evolving the data to fixed Q^2. The data[16] indicates that such assumptions are reasonable within the present accuracy of the data. However future high precision measurements may require more sophisticated techniques (such as QCD evolution of the structure functions [1]).

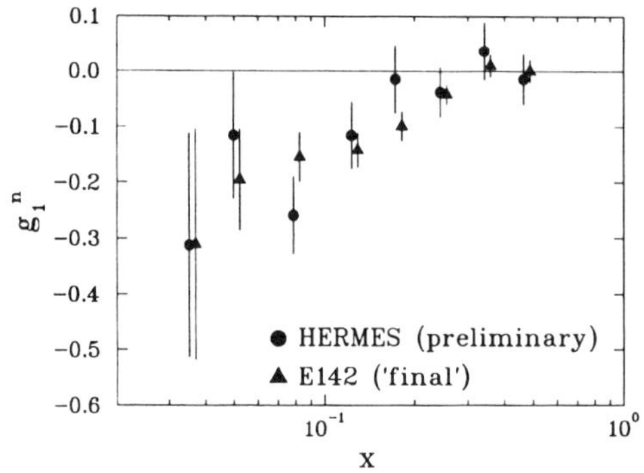

Fig. 3. SLAC E142[13–14] and HERMES[15] data for g_1^n from ^{3}He.

Second, the data is necessarily taken over a finite range in x, such that an integral over all x requires extrapolation. For the large x region this is straightforward, since g_1 is proportional to the quark distributions it must approach zero as $x \to 1$. However the low x region is problematic, as there is no clear dependence expected. Previously, simple Regge type models have been used, but these have been recently called into question by new next-to-leading-order QCD calculations[1], that suggest a significantly different behavior.

For the Bjorken sum (see Fig. 4) the data do not agree with the simple prediction without QCD corrections. However when the QCD corrections are included to fourth order[18] (see Fig. 4) the agreement is excellent. In fact by assuming that the Bjorken sum rule is valid, a competitively precise value for the strong coupling constant of QCD (α_s) has recently been extracted[19].

A comparison of the data with the prediction from the Ellis-Jaffe sum rule (including QCD corrections to third order[18]) is shown in Fig. 5 for p, d, and ^{3}He targets. In all cases the sum rule is clearly violated, suggesting a suppression in the quark contribution to the nucleon spin. However the simple constituent quark model predictions for the quark spin contribution may still be valid if the contribution from the gluons in the form of the axial anomaly is large[1].

Direct information on the contribution of the valence quarks has recently come from semi-inclusive hadron asymmetries from SMC[20]. These new data, while require some assumptions about the form of quark fragmentation, are the first information on the valence quark contribution to the nucleon spin.

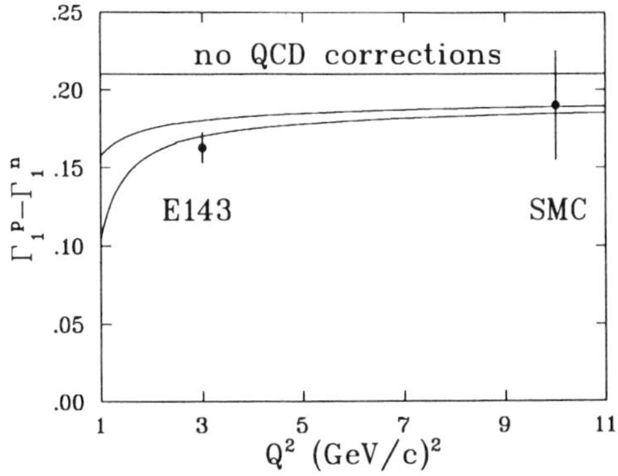

Fig. 4. Comparison of Bjorken Sum Rule with E143 and SMC.
The curves show the range (due to uncertainties in α_s) of the higher order QCD predictions.

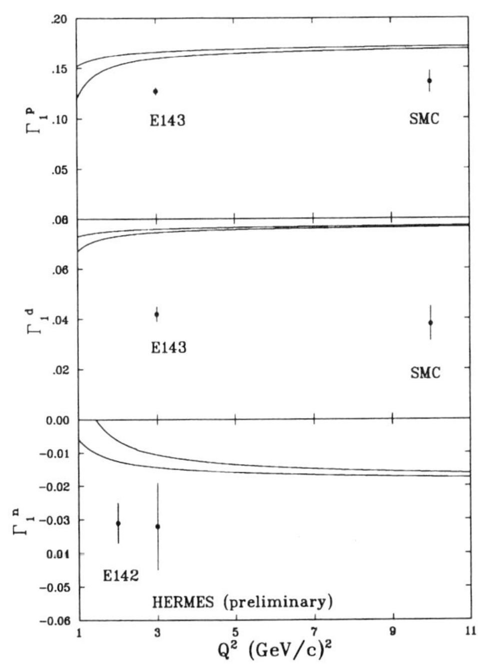

Fig. 5. Comparison of Ellis-Jaffe Sum Rule with E142, E143, SMC, and HERMES.

4 Future

The next five years should see a dramatic increase in the amount of data. SLAC and DESY will make new measurements for p, d, and ^{3}He targets. Precise semi-inclusive data is expected from HERMES. CEBAF[21] has an extensive program of measurements planned at large x and low Q^2. A new CERN proposal (COMPASS)[22] could extend the high Q^2 data and provide information on semi-inclusive charm production, that may shed light on the contribution of gluons to the nucleon's spin. RHIC[23] (via polarized Drell-Yan and direct photons) also has potential to provide direct information on the gluon contribution. In addition an improved theoretical understanding my be forthcoming with further improvements in the most advanced QCD calculations.

References

1. S. Forte, contribution to these proceedings.
2. M. J. Alguard et al.,*Phys. Rev. Lett.* **37**, 1261 (1976),*Phys. Rev. Lett.* **41**, 70 (1976).
3. G. Baum et al.,*Phys. Rev. Lett.* **51**, 1135 (1983),*Phys. Rev. Lett.* **45**, 2000 (1980).
4. J. Ashman et al.,*Phys. Lett.* B **206**, 364 (1988).
5. J. Ashman et al.,*Nucl. Phys.* B **328**, 1 (1989).
6. B. Adeva et al.,*Phys. Lett.* B **302**, 533 (1993).
7. D. Adams et al.,*Phys. Lett.* B **329**, 399 (1994).
8. B. Adeva et al.,*Phys. Lett.* B **320**, 400 (1994).
9. D. Adams et al.,*Phys. Lett.* B **357**, 248 (1995).
10. A. Witzmann, proceedings of the DIS96 Conference, Rome, (1996).
11. K. Abe et al.,*Phys. Lett.* B **74**, 346 (1995).
12. K. Abe et al.,*Phys. Rev. Lett.* **75**, 25 (1995).
13. P. L. Anthony et al.,*Phys. Rev. Lett.* **71**, 959 (1993).
14. P. L. Anthony et al.,SLAC-PUB-95-7103, 1996.
15. M. Vetterli, Proceedings of the DIS96 Conference, Rome, (1996).
16. K. Abe et al.,*Phys. Lett.* B **364**, 61 (1995).
17. K. Abe et al.,*Phys. Rev. Lett.* **76**, 587 (1996).
18. R. D. Ball, et al.,*Phys. Lett.* B **378**, 255 (1996).
19. J. Elliss and M. Karliner, *Phys. Lett.* B **341**, 397 (1995).
20. B. Adeva et al.,*Phys. Lett.* B **369**, 93 (1996).
21. CEBAF Experiments 89-042, 91-023, 93-009, 94-010, (www.cebaf.gov).
22. COMPASS Proposal, CERN/SPSLC 96-14 (1996).
23. Relativistic Heavy Ion Collider (RHIC) Project (www.rhic.bnl.gov).

POLARIZED STRUCTURE FUNCTIONS: A STATUS REPORT

STEFANO FORTE
INFN, Sezione di Torino
via P. Giuria 1, I-10125 Torino, Italy

We draw a balance of the current knowledge on polarized structure functions measured in deep-inelastic scattering. We discuss how measurements of the x and Q^2 dependence of the structure function g_1 and the scaling of its moments test perturbative QCD at next-to-leading order and beyond. We summarize the knowledge of polarized parton distributions which can be achieved with current data, in particular the determination of the first moment of the quark and gluon distributions and the axial charge of the nucleon. We examine what can in principle be learnt and briefly discuss what future experiments could teach us on the polarized structure of the nucleon.

1. From structure functions to polarized partons.

Structure functions measured in deep-inelastic scattering provide the cleanest access to the structure of the nucleon, as probed in hard processes, and understood in perturbative QCD. In the polarized case only very recently have theory and experiment progressed to the point that it is possible to extract from experiment physically meaningful information, as given by nucleon matrix elements of quark and gluon operator, and compare the experimental results with their predicted NLO perturbative behavior, as given by the x and Q^2 dependence embodied in the Altarelli-Parisi equations.

Here we will chart the current status of this knowledge, by summarizing how recent data test perturbative QCD, reviewing what we have learnt on the polarized parton structure of the nucleon, and discussing what future experiments could teach us. In Sect. 2 we will review recent developments related to the study of the Bjorken sum rule, which singles out the hard structure of perturbative QCD. In Sect. 3 we will discuss the behavior of the structure function $g_1(x, Q^2)$ in next-to-leading order (NLO) in the (x, Q^2) plane, with special regard to the small-x region. In Sect. 4 we will review the current knowledge of polarized parton distributions, and in particular their first moments, which are related to the parton interpretation of the nucleon spin. In Sect. 5 we will finally compare the information which could be in principle obtained on polarized partons with current and future experimental prospects.*

* For a more detailed introduction to the aspects of the theory of the polarized structure

2. The Bjorken sum rule.

The nonsinglet first moment of the structure function g_1 has the peculiar feature of being related to the matrix element of a conserved current through a nontrivial coefficient function:

$$\Gamma_1^{\rm NS}(Q^2) \equiv \int_0^1 dx\, g_1^{\rm NS}(x,Q^2) = \frac{1}{2} C_{\rm NS}[\alpha_s(Q^2)] a_{\rm NS}. \tag{2.1}$$

This implies that its measurement probes the hard part of the total polarized inclusive nonsinglet γ^*-p cross section, which is proportional to the coefficient function $C_{\rm NS}(Q^2)$, up to a scale-independent coefficient, the nucleon matrix element of the nonsinglet axial current $a_{\rm NS}$,† which incorporates all the soft, target-dependent physics. Hence, determining $C_{\rm NS}(Q^2)$ at various scales probes the perturbative computation of this cross-section, which is under theoretical control and indeed has been accomplished[4] up to ordet α_s^3.

In the particular case of the isotriplet combination the current matrix element can in turn be related, using isospin algebra, to that which governs nucleon β decay, and thus expressed in terms of the corresponding decay constant g_A:

$$a^{I=1} = \frac{1}{6}\langle p,s|j_5^{\mu\, I=1}|p,s\rangle \frac{s_\mu}{M} = \frac{1}{6} g_A \tag{2.2}$$

where M and s^μ are the nucleon mass and spin and the axial current is $j_i^\mu = \bar\psi\gamma^\mu\gamma_5\lambda_i\psi$, λ_i being an isospin matrix. This gives an absolute prediction for the isotriplet first moment (Bjorken sum rule): testing the normalization of this prediction tests isospin symmetry in this channel, and measuring its scale dependence tests perturbative QCD, in particular the size and running of the strong coupling $\alpha_s(Q^2)$.

The predicted scale dependence of the sum rule (with[8] $\alpha_s(M_Z) = 0.119 \pm 0.116$) is shown in Fig. 1, which demonstrates its sensistivity both to the value of the strong coupling, and to higher order corrections. Notice the jump at the charm threshold, which is due to the explicit n_f dependence of the coefficient function at order α_s^2 and beyond (the discontinuity would be smoothened in a more refined treatment of thresholds). Recently, both the the SMC[5] and

---

function $g_1(x,Q^2)$ relevant to this paper see e.g. ref. 1, whose notation and conventions we will follow; a general discussion of the phenomenology of g_1 is e.g. in ref. 2. For a comprehensive introduction to the problems related to the determination of the first moment of g_1 (the "spin crisis") see ref. 3.

† Strictly speaking, $a_{\rm NS}$ is only scale independent far from heavy quark thresholds, otherwise it acquires a scale dependence due to the appearance of a new heavy quark contribution as the threshold is passed.

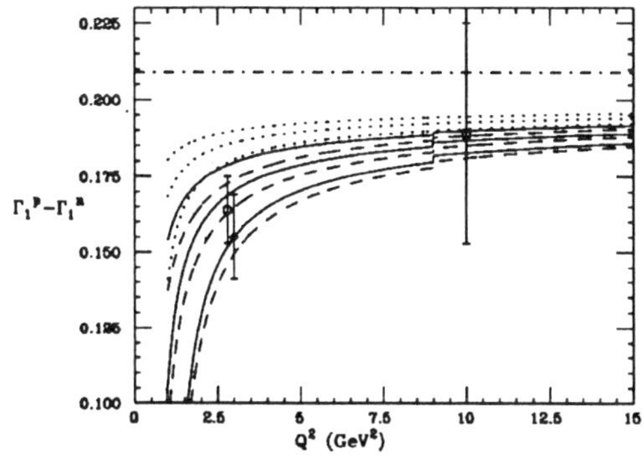

Fig. 1: Scale dependence of the Bjorken sum rule up to order α_s (dotted), α_s^3 (solid) and with higher twist corrections (dashed). The three sets of curves correspond to $\alpha_s(M_Z) = 0.119 \pm 0.006$. The data points are from ref. 5 (square), 6 (diamond) and the global average of ref. 7 (circle) (slightly offset from Q^2=3 GeV^2 to improve readability).

the E142-E143 collaboration[6] have released preliminary determinations of the isotriplet first moment, repectively obtained 10 GeV^2 combining proton and deuteron data, and at 3 GeV^2 also including ^{3}He data; the results are manifestly consistent with the QCD prediction. Apparent from Fig. 1 is also the much greater sensitivity of the sum rule to higher order corrections and the value of α_s when the scale is low. It follows that a measurement of equal accuracy leads a priori to a more stringent test of the sum rule (or, equivalently, to a better determination of α_s) if performed at a low scale, but it is then affected by a larger theoretical uncertainty related to lack of knowledge of higher order corrections. In fact, as we will discuss in Sect. 3 and 4, the experimental values of the first moment are themselves obtained by evolving to a common scale data taken at several values of Q^2: this involves an error which is likely to be underestimated at low scales, so that the apparent greater precision of the low Q^2 data point in fig. 1 is in fact in part due to the neglect of this systematics. Be that as it may, if all data are evolved to a common scale and combined, one may arrive at a determination of the sum rule which, if taken at face value, is very precise, and would determine α_s to competitive accuracy,[7] as clearly shown in Fig. 1.

The perturbative computation of $C_{\rm NS}(Q^2)$ is in fact extended to high enough order that one may wonder whether the asymptotic nature of perturbation theory in QCD may already manifest itself. Specifically, the asymptotic series for $C_{\rm NS}(Q^2)$ can be summed by the Borel method but only up to an ambiguity of order $\frac{1}{Q^2}$: one can then try to estimate the size of this ambiguity, either by approximate computation of the high-order behavior of the series,[9] or on the basis of the terms which are already known (extrapolated with the method of Padé approximants).[10] This ambiguity must cancel against an equal and opposite ambiguity in the contribution $c_{\rm HT}$ to the sum rule from higher twist operators (such as $\bar\psi\gamma^\mu\gamma_5\psi F_{\mu\nu}$): $\Gamma^{I=1} = \frac{1}{6}\left[C_{\rm NS}(Q^2)g_A + \frac{C_{\rm HT}}{Q^2} + O(1/Q^4)\right]$. The estimates of ref.s 9,10 lead to a value of the ambiguity which is of the same size as the contributions of such operators computed with QCD sum rule methods, which give $C_{\rm HT} \approx -0.01$.[11] This suggests that even though the sum rule estimates are of the right order of magnitude, their actual size is entirely uncertain. The logarithmic corrections to the scale dependence of the twist four, spin one operators which contribute to the first moment of g_1 have also been computed recently.[12].

3. Parton distributions at next-to-leading order

The moments of $g_1(N,Q^2) = \int_0^1 dx\, g_1(x,Q^2)$ generically depend on scale both because of the α_s dependence of coefficient functions, and because of the scale dependence of the matrix elements of quark and gluon operators to which they are related by

$$\begin{aligned} g_1(N,Q^2) =&\frac{\langle e^2\rangle}{2}[C_{\rm NS}(N,\alpha_s)\Delta q_{\rm NS}(N,Q^2)+ \\ &C_{\rm S}(N,\alpha_s)\Delta\Sigma(N,Q^2) + 2n_f C_g(N,\alpha_s)\Delta g(N,Q^2)], \end{aligned} \tag{3.1}$$

where $\Delta q_{\rm NS} = \sum_{i=1}^{n_f}\left(\frac{e_i^2}{\langle e^2\rangle} - 1\right)(\Delta q_i + \Delta\bar q_i)$ and $\Delta\Sigma = \sum_{i=1}^{n_f}(\Delta q_i + \Delta\bar q_i)$ (with $\langle e^2\rangle = \frac{1}{n_f}\sum_{i=1}^{n} e_i^2$) are respectively the nonsinglet and singlet combinations of moments of the quark distributions. The scale dependence of the polarized quark and gluon distributions is in turn governed by the evolution equations

$$\begin{aligned} \frac{d}{dt}\Delta q_{\rm NS} &= \frac{\alpha_s(t)}{2\pi}\gamma^N_{\rm NS}\Delta q_{\rm NS} \\ \frac{d}{dt}\begin{pmatrix}\Delta\Sigma\\ \Delta g\end{pmatrix} &= \frac{\alpha_s(t)}{2\pi}\begin{pmatrix}\gamma^N_{qq} & 2n_f\gamma^N_{qg}\\ \gamma^N_{gq} & \gamma^N_{gg}\end{pmatrix}\begin{pmatrix}\Delta\Sigma\\ \Delta g\end{pmatrix}. \end{aligned} \tag{3.2}$$

The physically relevant moments of quark and gluon distributions can thus be extracted from the measurement of the moments of g_1 at a pair of different scales. In fact, however, the moments of g_1 cannot be determined directly because the structure function can only be measured in a finite range of x ($x \to 0$ corresponds to the energy of the γ^*-nucleon collision going to infinity); furthermore, present-day experiments only determine g_1 along a curve $Q(x)$ (rather than at a fixed scale) with Q increasing with x. This implies that the moments of g_1 can only be extracted indirectly, through an analysis of its x and Q^2 dependence: all the available experimental information is then summarized in a set of polarized parton distributions at a reference scale, from which g_1 at all x and Q^2 are found by solving the evolution equations and taking (and inverting) moments. The potential accuracy of such an analysis has substantially improved recently since the complete set of anomalous dimensions γ has now been determined up to NLO.[13]

Given the need to extrapolate the data to $x = 0$ in order to compute moments, it is particularly interesting to study the implications of perturbative evolution for the small x behavior of g_1, which follows from the behavior of the anomalous dimensions γ eq. (3.2) around their rightmost singularity in N space. This is located at $N = 0$ and is on general grounds of the form[14] $\alpha_s\gamma \underset{N\to 0}{\sim} (\alpha_s/N)^{2n+1}$ at n-th perturbative order. Already at LO this implies that any starting parton distributions leads to a growth of $|g_1(x,Q^2)|$ at small x and large Q^2;[15] in fact in this limit the ratio of Δg and $\Delta\Sigma$ is fixed and negative, so that $g_1 < 0$ asymptotically at small x and large Q^2 for any reasonable input parton distribution.[16] The generic form of the rise at n-th perturbative order is then

$$\Delta f \sim \frac{1}{(\xi\zeta)^{1/4}} e^{2\gamma_f\sqrt{\xi\zeta}} \left(1 + \sum_{i=2}^{n} \epsilon_f^i \left(\sqrt{\frac{\xi}{\zeta}}\right)^{2n-1} \alpha_s^n\right), \tag{3.3}$$

where $\xi \equiv \ln\frac{x_0}{x}$, $\zeta \equiv \ln\frac{\alpha_s(Q_0^2)}{\alpha_s(Q^2)}$, and Δf denotes either the nonsinglet quark distribution, or the two linear combinations of the gluon and singlet quark distributions which correspond to eigenvectors of the small N anomalous dimension matrix. Notice that this rise contradicts the Regge expectation[17] that g_1 should at most be constant at small x.

The coefficients ϵ_f in eq. (3.3) can be determined at NLO;[18] beyond it they have been obtained at fixed coupling both in the nonsinglet[14,19] and singlet[20] case on the basis of the analysis of the leading double logarithmic singularities. These results (unlike their unpolarized counterparts) are not supported by suitable factorization theorems, even though they do agree with the available NLO computations. If correct, they lead to the prediction that the series in eq. (3.3) exponentiates, thus leading to a power-like rise of $g_1 \sim x^{-\lambda}$. The value of λ cannot be computed precisely since it depends on α_s, which is assumed to be constant in these computations; taking $\alpha_s \sim 0.2$ (which corresponds to a scale of roughly 10 GeV^2) would lead to $\lambda \approx 0.5$ in the nonsinglet and $\lambda \gtrsim 1$ in the singlet case, implying that the first moment of g_1 would then diverge. This casts some doubts on the applicabilty of the analysis, at least in its naive form. However, it suggests that a strong rise of g_1, which might be of considerable phenomenological relevance, should be generated by perturbative QCD. The all order analysis confirms the LO result that this rise will preserve the sign of the nonsinglet, but lead to a negative singlet contribution (which asymptotically dominates g_1 at small x) even if g_1 is positive at the starting scale.

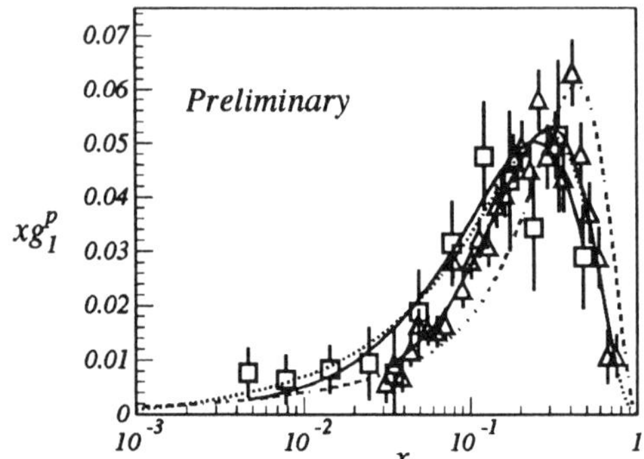

Fig. 2: Plot of xg_1^p. The squares are SMC data (preliminary)[21,5] and the diamonds E143 data.[22] The curves are the results of a NLO fit[26] to all[22–25,5] available data: long dashes: Q^2=1 GeV^2, short dashes: Q^2=10 GeV^2, solid: $Q^2 = Q^2(x)$ corresponding to the two data sets.

4. Phenomenology of g_1.

The scaling violations displayed by current data for g_1^p are shown in Fig. 2: what might appear as a systematic discrepancy of the two data sets[21,22] is in fact a consequence of the fact that in each x bin the SMC data are taken at a higher value of Q^2 than the E143 data. These scaling violations are largely due to scale dependence of the asymmetry $A_1 = g_1/F_1$, i.e. they would be much smaller if the asymmetry were scale independent. Thus, even though the individual experiments, because of their limited kinematic coverage, cannot measure[25,5] the scale dependence of the asymmetry directly, the data do show evidence for this scale dependence when the two data sets are combined.

4.1 Parton distributions

The observation of sizable scaling violations, together with the availability of proton[21,22] and deuteron[23,24] data which allow disentangling the singlet and nonsinglet contributions, make a full NLO determination of polarized parton distributions possible.[18,27,28] A detailed study of the shape of polarized parton distributions[28] shows (see Fig. 3) that while it is possible to separate out valence (defined as $\Delta\bar{q} - \Delta q$) and sea components (assumed to be SU(3) flavor symmetric) the gluon distribution is only constrained by present-day data in the small x region ($x \lesssim 0.03$) where scaling violations driven by it are largest: otherwise the data are compatible with a variety of possible large x behaviors of Δg and do not even constrain its sign when $x \gtrsim 0.1$.

The small x behavior of the parton distributions is particularly interesting. The proton data (Fig. 4a) display a marked rise at small x (and low Q^2, where the small x data are taken): since g_1^p at small x is positive, this appears to be due to a rise of the nonsinglet contribution, because a rising singlet would produce substantial scaling violations which turn it into a rapid drop. Indeed,

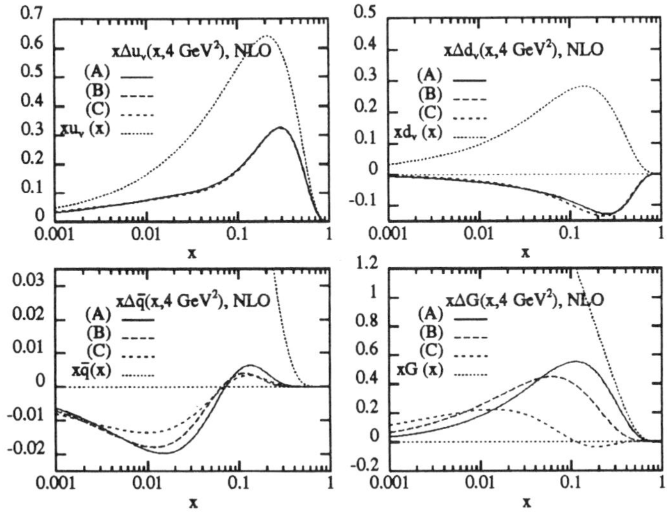

Fig. 3: Polarized parton distributions (from Ref. 28.). The sets A-C correspond to different choices for the large x behavior of Δg; the dotted curves are unpolarized distributions.

the data for the deuteron, which is almost pure singlet (the octet contribution to it is quite small) do not display such a rise (Fig. 4b). This then implies[18,28] an effective behavior $\Delta q_{\rm NS} \sim \frac{1}{\sqrt{x}}$ in the measured region, in contradiction to Regge expectation, and reminescent of the predictions[14,19] based on the summation of double logs discussed in Sect. 2.* Much more precise data at smaller values of x will be required to elucidate this issue.

4.2 First moments

The NLO analysis can in particular be used[18] to give a precision determination of the first moment of g_1, which is of primary theoretical interest because of its relation to the singlet axial current:

$$\Gamma_1^{\rm S}(Q^2) \equiv \int_0^1 dx\, g_1^{\rm S}(x,Q^2) = \frac{1}{2} C_{\rm S}[\alpha_s(Q^2)] a_0(Q^2), \tag{4.1}$$

where a_0 is the singlet axial charge $a_0(Q^2) = \langle p,s|j_5^\mu|p,s\rangle \frac{s_\mu}{M}$. Unlike its non-singlet counterpart eq. (2.2), $a_0(Q^2)$ depends on scale (see Fig. 5) because the classical conservation of the singlet axial current is spoiled by the axial anomaly in the quantized theory. Its nucleon matrix element measures a

* Interestingly, these predictions also imply that the small x behavior of the polarized and unpolarized nonsinglet should be similar; this also appears to agree with the data.

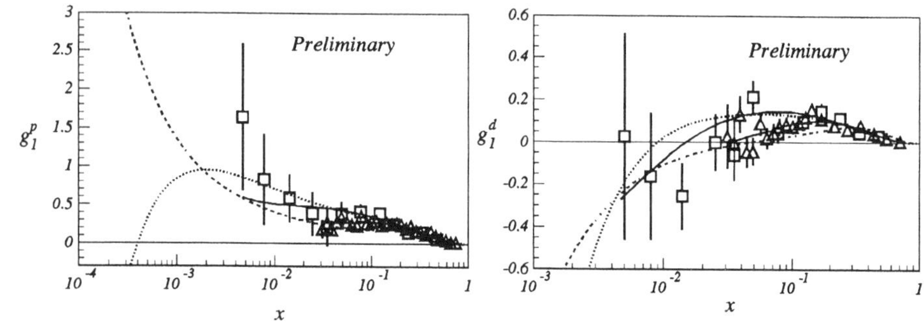

Fig. 4: Plot of g_1 for (a) proton and (b) deuteron. The deuteron data are from Ref. 5 (squares, preliminary) and from Ref. 24 (triangles); the proton data and curves are as in Fig. 2.

nontrivial combination eq. (3.1) of the first moments of the quark and gluon distributions (the total quark and gluon spin), which at LO simply reduces to $a_0(Q^2) = \Delta\Sigma(1, Q^2)$. The first moment is usually extracted from the data by assuming Q^2 independence of the asymmetry $A_1 = g_1/F_1$ to evolve all data to a common scale,[29] and extrapolating the data to $x = 0$ on the basis of the Regge expectation.[17] The axial charge can then be obtained directly by using SU(3) symmetry to separate off the nonsinglet component, and the perturbative expression of $C_S(\alpha_s)$ (which is known[30] up to order α_s^2). The results thus found from the published (or preliminary) values of the first moment, evolved to a common scale of 5 GeV^2, are collected in the table.*

Neither of these evolution and extrapolation procedures is consistent with perturbative QCD, as discussed in Sect. 2: the information contained in the data can instead be consistently extracted from a global NLO fit. In such case, the primary quantities which enter the analysis are the first moment of the quark and gluon distribution, from which the first moment of g_1 is then extracted using eq. (3.1), and the singlet charge using eq. (4.1). The main result of such an analysis[18] is that the naive procedure systematically underestimates somewhat the value of a_0, and substantially underestimates the theoretical uncertainty on it. This is partly due to the fact that the small x behavior is overconstrained by the Regge ansatz, and partly to the fact that corrections due to perturbative evolution (which are only known up to NLO) are potentially large. In the NLO analysis the small x behavior is fitted to the data and constrained by assuming a reasonably smooth form of the starting

* We used the $O(\alpha_s^3)$ expression[4] of the nonsinglet and the $O(\alpha_s^2)$ expression[30] of the singlet coefficient function, and the values of the triplet and octet charges from Ref. 31.

Ref.	Targ.	$a_0(5\ \mathrm{GeV}^2)$	Ref.	Targ.	$a_0(5\ \mathrm{GeV}^2)$
21	p	0.27 ± 0.15	6	p	0.18 ± 0.09
22	p	0.25 ± 0.10	5	d	0.23 ± 0.09
23	d	0.19 ± 0.11	6	^{3}He	0.41 ± 0.12
24	d	0.28 ± 0.06	32	^{3}He	0.43 ± 0.22

The singlet axial charge a_0 extracted from the values of the first moment Γ_1 published by the experimental collaborations, and evolved to a common scale. The data in the last column are all preliminary.

parton distributions, while higher order corrections to evolution are estimated by varying the renormalization and factorization scale. Using all the published data[21-24] leads to the value[18]

$$a_0(5\ \mathrm{GeV}^2) = 0.15\,^{+0.16}_{-0.11} \tag{4.2}$$

of the axial charge, to be compared to those of the table.

Experimental results for the singlet axial charge can be compared to expectations based on the quark model. In particular, the Zweig rule suggests that the strange quark component of the nucleon should be small: this then leads to the prediction that the singlet and the octet axial charge should be approximately equal to each other (Ellis-Jaffe sum rule[33]). Because the octet axial charge is scale independent to all orders, while the singlet charge, starting at NLO, is not, this prediction can only be exactly satisfied at one particular scale. However, the scale dependence of a_0 is only rather slight even at moderately large scales (Fig. 5b), so that the Ellis-Jaffe prediction can be considered to be an approximate one, which the result eq. (4.2) would then strongly contradict (the octet charge is[31] $a_8 = 0.579 \pm 0.025$). However, if one were to boldly extrapolate the perturbative scale dependence of a_0 outside the perturbative region proper (Fig. 5a), then the sum rule could be satisfied by only assuming it to be valid at a "hadronic" low scale of order of several hundreds MeV.

However, quark model predictions should apply to partonic observables, i.e. in this case to the first moments of the polarized singlet quark and gluon distributions. Now, the definition of the first moment of the quark distribution is entirely scheme dependent. This follows from the fact that $\alpha_s \Delta g(1,Q^2)$ is scale independent at LO:[34] if one performs a scheme change $\Delta\Sigma'(1,Q^2) = [1+\alpha_s z_{qg}\Delta g(1,Q^2)]\Delta\Sigma(1,Q^2)$ then $\Delta\Sigma'(1,Q^2) - \Delta\Sigma(1,Q^2)$ is asymptotically constant, so even asymptotically the ambiguity on $\Delta\Sigma$ is of the same order as its size.* It is then possible in particular to choose a scheme

* Notice that this is not the case for $\Delta g(1,Q^2)$ itself: if $\Delta g'(1,Q^2) = [1+\alpha_s z_{gg}]\Delta g(1,Q^2)$ then even though $\Delta g'(1,Q^2) - \Delta g(1,Q^2)$ is asymptotically constant Δg itself diverges (as α_s^{-1}) in the same limit, so the relative ambiguity vanishes asymptotically.

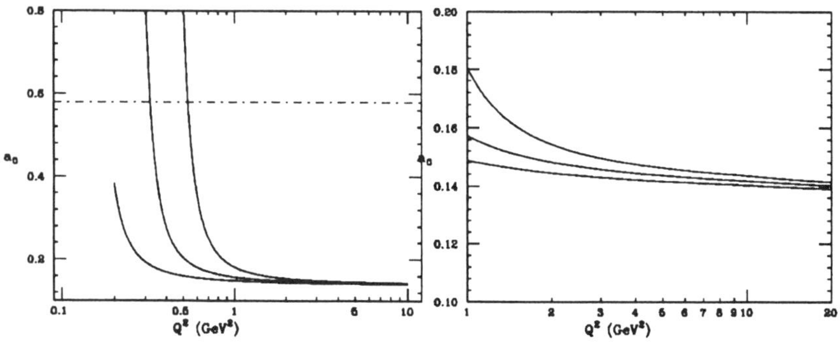

Fig. 5: Scale dependence of the axial charge $a_0(Q^2)$ eq. (4.2) computed at order α^2 at low (a) and intermediate (b) scales. The three curves correspond to $\alpha_s(M_Z) = 0.119 \pm 0.006$. The dot-dashed line is the value of the octet charge a_8.

where $\Delta\Sigma(1)$ is scale independent. In this scheme one can meaningfully compare the singlet quark distributions with the Ellis-Jaffe prediction since then the first moments of all polarized quark distributions (and linear combinations thereof) are separately scale independent.[34]

As the NLO analysis actually derives the axial charge from the polarized quark and gluon distributions, and specifically their first moments, these are also determined in the process. Using a scheme where $\Delta\Sigma(1)$ is scale independent leads to[18]

$$\Delta\Sigma(1) = 0.5 \pm 0.1, \qquad \Delta g(1, 5\,\mathrm{GeV}^2) = 2.6 \pm 1.4. \tag{4.3}$$

An analysis[28] with a different scheme choice for the quark leads essentially to the same result for $\Delta g(1, Q^2)$. This value of $\Delta\Sigma(1)$ agrees with the Ellis-Jaffe prediction; the size of $\Delta g(1, Q^2)$ may appear to be large, but is in fact natural recalling that it scales as $1/\alpha_s$: if one were to use NLO evolution down to very low scales the value eq. (4.3) would corresponds to vanishing of $\Delta g(1)$ around 700 MeV.

5. What can we learn?

The present generation of polarized experiments has brought our understanding of polarized parton distributions from the parton model to the level where it makes sense to use QCD at NLO. The data have shown evidence for scaling violations (Fig. 2), have given us a first crude understanding of the shape of parton distributions (Fig. 3), and have lead to a determination of the singlet axial charge and the first moment of the gluon distribution [eq.s (4.2),(4.3)] which, given the uncertainties involved, should be considered surprisingly precise.

While the determination of the first moment of the gluon distribution is dominated by the statistics and will thus improve somewhat as more abundant data in the currently explored kinematic region become available (specifically from SMC[5] and E155[6]), the determination of the singlet axial charge is already dominated by the systematics related to higher order corrections, as well as by the uncertainty in the small x behavior. Only the availability of data with a more complete coverage of the (x, Q^2) plane (such as those from the proposed GSI collider[35] or from HERA with a polarized proton beam[36]) would allow a substantial improvement here, as well as a determination of higher moments and in general of the x dependence of parton distributions. The study of scaling violations at small x in particular should turn out to be very fruitful, because it could confirm the non-Regge behavior suggested by current data (and seen[37] in unpolarized singlet structure functions at HERA), show the dramatic change of sign of g_1^p (see Fig. 4a) predicted by the QCD evolution equations as the scale is raised (thereby leading to a very precise determination of the polarized gluon distribution[38]), and possibly show evidence of the double logarithmic effects which characterize the polarized small x behavior.[19,20] On the theoretical side, there is room for improvement from a better understanding of double logs at small x, consistent with the running of α_s, and from a more detailed understanding of higher twist effects, both kinematical[39] (such as related to target mass effects) and dynamical,[11] which have been so far studied systematically only at the level of first moments.

The theory and phenomenology which we have discussed so far are based on fully inclusive measurements of longitudinal polarization with lepton beams and nucleon (proton and deuteron, actually) targets. These allow (at leading twist) only the determination of the combinations of polarized parton distributions which contribute to g_1 eq. (3.1). Firstly, this only measures the C-even combination $q + \bar{q}$ of polarized quark distributions: the "valence" $q - \bar{q}$ component can only be disentangled[27,28] by making specific assumptions on the symmetry of the sea. Furthermore, whereas comparing proton and deuteron data allows to separate off the isotriplet component, a further flavor decomposition can only be done at the level of first moments by using SU(3) symmetry and data from hyperon β decays, while for the full x dependence it again requires[27,28,18] assumptions on the symmetry of the sea.

For all moments beyond the first the singlet quark and gluon contributions can then separated by the observation of scaling violations. The case of the first moment is special because whereas the gluon is again determined by scaling violations the quark is completely ambiguous (see Sect. 3), and can thus only be determined within a specific scheme choice. It is important to notice that this is an inevitable consequence of the scaling of $\Delta g(1, Q^2)$, which allows an asymptotically nonvanishing redefinition of the singlet quark, and does not depend on the specific process. Thus, a determination of $\Delta\Sigma(1, Q^2)$ from any hard process will be subject to the same ambiguity, and there is no piece of data which could eliminate it, even in principle. Notice however that the singlet valence component, which decouples from the gluon, is not subject to this ambiguity.

A simple way of widening the source of available information is to consider nuclear targets. In particular, ^{3}He targets provide a direct measurement of the neutron structure function, to the extent that only the neutron carries the nuclear spin. Corrections due to higher spin components in the nuclear wave function[40] have been studied and seem under theoretical control (Fermi motion effects appear[41] to be negligible). On the contrary, there are only preliminary studies[42] (in the nonsinglet sector only) of shadowing effects, which suggest that they may be quite large (up to 20% around $x = 0.1$ where g_1 is sizable): a systematic investigation of these effects is required if the ^{3}He experiments (specifically HERMES[32] and E154[6]) are to produce precise determinations of the neutron polarized structure function.

More ambitious programs (perhaps more theoretically than experimentally) involve going beyond fully inclusive measurements. Tagging hadrons in the final state of a DIS event leads to information on the struck current, provided one can separate out the fragmentation of the current itself (as described by fragmentation functions) from the fragmentation of the target (parametrized by fracture functions[43]). In principle, this allows disentangling of the individual flavor components, and also of the valence and sea contributions: however, beyond LO there is no simple way of associating simple observables (such as asymmetry ratios) to individual observables (such as the isotriplet valence component), other than by making assumptions, for instance about the symmetry of the sea, which is what one would want to test in the first place. However, these data can be used in a global NLO analysis (the general framework of which can be already set up[44]) where they may significantly constrain the final result. Such experiments are currently underway at HERMES,[45] while an ambitious program has been proposed at CERN.[46] An alternative method to disentangle valence and sea components for individual flavors is to use charged-current events.[47] Even though this would be feasible at HERA with a polarized proton beam, it does not appear to be competitive.

A separate class of exclusive or semi-inclusive experiments can be used to measure the polarized gluon distributions in processes where it contributes at leading order (such as heavy quark production by photon-gluon fusion). It is then possible to evade the tradeoff of inclusive DIS, where the gluon is measured by scaling violations, which are larger at low scales where however scheme ambiguities are also large. The cleanest measurement here is possibly heavy quark photoproduction, where since only one large scale (the quark mass) is present the cross section can be reliably computed in perturbation,[48] so that a study of the p_T dependence of the production asymmetry can lead to a sensitive determination of the gluon distribution.[49]

As more precise data become available, it will also be possible to take the theoretically very ambitious step of going beyond longitudinal polarization. So far, the structure function g_2 has been considered a higher twist background to the measurement of g_1, and recent determinations[50] of it have essentially achieved the aim of reducing this background. A study of the x and Q^2 dependence of g_2 as might be possible in a dedicated experiment would open Pandora's box of twist-3 operators and their mixing,[51] and lead us to leave the safe context of polarized parton distributions defined by twist two operators.

Acknowledgements: I thank J. Domingo and the organizing committee for their hospitality in Williamsburg, M. Anselmino, R. Ball and G. Ridolfi for several discussions, and the SMC collaboration and especially A. Deshpande and J. Lichtenstadt for illuminating conversations on their data.

References

1. S. Forte, in "Frontiers in Strong Interactions", proc. of the 7th Rencontres de Blois, M. Haguenauer, ed. (Editions Frontières, Paris, 1996), hep-ph/9511345.
2. S. Forte, in "Radiative Corrections: Status and Outlook", B. F. L. Ward, ed. (World Scientific, Singapore 1995), hep-ph/9409416.
3. G. Altarelli, in "The challenging questions", Proc. of the 1989 Erice School, A. Zichichi, ed. (Plenum, New York, 1990).
4. S. G. Gorishny and S. A. Larin, *Phys. Lett.* **B172** (1986) 109; S. A. Larin and J. A. M. Vermaseren, *Phys. Lett.* **B259** (1991) 345.
5. A. Witzmann, for the SMC Collab., talk at "DIS 96", Rome, April 1996.
6. S. Rock, for the E142 and E143 Collab., talk at "DIS 96", Rome, April 1996.
7. J. Ellis and M. Karliner, *Phys. Lett.* **B341** (1995) 397; lectures at the 1995 Erice school, hep-ph/9601280.
8. G. Altarelli, in "DIS 95", J. F. Laporte and Y. Sirois, eds. (Ecole Polytechnique, Paris 1996).
9. M. Meyer-Herrmann et al., hep-ph/9605229.
10. J. Ellis et al., *Phys. lett.* **B366** (1996) 268.
11. L. Mankiewicz, E. Stein and A. Schäfer, hep-ph/9510418.
12. H. Kawamura et al., hep-ph/9603338.
13. R. Mertig and W. L. van Neerven, *Zeit. Phys.* **C70** (1996) 637; W. Vogelsang, *Phys. Rev.* **D54** (1996) 2023.
14. R. Kirschner and L. Lipatov, *Nucl. Phys.* **B213** (1983) 122.
15. M. A. Ahmed and G. G. Ross, *Phys. Lett.* **B56** (1975) 385; M. B. Einhorn and J. Soffer, *Nucl. Phys.* **B74** (1986) 714; A. Berera, *Phys. Lett.* **B293** (1992) 445.
16. R. D. Ball, S. Forte and G. Ridolfi, *Nucl. Phys.* **B 444** (1995) 287
17. R. L. Heimann, *Nucl. Phys.* **B64** (1973) 429.
18. R.D. Ball, S. Forte and G. Ridolfi, *Phys. Lett.* **B378** (1996) 255.
19. J. Bartels, B. I. Ermolaev and M. G. Ryskin, *Z. Phys.* **C70** (1996) 273.
20. J. Bartels, B. I. Ermolaev and M. G. Ryskin, hep-ph/9603204.
21. SMC Collaboration, *Phys. Lett.* **B329** (1994) 399.
22. E143 Collaboration, *Phys. Rev. Lett.* **74** (1995) 346.
23. SMC Collab., *Phys. Lett.* **B357** (1995) 248.
24. E143 Collab., *Phys. Rev. Lett.* **75** (1995) 25.
25. E143 Collab., *Phys. Lett.* **B364** (1995) 61.

26. R. D. Ball, A. Deshpande, S. Forte, V. Hughes, J. Lichtenstadt and G. Ridolfi *unpublished.*
27. M. Glück et al., *Phys. Rev.* **D53** (1996) 4775.
28. T. Gehrmann and W. J. Stirling *Phys. Rev.* **D53** (1996) 6100.
29. J. Ellis and M. Karliner, *Phys. Lett.* **B313** (1993) 131.
30. S. A. Larin, *Phys. Lett.* **B334** (1994) 192.
31. F. E. Close and R. G. Roberts, *Phys. Lett.* **B336** (1994) 257.
32. D. De Schepper, for the HERMES Collab., talk at "DIS 96", Rome, April 1996.
33. J. Ellis and R. L. Jaffe, *Phys. Rev.* **D9** (1974) 1444.
34. G. Altarelli and G. G. Ross, *Phys. Lett.* **B212** (1988) 391.
35. D. von Harrach, V. Metag and A. Schäfer, workshop on long-term perspectives at GSI.
36. J. Lichtenstadt, talk at "DIS 96", Rome, April 1996.
37. A. de Roeck, Cargèse lectures, preprint DESY-95-025; S. Forte and R. D. Ball, *Acta Phys. Pol.* **B26** (1995) 2097
38. R. D. Ball et al., in proc. of the workshop on future physics at HERA.
39. H. Kawamura and T. Uematsu, *Phys. Lett.* **B343** (1995) 346.
40. C. Ciofi degli Atti et al, *Phys. Rev.* **C48** (1993) 968
41. W. Melnitchouk, G. Piller and A. W. Thomas, *Phys. Lett.* **B346** (1995) 165.
42. L. Frankfurt, V. Guzey and M. Strikman, hep-ph/9602301.
43. L. Trentadue and G. Veneziano, *Phys. Lett.* **B323** (1994) 201.
44. D. de Florian et al., hep-ph/9603302.
45. K. Ackerstaff, for the HERMES Collab., talk at "DIS 96", Rome, April 1996.
46. COMPASS proposal, preprint CERN/SPSLC 96-14.
47. M. Anselmino, P. Gambino and J. Kalinowski, *Z. Phys.* **C64** (1994) 267.
48. M. Glück, E. Reya and W. Vogelsang, *Nucl. Phys.* **B351** (1991) 579.
49. S. Frixione and G. Ridolfi, hep-ph/9605209.
50. SMC Collaboration, *Phys.* Lett. **B336** (1994) 125; E143 Collaboration, *Phys. Rev. Lett.* **76** (1996) 587.
51. For a review see R. L. Jaffe, hep-ph/9602236.

STRUCTURE FUNCTIONS AND GLUON DENSITY

Antje Brüll
MPI für Kernphysik Heidelberg, Postfach 103980
D-69029 Heidelberg, Germany

Abstract

The status of unpolarised nucleon structure function measurements is reviewed. Good agreement between the results of the high statistics charged lepton scattering experiments at SLAC, by the BCDMS collaboration and the more recent measurements by the NMC collaboration is observed. This leads to a precise knowledge of the structure functions $F_2^p(x, Q^2)$ and $F_2^d(x, Q^2)$ in the wide kinematic range of $0.006 < x < 0.9$ and $0.5 < Q^2 < 200$ GeV^2. A parametrisation of these structure functions and their uncertainties is presented. New data from the HERA collider experiments extend the kinematic range to even lower values of x and cover Q^2 values up to 5000 GeV^2.

The comparison of the unpolarised structure function measurements with perturbative QCD allows one of the most precise determinations of the strong coupling constant α_S. From a QCD analysis of the low x NMC + HERA data the gluon distribution of the nucleon is determined down to $x \sim 10^{-4}$ and shows a steep rise of the gluon density at low x.

Deep inelastic scattering experiments at different incident energies allow the determination of $R(x, Q^2)$, the ratio of longitudinally to transversely polarised virtual photon absorption cross sections. First preliminary results from the NMC measurements at low x are presented.

New precise data on various nuclei allow a detailed study of the x, Q^2 and A dependence of nuclear effects on structure functions. An apparent disagreement between the NMC and recently published E665 data on F_2^A/F_2^D is shown to disappear when calculating the nuclear ratios with respect to carbon instead of deuterium.

1 Introduction

Deep inelastic lepton nucleon scattering provides a unique possibility to study the internal structure of the nucleon. The measurement of the nucleon structure function $F_2(x, Q^2)$ determines the momentum distributions of the quarks in the nucleon, essential for predictions of reaction rates at future high energy colliders.

The Q^2 dependence of F_2 can be compared to the predictions of perturbative QCD and thus allows the determination of the strong coupling constant, the only free parameter in the theory. From data at low values of x, where the Q^2 evolution of F_2 is driven by the Q^2 evolution of the gluons, the gluon distribution can be determined.

Nuclear effects in structure functions were first observed by the EMC collaboration in the comparison of the structure functions of deuterium and iron. Using a complementary target setup the NMC measurements allow a determination of these effects with very low systematic uncertainties and over a wide kinematic range.

In the deep inelastic scattering of a charged lepton on a nucleon the differential cross section for the one-photon exchange can be written in terms of the nucleon structure function, $F_2(x, Q^2)$, and the ratio of longitudinally to transversely polarised virtual photon absorption cross sections, $R(x, Q^2)$, as:

$$\frac{d^2\sigma(x,Q^2)}{dxdQ^2} = \frac{4\pi\alpha^2}{Q^4} \cdot \frac{F_2(x,Q^2)}{x} \cdot \left\{1 - y - \frac{Q^2}{4E^2} + \frac{y^2 + Q^2/E^2}{2\,(1 + R(x,Q^2))}\right\}, \quad (1)$$

where α is the fine structure constant, $-Q^2$ the four-momentum transfer squared, E the energy of the incident lepton and m the lepton mass. The two scaling variables x and y are defined as $x = Q^2/2M\nu$ and $y = \nu/E$, where ν is the virtual photon energy in the laboratory frame and M is the proton mass. This one-photon exchange cross section is obtained from the measured differential cross section by correcting for higher order electroweak effects.

In the quark-parton model the variable x is equal to the fraction of the nucleon momentum carried by the parton on which the scattering has occured. The structure function F_2 is represented by the sum of all quark distributions, in case of charged lepton scattering weighted with the squares of the quark charges. The second structure function R is identical to zero. All quark distributions and consequently all structure functions depend only on the variable x (Bjorken scaling).

In the framework of QCD where strong interactions between quarks are taken into account an additional dependence of the structure functions on Q^2 is introduced. In leading order this scaling violation is due to processes of gluon radiation and quark pair production and leads to a logarithmic Q^2 dependence of F_2. Quantitatively it is described by the QCD evolution equations (*GLAP equations*) which predict the parton distributions at all values of Q^2 provided their x dependence is known at some value Q_0^2. In addition to the logarithmic Q^2 evolution predicted by perturbative QCD, non-logarithmic contributions to the Q^2 dependence of F_2 come from target mass effects and the interaction of the struck quark with the spectator quarks (higher twist effects). At a given x, they both behave like power series in $1/Q^2$ and may thus become important at low values of Q^2.

2 Structure function measurements

Since the first measurement of deep inelastic lepton nucleon scattering at the Stanford Linear Accelerator Center (SLAC) at the end of the 60's, a large number of experiments using different lepton beams and different targets have been performed to measure the structure functions of the nucleon.

The most precise results have been obtained with the SLAC electron beam, the high energy muon beam at CERN and the neutrino beams at CERN and Fermilab. High statistics experiments with hydrogen and deuterium targets, which give the most direct access to the nucleon structure functions F_2^p and F_2^n, could be done only with charged lepton beams. To compensate for the small cross section high statistics $\nu - N$ scattering experiments were using nuclear targets of higher density (usually iron). Since 1992 measurements with colliding positron and proton beams at much higher center of mass energy are performed at the HERA accelarator at DESY. Table 1 gives an overview over the most important unpolarised structure function measurements.

Experiment	Location	Reaction	kinematic range	
SLAC	Stanford	$e - N$ $e - A$	$0.07 < x < 0.9$,	$0.5 < Q^2 < 20$
BCDMS	CERN	$\mu - N$	$0.07 < x < 0.8$,	$7 < Q^2 < 200$
NMC	CERN	$\mu - N$ $\mu - A$	$0.006 < x < 0.6$, $5 \cdot 10^{-4} < x < 0.6$,	$0.5 < Q^2 < 75$ $0.1 < Q^2 < 150$
E665	Fermilab	$\mu - N$ $\mu - A$	$5 \cdot 10^{-4} < x < 0.3$,	$0.1 < Q^2 < 250$
CCFR	Fermilab	$\nu - Fe$	$0.015 < x < 0.7$,	$5 < Q^2 < 400$
H1 ZEUS	DESY	$e - p$	$5 \cdot 10^{-5} < x < 0.3$,	$1 < Q^2 < 5000$

Table 1: Measurements of unpolarised structure functions

Recently the analysis of all NMC structure function measurements was completed[1]. These data were collected at four different beam energies between 90 and 280 GeV and thus cover with high precision a wide kinematic range. As the earlier published NMC data[2] on F_2^p and F_2^d they are found in very good agreement with the results obtained at SLAC[6] and by the BCDMS collaboration[4]. Therefore the NMC, SLAC and BCDMS data sets were used to parametrise the nucleon structure functions. In the fits the individual data points were weighted with their statistical errors and the relative normalisation of the various data sets was fitted. Optimal agreement between the experiments was found with changes in normalisation inside the uncertainties given by the experiments (i.e. below 2%). An additonal free parameter was included to account for a possible miscalibration of the scattered muon energy in the BCDMS data and was determined from the fit to be +0.2% for both the proton and the deuteron data. The result of the fit for the deuteron is shown in fig. 1 together with the final NMC data and the SLAC and BCDMS results. In the figure, the points have been renormalised according to the fitted normalisation parameters and the BCDMS data have also been adjusted for the energy miscalibration mentioned above. The broken curves show the uncertainty in the parametrised structure function. To determine these uncertainty the systematic errors and their correlations were taken into account. The total errors on F_2^p and F_2^d in the range of the fixed target experiments ($0.005 < x < 0.9, 0.5 < Q^2 < 200$ GeV^2) are between 1.5% and 5% and include the normalisation uncertainty of the parametrisation.

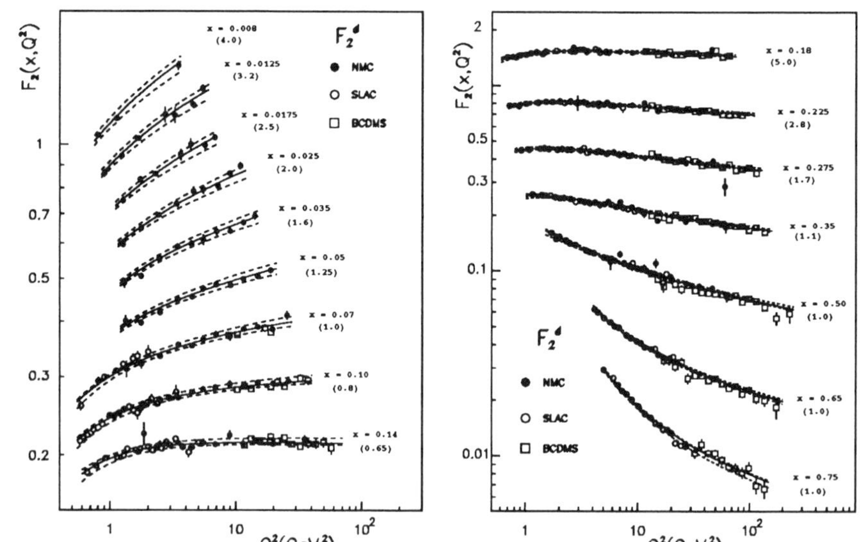

Figure 1: The deuteron structure function F_2^d and its uncertainty

New data on F_2^p and F_2^d from the fixed target experiment E665 at Fermilab[5] extend to very low x (0.0004) and are found in reasonable agreement with the NMC results. Structure functions measured at the HERA $e-p$ collider by the H1[6] and ZEUS[7] experiments have now reached the impressiv precision of about 5% over the large kinematic range of $10^{-4} < x < 0.2$ and $2 < Q^2 < 1000$ GeV^2. These data extend the measured x range towards much lower values and show a steep rise of the structure function at low x. Good agreement with the the results from fixed target experiments is observed in the whole Q^2 range (see fig. 2).

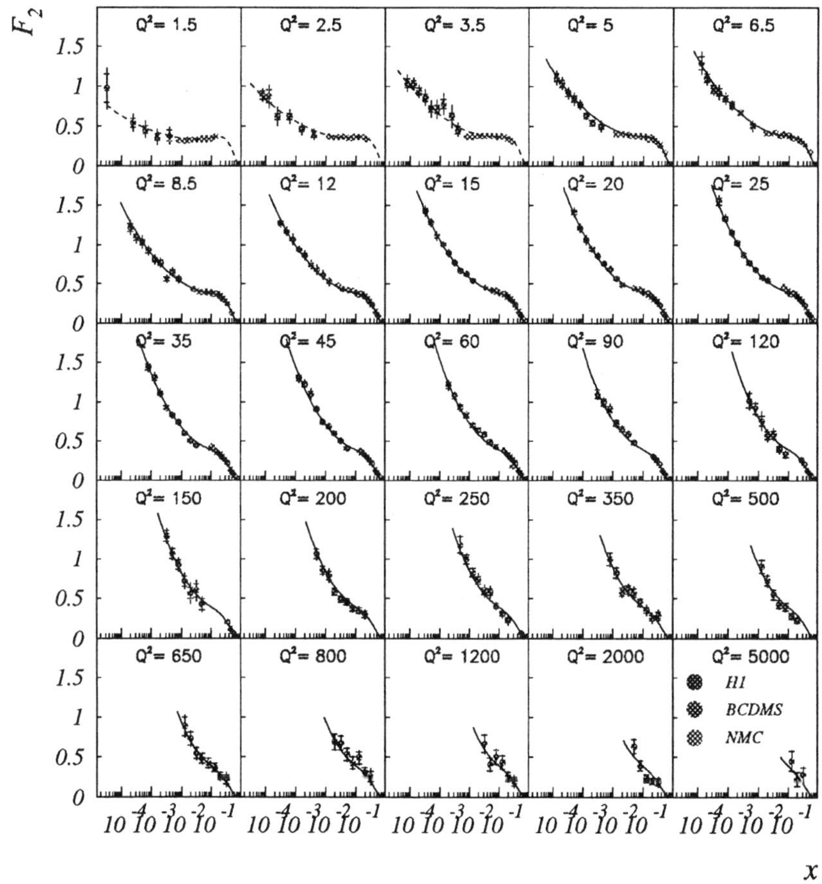

Figure 2: F_2^p measured by H1 (similar results exist from ZEUS)

3 Comparison with QCD

Although the x dependence of the parton distributions Φ_i cannot be calculated by perturbative methods, their evolution with Q^2 is given by perturbative QCD. Provided $\Phi_i(x)$ is known at any arbitrary value Q_0^2, it can be calculated via the evolution equations for all values of Q^2. For a quark distribution q_i and the gluon distribution g these evolution equations (*Gribov-Lipatov-Altarelli-Parisi equations*) take the following form:

$$\frac{dq_i(x,Q^2)}{dlnQ^2} = \frac{\alpha_S(Q^2)}{2\pi}\int_x^1 \frac{dz}{z}\left[q_i(z,Q^2)P_{qq}(\frac{x}{z}) + g(z,Q^2)P_{qg}(\frac{x}{z})\right] \quad (2)$$

$$\frac{dg(x,Q^2)}{dlnQ^2} = \frac{\alpha_S(Q^2)}{2\pi}\int_x^1 \frac{dz}{z}\left[\sum_{i=1}^{n_f}\left(q_i(z,Q^2)+\overline{q_i}(z,Q^2)\right)P_{gq}(\frac{x}{z}) + g(z,Q^2)P_{gg}(\frac{x}{z})\right] \quad (3)$$

Here the P_{ij} are the *spiltting functions* that can be calculated within perturbative QCD as power series in $\alpha_S(Q^2)$.

The principle of all comparisons of structure function measurements with the expectation of perturbative QCD is thus an iterative procedure: for one (in principle arbitray) value of Q_0^2 the x dependences of the parton distributions are parametrised, a value for α_S is choosen and the structure functions are calculated for all x, Q^2 values using the evolution equations. The value of α_S and the parameters of the parton distributions are varied until optimal agreement with the data is obtained.

The most precise determinations of α_S come from the QCD analysis of the nonsinglet structure function xF_3 by the CCFR collaboration[8] and the analysis of the combined SLAC and BCDMS F_2 measurements[9]. Both analysis were done in second order α_S in the $\overline{\text{MS}}$ renormalisation and factorisation schemes. They lead to values of α_S of

$$\text{CCFR:} \quad \alpha_S(M_Z^2) = 0.111 \pm 0.004\ (exp) \pm 0.004\ (theory)$$

$$\text{SLAC+BCDMS:} \quad \alpha_S(M_Z^2) = 0.113 \pm 0.003\ (exp) \pm 0.004\ (theory)$$

At low values of x the evolution of structure functions is dominated by the gluons. A QCD analysis of such data thus allows a determination of the gluon density. The present knowledge of $xg(x)$ is shown in fig. 3 where $xg(x)$ determined from the NMC data[10] and the result from the H1 experiment[11] for two different values of Q^2 are presented.

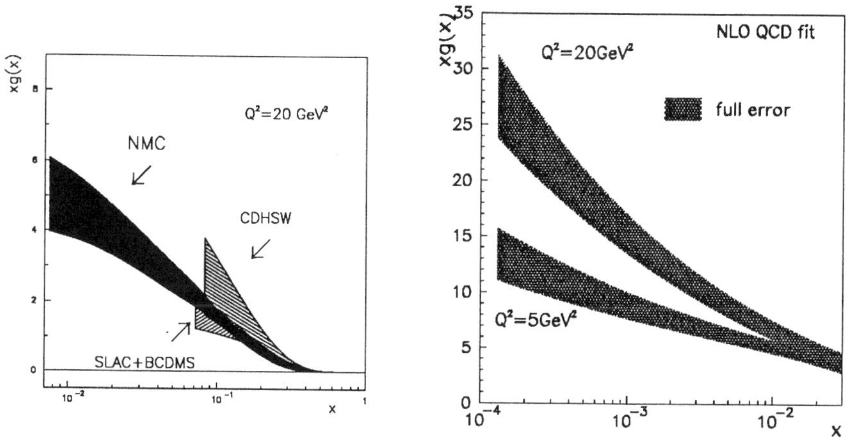

Figure 3: The gluon distribution determined by fixed target experiments and by HERA

4 Measurements of $R(x, Q^2)$

The final NMC data set includes data from four beam energies and two different triggers and spans a wide kinematical range. It thus allows a measurement of R, the ratio of longitudinally to transversely polarised virtual photon absorption cross sections, for $0.002 < x < 0.12$. The extracted values of R are shown in fig. 4 in comparison with the results of earlier measurements at comparable values of Q^2. They are compatible with the prediction of perturbative QCD and represent the first precise measurement of R at low values of x.

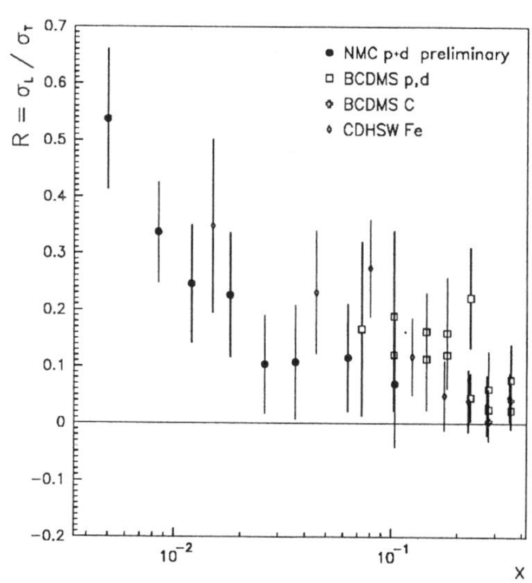

Figure 4: Measurements of $R(x)$

5 Nuclear effects

Since the original discovery by the EMC which observed a significant difference between the structure functions of free and bound nucleons the experimentel knowledge on nuclear effects in deep inelastic scattering has steadily increased. Expecially the NMC[12] and SLAC[13] measurements on nuclei between ^{4}He and ^{207}Pb have determined the x and A dependence of the "EMC effect" with good precision. In a NMC measurement on tin and carbon at different beam energies and with high statistics[14], for the first time a significant positiv Q^2 dependence was observed for $0.01 < x < 0.08$ (see fig. 5a).

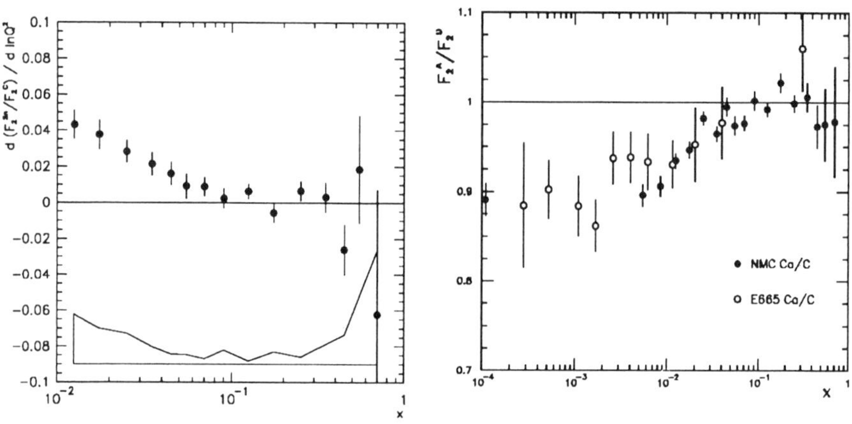

Figure 4: (a) Q^2 dependence of F_2^{Sn}/F_2^C (b) Comparison of NMC and E665 results

Recently the E665 collaboration published results[15] for the ratios F_2^C/F_2^D, F_2^{Ca}/F_2^D and F_2^{Pb}/F_2^D which showed a significant discrepancy to the earlier results obtained by NMC[12]. Fig. 5b shows the good agreement between the two experiments if one plots the ratios with respect to carbon instead of the deuteron. Therefore, the most likely explanation for the observed discrepancy might be an imperfect acceptance cancellation in the E665 nuclear data compared to the deuteron ones.

1. NMC, P. Amaudruz et al., preprint CERN-PPE/95-138, submitted to Phys. Lett.
2. NMC, P. Amaudruz et al., *Phys. Lett.* **B295** (1992) 159.
3. SLAC, L. W. Whitlow et al., *Phys. Lett.* **B282** (1992) 475.
4. BCDMS, A. C. Benvenuti et al., *Phys. Lett.* **B223** (1989) 485.
BCDMS, A. C. Benvenuti et al., *Phys. Lett.* **B237** (1990) 592.

5. E665, preprint Fermilab-Pub-95/396-E, subm. to Phys. Rev. D
A. V. Kotwal, Ph.D. Thesis, Harvard University (1995).
6. H1, S. Aid et al., preprint DESY 96-039, subm. to Nucl. Phys. B
7. ZEUS, M. Derrick et al., preprint DESY 96-076, subm. to Z. Phys. C
8. CCFR, P. Z. Quintas et al., *Phys. Rev. Lett.* **71** (1993) 1307.
9. M. Virchaux and A. Milsztajn, *Phys. Lett.* **B274** (1992) 221.
10. NMC, M. Arneodo et al., *Phys. Lett.* **B309** (1993) 222.
11. H1, presented at the Workshop on Deep Inelastic Scattering, Rome, April 1996.
12. NMC, P. Amaudruz et al., *Nucl. Phys.* **B441** (1995) 3.
NMC, M. Arneodo et al., *Nucl. Phys.* **B441** (1995) 12.
13. SLAC-E139, J. Gomez et al., *Phys. Rev.* **D49** (1994) 4348.
14. NMC (A. Mücklich), presented at the Workshop on Deep Inelastic Scattering and QCD, Paris, April 1995.
15. E665, M. R. Adams et al., *Z. Phys.* **C67** (1995) 403.

NEUTRAL AND CHARGED CURRENT PROCESSES IN POSITRON PROTON COLLISIONS AT HERA

Peter Truöl
Physik-Institut, Universität Zürich, CH-8057 Zürich, Switzerland

Both charged and neutral electroweak current induced events in new domains of the kinematical variables Bjorken-x and momentum transfer Q^2 have been recorded by the two collaborations at the HERA electron (or positron) proton storage ring. New results from the 1994 and partly also the 1995 run with an integrated luminosity of 5 pb^{-1} each are presented with emphasis on $(e^{\pm}, \overline{\nu}_e(\nu_e))$ cross sections, quark substructure, the gluon density in the proton extracted from structure function data, and diffractive processes.

In this talk I will report on the latest data obtained by the two major experiments at the HERA electron proton storage ring. I am therefore trying to represent as unbiased as possible the ZEUS- and my own H1-collaboration. The title chosen for this talk reflects the fact that most of the data shown stem from the 1994 and 1995 positron runs. I will start by briefly summarizing the accelerator status, then introduce the basic processes and experimental observables. More details on the detectors, the kinematical acceptances and the experimental challenges are given in the result sections. The latter will deal with three topics – charged and neutral currents at high momentum transfer, which can be contrasted with standard model predictions to exhibit possibly exotic phenomena, measurement of the structure function F_2 in new kinematical domains, and diffractive processes.

1 The HERA $(e^{\pm}, p)$–Collider

HERA, the first electron proton storage ring is just starting its fifth year of running. In 1995, 820 GeV protons collided with 27.5 GeV positrons, close to the design energy of $E_e = 30$ GeV. Positron fills last longer in the ring, which is why they are preferred over electrons at the moment. The circulating currents ($I_e = 36$ mA, $I_p = 73$ mA) have reached about half their design values, and so has the luminosity ($\mathcal{L} = 7\ \mu\mathrm{b}^{-1}\mathrm{s}^{-1}$). With the planned improvements of the electron radio frequency and transverse feedback system, as well as of the proton linac used for injection, the design luminosity of $\mathcal{L} = 15\ \mu\mathrm{b}^{-1}\mathrm{s}^{-1}$ should be reached in 1998. Then we should be able to collect data equivalent to an integrated luminosity of $\mathcal{L}_0 = 35\ \mathrm{pb}^{-1}$ per year, an estimate for which we assume that the average current will be half the peak current, that the machine will be available half the time during a third of the year. Last year both experiments were delivered $\mathcal{L}_0 = 10$ pb^{-1} and about half of that went to the storage media in each experiment, a fraction which can certainly be improved. However compared to the situation at the time of the last PANIC conference in Perugia, where we were only able to show first indications of HERA life [1] from $\mathcal{L}_0 = 25$ nb^{-1}, the progress is remarkable.

2 Introduction to (e,p)–Collision Physics

At higher energy the standard topology in a neutral current electroweak interaction (Figure 1a) is that of a final state jet following the struck quark, the remnant diquark system producing a jet or jets in forward (proton beam) direction and the scattered electron balancing kinematically the hadronic recoil. We are entering at HERA the regime where Z^0 and photon exchange are equally important. The charged current process (Figure 1b) is mediated by a charged vector boson, for the positron case a W^+ interacting with a d–quark. The final state neutrino escapes of course undetected. QED radiative corrections and QCD gluon radiation must be considered in a complete analysis, too. A typical event observed in the H1-detector (Figure 2) illustrates this topology.

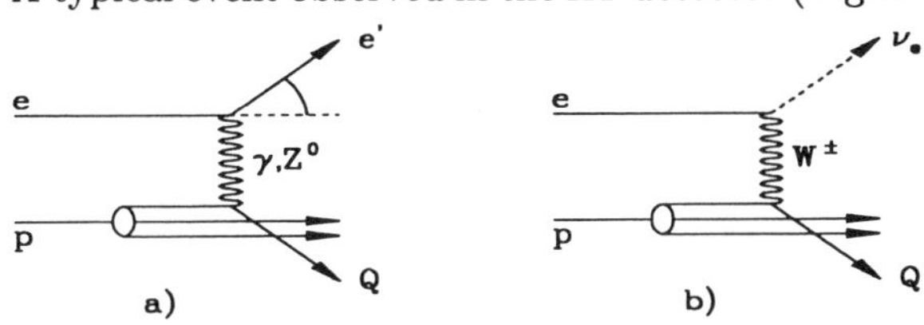

Figure 1: Neutral (a) and charged (b) current ep interaction.

Our collider experiments differ from fixed target experiments mainly because both scattered electron and hadronic recoil are detected over the full solid angle. This provides a number of advantages. The reconstruction of kinematical variables can be crosschecked (see Section 4), missing particles can be identified, e. g. in diffractively produced events (see Section 5) and charged current events (see Section 3), and a detailed spectroscopy the hadronic final state is possible. The luminosity is measured in both experiments using bremsstrahlung events detected in small angle electron and photon taggers.

3 Charged and Neutral Currents at High Momentum Transfer – the Standard Model and Beyond

In the discussion of data from HERA four kinematical variables are commonly used to describe the process $e(k) + p(P) \to e'(k')X(p_X)$ at fixed center of mass energy $\sqrt{s} \sim 300$ GeV. Besides the four momentum transfer Q these are the Bjorken scaling variable x, the boson proton center of mass energy W, and

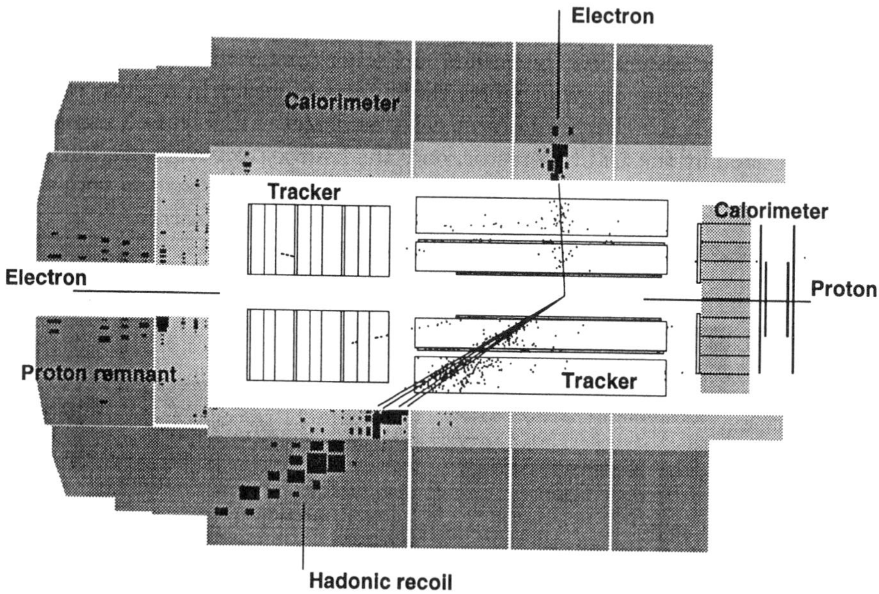

Figure 2: A high momentum transfer deep inelastic scattering event observed in the H1-detector. Electrons are entering from the left, protons from the right. The electron leaves the interaction point with an angle θ_e relative to the proton direction, the stream of hadrons with an average angle θ_h. The size of the squares illustrates the energy deposited in the electromagnetic and the hadronic calorimeter, the lines the direction of the charged particle tracks. In the transverse bending plane, current jet and electron emerge back to back (see also Figure 3). The H1-detector is based on a liquid argon calorimeter embedded into a large superconducting coil. On the outside and not shown here is the iron flux return yoke instrumented with muon chambers. The resolution is $\sigma/E = 50\%/\sqrt{E}$ for hadrons (Fe-absorber plates) and $\sigma/E = 11\%/\sqrt{E}$ for electrons (Pb).

the fraction y of the maximum energy transfer (ν, ν_{max}) in a frame where the proton is at rest

$$Q^2 \equiv -q^2 = -(k-k')^2,\ x = \frac{-q^2}{P \cdot q},\ W^2 = (q+P)^2,\ y = \frac{q \cdot P}{q \cdot k} \equiv \frac{\nu}{\nu_{max}}$$

The energy required in a fixed target experiment equivalent to HERA is $\nu_{max} = 48$ TeV. The following relations hold: $W^2 = ys(1-x) \approx ys$, $Q^2 = xys$, $Q^2_{max} = s$. Instead of the scattering angle often the pseudorapidity variable η is used: $\eta = -\log\tan\theta/2$. Compared to existing data from fixed target experiments the accessible range in Q^2, s and x is extended by two orders of magnitude. A spatial resolution $\Delta = Q^{-1}_{max} = 10^{-18}$ m can be reached. In the photoproduction regime ($Q^2 \approx 0$, not covered in this talk) we measure cross sections in the energy region $100 \leq W_{\gamma,p} \leq 200$ GeV. For the reconstruction of the kinematical quantities either the energy and angle of the scattered electron or of the hadronic system may be used or any other combination of two of these observables, depending on what yields the highest accuracy in a given region of phase space. In the charged current case, as depicted in Figure 3 using an event registered by the ZEUS detector, only the hadrons are available, but the corresponding kinematical resolution can be deduced by comparison with the neutral current sample, disregarding and comparing to the electron.

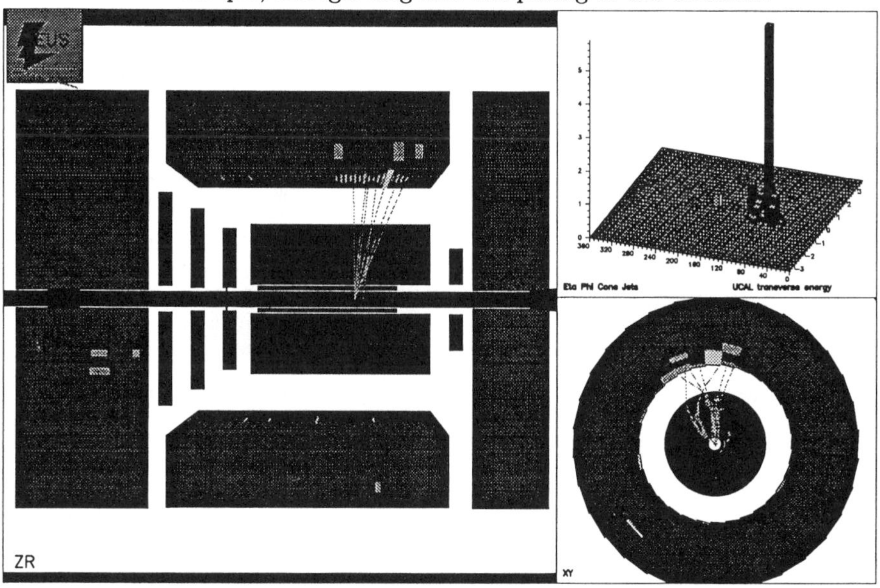

Figure 3: A moderate momentum transfer charged current $(e^+, \overline{\nu}_e)$ observed in the ZEUS-detector and tagged by a large value of missing transverse momentum. The neutrino escapes unseen. Q^2, x and y are reconstructed from the energy E_h and the angle θ_h of the stream of hadrons balancing the neutrino quantities. The ZEUS-detector is based on a uranium scintillator calorimeter surrounding a smaller superconducting coil. The muon system is on the outside is not shown. The resolution is $\sigma/E = 35\%/\sqrt{E}$ for hadrons and $\sigma/E = 17\%/\sqrt{E}$ for electrons. The right plot shows the transverse energy measured with the calorimeter for a given bin in pseudorapidity and azimuthal angle ϕ.

The cross section data [2,3] at high Q^2 are shown in Figure 4. Standard model predictions describe the data very well, nearly independent of how the parton densities are extrapolated from the low Q^2 data. Limits on fermion substructure can be set assuming either a classical charge form factor

$$\frac{d\sigma}{dQ^2} = \left(\frac{d\sigma}{dQ^2}\right)_{SM} f_e(Q^2) f_q(Q^2) \quad \text{with} \quad f_{e/q} = 1 - \frac{1}{6} R^2_{e/q} Q^2$$

or introducing a Lagrangian for an exotic extension of the type

$$L = \frac{g^2}{4\pi}\frac{1}{\Lambda^2}\left(e_{L/R}\gamma_\mu \overline{e}_{L/R}\right)\left(q_{R/L}\gamma^\mu \overline{q}_{R/L}\right)$$

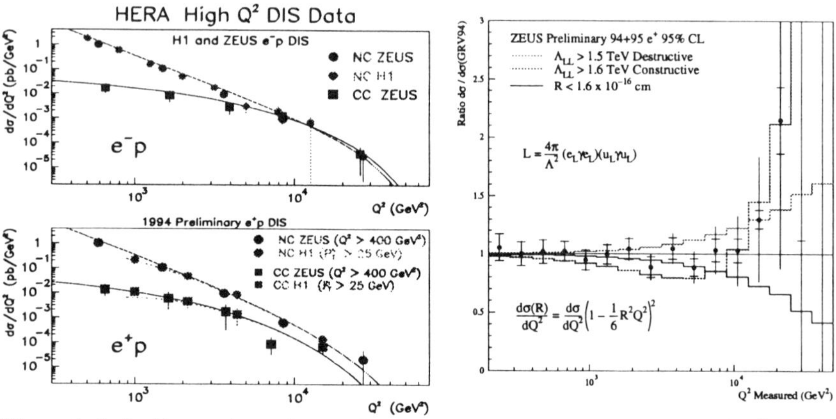

Figure 4: Left: Comparison of neutral and charged current cross section $d\sigma/dQ^2$ at high Q^2 to standard model predictions. Right: Ratio of the measured neutral current cross section to standard model predictions (inner [outer] error bars systematical [statistical]). The curves indicate expectations from compositeness (dashed) or quark substructure assuming a classical charge form factor dependence (solid).

Since $(g-2)$ and (e^+, e^-) experiments already put stringent limits on R_e the HERA data are sensitive to R_q, namely $R_q < 2.6$ am (10^{-18} m) (H1 [4]) and $R_q < 1.6$ am (ZEUS [5]). Assuming $g^2/4\pi = 1$ the limits for the mass scale of composite objects formed by lepton and quark constituents depend on whether the exotic admixture interferes destructively or constructively with the standard model currents, on the chirality and lastly on the quark species [4,5,2]. Typical values are 1 TeV $< \Lambda <$ 2 TeV, comparable to limits by Tevatron- and LEP-experiments (see Figure 4).

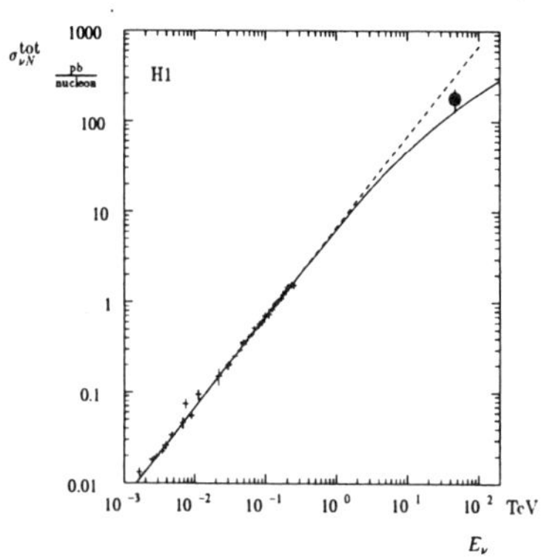

Electroweak unification leads to the approximate equality of neutral and charged current cross sections for momentum transfers $Q^2 \approx M_W^2 \approx M_Z^2$. This is indeed observed [6,7] as Figure 4 demonstrates, too.

Figure 5: The measured (e^-, ν_e) cross section ($p_{T,miss} > 25$ GeV) converted into an equivalent energy (ν, N) cross section and compared to the low energy data. The dashed curve corresponds to a linearly rising cross section, the full curve includes the W propagator effect.

The charged current cross section in the standard model is given by

$$e^- : \quad \frac{d^2\sigma^-}{dx dQ^2} = \frac{G_F^2}{2\pi}\left(\frac{M_W^2}{Q^2 + M_W^2}\right)^2 \left[(u+c) + (1-y)^2(\bar{d} + \bar{s} + \bar{b})\right]$$

For the quark densities $u, d, s, \ldots$ the charge conjugate has to be taken for the e^+ case, where the coupling is mainly to the $d-$quark. At lower energies the (ν, N) cross sections rise linearly with the laboratory neutrino energy E_ν (Figure 5). It is well known, that unitarity would be violated, if this behavior continues forever. The propagator term from the exchange of an intermediate vector boson W will prevent this. The predicted deviations from linearity at higher energies have now been directly seen in the HERA electron capture data [8], equivalent to $E_\nu = 48$ TeV. Combining the results from the analysis of both collaborations [9] (about 300 events) one obtains $M_W = 81.5 \pm 6.0$ GeV. First results for differential distributions in x and y are available, too, and start to constrain the parton densities [9].

4 Structure Functions in Neutral Current Data

From deep inelastic $e^\pm$ scattering on protons (all unpolarised) one determines mainly one structure function

$$\frac{d^2\sigma}{dxdQ^2} = \frac{2\pi\alpha^2}{xQ^4}\left[2(1-y) + \frac{y^2}{1+R}\right] F_2(x, q^2)$$

F_2 is the charge weighted sum of the parton densities within the proton, and R is related to the ratio of the total photoabsorption cross sections for longitudinally and transversely polarised virtual photons

$$R = \frac{\sigma^L(\gamma^* p)}{\sigma^T(\gamma^* p)} \equiv \frac{F_L}{2xF_1} = \frac{F_2 - 2xF_1}{2xF_1}$$

The contributions from the second term containing R typically amount to a few % except at high y, and one can safely neglect this term or use theoretical input for R (see below). The large data set of both collaborations [10,11] spanning the region $3 \times 10^{-5} < x < 0.32$ and $1.5\ \text{GeV}^2 < Q^2 < 5000\ \text{GeV}^2$ is shown in Figure 6. To reach the low x region either events from dedicated runs with an interaction vertex shifted in the proton direction or with initial state photon radiation were used. Data from a small beam pipe calorimeter were used by the ZEUS collaboration [12] to reach the transition region to photoproduction. Using both the hadron and electron information in the reconstruction a remarkable precision has already been reached, e. g. the systematic errors (values from the H1 experiment [10]) could be kept below 5 % at the lower Q^2 end, often larger than the statistical precision. The luminosity is known to an accuracy of 2 % and all efficiencies are determined from data. The electron energy scale is known to 1 % in the backward region and to 3 % in the central region (liquid argon calorimeter), the hadronic energy scale to 4 %. Both experiments agree with each other and join smoothly now the fixed target data [13,14,15].

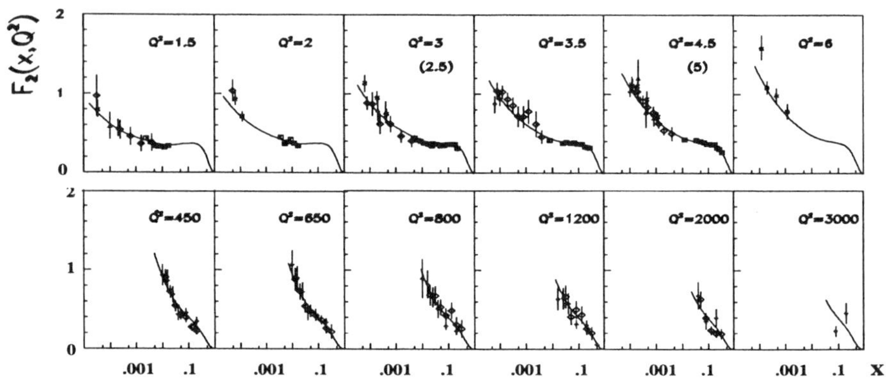

Figure 6: Excerpt of H1 [10] (open circles), ZEUS [11] (full symbols), and fixed target [13,14,15] data on F_2, plotted in function of x for different bins in Q^2. The solid curve indicates the QCD-fit performed by the ZEUS-collaboration.

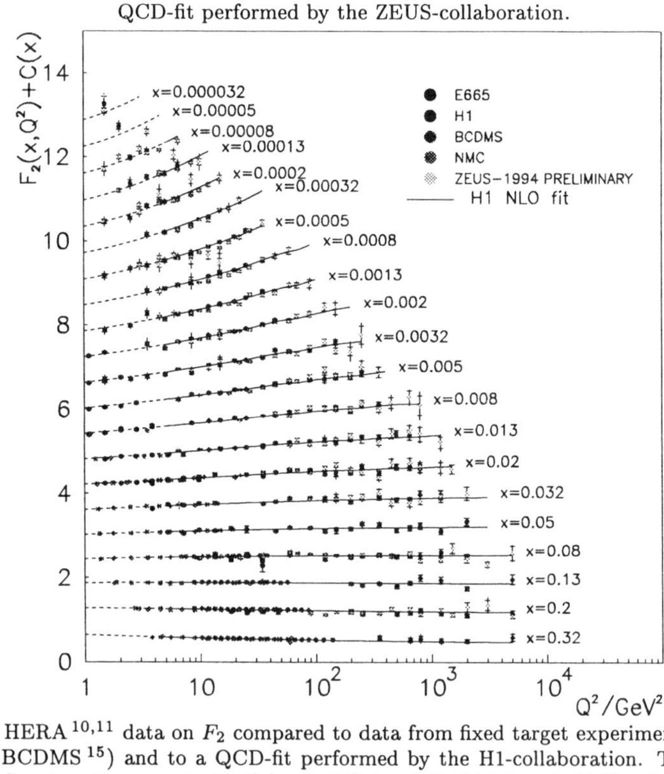

Figure 7: HERA [10,11] data on F_2 compared to data from fixed target experiments (E665 [13], NMC [14], BCDMS [15]) and to a QCD-fit performed by the H1-collaboration. The data are offset in x by a constant $C(x) = 0.6(i(x) - 0.4), i(0.32) = 1, i(0.2) = 2, \ldots$.

The data show a strong increase of F_2 with decreasing x linked to a rising gluon density at low x

$$F_2 \propto x^{-\lambda} \ \ (x < 0.1) \Leftrightarrow F_2 \propto W^{2\lambda} \ \ (Q^2 \approx xW^2) \ \text{with} \ 0.2 < \lambda < 0.4$$

This is one of the central results of the HERA experiments. All parametrisations of parton densities, which assume a flat gluon density at low x, though all compatible with the older fixed target data, are now ruled out. Speculations about what happens at even lower x, when the gluon density grows so strongly, that gluons will overlap, loose there identity or even form hot spots are still kept alive by the data. When plotted versus Q^2 for fixed bins of x (Figure 7), the data exhibit scaling violations (i. e. rise with Q for $x < 0.01$). Since the logarithmic derivative of F_2 is connected with the gluon density, the increasing slope again points directly to the rising gluon density $g(x, Q^2)$

$$\frac{\partial F_2}{\partial \ln Q^2} \propto \alpha_s x g(x, q^2)$$

Only at very low Q^2 (Figure 8) the rise of F_2 with decreasing x levels out, as was deduced from the photoproduction cross sections [16].

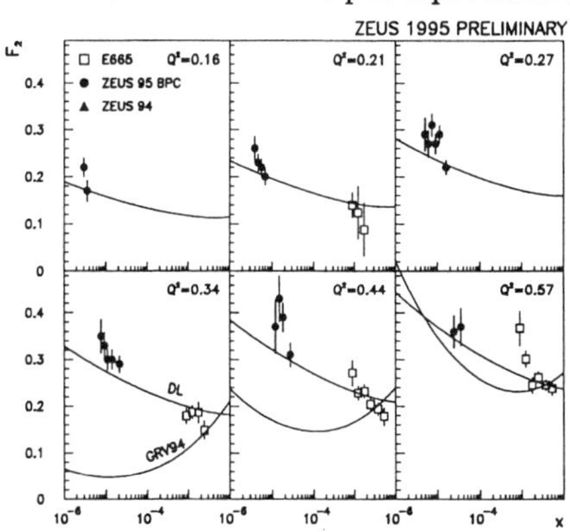

Figure 8: ZEUS low Q^2 and low x data from the beam pipe calorimeter indicating the transition to the photoproduction regime at Q^2. Upper curve derived from fit to photoproduction data [16], lower curve extrapolation of QCD fit to deep inelastic data [17].

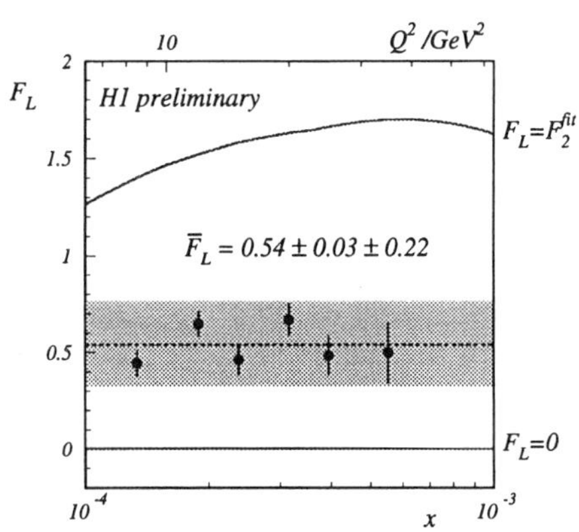

Figure 9: The longitudinal structure function F_L determined the H1 data for $y \geq 0.6$, subtracting the transverse contribution from an evolution of the low y ($y < 0.4$) data.

To obtain the gluon density one can subject the HERA and the fixed target data to a next-to-leading order QCD-fit. Figure 10 shows as an example the result obtained by the H1-collaboration [10], which started the evolution at $Q^2 = 4$ GeV^2 with the distributions given in the caption. This fit was also used to evolve F_2 from low values of y to high values of y, subtract the transverse part and determine F_L for the first time at HERA (Figure 9). The data are seen to agree well with QCD predictions using the structure function parametrisation of Glück, Reya and Vogt [17], which starts from sea quark like parton densities at $Q^2 = 1.5$ GeV^2. The latter ansatz also fits well all the available F_2 data.

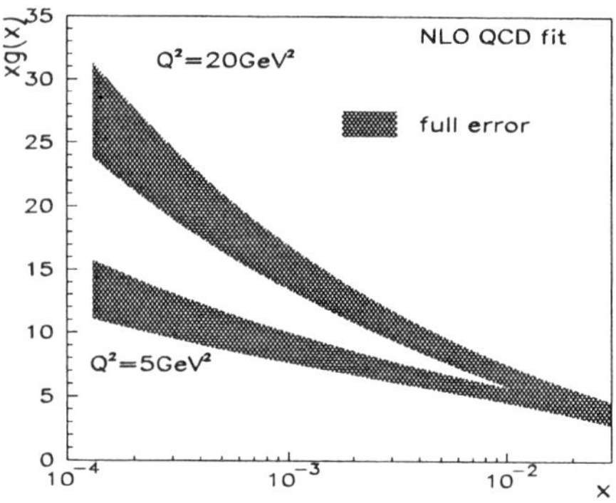

Figure 10: The gluon density $xg(x)$ extracted from the QCD fit to the H1 and fixed target data on F_2 with the systematic correlations between the different data points taken into account. The error band does not include the uncertainty in the QCD scale parameter $\Lambda \approx$ 200 MeV. $\delta\Lambda = 65$ MeV changes the $xg(x)$ by 9 %. The gluon density and the sea quark density $S(x)$ are strongly related. At low x one finds with three flavors $F_2 = (11/18)xS$, $xf \approx= A_f x^{B_f}$ with $f = S$ or $f = g$ and $A_S \approx A_g \approx -0.20$.

5 Diffractive Processes

My last subject is diffractive scattering. This an old subject at this conference series, starting in the sixties with data from the CERN SC on pion scattering of nuclei [18] exhibiting optical black disc diffraction patterns, and continuing to the impressive data on electron scattering of nuclei [19] spanning many orders of magnitude in cross section and revealing with increasing momentum transfer more and more of the nuclear density near the center. The experimental evidence which forced us to think about this subject at HERA [20] is shown in Figure 11.

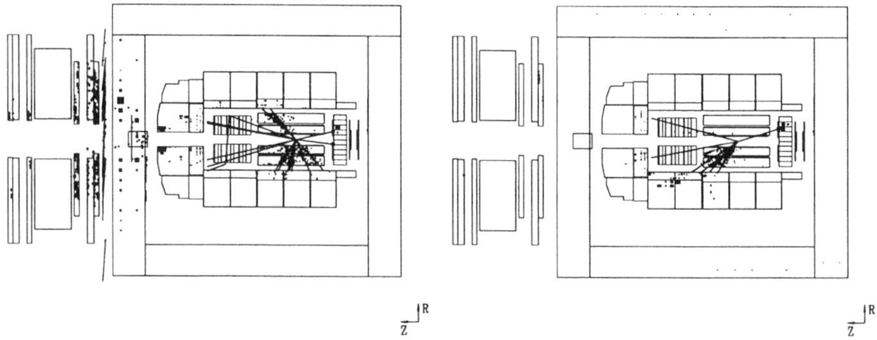

Figure 11: Left: standard deep inelastic scattering event, right: diffractive deep inelastic scattering event (rapidity gap event).

In a standard deep inelastic scattering event one observes the scattered electron (here in the backward electromagnetic calorimeter), the hadrons indicating the current quark (here in the central, hadronic part of the liquid argon calorimeter) and in forward direction the remnants of the proton (here producing a lot of activity in the forward liquid argon and the plug calorimeter, and the muon chambers in the instrumented iron and the forward muon spectrometer). About 10 % of the events show a different signature. While electron and hadronic recoil are present, too, there is no hadronic activity at all visible in forward direction, more precisely between 2°, where our acceptance starts, and 80°, where the first track is seen, the detector is empty. Translating this observation into the high energy physics jargon, there is a gap in hadronic activity between the maximum value of rapidity, which can be seen in the experiment ($\eta = 3.4$) and the rapidity of the first hadron ($\eta = 0$). Though such an event topology reminds one immediately of diffractive hadron-hadron scattering, say πp or pp, where the target is left more or less intact, and in our case escapes unseen through the beam pipe, one must experimentally prove, that these events are indeed of diffractive nature. A conceptually easy, but experimentally challenging way of doing this, is to actually fill the hole in forward direction, and to measure what is coming out. Both ZEUS and H1 use tracking devices located in so called roman pots for this, which can be driven into the vacuum system close to the circulating proton beam, once the tuning phase for luminosity optimisation is over. The signal seen in these detectors (Figure 12a) indicates, when the proton stays intact or is only mildly excited. By studying the decrease of the cross section with angle or with the hadronic momentum transfer t, as was done by the ZEUS-collaboration [21], one observes directly the $\exp(-bt)$–dependence characteristic of diffractive scattering (Figure 12b).

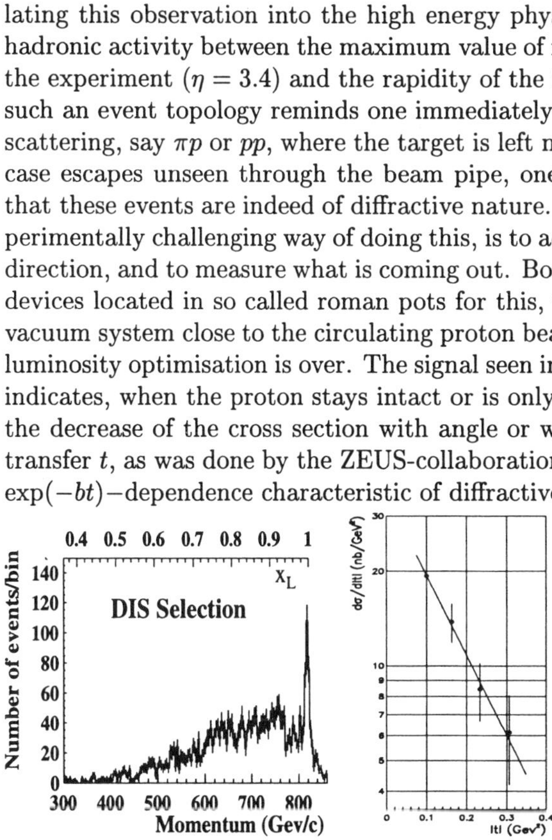

Figure 12: Left (a): Momentum spectrum measured in the ZEUS leading proton spectrometer equipped with silicon microstrip detectors. Right (b): the measured cross section in function of the hadronic momentum transfer t.

To introduce the more sophisticated method let us recall what produces the forward activity in the normal events referring to Figure 13. Both the struck quark and the remaining diquark are coloured objects and in the strong colour field in between them is the area where the final state hadrons are produced and where the colour flow is neutralized.

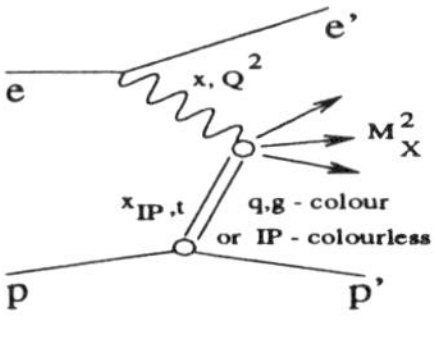

Figure 13: Normal and diffractive deep inelastic scattering lead to the exchange of different objects.

Hence with our rapidity gap events we are forced to consider a situation, where the colourless three-quark system remains intact and emits another colourless object, which then interacts with the photon and produces the hadronic system X seen in the detector. To prove, that we are dealing with a diffractive process, we can now use the full information provided by the detector and reconstruct the kinematics. This procedure also sheds light on the nature of the exchanged object.

Instead of Q^2, W, the mass M_X and t it has been found more convenient to use $Q^2, x_{I\!P}, \beta$ and t

$$x_{I\!P} = \frac{Q^2 + M_X^2}{Q^2 + W^2} \equiv 1 - x_L = 1 - \frac{|\vec{P'}|}{|\vec{P}|} \qquad \beta = \frac{Q^2}{Q^2 + M_X^2}$$

$x_{I\!P}$, i. e. the fraction of the protons momentum carried by the exchanged object, can be determined either from the mass of hadrons seen in the detector M_X and Q^2 determined from the momentum of the scattered electron or from a forward proton spectrometer measuring $\vec{P'}$. Assuming, that the exchanged object consists of several smaller parts, e. g. q, $\overline{q}$ or g, and only some of them are actually interacting with the virtual photon, then β is equivalent to the fraction of the exchanged momentum carried by the interacting quark or gluon.

It is worthwhile emphasizing at this point, what HERA brings us new here. The momentum transfer Q^2 can be varied over a wide range, which allows us to look deeper into the structure of the exchanged object, and to determine whether it is a common hadron, made of many gluons or of high-mass $q\overline{q}$–pairs. In terms of these four variables one measures

$$F_2^{D(3)}(Q^2, \beta, x_{I\!P}) = \int \frac{d^4\sigma}{dQ^2 d\beta dx_{I\!P} dt} \left[\frac{4\pi\alpha^2}{xQ^2}(1 - y + \frac{y^2}{2})\right]^{-1} dt$$

The results for $F_2^{D(3)}$ are displayed in Figure 14. Apart from the impressive amount of data, it immediately strikes the eye, that the cross section in all bins with more than two data points falls off rapidly as $1/x_{I\!P}^n$, with n very close to one. This is the behavior characteristic of and hence proof for a diffractive process. To investigate further the nature of the exchanged object we average over $x_{I\!P}$ and investigate the quantity (Figure 14)

$$\tilde{F}_2(\beta, Q^2) = \int_{0.003}^{0.05} F_2^{D(3)}(\beta, Q^2, x_{I\!P}) dx_{I\!P}$$

Now remembering that β is the fraction of the diffractively exchanged momentum carried by the parton, which then interacts with the current, we can use

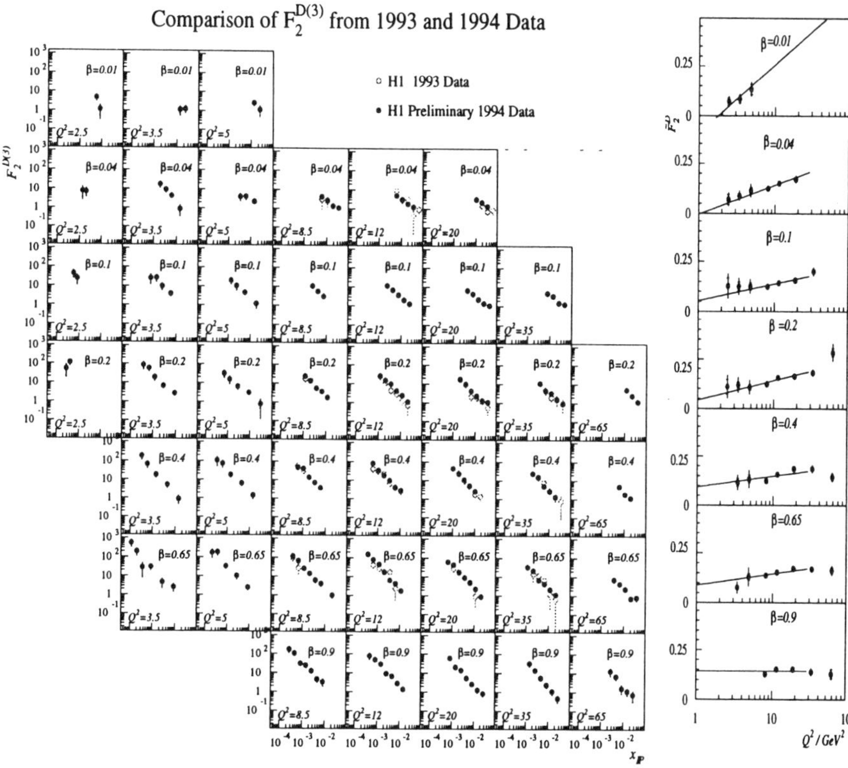

Figure 14: Left: The diffractive contribution [22] to the proton structure function F_2 plotted versus $x_{\mathbb{P}}$ for different β and Q^2. Right: The structure function of the diffractive exchange $\tilde{F}_2^{D(2)}(\beta, Q^2)$ plotted versus Q^2 for different bins in β.

the results for the proton structure function F_2 we discussed before as a guide and regard $\tilde{F}_2(\beta, Q^2)$ as the structure function of the exchanged object. For the proton we observed, that F_2 is independent of Q^2 for large and medium x, and rises more and more with Q^2, the lower x (or β as we call it here) becomes. This rise with Q^2 indicated, that at low $x(\beta)$ most of the partons found are gluons, as was apparent from the gluon density shown in Figure 10 above. Now here we see, that $\tilde{F}_2$ nearly never becomes flat, even up to $\beta = 0.65$. This is evidence that the exchanged object is predominantly of gluonic nature. If one is daring enough one can go beyond this observation and treat the measured quantity as if it was indeed the structure function of the pomeron, the traditional name for the diffractive exchange, make a full QCD evolution leading order fit to the data, and come back with a parton density. H1 has done this[22], and found that the fraction of the exchanged momentum carried by the gluons varies between 90 % (at 1 GeV/c^2) and 80 % (at 100 GeV/c^2). For low Q^2 the gluon density is strongly peaked near $\beta \approx 1$ indicating a leading gluon mechanism [23].

References

1. T. Naumann, Proc. XIII. Int. Conf. on Part. and Nuclei, Perugia (1993), A. Pascolini ed., World-Scientific (Singapore), p. 354; K. R. Long, ibid. p. 357.
2. D. Williams, in reference [24].
3. S. Aid et al., *Phys. Lett.* B **379**, 319 (1996).
4. S. Aid et al., *Phys. Lett.* B **353**, 578 (1995).
5. N. Brummer (ZEUS), Proc. Conf. on Results and Perspectives in Particle Physics, La Thuile, Italy (1996), in print.
6. M. Derrick et al., *Phys. Rev. Lett.* **75**, 1066 (1995); DESY-Report 96-104, *Z. Phys.* C , (1996), in print.
7. T. Ahmed et al., *Z. Phys.* C **67**, 565 (1995).
8. S. Aid et al., *Phys. Lett.* B **324**, 241 (1994).
9. A. Pieuchot, in reference [24].
10. S. Aid et al., *Nucl. Phys.* B **470**, 3 (1996).
11. M. Derrick et al., *Z. Phys.* C **65**, 379 (1995), *Z. Phys.* C **69**, 607 (1996).
12. Q. Zhu, in reference [24].
13. M. R. Adams et al., FNAL-Pub-95/396-E (1995).
14. M. Arneodo et al., *Phys. Lett.* B **364**, 107 (1995).
15. A. C. Benvenuti et al., *Phys. Lett.* B **223**, 485 (1989).
16. A. Donnachie, and P. V. Landshoff, *Z. Phys.* C **61**, 139 (1994).
17. M. Glück, E. Reya and A. Vogt, *Z. Phys.* C **67**, 433 (1995).
18. J. P. Stroot, Proc. IV. Int. Conf. High-Energy Physics and Nuclear Structure, Dubna (1971), V. P. Dzhelepov ed., p. 221.
19. I. Sick, Proc. VI. Int. Conf. High-Energy Physics and Nuclear Structure, Santa Fe (1975), AIP-Conf. Proc. 26, p. 388.
20. M. Derrick et al., *Phys. Lett.* B **315**, 481 (1993); T. Ahmed et al., *Nucl. Phys.* B **429**, 477 (1994).
21. E. Barberis, in reference [24].
22. M. Klein, in reference [24].
23. F. E. Close and J. R. Forshaw, *Phys. Lett.* B **369**, 33 (1996).
24. Proc. Int. Workshop on Deep Inelastic Scattering and Related Phenomena, Roma (1996), World-Scientific (Singapore), in print.

Solar Neutrino Experiments

K. NAKAMURA
National Laboratory for High Energy Physics (KEK)
Tsukuba, Ibaraki 305, Japan

Results from the four solar-neutrino experiments are reviewed. Updated results are consistent with the previous results, and all the experiments observe less neutrino fluxes than expected from the standard solar model calculations. Two solar neutrino problems are identified: one is the deficit of ^{8}B solar neutrinos and the other is the strong suppression of ^{7}Be solar neutrinos. Related to the ^{8}B solar-neutrino problem, better determination of S_{17} is needed. However, it seems that the ^{7}Be solar-neutrino problem makes the astrophysical solution unlikely. On the other hand, neutrino oscillations, both in matter and in vacuum, provide possible solutions. Next-generation solar-neutrino experiments are awaited to solve these problems. Future of the solar neutrino experiments is reported with emphasis placed on the commencement of SuperKamiokande and the present status of the SNO construction, and their prospects in resolving the solar-neutrino problems.

1 Introduction

Solar neutrinos have been observed by four experiments, Homestake chlorine experiment, Kamiokande νe scattering experiment, and SAGE and GALLEX gallium experiments. All these experiments have observed less solar neutrinos than expected from the calculations based on the standard solar models (SSMs). In reference to the SSM calculations by Bahcall and Pinsonneault in 1995 [1] (BP95), chlorine experiment observes $\sim$ 27 %, Kamiokande $\sim$ 42 %, and combined gallium experiments $\sim$ 51 % of the predictions.

Two problems have been identified so far. As can be seen from Fig. 1, Kamiokande purely observes ^{8}B solar neutrinos. This leads to the famous solar neutrino problem, *i. e.*, the deficit of the ^{8}B solar-neutrino flux. Analyzing both chlorine and Kamiokande results, it follows that there is strong suppression of the contribution from ^{7}Be solar neutrinos to the chlorine capture rate. [2–7] If the suppression factor R is defined as the ratio of the observed flux to the SSM prediction, $R(\phi_\nu(^7\mathrm{Be})) \ll R(\phi_\nu(^8\mathrm{B}))$. This conclusion is also supported by the results of the low counting rates observed by the gallium experiments. The strong suppression of the ^{7}Be solar neutrinos is the other solar neutrino problem, and it makes any astrophysical arguments to solve the solar-neutrino problems untenable simply because ^{8}B is produced from ^{7}Be in the pp chain of the solar-energy generation.

On the other hand, the resonant neutrino oscillation in matter, MSW mechanism, provides a promosing solution to the solar-neutrino problems. Two

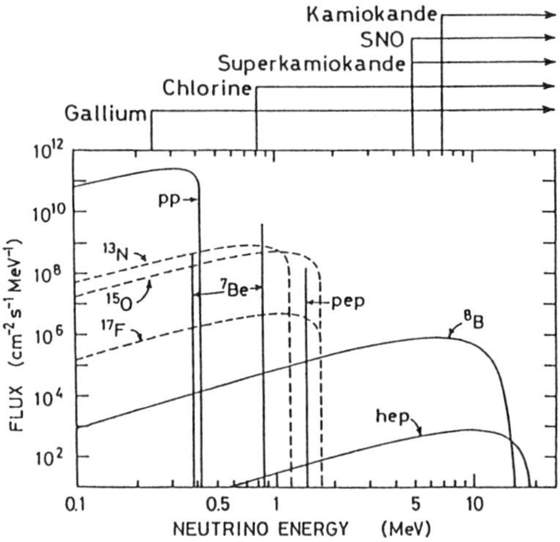

Figure 1: The solid curves show the energy spectra of solar neutrinos from the pp chain and the dashed curves show those from the CNO cycle. The detectable energy regions in various solar neutrino experiments are also shown.

regions in the $\sin^2 2\theta$ *vs.* Δm^2 parameter plane, small and large mixing-angle solutions, can explain all the aspects of the observed results. Yet there is another solution, vacuum neutrino oscillations with a large value of $\sin^2 2\theta$ and an extremely small value of Δm^2.

2 Present Status of the Solar Neutrino Experiments

2.1 Updated Results

The Homestake experiment,[8] SAGE,[9] and Kamiokande[10] updated their results before this conference. These results are summarized in Table 1 and compared with the BP95 SSM calculations [1] with and without diffusion of helium and metal. Shortly after the conference, GALLEX also updated the result. [12] For completeness, I include the updated GALLEX result in Table 1, though it was not discussed in the conference.

The Homestake result [8] has been obtained using the data up to Run 133, covering the observation period 1970.3 - 1994.4. The GALLEX result [12] represents the combined 53 runs from May 1991 to October 1995, which include the recent 14 runs of GALLEX III. These 14 runs give an about 1.5σ lower capture rate, $53.9 \pm 10.6 \pm 3.1$ SNU, than the combined result from 53 runs.

Table 1: Recent solar-neutrino results compared with the BP95 SSM calculations. [1] BP95(Best) is calculated with helium and metal diffusion and BP95(ND) is calculated without diffusion. The Homestake, GALLEX, and SAGE results and the corresponding calculations are given in terms of neutrino capture rates in units of SNU, and the Kamiokande result and the corresponding calculations are given in terms of the ^{8}B solar-neutrino flux in units of $10^6 cm^{-2}s^{-1}$.

Experiment	Result	BP95(Best)	BP95(ND)
Homestake	$2.54 \pm 0.14 \pm 0.14$	$9.3^{+1.2}_{-1.4}$	$7.0^{+0.9}_{-1.0}$
GALLEX	$69.7 \pm 6.7^{+3.9}_{-4.5}$	137^{+8}_{-7}	126^{+6}_{-6}
SAGE	72^{+12+5}_{-10-7}	137^{+8}_{-7}	126^{+6}_{-6}
Kamiokande	$2.80 \pm 0.19 \pm 0.33$	$6.62(1.00^{+0.14}_{-0.17})$	$4.85(1.00^{+0.14}_{-0.17})$

The SAGE group reanalyzed the SAGE II data using the pulse form from proportional counters, having been measured with a wave-form digitizer, for event selection. In this way, they were able to observe the L peak in addition to the K peak. The result shown in Table 1 is obtained by combining SAGE I and II, the observation period being early 1990 - October 1993. The Kamiokande solar-neutrino observation, began in January 1987, has come to an end in February 1995 because of the heavy work in constructing SuperKamiokande. The result shown in Table 1 is the average $\phi_\nu(^8B)$ in this entire observation period. The events with recoil electron energy > 7 MeV were analyzed. In addition to the average flux, the Kamiokande group updated the results of (i) the recoil electron energy distribution which is consistent with the β decay spectrum of ^{8}B, (ii) $\phi_\nu(^8B)$ as a function of time, showing no evidence of flux time variation, and (iii) daytime flux (2.70 ± 0.27) and nighttime flux ($2.87^{+0.27}_{-0.26}$), each in units of 10^6 cm^{-2} s^{-1}, showing no significant difference.

The updated results from all the four experiments are statistically consistent with the previous results. In particular, all the experiments observe less neutrino fluxes than expected from the SSM calculations.

A very important aspect of the solar-neutrino experiments is the calibrations using intense man-made neutrino sources by GALLEX and SAGE. The ^{51}Cr source produces 751 keV and 431 keV neutrinos with 90% and 10% probabilities, respectively, and is suited for the calibratin of the gallium experiments. The first GALLEX calibration result was 100 ± 10 % of expectation. [11] A preliminary result from the second GALLEX calibration [12] is 83 ± 10 % of the expectation, and the combined result of the two calibrations is (92 ± 8) %.

The SAGE result [9] is 0.95 ± 0.11. Both GALLEX and SAGE demonstrated the overall validity of the methods employed by these gallium experiments. It should be noted, however, that the demonstration of the validity might not be complete, because the capture of 751 keV neutrinos produce $\sim$ 520 keV electrons, while the capture of pp neutrinos involve lower-energy electrons.

2.2 Solar Neutrino Problems

As has been described in Sec. 1, there are two solar-neutrino problems, (i) deficit of the ^{8}B solar neutrinos, $R(\phi_\nu(^8\mathrm{B})) \sim 0.5$ and (ii) strong suppression of ^{7}Be solar neutrinos, *i.e.*, $R(\phi_\nu(^7\mathrm{Be})) \ll R(\phi_\nu(^8\mathrm{B}))$. These problems still persist for the updated results of the solar-neutrino experiments.

The ^{8}B solar-neutrino flux is proportional to an astrophysical S factor for the reaction $^7\mathrm{Be} + \mathrm{p} \rightarrow {}^8\mathrm{B} + \gamma$, S_{17}. Related to the solar-neutrino problem (i), there exists the S_{17} problem. BP95 [1] and SSM calculations by Turck-Chièze and Lopes[13] use $S_{17} = 22.4 \pm 2.1$ eV·b obtained by the extrapolation of Johnson *et al.* [14] However, the different data sets used by Johnson *et al.* have mutually inconsistent normalization. Barker and Spear [15] discarded the data sets showing consistently higher cross sections and obtained $S_{17} \sim 17$ eV·b. From a recent experiment of Coulomb dissociation $^{208}\mathrm{Pb}(^8\mathrm{B},\ ^7\mathrm{Be\ p})^{208}\mathrm{Pb}$, Motobayashi *et al.* deduced[16] $S_{17} = 16.7 \pm 3.2$ eV·b. In the direct capture reaction E2 contribution is negligible, while in Coulomb dissociation it is non-negligible and strongly dependent on the model of the ^{8}B structure. Nevertheless, analyses[17] of the Motobayashi *et al.* data give similar, or suggest even lower, values of S_{17} deduced by Motobayashi *et al.*[16] If $S_{17} = 17$ eV·b is used instead of 22.4 eV·b, the SSM flux of the ^{8}B solar neutrinos will be reduced by 25%. Though this alone would not completely solve the solar-neutrino problem (i), better determination of S_{17} is clearly very important.

It should be noted that the SSM of Dar and Shaviv [18], which includes the element diffusion, predicts a significantly low $\phi_\nu(^8\mathrm{B})$ of 2.49×10^6 cm^{-2} s^{-1}, which is consistent with the Kamiokande result. If this is the case, the solar-neutrino problem (i) disappears. Dar and Shaviv adopted $S_{17} = 17$ eV·b. They also made their own extrapolation to obtain $S_{33} = 5.6$ MeV·b and $S_{34} = 0.45$ keV·b, in contrast to the values used in BP95, $S_{33} = 4.99$ MeV·b and $S_{34} = 0.524$ keV·b. Since $\phi_\nu(^8\mathrm{B}) \sim S_{17} S_{34} S_{33}^{-1/2}$, the use of these astrophysical S factors reduces $\phi_\nu(^8\mathrm{B})$ to $\sim$ 60% of the BP95 flux. Further reduction of $\phi_\nu(^8\mathrm{B})$ seems to result mainly from different methods in dealing with the element diffusion and different equation of states used. A lesson to non-specialists in SSM may be that SSM might still involve quite large ambiguities. What nuclear physicists can contribute here is to provide better data to determine S_{33} and S_{34} as well as to examine the extrapolation method.

It seems almost impossible to solve the other solar-neutrino problem, strong suppression of $\phi_\nu(^7\mathrm{Be})$, astrophysically, as demonstrated in Refs.[2,4,19–23] Not only SSMs including that of Dar and Shaviv[18] but also non-standard solar models are unable to explain this.

Resonant neutrino oscillations in matter between two flavors, on the other hand, provide a natural and elegant solution to both of the solar neutrino problems without any *a priori* assumptions or fine tuning. The essential point is that this mechanism causes energy-dependent suppression factor (or survival probability) of electron neutrinos emitted by the sun. Oscillations into both active and sterile neutrinos are considered. Several authors made extensive MSW anlyses using all the existing data and ended up with similar results. Two solutions called the small and large mixing-angle solutions are known. In a recent analysis, Hata and Haxton [23] used the solar-neutrino data as of 1994 and BP95 SSM [1]. They incorporated the theoretical uncertainties in their analysis. For oscillations into active neutrinos, the small mixing-angle solution has central values of $\Delta m^2 \sim 6 \times 10^{-6}$ eV^2 and $\sin^2 2\theta \sim 7 \times 10^{-3}$ and the large mixing-angle solution has central values of $\Delta m^2 \sim 2 \times 10^{-5}$ eV^2 and $\sin^2 2\theta \sim 0.6$. Although acceptable, the large mixing-angle solution fits the data with a larger value of χ^2_{min} than the small mixing-angle solution. See, also, more recent analysis by Bahcall and Krastev. [25] For oscillations into sterile neutrinos, only the small mixing-angle solution is acceptable. [24]

Yet another solution to the solar-neutrino problems is provided by two-flavor neutrino oscillations in vacuum. [25,26] With BP95 SSM fluxes, [1] disconnected regions of the acceptable solution exist [25] at $5 \times 10^{-11} \lesssim \Delta m^2 \lesssim 10^{-10}$ eV^2 and $\sin^2 2\theta \gtrsim 0.65$. With SSM fluxes, no acceptable solution exists for vacuum neutrino oscillations into sterile neutrinos. [26]

3 Future of the Solar-Neutrino Experiments

3.1 New Experiments

The most recent and the most significant news of the solar-neutrino experiments is the start-up of the SuperKamiokande experiment. SuperKamiokande is a 50,000-ton imaging water-Cherenkov detector, located near the Kamiokande site. Its fiducial mass of 22,000 tons is viewed by 11,146 Hamamatsu 20″ photomultipliers. It has more than 2 m thick water-Cherenkov anticounter layer. The construction of SuperKamiokande started in December 1991, and it was almost completed by the end of 1995 when water-fill started. After a few months of test runs, the formal data-taking of SuperKamiokande started on April 1, 1996 at 0:00. At present, data-taking efficiency is about 70 %, and the peak of the solar-neutrino signal has already been observed in the angular distribution of low-energy electron events with respect to the sun-earth direction. The present trigger threshold is $\sim$ 5.5 MeV, but this will be lowered in future. The goal of the solar-neutrino observation in SuperKamiokande is the measurement of recoil electrons as low as 5 MeV with $\gtrsim$ 90 % efficiency. For that purpose, reduction of the Rn level in water to $\sim$ 0.005 Bq/m^3 is needed. Now the Rn level is $\sim$ 0.05 Bq/m^3. If this is achieved, the expected solar-neutrino event rate is $\sim$ 30/day assuming $R(\phi_\nu(^8\text{B}))\sim 0.5$.

Another important future solar-neutrino experiment is Sudbury Neutrino Observatory (SNO). It will consist of 1,000 tons of D_2O surrounded by 7,300 tons of H_2O, and observe ^{8}B solar neutrinos. The important feature of SNO is the capability of observing neutral-current events, $\nu d \rightarrow pn\nu$, the signal being γ-rays ($\Sigma E_\gamma = 8.6$ MeV) from neutron capture by ^{35}Cl dissolved in heavy water or output from discrete ^{3}He neutron counters installed in heavy water. If the SSM flux is assumed, the event rate will be 9,750/year. In addition, charged-current events, $\nu d \rightarrow ppe^-$, and νe scattering events will be observed with event rates of 2,800 and 1,100/year, respectively. The status of the SNO construction as of April 1996 is the following. The upper half of the photomultiplier support structure was complete, with about 5,000 photomultipliers in place with cables, and the upper half of the acrylic vessel for haevy water was constructed. After construction of the lower half of the detector, water-fill is expected early in 1997. The first run is planned to be conducted with pure heavy water, and after about 8 months decision will be made on the choice between adding $MgCl_2$ and installing ^{3}He detector array for the measurement of neutral-current events. Since these techniques have different systematics, it is considered to run with each of these techniques for some time.

At Homestake, an iodine experiment is being prepared as a new project. Technically, the iodine experiment is very similar to the chlorine experiment. Solar neutrinos are captured by ^{127}I to produce ^{127}Xe with an effective threshold of 789 keV. ^{127}Xe decays by electron capture with a half life of 36.4 days. With the SSM fluxes, a capture rate of 36.4 SNU is expected with 18.4 SNU and 13.0 SNU from ^{8}B and ^{7}Be solar neutrinos, respectively.[27] Thus, this experiment has larger sensitivity to $\phi_\nu(^7\text{Be})$ than the chlorine experiment. However, there are rather large ambiguities in the neutrino capture cross section of ^{127}I, and calibration experiments are needed. It is planned to start the iodine experiment late in 1996 with 235 tons of a solution of NaI in water, in which 100 tons of ^{127}I is contained. In future, it is planned to increase ^{127}I to 1,000 tons.

At Gran Sasso, there are activities for new solar-neutrino projects, Borexino, ICARUS, and GNO (Gallium Neutrino Observatory). Borexino is a highly purified liquid scintillation detector with a fiducial mass of 100 tons. With a threshold as low as 250 keV, it will be sensitive to ^{7}Be solar neutrinos. If SSM with no neutrino oscillations is assumed, an event rate of 50/day is expected. The final installation of Borexino is expected in 1998 - 1999. ICARUS is a liquid-argon time projection chamber. With a 5-MeV threshold, this detector is sensitive to ^{8}B solar neutrinos. A 600-ton detector has been proposed, and if approved, construction could be started in 1997. GNO is an extended version of the gallium experiment proposed in three stages: GNO30 with present GALLEX by purchasing 30 tons of gallium now in use (< 1999), GNO60 with a total of 60 tons of gallium in hydrochloric solution (1999 - 2001), and GNO100 with a total of 100 tons of gallium, possibly in metallic form (> 2001).

3.2 Prospects

If the MSW mechanism is responsible for the solar-neutrino problems, the high-statistics of SuperKamiokande would allow to identify the true solution from the two possibilities in a few years. The small mixing-angle solution would be positively identified by the distortion of the electron energy spectrum and the large mixing-angle solution by the day-night flux asymmetry. [28] The same is true for SNO. In addition, SNO can measure the ^{8}B solar-neutrino flux with both charged-current (CC) and neutral-current (NC) interactions. If the CC flux is smaller than the NC flux, undoubtedly it signifies neutrino oscillations either in matter or in vacuum. If the vacuum oscillation is the true solution, a positive identification would need the measurement of the ^{7}Be solar neutrinos by BOREX. If the CC flux is identical with the NC flux, it means either the standard neutrinos (in this case, other mechanisms must be sought for the cause of the solar-neutrino problems) or neutrino oscillations into sterile neutrinos. In the latter case, the only possibility is the MSW small mixing-angle solution as seen in subsect. 2.2 and will be positively identified by the distortion of the electron-energy spectrum in SuperKamiokande and in SNO.

4 Conclusions

The updated results of the four solar neutrino experiments have been presented. They are consistent with the previous results and show solar-neutrino problems, $R(\phi_\nu(^8\mathrm{B})) \sim 0.5$ and $R(\phi_\nu(^7\mathrm{Be})) \ll R(\phi_\nu(^8\mathrm{B}))$. The role of the next-generation solar-neutrino experiments is to solve these problems. With regard to this, a recent big news is the start-up of SuperKamiokande. It will open a new era of the solar-neutrino observations together with SNO and other future experiments.

Acknowledgments

I wish to thank Dr. Y. Totsuka, Dr. R. Davis, Dr. K. Lande, Dr. T. Kirsten, Dr. V.N. Gavrin, and Dr. A. McDonald for useful information.

References

1. J.N. Bahcall and M.H. Pinsonneault, *Rev. Mod. Phys.* **67**, 781 (1995).
2. N. Hata, S. Bludman, and P. Langacker, *Phys. Rev.* D **49**, 3622 (1994).
3. V. Castellani *et al.*, *Phys. Lett.* B **324**, 425 (1994).
4. X. Shi, D.N. Schramm, and D.S.P. Dearborn, *Phys. Rev.* D **50**, 2414 (1994).
5. W. Kwong and S.P. Rosen, *Phys. Rev. Lett.* **73**, 369 (1994).
6. J.N. Bahcall, *Phys. Lett.* B **338**, 276 (1994).
7. S. Parke, *Phys. Rev. Lett.* **74**, 839 (1995).
8. B.T. Cleveland *et al.*, *Astrophys. J.*, submitted.
9. V.N. Gavrin, talk presented at the 7th Int. Conf. on Neutrino Phys. and Astrophys., Helsinki, 1996.
10. Y. Fukuda *et al.*, *Phys. Rev. Lett.* **77**, 1683 (1996).
11. P. Anselmann *et al.*, *Phys. Lett.* B **342**, 440 (1995). The result has been updated by Ref. [12]
12. W. Hampel *et al.* *Phys. Lett.* B, submitted.
13. S. Turck-Chièze and I. Lopes, *Astrophys. J.* **408**, 347 (1993).
14. C.W. Johnson *et al.*, *Astrophys. J.* **392**, 320 (1992).
15. F.C. Barker and R.H. Spear, *Astrophys. J.* **307**, 847 (1986).
16. T. Motobayashi *et al.*, *Phys. Rev. Lett.* **73**, 2680 (1994).
17. M. Gai and C.A. Bertulani, *Phys. Rev.* C **52**, 1706 (1995); R. Shyam, I.J. Thompson, and A.K. Dutt-Mazumder, *Phys. Lett.* B **371**, 1 (1996).
18. A. Dar and G. Shaviv, *Astrophys. J.* **468** (1996), in press.
19. S.A. Bludman *et al.*, *Phys. Rev.* D **47**, 2220 (1993).
20. V. Castellani *et al.*, *Phys. Rev.* D **50**, 4749 (1994).
21. V. Berezinsky, G. Fiorentini, and M. Lissia, *Phys. Lett.* B **341**, 38 (1994).
22. G.L. Fogli and E. Lisi, *Astropart. Phys.* **3**, 185 (1995).
23. N. Hata and W. Haxton, *Phys. Lett.* B **353**, 422 (1995).
24. N. Hata and P. Langacker, *Phys. Rev.* D **50**, 632 (1994).
25. J.N. Bahcall and P.I. Krastev, *Phys. Rev.* D **53**, 4211 (1996).
26. P.I. Krastev and S.T. Petcov, *Phys. Rev. Lett.* **72**, 1960 (1994); *Phys. Rev.* D **53**, 1665 (1996).
27. J. Engel, P.I. Krastev, and K. Lande, *Phys. Rev.* C **51**, 2837 (1995).
28. See, for example, Ref. [24]

Current Status of J^{PC}-exotic Mesons and Glueballs

S. U. Chung

Brookhaven National Laboratory, Upton, NY 11973

Abstract

An exotic meson with $J^{PC} = 1^{-+}$ has been studied by three groups, in the vicinity of the $a_2(1320)$ meson in the decay channel $\pi\eta$. In the first part of this note, the current status and future prospects of this exotic meson, especially on the new result coming from the data of BNL-E852 Collaboration, are given. In the second part, a brief summary of the scalar and tensor glueball candidates is given.

1 Introduction

Although the quark model has been successful in describing and classifying hadrons, there is mounting evidence that the simple picture of a meson consisting of a quark and an anti-quark must be extended to include multi-quark states and those containing valence gluons. J^{PC}-exotic mesons, i.e. the states belonging to $J^{PC} = 0^{+-}$, 1^{-+}, 2^{+-} and 0^{--}, cannot be quarkonia. The current status and future prospects of a $\pi\eta$ state with $J^{PC} = 1^{-+}$ are given.

A short comment is in order on notations for the partial waves commonly encountered in the literature. A pure wave coupling to a system of two pseudoscalars can be completely specified by the orbital angular momentum (L), its z-component (M) and the reflectivity quantum number (ϵ). The latter is usually chosen such that it coincides with the naturality of the exchanged particles (Reggeons) in πN or $K N$ production processes. The spin J is simply L and the parity P is given by $(-)^L$. The usual notation is $|J^{PC} M^{\epsilon}\rangle$ or $|\epsilon\, LM\rangle = L_{M\epsilon}$. Here C is the charge conjugation. If $M = 0$, there is only one ϵ, i.e. $\epsilon = -1$. In addition, one makes the usual assumption that the partial waves for $M > 1$ are negligible, so that M can be dropped from the subscript above. One then merely deals with the following three partial waves, without ambiguity: L_0, L_+ and L_-, where the notations S, P, D, F are used for $L = 0$, 1, 2, 3, etc.

In this note, an J^{PC}-exotic meson is denoted following the standard convention in use, i.e. the name is determined by the combination $\{P, C\}$ of the meson. Hence, an isovector $J^{PC} = 1^{-+}$ state is denoted π_1, while its isoscalar partners are given the names η_1 and η_1'. It should be noted that the Particle Data Group[1] suggests a different convention for exotic mesons.

2 Observation of the $\pi_1(1405)$ meson

The $\pi_1(1405)$ state, called the $M(1405)$ meson, was originally reported by the GAMS group[2], seen in the reaction $\pi^- p \to \pi^0 \eta n$ at 100 GeV/c. The forward-backward asymmetry, obtained from the cosine of the Jackson angle in the $\pi^0\eta$ system, shows a substantial non-zero swing in the $a_2(1320)$ region. This can happen only if there is a strong interference effect between even-odd waves in the $a_2(1320)$ region. After a partial-wave analysis of the $\pi^0\eta$ system, the GAMS group reported that both the D_0 and the P_0 are resonant in the vicinity of the $a_2(1320)$. Because the $a_2(1320)$ and the M are seen to be produced via unnatural-parity exchange, it could be concluded that the $b_1(1235)$ exchange is the dominant one. This also indicates that the M should decay into the $b_1(1235)\pi$ final state.

There exist two additional pieces of evidence regarding the $M/\pi_1(1405)$ meson. The VES experiment[3] studied the reaction $\pi^- N \to \pi^- \eta N$ at 37 GeV/c. The $M/\pi_1(1405)$ signal is present but small with large errors in the $\pi\eta$ channel. The E179 collaboration[4] at KEK studied the $\pi^-\eta$ system in the reaction $\pi^- p \to \pi^- \eta p$ at 6.3 GeV/c. As seen in fig. 1, both D_+ and P_+ waves peak at the $a_2(1320)$ mass region; therefore, ρ- or f_2-exchange processes seem to play a central role in the production of these waves. Since the phase variation between these waves seems constant throughout the $a_2(1320)$ mass region, one must conclude that both are resonant. They have determined that $m(a_2) = 1325.1 \pm 5.1$ MeV and $\Gamma(a_2) = 126.2 \pm 14.2$ MeV for the $a_2(1320)$, and $m(\pi_1) = 1323.1 \pm 4.6$ MeV and $\Gamma(\pi_1) = 143.2 \pm 12.5$ MeV for the $M/\pi_1(1405)$.

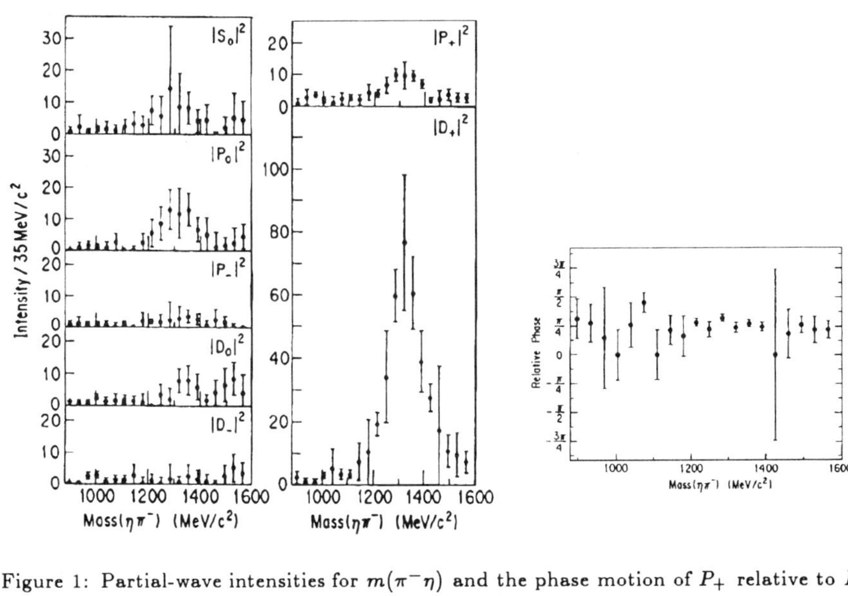

Figure 1: Partial-wave intensities for $m(\pi^-\eta)$ and the phase motion of P_+ relative to D_+ from the KEK data

3 New result on the $J^{PC} = 1^{-+}$ $\pi\eta$ state

The BNL-E852 Collaboration[5] is in the process of carrying out a study of the reactions $\pi^- p \to \eta\pi^- p$ and $\pi^- p \to \eta\pi^0 n$ at 18 GeV/c, where the η decay is observed in the channels $\gamma\gamma$ or $\pi^+\pi^-\pi^0$. The data presented in this note are based on 45,000 events fitting the reaction $\pi^- p \to \eta\pi^- p$ with $\eta \to \gamma\gamma$. The mass spectrum of $\eta\pi^-$ is shown in Fig. 2, and the dotted histogram corresponds to non-η background in the data. It is seen that the $a_2(1320)$ peak is most prominent, with relatively pure η events in this region.

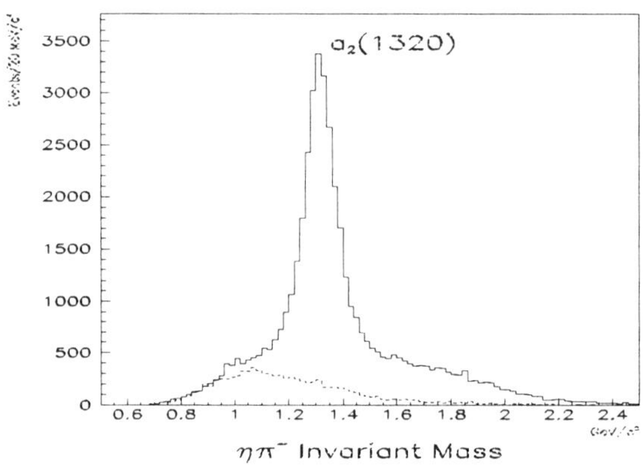

Figure 2: $m(\pi^-\eta)$ spectrum in 20-MeV bins from the BNL-E852 data

The partial-wave analysis has been done on the data, but the results are highly preliminary and are not presented here; however, one can glean a lot from a study of the structures in the unnormalized moments,

$$H(LM) = \sum_{i}^{n} d^L_{M0}(\theta_i) \cos M\phi_i \tag{1}$$

where n is the number of raw experimental events in a given $m(\eta\pi^-)$ bin and (θ, ϕ) describes the orientation of the η in an $\eta\pi^-$ rest frame (the Jackson frame). One notes that $H(00)$ is merely the raw experimental mass distribution. In the limit of perfect acceptance, the experimental moments of interest are given, in terms of the partial-waves P and D,

$$\begin{aligned} H(30) &= \frac{3}{7\sqrt{5}}(\sqrt{3}P_0D_0 - P_-D_- - P_+D_+) \\ H(32) &= \frac{1}{7}\sqrt{\frac{3}{2}}(P_-D_- - P_+D_+) \\ H(40) &= \frac{2}{7}D_0^2 - \frac{4}{21}(D_-^2 + D_+^2) \\ H(42) &= \frac{\sqrt{10}}{21}(D_-^2 - D_+^2) \end{aligned} \tag{2}$$

with a shorthand notation in which A^2 stands for $|A|^2$ and AB for $2\,\mathrm{Re}\{AB^*\}$.

In Fig. 3 are shown the experimental moments $H(30)$, $H(32)$, $H(40)$ and $H(42)$. As expected, one sees that the $a_2(1320)$ peak is dominated by $H(40)$

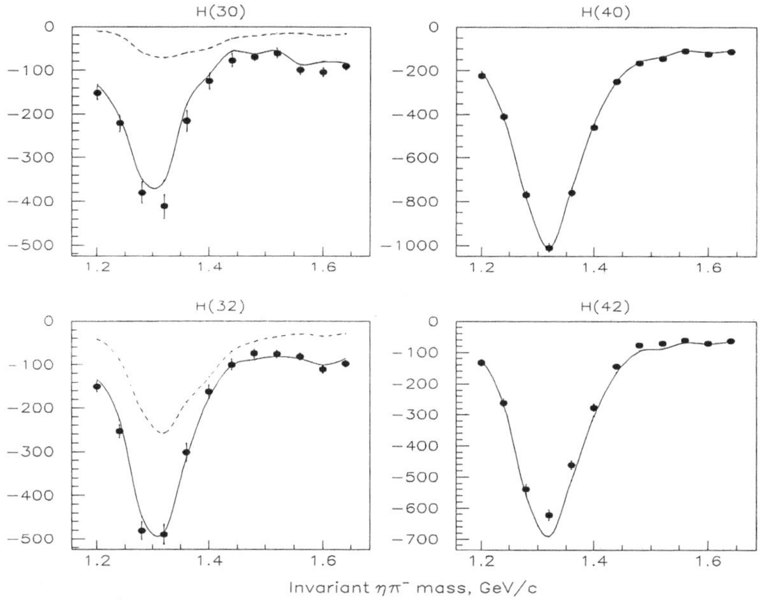

Figure 3: Experimental $H(LM)$ moments as a function $m(\pi^-\eta)$ from the BNL data

and $H(42)$ moments; as they are both large and negative, one concludes that the D_+ wave must be dominant in the a_2 region. The crucial observation is that the experimental moments $H(30)$ and $H(32)$ are both large in the a_2 region as well and have the same sign (negative). From (2), one finds that the only possible solution requires presence of a substantial P_+ wave in the a_2 region. A study of the finite acceptance of the apparatus reveals that it cannot account for the structures seen in the $H(30)$ and $H(32)$ moments—a more quantitative measure is given in the next paragraph. The next crucial feature of the moments is that the mass and the width of the peaks in all the moments $H(30)$, $H(32)$, $H(40)$ and $H(42)$ are very much like those of the $a_2(1320)$ itself. From this one must conclude that the P_+ wave is resonant, as is the a_2 of course, and that the two phases track with each other throughout the a_2 region.

A full likelihood analysis of the data has been performed, taking into account the finite acceptance of the apparatus in the fits. One finds that S-, P- and D-waves are sufficient to give acceptable fits to the data below 1.8 GeV. The required waves are S_0, P_0, P_-, P_+, D_0, D_- and D_+. Of these, only the P_+ and D_+ waves are significant, as expected, in the a_2 region. It can be shown that the eightfold ambiguities endemic for a two-pseudoscalar system are confined to the waves of unnatural-parity exchange ($\epsilon = -1$) and there are no ambiguities involving the two waves, D_+ and P_+, with natural-parity exchange. The results of a fit—still preliminary—are given in Fig. 3 as solid lines; it is seen that the fit describes well the structures in the a_2 region. The dotted lines in $H(30)$ and $H(32)$ result from the leakage of the D_+ wave, resulting evidently from finite acceptance of the apparatus alone. A preliminary result of the likelihood fit has been presented by S. Taegar[13] in the parallel session; one sees, most clearly and dramatically, that a substantial P_+ wave is required in the a_2 region in this data and that the phase difference between between P_+ and D_+ is nearly constant throughout the a_2 region.

4 Current Status of Glueballs

The scalar ($J^{PC} = 0^{++}$) glueball is predicted from the lattice QCD[14] at a mass 1.55 ± 0.05 or 1.74 ± 0.07 GeV, while the tensor ($J^{PC} = 2^{++}$) glueball is predicted, again from the lattice QCD, at a mass 2.27 ± 0.10 or 2.36 ± 0.13 GeV. In this section, a brief summary is given of the experimental candidates.

The Crystal Barrel Collaboration[6] observed a state $f_0(1500)$ with unusual decay properties, i.e.

$$\pi\pi \ : \ \eta\eta \ : \ \eta\eta' = 3 \ : \ 0.70 \pm 0.27 \ : \ 1.00 \pm 0.46$$

in the final states $\pi^0\pi^0\pi^0$ (see Fig. 4) $\pi^0\eta\eta$ and $\pi^0\eta\eta'$. Amsler and Close[7] argue that this state is mostly a scalar glueball mixed with some $n\bar{n}$ where

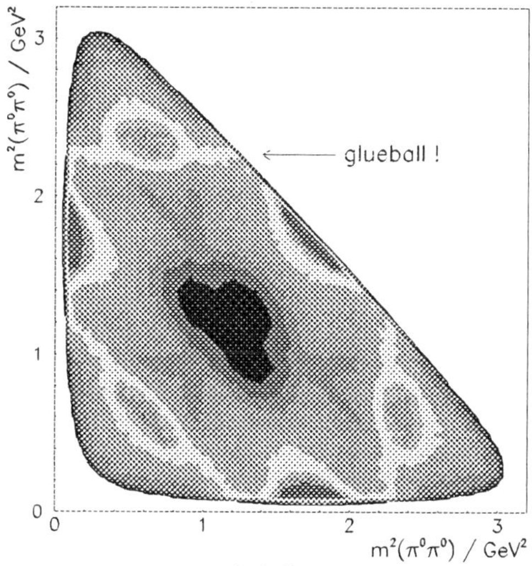

Figure 4: Dalitz plot for $p\bar{p} \to \pi^0\pi^0\pi^0$ from the Crystal Barrel data

$n = \{u, d\}$ and assume that the $K\bar{K}$ decay would be suppressed. Central to their argument is that the ground-state 0^{++} nonet consists of the states $a_0(1450)$ and $f_0(1370)$, both reported by the Crystal Barrel Collaboration[8]. This scenario requires that an additional scalar state should exist in the mass range 1.6 to 1.8 GeV, predominantly composed of $s\bar{s}$. More recently the Crystal barrel group[9] reported observation of the $f_0(1500)$ in the final state $\pi^0 K_L K_L$ and the relative branching ratio is quoted as

$$\frac{f_0(1500) \to K_L K_L}{f_0(1500) \to \eta\eta} = 0.76 \pm 0.27$$

In the limit of perfect $SU(3)$ flavor symmetry, a glueball should have a decay pattern given by

$$\pi\pi \ : \ \eta\eta \ : \ \eta\eta' \ : \ K\bar{K} = 3 \ : \ 1 \ : \ 0 \ : \ 4$$

This would imply that $f_0(1500)$ is a complex object—presumably more so than that envisaged by Amsler and Close—consisting not only of a scalar glueball and $n\bar{n}$ but also of $s\bar{s}$. The status of $f_0(1500)$ as mostly a glueball remains uncertain until a more clear picture emerges on all the members of the scalar nonet.

According to the PDG review[1], the ground-state scalar nonet consists of $a_0(980)$, $f_0(980)$ and $f_0(1300)$, at variance with the view proposed by Amsler and Close. As a matter of fact, Godfrey and Isgur[10] obtained $a_0(1090)$,

$f_0(1090)$ and $f_0'(1360)$ as members of the ground-state scalar nonet. But their predictions for the first radially excited scalar nonet are $a_0(1780)$, $f_0(1780)$ and $f_0'(1990)$. Once again, it is difficult to place $f_0(1500)$ in this scheme, leading to the conclusion that it must have a large admixture of non-quarkonium states. Its decay pattern confirms in addition the exotic character of this state. Finally, it should be pointed out that there exists another glueball candidate $f_0(1590)$, previously seen by GAMS [11] and VES[12], which may or may not be the $f_0(1500)$ seen in the Crystal Barrel data.

Some time ago, Lindenbaum[15] and his group argued that one or more of the three $\phi\phi$ states, with masses/widths 2.01/0.20, 2.30/0.15 and 2.34/0.32 GeV, could be the tensor glueball because they were produced prominently in an OZI-violating reaction, $\pi^- p \to \phi\phi n$. More recently, the JETSET Collaboration[16] studied the production reaction $p\bar{p} \to \phi\phi$ near the $\phi\phi$ threshold. They find a strong 2^{++} signal near the threshold, but they report merely a single peak at mass 2.20 ± 0.01 GeV with width 0.09 ± 0.02 GeV. Further analyses by the JETSET group on a more complete data sample should be very helpful in clarifying the $\phi\phi$ states between 2.0 and 2.3 GeV.

There exists another 2^{++} glueball candidate, called $\xi(2230)$, which could turn out be the sought-after tensor glueball. It was first seen in the $K\bar{K}$ system in the process $J/\psi \to \gamma K\bar{K}$ by the Mark III Collaboration[17]. Although its spin J was not determined because of limited statistics, this state could be the tensor glueball—it is produced in a glue-rich channel with a very narrow width (less than 30 MeV), making it unlikely to be an ordinary quarkonium[18]. More recently, the BES Collaboration[19] observed $\xi(2230)$ in the decay channels $K\bar{K}$, $\pi\pi$ and $p\bar{p}$(see Fig. 5). However, due to limited statistics, the results of their spin-parity analysis was not given in the publication, although a preliminary result[20] favored $J = 2$ over $J = 4$. Recently, an analysis of the decay $J/\psi \to \gamma p\bar{p}$ was performed on the Mark III data[21]. The $\xi(2230)$ is not observed in the $p\bar{p}$ system, but the upper limit for its quoted branching ratio is some 7 times larger than that obtained by the BES Collaboration. So there is no apparent conflict; nevertheless, absence of the signal in the Mark III data is unsettling. One needs evidently a high-statistics study of this state in the J/ψ radiative decays and other production channels, before one could attempt to identify $\xi(2230)$ as the tensor glueball. In any event, this narrow state cannot be Lindenbaum's $\phi\phi$ states between 2.0 and 2.3 GeV; a $\phi\phi$ state has in fact been seen in the J/ψ radiative decay[22], but its likely J^{PC} assignment favored 0^{-+}.

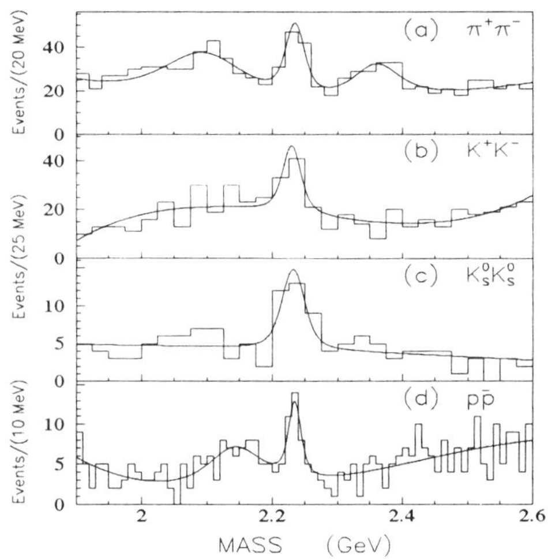

Figure 5: Fitted mass spectra from J/ψ radiative decays (BES Collaboration)

5 Discussions and Future Prospects

The KEK group sees a state at mass 1323 MeV and width 143 MeV in the $\pi\eta$ channel, produced via natural-parity exchange. This result is in conflict with an earlier GAMS result which showed a 1^{-+} state with unnatural-parity exchange. (Sadovsky[23] emphasized that the analysis technique, originally employed by the GAMS group, was unreliable.) The VES group also studied the $\pi\eta$ channel but they report no significant 1^{-+} state in the 1.3–1.4 GeV region. Now the new results from the BNL-E852 Collaboration tend to confirm those of the KEK data. It should be noted that the BNL data at 18 GeV/c is three times the KEK data (on the same reaction but at a much lower beam energy of 6.3 GeV/c). A definitive statement from the BNL Collaboration is expected soon, since their current analysis is based on 1/4 of the total data sample they have taken in 1994 and 1995.

In addition to the $\pi_1(1320)$ covered in this note, the VES group[3] claims a 1^{-+} candidate at 1620 MeV in the decay channels $\pi\rho$ and $\pi\eta'$, produced via natural-parity exchange. The BNL-E818 Collaboration[24] reported observation of a 1^{-+} $f_1(1285)\pi$ state in the 1.9–2.0 GeV mass region. All of these are in need of confirmation.

It is hoped that the situation will soon be very much clarified by the impending new results from partial-wave analyses on the BNL-E852 data, which consist of some 10^9 triggers collected over two years. In addition to $\eta\pi$, the decay channels under study include $\eta'\pi$, $\eta\eta$, $\eta\pi\pi$, $(\pi\pi\pi)^-$, $f_1(1285)\pi$, $\omega\pi$, $b_1(1235)\pi$ and $\omega\eta$. The E852 Collaboration is planning to take more data in 1997 with a segmented Cerenkov Counter for π/K separation. Because the gluons couple equally to all flavors, a comprehensive study of gluonic hadrons necessarily involves $s\bar{s}$ final states. The decay channels of interest include $K\bar{K}\pi$, $K\bar{K}\pi\pi$, $K\bar{K}\eta$, $K\bar{K}\omega$, etc.

A scalar (tensor) glueball is expected in the mass range 1.5 to 1.8 GeV (2.2 to 2.4 GeV). There are several experimental candidates; two new entries are $f_0(1500)$ for the scalar glueball and $\xi(2230)$ for the tensor glueball. However, the task of assignment is complicated by our incomplete knowledge of the scalar and tensor nonets, in the ground-states and in the first radial excitations. Until they are fully sorted out, from the studies of a variety of production and formation processes, the glueball candidates will remain as such.

The relevant experiments, both on-going and planned throughout the world, should shed light on the missing and uncertain members of the quarkonium family and those that lie beyond, in the near future. New results are expected from the analyses being carried out on the IHEP data taken by the VES and the GAMS group. Further new information should be forthcoming from the massive statistics being accumulated by the Crystal Barrel and Obelix collaborations at LEAR, on the $\bar{p}p$ and $\bar{p}n$ annihilations at rest and in flight. The WA76 collaboration with the Omega Spectrometer at CERN, and its extension WA91, have been collecting data since 1991, for a study of exotic mesons in the central production at the SPS. Towards the end of this decade and at the beginning of the next millennium, one can look forward to new experiments such COMPASS planned at CERN SPS, Japan Hadron Facility with 50 GeV protons at KEK, Beijing τ-charm factory, and upgraded AGS and RHIC at BNL.

Acknowledgments

The author wishes to thank the organizers for inviting him to speak in PANIC96, an excellent international conference devoted to QCD.

References

1. Review of Particle Properties, Particle Data Group, Phys. Rev. **D50**, Part I (1994).

2. D. Alde, *et al.*, Phys. Lett. **B205**, 397 (1988).
3. G. M. Beladidze, *et al.*, Phys. Lett. **B313**, 276 (1993).
4. H. Aoyagi, *et al.*, Phys. Lett. **B314**, 246 (1993).
5. The BNL-E852 collaboration consists of BNL, IHEP (Protvino), Indiana, U. Mass.(Dartmouth), Moscow State, Northwestern (Evanston), Notre Dame, RPI (Troy, NY). The group is currently carrying out a systematic study of the exotic mesons produced in $\pi^- p$ interactions at 18 GeV/c at the existing BNL Multi-Particle Spectrometer (MPS), augmented by charged-particle and photon detectors around the target and at the downstream end of the MPS magnet.
6. C. Amsler, *et al.*, Phys. Lett. **B355**, 425 (1995); Phys. Lett. **B340**, 259 (1994).
7. C. Amsler and F. E. Close, Phys. Rev. D**53**, 295 (1996); Phys. Lett. **B353**, 385 (1995).
8. C. Amsler, *et al.*, Phys. Lett. **B333**, 277 (1994);
9. S. Spanier, 'Observation of $f_0(1500)$ decay into $K_L K_L$,' presented at PANIC96, Williamsburg, VA (May 1996).
10. S. Godfrey and N. Isgur, Phys. Rev. D**32**,189 (1985).
11. D. Alde, *et al.*, Phys. Lett. **B269**, 485 (1986).
12. G. M. Beladidze, *et al.*, Sov. J. Nucl. Phys. **55**, 1535 (1992).
13. S. A. Taegar, 'Partial-wave Analyses of the $\eta\pi^-$ and $\eta'\pi^-$ systems for reactions $\pi^- p \to \eta\pi^- p$ and $\pi^- p \to \eta'\pi^- p$ at 18 GeV/c,' presented at PANIC96, Williamsburg, VA (May 1996).
14. G. S. Bali, *et al.*, Phys. Lett. **B309**, 378 (1993); H. Chen, *et al.*, Nucl. Phys. (Proc. Suppl.) **B34**, 357 (1994).
15. A. Etkin, *et al.*, Phys. Lett. **B201**, 568 (1988).
16. A. Palano, Proc. Hadron95, Manchester, England (July 1995), p. 171.
17. R. M. Baltrusaitis, *et al.*, Phys. Rev. Lett. **56**, 107 (1986).
18. H. G. Blundell and S. Godfrey, Phys. Rev. D**53**, 3700 (1996).
19. J. Z. Bai, *et al.*, Phys. Rev. Lett. **76**, 3502 (1996).
20. S. Jin, Proc. Workshop on Hadron Physics at e^+e^- Colliders, IHEP, Beijing (October 1994), p. 77.
21. T. H. Burnett, A. Guy and P. M. Mockett, Proc. Hadron95, Manchester, England (July 1995), p. 457.
22. J. Z. Bai, *et al.*, Phys. Rev. Lett. **65**, 1309 (1990).
23. S. A. Sadovsky, 'On the Choice of Minimization functional in the partial-wave analysis,' IHEP preprint 91-15, IHEP/Protvino, 1991.
24. J. H. Lee, *et al.*, Phys. Lett. **B323**, 227 (1994).

QCD FROM A LARGE N PERSPECTIVE

ANEESH V. MANOHAR
Department of Physics 0319, University of California at San Diego, 9500 Gilman Drive, La Jolla, CA 92093, USA

The $1/N_c$ expansion of QCD can be used to calculate properties of nucleons and Δ baryons, such as masses, magnetic moments, and pion couplings. The predictions of the $1/N_c$ expansion are in excellent agreement with the experimental data. The $1/N_c$ expansion also provides an understanding of the relation between the quark model, the Skyrme model, and QCD. This talk reviews some of the recent developments.

1 Introduction

Many features of the baryon sector of QCD have recently been understood both qualitatively and quantitatively using the $1/N_c$ expansion.[1-6] Here N_c is the number of colors, which is three in the real world. The main results are:

- There is a new symmetry in the baryon sector of QCD in the large N_c limit. For two light flavors, the symmetry is a contracted $SU(4)$ symmetry which connects the $u\uparrow$, $u\downarrow$, $d\uparrow$, $d\downarrow$ in a baryon. The $SU(4)_c$ symmetry relates the nucleon and Δ, which belong to a single irreducible representation of $SU(4)_c$.

- One can compute the higher order $1/N_c$ corrections in a systematic way.

- Many results previously obtained in models such as the quark model or the Skyrme model can be proved to be true in QCD to order $1/N_c$ or order $1/N_c^2$. The $1/N_c$ expansion explains why some model predictions work better than others; results obtained to order $1/N_c^2$ work about three times better than those obtained to order $1/N_c$. The $1/N_c$ expansion provides a unified understanding of different hadronic models, and more importantly, allows one to compute some quantities (e.g. baryon masses, pion couplings, magnetic moments) directly from QCD without any model assumptions.

- The $1/N_c$ expansion is intimately connected with baryon chiral perturbation theory. It explains why the Δ must be included to have a consistent perturbative chiral expansion for baryons. Including the Δ converts baryon chiral perturbation theory from a strong coupling expansion in powers of N_c to a weak coupling expansion in powers of $1/N_c$.

- The sigma term puzzle, and the success of various baryon mass formulæ such as the Gell-Mann–Okubo formula can now be understood
- The $1/N_c$ expansion provides some information on the nucleon-nucleon potential. (See the talk by D.B. Kaplan in this proceedings.)

2 Large N_c Counting Rules

Large N_c QCD[7] is a non-trivial theory with confinement and chiral symmetry breaking. The physical states are mesons and baryons. The $1/N_c$ counting rules in the meson sector are well-known.[7,8] The pion decay constant f_π, which is like a one-meson amplitude, is of order $\sqrt{N_c}$. Every additional meson in a vertex brings a suppression factor of $1/\sqrt{N_c}$. Thus, three-meson amplitudes are of order $1/\sqrt{N_c}$, four-meson amplitudes are of order $1/N_c$, etc. The counting rules imply that while large N_c QCD is a strongly interacting theory in terms of quarks and gluons, it is equivalent to a weakly interacting theory of mesons (and, as we will soon see, baryons). Loop corrections in the meson theory are suppressed by powers of $1/N_c$, which is consistent with a semiclassical expansion in powers of $\hbar/N_c$.

Large N_c baryons are more difficult to study than mesons, because the number of quarks in the baryon is N_c. The large N_c counting rules for baryon amplitudes were given by Witten.[9] The baryon matrix elements of any quark bilinear such as $\bar{\psi}\psi$ are $\leq \mathcal{O}(N_c)$, because the operator can be inserted on N_c possible quark lines. One finds an inequality, rather than an equality, because it is possible that there is a cancellation between the various diagrams. These possible cancellations will turn out to be very important in the analysis that follows. I will assume that the baryon mass M and the axial coupling g_A are both of order N_c, i.e. that there is no cancellation for these two quantities. All models of the baryon, such as the Skyrme model, the non-relativistic quark model, or the bag model, satisfy this assumption. For example, in the non-relativistic quark model $g_A = (N_c + 2)/3$, which gives the well-known value of $5/3$ for $N_c = 3$.

With these assumptions, the pion-nucleon coupling is of the form

$$(g_A/f_\pi)\ \bar{N}\gamma_\mu\gamma_5\tau^a N\ \partial_\mu\pi^a,$$

and is of order $\sqrt{N_c}$, since $g_A \sim N_c$ and $f_\pi \sim \sqrt{N_c}$. Recoil effects are of order $1/N_c$, and can be neglected. This allows one to simplify the expression for the nucleon axial current. The time component of the axial current between two nucleons at rest vanishes. The space components of the axial current between

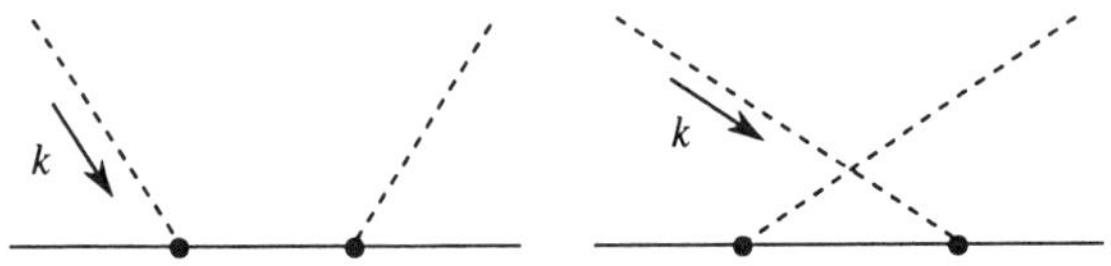

(a) (b)

Figure 1: The leading contribution to πN scattering to order E in the large N_c and chiral limits.

nucleons at rest can be written as

$$\langle N|\,\bar{\psi}\gamma^i\gamma_5\tau^a\psi\,|N\rangle = gN_c\,\langle N|\,X^{ia}\,|N\rangle\,, \tag{1}$$

where $\langle N|\,X^{ia}\,|N\rangle$ and g are of order one. The coupling constant g has been factored out so that the normalization of X^{ia} can be chosen conveniently. X^{ia} is an operator (or 4×4 matrix) defined on nucleon states $p\uparrow$, $p\downarrow$, $n\uparrow$, $n\downarrow$, which has a finite $N_c\to\infty$ limit.

3 Consistency Conditions

One can use the qualitative $1/N_c$ counting rules to obtain consistency conditions for various baryonic quantities, such as the axial couplings X^{ia}. These consistency conditions can be solved to obtain results for baryons that are in good agreement with the experimental data at $N_c = 3$.

Consider pion-baryon scattering at fixed energy E (as $N_c\to\infty$) in the chiral limit. The leading contribution is from the pole graphs in Fig. 1, which contribute at order E provided the intermediate state is degenerate with the initial and final states. Otherwise, the pole graph contribution is of order E^2. In the large N_c limit, the pole graphs are of order N_c. There is also a direct two-pion-N coupling that contributes at order E, which is of order $1/N_c$ in the large N_c limit and can be neglected.

The pion-nucleon scattering amplitude for $\pi^a(q)+N(k)\to\pi^b(q')+N(k')$ from the pole graphs is

$$-i\;q^i q'^j\,\frac{N_c^2 g^2}{f_\pi^2}\left[\frac{1}{q^0}X^{jb}X^{ia}-\frac{1}{q'^0}X^{ia}X^{jb}\right], \tag{2}$$

where the amplitude is written in the form of an operator acting on nucleon states. Both initial and final nucleons are on-shell, so $q^0 = q'^0$. The product of the X's in Eq. 2 then sums over the possible spins and isospins of the intermediate nucleon. Since $f_\pi \sim \sqrt{N_c}$, the overall amplitude is of order N_c, which violates unitarity at fixed energy, and also contradicts the large N_c counting rules of Witten. Thus $N_c \to \infty$ QCD with a $I = J = 1/2$ nucleon multiplet interacting with a pion is an inconsistent field theory. There must be other states that cancel the order N_c amplitude in Eq. 2 so that the total amplitude is order one, consistent with unitarity. One can then generalize X^{ia} to be an operator on this degenerate set of baryon states, with matrix elements equal to the corresponding axial current matrix elements. With this generalization, the form of Eq. 2 is unchanged, and we obtain the first consistency condition for baryons,[1]

$$\left[X^{ia}, X^{jb}\right] = 0 \,. \tag{3}$$

This consistency condition implies that the baryon axial currents are represented by a set of operators X^{ia} which commute in the $N_c \to \infty$ limit, a result also derived by Gervais and Sakita. [10] There are additional commutation relations,

$$\begin{aligned} \left[J^i, X^{jb}\right] &= i\,\epsilon_{ijk}\, X^{kb}, \\ \left[I^a, X^{jb}\right] &= i\,\epsilon_{abc}\, X^{jc}, \end{aligned} \tag{4}$$

since X^{ia} has spin one and isospin one.

The algebra in Eqs. 3 and 4 is a contracted SU(4) algebra. To see this, consider the algebra of operators in the non-relativistic quark model, which has an $SU(4)$ symmetry. The operators are

$$J^i = q^\dagger \frac{\sigma^i}{2} q, \quad I^a = q^\dagger \frac{\tau^a}{2} q, \quad G^{ia} = q^\dagger \frac{\sigma^i \tau^a}{4} q, \tag{5}$$

where J^i is the spin, I^a is the isospin, and G^{ia} are the spin-flavor generators. The commutation relations involving G^{ia} are

$$\begin{aligned} \left[G^{ia}, G^{jb}\right] &= \frac{i}{4}\,\epsilon_{ijk}\delta_{ab}\, J^k + \frac{i}{4}\,\epsilon_{abc}\delta_{ij}\, I^c, \\ \left[J^i, G^{jb}\right] &= i\,\epsilon_{ijk}\, G^{kb}, \\ \left[I^a, G^{ib}\right] &= i\,\epsilon_{abc}\, G^{jc}, \end{aligned} \tag{6}$$

The algebra for $N_c \to \infty$ baryons in QCD is given by taking the limit

$$X^{ia} \equiv \lim_{N_c \to \infty} \frac{G^{ia}}{N_c}, \tag{7}$$

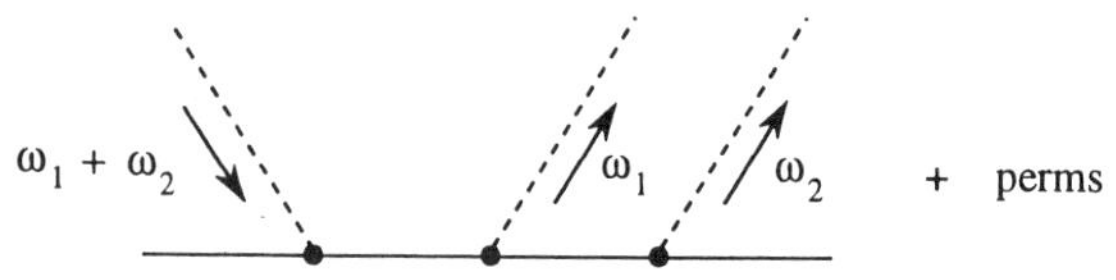

Figure 2: The leading contribution to $\pi N \to \pi\pi N$ in the large N_c and chiral limits.

The $SU(4)$ commutation relations Eq. 6 turn into the commutation relations Eqs. 3–4 in the $N_c \to \infty$ limit. The limiting process Eq. 7 is a Lie algebra contraction.

We have just proved that the the $N_c \to \infty$ limit of QCD has a contracted $SU(4)$ symmetry in the baryon sector. The unitary irreducible representations of the contracted Lie algebra can be obtained using the theory of induced representations, and can be shown to be infinite dimensional. The simplest irreducible representation is a tower of states with $I = J = 1/2, 3/2, etc.$, which is the set of states of the Skyrme model, or the large N_c non-relativistic quark model. More complicated irreducible representations correspond to baryons containing strange quarks.

The pion-baryon coupling matrix X^{ia} can be completely determined (up to an overall normalization g), since it is a generator of the $SU(4)_c$ algebra. It is easy to show that the $N_c \to \infty$ QCD predictions for the pion-baryon coupling ratios are the same as those obtained in the Skyrme model or non-relativistic quark model[11–13] in the $N_c \to \infty$ limit, because both these models also have a contracted $SU(4)$ symmetry in this limit. In the Skyrme model, the axial current in the $N_c \to \infty$ limit is $X^{ia} \propto \mathrm{Tr} A \tau^i A^{-1} \tau^a$. The X's commute (and so form part of an $SU(4)_c$ algebra), since A is a c-number. We have already seen how the quark model algebra reduces to $SU(4)_c$ in the large N_c limit.

4 $1/N_c$ Corrections

What makes the $1/N_c$ expansion for baryons interesting is that it is possible to compute the $1/N_c$ corrections. This allows one to compute results for the physical case $N_c = 3$, rather than for the strict $N_c = \infty$ limit, which is only of formal interest.

The $1/N_c$ corrections to the axial couplings X^{ia} are determined by considering the scattering process $\pi^a + N \to \pi^b + \pi^c + N$ at low energies. The baryon pole graphs that contribute in the $N_c \to \infty$ limit are shown in Fig. 2. The axial coupling X^{ia} can be expanded in a series in $1/N_c$,

$$X^{ia} = X_0^{ia} + \frac{1}{N_c} X_1^{ia} + \ldots \tag{8}$$

The amplitude for pion-nucleon scattering from the diagrams in Fig. 2 is proportional to

$$N_c^{3/2} \left[X^{ia}, \left[X^{jb}, X^{kc}\right]\right],$$

and violates unitarity unless the double commutator vanishes at least as fast as $N_c^{-3/2}$, so that the amplitude is at most of order one. (In fact, one expects that the double commutator is of order $1/N_c^2$ since the corrections should only involve integer powers of $1/N_c$. This result also follows from the $N_c \to \infty$ counting rules which imply that each additional pion gives a factor of $1/\sqrt{N_c}$ in the amplitude.) Substituting Eq. 8 into the constraint gives the consistency condition

$$\left[X_0^{ia}, \left[X_1^{jb}, X_0^{kc}\right]\right] + \left[X_0^{ia}, \left[X_0^{jb}, X_1^{kc}\right]\right] = 0, \tag{9}$$

using $\left[X_0^{ia}, X_0^{jb}\right] = 0$ from Eq. 3. The only solution to Eq. 9 is that X_1^{ia} is proportional to X_0^{ia}.[1] This can be verified by an explicit computation of X_1^{ia} using reduced matrix elements, or by using group theoretic methods.[3,6] Thus we find that

$$X^{ia} = \left(1 + \frac{c}{N_c}\right) X_0^{ia} + \mathcal{O}\left(\frac{1}{N_c^2}\right), \tag{10}$$

where c is an unknown constant. The first correction to X^{ia} is proportional to the lowest order value X_0^{ia}, so the $1/N_c$ correction to the axial coupling constant ratios vanishes.[1] We have now shown that the predictions for the ratios of pion couplings, such as $g_{\pi NN}/g_{\pi N\Delta}$ or $g_{\pi NN}/g_{\pi\Delta\Delta}$ are equal to the prediction in the Skyrme model up to corrections of order $1/N_c^2$. The Skyrme model results for the *ratios* of couplings are known to agree with experiment to about 10% accuracy.[14]

At order $1/N_c$, the baryons in an irreducible representation of the contracted $SU(4)$ Lie algebra are no longer degenerate, but are split by an order $1/N_c$ mass term ΔM. The intermediate baryon propagator in Eq. 2 should be replaced by $1/(E - \Delta M)$. The energy E of the pion is order one, whereas ΔM is of order $1/N_c$, so the propagator can be expanded to order $1/N_c$ as

$$\frac{1}{E - \Delta M} = \frac{1}{E} + \frac{\Delta M}{E^2} + \ldots \, . \tag{11}$$

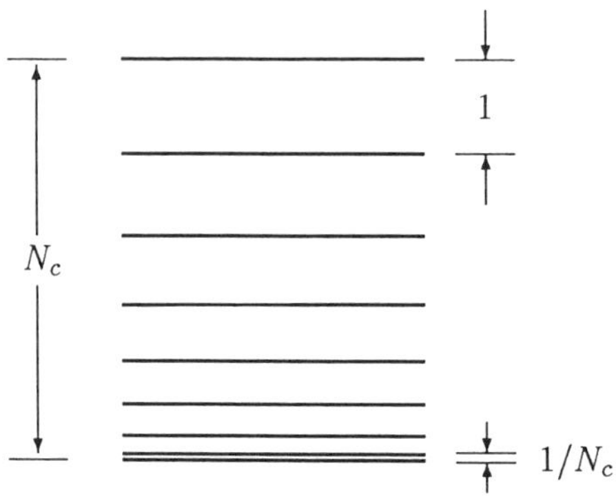

Figure 3: The baryon mass spectrum including the J^2/N_c term.

Including the $1/N_c$ corrections to the propagator does not affect the derivation of Eq. 3, as the two terms in Eq. 11 have different energy dependences. The first term leads to the consistency condition Eq. 3 and the second gives the consistency condition on the baryon masses,[2,3]

$$\left[X^{ia}, \left[X^{jb}, \left[X^{kc}, \Delta M\right]\right]\right] = 0. \tag{12}$$

This constraint can be used to obtain the $1/N_c$ corrections to the baryon masses. The constraint Eq. 12 is equivalent to a simpler constraint obtained by Jenkins using chiral perturbation theory[2]

$$\left[X^{ia}, \left[X^{ia}, \Delta M\right]\right] = \text{constant}. \tag{13}$$

The solution of Eq. 12 or 13 is that the baryon mass splitting ΔM must be proportional to $J^2/N_c = j(j+1)/N_c$, where j is the spin of the baryon.[2] This is precisely the form of the baryon mass splitting in the Skyrme model.

The structure of the $1/N_c$ corrections shows that the expansion parameter is J/N_c, where J is the spin of the baryon. For example, the baryon mass spectrum including the J^2/N_c mass splitting has the form shown in Fig. 3. The correction terms are only small near the bottom of the (infinite) baryon tower. For this reason, the $1/N_c$ expansion will only be considered for baryons with spin J held fixed as $N_c \to \infty$.

5 Results

It is relatively simple to obtain consistency conditions such as Eq. 3,9 or 12 using the $1/N_c$ expansion. Solving these equations is straightforward, but will not be described here. The general solution for any physical quantity $\mathcal{M}$ has the form of a series expansion,

$$\mathcal{M} = N_c^r \, \mathcal{P}\left(X^{ia}, \frac{J^k}{N_c}, \frac{I^a}{N_c}\right), \tag{14}$$

where $\mathcal{P}$ is a polynomial in its arguments, with each operator appearing with a coefficient $c\,(1/N_c)$ which is order one as $N_c \to \infty$. The overall power of r is given by the N_c power counting rule for the quantity $\mathcal{M}$ being considered. All the terms in the polynomial $\mathcal{P}$ must have the same spin and isospin as $\mathcal{M}$. In addition, there are a set of operator identities that allow one to eliminate many terms in $\mathcal{P}$ with contracted spin or isospin indices. For two flavors, these identites are[3]

$$\begin{aligned} X^{ia}X^{ib} &= \delta^{ab}, & X^{ia}X^{ja} &= \delta^{ij}, \\ X^{ia}J^{i} &= -I^{a}, & X^{ia}I^{a} &= -J^{i}, \\ \epsilon_{ijk}X^{ia}X^{jb} &= \epsilon_{abc}X^{kc}, & \epsilon_{abc}X^{ia}X^{jb} &= \epsilon_{ijk}X^{kc}. \end{aligned} \tag{15}$$

For example, if $\mathcal{M}$ is the baryon mass, $r = 1$ since the baryon mass is of order N_c. The general expansion has the form

$$M = N_c\left(c_0 + c_1\frac{J^2}{N_c^2} + c_2\frac{J^4}{N_c^4} + \ldots\right). \tag{16}$$

All other terms in $\mathcal{P}$ can be eliminated using Eq. 15. At present, it is not known how to compute the coefficients c_i.

One can obtain relations among baryon quantities by working to a given order in $1/N_c$, neglecting all higher order terms in $\mathcal{P}$. The accuracy of the relations is determined by the order in $1/N_c$ of the neglected operators. As a trivial example, consider the baryon masses in the two flavor case. The physical states are the N and Δ, with $I = J = 1/2$ and $3/2$, respectively. To lowest order, one finds that

$$M_N = N_c c_0 + \mathcal{O}\left(1/N_c\right), \quad M_\Delta = N_c c_0 + \mathcal{O}\left(1/N_c\right), \tag{17}$$

and the prediction

$$M_N = M_\Delta + \mathcal{O}\left(1/N_c\right). \tag{18}$$

This relation is a homogenous relation among the baryon masses. It is convenient to write Eq. 18 in the form

$$\frac{M_\Delta - M_N}{\left(M_\Delta + M_N\right)/2} = \mathcal{O}\left(1/N_c^2\right), \tag{19}$$

where the numerator is the difference of the two sides of Eq. 18, and the denominator is the average of the two sides. The form Eq. 19 does not depend on the overall scale of Eq. 18. Including the first correction gives

$$M_N = N_c c_0 + c_1 \frac{3}{4N_c} + \mathcal{O}\left(1/N_c^3\right), \quad M_\Delta = N_c c_0 + c_1 \frac{15}{4N_c} + \mathcal{O}\left(1/N_c^3\right) \quad (20)$$

In this simple example, there is no relation including the $1/N_c$ correction because two masses are given in terms of two parameters. However, one obtains many non-trivial predictions when the results are extended to three flavors.

The baryon masses have been analyzed in a simultaneous expansion in $SU(3)$ breaking (due to m_s), isospin breaking, and $1/N_c$ by Jenkins and Lebed.[15] They obtain a number of relations of the form Eq. 19. Their relations for isospin averaged masses (such as $(M_P + M_N)/2$) are shown graphically in Fig. 4. What is plotted are mass splittings of the form Eq. 19; the error bars are the experimental errors on the measured baryon masses. The relations are valid to some order in ϵ and $1/N_c$, where $\epsilon \sim 0.3$ is a measure of $SU(3)$ breaking. It is clear from Fig. 4 that the $1/N_c$ expansion explains the hierarchy of the baryon mass relations. For example, there are three relations which should be true up to corrections of order ϵ. One of them is order ϵ/N_c, one of order ϵ/N_c^2 and one of order ϵ/N_c^3. The hierarchy between these relations is obvious from Fig. 4. An $SU(3)$ analysis alone does not explain why some of the order ϵ relations work better than others. Similarly, the $1/N_c$ expansion explains the hierarchy in the two ϵ^2 relations. One can also see that the $1/N_c^2$ mass relation, which is not suppressed by $SU(3)$ breaking, works to about 11%. One can understand the baryon mass spectrum to a fractional error of 0.001 (i.e. ~ 1 MeV) in *the isospin averaged* masses using $1/N_c$ and $SU(3)$ symmetry.

A similar analysis can be done for the baryon magnetic moments.[16,17] The $1/N_c$ relations for the isovector and isoscalar magnetic moments are shown in Fig. 5. Most of the magnetic moment relations derived using the quark model can now be obtained directly from QCD, to some order in $1/N_c$. The $1/N_c$ expansion also explains why some relations work better than others; something which is left unexplained by the quark model. There is also one new relation. Relation 7, which relates the $\Delta^+ \to p$ transition moment to $p - n$, is a $1/N_c^2$ relation (which is also true in the quark model) which works to 30%. This is one instance in which the $1/N_c^2$ correction is substantial. All the $1/N_c$ relations are homogeneous, and do not set the scale of the magnetic moments. In the quark model, one can get a good estimate of the size of the magnetic moments from the constituent quark mass. This is not explained by the $1/N_c$ expansion.

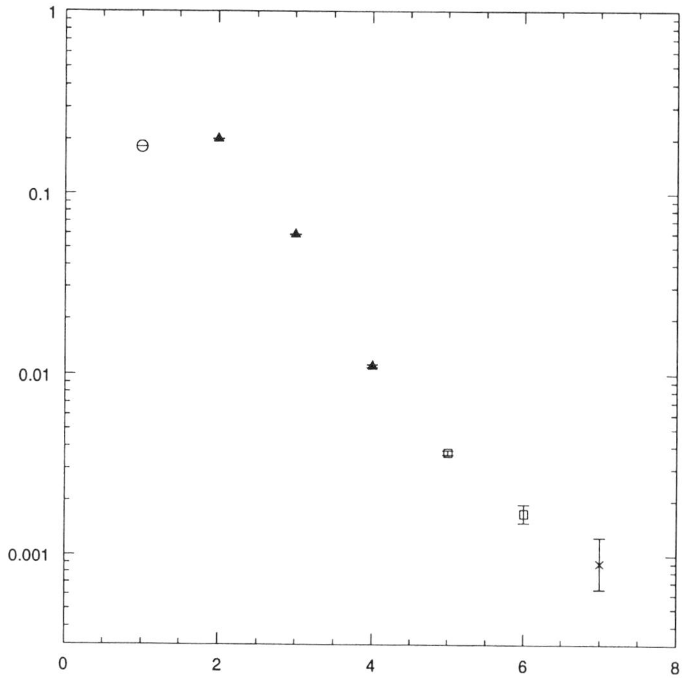

Figure 4: Plot of the isospin averaged mass relations of Jenkins and Lebed.[15]. The error bars are the experimental errors on the measured baryon masses. In terms of the $SU(3)$ breaking parameter $\epsilon \sim 0.3$, and $1/N_c$, the open circle is a relation of order $1/N_c^2$, the three solid triangles are relations of order ϵ/N_c, ϵ/N_c^2, and ϵ/N_c^3, the open squares are relations of order ϵ^2/N_c^2, and ϵ^2/N_c^3 and the $\times$ is a relation of order ϵ^3/N_c^3. The $1/N_c$ hierarchy is obvious.

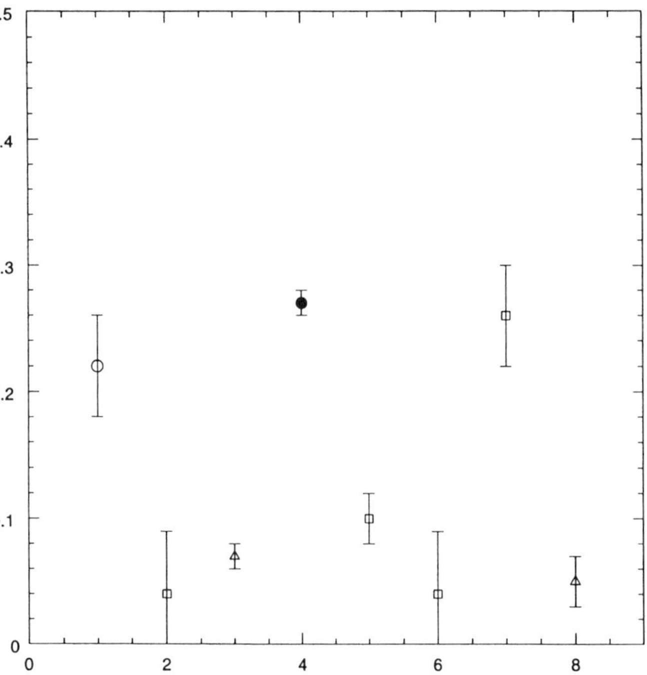

Figure 5: Plot of the magnetic moment relations. The error bars are the experimental errors on the measured magnetic moments. The open points are relations which are true in $1/N_c$ and in the non-relativistic quark model, the filled point is a relation which is true in $1/N_c$, but has no quark model analog. The circles are relations of order $1/N_c$, the squares of order $1/N_c^2$, and the triangles of order ϵ/N_c.

6 Other Results

There are many other results that have been obtained using the $1/N_c$ expansion, some of which are summarized here.

- One loop corrections to the baryon axial currents are large, and disagree with experiment, unless one includes intermediate Δ states in loops.[18] There are large cancellations between N and Δ intermediate states. The best fit values for the axial couplings when Δ's are included are close to $SU(6)$ values.[18] The $1/N_c$ explains all of these results. One can show that the one-loop corrections to the axial currents have the structure

$$N_c\,[X,[X,X]] \tag{21}$$

  Eq. 21 is a commutator that is constrained by the large N_c consistency conditions, and is of order $1/N_c^2$. Thus including intermediate Δ states produces a cancellation to two orders in $1/N_c$, so the one-loop correction is of order $1/N_c$. This cancellation converts an expansion in N_c to a semiclassical expansion in $\hbar/N_c$. If intermediate Δ states are not included, the one-loop correction is of order N_c times the tree level result, since the cancellation conditions on X do not hold in the nucleon sector alone. Baryon chiral perturbation theory without a Δ is a strong coupling expansion in powers of N_c. The strongly coupled theory presumably dynamically generates a Δ.

- One loop corrections to the baryon masses are large ($\sim$ 500 MeV). Most of the correction is $SU(3)$ singlet or octet, so the large non-linearities do not violate the Gell-Mann–Okubo formula.[19] This resolves the sigma-term puzzle.[20]

- Certain Skyrme model results on baryons containing heavy quarks using the bound state approach to strangeness of Callan and Klebanov follow from large N_c alone.[2,21,22]

- The chiral bag model arises naturally in the $N_c \to \infty$ limit.[23,24]

- One can compute the F/D ratio and an equal spacing rule for the axial couplings in β decay.[3]

- One understands the s-wave hyperon non-leptonic decay amplitudes.[25,6]

- One can analyze $SU(3)$ breaking in the baryon axial vector currents.[26] One finds that $\Delta u + \Delta d - 2\Delta s$ is shifted from its $SU(3)$ symmetric value of 0.6 to 0.3 ± 0.1. This affects the interpretation of the recent experiments on the g_1 structure function of the nucleon.

7 Conclusions

The $1/N_c$ expansion shows that most of the spin-flavor structure of baryons can be understood using contracted $SU(4)$ symmetry. It provides a unifying symmetry which connects the various quark models (non-relativistic, bag), the Skyrme model, and QCD. The $1/N_c$ expansion explains many results found earlier in baryon chiral perturbation theory in a systematic way, and it is possible to combine baryon chiral perturbation theory with the $1/N_c$ expansion of QCD.[27]

Acknowledgments

This work was supported in part by the Department of Energy grant DOE-FG03-90ER-49546, by a PYI award PHY-8958081 from the National Science Foundation, and by a "Profesor Visitante IBERDROLA de Ciencia y Tecnología" position at the Departmento de Física Teórica of the University of València. I would like to thank the Benasque Center for Physics for hospitality while this manuscript was written.

References

1. R. Dashen and A.V. Manohar, *Phys. Lett.* B **315**, 425 (1993), *Phys. Lett.* B **315**, 438 (1993).
2. E. Jenkins, *Phys. Lett.* B **315**, 431 (1993), *Phys. Lett.* B **315**, 441 (1993), *Phys. Lett.* B **315**, 447 (1993).
3. R. Dashen, E. Jenkins, and A.V. Manohar, *Phys. Rev.* D **49**, 4713 (1994).
4. C. Carone, H. Georgi, and S. Osofsky, *Phys. Lett.* B **322**, 227 (1994).
5. M. Luty and J. March-Russell, *Nucl. Phys.* B **426**, 71 (1994).
6. R. Dashen, E. Jenkins, and A.V. Manohar, *Phys. Rev.* D **51**, 3697 (1995).
7. G. 't Hooft, *Nucl. Phys.* B **72**, 461 (1974).
8. S. Coleman, in *Aspects of Symmetry* (Cambridge University Press, Cambridge, 1989).
9. E. Witten, *Nucl. Phys.* B **160**, 57 (1979).
10. J.-L. Gervais and B. Sakita, *Phys. Rev.* D **30**, 1795 (1984).
11. A.V. Manohar, *Nucl. Phys.* B **248**, 19 (1984).
12. G. Karl and J.E. Paton, *Phys. Rev.* D **30**, 238 (1984).
13. M. Cvetic, and J. Trampetic, *Phys. Rev.* D **33**, 1437 (1986).
14. G.S Adkins, C.R. Nappi, and E. Witten, *Nucl. Phys.* B **228**, 552 (1983).
15. E. Jenkins and R.F. Lebed, *Phys. Rev.* D **52**, 282 (1995).
16. E. Jenkins and A.V. Manohar, *Phys. Lett.* B **335**, 452 (1994).

17. M. Luty, J. March-Russell, and M. White, *Phys. Rev.* D **51**, 2332 (1995).
18. E. Jenkins and A.V. Manohar, *Phys. Lett.* B **255**, 558 (1991), *Phys. Lett.* B **259**, 353 (1991).
19. E. Jenkins, *Nucl. Phys.* B **368**, 190 (1992), E. Jenkins and A.V. Manohar, *Phys. Lett.* B **281**, 336 (1992).
20. R.L. Jaffe, *Phys. Rev.* D **21**, 3215 (1980).
21. C. Callan and I. Klebanov, *Nucl. Phys.* B **262**, 365 (1985), C. Callan, K. Hornbostel, and I. Klebanov, *Phys. Lett.* B **202**, 269 (1988).
22. E. Jenkins, A.V. Manohar, and M.B. Wise, *Nucl. Phys.* B **396**, 27 (1993), *Nucl. Phys.* B **396**, 38 (1993), E. Jenkins and A.V. Manohar, *Phys. Lett.* B **294**, 273 (1992), Z. Guralnik, M.E. Luke, A.V. Manohar, *Nucl. Phys.* B **390**, 474 (1993).
23. N. Dorey, J. Hughes, and M. Mattis, *Phys. Rev. Lett.* **73**, 1211 (1994).
24. A.V. Manohar, *Phys. Lett.* B **336**, 502 (1994).
25. E. Jenkins, *Nucl. Phys.* B **375**, 561 (1992).
26. J. Dai, R.F. Dashen, E. Jenkins, and A.V. Manohar, *Phys. Rev.* D **53**, 273 (1996).
27. E. Jenkins, *Phys. Rev.* D **53**, 2625 (1996).

Rare Decays

Toshio Numao
TRIUMF, 4004 Wesbrook Mall, Vancouver, Britich Columbia, V6T 2A3, Canada

The present status of rare decays of mesons and leptons are reviewed in terms of CP violation and lepton flavor violation. Neutrinoless double beta decays are also discussed in terms of lepton number violation.

1 Introduction

Rare decays have played a key role in the development of the standard model. The decay $\pi^+ \to e^+\nu$, which is helicity-suppressed to the level of 10^{-4}, provided a confirmation of the V–A nature of the weak interaction.[1] It is now studied in terms of e-μ universality in weak interactions at a level of 0.2 %.[2] CP violation was observed [3] by the decay mode $K_L^0 \to \pi^+\pi^-$ at the level of 10^{-3}. This mode in conjunction with similar modes is being studied by several groups [4] at a precision level of 10^{-4}; the double ratio of K_L^0 and K_S^0 decays to charged and neutral pions is expected to differ slightly from unity by $6Re(\frac{\epsilon'}{\epsilon})$, where $\frac{\epsilon'}{\epsilon}$ is a ratio of direct (by a CP violating decay amplitude) and indirect (CP violation by state-mixing) components. The suppression of the decay $K_L^0 \to \mu^+\mu^-$ [5] indicated the absence of flavor changing neutral current which suggested the existence of the charm quark. This mode is now being studied [6] in terms of V_{td} . Studies of these classic rare decays can now be categorized as "precision experiments". In the present talk, suppressed weak decays with accumulated events less than 50 are considered to be "rare decays".

There are other types of rare decays which have never been observed, perhaps because they are prohibited in the standard model. In the following chapters, some unanswered questions in the standard model are addressed in conjunction with studies of rare decays. Topics include "allowed" rare decays to investigate the origin of CP violation, forbidden decays to study the problem of the multiple quark-lepton generations, and double β ($\beta\beta$) decays to study the neutrino mass problem.

2 Allowed Rare Decays

The three-generation standard model has incorporated CP violation by including a complex phase in the off-diagonal corners of the Cabibbo-Kobayashi-

Maskawa (CKM) weak quark-mixing matrix. The least known elements are the two corner off-diagonal elements, V_{td} and V_{ub} . Orthogonality between the first and the third columns of the matrix relates these two matrix elements (assuming the diagonal elements to be nearly unity):

$$V_{ub}^{*} + V_{cd} \cdot V_{cb}^{*} + V_{td} \sim 0. \tag{1}$$

The above equation is usually illustrated by a unitarity triangle in the complex plane. The bottom side $V_{cd} \cdot V_{cb}^{*}$ is well determined from the Cabibbo angle and precision measurements of $B \rightarrow D$ decays. The height of the triangle is the imaginary part of V_{td} or V_{ub} , and is responsible for CP violation. The question of CP violation is whether the triangle is flat or not. Since the absolute value of V_{ub} may reasonably be determined with the development of theories and experiments, the next step is to measure the height, the length of the remaining side $|V_{td}|$ and the three angles, α, β and γ; multiple constraints on this triangle are essential elements of testing the standard-model origin of CP violation, or of finding a clue to physics beyond the standard model.

Since B decays play a significant role in the determination of the triangle, rare B decays will be briefly discussed to show the status of CP-related studies. The time-dependent yield of the decay $B^0(\overline{B^0}) \rightarrow \psi K_S^0$ measures $sin2\beta$ in the triangle without effects of competing diagrams. About 40 events were observed by the CDF group.[7] In order to determine the angle β at a level of 10 %, about 10^3 tagged events (B^0 or $\overline{B^0}$) are required. The observation of CP violation in this mode is the primary goal of many B-factories.
The time dependence of the decay $B^0(\overline{B^0}) \rightarrow \pi^+\pi^-$ (sensitive to $sin2\alpha$) is another CP violating observable to be studied at B-factories. There may be a significant contribution from another CP violating diagrams with a loop that complicates the extraction of the CP phase (the so-called penguin pollution), and this may require other measurements of similar decay modes for isospin or SU(3) analyses.[8] A few events have been observed [9] at the level of 10^{-5}, but because of the difficulty in doing particle identification between pions and kaons the branching ratios were only quoted as a sum of $B^0(\overline{B^0}) \rightarrow \pi^+\pi^-$ and $B^0(\overline{B^0}) \rightarrow K^+\pi^-$.
When the amplitude of the decay $B \rightarrow \rho(\omega)\gamma$ is compared with a similar decay $B \rightarrow K^*\gamma$, the ratio provides $|\frac{V_{td}}{V_{ts}}|$. The decay $B \rightarrow \rho(\omega)\gamma$ is expected occur at a branching ratio around 10^{-6} and the present limit [10] for $B \rightarrow \rho\gamma$ is 3.1×10^{-5}. However, there is a theoretical uncertainty in the corrections up to the size of the main diagrams.

The decay $K^+ \to \pi^+\nu\bar{\nu}$ proceeds via second order weak interactions and is sensitive to $|V_{td}|$. The uncertainties from the diagrams with a virtual charm quark and by long distance interactions have been estimated to be ~7 %.[11] The expected branching ratio is $0.5 - 3 \times 10^{-10}$ depending on the value of V_{td}. Experimentally, the search is done by looking for a single pion track with no other activity in the detector system. Two major background sources are from $K^+ \to \mu^+\nu$ and $K^+ \to \pi^+\pi^0$. Therefore, excellent π-μ identification and photon vetoing capability to reject events with photons from π^0 decays are required in the experiment. Using stopped K^+'s, the E787 group at BNL measures the momentum, kinetic energy and the range of the decay particle for particle identification. The tracking region is almost entirely covered by an electro-magnetic calorimeter. In addition, the pulse shape in the stopping counter is measured to positively identify the decay sequence, the decay $\pi^+ \to \mu^+\nu$ followed by the decay $\mu^+ \to e^+\nu\bar{\nu}$. The present upper limit for this mode is 2.4×10^{-9} (90 % c.l.).[12] Along with the upgrade of the accelerator, the group has upgraded the detector to reach a sensitivity below 10^{-10}.

The decay $K_L^0 \to \pi^0\nu\bar{\nu}$ is theoretically clean and involves only direct CP-violation. The expected branching ratio $1 - 5 \times 10^{-11}$ is proportional to the square of the imaginary part of V_{td} (the height of the triangle). Since all the particles involved in this decay mode are neutral, any additional kinematic constraints are very important. The present limit, 5.8×10^{-5}, comes from the Fermilab experiment E799 [13] in which the decay products of $\pi^0 \to e^+e^-\gamma$ ($BR = 1\%$) were measured to get the vertex but a significant part of the acceptance was lost. The group is going to push the limit to 10^{-8}, but still three orders of magnitude away from the prediction. Dedicated experiments with a sensitivity $10^{-11} - 10^{-12}$ are now being proposed with new ideas to get additional constraints such as using a "pencil" beam and making measurements of time-of-flight of neutral kaons.

The decays $K_L^0 \to \pi^0 e^+e^-$ and $K_L^0 \to \pi^0\mu^+\mu^-$ are also sensitive to direct CP violation with an expected branching ratio in the range $10^{-11} - 10^{-12}$. These decay modes, however, receive contributions from indirect CP violation and two-photon diagrams (CP conserving) at a level of $10^{-11} - 10^{-13}$. In addition, the background from the decay $K_L^0 \to \ell^+\ell^-\gamma\gamma$ at the level of 10^{-11} is unavoidable.[14] The present limits are $4.3(5.1) \times 10^{-9}$ (90 %) [15] for the decay $K_L^0 \to \pi^0 e^+e^-$ ($K_L^0 \to \pi^0\mu^+\mu^-$), and will be improved by the on-going experiments [16] to a sensitivity of 10^{-11}.

The decays $K^+ \to \pi^+\mu^+\mu^-$ and $K^+ \to \pi^+\gamma\gamma$ have been observed recently by the E787 group at levels consistent with Chiral Perturbation Theory (ChPT) predictions, and the results were presented in this conference.[17] One event was recently reported for the decay $K_L^0 \to e^+e^-\mu^+\mu^-$.[18] These decays provide important information for ChPT parameters to estimate CP conserving and indirect CP violating "background" in the decay $K_L^0 \to \pi^0 e^+e^-$.

3 Lepton Flavor Violating Decays

The old question of "What is the muon" is still around us, but it is now phrased in more general terms: "Why are there three generations?" Deeper understanding of the generation problem may be found in searches for transitions between generations, namely Lepton Flavor Violation (LFV) decays. Many extensions of the standard model including Supersymmetry models[19] and Grand Unified Theories (GUT) predict LFV decays. Searches for LFV decays such as $\mu^+ \to e^+\gamma$ and μ-e conversion in the field of a nucleus ($\mu^- A \to e^- A$) have been performed, and in the 60's kaon experiments joined the LFV search. Recently, τ, D, B and Z^0 experimental searches are done to look for LFV decays, but the sensitivities they can reach are still six orders of magnitude worse than those of μ's and K's.

There is no straight-forward answer for what is the best LFV mode to search for since most answers are highly model dependent. However, according to Marciano,[20] some general remarks may be possible though model dependence can not totally be removed from the comparison. For the case of the purely leptonic decays $\ell' \to \ell e^+e^-$ (assuming Z diagram dominance) and $Z^0 \to \ell'^+\ell^-$, there is a simple relation:

$$BR(Z^0 \to e^+e^-)/BR(\ell' \to \ell e^+e^-) \sim 1/6. \tag{2}$$

The present μ and τ decay limits set very tight constraints on the corresponding Z decays. The relation between B and K decays can be represented by

$$\frac{BR(K_L^0 \to \mu^\pm e^\mp)}{BR(B^0 \to \mu^\pm e^\mp)} \sim \frac{\tau_K m_K}{\tau_B m_B}\frac{C_K^2}{C_B^2} \sim 10^4, \tag{3}$$

where C_K and C_B are effective coupling constants for K and B systems, respectively, including the mixing angles. In this sense, kaon decays are favored significantly, especially when the "beam intensity" factor of 10^6 is taken into account. However, for higgs couplings the mass ratio may be significant, making B decays attractive.

Among muon decays, if the photon diagram is dominant the rate for the decay $\mu^+ \to e^+e^+e^-$ is roughly α/π of $\mu^+ \to e^+\gamma$ due to electro-magnetic

Processes	B.R.	References
$\mu^- A \rightarrow e^- A$	8.4×10^{-13}	PANIC '96
$\mu^+ \rightarrow e^+\gamma$	4.9×10^{-11}	PR D38 2077 (1988)
$\mu^+ e^- \rightarrow \mu^- e^+$	7.9×10^{-9}	WEIN '95
$K_L^0 \rightarrow \mu^\pm e^\mp$	3.3×10^{-11}	PRL 70 1049 (1993)
$K_L^0 \rightarrow \pi^0\mu^\pm e^\mp$	3.2×10^{-9}	23rd INS sympo. (1995)
$K^+ \rightarrow \pi^+\mu^+ e^-$	2.1×10^{-11}	PRL 64 165 (1990)
$\tau \rightarrow \mu\gamma$	4.2×10^{-6}	PRL 70 138 (1993)
$D^0 \rightarrow \mu^\pm e^\mp$	1.9×10^{-5}	CLNS 96-1389 (1996)
$B^0 \rightarrow \mu^\pm e^\mp$	5.9×10^{-6}	PR D49 5701 (1994)
$Z^0 \rightarrow \mu^\pm e^\mp$	1.7×10^{-6}	ZP C67 555 (1995)

Table 1: A summary of typical LFV decay limits.

coupling. The branching ratio of the mode $\mu^- A \rightarrow e^- A$ is expected to be $N \cdot \frac{\alpha}{\pi}$ of $\mu^+ \rightarrow e^+\gamma$ because there are N nucleons contributing to the process. However, again this is just a general statement and actual cases are model-dependent. For example, if the process goes though a leptoquark, $\mu^+ \rightarrow e^+\gamma$ and $\mu^+ \rightarrow e^+e^+e^-$ are significantly suppressed.

The signature of neutrinoless $\mu - e$ conversion in the field of an atomic nucleus ($\mu^- A \rightarrow e^- A$) is a single electron at the energy $E_e \sim m_\mu - \mathrm{B}_\mu$ where B_μ is the binding energy of the muonic atom. The major background source comes from electrons from μ decay in orbit, in which the electron energies could reach the maximum available energy for the $\mu^- A \rightarrow e^- A$ reaction although the spectrum sharply drops at the end-point. Good energy (momentum) resolution is essential in this experiment to separate the background. In this conference, the SINDRUM-II group at PSI reported a new limit 8.4×10^{-13} (90 % c.l.) in titanium. The group may push the sensitivity by another order of magnitude after the installation of a $\pi \rightarrow \mu$ converter to increase the beam intensity while reducing pion contamination in the beam.

The MEGA group at Los Alamos finished data taking for the $\mu^+ \rightarrow e^+\gamma$ search. They accumulated data corresponding to 10^{14} muon decays. The difficulty of this experiment is to get good energy and angle resolutions to suppress background arising from chance coincidences and the physical process, $\mu \rightarrow e\nu\bar{\nu}\gamma$. The goal is to reach a sensitivity level below 10^{-12}. In the search for muonium-antimuonium conversion ($\mu^+e^- \rightarrow \mu^-e^+$) , a thermal muonium drifting out from the target converts to an anti-muonium in vacuum. An electron from the decay $\mu^- \to e^- \nu \overline{\nu}$ is detected with the SINDRUM-I detector while a very slow positron is accelerated by an electric field and detected. The present limit 0.018 G_F [21] may be improved by another order of magnitude. The BNL experiment E871, search for $K_L^0 \to \mu^\pm e^\mp$, is taking data and expects to improve the current limit 3.3×10^{-11} by another order of magnitude. In this experiment, the K_L^0 beam is stopped immediately after the decay volume to reduce the counting rates of the downstream counters. Also, the number of track measurements has been increased to cope with random hits due to high counting rates. Another BNL experiment E865 is pushing down the limit for the decay $K^+ \to \pi^+ \mu^+ e^-$ by a factor of 70. This group also went through a major upgrade to clean up the beam halo and to gain more acceptance. A similar neutral decay mode $K_L^0 \to \pi^0 \mu^\pm e^\mp$ is being studied by the E799 group at Fermilab.

4 Double Beta ($\beta\beta$) Decays

Neutrinoless $\beta\beta$ decay is forbidden in the standard model because it violates the lepton number conservation law; in order to absorb the virtually emitted neutrino in the "first" decay, neutrinoless modes require $\nu = \overline{\nu}$, which means that the neutrino should be a majorana particle. It also means that there should be some mechanism to deal with the wrong helicity of the virtual neutrino. This can be achieved by introduction of neutrino mass, introduction of Majoron, and/or introduction of right-handed currents to accept the wrong helicity.

In neutrinoless $\beta\beta$ decays except the Majoron mode, the sum of two electron energies is almost equal to the available Q-value. When the source can not be used as a detector, e.g. in ^{82}Se and ^{100}Mo, thin isotopically enriched foils are sandwiched by detectors to measure two electrons. Since the invention of Ge detector, $\beta\beta$ decay searches in ^{76}Ge have taken the leading role. The tightest lower limit for neutrinoless $\beta\beta$ decays, $T_{1/2} > 6.4 \times 10^{24} y$ based on 12.5 $kg \cdot y$ data,[22] comes from the Heidelberg-Moscow collaboration using several 2-kg enriched (86 %) ^{76}Ge detectors. Figure 1 shows the present status of neutrinoless $\beta\beta$ decay limits.[22,23,24] In many sensitive nuclei such as ones in Fig. 1, direct measurements now surpass geo-chemical ones.

For the neutrino mass limit, the half-life is connected to a neutrino mass through nuclear matrix elements, which are reasonably estimated by shell model calculations for light nuclei up to $A \sim 100$. However, truncation of

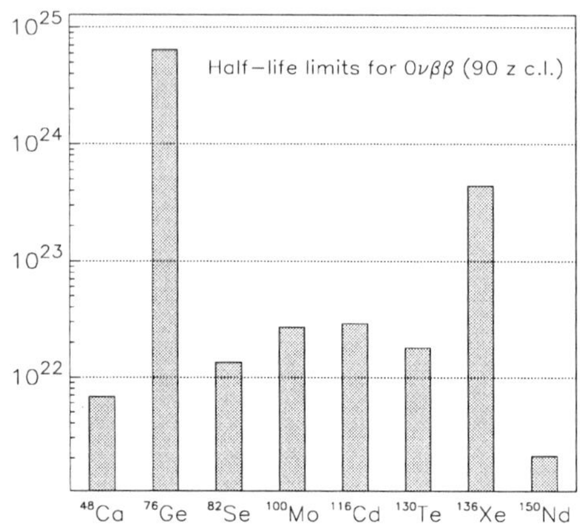

Figure 1: Life-time limits of $0\nu\beta\beta$ decays

the shell model space significantly affects the results for heavier nuclei. On the other hand, the Quasi-particle Random Phase Approximation (QRPA) provides reasonable estimations for heavier nuclei although there are some uncertainties due to unknown strengths of particle-particle interactions. These calculations are confirmed by two-neutrino $\beta\beta$ decays, ordinary β decays and B(E2) values, and then applied to neutrinoless $\beta\beta$ decays. Two-neutrino $\beta\beta$ decays in ^{48}Ca were measured with a Time Projection Chamber (TPC) by the University of California Irvine group at $5.3\ ^{+3.5}_{-1.5} \times 10^{19} y$.[25] This is in good agreement with shell model calculations, confirming the validity around the closed shell region. Measurements of other two-neutrino $\beta\beta$ decays are also in reasonable agreement with calculations.

The tightest majorana neutrino mass limit $m_\nu < 0.6$ eV comes from ^{76}Ge. New neutrinoless $\beta\beta$ limits for ^{100}Mo (for the mass mechanism, $T_{1/2} > 5.2 \times 10^{22}$ (68 % c.l.)) were presented in this conference by the Osaka group.[23] The corresponding mass limit (90 %) is 3.0 eV. A search in ^{100}Mo may be significantly improved by the coming of the NEMO-III detector which utilizes 10 kg of enriched isotope. The Gotthard TPC experiment [24] also reported their latest results at this conference. The lower half-life limit (90 %) of ^{136}Xe is now $4.4 \times 10^{23} y$, which corresponds to $m_\nu < 2.2$ eV.

5 Conclusion

The standard-model origin of CP violation will soon be extensively tested by rare decays of K and B mesons. So far, there is no evidence of physics beyond the standard model in rare decays, but improvement by another order of magnitude may lead to a breakthrough.

The author wishes to thank Drs. W. Bertl, M. Cooper, P. Wintz, H.V. Klapdor-Kleingrothaus, J-L. Vuilleumier, H. Ejiri and M. Moe for providing information, and Dr. D. Bryman for reading the manuscript.

References

1. T. Fazzini, *et al., Phys. Rev. Lett.* 1 247 (1958).
2. D.I. Britton *et al., Phys. Rev.* D49 28 (1994).
3. J.H. Christenson *et al., Phys. Rev. Lett.* 13 138 (1964).
4. E832 at Fermilab, NA48 and CPLEAR at CERN, and CLOE at DAΦNE.
5. W.C. Carithers *et al., Phys. Rev. Lett.* 30 1336 (1973).
6. E871 group at Brookhaven National Laboratory.
7. F. Abe *et al., Phys. Rev. Lett.* 76 2015 (1996).
8. M. Gronau and J.L. Rosner, *Phys. Rev. Lett.* 76 1200 (1996).
9. D.M. Asner *et al., Phys. Rev.* D53 1039 (1996).
10. M. Athanas *et al.,* Int. Conf. HEP, Glasgow, 1994, CLEO Conf 94-2.
11. G. Buchalla and A.J. Buras, *Nucl. Phys.* B412 106 (1994).
12. S. Adler *et al., Phys. Rev. Lett.* 76 1421 (1996).
13. M. Weaver *et al., Phys. Rev. Lett.* 72 3758 (1994).
14. B.H. Greenlee, *Phys. Rev.* D42 3724 (1990).
15. D.H. Harris *et al., Phys. Rev. Lett.* 71 3914 and 3918 (1993).
16. E162 at KEK and E799 at Fermilab.
17. J. Haggerty, PANIC '96.
18. P. Gu *et al., Phys. Rev. Lett.* 76 4312 (1996).
19. R. Barbieri, L. Hall and A. Strumia, *Nucl. Phys.* B445 219 (1995).
20. W. Marciano, Rare Decay Symposium, Vancouver, p1, 1988.
21. W. Bertl *et al.,* WEIN Yymposium, Osaka, p117, 1995.
22. H.V. Klapdor-Kleingrothaus, WEIN Symposium, Osaka, p174, 1995; Phys. Lett. B356 450 (1995), and references therein.
23. N. Kudomi *et al.,* PANIC '96 Abstract p480.
24. A. Paic, PANIC '96 Abstract p487 and private communication.
25. A. Balysh *et al.,* UCI-Neutrino 95-11; TAUP'95, Toledo, Spain, 1995.

LATTICE RESULTS IN HADRON PHENOMENOLOGY

G. MARTINELLI
Dip. di Fisica, Università degli Studi di Roma "La Sapienza"
and INFN, Sezione di Roma, P.le A.Moro 2, 00185, Rome, Italy

By taking as an example the calculation of the moments of the structure functions in deep inelastic scattering, the main features, advantages and limitations of the lattice approach in the computation of hadronic quantities are reviewed. The Bjorken sum rule is also considered, for discussing the theoretical aspects related to the calculation of power (higher-twist) corrections in effective theories.

1. Introduction

In the last years many important progresses have been made in lattice QCD, thanks to a great theoretical effort in understanding the lattice quantum field theory and to an extraordinary increase in the available computer power. Almost all aspects of particle physics are currently studied on the lattice and, in some important cases, particularly for weak hadronic decays, the results have already had a significant impact on phenomenological studies.

In this talk I will discuss some recent calculations of hadronic matrix elements which are relevant in Deep Inelastic Scattering (DIS). This is a prototype a wide class of physical processes where the dynamics is controlled by a scale μ which is much larger than Λ_{QCD}, and an "effective theory" can (must) be introduced. Besides DIS, where the large scale is given by the virtuality of the photon which probes the nucleon ($\mu^2 = -q^2$), this class includes among the others $e^+ e^-$ annihilation into hadrons at large center of mass energies ($\mu^2 = E^2_{c.m.}$) or weak decays of hadrons containing a heavy quark Q ($\mu^2 = m_Q^2$). In all these cases the theoretical calculation can usually be divided in two parts: the calculation of the effects of strong interactions at short distances, which can be done in perturbation theory, and the calculation of the hadronic matrix elements of local operators. The latter contain the effects of strong forces in the infrared region where perturbation theory cannot be applied. It is then necessary to find a non-perturbative method/approximation to compute the matrix elements. To this purpose, several different methods have been developed in the past: chiral Lagrangians, $1/N_c$ expansion, QCD sum rules, quark models, etc. Lattice simulations provide an opportunity for the evaluation of these matrix elements *from first principles and with no model parameters or assumptions.* Moreover there are no free parameters once the quark masses and a fundamental scale, e.g. the string tension, have been fixed.

It will take some time until the approximations present in these calculations can be brought fully under control. It is very satisfying to note, however, that during the last few years a large number of fundamental quantities have been computed non-perturbatively, and that it has often been possible to reduce significantly the systematic uncertainties in these computations. In this respect the lattice approach is unique. In the course of this review a particular attention will be paid to a discussion of the systematic errors present in the lattice calculations, and of the progresses made in reducing these errors.

In effective theories where an expansion in inverse powers of a large scale μ is used, we are often confronted to the problem of controlling power corrections. This is particularly important in the Heavy Quark Effective Theory (HQET) [1], for which the large scale, typically the mass of the b-quark, cannot be varied, or in cases where higher-twist effects are important, as for example for the Gross-Llewellyn Smith [2] or the Bjorken [3] sum rules. In these cases the computation of a physical quantity to the wanted accuracy, i.e. beyond the lowest order in the large scale expansion, requires both the non-perturbative computation of the matrix elements of higher-dimensional operators, and the perturbative calculation of the Wilson coefficient functions, to a sufficiently high order in perturbation theory for the cancellation of "renormalon ambiguities" to be under control [4,5]. This is true not only for the lattice method but for all the non-perturbative approaches which are used to compute the hadronic matrix elements. I will discuss below the problems related to the calculation of power (higher-twist) corrections by considering as a specific example the Bjorken sum rule.

The plan of this talk is the following: in section 2. a very short introduction to lattice QCD will be given; in section 3. the limitations and systematic errors present in the calculation of hadronic matrix elements, and the methods used to reduce them will be discussed; in section 4. some recent lattice results for matrix elements of operators relevant for the unpolarized and polarized structure functions will be presented; power corrections will be discussed in section 5., followed by the conclusion in section 6..

2. The lattice approach

All physical quantities (masses, cross-sections, decay amplitudes, etc.) of a quantum system described by an action S can be derived from T-products of suitable fields Φ of the theory. In the functional formulation of the theory, the T-products can be written as

$$\langle O(\Phi)\rangle = \frac{\int d[\Phi] e^{iS(\Phi)}\Phi(x_1)\ldots\Phi(x_n)}{\int d[\Phi] e^{iS(\Phi)}} . \tag{1}$$

Thus the problem is reduced to the calculation of the functional integral of eq. (1).

In order to compute (1) in numerical simulations, the following steps are needed:

1. An ultra-violet cut-off is introduced by replacing the continuum space-time by a mesh of discrete lattice points. The introduction of an ultra-violet cut-off is a necessary step in any renormalible theory, such as QCD, in order to eliminate the infinities from physical quantities, which at the end of the calculations should be finite and independent of the choice of the cut-off.

2. An infrared cut-off is introduced by working on a finite space-time volume; in this way the integral (1) is performed on a finite number of variables; this is unavoidable in numerical calculations.

3. After steps 1. and 2., the numerical integral is, at least in principle, feasible, because only a finite number of integration variables is involved. In practice, however, we have to use some kind of "important sampling" integration technique to make it affordable. This can be achieved only by rotating the Minkowsky space-time into a four dimensional Euclidean space. [6].

The three steps above are at the origin of some systematic effects or limitations, to be discussed below, in the calculation of physical quantities.

For lack of space, I will describe the calculation of the hadronic masses and matrix elements only through an explicit, simple example. Let us consider the T-product of two axial currents with the quantum number of a (charged) pion[a]

$$G(t) = \sum_{\vec{x}} \langle A_0(\vec{x},t) A_0^\dagger(\vec{0},0)\rangle \ , \tag{2}$$

where $A_0(\vec{x},t) = \bar{u}(\vec{x},t)\gamma_0\gamma_5 d(\vec{x},t)$. By inserting a set of (one-particle) states in eq. (2) we have

$$\begin{aligned} G(t) &= \sum_n \frac{|\langle 0|A_0|n\rangle|^2}{2M_n} e^{-M_n t} \\ &\to \frac{|\langle 0|A_0|\pi\rangle|^2}{2M_\pi} e^{-M_\pi t} \ , \end{aligned} \tag{3}$$

as $t \to \infty$. Thus, by studying the behaviour of the two-point function at large time-distances, it is possible to extract the pion mass M_π and its decay

[a] After the rotation to the Euclidean, T-products can be written as correlation functions.

constant f_π, $\langle 0|A_0|\pi\rangle = f_\pi M_\pi$. Similarly, by studying the time dependence of three-point functions, it is possible to extract more complicated matrix elements, such as those which allow the determination of the electro-magnetic form factors, the moments of the structure functions, radiative, semi-leptonic or non-leptonic weak amplitudes and mixings, etc.

An observation is necessary at this point. There is no difficulty in studying amplitudes which involve only one-particle states, such as electro-magnetic space-like form factors, or matrix elements of the local operators which appear in the Wilson Operator Product Expansion (OPE) in DIS processes. This is not true, however, for processes involving more than one particle in the final state, as a consequence of the use of the Euclidean space-time instead of the Minkowsky one. The Maiani-Testa no-go theorem [8] demonstrates that, since *in-* and *out*-states are indistinguishable in the Euclidean, from the study of the correlation functions one obtains an average of the *in-* and *out-* amplitudes instead of the physical one, thus losing all information about final state interactions. For example, in the calculation of the decay of a D-meson into two light mesons M_1 and M_2, mediated by the weak Hamiltonian $\mathcal{H}^W$ one gets

$$A^W = \frac{1}{2}\Big({}_{out}\langle M_1 \; M_2|\mathcal{H}^W|D\rangle +_{in} \langle M_1 \; M_2|\mathcal{H}^W|D\rangle\Big) \; , \tag{4}$$

instead of ${}_{out}\langle M_1 \; M_2|\mathcal{H}^W|D\rangle$ only. Moreover, at large time-distances, the correlation functions from which the amplitudes are usually extracted, cf. eq. (3), are dominated by the off-shell matrix element $\langle M_1(\vec{p}_1 = 0) \; M_2(\vec{p}_2 = 0)|\mathcal{H}^W|D\rangle$, instead of the physical amplitude. The Maiani-Testa theorem prevents the calculation of time-like electro-magnetic form factors and strong and weak decays in two or more hadron final states. Its extension can be used to show that it also limits the possibility of computing inclusive cross-sections and decay rates, see also below. A possible way out of this probl em, under mild assumptions, although at the price of introducing some model dependence, has been recently proposed in ref. [9].

3. Systematic errors in lattice calculations

An important systematic error comes from the fact that most of the results from numerical simulations of lattice QCD have been obtained in the so called "quenched approximation", i.e. by neglecting "sea quark" loops. This was found to have a small effect for the proton to ρ mass ratio, $\sim 10\%$ (in the quenched case one finds $M_\mathcal{N}/M_\rho = 1.2\pm0.1$), but it is expected to be important for the nucleon σ-term or the polarized structure functions. The systematic error induced by this approximation on the different quantities is not known a priori, although is some cases it can be roughly estimated [10,11]. Preliminary results obtained in the "unquenched" theory, though at values of the quark masses significantly larger than the physical ones (typically of the order of the strange quark mass) are already available. According to these studies, for most of the quantities the quenched and unquenched cases give very similar results (within 5–10%) upon a suitable shift of the renormalized strong coupling constant at a given scale.

Other systematic effects come from steps 1. (introduction of an ultra-violet cut-off) and 2. (introduction of an infrared cut-off), cf. sec. 2.. Another source of uncertainty is due to the renormalization procedure necessary to relate the matrix elements of lattice bare operators to the corresponding physical amplitudes. I discuss now these three points in turn.

- Discretization errors: by working with a finite lattice spacing a, corresponding to an ultra-violet cut-off $\sim a^{-1}$, we expect a distortion of the mass spectrum and an error on the value of the matrix elements of local operators of $O(Ma)$, where M is a mass-scale of order Λ_{QCD}. To give an example, for the (light-quark) vector and axial-vector currents, the systematic errors on the matrix elements have been measured and found as large as 25%.

  Numerical simulations are usually performed at relatively large lattice spacings, typically $a^{-1} \sim$ 2–4 GeV, and one should bear in mind that, in order to reduce the lattice spacing of a factor 2, one needs at least a factor of 32 in computer time and memory. A simple and theoretically attractive possibilty to reduce these effects is to apply the Symanzik improvement programme, where the $O(a)$ cut-off effects for on-shell quantities are cancelled by adding local counter-terms to the lattice action and to the composite operators of interest. This programme has been pursued by several groups in the recent past [12,13], using the so-called tree-level improved SW-Clover action [14]. It has been shown that the use of the SW-Clover action reduces sizeably discretization errors, e.g. from 25% to less than 5% in the matrix elements of the light quark currents [16]. A further important progress has recently been made by the group of Lüscher, which was able to extend the tree-level improvement proposed in ref. [14] to an all-order non-perturbative improvement thus reducing discretization errors to $O(M^2a^2)$ [15]. This is expected to be very effective in heavy quark physics, for which discretization errors of still about 15% were measured in spite of the use of a tree-level improved action.

- Finite-volume effects: in most of the present quenched calculations the error induced by the necessity of working on a finite volume is very small. Indeed, even though it is still not possible to work at values of the quark masses as small as the u and d-quark masses, the dependence of most of the quantities on m_q is very smooth. Thus one can work with quark masses as small as 50 MeV, with very small finite volume effects, and then extrapolate to the physical point with negligible errors. Preliminary studies seem to indicate that finite volume effects can be larger in the unquenched case. The values of the quark masses at which it is possible to work at present is, however, still rather large, because of computer-time limitations. Thus, also in this case, finite-size effects are smaller than other systematic errors, and not a real problem at the level of accuracy that can be presently reached.

- Renormalization of lattice operators: suitable renormalization conditions have to be imposed in order to obtain physical amplitudes from the matrix elements of the bare lattice operators $\hat{O}(a)$. Schematically we have

$$\langle \hat{O} \rangle_{\rm phys} = Z_{\hat{O}} \langle \hat{O}(a) \rangle \ . \tag{5}$$

We ought to know the renomalization constant (more in general the renormalization matrix) $Z_{\hat{O}}$ in order to make physical predictions (this is a necessary step in all theoretical approaches and not peculiar to lattice calculations). In principle, for finite or logarithmically divergent operators, one can compute $Z_{\hat{O}}$ in perturbation theory [b]. Lattice perturbation theory however behaves rather badly due to the presence of tadpole-like diagrams which are typical of the lattice case. The inaccuracy of lattice perturbation theory has also been demonstrated explicitely in several cases where a non-perturbative determination of the renormalization constants was possible. Several remedies to this problem have been proposed in the recent years: boosted perturbation theory [19], non-perturbative renormalization of the operators using the continuum Ward Identities [17,20], non-perturbative renormalization on quark and gluon states in a fixed gauge [21]. Boosted perturbation theory is in a sense the extension to lattice perturbation theory of the successful BLM proposal in the continuum [22]. The uncertainties of boosted perturbation theory are larger than those of the non-perturbative methods, and one finds the same problems as standard perturbation theory for power-divergent operators, where a non-perturbative approach is compulsory. For this reason, in order to eliminate the systematic error coming from the renormalization of lattice operators, it is certainly more convenient

[b] This is not possible instead in the case of power-divergent operators [5,17,18], such as those which appears when considering higher-twist effects, see below.

to use the Ward identity method[17,20] and the non-perturbative approach of ref. [21 c].

4. Structure functions on the lattice

In inclusive deep inelastic processes, the hadronic tensor $W^{\mu\nu}$ is computed from the forward matrix element of the T-product of two electro-magnetic (or weak) currents,

$$W^{\mu\nu} = \mathrm{Im}\left[2i \int d^4x\, e^{iq\cdot x}\, \langle \mathcal{N}|T\{J^\mu(x)\,,\, J^\nu(0)\}|\mathcal{N}\rangle\right] \,, \tag{6}$$

which can also be written as

$$\begin{aligned} W^{\mu\nu} &= \mathrm{Im}\Big[2i \sum_X \int dt\; e^{i(p^0_{\mathcal{N}}+q^0-E_X)t} \\ &\quad \times (2\pi)^3\, \delta^3(\vec{p}_{\mathcal{N}}+\vec{q}-\vec{p}_X)\langle \mathcal{N}|J^\mu(0)|X\rangle\langle X|J^\nu(0)|\mathcal{N}\rangle\Big] \\ &\sim \sum_X (2\pi)^4\delta^4(p_{\mathcal{N}}+q-p_X)\langle \mathcal{N}|J^\mu(0)|X\rangle\langle X|J^\nu(0)|\mathcal{N}\rangle \,. \end{aligned} \tag{7}$$

In equations (6) and (8), $W^{\mu\nu}$ is the standard hadronic tensor, defined in terms of two structure functions (three for neutrino DIS)

$$\begin{aligned} W^{\mu\nu} &= \left(-g^{\mu\nu} + \frac{q^\mu q^\nu}{q^2}\right) W_1(Q^2,x) \\ &+ \left(p^\mu_{\mathcal{N}} - \frac{(p_{\mathcal{N}}\cdot q)q^\mu}{q^2}\right)\left(p^\nu_{\mathcal{N}} - \frac{(p_{\mathcal{N}}\cdot q)q^\nu}{q^2}\right) W_2(Q^2,x)\,; \end{aligned}$$

q is the four momentum carried by the virtual photon; $Q^2 = -q^2$ and $x = Q^2/2(p_{\mathcal{N}}\cdot q)$; $p_{\mathcal{N}}$ and $p_X \equiv (E_X, \vec{p}_X)$ are the nucleon and final state hadron momenta respectively.

It is illusory to imagine that, even in the far future, when the lattice spacing a will become very small ($\Lambda_{QCD} \sim 1/M_{\mathcal{N}} \ll 1/|x| \ll 1/a$) and the computer resources will be powerfull enough, it will be possible to compute directly the T-product of eq. (6) on the lattice. The reason is that, by working in an Euclidean space-time, as unavoidable in numerical simulations, cf. sec.

---

[c] There is still some error in these procedures which is connected with the discretization. This is, however, common with boosted perturbation theory and I am assuming for this discussion that dicretization errors are negligible.

2., eq. (8) becomes

$$W^{\mu\nu} = \mathrm{Im}\Big[2i \sum_X \int dt \; e^{(p^0_{\mathcal{N}}+q^0-E_X)t} \times (2\pi)^3 \, \delta^3(\vec{p}_{\mathcal{N}}+\vec{q}-\vec{p}_X)\langle\mathcal{N}|J^\mu(0)|X\rangle\langle X|J^\nu(0)|\mathcal{N}\rangle\Big] . \quad (8)$$

Thus, at large time-distances, the integral over t is dominated by the states with the smallest energy E_X, and not by those which satisfy the energy conservation condition $E_X = p^0_{\mathcal{N}} + q^0$. This is similar to what happens in the case of exclusive decays, as discussed in ref.[8]. The same problem is also encountered in inclusive decays of heavy flavours on the lattice, or in other analytic approaches which study inclusive cross-sections or decay rates in the Euclidean [23]. The only possible solution to this problem is to model the energy dependence of the integrand in eq. (9), and then make an analytic continuation of the model to Minkowsky, analogously to what has been proposed in ref. [9] for exclusive decays. This will unavoidably introduce a certain model dependence in the final results.

The above problem can, fortunately, be avoided in the short-distance (large momentum transfer) limit, where we can expand the T-product of the two currents using the Wilson OPE. Schematically, we have

$$\lim_{|x|\to 0} T\{J(x), J(0)\} = \sum_i C_i\left(\frac{|x^2|}{\mu^2}, g(\mu)\right) O^i(0) \, , \quad (9)$$

where C_i are the Wilson coefficients, $O^i(0)$ local operators and μ is the scale at which the operators are renormalized.

The operators of the OPE are usually classified in terms of "twist = dimension – spin". The operator product expansion is an expansion in inverse powers of a large scale, higher and higher twists corresponding to higher and higher powers of Λ_{QCD}/Q. The matrix elements of the leading-twist operators are directly related to the moments of the parton distributions functions; the matrix elements of higher-twist operators correspond usually to multi-parton distributions [24].

In the unpolarized case, the leading-twist contribution to the moments of the structure functions can be written as ($W_{1,2}$ are related via trivial kinematical factors to $F_{1,2}$)

$$\begin{aligned} 2\int_0^1 dx \; x^{n-1} F_1(x,Q^2) &= \sum_i C^i_{1,n}\left(\mu^2/Q^2, g(\mu)\right) v^i_n(\mu) \\ \int_0^1 dx \; x^{n-1} F_2(x,Q^2) &= \sum_i C^i_{2,n}\left(\mu^2/Q^2, g(\mu)\right) v^i_n(\mu) \, , \end{aligned} \quad (10)$$

Observable	Value in the chiral limit
$\langle x\rangle^u$	$0.45 \pm 0.03 \pm 0.02$
$\langle x\rangle^d$	$0.19 \pm 0.01 \pm 0.02$
$\langle x^2\rangle^u$	0.10 ± 0.02
$\langle x^2\rangle^d$	0.04 ± 0.01
$\langle x^3\rangle^u$	0.02 ± 0.01
$\langle x^3\rangle^d$	-0.001 ± 0.007
Δu	0.83 ± 0.07
Δd	-0.24 ± 0.02
a_2^u	0.15 ± 0.03
a_2^d	-0.05 ± 0.01
d_2^u	-0.23 ± 0.02
d_2^d	0.040 ± 0.007

Table 1: Results for the polarized and unpolarized structure functions of the proton. All the number refer to operators renormalized at the scale $\mu = a^{-1} \sim 2$ GeV in the $\overline{MS}$ renormalization scheme.

where $C^i_{1,n}\left(\mu^2/Q^2, g(\mu)\right)$ and $C^i_{2,n}\left(\mu^2/Q^2, g(\mu)\right)$ are the Wilson coefficients and $v^i_n(\mu)$ the matrix elements of local quark and gluon operators renormalized at the scale μ. In lattice calculations, only the matrix elements of valence-quark operators have been considered to date. In this case we have

$$\frac{1}{2}\sum_{\vec{s}}\langle\vec{p},\vec{s}|\mathcal{O}^i_{\{\mu_1,\ldots,\mu_n\}}(\mu)|\vec{p},\vec{s}\rangle = 2\, v^i_n(\mu)\,[p_{\mu_1}\cdots p_{\mu_n} - \text{traces}]$$

$$\mathcal{O}^i_{\{\mu_1,\ldots,\mu_n\}} = \left(\tfrac{i}{2}\right)^{n-1} \bar{\psi}_i\gamma_{\mu_1}D_{\mu_2}\ldots D_{\mu_n}\psi - \text{traces}\ ,\quad (11)$$

with $i = u$ or d; $\{\ldots\}$ denotes symmetrization with respect to the Lorentz indices μ_k; $\vec{p}$ and $\vec{s}$ are the nucleon momentum and spin and the normalization of the states is given by $\langle\vec{p},\vec{s}|\vec{k},\vec{t}\rangle = (2\pi)^3\, 2E_{\vec{p}}\ \delta(\vec{p}-\vec{k})\ \delta_{\vec{s},\vec{t}}$.

The $v^i_n(\mu)$'s have a parton model interpretation

$$v^i_n(\mu) = \langle x^{n-1}\rangle^i\ , \quad (12)$$

where x is the fraction of the nucleon momentum carried by the valence quarks at the scale μ, i.e. $\langle x^{n-1}\rangle^u = \int_0^1 dx\ x^{n-1}\ u(x)$ and $\langle x^{n-1}\rangle^d = \int_0^1 dx\ x^{n-1}\ d(x)$.

In table 1, recent results obtained in ref. [25] are reported. They were obtained in the quenched approximation, by using the fermion Wilson action at $\beta = 6/g^2 = 6.0$, corresponding to an inverse lattice spacing $a^{-1} \sim 2$ GeV, on a $16^3 \times 32$ lattice volume. The quark operators were renormalized both perturbatively and non-perturbatively using the method of ref. [21]. The results for the non-perturbative renormalization constants were found to agree, within 20% with those obtained using boosted perturbation theory. Thus the main systematic errors are expected to come from discretization effects, which could be reduced by using an improved lattice fermion action, and from the quenched approximation. The size of the lattice used in ref. [25] is not very large (in physical units), compared to modern lattice calculations, and could be at the origin of a some further systematic effect. In the table the systematic errors on $\langle x\rangle^{u,d}$ are my interpretation of the different results for the average momentum of the quarks obtained by using different lattice operators to compute this quantity.

The main conclusions of the study of ref. [25] are the following:

- $\langle x^{n-1}\rangle^{u,d}$ shows a stronger dependence on the quark mass m_q for larger values of n. The lowest moment, $n=2$, is pratically independent of m_q.
- There is some indication that the d-quark has a softer distribution than the u-quark.
- Valence quarks carry significantly more momentum, $\langle x\rangle^u + \langle x\rangle^d \sim 0.64$ than what is found in phenomenological analyses of deep inelastic data[30].

  It is not surprising that valence quarks carry more momentum in the quenched approximation than in the unquenched case, since sea-quarks are absent in this limit. The size of the effect looks, however, too large, and may indicate the presence of other systematic errors in the lattice results. The calculation should be repeated by using an improved fermion lattice action, on a larger volume, in order to reduce the systematic errors other than that due to the quenched approximation. Of course, the unquenched calculation should also be attempted.

In the polarized case, let us consider the following momenta

$$\begin{aligned} 4\int_0^1 dx\, x^n g_1(x,Q^2) &= \sum_i E^i_{1,n}\left(\mu^2/Q^2, g(\mu)\right) a^i_n(\mu) \\ 4\int_0^1 dx\, x^n g_2(x,Q^2) &= \frac{n}{n+1}\sum_i \Big(E^i_{2,n}\left(\mu^2/Q^2, g(\mu)\right) d^i_n(\mu) \\ &\quad -E^i_{1,n}\left(\mu^2/Q^2, g(\mu)\right) a^i_n(\mu)\Big)\,, \end{aligned} \tag{13}$$

where

$$\langle \vec{p}, \vec{s}|O^{5,i}_{\{\sigma,\mu_1,\ldots,\mu_n\}}(\mu)|\vec{p}, \vec{s}\rangle = \frac{1}{n+1}\, a^i_n(\mu)\left[s_\sigma p_{\mu_1}\cdots p_{\mu_n} - \text{traces}\right]$$

$$\mathcal{O}^i_{\sigma,\mu_1,\ldots,\mu_n} = \left(\frac{i}{2}\right)^{n-1} \bar{\psi}_i \gamma_\sigma \gamma_5 D_{\mu_1} \ldots D_{\mu_n} \psi - \text{traces}, \quad (14)$$

and

$$\langle \vec{p}, \vec{s} | \mathcal{O}^{5,i}_{[\sigma,\{\mu_1],\ldots,\mu_n\}}(\mu) | \vec{p}, \vec{s} \rangle = \frac{1}{n+1} d^i_n(\mu) \Big[(s_\sigma p_{\mu_1} - s_{\mu_1} p_\sigma) p_{\mu_2} \cdots p_{\mu_n} + \ldots - \text{traces} \Big] . \quad (15)$$

[...] denotes untisymmetrization with respect to the Lorentz indices. Notice that for $n = 0$ one is simply computing the forward matrix element of the u- or d axial-vector currents on the proton. In the parton language this corresponds to

$$a^u_0 = 2\Delta u \quad a^d_0 = 2\Delta d , \quad (16)$$

where Δu and Δd determine the fraction of the nucleon spin carried by the u and d quarks respectively and the axial-vector form factor is given by $g_A = \Delta u - \Delta d$. The structure function g_2 consists of two contributions: a^i_n is the so-called Wandzura-Wilczek term [26], which corresponds to a twist-two operator, whereas d^i_n corresponds to a twist-three one. A further twist-three contribution, proportional to the quark mass, was not considered in ref. [25]. Results for these quantities are also given in table 1. From this table one obtains

$$g_A = \Delta u - \Delta d = 1.07(9) \quad (17)$$

in good agreement with previous lattice determinations of the same quantity, $g_A = 1.10(8)$ and $g_A = 1.18(7)$ from ref. [27], but lower than the experimental value $g_A = 1.26$.

In ref. [25], they also found

$$\int_0^1 dx\ x \left(g_1^p(x, Q^2) - g_1^n(x, Q^2) \right)_{Q^2 \simeq 4\text{GeV}} = 0.174(25) , \quad (18)$$

to be compared with the experimental value of ~ 0.163 [28] and

$$\int_0^1 dx\ x^2 \left(g_1^p(x, Q^2) - g_1^n(x, Q^2) \right)_{Q^2 \simeq 4\text{GeV}} = 0.016(5) . \quad (19)$$

The results of ref. [25] suggest for d_2 values much larger than recent estimates obtained by using QCD sum rules [29].

5. Power corrections

In this section, I discuss the problem of controlling power corrections in effective theories, by considering as an example the Bjorken sum rule[3]

$$\int_0^1 dx\,\left(g_1^p(x,Q^2)-g_1^n(x,Q^2)\right)=\frac{1}{6}C_2(\alpha_s(Q^2))\langle O_2^\mu\rangle-\frac{4}{27}C_4(\alpha_s(Q^2))\frac{\langle O_4^\mu\rangle}{Q^2}+..,\tag{20}$$

where the operator O_2^μ is the leading-twist operator given in eq. (14), with one Lorentz-index. O_2^μ corresponds to the axial-vector current, i.e. $\langle O_2^\mu\rangle = g_A$. The first higher-twist correction is given by the dimension 5 operator $O_4^\mu = \bar{u}g\tilde{G}^{\mu\sigma}\gamma_\sigma u - \bar{d}g\tilde{G}^{\mu\sigma}\gamma_\sigma d$. C_2 and C_4 are the Wilson coefficients entering in the OPE.

Equation (20) is typical of all effective theories. The matrix element of the leading-twist operator is known either because it has been measured from some other physical quantity, like in the present case, or because it is known a priori. For example, in e^+e^- annihilation into hadrons, the leading term is the matrix element of the identity operator on the vacuum and in the HQET it is the forward matrix element of the scalar density $h^\dagger h$, which is a conserved current, i.e. $\langle H|h^\dagger h|H\rangle = 1$. Let us assume also that α_s is known because it has been measured in some other process (alternatively α_s is the quantity that we want to extract from the Bjorken sum rule). Thus, in order to predict the l.h.s. of eq. (20), we need a non-perturbative determination of the matrix element of O_4^μ, and a perturbative calculation of the Wilson coefficients C_2 and C_4. This is not the end of the story, however. Indeed, if we use dimensional regularization, the perturbative series of both the Wilson coefficients, and of the matrix elements of higher twist operators, are divergent and moreover not Borel summable, due to the presence of renormalon singularities[4]. For this reason, the "sum" of the perturbative series of C_2

$$\begin{aligned}C_2(\alpha_s(Q^2)) &= 1-\frac{\alpha_s(Q^2)}{\pi}-3.58\left(\frac{\alpha_s(Q^2)}{\pi}\right)^2\\ &- 20.21\left(\frac{\alpha_s(Q^2)}{\pi}\right)^3+\ \dots\ +C_2^{n+1}\left(\frac{\alpha_s(Q^2)}{\pi}\right)^{n+1},\end{aligned}\tag{21}$$

$$C_2^{n+1}\sim n!n^\gamma\left(\frac{\beta_0}{4\pi}\right)^n$$

is not unique, the ambiguity being of $O(1/Q^2)$, or higher. This ambiguity is cancelled by the corresponding one in the matrix element of O_4^μ. This implies that renormalization schemes based on dimensional regularization do not define higher-twist operators unambigously, and alternative definitions must be used.

A possibility is to introduce a hard ultra-violet cut-off Λ, as for example by computing the matrix elements of the operators on the lattice [d]. In this case we have $\Lambda \sim a^{-1}$. It is natural then to expand the Bjorken sum rule in terms of the bare operators $O_2^\mu(\Lambda)$ and $O_4^\mu(\Lambda)$.

With a hard cut-off the operator O_4^μ mixes with O_2^μ, with a mixing coefficient which diverges as Λ^2

$$\langle O_4^\mu(\Lambda)\rangle = D_2(\alpha_s(\Lambda))\Lambda^2\langle O_2^\mu\rangle_0 + D_4(\alpha_s(\Lambda))\langle O_4^\mu\rangle_0\,, \tag{22}$$

where $\langle O_{2,4}^\mu\rangle_0$ are the three-level matrix elements. Thus we have

$$\begin{aligned}\int_0^1 dx\left(g_1^p(x,Q^2) - g_1^n(x,Q^2)\right) &= \frac{1}{6}C_2(\alpha_s(Q^2))\langle O_2^\mu(\Lambda)\rangle \\ &- \frac{4}{27}C_4(\alpha_s(Q^2))\frac{\langle O_4^\mu(\Lambda)\rangle}{Q^2} + \dots\,,\end{aligned} \tag{23}$$

where C_2 has the form [e]

$$C_2(\alpha_s(Q^2)) = c_2(\alpha_s(Q^2)) + \tilde{c}_2(\alpha_s(Q^2))\frac{\Lambda^2}{Q^2}\,, \tag{24}$$

where $c_2(\alpha_s(Q^2)) = \sum_k c_2^k \alpha_s(Q^2)^k$ and similarly for $\tilde{c}_2$. One can show that the renormalon singularities disappear by using a hard cut-off [5,31], and that the perturbative series of the Wilson coefficients, as well as those of the operator matrix elements, are well behaved and free of renormalon ambiguities. The price to pay is that the matrix element of O_4^μ is quadratically divergent, its divergence beying cancelled by the corresponding term, $\tilde{c}_2$ in eq. (24), in the Wilson coefficient of the leading twist operator.

Being divergent as $\Lambda^2 \sim 1/a^2$, the matrix element of O_4^μ cannot be considered as the "physical" power correction to the Bjorken sum rule, because this is expected to be of order Λ_{QCD}^2. On the other hand, if one insists in defining a finite operator by subtracting its divergent part computed in perturbation theory, $O_4^\mu(\Lambda) \to O_4^\mu(\Lambda) - D_2(\alpha_s(\Lambda))\Lambda^2 O_2^\mu(\Lambda)$, then the renormalon ambiguity is reintroduced [5]. As discussed in ref. [5], one can envisage a non-perturbative procedure which allows to define a subtracted operator $\bar{O}_4^\mu$ which is finite and free of renormalon ambiguities. This, however, does not change the main problem.

---

[d] A different possibility which, however, leads to the same conclusions, is discussed in ref. [5].

[e] For simlicity, I omit $(\Lambda/Q)^2$ as argument of the Wilson coefficients.

As mentioned above, the perturbative series for C_2 is free of ambiguities of $O(1/Q^2)$. However, this arises as a cancellation of the renormalon singularities in the two series with coefficients c_2^k and $\tilde{c}_2^k$ as k becomes large. Each of these two series has a renormalon ambiguity of $O(1/Q^2)$. The cancellation of this ambiguity occurs between contributions which, in each order of perturbation theory, are of different order in $1/Q$, cfr. eq. (24). The low-order terms in the two series can be very different from each other, but the high order behaviour is governed by the same renormalon singularity and is the same. In order to predict the Bjorken sum rule, we have to control the two series up to a precision of better than $O(\Lambda^2_{QCD}/Q^2)$. A related practical problem is the cancellation of the power divergences of $O(\Lambda^2)$ in the coefficient function C_2 with those in the matrix element $\langle O^\mu \rangle$ in eq. (23). C_2 has to be computed to sufficiently high order so that this cancellation also occurs with a precision of better than $O(\Lambda^2_{QCD}/Q^2)$. Notice that in most of the cases, only the first order coefficient, corresponding to $\tilde{c}_2^1$ is known.

In ref. [5] several examples of power corrections in effective theories were studied, and the question of the precision that can be achieved was addressed. The conclusion of this analysis was that in all the example considered, it is not possible to get results with sufficient accuracy unless one is able to control perturbation theory at higher orders than those available at present. I believe that this also holds true in the case of the Bjorken sum rule.

6. Conclusion and outlook

In this talk, I have tried to present the main features of the lattice approach to the calculation of physical hadronic quantities. In order to be specific, I have focussed the discussion on DIS processes, including power corrections, by using the example of the Bjorken sum rule.

I have emphasized mostly the problems and limitations of the method, rather than its successes, to indicate those directions where future efforts shold be focused in order to reduce the present systematic errors and uncertainties.

To avoid a pessimistic impression, I want to conclude by listing some of the physical quantities which have been predicted on the lattice with "good" accuracy. For "good" accuracy, I mean that either they have been confirmed by experiments, or they agree with results obtained by different methods, or (but this is weaker) that we believe that the systematic errors are sufficiently under control. These quantities include

- Meson decay constants: f_{D_s} has been predicted first in 1988 [32], long before its first experimental measurement; its theoretical value has changed very little in the years, and the present result from the lattice, $f_{D_s} = 235 \pm 15$ MeV [33] is in striking agreement with the experimental world average $f_{D_s} = 241 \pm 20 \pm 30$ MeV [34]. Also f_D, f_B and f_{B_s} have been predicted [33–36].

- Semileptonic form factors of heavy mesons: there is a good agreement between lattice calculations and experimental measurements of the form factors in $D \to K^*$ and K semileptonic decays (f^+, V, A_1 and A_2) [33]. Similar results, but with large uncertainties due to the extrapolation in the heavy quark mass and momentum transfer, exist also for $B \to \pi$ and ρ semileptonic decays [33,35]. They have been used in the experimental analyses in order to extract $|V_{ub}|$, with results consistent with the data and with other theoretical estimates [36].

- $\Delta F = 2$ amplitudes: the quenched lattice calculation of the kaon B-parameter B_K relevant for K^0–$\bar{K}^0$ transistions has already reached an accuracy of about 5% and results for this quantity also exist in the unquenched case [37]. There are also several results for the B parameters relevant for $B^0_{d,s}$–$\bar{B}^0_{d,s}$ mixings [35]. These results have been used in several theoretical analyses of CP violation in the standard model and beyond [38].

- Other quantities: several other quantities important to particles physics have also been computed more or less accurately on the lattice. Among these quantities, let me recall the strong coupling constant α_s, the critical temperature of the deconfining and chiral phase transitions, the masses of the strange, charmed and bottom quarks, the nucleon σ-term, the glueball spectrum and many others.

In conclusion, the accuracy of the results obtained for hadronic quantities in lattice QCD is improving constantly in time, many of these results have had an important phenomenological impact, and, in many cases, the systematic errors have already been reduced to an acceptable level. In some other cases, however, much hard work is needed to fully control these errors.

References

1. For a review see M. Neubert, Phys. Rep. 245 (1994) 259.
2. D.J. Gross and C.H. Llewellyn Smith, Nucl. Phys. B14 (1969) 337.
3. J. Bjorken, Phys. Rev. 148 (1966) 1467.
4. See for example A.H.Mueller, Nucl. Phys. B 250 (1985) 327; Phys. Lett. B308 (1993) 355; Proceedings of the Workshop “QCD: 20 years Later”,

Aachen, June 1992 eds. P.M.Zerwas and H.A.Kastrup (World Scientific, Singapore, 1993) p. 162.

5. G. Martinelli and C.T. Sachrajda, Phys. Lett. B354 (1995) 423, to appear in Nucl. Phys.B. and CERN-TH/96-117 hep-ph/9605336.
6. For a general discussion of numerical simulations of quantum field theories, see for example G. Parisi, "Field theory, disorder and simulations", World Scientific, Singapore (1992) 503 p.
7. More details on the lattice approach can be found in ref.[6] or in G. Martinelli, in Cargese 1989, Particle Physics, p. 239.
8. L. Maiani and M. Testa, Phys. Lett. B245 (1990) 585.
9. M. Ciuchini, E. Franco, G. Martinelli and L. Silvestrini, Phys. Lett. B380 (1996) 353.
10. J.N. Labrenz and S.R. Sharpe, Nucl. Phys. B34 (Proc. Suppl.) (1994) 335 and UW-PT-96-07 hep-lat/9605034; S.R. Sharpe and Y. Zhang, Phys. Rev. D53 (1996) 5125.
11. C.W. Bernard and M.F.L. Golterman, Nucl. Phys. B30 (Proc.Suppl.) (1993) 217, Phys. Rev. D49 (1994) 486 and Nucl. Phys. B34 (Proc. Suppl.) (1994) 331; M.F.L. Golterman, WASH-U-HEP-94-62 hep-lat/9405002 and Acta Phys. Polon. B25 (1994) 1731.
12. G. Heatlie et al., Nucl. Phys. B352 (1991) 266; APE Collaboration, Phys. Lett. B345 (1995) 513; Nucl. Phys. B437 (1995) 641; Phys. Lett. B326 (1994) 295; Nucl. Phys. B413 (1994) 461.
13. UKQCD Collaboration, Phys. Lett. B284 (1992) 377; Nucl. Phys. B407 (1993) 331; Phys. Rev. D49 (1994) 474 and 1594; Phys. Rev. D51 (1995) 4905; Phys. Rev. D52 (1995) 5067; Phys. Rev. D54 (1996) 3619 and 3526.
14. B. Sheikoleslami and R. Wohlert, Nucl. Phys. B259 (1985) 572.
15. Karl Jansen et al., Phys. Lett. B372 (1996) 275; M. Luscher, S. Sint, R. Sommer and P. Weisz, DESY-96-086, hep-let/9605038; M. Luscher et al., DESY-96-157, hep-lat/9608049.
16. G. Martinelli, C.T. Sachrajda and A. Vladikas, Nucl. Phys. B358 (1991) 212.
17. M. Bochicchio et al., Nucl. Phys. B262 (1985) 331.
18. L. Maiani, G. Martinelli, C.T. Sachrajda, Nucl. Phys. B368 (1992) 281.
19. P. Lepage and P.B. Mackenzie, Phys. Rev. D48 (1993) 2250.
20. L. Maiani and G. Martinelli, Phys. Lett. B178 (1986) 265; G. Martinelli, S. Petrarca, C.T. Sachrajda and A. Vladikas, Phys. Lett. B311 (1993) 241; D.S. Henty, R.D. Kenway, B.J. Pendleton and J.I. Skullerud, Phys. Rev. D51 (1995) 5323.
21. G. Martinelli, C. Pittori, C.T. Sachrajda, M. Testa and A. Vladikas,

Nucl. Phys. B445 (1995) 81.
22. S. Brodsky, P. Lepage and P.B. Mackenzie, Phys. Rev. D28 (1983) 228.
23. B Blok and I. Halperin, TECHNION-PH-96-14, hep-ph/9605441.
24. R.K. Ellis, W. Furmanski and R. Petronzio, Nucl. Phys. B207 (1982) 1, Nucl. Phys. B212 (1983) 29 and refs. therein.
25. M. Gockeler et al., (Nucl. Phys. B42 (Proc. Suppl.) (1995) 337; Nucl. Phys. B472 (1996) 309; Phys. Rev. D53 (1996) 2317; DESY-96-151, hep-lat/9608046.
26. S. Wandzura and F. Wilczek, Phys. Lett. B72 (1977) 195.
27. K.F. Liu et al., Phys. Rev. D49 (1994) 4755.
28. J. Ellis and M. Karliner, Phys. Lett. B341 (1995) 397; D. Adams et al., Phys. Lett. B329 (1994) 399; K. Abe et al., Phys. Rev. Lett. 74 (1995) 346.
29. I.I. Balitsky, V.M. Braun and A.V. Kolesnichenko, Phys. Lett. B242 (1990) 245; E. Stein et al., Phys. Lett. B343 (1995) 369.
30. see for example A.D. Martin, W.J.Stirling and R.G. Roberts, Phys. Rev. D47 (1993) 867; for a recent upgrade of the parton distribution functions see also Phys. Lett. B381 (1996) 353 and refs. therein.
31. I.I.Bigi, M.A.Shifman, N.G.Uraltsev and A.I.Vainshtein, Phys. Rev.D 50 (1994) 2234,
32. M.B. Gavela, L. Maiani, G. Martinelli and O. Pène, Phys. Lett. B206 (1988) 113 and L. Maiani, Nucl. Phys. B9 (Proc. Suppl.) (1989) 193; C. Bernard, T. Draper, G. Hockney and A. Soni, Phys. Rev. D38 (1988) 3540.
33. G. Martinelli, talk given at 6th International Symposium on Heavy Flavor Physics, Pisa, Italy, 6-9 Jun 1995, to appear in the Proceedings.
34. J. Richman, rapporteur talk given at the XXVIII International Conference on High Energy Physics, June 26-30 1996, Warsaw (Poland), to appear in the Proceedings.
35. J. Flynn, rapporteur talk given at Latt96, 14th International Symposium on Lattice Field Theory, St. Louis, 4-8 June 1996, to appear in the Proceedings.
36. L. Gibbons, rapporteur talk given at the XXVIII International Conference on High Energy Physics, June 26-30 1996, Warsaw (Poland), to appear in the Proceedings; see also the rapporteur talk by J. Flynn at the same conference.
37. S. Sharpe, rapporteur talk given at Latt96, 14th International Symposium on Lattice Field Theory, St. Louis, 4-8 June 1996, to appear in the Proceedings.
38. see e.g. E. Franco et al., Z. Phys. C68 (1995) 239 and refs. therein.

STRANGENESS, CHARM AND BEAUTY IN HADRONS

E.A. CHUDAKOV
University of Heidelberg, Philosophenweg 12 D-6900 Heidelberg, Germany

The talk is dedicated to spectroscopy of particles with heavy flavours. A general overview of the current picture is given and several recent experimental results in this field are discussed.

1 Introduction

A quark is called heavy if its mass is much larger than the scale of QCD $\Lambda_{QCD}\approx$ 200 MeV. Top ($\approx$ 170 GeV), beauty ($\approx$ 4.5 GeV) and to a certain extent charm ($\approx$ 1.5 GeV) may thus be considered as heavy. Top is so heavy that it decays before it can hadronise and therefore there is no top spectroscopy to study. For charm and beauty the quark model predicts rich spectroscopy. Since the size of a heavy quark is small in comparison with the QCD scale $1/\Lambda_{QCD}$ it looks point-like for the "light" quarks of the same hadron. Therefore the properties of two particles which are different only by the flavours of their heavy quarks but have the same structure of light quarks (say $\Lambda_c$and $\Lambda_b$which are different only by the heavy quarks c or b) should be similar (except for their masses, of course). This flavour independence may be compared with the isotopic independence of chemical properties of atoms.
Spin-dependent interactions play an important role in hadron spectroscopy adding a mass term for the interaction of two quarks of $\Delta M_{1,2} \propto \frac{<\mathbf{S_1}\cdot\mathbf{S_2}>}{m_1 \cdot m_2}$. In the light quark sector spin-spin interaction provides mass splittings between multiplettes, for example about 300 MeV between proton and Δ. For a very heavy quark its spin interaction vanishes. Therefore the mass of a heavy hadron does not depend on spin of its heavy quark. One may conclude that a new symmetry appears for particles with heavy quarks, namely on its *flavour* and on its *spin*, called Heavy Quark Symmetry (HQS). This symmetry may simplify both spectroscopy of heavy particles and their decay properties.

2 Calculations of Masses

2.1 Early Phenomenology

The early phenomenological approach to particle masses using the "naive" quark model and the concept of colour happened to be surprisingly successful. The mass of a hadron is described as a sum of the constituent masses of all valence quarks and a sum of spin-spin interaction terms for all quark pairs times a factor of 1 for mesons of 1/2 for baryons coming from colour SU_3:

$$M = \sum_i m_i + a_{SS} \sum_{j>i} \frac{<\mathbf{S}_i \mathbf{S}_j>}{m_i m_j} \times \begin{cases} 1 & \text{mesons} \\ \frac{1}{2} & \text{baryons} \end{cases} \tag{1}$$

The colourmagnetic spin-spin interaction term is introduced by analogy with atomic physics and is also called "hyperfine". The free parameters are the constituent masses and the factor for the spin-spin interaction. There are more particles than parameters and using the measured masses of pseudoscaler and vector mesons one obtains the following set of parameters: $m_u = m_d \approx$ 310 MeV, $m_s \approx$ 480 MeV, $a_{SS} \approx$ 160 MeV, for $\frac{1}{2}^+$ and $\frac{3}{2}^+$ baryons the results are slightly different. One can calculate the masses of ground state hadrons with an accuracy of better than 20 MeV. One should also point out that the same constituent quark masses described successfully the magnetic momenta of baryons. This simple ansatz however does not describe excited states and is not theoretically satisfying.

2.2 Potential Models

Still, one can not calculate rigorously the masses of hadrons within QCD, apart from lattice calculations. Many potential models[1,2,3,4] were developed which use the quark model with incorporated "correct" QCD elements, a phenomenological approach to confinement motivated by QCD, and relativistic effects included. The hamiltonian used consists of: free quark hamiltonian, one gluon exchange and confinement terms ($\alpha_s/r - br - c$), spin-spin "hyperfine" term, spin-orbital term, and annihilation term (for isoscalers). Some problems have to be specially treated, like $\alpha_s(Q^2)$ behaviour at $Q^2 \to 0$ and relativistic corrections.

In one of the models [2] mesons and baryons up to high orbital excitations were treated with the same set of parameters. The estimated sources of errors are relativistic corrections and possible interaction with final states. The expected accuracy is about 10 MeV/c^2, while the observed accuracy is about 20 MeV/c^2 (apart from $\Lambda(1405)S_{01}$). Potential models have a solid theoretical basis in comparison with the naive model and provide a high prediction power for excited hadrons.

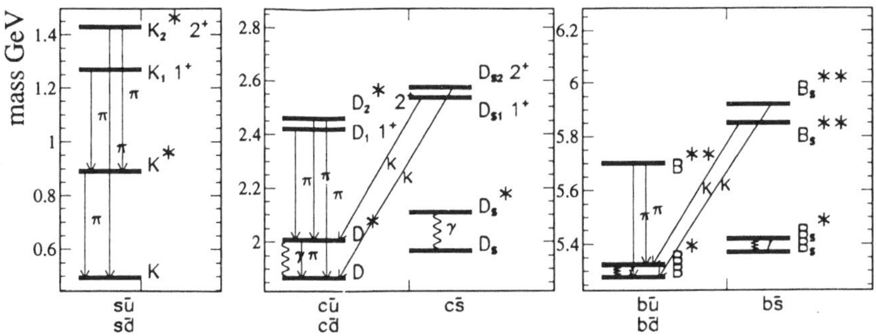

Figure 1: Masses and dominant decays of charm and beauty mesons. Strange mesons are shown for comparison at the left plot. The vertical scales are identical with the exception of an offset.

3 Overview of the Data

3.1 Mesons

Experimentally established charm and beauty mesons are presented on Fig. 1. All ground states for pseudoscaler and vector mesons have been found. The splittings between pseudoscaler and vector mesons become smaller for heavier quarks, as expected from Eq. 1. Orbitally excited charm and beauty mesons also have been observed. HQS [5] predicts a certain pattern for the masses and decay properties of the mesons with L=1, which can exist as 0^+,1^+ and 2^+ states. Charmed mesons can decay: $D^{**}(J^P) \rightarrow D(0^-)\pi(0^-)$, $D(1^-)\pi(0^-)$. The spin of the heavy quark is decoupled from the spins and orbital momenta of the rest. Therefore in the decays the spin of the heavy quark (S_Q) and the full spin of the light quarks (S_l) are conserved separately.

Spin-parity conservation (Table 1) allows decays of $S_l = \frac{1}{2}$ mesons via S-

J^P	S_l	decay	Waves allowed		
			J^P	S_l	HQS
0^+	$\frac{1}{2}$	$D\pi$	S	$0+\frac{1}{2}$	S
1^+	$\frac{1}{2}$	$D^*\pi$	S	$0+\frac{1}{2}$	S
			D	$2+\frac{1}{2}$	×
1^+	$\frac{3}{2}$	$D^*\pi$	S	$0+\frac{1}{2}$	×
			D	$2+\frac{1}{2}$	D
2^+	$\frac{3}{2}$	$D\pi$	D	$2+\frac{1}{2}$	D
		$D^*\pi$	D	$2+\frac{1}{2}$	D

decay	Γ estim	Γ meas
D^{*0}		<2 MeV
D^{*+}		<0.13 MeV
D_1	15. MeV	20 MeV
D_2^*	35. MeV	20 MeV
D_{s1}	5. MeV	<2.3 MeV
D_{s2}^*	10. MeV	16 MeV
B_1	40. MeV	<80 MeV
B_{s2}	5. MeV	<20 MeV

Table 1: Left: Selection rules for decays of L=1 D-mesons. Right: Measured and predicted decay widths of excited mesons.

waves while $S_l = \frac{3}{2}$ 2^+ mesons can decay only via D-waves. $S_l = \frac{3}{2}$ 1^+ meson can decay both via S and D-waves. HQS imposes an additional selection rule forbidding its decay via S-wave. Therefore both $S_l = \frac{3}{2}$ mesons should have close masses and be relatively narrow, while $S_l = \frac{1}{2}$ mesons should be wide.

The narrow charmed mesons $D_1(2420)^0$ and $D_2^*(2460)^0$ have been found and their quantum numbers identified[6]. In the beauty sector the narrow states have been seen but could not yet been resolved. The wide ones have not been found yet. Charm–strange and beauty–strange mesons should have a similar pattern (fig. 1)

3.2 Baryons

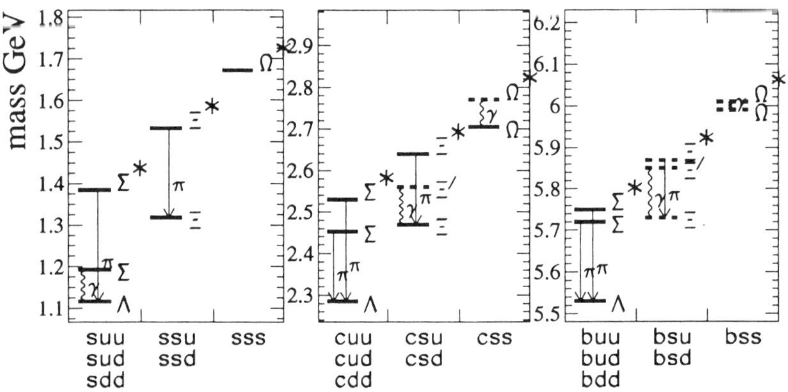

Figure 2: Masses and dominant decays of $\frac{1}{2}^+$ and $\frac{3}{2}^+$ charm and beauty baryons. Strange baryons are shown for comparison. $\frac{3}{2}^+$ states are denoted by asterisks.

Charm and beauty baryons are presented on Fig. 2. The states still not found experimentally are denoted by dashed lines. If the mass is measured its value is used, otherwise the mass of the closest observed particle is taken and the predicted mass splitting is added. The left column present Λ - Σ type particles. The Λ - Σ splitting increases with the mass of the heavy quark while the Σ - Σ^* splitting decreases, since Σ and Σ^* are different by the spin of the heavier quark and for very heavy quarks these states should degenerate. Indeed, in the beauty sector they are very close and about 200 MeV higher than the Λ state, as it was predicted by calculations. Starting with charm, Σ-like states have enough mass to decay strongly. Ξ $\frac{1}{2}^+$ states split for charmed and beauty particles, in the same way as Λ and Σ do. For charm the splitting is expected to be about 100 MeV, so the higher state should decay radiatively. The $\frac{3}{2}^+$

state decays strongly to the ground one. In the beauty sector the $\frac{1}{2}^+$ higher and $\frac{3}{2}^+$ states should converge to a mass about the pion decay threshold, so it remains to be seen if they decay strongly or radiatively. And finally, the splitting between $\frac{1}{2}^+$ and $\frac{3}{2}^+$ Ω - type states are expected to be about 60 MeV for charm and 20 MeV for beauty, so the $\frac{3}{2}^+$ states should decay radiatively. Two orbitally excited charmed baryons have been found, identified as $\Lambda_c(2593)\frac{1}{2}^-$ and $\Lambda_c(2625)\frac{3}{2}^-$, similar to the hyperons $\Lambda(1405)S_{01}$ and $\Lambda(1520)D_{03}$.

4 Recent Experimental Results

5 Experimental Methods

The experimental signatures of heavy particles include the observation of peaks in appropriate mass spectra of decay products, detection of a separation between the production and decay points and the detection of special decay products, like leptons from semileptonic decays. Obtaining of large samples of fully reconstructed particles is not easy since the production cross–sections of charm is low and a typical hadronic decay mode has a probability of a few percent. For beauty the branching ratios are lower and beauty decays predominantly to charm which also has to be detected.
Heavy quarks can be produced in all types of interactions. In hadronic interactions at $\sqrt{s} \sim 30$ GeV (fixed target at CERN and FNAL) charm is produced in about 10^{-3} of interactions. The highest statistics obtained is about 10^5 of fully reconstructed charmed particles (E791 at FNAL). At $\sqrt{s} \sim 1800$ GeV (FNAL) beauty is produced at the same rate of 10^{-3}. CDF has reconstructed about 10^3 beauty particles. In photoproduction at $E_\gamma \sim 200$ GeV charm is produced at a rate of 10^{-2}. E687 at FNAL has reconstructed about 10^5 charmed particles. In neutrino interactions charm is produced at a rate of about 10^{-1} providing small (~ 100) but clean samples used to study excited states. And finally, e^+e^- interactions provide a clean and abundant source of both beauty and charm with statistics of about 10^5 of fully reconstructed charmed particles at CLEO and about 10^3 beauty particles at LEP.
For the discussion of the recent experimental contributions to heavy quark spectroscopy several results have been selected which illustrate different experimental methods applied to different types of heavy particles.

5.1 Ground States of Charm

Until recently Ω_c^0 (css) has been the least known ground state charmed particle, rated only two PDG-stars [8]. It decays weakly predominantly to S=-3 final states and is difficult to find because of low production rates and its very short lifetime. Two new measurements appeared recently, both from fixed target experiments. The first measurement was done by E687 [9] photoproduction experiment at FNAL running at $E_\gamma \sim 200$ GeV. Two decay channels were observed (Table 2 and Figure 3). The other experiment - WA89 [10] at CERN used a Σ^- beam at 340 GeV/c. Six decay modes were observed (Table 2). Both experiments measured the lifetime of Ω_c^0.

exper.	decay	events	mass MeV/c^2
E687	$\Omega^-\pi^+$	10/4	2706±4
	$\Sigma^+K^-K^-\pi^-$	42/30	2700±3
WA89	$\Omega^-\pi^+$	20/10	2723±7
	$\Omega^-\pi^+\pi^-\pi^-$	20/10	2714±7
	$\Xi^-K^-\pi^+\pi^+$	60/80	2707±5
	$\Xi^-K_S^0\pi^+$	26/28	2695±4
	$\Lambda K^-K^-\pi^+\pi^+$	30/31	2704±4
	$\Lambda K^-K_S^0\pi^+$	40/34	2705±4
			2704±2.5

Table 2. New observations of Ω_c^0

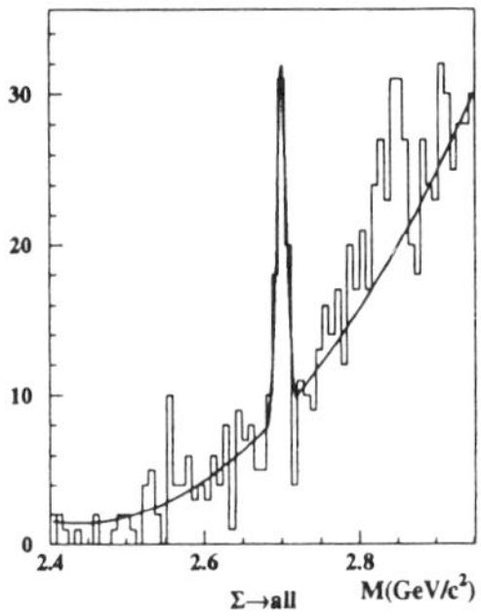

Figure 3. $\Omega_c^0 \to \Sigma^+K^-K^-\pi^-$, E687

5.2 *Excited States of Charm*

CLEO has accumulated a luminosity about 5 pb^{-1} at $\Upsilon(4S)$ (10.6 GeV) and off-resonance. These data were used for the search for $\Xi_c^*(\frac{3}{2}^+)$ decays to $\Xi_c\pi$. Using different decay modes of the daughter particles they selected about 300 candidates of Ξ_c^0 and about 300 of Ξ_c^+. In the distributions for $\Delta M = M(\Xi_c\pi) - M(\Xi_c)$ they observe narrow peaks for both charge combinations, interpreted as the first observation of Ξ_c^* [11]:

mode	ΔM MeV/c^2	events	Γ MeV/c^2
$M(\Xi_c^+\pi^-) - M(\Xi_c^+)$	178.2±0.5±1.0	55±12	<5.5 90% CL
$M(\Xi_c^0\pi^+) - M(\Xi_c^0)$	174.3±0.5±1.0	34±10	<3.1 90% CL

5.3 *Ground States of Beauty*

Masses of B_s^0 and Λ_b^0 have been recently re-measured at LEP [12,13,14] and CDF [15]. LEP experiments observed 3-5 fully reconstructed particles in $D_s\pi$ or $\Lambda_c^+\pi$ decay modes, based on statistics of about 3 10^6 of Z^0 hadronic decays per experiment.

	Λ_b^0	B_s^0
PDG-94	5641 ± 50 MeV/c^2	5375 ± 6 MeV/c^2
	(ISR,UA1)	(DELPHI,CDF,ALEPH)
New	$5668 \pm 16 \pm 8$ MeV/c^2 (DELPHI)	$5374 \pm 16 \pm 2$ MeV/c^2 (DELPHI)
	$5614 \pm 21 \pm 4$ MeV/c^2 (ALEPH)	$5370 \pm 2.3 \pm 1.3$ MeV/c^2 (CDF)
Mean	5645 ± 13 MeV/c^2	5372 ± 2.4 MeV/c^2

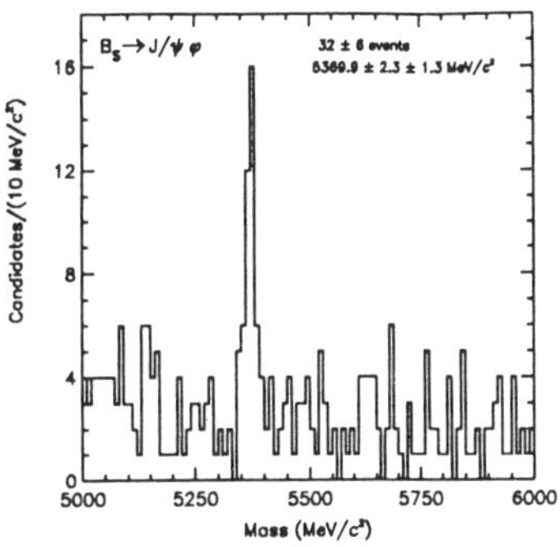

Figure 4: B_s^0 from CDF.

CDF has accumulated an integrated luminosity of 100 pb^{-1}, though only about 20% were used so far for the B_s^0 search. In order to reduce the background they looked for a relatively rare decay mode $B_s^0 \rightarrow J/\psi + \phi$, $J/\psi \rightarrow \mu^+\mu^-$ and applied strong cuts on the P_T of secondary particles ($>$ 2 GeV/c). CDF observes 32±6 B_s^0 events (Figure 4), improving the error on the mass, compared to the PDG-94 value.

5.4 Excited States of Beauty

Recent measurements of excited beauty states were performed at LEP[16,17,18,19]. 22% of all Z^0 hadronic decays go to beauty. Instead of looking for "exclusive decays" (fully reconstructed ones) LEP experiments look for "inclusive decays", tagging observed jets as beauty decay candidates. Different tagging techniques are used by the four LEP experiments.

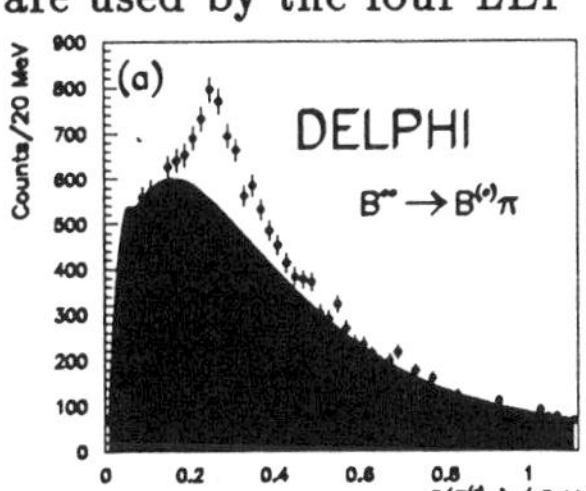

Figure 5: B** from DELPHI.

About 10^5 candidates per experiments can be selected, with purity better than 95%. Their momentum is measured with accuracy of about 10%. Mass spectra $Q{=}M(b-jet+\pi){-}M(b-jet){-}M(\pi)$ (Figure 5) show peaks at $Q{\approx}285$ MeV/c^2 with a width of $\sigma(Q){\approx}$ 60 MeV/c^2, interpreted as unresolved combination of orbitally exited B-mesons. Using the same method B_s^{**}, Σ_b and Σ_b^* were found.

6 Conclusion

In the last 3 years new experimental data were obtained on spectroscopy of particles with heavy flavours. In the charm sector one particle was found (Ξ_c), two other confirmed (Ω_c^0, Λ_c^*) and quantum numbers of several excited states were determined. One $\frac{1}{2}^+$ and two $\frac{3}{2}^+$ baryons are still missing. In the beauty sector a number of new states have been found (B^{**}, B_s^{**}, Σ_b and Σ_b^*). All beauty–strange baryons are still missing. Spectroscopy of particles with heavy quarks looks well described by models, to an accuracy of about 20 MeV/c^2. In the next 3 years one may expect new results in the charm sector coming from new fixed target experiments at FNAL (E781 and E831) which are going to reach statistics of 10^6 reconstructed particles.

References

1. A. De Rujula, H. Georgi and S. L. Glashow, *Phys. Rev.* D **12**, 147 (1975)
2. K. Maltman and N. Isgur, *Phys. Rev.* D **22**, 1701 (1980); S. Capstick and N. Isgur, *Phys. Rev.* D **34**, 2809 (1986); N. Isgur and M. Wise, *Phys. Lett.* B **232**, 113 (1989); N. Isgur and M. Wise, *Phys. Lett.* B **237**, 527 (1990); N. Isgur and M. Wise, *Nucl. Phys.* B **348**, 276 (1991);
3. W-Y. P. Hwang and D. P. Lichtenberg, *Phys. Rev.* D **35**, 3526 (1987);
4. C. Itoh et al., *Phys. Rev.* D **40**, 3660 (1989);
5. N. Isgur and M. Wise, *Phys. Rev. Lett.* **66**, 1130 (1991);
6. ARGUS collab., H. Albrech at al., *Phys. Lett.* B **232**, 398 (1989); CLEO collab., P. Avery et al., *Phys. Lett.* B **331**, 236 (1994); E687 collab., P.L. Frabetti et al., *Phys. Rev. Lett.* **72**, 324 (1994);
7. ARGUS collab., H. Albrech at al., *Phys. Lett.* B **317**, 227 (1993); CLEO collab., P. Avery et al., *Phys. Rev. Lett.* **74**, 3331 (1995); E687 collab., P.L. Frabetti et al., *Phys. Rev. Lett.* **72**, 961 (1994);
8. Particle Data Group, L. Montanet et al., *Phys. Rev.* D **50**, 1 (1994);
9. E687 collab., P.L. Frabetti et al., *Phys. Lett.* B **300**, 190 (1993); *Phys. Lett.* B **338**, 106 (1994);
10. WA89 collab., M.I. Adamovich et al., *Phys. Lett.* B **358**, 151 (1995);
11. CLEO collab., L. Gibbons et al., CLEO 96-4 (1996);S. Timm, contribution to PANIC96.
12. DELPHI collab., P. Abreu et al., *Phys. Lett.* B **324**, 500 (1994);
13. DELPHI collab., P. Abreu et al., *Phys. Lett.* B **374**, 351 (1996);
14. ALEPH collab., P. Spagnolo et al., CERN PPE/96-028 (1996), sumb. to Physics Letters B.
15. CDF collab., F. Abe et al., *Phys. Rev.* D **53**, 3496 (1996);
16. DELPHI collab., P. Abreu et al., *Phys. Lett.* B **345**, 598 (1995);
17. OPAL collab., R. Akers et al., *Z. Phys.* C **66**, 19 (1995);
18. ALEPH collab., D. Busculic et al., *Z. Phys.* C **69**, 393 (1996);
19. M. Feindt, CERN-PPE/95-139 (1995); I.J. Kroll, FERMILAB-CONF-96-032 (1996);

Strange Quark Currents in the Nucleon

W. Korsch
for the SAMPLE Collaboration
Kellogg Radiation Laboratory, California Institute of Technology, Pasadena, CA 91125, USA

During the last years many experiments in nuclear and particle physics were dedicated to studying the structure of the nucleon. Besides the electromagnetic form factors of the proton and especially the neutron, recently many experiments have been trying to extract information on the axial- and pseudo-scalar currents in the nucleon. The SAMPLE experiment at MIT-Bates is trying to isolate the strange magnetic moment of the proton by means of parity violating electron-proton scattering. The goal is to restrict G_M^s to an overall uncertainty of ± 0.22.

1 Introduction

Understanding the structure of the nucleon has become a major effort in experimental and theoretical nuclear physics within the last decade or so. Different theoretical models are able to describe certain properties of the nucleon and other baryons quite successfully, e.g. in the constituent quark model all baryons are made of three spin 1/2 Dirac particles bound by a non-relativistic potential. This model is very powerful in reproducing the magnetic moments and the excitation spectra of the nucleon. However, in the framework of QCD, the fundamental theory of strong interactions, the structure of the nucleon is more complicated. The three valence quarks are bound by gluons (spin 1) which do interact with themselves and can fluctuate into quark-antiquark ("sea" quarks, spin 1/2)) pairs. These $q\bar{q}$ pairs can be $u\bar{u}$, $d\bar{d}$, $s\bar{s}$, or even heavier quark pairs. Such a picture naturally raises many questions. For example: How do the different constituents contribute to the nucleon's mass, momentum, and spin? How do they contribute to the magnetic moment? How can we describe e.g. the vector- and axial-vector current distributions in the nucleon in terms of its constituents? Especially the strange quark contributions are of interest, since these quarks are the lightest next to the up and down quarks. Several experiments in very different energy regimes tried to extract the contribution of the strange (and anti-strange) quarks to different properties of the nucleon. One possible indication for a sizable amount of strangeness in the nucleon comes from the disagreement in the value of the σ-term, $\langle p \mid m_u u\bar{u} + m_d d\bar{d} \mid p \rangle$, when extracted from the baryon mass spectrum or from $\pi N \rightarrow \pi N$ scattering. The estimate for σ from the mass spectrum is about 25 MeV, whereas the extraction of σ from πN scattering favors a value of about 45 MeV.[1] About 10 MeV of discrepancy can be assigned to $SU(3)_f$ breaking effects and the residual 10 MeV are accounted for when about 10-15% of the nucleons rest mass is carried by strange quarks.

Further indications for strangeness in the nucleon are extracted from deep-inelastic $\nu(\bar{\nu})$-N scattering. These scattering processes allow access to the s- and $\bar{s}$ probability distributions in the nucleon. The momentum carried by strange-quarks (and anti-quarks) can be extracted from these distributions. The contribution to the overall proton momentum amounts to 4.1±0.4%. [2]

Recent measurements of the spin structure functions of the proton in deep-inelastic polarized lepton-nucleon scattering revealed a violation of the Ellis-Jaffe sum-rule on the 2 sigma level. [3] A consequence of this result is that about −10 % of the spin of the proton is carried by strange quarks (with an orientation antiparallel to the proton spin). Here, one assumes that $SU(3)_f$ is a good symmetry and the coupling constants F and D in semileptonic hyperon decays are the same for the whole baryon octet.

All these different results indicate a quite sizable amount of strangeness in the nucleon. The SAMPLE experiment tries to extract further information about strange quarks in the nucleon by measuring the parity violating asymmetry in elastic polarized electron-proton scattering at a low energy. The measured asymmetry can be related to the strange vector matrix element $\langle p \mid s\gamma_\mu\bar{s} \mid p\rangle$ or, since the average four-momentum transfer Q^2 is only 0.1 GeV^2/c^2, to the anomalous magnetic moment induced by the strange quarks. [4]

2 The SAMPLE Experiment

A different way of accessing strange quark contributions to the nucleon is parity violating lepton nucleon scattering. Parity violation is induced by the axial part of the neutral current. In the standard model the electromagnetic and axial-currents of the nucleon can be written in terms of the individual quark currents: [5]

$$J_\mu^{em} = \sum_f Q_f \cdot q_f\gamma_\mu\bar{q}_f$$

$$J_\mu^{n.c.} = \sum_f g_v \cdot q_f\gamma_\mu\bar{q}_f \qquad J_{\mu_5}^{n.c.} = \sum_f g_a \cdot q_f\gamma_\mu\gamma_5\bar{q}_f$$

Here, the Q_f are the electric charges of quarks with flavor f, g_v and g_a are the neutral current vector and axial-vector coupling constants. In low- and intermediate energy lepton-nucleon scattering these currents are typically described in terms of form factors. The electromagnetic Pauli and Dirac form factors are often expressed in terms of the Sachs form factors, G_p^M, G_p^E,

G_n^M, and G_n^E. Therefore, performing a flavor decomposition in terms of form factors, the electromagnetic current distributions can be written as follows:

$$G_x^{T=1} = \frac{1}{2}(G_x^p - G_x^n) \quad G_x^{T=0} = \frac{1}{2}(G_x^p + G_x^n)\ , \quad x = E, M$$

Equivalently one can rewrite the neutral vector current distributions

$$G_x^{nc} = 2(1 - 2sin^2\theta_w)G_x^{T=1} + (-4sin^2\theta_w)G_x^{T=0} + (-1)G_x^s\ , \quad x = E, M$$

and the axial- and pseudo-scalar current distributions

$$G_y^{nc} = (-2)G_y^3 + (0)G_y^8 + (1)G_y^s\ , \quad y - A, P$$

Here, we separated out the individual contributions coming from the strange quarks, $G_{x,y}^s$. G_y^3 and G_y^8 are the SU(3)$_f$ triplet- and octet-current parts. The '0' in front of G_y^8 is supposed to indicate that there is no contribution from this current on tree level, although higher order effects make this contribution finite and therefore strange quarks will contribute to this term as well. The parity violating asymmetry in $\vec{e}$-p scattering is induced by the interference term between the electromagnetic and the weak neutral current interaction. The asymmetry can be written as

$$A_{LR} = \frac{d\sigma_L - d\sigma_R}{d\sigma_L + d\sigma_R} = \frac{M^{em}M_{pv}^{nc}}{M^{em2}} = \frac{-G_F \mid Q^2 \mid}{2\pi\alpha\sqrt{2}}\left(\frac{(ej^\mu J_\mu^{cm})(g_V^l j^\mu J_{\mu_5}^{nc} + g_A^l j^{\mu_5} J_\mu^{nc}}{L^{\mu\nu}W_{\mu\nu}}\right)$$

Here, e, g_V^l, and g_A^l are the electric charge and the electroweak coupling constants, the $j's$ are the leptonic currents, the $J's$ the hadronic currents, and the $L^{\mu\nu}$ and $W_{\mu\nu}$ are the corresponding current tensors. This asymmetry can be expressed completely in terms of the above defined form factors.

$$A_{LR} = \frac{-G_F \mid Q^2 \mid}{4\pi\alpha\sqrt{2}}\left(\xi_V^p + \frac{\epsilon G_E^p(\xi_V^n G_E^n + \xi_V^0 G_E^s) + \tau G_M^p(\xi_V^n G_M^n + \xi_V^0 G_M^s)}{\epsilon(G_E^p)^2 + \tau(G_M^p)^2} - \right.$$
$$\left. \frac{(1 - 4sin^2\theta_w)\sqrt{1-\epsilon^2}\sqrt{\tau(1+\tau)}G_M^p G_A^p}{\epsilon(G_E^p)^2 + \tau(G_M^p)^2}\right),$$

ϵ=$[1+2|q^2/Q^2 \mid \tan^2(\theta/2)]^{-1}$ and τ=$|Q^2 \mid/4m_p^2$. The ξ's are electroweak coupling constants with radiative corrections included.[6] The goal of the SAMPLE experiment is to extract the strange magnetic form factor using elastic electron proton scattering at backward angles at a beam energy of 200 MeV. The scattering angle covers 130° to 160° with an average Q^2 of about 0.1 GeV^2/c^2. Under these conditions the asymmetry can be approximated by

$$A_{LR} \approx \frac{G_M^p G_M^{nc} - 0.20 G_M^p G_A^p}{(G_M^p)^2} \propto -\frac{1}{4}\frac{G_M^n}{G_M^p} - \frac{1}{4}\frac{G_M^s}{G_M^p} - 0.20\frac{G_A^p}{G_M^p}\ .$$

The expected asymmetry in the SAMPLE kinematics is about $-7 \cdot 10^{-6}$, if one assumes that the strange quarks have no contribution. There are many model predictions for the strange magnetic moment, ranging from about +0.4 to −0.75 (see e.g. reference[7] and references therein). For $G_M^s = -0.3$, a typical value predicted by several models, the additional contribution coming from the strange quarks is about $-1 \cdot 10^{-6}$. This amounts to a 72% contribution of the asymmetry coming from G_M^n, about 13% from G_A^p, and about 15% from G_M^s. Therefore, in order to constraint G_M^s to better than ±0.3 a measurement of the parity violating asymmetry of better than 13% is necessary. The SAMPLE experiment tries to extract the strange magnet moment with an accuracy of ±0.22.

2.1 The Experimental Setup

The SAMPLE experiment is designed to have maximum sensitivity to the parity violating asymmetry at backward angles and therefore its contributions from the magnetic moment of the nucleon.

As pointed out in the previous section, the expected asymmetry is on the order of $-7 \cdot 10^{-6}$ and the goal is to measure this number with an overall uncertainty of about 10%. This clearly means that high statistics and an extremely good understanding of the whole experimental setup, including polarized electron source, accelerator, and detector, is needed.

A source for polarized electrons in combination with a Wien filter, acting as a spin rotator, injects polarized electrons into the accelerator. At a beam energy of 200 MeV the electron beam gets bent (36.5°) into the Bates experimental North Hall. At the end of the beamline the electrons get scattered off a high power 40 cm long liquid hydrogen target and detected in a large acceptance air Čerenkov detector. The Čerenkov detector is axially symmetric around the beam pipe and consists of 10 concave mirrors covering a total solid angle of about 1.5 sr. The Čerenkov light gets focussed onto ten 8″ photomultiplier tubes. These tubes are mounted in lead cylinders as indicated in figure 1. The MIT-Bates linac delivers a pulsed beam with a duty factor of 1%. The peak current of polarized electrons is 4 mA with a polarisation of about 35 %. The repetition rate is 600 Hz and the pulse length is about 15 μs. The high peak current causes high instantaneous rates in the phototubes which makes it impossible to count individual pulses. The signals are integrated over the length of a beam pulse and then normalized to the charge. In front of each of the lead cylinders a thin remotely controlled shutter is mounted which can be closed to measure the background. A proper understanding of the background is a major task of this experiment. Possible background sources like

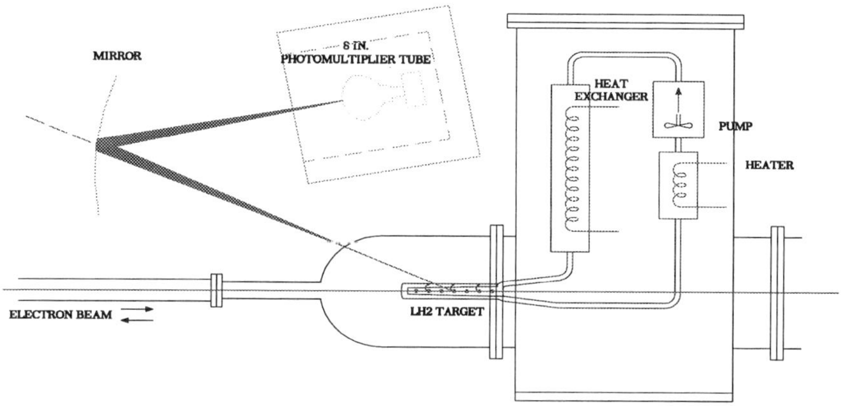

Figure 1: Schematic drawing of the SAMPLE detector and target.

e.g. electromagnetic showers from beam halo, scintillation in the air generated by electrons which were not back-scattered from the target, X-rays, and other sources of light have to be identified and minimized. Therefore, a large fraction of the beamtime is dedicated to background studies (shutter closed runs with full and empty target, low peak current runs using coincidence setups between scintillation detectors and the Čerenkov light, energy spectra in NaI detectors from the elastically scattered electrons off the target protons).

The power deposited in the liquid target due to energy loss is ≈ 500 W. A possible concern is (local) target boiling when the beam passes through the target. If target boiling occurs, the width of the signal will increase. No indications of target boiling were found at a $<1\%$ level of accuracy as the peak beam current was varied from 0.4 mA to 4 mA.

The main challenge of parity experiments is to keep the systematic effects well under control. An important feature in the SAMPLE experiment is the option of flipping the helicity of the electrons rapidly. The electrons are polarized by means of laser optical pumping in a bulk GaAs crystal. The laser system consists of an Ar-ion laser followed by a Ti-sapphire laser. A Pockels cell is used to flip the light helicity at a rate of 600 Hz which is perfectly matched to the pulse structure of the accelerator. These 600 pulses per second are subdivided in 10 'timeslots', meaning that a quasi-random helicity pattern is assigned to the first 10 pulses, for the following 10 pulses the helicity pattern is inverted, the next 10 timeslots will have a new quasi-random structure. The asymmetries are extracted by forming pulse pair asymmetries of two corresponding pulse pairs, i.e. pulse pair 1-10, 2-11, ..., this ensures that 60 Hz noise gets suppressed. Another important systematic source is the effect of beam loading. This effect is caused by the fact that pulses with different charge will have slightly different energies. Beam loading is a feedback of the accelerated charge to the accelerator RF. This effect is 2.3 MeV/mA for the MIT-Bates accelerator. Any helicity correlated change in energy is associated with a parity conserving change in the scattering cross-section causing a false asymmetry in the measurement. In order to minimize this effect a feedback loop which had been installed during the first parity experiment at MIT-Bates[8] measures the charge at the end of the accelerator and feeds its signal back to the laser Pockels Cell. Any helicity correlated charge differences are compensated by adjusting the Pockels Cell accordingly every few minutes. In addition to this feedback loop pulse-to-pulse energy fluctuations are suppressed by the Bates energy compression system (ECS). The compression can be as large as a factor of ten. The whole beam line is equipped with many beam postion and current monitors to ensure that the beam gets transported to the target without any losses. Another important source of systematic errors is the fact that the response of the detector system is a function of the beam position and angle as measured just upstream of the detector. If the positions or angles change in a helicity correlated way the signal may contain false asymmetries. The measured asymmetry must therefore be corrected in the following way:

$$A_c = A_m - \frac{1}{S}\Sigma\frac{\partial S}{\partial \alpha_i}\delta\alpha_i^{LR},$$

here, S is the normalized detector yield, α_i is the beam position, angle, or energy, and $\delta\alpha_i^{LR} = \alpha_i^L - \alpha_i^R$ is the helicity correlated difference in the beam parameter. Since the detector system is designed to be axially symmetric around the beam pipe and target these effects should be small. The response of the detector can either be studied by varying the beam parameters in a controlled way (rapid pulsing of steering coils and energy variations with a fast RF phase-shifter) or by using the intrinsic motion of the beam. The SAMPLE experiment makes use of the intrinsic beam motion and corrects the data using a formulism very similar to multiple regression in data analysis. Typically the measured slopes (responses) are less than 1% per mm.

2.2 First Results

First data were taken in fall 1995. Figure 2 summarizes the main result of this run. The left top figure shows that the charge asymmetry measured at

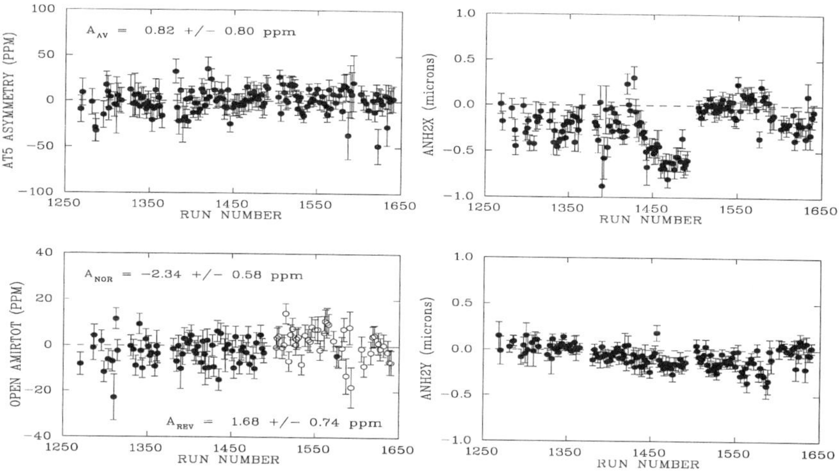

Figure 2: Results from the Fall 1995 run. All these plots are preliminary and the error bars are statistical only. Top-left: charge asymmetry measured at the end of the accelerator, top-right: x-position asymmetry, bottom-right: y-position asymmetry, bottom-left: raw asymmetry for all mirrors (shutters open).

the end of the accelerator is on average consistent with zero. Further the position "asymmetries" in x and y (x:top-right, y:bottom-right) are sometimes as large as about 600 nm. These asymmtries enter directly in the evaluation of the errors and will be improved in future runs. The figure on the bottom-left shows the raw mirror pulse-pair asymmetries when the shutters in front of the phototubes are open. All 10 mirrors are added together. Each data point is the result of about one half hour run. The corrections due to correlations with beam parameters have not been applied. The result is inconsistent with zero and the pronounced structure in the position asymmtries is not reflected in these data. This shows that the position sensitivity is strongly suppressed. It is further noticable that the asymmetry changes sign when the sign of the polarization is reversed. This "slow" reversal of the helicity was achieved by a manual rotation of a $\lambda/2$ plate in front of the Pockels Cell. The magnitude of the asymmetry is within errors symmetric around zero which shows that there is no significant offset. The data shown in the figure represent about 5% of the total expected. SAMPLE plans to extract G_M^s with the following accuracy: ±0.117 (stat.), ±0.128 (syst.), ±0.135 (theo., electroweak radiative corrections). Clearly, a large contribution is associated with radiative corrections to the axial current. These corrections are presently not easily calculable. This theoretical uncertainty can be strongly suppressed by measuring the parity violating asymmetry in quasielastic scattering from deuterium.[9] The proton and deuteron asymmetries can be parameterized in the following way, if the "static approximation" for the deuteron is assumed:

$$A_d = [\frac{49}{\sigma_d}][-\frac{G_F Q^2}{\pi\alpha\sqrt{2}}] \cdot [1 + 0.22(1 + R_A^{T=1})G_A^3 - 0.10 G_M^s] \quad (1)$$

$$A_p = [\frac{26}{\sigma_p}][-\frac{G_F Q^2}{\pi\alpha\sqrt{2}}] \cdot [1 + 0.24(1 + R_A^{T=1})G_A^3 - 0.61 G_M^s] \quad (2)$$

Taking the ratio of these two asymmetries gives:

$$\frac{A_d}{A_p} \propto 1 + 0.02(1 + R_A^{T=1})G_A^3 - 0.51 G_M^s \quad (3)$$

This clearly reduces the contribution of the axial radiative corrections and should allow an extraction of G_M^s which is dominated by statistical and experimental systematic errors.

SAMPLE is the first experiment which tries to extract a strange form factor of the nucleon. Several other experiments at MAMI and TJNAF will continue this program in the future and extract the electric and magnetic strange form factors for Q^2 values up to about 0.7 GeV^2/c^2.

This work was supported by NSF grants PHY-9420470 (Caltech), PHY-9420787 (Illinois), PHY-9457906/PHY-9229690 (Maryland), PHY-9208119 (RPI), and DOE cooperative agreement DE-FC02-94ER40818 (MIT-Bates).

References

1. J.Gasser, H. Leutwyler, and M.E. Sainio, *Phys. Lett.* B **253**, 252 (1991).
2. A.O. Bazarko *et al.*, e-print # hep-ex/9406007 (1994); R.L. Jaffe, private communication.
3. B. Filippone, these proceedings.
4. R.D. McKeown, *Phys. Lett.* B **219**, 140 (1990).
5. D.H. Beck, *Phys. Rev.* D **39**, 3248 (1989).
6. M.Musolf, T.W. Donnelly, J. Dubach, S.J. Pollock, S. Kowalski, and E.J. Beise, Phys. Rep. **239**, 1 (1994).
7. H.-W. Hammer, U.-G. Meissner, D. Drechsel, *Phys. Lett.* B **367**, 323 (1996); H.-C. Kim, T. Watabe, K. Goeke, e-print # hep-ph/9606440 (1996).
8. P.A. Souder *et al.*, *Phys. Rev. Lett.* **65**, 694 (1990).
9. Bates-experiment # 94-11 (M. Pitt and E. Beise, co-spokespersons).

RECENT EXPERIMENTS ON THE ELECTRIC FORM FACTOR OF THE NEUTRON

F. KLEIN
Physikalisches Institut, Universität Bonn,
Nussallee 12, D–53115 Bonn, Germany

*for the **A3–COLLABORATION at MAMI***
Institut für Kernphysik, Universität Mainz, D-55099 Mainz
Institut für Physik, Universität Mainz, D-55099 Mainz
Department of Physics & Astronomy, University of Glasgow, Glasgow, U.K.
Physikalisches Institut, Universität Tübingen, D-72076 Tübingen
Ecole Normale Supérieure, Paris, France
Physikalisches Institut, Universität Bonn, D-53115 Bonn

Form factors for elastic electron scattering parametrise the collective response of a struck object and provide information vital for the understanding of its ground state. The quality of existing data for the neutron is much behind that for the proton; it will now benefit from coincidence experiments at modern cw accelerators. For the electric form factor the scattering of high energy polarised electrons from polarised targets or the spin transfer to the recoiling neutron from unpolarised targets now exhibits its full potential in running experiments and delivers first results.

1 Introduction

Protons and neutrons, generically termed nucleons, are the constituents of atomic nuclei and as such the carriers of the mass which makes up our world and ourselves. Characteristic properties as magnetic moments, radii and the occurrence of resonances give ample evidence for their internal structure which in our belief should ultimately be derivable from the strong colour force between their constituents, the quarks. Elastic form factors are a measure of the ability of a composite system to embed an impuls experienced in a scattering (spacelike) or annihilation/creation (timelike) process into the ground state wave function of its constituents. Proton form factors have been extensively measured in single arm experiments through the classical Rosenbluth separation of the cross section for elastic electron scattering. The neutron is much less accessible in this way thus leaving a big gap in the test and understanding of the nucleon structure. Recent experimental achievements can now improve on this deficiency; this is the subject of the third section following a summary of the formalism and the general status of elastic form factors in electron scattering.

2 Nucleon Form Factors in Electron Scattering

The interaction of high energy electrons with the quanta of the electromagnetic field is fully understood and described in the frame of quantum electrodynamics. For spin=1/2 objects with an unknown internal structure like nucleons the charge–current–density is modified by two scalar functions $F_1(Q^2)$ and $F_2(Q^2)$, called the Dirac and Pauli form factors respectively which only depend on the square of the four-momentum transfer Q^2 received in the scattering process. Linear combinations of them make up the magnetic and electric Sachs form factors

$$G_M(Q^2) = F_1(Q^2) + F_2(Q^2), \qquad G_E(Q^2) = F_1(Q^2) - \frac{Q^2}{4M_p^2} * F_2(Q^2) \quad (1)$$

which can be interpreted as the Fourier transforms of the distribution of magnetism and charge in a frame with no energy transfer. With these the cross section for elastic electron scattering can be cast into the Rosenbluth–form

$$\frac{d\sigma}{d\Omega} = \sigma_{Mott} * \left(\frac{G_E^2 + \tau \cdot G_M^2}{1+\tau} + 2\tau \cdot G_M^2 \cdot tg^2(\theta/2) \right), \qquad \tau = \frac{Q^2}{4M_p^2} \quad . \quad (2)$$

Due to the different angular dependence G_E and G_M can, in principle, be separated in the elastic cross section taken at different scattering angles and constant momentum transfers.

At zero momentum transfer the slope of any form factor with respect to Q^2 defines a mean squared radius of an associated distribution. In this way the electric proton radius has been determined to be $< r_{E,p}^2 >^{1/2} = 0.862 \pm 0.012\, fm$ [1]. The decomposition of the G_E–radius into F_1– and F_2–radii illustrates a fact which has already been pointed out by Foldy [2] when first experimental results became available[3,4]:

$$< r_E^2 > := -6 \cdot \frac{dG_E(0)}{dQ^2} = -6 \cdot \frac{dF_1(0)}{dQ^2} + \frac{3}{2M^2} * F_2(0) \quad (3)$$

$$< r_E^2 >_{proton} \simeq 0.56 + 0.184\, fm^2$$

$$< r_E^2 >_{neutron} \simeq \pm\ ? + -0.126\, fm^2$$

Contrary to the proton for the neutron by far the largest contribution to $< r_E^2 >$ originates from the absolute value of $F_2(0)$ (not its slope!) i.e. the anomalous magnetic moment. The Dirac– or F_1–radius is there only small and even positive according to the most recent determination $< r_E^2 >_{neutron} \simeq -0.113 \pm 0.003 \pm 0.004\, fm^2$ from a measurement of the neutron transmission through a lead target [5].

Our empirical knowledge about the form factors of the proton (in the spacelike region) comes from single arm electron scattering through application of the Rosenbluth–separation. Due to the kinematical factor τ in front of G_M at low momentum transfers ($Q^2 < 0.5\ GeV^2/c^2$) $G_{E,p}$ dominates the cross section while at higher Q^2 its contribution dies out (typ. 5 % at $Q^2 \simeq 9\ GeV^2/c^2$). A Rosenbluth separation was made[6,7] up to 9 GeV^2/c^2 while for higher Q^2 up to 30 GeV^2/c^2 $G_{M,p}$ was directly taken from the cross section with the small $G_{E,p}$-contribution subtracted[8].

Neutron form factors raise more difficulties: as a target of sufficient density it is only available in nuclei where it underlies the Fermi motion and is necessarily accompanied by protons. In this respect the deuteron was usually chosen to extract the en–cross section from the quasielastic peak in single arm experiments by subtraction of the dominating proton contribution, with correspondingly large errors (for a compilation see [9]). Now, at high duty cycle accelerators the occurrence of an en–scattering can be explicitly tagged by detection of the outgoing neutron in coincidence experiments[10,11]. Particular emphasis is given to the control of nuclear effects by simultaneous measurement of $D(e,e'p)$ and to the calibration of the neutron detection efficiency and promise sharper limits on this important quantity[12,13 a].

The electric neutron form factor $G_{E,n}$ is most unfavourable; its smallness compared to $G_{M,n}$ renders a Rosenbluth separation very difficult. From recent measurements of the response functions R_L and R_T for quasielastic deuteron break up in the range $Q^2 \simeq 1.75 - 4\ GeV^2/c^2$ low values for $G_{E,n}$ have been obtained[16]; they demand, despite their large uncertainties, a cancellation in $F_{1,n} - \tau * F_{2,n}$ ($\equiv G_{E,n}$) implying that the Dirac form factor $F_{1,n}$ is different from zero in this regime. For a long time the most precise data for the intermediate momentum transfer range came from the study of the elastic deuteron structure function $A(Q^2)$[17,18] which provides a higher sensitivity on $G_{E,n}$ through the square of the isoscalar form factor $(G_{E,n} + G_{E,p})^2$. Unfortunately the necessary unfolding of the deuteron wavefunction etc. introduces a substantial uncertainty in the extracted $G_{E,n}$ (fig. 2).

3 Polarisation Experiments

In the search for a sensitive observable with little model dependence Arnold, Carlson and Gross[19] suggested to exploit the spin transfer in $n(\vec{e},e'\vec{n})$ from electrons with helicity P_l to neutrons where the interference between electric and magnetic form factor shows up in the transverse polarisation P_x of the

[a] Also polarisation experiments on $G_{M,n}$ have been reported[14,15]

recoiling neutron with respect to $\vec{q}$ according to

$$P_x = -P_l * \frac{\sqrt{2\tau\epsilon \cdot (1-\epsilon)} \cdot G_E \cdot G_M}{\epsilon \cdot G_E^2 + \tau \cdot G_M^2}, \qquad (4)$$

with θ denoting the electron scattering angle and $\epsilon = [1+2\cdot(1+\tau)tg^2\theta/2]^{-1}$ the polarisation parameter of the virtual photon. More generally the asymmetry of the cross section for $\vec{e}-\vec{n}$ scattering depends on the transverse and longitudinal polarisation P_x and P_z of the neutron, either in the initial or in the final state [20]:

$$A = -P_l * \frac{\sqrt{2\tau\epsilon \cdot (1-\epsilon)} \cdot G_E \cdot G_M \cdot P_x + \tau\sqrt{1-\epsilon^2} \cdot G_M^2 \cdot P_z}{\epsilon \cdot G_E^2 + \tau \cdot G_M^2} \qquad (5)$$

For a neutron polarisation P_x perpendicular to the momentum transfer $\vec{q}$ the asymmetry is proportional to $G_E \cdot G_M$ while for $\vec{P} \parallel \vec{q}$ it is essentially independent of the form factors and can be used for calibration purposes.

Several ways to approximate the ideal case of the $n(\vec{e}, e'\vec{n})$– or $\vec{n}(\vec{e}, e')$–reaction with feasible targets have been investigated. Polarised $^3\vec{H}e$ is assumed to carry a high n–polarisation of about 87 % in quasielastic scattering [21]. Its magnitude will even cancel in the ratio A_x/A_z of asymmetry measurements. However, in the inclusive $^3\vec{H}e(\vec{e}, e')$–process the proton contribution to the asymmetry, even in quasielastic kinematics, is difficult to control [22] and makes the extraction of $G_{E,n}$ somehow critical. A theoretical work [23] suggested a more sophisticated choice of the polarisation angle $\beta \neq 90°$ to minimise the proton contribution. A recent measurement of the spin asymmetries in $^3\vec{H}e(\vec{p}, 2p)$ and $^3\vec{H}e(\vec{p}, np)$ quasielastic scattering using a PWIA–analyses [24] supports the reliability of theoretical calculations at a level of 20 %.

The exclusive reaction $^3\vec{H}e(\vec{e}, e'n)$ avoids the proton contribution provided that final state interactions can be neglected. Theoretical studies [25] suggest that in collinear kinematics this condition holds for momentum transfers $Q^2 > 0.3\ GeV^2/c^2$.

For the spin transfer in $D(\vec{e}, e'\vec{n})$ Arenhövel [26] confirmed a persisting sensitivity to $G_{E,n}$ and little dependence on the chosen deuteron wave function for quasifree kinematics[b]. The transverse polarisation P_x of the recoiling neutrons can be analysed through the ϕ–asymmetry in a second np–scattering with an analysing power $A(\theta_{n'}, T_n)$ of about 50 % at $\theta_{n'} \simeq 30°$ and energies of interest here [27]:

$$I(\theta_{n'}, \phi) = I_0 * (1 + A(\theta_{n'}, T_n) * cos\phi) \qquad (6)$$

---

[b]This arguement also applies to the cross section asymmetry in $\vec{D}(\vec{e}, e'n)$.

It has been demonstrated [28] that the np–scattering in the detection process itself can be used for the spin analysis with great advantage. In a first experiment already done at Bates [29] several horizontally installed detector bars were followed by two other sets of detectors positioned at the angle of maximum analysing power. Unfortunately, the low duty factor of the accelerator made this coincidence experiment very difficult. Recent experimental studies indicate that also at neutron energies above 400 MeV the effective analysing power is still large [30]. This will preserve the potential of experiments using recoil polarimetry [31] at higher momentum transfer in comparison with those using polarised solid state targets in the reaction $\vec{D}(\vec{e}, e'n)$ [32].

3.1 $G_{E,n}$-Experiments at MAMI

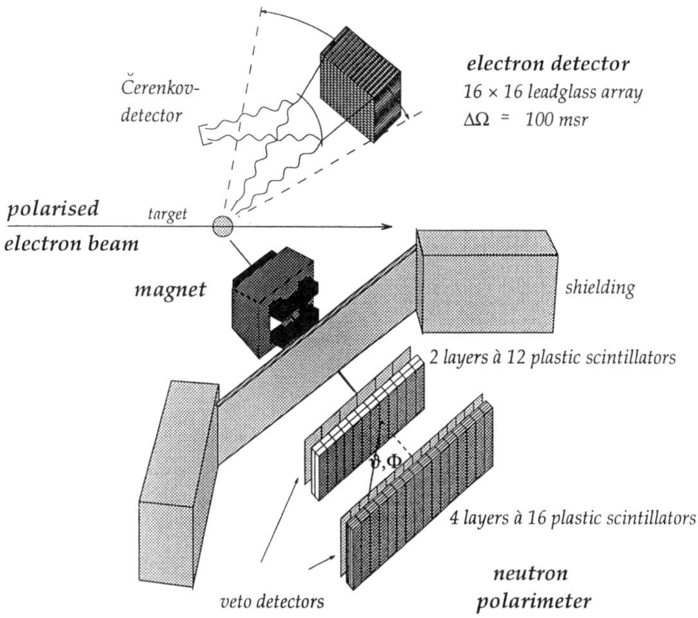

Figure 1: set up for the $G_{E,n}$ experiments at MAMI, magnet only used for $D(\vec{e}, e'\vec{n})$-experiment

At MAMI $^3\vec{H}e(\vec{e}, e'n)$- and $D(\vec{e}, e'\vec{n})$-experiments have been performed with one common large solid angle detector (fig. 1). It consists of a Pb–glass array with a Čerenkov detector for the electrons and two well shielded walls of neutron detectors. The source of polarised electrons is based on photoelectron emission of GaAsP [33,34,35]. The spin precession in MAMI is compensated with the help of a spin rotator inserted in the beamline before accelerator injection [36]. It allows to preset the spin of the electrons to any desired direction and is thus suited for various types of experiments. With a Mott polarimeter the electron polarisation can be determined before injection at an energy of 100 keV and at 855 MeV by Møller scattering from a magnetized iron foil in a polarimeter similar to the one described in ref.[37] but detecting the Møller electrons in coincidence [38,39]. The degree of polarisation increased from initially 30 % to 50 % with beam currents of 5 – 10 μA at the $^3\vec{H}e$–target. Meanwhile, with a strained crystal a beam polarisation of about 75 % at currents of 2 μA is routinely achieved.

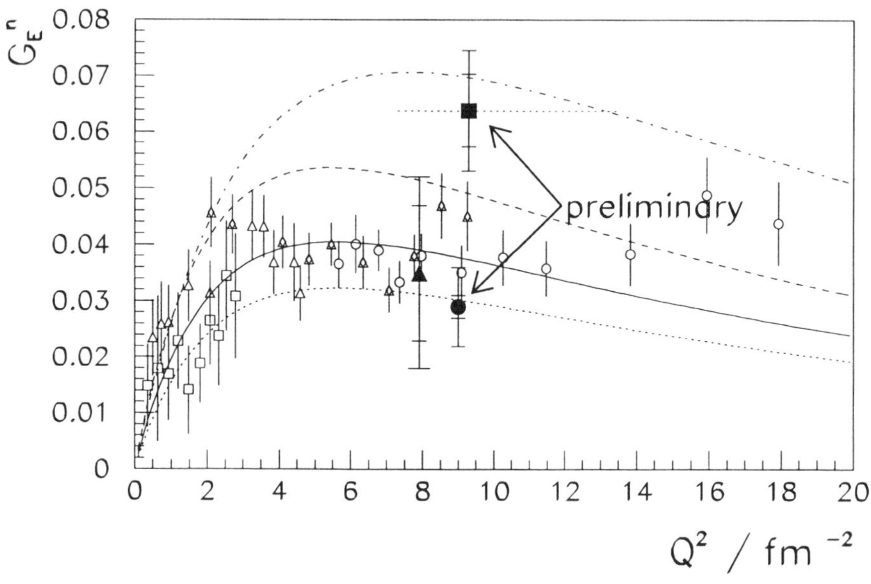

Figure 2: $G_{E,n}$ data from [Pl90] with curves for different potentials in the extraction; full points: from the A3–collaboration (triangle: $^3\vec{H}e(\vec{e}, e'n)$-data point [40], circle: new $^3\vec{H}e(\vec{e}, e'n)$-data, box: new $D(\vec{e}, e'\vec{n})$-data, both preliminary!

$^3\vec{H}e(\vec{e}, e'n)$–experiments

In the $^3\vec{H}e(\vec{e}, e'n)$-experiment the neutron detector was operated as a time–of–flight device. ^{3}He gas was spinpolarised by metastability exchange scattering with optically pumped metastable ^{3}He atoms in their $1s2s^3S_1$ state[41] and compressed to 1 bar in the target cell where a polarisation of 50 % was maintained. The result of an exploratory run is described in more detail elsewhere [40] and plotted in fig. 2 along with the data from[18]. The preliminary result of the full run averaged over $Q^2 = 7 - 12\ fm^{-2}$ is also shown in fig. 2. A new experiment to access higher momentum transfers ($Q^2 \simeq 0.6\, GeV^2/c^2$) based on a magnetic electron spectrometer and a higher target density is in preparation at MAMI [42].

$D(\vec{e}, e'\vec{n})$–experiments

In this experiment the polar and azimuthal angles of the second (analysing) neutron scattering $\theta_{n'}$ and $\phi_{n'}$ were reconstructed from the hit positions in the front and rear detector planes. With the number of events $N^{\pm}(\phi)$ for $\pm$ helicity states of the beam the ϕ–asymmetry $A(\phi)$ was determined through the ratio

$$\frac{1 - A(\phi)}{1 + A(\phi)} = \sqrt{\frac{N^+(\phi) * N^-(\phi + \pi)}{N^-(\phi) * N^+(\phi + \pi)}} \tag{7}$$

in which the detector efficiencies and the beam fluxes cancel. The determination of the desired recoil polarisation P_x from the amplitude of the asymmetry curve requires the knowledge of the effective analysing power of the polarimeter which depends crucially on the kinematical cuts applied to the n–scattering in the first detector plane.

To circumvent this problem the redundancy in the longitudinal neutron polarisation has been exploited. On their way L through a magnet with a vertical field $\vec{B}$ (i.e. perpendicular to the scattering plane) the neutrons undergo a spin precession due to their anomalous magnetic moment $\mu_n = -1.91 \cdot \mu_N$ by the angle

$$\chi \quad = \quad 0.61\,\frac{rad}{T \cdot m} * \int_L B \cdot dl/\beta_n \ . \tag{8}$$

After rotation by χ the transverse polarisation P_T receives contributions from both, the term of interest a_E containing $G_{E,n} \cdot G_{M,n}$ and a term a_M originating from $G^2_{M,n}$. Thus, the resulting asymmetry curve for the second scattering is a superposition of two contributions

$$A(\chi) \;=\; k * \{a_E \cdot cos\chi \;\; + a_M \cdot sin\chi\} \tag{9}$$

with the amplitudes $a_{E,M}$ and the calibration factor k. The ratio a_E/a_M or the angle of zero crossing χ_0, respectively, is directly related to the ratio of the electric and magnetic form factors

$$tg\,\chi_0 \equiv \frac{a_E}{a_M} = \frac{1}{\sqrt{\tau \;+\; \tau \cdot (1+\tau) \cdot tg^2\vartheta/2}} * \frac{G_E}{G_M} \;. \tag{10}$$

and depends neither on the analysing power of the polarimeter nor on the polarisation of the electron beam. Asymmetry data for various precession

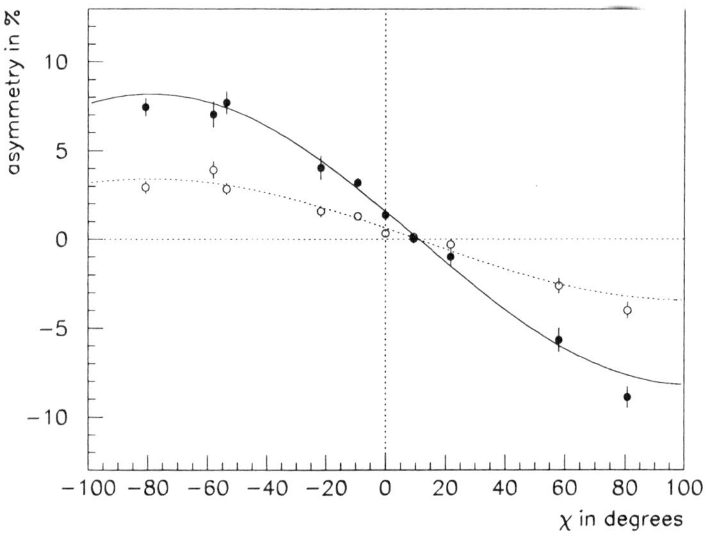

Figure 3: Reconstructed ϕ–asymmetries for various precession angles $\chi \propto \int_L B \cdot dl/\beta_n$, averaged over detector acceptance, for 2 different cuts on secondary scattering.

angles $\chi \propto \int_L B \cdot dl/\beta_n$ are shown in fig. 3. Kinematical cuts in the second scattering of the neutrons change the amplitude (i.e. the effective analysing power) but not the zero–crossing χ_0. The preliminary value for $G_{E,n}$ calculated from eq. 10 in simple average over the full detector acceptance is higher than others as shown in fig. 2.

Conclusion

Elastic form factors contain condensed information about the nucleon ground state. Data exist for proton and neutron in the timelike and spacelike region with great variance in precision. For the neutron our knowledge is generally unsatisfactory and takes much away from the potential constraints of isospin symmetry in tests of nucleon models. New cw accelerators now make coincidence experiments with explicit tagging of genuine en–interactions feasible. This technique has already delivered first results on $G_{M,n}$ which plays a central role for the normalisation of experiments on strange and electric form factors.

Among electromagnetic form factors $G_{E,n}$ is the least constrained. Only the radius is precisely known and fixes its slope with respect to Q^2 at zero momentum transfer, indicating a small value for the Dirac form factor $F_{1,n}$. Contrary, at high momentum transfers the observed smallness of $G_{E,n}$ demands a finite value for $F_{1,n}$. New polarisation experiments will now improve on the sensitivity and model dependence. Data have been taken for ${}^3\vec{H}e(\vec{e}, e'n)$ and $D(\vec{e}, e'\vec{n})$ at $Q^2 \simeq 9\ fm^{-2}$ with statistical errors of typically 10 %. The first preliminary results indicate a trend for the $D(\vec{e}, e'\vec{n})$-analysis towards a higher value. If it persists in the final analysis these experiments will also give access to nuclear effects in spin physics. Fortunately enough we can soon expect more and independent results on this delicate and important quantity from coming experiments at the newly commissioned Thomas Jefferson National Accelerator Facility.

Acknowledgements

This work has been supported by the Deutsche Forschungsgemeinschaft within the Sonderforschungsbereich 201.

References

1. G. G. Simon et al., *Nucl. Phys.* A **333**, 381 (1980)
2. L. Foldy, *Phys. Rev.* **87**, 693 (1952), *Rev. Mod. Phys.* **30**, 471 (1958)
3. E. Fermi and L. Marshal, *Phys. Rev.* **72**, 1139 (1947)
4. Havens, Rabi and Rainwater, *Phys. Rev.* **72**, 634 (1947)
5. S. Kopecki et al., *Phys. Rev. Lett.* **74**, 2427 (1995)
6. P. Bosted et al., *Phys. Rev. Lett.* **68**, 3841 (1992)
7. A. F. Sill et al., *Phys. Rev.* D **48**, 29 (1993)
8. R. G. Arnold et al., *Phys. Rev. Lett.* **57**, 174 (1986)
9. I.Sick, *Proc. 6th Miniconf. on electron scattering*, p.193, NIKHEF 1989

10. H. Anklin et al., *Phys. Lett.* B **336**, 313 (1994)
11. E. Bruins et al., *Phys. Rev. Lett.* **75**, 21 (1995)
12. J. Jourdan, *contrib. paper to this conference, PANIC96 abstracts p. 51*
13. W. Brooks, *contrib. paper to this conference, PANIC96 abstracts p. 12*, CEBAF Exp. proposal E–94–017 and extension
14. Markowitz et al., *Phys. Rev.* C **48**, R5 (1993)
15. H. Gao et al., *Phys. Rev.* C **50**, R546 (1994)
16. A. Lung et al., *Phys. Rev. Lett.* **70**, 718 (1993)
17. S. Galster et al., *Nucl. Phys.* B **32**, 221 (1971)
18. S. Platchkov et al., *Nucl. Phys.* A **510**, 740 (1990)
19. R. G. Arnold et al., *Phys. Rev.* C **23**, 363 (1981)
20. T. W. Donnelly et al., *Ann. of Phys.* **169**, 247 (1986)
21. J. Friar, *Phys. Rev.* C **42**, 2310 (1990)
22. R. W. Schulze and P. U. Sauer, *Phys. Rev.* C **48**, 38 (1993)
23. C.Ciofi degli Atti et al., *Phys. Rev.* C **51**, 1108 (1995)
24. R. G. Milner et al., *Phys. Lett.* B **379**, 67 (1996)
25. J. M. Laget, *Phys. Lett.* B **273**, 367 (1991) (1980),
26. H. Arenhövel, *Phys. Lett.* B **199**, 13 (1987); *Z. Phys.* A **331**, 509 (1988)
27. R. A. Arndt et al., *Phys. Rev.* C **15**, 1002 (1977) and SAID–Program, Phase shift solutions SP84, unpublished
28. T. N. Taddeucci et al., *Nucl. Instrum. Methods* A **241**, 448 (1985)
29. T. Eden et al., *Phys. Rev.* C **50**, R1749 (1994)
30. R.Madey, *private communication during this conference*
31. T. Eden et al., *contrib. paper to this conference, PANIC96 abstracts p. 74*
32. CEBAF experiment proposal E–93–026
33. K. Aulenbacher, Proc. Workshop on Photocathodes for polarized e– sources for accelerators, SLAC–report 432, p. 1
34. W. Hartmann et al., *Nucl. Instrum. Methods* A **286**, 1 (1990)
35. H. G. Andresen et al., *Proceedings of "High Energy Spin Physics"*, Bonn 1990, (Meyer, Steffens, Thiel, eds.), Vol. 2, p. 12
36. K. H. Steffens et al., *Nucl. Instrum. Methods* A **325**, 378 (1993)
37. B. Wagner et al., *Nucl. Instrum. Methods* A **294**, 541 (1990)
38. M. Ertel et al., Proceedings of "High Energy Spin Physics", Bonn 1990, (Meyer, Steffens, Thiel, eds.), Vol. 2, p. 43
39. D.Eyl et al., *Z. Phys.* A **352**, 211 (1995)
40. M. Meyerhoff et al., *Phys. Lett.* B **327**, 201 (1994)
41. G. Eckert et al., *Nucl. Instrum. Methods* A **320**, 53 (1992)
42. Measurement of the Electric Form Factor of the Neutron via $^3\vec{He}(\vec{e}, e'n)$, proposal to the joint MAMI/ELSA PAC, Mainz 1995

Structure and Motion of Nucleons in Nuclei

W. U. Boeglin
Physics Department
Florida International University
University Park
Miami FL 33199

Abstract

The motion of nucleons in nuclei is closely related to the nucleons wave function which depends on the nucleon-nucleon force. The short range part of the nucleon nucleon force is expected to give rise to short range correlations. Results of the investigation of the effect of the interaction between nucleons on their structure and on their wave functions are discussed.

1 Introduction

The question of how nucleons move in nuclei and whether their internal structure is affected by the nuclear medium is one of the fundamental problems in nuclear physics. The wave function of nucleons in nuclei depends on the structure of the underlying nucleon nucleon (NN) interaction. The goal of measuring high momentum components in nuclei is the search for effects from the short range part of the NN interaction. The nuclear binding might affect the internal structure of nucleons. A possibility to determine changes of the bound nucleon structure is to compare experimental data with theoretical model predicitons, which use free nucleon form factors. The remaining discrepancies between experimental data and the models are then explained by modifications of the nucleon form factors. The data discussed are obtained from quasi-free inclusive and semi-exclusive electron scattering experiments.

2 Inclusive Electron Scattering

The scattered electron transfers momentum $\vec{q}$ and energy ω to the nucleus (nucleon). Depending on the energy transferred (ω), the nucleus can either recoil without excitation (elastic electron scattering) or low lying bound states can be excited. At increasing values of ω, the giant resonances are excited, and at around a value of $\vec{q}\,^2/2M_p$, nucleons are knocked out and one assumes the electrons scatter incoherently off the individual nucleons.

2.1 Longitudinal/Transverse Separations

In general the inclusive cross section can be written as follows:

$$\frac{d^3\sigma}{d\omega d\Omega_e} = \sigma_{Mott}\left(v_L R_L + v_T R_T\right),$$

where σ_{mott} is the mott cross section, which describes the scattering of a point particle with spin 0. R_L and R_T are the longitudinal and the transverse response functions. By varying the incident electron energy and the scattering angle, so as to keep q and ω constant, the longitudinal and the transverse response functions can be separated (Rosenbluth separation). If the cross sections obtained for a variety of momentum transfers show y-scaling behavior, one can be certain that the reaction mechanism consists of scattering of individal nucleons. In the simplest approximation this should also be applicable to the separated response functions. Figure 1 shows the experimental scaling functions $F_L(y)$ and $F_T(y)$ [1] Both response functions clearly show scaling, thereby

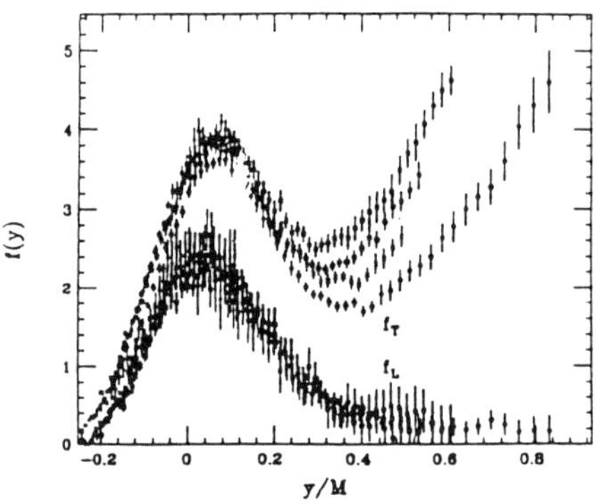

Figure 1: The two scaling functions $F_L(y)$ and $F_T(y)$ from ^{12}C [1].

indicating quasi-free scattering, but the two scaling functions are not identical. The longitudinal resonse seems to be suppressed. The same supression has been observed in many other nuclei ranging from ^4He up to uranium [4,5]. The calculation of these response functions is however very sensitive to details of the nuclear models used and the transverse response function is sensitive conrtibutions from twobody currents such as meson exchange currents. Consequently a better, more model independent quantity, has been investigated, namely the coulomb sum rule.

2.2 Coulomb Sum Rule

The coulomb sum rule can be written as follows:

$$S_L(q) = \frac{1}{Z}\int_{\omega}^{\infty} \mathrm{d}\omega \frac{R_L(q,\omega)}{\tilde{G}_E^2(Q^2)}$$

Where $\tilde{G}_E^2(Q^2)$ is an effective nucleon electric form factor containing neutron and proton contributions. In the Fermi Gas model one obtains $S_L(q) \to 1$ for $q \to \infty$. The longitudinal response is not affected by meson exchange currents to first order so one expects the impulse approximation to be valid. The experimental values obtained for S_L have typically been around 0.6 [4]

This supression of the longitudinal response could be explained by a modification of the nucleon form factors which uses an increased charge but an unchanged magnetic radius of the nucleon [2,3]. Recently however new data and a new analysis of the quasi free scattering on ^{12}C and ^{56}Fe are in good agreement with new and improved theoretical calculations[5] which do not require any modified nucleon form factors (figure 2). From these recent studies of quasi

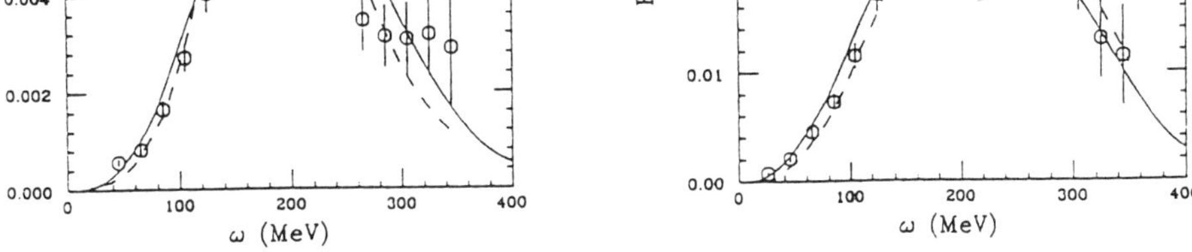

Figure 2: The separated response functions which agree well with theoretical calculations which do not need modified nucleon form factors[5].

free scattering, one can conclude that no modifications of the bound nucleons could be observed. The coulomb sum rule agrees well within the error bars with recent theoretical values[5].

3 (Semi-)Exclusive Scattering: (e,e'p)

While inclusive quasi-free scattering is averaged and summed over all possible inital and final states, the observation of the ejected proton in the (e,e'p) reaction permits a more detailed study of bound protons.

The (e,e'p) reaction can be schematically understood as follows. The virtual photon is absorbed by a bound nucleon having a certain initial momentum p_i, which subsequently leaves the nucleus with a kinetic energy T_p. The residual system may remain in its ground state or in an excited state characterized by the missing energy ($\epsilon_m = \omega - T_p - T_r$) and has a recoil momentum $\vec{p_{miss}} = \vec{q} - \vec{p_f}$. Here T_r is the kinetic energy of the recoiling system.

Assuming one photon exchange, the cross section for a coincidence reaction where one particle is detected in coincidence with the scattered electron can be written as follows[6], [7]:

$$\frac{\mathrm{d}^6\sigma}{\mathrm{d}\omega\mathrm{d}\Omega_e\mathrm{d}T_p\mathrm{d}\Omega_p} = \sigma_{Mott}\left(v_L R_L + v_T R_T + v_{LT} R_{LT}\cos\phi + v_{TT} R_{TT}\cos 2\phi\right)$$

The function R_i and R_{ij} are response functions and consist of combinations of matrix elements of the electromagnetic currents in the system investigated. The simplest approximation to describe this reaction, assumes the nucleon to be free after it has absorbed the virtual photon. It leaves the nucleus without any further interaction and can be described by a plane wave. The struck nucleon is detected in coincidence with the scattered electron. These are the assumptions of the Plane Wave Impulse Approximation (PWIA). In this approximation, the recoil momentum of the residual system is antiparallel to the initial momentum of the struck nucleon. The cross section can be factorized as follows:

$$\frac{\mathrm{d}^6\sigma}{\mathrm{d}\omega\mathrm{d}\Omega_e\mathrm{d}\Omega_p\mathrm{d}T_p} = K \cdot \sigma_{ep} \cdot S(E, p_i)$$

Here σ_{ep} describes the elementary electron proton (off shell) cross section for scattering an electron off a moving (bound) nucleon[9]. The factor K is a kinematical factor (in this case $K = p_f E_p$), and $S(E, p_i)$ is the spectral function which describes the probability of finding a proton with the separation energy E and an initial momentum p_i. For a bound hole state α of the residual system, the spectral function can be written as follows: $S(E, p_i) = \rho_\alpha(p_i)\delta(E - E_\alpha)$, where E_α is the separation energy of a proton from the state α and ρ_α the momentum density.

In the Independent Particle Shell Model (IPSM) ρ_α corresponds to the Fourier transform of the hole wave function. The spectroscopic factor is given by $\int \rho_\alpha^{exp} d^3p_{miss}/2J_\alpha + 1$, where $2J_\alpha + 1$ is the calculated occupation number.

When a proton is knocked out, in reality the emerging nucleon does interact with the residual nucleus, and its wave function is not a plane wave any more. In this case $\vec{p}_f \neq -\vec{p}_{miss}$, and the spectral function has to be replaced by the distorted spectral function $S_{dist}(E, \vec{p}_{miss}, \vec{p}_f)$. The Final State Interaction (FSI) for a heavy nucleus is described by means of an optical potential. This approach corresponds to the Distorted Wave Impulse Approximation (DWIA) [7].In both approximations the electron waves have been described by plane waves which is not valid for heavy nuclei, this problem has been discussed in reference [8]. The spectroscopic factors determined for a variety of nuclei were reduced by typically 35% compared to Hartree-Fock and random phase approximation calculations [11,12].

3.1 Probing the nucleon form factors

At MIT-Bates a series of experiments have been performed on ^{12}C where the $\vec{q}$-dependence of the spectroscopic factors has been investigated for momentum transfers between 400 MeV/c and 800 MeV/c [13]. A modification of the nucleon form factors should result in a different $\vec{q}$-dependence of the spectroscopic factors. Within the error bars of the experiments (about 15%) no deviations could be observed.

The ratio between the transverse and the longitudinal response functions of the (e,e′p) reaction in parallel kinematics of ^{12}C has been investigated [14]. The ratio $R_G = \frac{2M_p}{Q}\sqrt{R_T/RL}$ in PWIA is equal to the ratio between the nucleons' magnetic and electric form factor. The observed deviation from this value could be explained as contributions from two-body currents and FSI [15, 16]. One can conclude, that within the error bars of the experiments no clear evidence for a modification of the nucleon form factors could be found in the coincidence experiments, however, it is clear that the reaction mechanism is considerably more complex than is asumed in the impulse approximation.

3.2 High Momentum Components

The observed reduction of the spectroscopic factors is attributed to the effect of short range correlations which should affect the high momentum part of the momentum distribution. The study of these kinematical regions has only recently become possible with the advent of electron accelerators which produce continous electron beams such as in Mainz (MAMI), at NIKHEF (Amps) or now at CEBAF. A few new results are presented here.

As part of a program to study the Deuteron the reduced cross section $\sigma_{red} = \sigma_{exp}/\kappa\sigma_{ep}^{cc1}$, which in PWIA corresponds to the momentum distribution, has been measured up to missing momenta of 930 MeV/c at momentum transfers between 600 MeV/c and 700 MeV/c. At missing momenta up to 350 MeV/c, the data follow the calculation using the Paris potential, while at larger missing momenta two-body current contributions such as meson exchange currents (MEC) and isobar currents (IC) become very important. It is in this region that considerable differences between the experimental data and calculations seem to exist. These large contributions can be understood by the fact that in order to measure these kinematics, large energy transfers (up to 600 MeV) are necessary which would inclusively correspond to the excitation of the delta. These data are being analysed further.

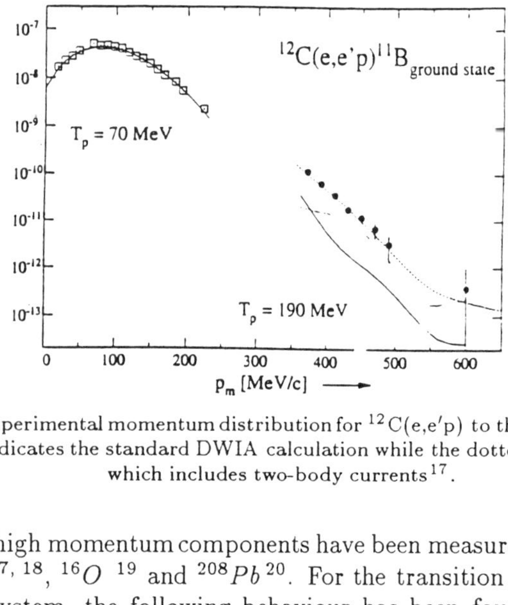

Figure 3: The experimental momentum distribution for $^{12}C(e,e'p)$ to the ground state of ^{11}B. The solid line indicates the standard DWIA calculation while the dotted line is a calculation which includes two-body currents[17].

Recently high momentum components have been measured in heavier nuclei such as ^{12}C [17, 18], ^{16}O [19] and ^{208}Pb [20]. For the transition to bound states in the residual system, the following behaviour has been found: In carbon the DWIA calculation which is fitted to low p_{miss} data could not reproduce the high p_{miss} data (figure 3). The inclusion of long range correlations improve the agreement. The best reproduction is obtained from a Hartree-Fock calculation which includes two-body currents.

At large missing energies the cross section in the continuum is much larger than the one for the bound state and therefore dominates the (e,e'p) cross section. These large energies are populated by two nucleon knock out (mostly np pairs) and are dominated by MEC. In oxygen the proton knock out from the 1p orbitals has been studied[19] for missing momenta up to 700 MeV/c. Here the low missing momentum region as well as the high p_{miss} data, which have been measured at high kinetic proton energies (180 to 250 MeV), could not be reproat a kinetic energy of 93 MeV. These low p_{miss} data agree well with results obtained at NIKHEF (figure 4). At large missing energies the same behavior as

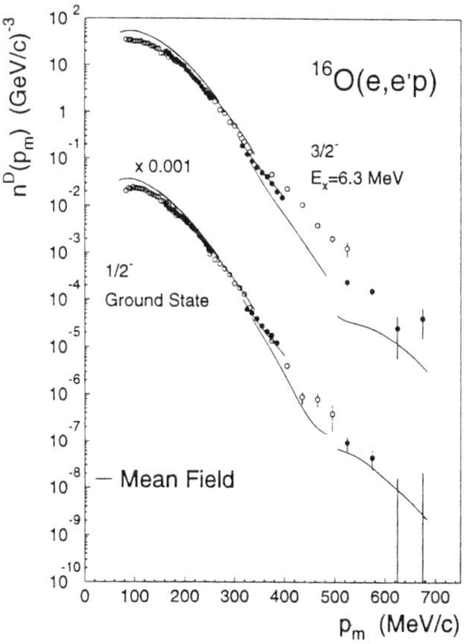

Figure 4: Reduced cross section (momentum distribution in PWIA) for proton knock out of the $1p_{1/2}$ and the $1p_{3/2}$ shells.

observed for carbon has been found. The analysis of the (e,e′p) experiment on ^{208}Pb to the bound state also shows the importance of long range correlations in the nucleon wave function. A recent analysis of the missing energy dependence of the momentum distribution indicates that the effect of long range correlations decreases and that the effect of short range correlations increases with increasing missing energy [21].

4 Summary and Conclusion

New data and a new analysis of inclusive quasi free scattering data on ^{12}C and ^{56}Fe fulfill the coulomb sum rule within the error bars of the experiment without the necessity of modified form factors. All exclusive data available so far, which aimed at the investigation of the form factors of bound nucleons, could be reproduced without modified nucleon form factors within their error bars (15-20 %). High momentum components observed in (e,e′p) reactions to bound states in the residual system seem to be strongly affected by long range correlations. At large missing momenta, the (e,e′p) cross section is dominated by large excitation energies. Modern calculations predict that short range correlations produce large missing energies at large missing momenta. The large missing energy regions seem to be populated by two-body knock out reactions mediated by two-body currents, namely MEC. New experiments are necessary with large momentum transfers and large incident energies to obtain further insight into the nature and importance of short range correlations.

1. J. M. Finn, R. W. Lourie and B. H. Cottman, *Phys. Rev.* **C29**, (1984) 2230
2. J. V. Noble, *Phys. Rev. Lett* **46** (1981) 412
3. P. J. Mulders, *Phys. Rev. Lett.* **54** (1985) 2560
4. P. J. Mulders, *Phys. Rep.* **185** (1990) 83
5. J. Jourdan, *Phys. Lett.* **B353** (1995) 189
6. A.E.L. Dieperink and T. de Forest jr., *Ann. Rev. Nucl. Sc.* **25** (1975)1; S. Frullani and J. Mougey *"Single-Particle Properties of Nuclei through (e,e′p) Reactions"*, *Adv.in Nucl. Phys.* **14** (1984)
7. S.Boffi, C.Giusti and F.D.Pacati, *Phys. Rep.* **226** (1993) 1
8. C.Giusti and F.D.Pacati, *Nucl.Phys.* **A473** (1987) 717; *Nucl.Phys.* **A485** (1988) 461
9. T. de Forest jr., *Nucl. Phys.* **A392** (1983) 232
10. H. W. L. Naus, S. J. Pollock, J. H. Koch and U. Oelfke, *Nucl. Phys* **A509** (1990) 717
11. L. Lapikás, *Nucl. Phys,* **A553** (1993) 297c
12. P.K.A. de Witt Huberts, *Journal of Physics* **C16** (1990) 507
13. L. B. Weinstein et. al. *Phys. Rev. Lett.* **64** (1990) 1646
14. G. van der Steenhoven et. al.,*Phys. Rev. Lett.* **57** (1986) 182, *Phys. Rev. Lett.* **58** (1987) 1727
15. T. D. Cohen, J. W. Van Orden and A. Pickelsheimer, *Phys. Rev. Lett.* **59** (1987) 1267, T. Suzuki, *Phys. Rev.* **C37** (1988) 549
16. M. Kohno, *Phys. Rev.* **C38** (1988) R584
17. L.J.H.M. Kester et. al., *Phys. Lett.* **B 366** (1996) 44
18. L.J.H.M. Kester et. al. *Phys. Lett.* **B344** (1995) 79
19. K.I. Bloqvist et.al. *Phys. Lett* **B344** (1995) 85
20. I. Bobeldijk et.al., *Phys. Rev. Lett.* **73** (1994) 2684
21. I. Bobeldijk et.al., *Phys. Lett.* **B353** (1995) 32

ELECTRON SCATTERING FROM A TENSOR-POLARIZED DEUTERIUM INTERNAL TARGET AT NIKHEF.

J. F. J. van den Brand,[2,3,6]

for the

91-12 COLLABORATION

M. Ferro-Luzzi,[1] M. Bouwhuis,[2] E. Passchier,[2] Z.-L. Zhou,[3] R. Alarcon,[4] M. Anghinolfi,[5] R. van Bommel,[2] T. Botto,[2] J. F. J. van den Brand,[2,3,6] M. Buchholz,[3] H.J. Bulten,[3] S. Choi,[4] J. Comfort,[4] S. Dolfini,[4] R. Ent,[7] C. Gaulard,[4] D. Higinbotham,[8] C. W. de Jager,[2] E. P. van Klaveren,[2] E. Konstantinov,[9] J. Lang,[1] D. J. de Lange,[2] M. A. Miller,[3] D. Nikolenko,[9] G. J. Nooren,[2] N. Papadakis,[2] I. Passchier,[2] H. R. Poolman,[2] S. G. Popov,[9] I. Rachek,[9] M. Ripani,[5] E. Six,[4] J. J. M. Steijger,[2] M. Taiuti,[5] O. Unal,[3] N. Vodanis,[2] H. de Vries.[2]

[1] *Institut für Teilchenphysik, Eidg. Technische Hochschule, CH-8093 Zürich, Switzerland*
[2] *NIKHEF, P.O. Box 41882, 1009 DB Amsterdam, The Netherlands*
[3] *Department of Physics, University of Wisconsin, Madison, Wisconsin 53706, USA*
[4] *Department of Physics, Arizona State University, Tempe, AZ 85287, USA*
[5] *Istituto Nazionale di Fisica Nucleare (INFN), I-16146 Genova, Italy*
[6] *Department of Physics and Astronomy, Vrije Universiteit, Amsterdam, The Netherlands*
[7] *CEBAF, Newport News, VA 23606, and Department of Physics, Hampton University, Hampton, VA 23668, USA*
[8] *Department of Physics, University of Virginia, Charlottesville, VA 22901, USA*
[9] *Budker Institute for Nuclear Physics, Novosibirsk, 630090 Russian Federation*

We report on measurements with a polarized deuterium target internal to the NIKHEF electron storage ring. The target and polarimetry techniques are described. Results are given for an absolute measurement of the tensor analyzing powers T_{20} and T_{22} in elastic electron-deuteron scattering at momentum transfer of 1.6 fm^{-1}. The techniques demonstrated have broad applicability to further measurements of spin-dependent electron scattering.

1. Introduction

Measurements of spin-dependent electron scattering have the potential to enhance our understanding of nucleon and nuclear structure. For example, spin observables in elastic, quasi-elastic, and deep-inelastic scattering from polarized deuterium are predicted to provide important information on the effects of D-wave components in the ground state of ^{2}H, the largely unknown charge form factor of the neutron, and the neutron spin structure functions. This has prompted development of both polarized ^{2}H targets for use with internal [1] or external beams [2] and polarimeters for measuring the polarization of recoiling hadrons [3]. Indeed the first round of measurements of spin-dependent e-^{2}H scattering has been carried out at Novosibirsk [4,5], Bonn [6], MIT-Bates [7,8], and SLAC [9].

Polarized internal gas targets in electron storage rings have the advantage that spin-dependent scattering from chemically and isotopically pure atomic species of high polarization can be realized. They offer rapid polarization reversal and flexible orientation of the nuclear spin direction by using low magnetic holding fields, a low thickness at high luminosity which allows for the detection of low-energy recoiling hadrons, and access to a broad kinematic range by using large acceptance detectors. For polarized deuterium one has the additional ability to reverse the tensor polarization, P_{zz}, at fixed vector polarization, P_z, and vice versa. Consequently, small systematic errors can be expected.

Since the physics issues in polarized elastic electron-deuteron scattering are well known, we restrict the discussion in section 2. to an overview of the target setup and the polarimetry techniques. The tensor analyzing powers T_{20} and T_{22} for elastic electron-deuteron scattering were measured at a momentum transfer of 1.6 fm^{-1} and the data are presented in section 3..

2. Tensor-Polarized Deuterium Target and Polarimetry

An overview of the experiment is given in Fig. 1. The deuterium target is located internal to the AmPS electron ring at NIKHEF. Nuclear-polarized deuterium gas is provided by an atomic beam source (ABS). This source[1] consists of an RF-dissociator, a cooled nozzle, two sextupole electromagnets and two RF transition units[10]. The sextupole magnets focus, according to the Stern-Gerlach principle, atoms with electron spin up (states 1,2,3) and reject the states with spin down (4,5,6). A medium field unit induces a 1-4 transition, while a strong field unit induces either a 3-5 transition or a 2-6 transition, resulting in a tensor polarization of -2 or +1, respectively.

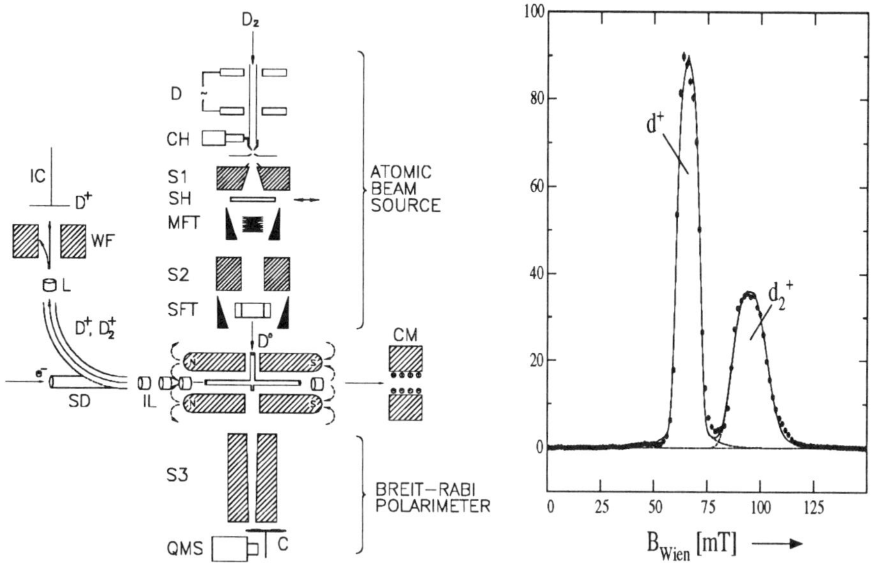

Figure 1: Left panel: Schematic outline of the atomic beam source, Breit-Rabi polarimeter, internal target and ion extraction system. All components, except the target holding field and the compensation magnet, are in the vacuum. D: RF-dissociator; CH: coldhead; S1,S2,S3: sextupole magnets; SH: shutter; MFT (SFT) medium (strong) field transition unit; IL: ion extraction lenses; SD: spherical deflector; WF: Wien filter; IC: ion collector; QMS: quadrupole mass spectrometer; C: chopper. Right panel: Ion current (nA) as a function of magnetic field in the Wien filter. The curves represent fits to the data.

The atomic beam from the ABS is injected into a T-shaped conductance limiter. In this way the atomic density is increased by two orders of magnitude compared to a free atomic beam. This so-called storage cell was cooled to 150 K to further increase the target density and polarization. The cell has been constructed from ultra-pure aluminum with a thickness of 25 μm. In order to minimize recombination and depolarization, it has been coated with a solution of PTFE3170 liquid Teflon in water. The storage cell (feed tube) has dimensions of 400 mm (130 mm) in length and 15 mm (12 mm) in diameter. Presently we obtain with this geometry a target thickness (viewed by the detectors) of 2×10^{13} at/cm^2. A small sample tube (4 mm diameter) is located opposite to the feed tube, and allows to extract a fraction (about 10%) of the injected gas for subsequent analysis by a Rabi polarimeter. This polarimeter consists of a 41 cm long permanent tapered sextupole magnet (entrance (exit) radius 3 mm (1.5 mm), tip field about 0.44 T) in front of a quadrupole mass spectrometer.

A chopper, located in between the magnet and the mass spectrometer, could be employed to subtract background gas. The Rabi polarimeter was used to continuously monitor the performance of the ABS during data taking.

Nuclear polarization was measured *in situ* by a polarimeter[11], which uses the well-known tensor analyzing power of the reaction $^3H(^2H,n)^4He$. The electron beam ionizes a fraction of the deuterium gas in the storage cell. These ions are extracted by a set of electrostatic lenses and an electrostatic spherical deflector, accelerated to 60 keV and transported to a tritiated titanium target. The 13 MeV recoiling neutrons are detected in two neutron detectors, located under 0° and 90°. The angle-dependent count rate for the reaction $^3H(^2H,n)^4He$ is given by $N(\theta) = N_0(1 + \frac{f}{4}P_{zz}(3\cos^2(\theta) - 1))$, where θ is the angle between the neutron momentum and the polarization axis, $f = 0.95$ for 60 keV deuterons, N_0 is the unpolarized count rate, and P_{zz} the polarization of the impinging deuterons. With this polarimeter, the tensor polarization of the target could be measured to a precision of 0.01 within seconds of beam time. Excellent separation of atomic (d^+) and molecular (d_2^+) ions was obtained by applying a Wien filter (see Fig. 1). The systematic error on P_{zz} measured with this polarimeter is estimated to be 2 %.

With the tritium polarimeter we measured the nuclear polarization of the *atoms* ($P_{zz}^+(D^+) = +0.664 \pm 0.019 \pm 0.017$ and $P_{zz}^-(D^+) = -1.215 \pm 0.036 \pm 0.023$). The target contains also molecules originating from imperfect dissociation, residual gas in the target chamber and ABS, and from recombination on the cell walls. We accurately determined the atomic fraction, by performing Wien scans of the ion current as a function of the magnetic field and by alternating between data from the ABS and from a buffer vessel, in which we could prepare arbitrary mixtures of deuterium (mass 4) and hydrogen (mass 2). The atomic fraction, $\eta = \frac{t_{D^+}}{t_{D^+}+2t_{D_2^+}}$ where t_i is the areal target thickness of the species, amounted to $\eta = 0.71 \pm 0.02$. Furthermore, the flexibility of our setup enabled us to study depolarization effects due to the spin-exchange mechanism, and due to electron-beam induced hyperfine transitions. We also found that the recombined molecules do retain part of the atomic polarization ($P_{zz}(D_2^+) = 0.81 \pm 0.33$). These issues are relevant for future measurements in the field (e.g. HERMES at DESY).

3. Tensor Analyzing Powers for Elastic e-d Scattering

Elastic electron scattering from the deuteron is completely described by the G_C, G_Q, and G_M form factors. Cross section measurements yield the structure functions $A(G_C,G_Q,G_M)$ and $B(G_M)$, which combined with $T_{20}(G_C,G_Q,G_M)$

allow the determination of these form factors. The present data set for elastic e-^{2}H scattering and the isoscalar charge form factor of the three-body system pose an interesting puzzle[12] for few-body theory: there exists *no* theoretical model capable of describing all measurements simultaneously. T_{20} has only a limited data set, and accurate measurements of this observable are required to address this issue. In addition, a measurement of T_{22} provides a stringent consistency check, since the value of T_{22} can be unambiguously determined from unpolarized elastic electron scattering.

We have measured the tensor analyzing powers T_{20} and T_{22}, with the Amsterdam Pulse Stretcher ring at NIKHEF with 565 MeV electrons. Several beam bunches were stacked into the ring, yielding currents up to 120 mA and a life time exceeding 15 minutes. Scattered electrons were detected in an electromagnetic calorimeter consisting of six layers of CsI(Tl) blocks covering a solid angle of 180 msr. Two plastic scintillators, one in front of the CsI(Tl) blocks, one sandwiched between the first two layers, provided the electron trigger. The total energy resolution obtained (about 8%) was sufficient to distinguish quasi-elastic events from inelastic events, in which a pion was produced. Two sets of wire chambers, one adjacent to the scattering chamber, one in front of the first trigger scintillator, were used for track reconstruction. The central angle of the electron detector was positioned at 35°. This resulted in a coverage in four-momentum transfer between $1.3 < q < 2.1$ fm^{-1} with a cross section and acceptance weighted average of $\bar{q} = 1.58$ fm^{-1}. The recoil deuterons were detected in a range telescope, consisting of 15 layers of 1 cm thick plastic scintillator preceded by 1 layer of 2 mm thickness. The detector was positioned at a central angle of 80°, and covered a solid angle of nearly 300 msr. The range telescope was preceded by two sets of wire chambers for track reconstruction. The minimum energy of the detected deuterons was 19 MeV. The left panel in Fig. 2 shows a time difference spectrum between electrons and hadrons for the event sample with and without cuts on angular correlations. It is seen that an unambiguous separation of the deuterons from protons was obtained by differences in time-of-flight, in energy loss in the scintillators, and by requiring kinematic correlations between electron and deuteron events. The right panel in Fig. 2 shows the reconstructed vertex distribution along the storage cell for ^{2}H(e,e$'$d) events. The expected triangular density distribution of the gas is observed, as illustrated by the curve obtained with a Monte Carlo technique. We also included in Fig. 2 the results from a run without gas flow from the ABS, established by inserting a shutter (shaded histogram). This unpolarized background amounts to 4% and was taken into account in the polarization determination.

The asymmetry $A = \sqrt{2}\frac{N^+ - N^-}{N^- \cdot P^+_{zz} - N^+ \cdot P^-_{zz}}$ was determined, where N^+ (N^-)

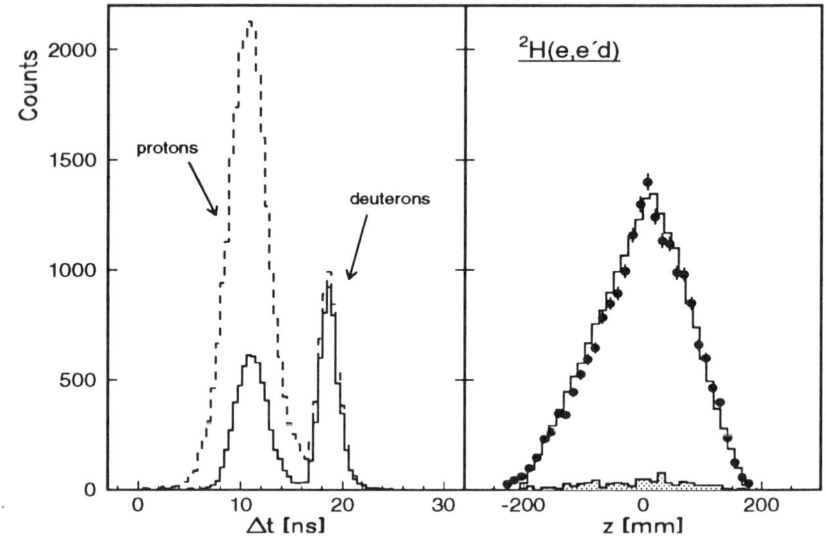

Figure 2: Left panel: time difference between electron and hadron for the event sample with (solid histogram) and without (dashed histogram) cuts on angular correlations. Right panel: reconstructed vertex position along the beam axis for the reaction ^{2}H(e,e$'$d). Both measurements with injected atomic beam (circles) and empty storage cell (shaded histogram) are shown. The curve indicates the Monte Carlo prediction.

corresponds to the number of events measured for P^+_{zz} (P^-_{zz}). The target spin vector was directed approximately parallel or perpendicular to the momentum transfer. The average spin angle was optimized for quasi-free scattering kinematics, resulting for elastic scattering in $\bar{\theta}^*_\parallel = 15°$ and $\bar{\theta}^*_\perp = 89°$. We found $A_\parallel = -0.317 \pm 0.028$ and $A_\perp = 0.210 \pm 0.019$. Since the spin dependent cross section is dominated by T_{20}, we expect the asymmetry to follow the spin-angular dependence of the associated Legendre function $P_2^0 = \frac{1}{2}(3\cos^2\theta^* - 1)$. The predicted change of sign in the tensor asymmetry is clearly observed when changing the spin angle. As a check on false asymmetries, we measured the asymmetry for unpolarized target gas and obtained $A_{unp} = 0.000 \pm 0.014$.

The 7% systematic error in T_{20} includes the uncertainties from the polarization measurement (6.6%) and the spin orientation angles (1.9%). Also included is the uncertainty from the correction for the small T_{21} and T_{22} contributions, which amounts to 0.7% and has been estimated from the existing data for $G_Q(Q^2)$ and $B(Q^2)$. The data for the two spin orientations also allow to extract T_{22}, resulting in $< T_{22} >= 0.022 \pm 0.019 \pm 0.003$. In Fig. 3 we compare the data for T_{20} and T_{22} with the results of other measurements and several state-of-the-art models [13,14,15,16,17].

We have included the results of experiments measuring T_{20} and T_{22} through

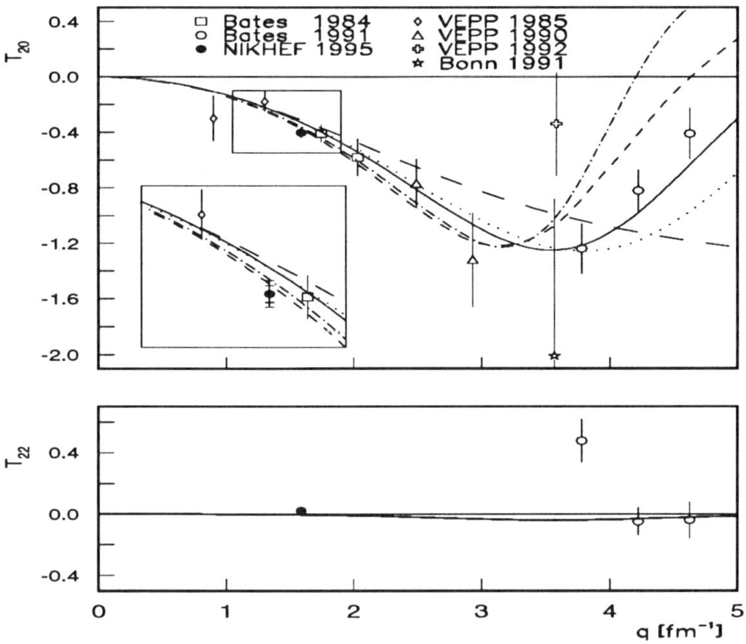

Figure 3: Data and theoretical predictions for T_{20} and T_{22}, as a function of momentum transfer. The curves represent various theoretical models: short-dashed for the NRIA including MEC from Ref. [13] by using the Argonne V14 potential; solid from Ref. [14] for the RIA including $\rho\pi\gamma$ and $\omega\sigma\gamma$ contributions; dotted for the RIA from Ref. [15] by using the Bonn Q potential; dot-dashed for the coupled-channels calculation of Ref. [16] for IA + MEC and model D′; long-dashed for the PQCD prediction of Ref. [17]. The inset shows the results in the q (T_{20}) range from 1.05 fm^{-1} (-0.55) to 1.90 fm^{-1} (-0.10). The heavy curve in the left bottom figure represents T_{22} calculated from a fit to the world data for $B(Q^2)$.

the recoil polarization technique [7,8]. The variation in theoretical predictions for T_{20} shows that precise measurements up to $q = 4$ fm^{-1} can distinguish between the various models shown. The experimental techniques established here are sufficient to obtain such data, with small statistical and systematic uncertainties, within a reasonable time frame.

Presently, we are upgrading both the ABS (by installing stronger sextupole magnets) and the electron storage ring (a smaller β function at the target region will enable the use of a cell with a smaller diameter. Furthermore, polarized electrons will be available in the fall of 1996). These modifications will allow us to measure polarization observables for hydrogen, deuterium, and ^3He with high precision. For example, we intend to measure T_{20} up to a momentum transfer of about 4 fm^{-1} in the spring of 1997.

Acknowledgements

This work was supported in part by the Stichting voor Fundamenteel Onderzoek der Materie (FOM), which is financially supported by the Nederlandse Organisatie voor Wetenschappelijk Onderzoek (NWO), the Swiss National Foundation, the National Science Foundation under Grants No. PHY-9316221 (Wisconsin), PHY-9200435 (Arizona State) and HRD-9154080 (Hampton), Nato Grant No. CRG920219, NWO Grant No. 713-119, and HCM Grant No ERBCHBICT-930606 and ERB4001GT931472.

References

1. Z.-L. Zhou *et al.*, Nucl. Instrum. Methods. **A378**, 40 (1996).
2. D.C. Crabb and D. Day, in *Proceedings of the 7th Workshop on Polarized Target Materials and Techniques, Bad Honnef, Germany, 1994* [Nucl. Instrum. Methods Phys. Res., Sect. A **9** (1995)].
3. S. Kox *et al.*, Nucl. Instrum. Methods. **A346**, 527 (1994); J.S. Réal, Thèse de l'Université de Grenoble 1, ISN 94-003, 1994, unpublished;
4. V.F. Dmitriev *et al.*, Phys. Lett. 157**B**, 143 (1985); B.B. Voitsekhovskii *et al.*, JETP Lett. **43**, 733 (1985).
5. R. Gilman *et al.*, Phys. Rev. Lett. **65**, 1733 (1990).
6. B. Boden *et al.*, Z. Phys. **C49**, 175 (1991).
7. M.E. Schulze *et al.*, Phys. Rev. Lett. **52**, 597 (1984).
8. I. The *et al.*, Phys. Rev. Lett. **67**, 173 (1991); M. Garçon *et al.*, Phys. Rev. C **49**, 2516 (1994).
9. K. Abe *et al.*, Phys. Rev. Lett. **75**, 25 (1995).
10. M. Ferro-Luzzi, Z.-L. Zhou, H.J. Bulten and J.F.J. van den Brand, Nucl. Instrum. Methods **A364**, 44 (1995).
11. Z.-L. Zhou, M. Ferro-Luzzi, J.F.J. van den Brand, H. J. Bulten *et al.*, accepted for publication in Nucl. Inst. Meth. **A** (June 1996).
12. H. Henning, J. Adam, Jr., P.U. Sauer, and A. Stadler, Phys. Rev. **C52**, R471 (1995).
13. R. Schiavilla and D.O. Riska, Phys. Rev. C **43**, 437 (1990).
14. E. Hummel and J.A. Tjon, Phys. Rev. C **42**, 423 (1990).
15. P.L. Chung, F. Coester, B.D. Keister, and W.N. Polyzou, Phys. Rev. C **37**, 2000 (1988).
16. P.G. Blunden, W.R. Greenberg, and E.L. Lomon, Phys. Rev. C **40**, 1541 (1989).
17. C.E. Carlson, Nucl. Phys. **A508**, 481c (1990).

Deuteron Photodisintegration: Preliminary Results from CEBAF

D.J. Abbott,[1] A. Ahmidouch,[2] C.S. Armstrong,[3] J. Arrington,[4] K.A. Assamagan,[5] O.K. Baker,[5] S.P. Barrow,[6] D.P. Beatty,[6] D.H. Beck,[7] S.Y. Beedoe,[8] E.J. Beise,[9] J.E. Belz,[10] C.W. Bochna,[7] P.E. Bosted,[11] E.J. Brash,[12,17] H. Breuer,[9] R.V. Cadman,[7] L. Cardman,[1] R.D. Carlini,[1] J. Cha,[5] N.S. Chant,[9] G. Collins,[9] C. Cothran,[13] W.J. Cummings,[14] S. Danagoulian,[8] F.A. Duncan,[9] J.A. Dunne,[1] D. Dutta,[15] T. Eden,[5] R. Ent,[1] B.W. Filippone,[4] T.A. Forest,[7] H.T. Fortune,[6] V.V. Frolov,[16] H. Gao,[7,*] D.F. Geesaman,[14] R. Gilman,[17] P.L.J. Gueye,[5] K.K. Gustafsson,[9] J-O. Hansen,[14] M. Harvey,[5] W. Hinton,[5] R.J. Holt,[7] H.E. Jackson,[14] C.E. Keppel,[5] M.A. Khandaker,[18] E.R. Kinney,[19] A. Klein,[20] D.M. Koltenuk,[6] A.F. Lung,[9] D.J. Mack,[1] R. Madey,[2,5] P. Markowitz,[21] K.W. McFarlane,[18] R.D. McKeown,[4] D.G. Meekins,[3] Z-E. Meziani,[22] M.A. Miller,[7] J.H. Mitchell,[1] H.G. Mkrtchyan,[23] R.M. Mohring,[9] J. Napolitano,[16] A.M. Nathan,[7] G. Niculescu,[5] I. Niculescu,[5] T.G. O'Neill,[14] B.R. Owen,[7] S. Pate,[24] D.H. Potterveld,[14] J.W. Price,[16] G.L. Rakness,[19] R. Ransome,[17] J. Reinhold,[14] P.M. Rutt,[17] G. Savage,[5] R.E. Segel,[15] N. Simicevic,[7] P. Stoler,[16] L. Tang,[5] B.P. Terburg,[7] D. Van Westrum,[19] W.F. Vulcan,[1] S. Williamson,[7] M.T. Witkowski,[16] S.A. Wood,[1] C. Yan,[1] B. Zeidman[14]

[*] *Speaker.* [1] *Thomas Jefferson National Accelerator Facility.* [2] *Kent State University.* [3] *College of William and Mary.* [4] *California Institute of Technology.* [5] *Hampton University.* [6] *University of Pennsylvania.* [7] *University of Illinois at Urbana-Champaign.* [8] *North Carolina A&T State University.* [9] *University of Maryland.* [10] *TRIUMF.* [11] *American University.* [12] *University of Regina.* [13] *University of Virginia.* [14] *Argonne National Laboratory.* [15] *Northwestern University.* [16] *Rennselaer Polytechnic Institute.* [17] *Rutgers University.* [18] *Norfolk State University.* [19] *University of Colorado at Boulder.* [20] *Old Dominion University.* [21] *Florida International University.* [22] *Temple University.* [23] *Yerevan Physics Institute.* [24] *New Mexico State University.*

CEBAF experiment E89-012 has measured the $d(\gamma,p)n$ differential cross section up to 4.0 GeV and the $d(\gamma,d)\pi^0$ cross section up to 3.2 GeV. We discuss the preliminary results of $d(\gamma,p)n$ at proton center-of-mass angles of 90° and 37°. The previous SLAC data at 90° are consistent with the constituent-counting-rule behavior at photon energies larger than 1.0 GeV. The CEBAF 90° data show good agreement with the previous SLAC data and continue to show the scaling behavior. The 37° data also show good agreement with the SLAC data and suggest the onset of the scaling behavior at photon energies above 3.0 GeV. Detailed analysis of $d(\gamma,p)n$ at all measured proton angles and of $d(\gamma,d)\pi^0$ is in progress.

Introduction

The deuteron is the simplest nuclear system and as such should be an excellent testing ground for investigating the transition from conventional nucleon-meson descriptions of a nuclear system to QCD descriptions in terms of quarks and gluons. Deuteron photodisintegration is a reaction that addresses the above question because large momentum transfers to the constituents can be obtained at photon energies of a few GeV [1].

For the exclusive two-body scattering process A+B → C+D at high energy and large transverse momentum, dimensional analysis predicts the following constituent counting rule for the differential cross-section [2]:

$$\frac{d\sigma}{dt} \propto s^{-(n-2)}, \tag{1}$$

where s and t are the Mandelstam variables, and n is the total number of elementary fields (photon, quark, etc.,) in the initial and final states. The constituent-counting-rule behavior has been observed in exclusive processes such as proton-proton scattering [3]. One of the very important questions in nuclear physics is whether nuclear reactions obey any scaling at high energy and large transverse momentum. For the $d(\gamma, p)n$ process, the constituent counting rule predicts:

$$\frac{d\sigma}{dt} \propto s^{-11}. \tag{2}$$

Fig. 1 shows the previous data for this reaction at $\theta_{c.m.} = 90^\circ$ of $s^{11}d\sigma/dt$ as a function of the photon energy. These data indicate scaling behavior starting at photon energies around 1 GeV, corresponding to proton transverse momenta of $\sim$ 1.0 GeV/c. The solid line represents Lee's meson-exchange calculation [6], which is a traditional calculation that reproduces the measured NN phase shifts up to 2.0 GeV and is also constrained by photomeson production data. Below 500 MeV the calculation gives a reasonable description of the data, but above 1.0 GeV the calculation disagrees with the data. The dashed line represents a reduced nuclear amplitude calculation [7] with a normalization factor chosen to agree with a data point at $E_\gamma = 0.8$ GeV. This curve falls below the high E_γ data and does not reach an asymptotic limit at these energies. Fig. 2 shows a similar plot of the SLAC NE17 data [4] at $\theta_{c.m.} = 37^\circ$. The experimental uncertainties preclude any conclusion with regard to the scaling behavior in the photon energy range below 3 GeV. In addition to Lee's meson-exchange calculation [6], Nagornyi *et al.* [8] have calculated the cross section and the angular distribution for the $d(\gamma, p)n$ reaction in the few GeV region based on the meson theory. However, their calculation can not describe the experimental data at the 90° and the 37° simultaneously.

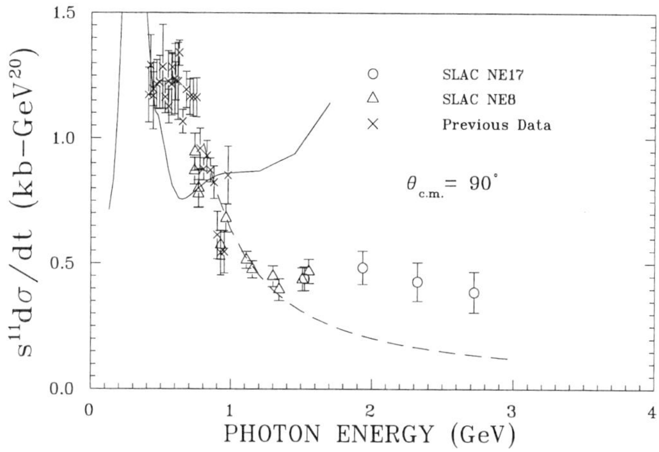

Figure 1: Existing data on deuteron photodisintegration as a function of the photon energy at $\theta_{c.m.} = 90°$. The open circles are the SLAC NE17 data [4], the open triangles are the SLAC NE8 data [5], and the crosses are all the other existing data. The solid line is a traditional meson-exchange calculation [6], and the dashed line is a reduced nuclear amplitude calculation [7].

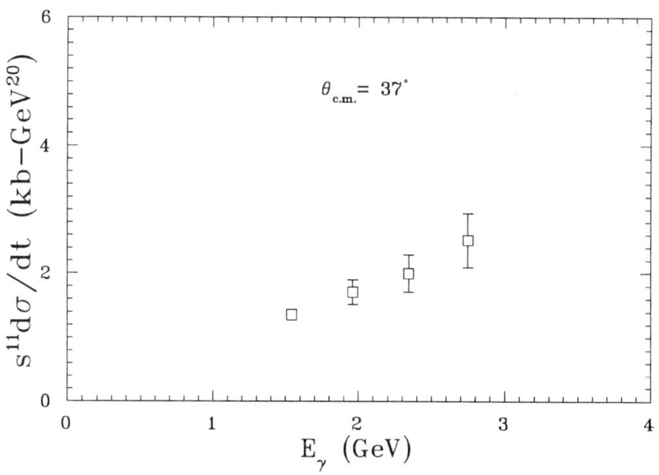

Figure 2: Existing data on deuteron photodisintegration as a function of the photon energy at $\theta_{c.m.} = 37°$ from the SLAC NE17 experiment [4].

Experimental Overview

In the present experiment (CEBAF E89-012) an untagged bremsstrahlung photon beam was incident on a cryogenic liquid deuterium target. The bremsstrahlung photons are produced by an electron beam incident on a copper radiator. Photoprotons are detected using a magnetic spectrometer. Since the $\gamma d \rightarrow pn$ reaction is a two-body process, the photon energy can be reconstructed by detecting the final state proton momentum and scattering angle. Events with pion production can be excluded by accepting only the protons with the highest momentum. The advantage of using an untagged bremsstrahlung photon beam near the end point is a photon flux that is several orders of magnitude higher than that of a tagged photon beam. This is very important for extending the measurement to higher energy because the cross section falls off very rapidly as the beam energy increases.

Experiment E89-012 was performed in Hall C at the Thomas Jefferson National Accelerator Facility (formerly CEBAF) in the spring of 1996. A 20 μA continuous (CW) electron beam at beam energies of 0.844 to 4.044 GeV in steps of 0.8 GeV was used. A 15-cm liquid deuterium target and a 6% copper radiator were employed for the experiment. The high momentum spectrometer (HMS) was used to detect protons at forward angles in the center-of-mass system and the short orbit spectrometer (SOS) was used for the backward angles. Both spectrometers have similar detector systems. Plastic scintillators were used to form the triggers and also to provide the time-of-flight information for particle identification. Drift chambers were employed for measuring the particle momentum and scattering angle. In addition, a gas Čerenkov counter and a shower counter were part of the detector system for particle identification. This experiment had good overlap with the previous SLAC measurements [4-5]. It not only extended the measurements to higher energies (4.0 GeV), but also had a more complete angular coverage. Although the $d(\gamma, d)\pi^0$ differential cross section was also measured in the present experiment, we discuss only the preliminary results from $d(\gamma, p)n$ here.

Data Analysis

As was discussed earlier the photon energy can be reconstructed from the two-body kinematics by detecting the final state proton momentum and angle. Thus, it is important to distinguish protons from pions and deuterons. For a proton momentum less than 2.7 GeV/c, a time-of-flight cut was sufficient to separate protons from pions and deuterons. For a proton momentum larger than 2.7 GeV/c, time-of-flight together with a gas Čerenkov counter provided the required proton/pion separation and the deuteron rejection. In addition to the particle identification, reasonable cuts on the reconstructed spectrometer quantities, both at the target and the focal planes, were applied.

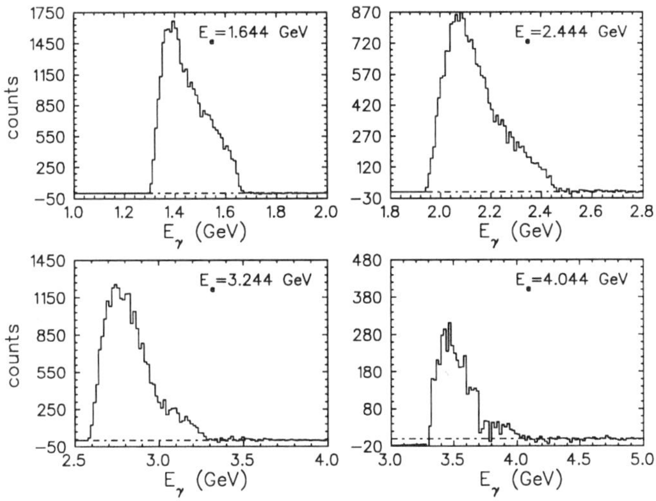

Figure 3: Proton spectra (background subtracted as discussed in the text) as a function of the reconstructed photon energy at a proton center-of-mass angle of 90°.

Background contributions from the target windows were removed by placing cuts on the reconstructed target position and subtracting the results obtained with an empty target cell of identical dimensions. The yield from electrodisintegration was measured by repeating the procedure without the radiator present. This background was subtracted from the photodisintegration yield with a correction factor taking into account the modification of the electron beam's flux and energy distribution by the radiator [9]. Fig. 3 shows the background-corrected proton spectra as a function of the reconstructed photon energy at four different beam energies for $\theta_{c.m.} = 90°$. The bremsstrahlung end-point which is determined by the electron beam energy is clearly visable in each case. The cut-off on the low energy side of the spectrum is caused by the spectrometer accepatnce.

Preliminary Results

To obtain the differential cross-section from the measurements, protons in the reconstructed photon energy bin of $E_0 - 85$ MeV $\leq E_\gamma \leq E_0 - 25$ MeV were selected for $\theta_{c.m.} = 37^\circ$, where E_0 is the electron beam energy. The energy bin size was chosen so that no protons from photopion production would be identified as protons from $d(\gamma, p)n$ process. Likewise for $\theta_{c.m.} = 90^\circ$, a reconstructed photon energy bin of $E_0 - 125$ MeV $\leq E_\gamma \leq E_0 - 25$ MeV was chosen. The bremsstrahlung photon flux was calculated with an estimated 3% uncertainty using the thick-target bremsstrahlung computer codes of Matthews and Owens[10], which were cross-checked by an independent code developed by Belz[9]. The spectrometer solid angle for the extended target was obtained from preliminary Monte Carlo simulations for each deuteron photodisintegration kinematics. The simulated acceptance agrees with the measurement to better than 10%. The liquid deuterium target density variation was measured to be less than 1%. The electron beam current was monitored continuously during the experiment using RF cavity monitors, and the beam current measurement was calibrated to 2% for a beam current of 20 μA. The absolute beam energy was measured to an accuracy of about 0.1%.

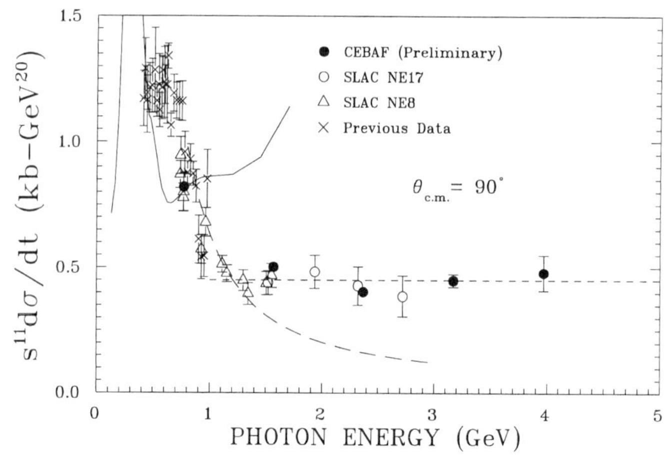

Figure 4: The preliminary CEBAF data together with the existing data as a function of the photon energy at $\theta_{c.m.} = 90^\circ$. The solid circles are the CEBAF data with statistical uncertainties only.

A proton absorption correction was applied to compensate for the proton loss going through the detector stack. This correction factor was measured for this experiment during the spectrometer checkout to be ($8\% \pm 1\%$). Corrections were also applied for the computer deadtime and the tracking efficiency.

Fig. 4 shows the preliminary CEBAF data at $\theta_{c.m.} = 90^\circ$ together with the previous data. The CEBAF data are in good agreement with the previous

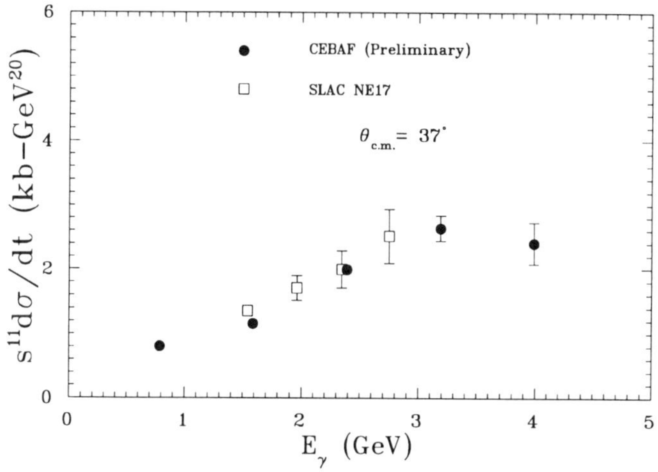

Figure 5: The preliminary CEBAF data together with the SLAC NE17 data [4] as a function of the photon energy at $\theta_{c.m.} = 37°$. The CEBAF data are shown with the statistical uncertainties only.

SLAC data in the overlapping region. The CEBAF data continue to show scaling behavior up to the highest beam energy of 4.0 GeV. The short-dashed line is the weighted average of the SLAC NE17 data [4] and the CEBAF data. Fig. 5 shows the preliminary CEBAF data at $\theta_{c.m.} = 37°$ together with the SLAC NE17 data [4]. The solid circles are the CEBAF data with statistical uncertainties only and are in good agreement with the SLAC NE17 data. The CEBAF data at 37° suggest the onset of the scaling behavior at a proton transverse momentum P_t of ~ 1.0 GeV/c, where $P_t{}^2 = \frac{1}{2} M_d E_\gamma \sin(\theta_{cm})^2$ with M_d the deuteron mass. The transverse momentum P_t of ~ 1.0 GeV/c corresponds to roughly the same P_t in the 90° case where the onset of scaling behavior is observed. The overall systematic uncertainty is dominated by the uncertainty in the tracking efficiency. For the CEBAF preliminary data at 90° and 37°, the overall systematic uncertainty is estimated to be 30%, but is expected to be $5 - 10\%$ after a more detailed analysis.

Summary

The CEBAF data show good agreement with the previous SLAC data in the overlapping region. The data at $\theta_{c.m.} = 90°$ continue to show scaling behavior and the data at 37° suggest the onset of scaling behavior at a proton transverse momentum of about 1.0 GeV/c. Measurements at higher energies at 37° would test whether scaling has set in at 3.0 GeV. It is surprising that the counting rule appears to work so well considering the fact that the momentum transfer to the individual quark is not above 1.0 $(\mathrm{GeV}/c)^2$. More detailed analysis of the angular distribution in $d(\gamma, p)n$ and of $d(\gamma, d)\pi^0$ is in progress.

Acknowledgments

We acknowledge the outstanding work from the staff of the accelerator division at the Thomas Jefferson National Accelerator Facility in delivering the high quality electron beam. This work is supported in part by the research grants from the U.S. Department of Energy and the U.S. National Science Foundation.

References

1. R.J. Holt, Phys. Rev. C **41**, 2400 (1990).
2. S.J. Brodsky and G.R. Farrar, Phys. Rev. Lett. **31**, 1153 (1973); V. Matveev *et al.*, Nuovo Cimento Lett. **7**, 719 (1973); G.P. Lepage and S.J. Brodsky, Phys. Rev. D **22**, 2157 (1980).
3. G. White *et al.*, Phys. Rev. D **49**, 58 (1994).
4. J.E. Belz *et al.*, Phys. Rev. Lett. **74**, 646 (1995).
5. J. Napolitano *et al.*, Phys. Rev. Lett. **61**, 2530 (1988); S.J. Freedman *et al.*, Phys. Rev. C **48**, 1864 (1993).
6. T.-S.H. Lee, Argonne National Laboratory Report No. PHY-5253-TH-88; T.-S.H. Lee, in *Proceedings of the International Conference on Medium and High Energy Nuclear Physics, Taipai, Taiwan, 1988* (World Scientific, Singapore, 1988), p.563.
7. S.J. Brodsky and J.R. Hiller, Phys. Rev. C **28**, 475 (1983).
8. S.I. Nagornyĭ, Yu.A. Kasatkin, and I.K. Kirchenko, Sov. J. Nucl. Phys. **55**, 189 (1992).
9. J.E. Belz, Ph.D. thesis, California Institute of Technology, 1994 (unpublished).
10. J.L. Matthews and R.O. Owens, Nucl. Instrum. Methods **111**, 157 (1973).

Quasi-free (e,e'p) reactions: the first look from CEBAF

D. Potterveld[1], D. Abbott[21], A. Ahmidouch[7], T. Amatuoni[24], C. Armstrong[23], J. Arrington[2], K. A. Assamagan[6], O. K. Baker[6], S. Barrow[17], K. Beard[6], D. Beatty[17], S. Beedoe[13], E. Beise[11], E. Belz[4], C. Bochna[9], H. Breuer[11], E. E. W. Bruins[10], R. Carlini[21], J. Cha[6], N. Chant[11], C. Cothran[22], W. J. Cummings[1], S. Danagoulian[13], D. Day[22], D. DeSchepper[10], J.-E. Ducret[20], F. Duncan[11], J. Dunne[21], D. Dutta[14], T. Eden[6], R. Ent[21], J. Fedchak[1], H. T. Fortune[17], V. Frolov[18], D. F. Geesaman[1], H. Gao[9], R. Gilman[19], P. Gueye[6], J. O. Hansen[1], W. Hinton[6], R. Holt[9], C. Jackson[13], H. E. Jackson[1], C. Jones[1], S. Kaufman[1], J. J. Kelly[11], C. Keppel[6], M. Khandaker[12], Wooyoung Kim[8], E. Kinney[4], A. Klein[16], D. Koltenuk[17], L. Kramer[10], W. Lorenzon[17], A. Lung[2], K. McFarlane[12], D. Mack[21], R. Madey[6,7], P. Markowitz[5], J. Martin[10], A. Mateos[10], D. Meekins[21], M. Miller[9], R. Milner[10], J. Mitchell[21], H. Mkrtchyan[24], R. Mohring[11], G. Niculescu[6], I. Niculescu[6], T. G. O'Neill[1], J. W. Price[18], J. Reinhold[1], C. Salgado[12], J. P. Schiffer[1], R. E. Segel[14], P. Stoler[18], V. Tadevosyan[24], L. Tang[6], B. Terburg[9], W. Turchinetz[10], D. van Westrum[4], Pat Welch[15], C. Williamson[10], S. Wood[21], C. Yan[21], Jae-Choon Yang[3], J. Yu[17], B. Zeidman[1], W. Zhao[10], B. Zihlmann[22]

[1] *Argonne National Laboratory, Argonne IL 60439* [2] *California Institute of Technology, Pasadena CA 91125* [3] *Chungnam National University, Taejon Korea* [4] *University of Colorado, Boulder CO 80309* [5] *Florida International University, University Park, FL 33199* [6] *Hampton University, Hampton VA 23668* [7] *Kent State University, Kent OH 44242* [8] *Kyungpook National University, Taegu, South Korea* [9] *University of Illinois, Champaign-Urbana IL 61801* [10] *Massachusetts Institute of Technology, Cambridge MA 02139* [11] *University of Maryland, College Park MD 20742* [12] *Norfolk State University, Norfolk VA 23504* [13] *North Carolina A & T, Greensboro NC 27411* [14] *Northwestern University, Evanston IL 60201* [15] *Oregon State University, Corvallis OR 97331* [16] *Old Dominion University, Norfolk, VA 23529* [17] *University of Pennsylvania, Philiadelphia PA 19104* [18] *Rensselaer Polytechnic Institute, Troy NY 12180* [19] *Rutgers University, New Brunswick NJ 08903* [20] *CE Saclay, Gif-sur-Yvette France* [21] *Thomas Jefferson National Accelerator Facility, Newport News VA 23606* [22] *University of Virginia, Charlottesville VA 22901* [23] *William and Mary, Williamsburg, VA 23187* [24] *Yerevan Physics Institute, Yeravan, Armenia*

Coincidence cross sections for $(e, e'p)$ quasi-elastic scattering were measured at CEBAF with high statistical precision for C, Fe, and Au targets for $0.6 < Q^2 < 3.3$ GeV^2. Missing energy and missing momentum distributions obtained from a preliminary analysis are in reasonable agreement with prior data from SLAC. The preliminary results are compared with a PWIA calculation to determine the nuclear transparency as a function of Q^2 and A. A Rosenbluth analysis to extract the longitudinal and transverse cross sections from these data is anticipated.

1 Introduction

The elastic scattering of electrons by a free proton is a well understood process, characterized by two electromagnetic form factors in the absence of polarization. When the scattering occurs from a nucleus, the process is more complex. The dominant reaction mechanism at modest Q^2 for kinematics near that of elastic proton scattering is quasi-elastic scattering, in which the electron scatters elastically from a moving, off-shell proton.[1] In the PWIA approximation, the cross section for this mechanism may be written as:

$$\frac{d^6\sigma}{dk'\, dE'_p\, d\Omega_{k'}\, d\Omega_{P'}} = P'\, E'_p\, \sigma_{ep}\, S(E, \vec{P}) \qquad (1)$$

where k' is the momentum of the scattered electron, P' and E'_p are the momentum and energy of the scattered proton, σ_{ep} is the elementary electron-proton cross section, and $S(E, \vec{P})$ is the probability density for striking a proton with initial energy E and momentum $\vec{P}$. The spectral function S is directly related to the nuclear wave function, and has normalization:

$$\int S(E, \vec{P})\, d^3\vec{P}\, dE = Z. \qquad (2)$$

An exclusive cross section measurement, in which E and P can be (approximately) determined from the final-state kinematics, can thus probe nuclear structure through the spectral function. Complicating this simple picture are effects beyond the PWIA, such as photon radiation, re-interaction of the emerging proton with the A-1 nucleons, and meson exchange currents. Exclusive measurements offer an opportunity to study these effects as well.

Previous work at SLAC[2] and BATES[3] focused on studying the transparency of the residual nucleons to the emerging proton. The SLAC data were compared with theoretical cross sections based on a spectral function derived from an independent particle shell model (IPSM) and including radiative effects and some corrections for short-range correlations in the nuclear wave function.[4] The ratios of experimental to theoretical cross sections $T = \sigma_{\rm exp}/\sigma_{\rm pwia}$ are plotted in Fig. 1. The BATES data were divided by the inclusive cross section after subtraction of the inelastic contribution; these ratios are also plotted in Fig. 1.

Absorption of the scattered protons results in $T < 1$. The ratio decreases with increasing A as one would expect. For $0.3 < Q^2 < 3.0$ GeV^2, T falls with increasing Q^2. This is consistent with the strong energy-dependence of the total proton-nucleon cross section for proton momenta below 1.5 GeV/c. Various authors[5,6,7,8,9,10,11] have attempted to calculate the Q^2 and A dependence of

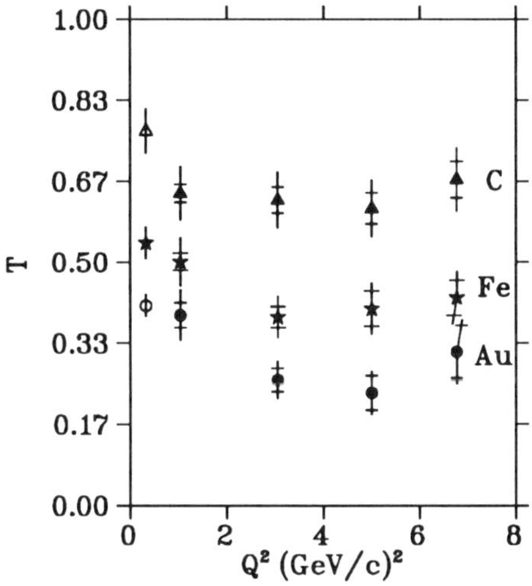

Figure 1: Transparency ratios for $(e, e'p)$ quasi-elastic scattering from C, Fe, and Au targets. The solid data points are the ratio of experimental to PWIA cross sections from Ref. 2. The open points are the ratio of exclusive to inclusive cross sections from Ref. 3. Statistical errors are indicated by horizontal ticks on the total error bars.

these data using a Glauber model of the proton absorption, or including more speculative effects such as color transparency—a proposed reduction in the final state interactions after a hard scattering that would cause an enhancement of the nuclear transparency at high Q^2. The SLAC data are consistent with some Glauber calculations, but cannot rule out the color transparency effect if its onset is slow enough in this kinematic range.

The present experiment studies proton propagation through nuclei in the Q^2 range $0.6 < Q^2 < 3.3$ $(\text{GeV/c})^2$. This is the region where the nuclear transparency is changing rapidly with Q^2, and any color transparency effect is expected to be small. By measuring high statistics data, including forward and backward angle data at the same Q^2 to separate the longitudinal and transverse response, we hope to improve our understanding of proton-nucleon interactions and to test the validity of the quasi-elastic reaction mechanism.

2 Experiment

This experiment (E91-013) was performed at CEBAF[a] in Hall C, using the High Momentum (HMS) and Short Orbit (SOS) spectrometers. A 20 μA

[a]Now the Thomas Jefferson National Accelerator Facility.

Table 1: Summary of measurements. The conjugate proton angles for elastic (e,p) scattering are shown in boldface.

Proton Kinetic Energy (MeV)	Electron Energy (MeV)	Electron Angle (degrees)	Proton Angles (degrees)	Q^2 $(GeV/c)^2$
350	845	78.5	27.8, **31.8**, 35.8, 39.8 43.8, 47.8	0.643
350	2445	20.5	35.4, 39.4, 43.4, 47.4 51.4, **55.4**, 59.4, 63.4 67.4, 71.4, 75.4	0.643
700	2445	32.0	31.0, 35.0, 39.0, **43.0** 47.0, 51.0, 55.0	1.283
970*	1645	80.0	**22.8**, 26.8, 30.8, 34.8	1.781
970	3245	28.6	33.5, 37.5, **40.5**, 44.5, 48.5, 52.5	1.784
1800	3245	50.0	**25.1**, 27.6, 30.1	3.305

* Only C and Fe data for these kinematics.

continuous (CW) beam of electrons was incident on solid targets of C, Fe, and Au, as well as a target of CH_2 plastic and an extended target of liquid hydrogen to check the absolute normalization relative to the well-known *ep* elastic cross section. Table 1 summarizes the measurements made. For each electron angle, cross sections were measured for a range of proton angles spanning the quasi-elastic peak. The HMS was used to detect the scattered electrons and the SOS was used to detect the scattered protons in coincidence, except for the data at a proton kinetic energy of 1.8 GeV, where the spectrometer roles were reversed. Typically, enough data were taken to ensure statistical uncertainties in the cross section of 1% or better.

The SOS consists of a quadrupole magnet followed by two dipoles in a configuration modelled after the MRS spectrometer of LAMPF. It features a short flight path (7.3 m) with large angular acceptance (7.7 msr) and momentum bite (40%), good momentum resolution (0.1%), and a maximum central momentum of 1.7 GeV. The HMS consists of three quadrupoles followed by a dipole, all superconducting with cold iron poles. It has a long flight path (23.2 m), comparable solid angle (6.8 msr), smaller momentum bite (18%), but better momentum resolution (0.05-0.1%), with a maximum central momentum of 7 GeV.

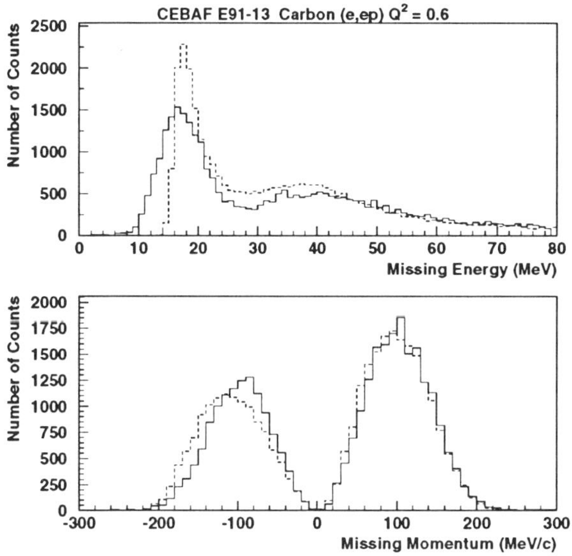

Figure 2: E_m and P_m distributions, not corrected for spectrometer acceptance.

In each spectrometer, four segmented planes of plastic scintillator were used to form a trigger and to provide time-of-flight particle identification. Two 6-plane drift chambers were used to measure the particle trajectories from which the scattering angle and momentum of the particle were calculated. Additional particle identification was provided by a gas threshold Cerenkov detector and a segmented lead glass shower array.

The solid angle of each spectrometer was defined by a 2-inch thick tungsten collimator with a large octagonal hole. These could be exchanged remotely with collimators featuring a grid of small holes. Extensive optics data were taken with these 'sieve-slit' collimators to calibrate the trajectory reconstruction coefficients for the scattering angles. Momentum reconstruction coefficients were determined by observing electrons scattered elastically from C and H nuclei at several spectrometer excitations.

Coincidence triggers were read out through fastbus hardware by the CODA data acquisition system.[12] After time-of-flight corrections, the coincidence timing resolution was typically 0.5 ns (fwhm), with a real to random ratio greater than 200. The calculation of absolute cross sections included corrections for

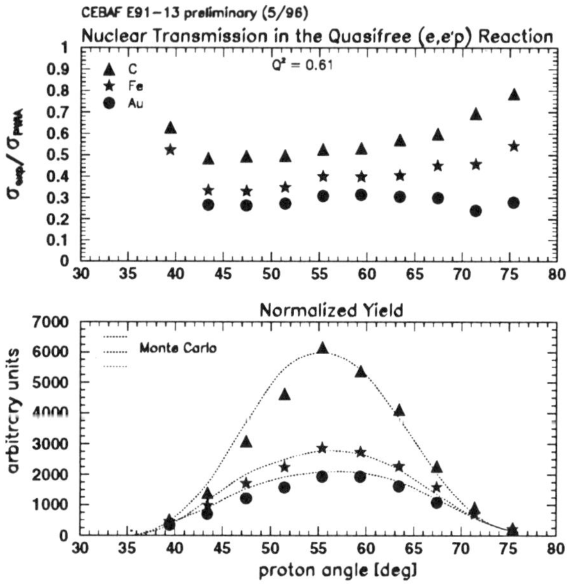

Figure 3: $T = \sigma_{\rm exp}/\sigma_{\rm pwia}$ for $(e, e'p)$ quasi-elastic scattering from C, Fe, and Au targets at $Q^2 = 0.6$ GeV^2, and the yield versus proton angle. The PWIA-based Monte Carlo results are shown as the dotted lines.

detector inefficiencies, computer dead-time, and proton losses from secondary scattering in the target and the spectrometer windows.

3 Preliminary results

The missing energy (E_m) and missing momentum (P_m) are defined as the difference between initial state and observed final-state quantities:

$$\begin{aligned} |P_m| &= |\vec{Q} - \vec{P}'| \\ E_m &= \omega - E'_p + M_p - T_{\rm A-1} \end{aligned} \qquad (3)$$

in which $\vec{Q}$ is the 3-momentum transfer, ω is the energy transfer, M_p is the proton mass, and T_{A-1} is the kinetic energy of the A-1 nucleons, with the convention that P_m is positive (negative) if $\vec{P}'$ is at a larger (smaller) angle with respect to the incident electron than $\vec{Q}$. E_m and P_m may be calculated

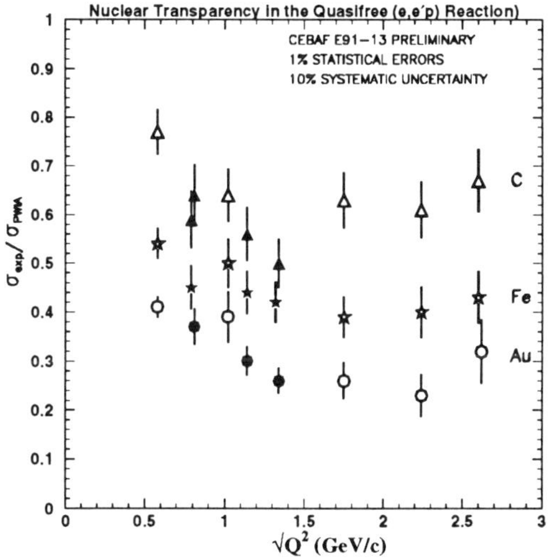

Figure 4: $T = \sigma_{\rm exp}/\sigma_{\rm pwia}$ for $(e, e'p)$ quasi-elastic scattering from C, Fe, and Au targets. Prior data[2,3] is shown as open symbols, and this experiment as solid symbols.

from the observed electron and proton momenta and angles. In Fig. 2 we plot the E_m and P_m distributions for the forward electron-angle C data at $Q^2 = 0.6$ GeV^2 as a solid line. The data have been cut with $|P_m| < 300$ MeV/c and $E_m < 80$ MeV. The scattering from p-shell and s-shell nucleons is clearly resolved into two peaks in E_m.

The dashed lines are the distributions calculated from a realistic Monte Carlo model[13] which includes radiative corrections, normalized to the same number of total events. The agreement is reasonable, especially in the tail region at large E_m where the cut is placed. The p-shell peak at $E_m \sim 20$ MeV appears to be slightly shifted and broadened relative to the Monte Carlo. Similar effects exist in the analysis of (e, p) elastic scattering from the hydrogen target, and are thought to originate in incorrect momentum and scattering angle reconstruction. The ongoing analysis of the spectrometer optics should improve the agreement with the Monte Carlo.

In Fig. 3, we show a comparison of experimental and Monte Carlo cross-sections for all the forward electron-angle, $Q^2 = 0.6$ GeV^2 data, as a function of proton angle. The same P_m and E_m cuts have been applied. In the top panel we plot the ratio of the cross sections, and in the bottom panel we plot the yields, with the Monte Carlo normalized to the average T of the corresponding data. All statistical uncertainties are smaller than the plotting symbols. One can see that there is essentially complete coverage of the proton angular distribution. The ratio of cross sections is fairly flat in the center, but rises at small and large proton angles. These angles correspond to large initial-state proton momenta that are underrepresented in the Monte Carlo (since the IPSM lacks the correlations from which they arise), leading to an increase in the ratio.

A 'central' value for the ratio of experimental to PWIA cross sections at each Q^2 is plotted versus $\sqrt{Q^2}$ in Fig. 4 for the partial data set analyzed as of May 1996. A correction for the wave function correlations has been included. These preliminary results contain a 10% systematic uncertainty, which is dominated by the uncertainty with which we understand the spectrometer acceptances. The systematic uncertainties should decrease substantially after all the optics data have been analyzed. We can see that there is fair agreement with the previous data. A full analysis of the entire data set, including a Rosenbluth separation of the longitudinal and transverse cross sections, is in progress.

Acknowledgments

We acknowledge the efforts of the TJNAF accelerator division in delivering high quality beam for this experiment. This work was supported in part by the U.S. Department of Energy, Nuclear Physics Division, under contract W-31-109-ENG-38, and by the National Science Foundation.

References

1. T. De Forest, *Nucl. Phys.* A **392**, 232 (1983).
2. T.G. O'Neill et al., *Phys. Lett.* B **351**, 87 (1995).
3. G. Garino et al., *Phys. Rev.* C **45**, 780 (1992).
4. T.G. O'Neill, California Institute of Technology Ph.D. thesis (1994).
5. G.R. Farrar et al., *Phys. Rev. Lett.* **61**, 686 (1988).
6. B.K. Jennings and G.A. Miller, *Phys. Rev.* D **44**, 692 (1991).
7. O. Benhar et al., *Phys. Rev. Lett.* **69**, 881 (1992).
8. L.L. Frankfurt et al., *Phys. Rev.* C **50**, 2189 (1994).
9. N.N. Nikolaev et al., *Nucl. Phys.* A **567**, 781 (1994).
10. A. Kohama, K. Yazaki, and R. Seki, *Nucl. Phys.* A **536**, 716 (1992).
11. A.S. Rinat and B.K. Jennings, *Nucl. Phys.* A **568**, 873 (1994).
12. W.A. Watson III et al., IEEE Trans. Nuc. Sci. **41**, 61.
13. N.C.R. Makins et al., *Phys. Rev. Lett.* **72**, 1986 (1994).

LOW ENERGY TESTS OF CHIRAL SYMMETRY

BARRY R. HOLSTEIN
Department of Physics and Astronomy, University of Massachusetts
Amherst, MA 01003

The present status of low energy tests of chiral invariance via chiral perturbation theory is reviewed, both in the meson and baryon sectors, and future prospects are discussed.

1 Introduction

When I was a student back in the 1960's, the goal of particle and nuclear physicists was to seek the fundamental laws of nature, and one of the requirements of such a "fundamental" law was that it be renormalizable. Well now it's 1996 and we've learned a few things in these three decades. One is that nonrenormalizable effective field theories can be just as if not more useful than their renormalizable siblings in certain situations. One of these is the case of QCD, where we have what we feel is a correct model of nature. However, it is written in terms of the "wrong" degrees of freedom (quarks and gluons rather than hadrons) and is impossible to solve because of its strong coupling and inherent nonlinearity. Much more useful in the arena of low energy physics is an effective Lagrangian, which is written in terms of experimental degrees of freedom—mesons and baryons—and which encodes the symmetries of the underlying QCD interaction—specifically for our purposes chiral symmetry, which exists in the limit in which the quark mass can be taken as vanishing. This is a program which was begun in the 60's with the effective two-derivative Lagrangian[1]

$$\mathcal{L}^{(2)} = \frac{F_\pi^2}{4}\text{Tr}D_\mu U D^\mu U^\dagger + \frac{F_\pi^2}{4}\text{Tr}2B_0 m(U+U^\dagger) \quad (1)$$

which describes the interaction of the Goldstone fields $\phi_i, i = 1..8$. Here $F_\pi = 92.4$ MeV is the pion decay constant, m is the quark mass matrix, B_0 is a phenomenological constant and

$$U = \exp\left(\frac{i}{F_\pi}\sum_{j=1}^{8}\lambda_j\phi_j\right) \quad (2)$$

is a nonlinear function of the fields which transforms as $LUR^\dagger$ under chiral rotations. When used at tree level this interaction is rather successful in predicting low energy interactions.[1] For example, expanding to fourth order in the

Coefficient	Value	Origin
L_1	0.65 ± 0.28	$\pi\pi$ scattering
L_2	1.89 ± 0.26	and
L_3	-3.06 ± 0.02	$K_{\ell 4}$ decay
L_5	2.3 ± 0.2	F_K/F_π
L_9	7.1 ± 0.3	$< r_\pi^2 >$
L_{10}	-5.6 ± 0.3	$\pi \to e\nu\gamma$

Table 1: Gasser-Leutwyler coefficients L_i and the means of determination.

fields one finds the well-known Weinberg predictions for $\pi\pi$ scattering lengths[2]

$$a_0^0 = \frac{7m_\pi}{32\pi F_\pi^2} \qquad a_0^2 = -\frac{m_\pi}{16\pi F_\pi^2} \tag{3}$$

which are confirmed by experiment. However, in order to go further and include loop effects one must include additional four-derivative pieces into the effective Lagrangian, with ten phenomenological constants $L_i, i = 1...10$ which can be determined experimentally as shown by Gasser and Leutwyler,[3] yielding values for these parameters as given in Table 1.

2 Mesons

As mentioned above, the one loop chiral expansion has been carried out in the cae of Goldstone boson interactions by many investigators. As emphasized by Weinberg,[4] this is basically an expansion in energy-momentum with a scale parameter $\Lambda_\chi \sim 1$ GeV, so that one is entitled to quit at one loop provided that energies are small compared to this scale. It is this for this reason that this is called chiral *perturbation* theory (χpt). Although ten seems at first like a large number of parameters, χpt is very predictive and the extent to which these predictions are valid is at some level a probe of the validity of QCD itself. This subject has been extensively reviewed in many places[5] and there is in general very good agreement between predicted and measured quantities as shown in Table 2 The one area here where there is a possible problem has to do with the required relationship between the charged pion polarizability α_E^π and the axial structure function h_A in radiative pion decay[8]

$$\alpha_E^\pi = \frac{\alpha h_A}{\sqrt{2} F_\pi m_\pi} \tag{4}$$

The chirally required value for the polarizability—2.8×10^{-4} fm^3—is at variance with the value found at Serpukov via radiative pion scattering[6] but not

Reaction	Quantity	Theory	Expt.
$\gamma \to \pi^+\pi^-$	$< r_\pi^2 >$ (fm^2)	0.44 (input)	0.44 ± 0.02
$\gamma \to K^+K^-$	$< r_K^2 >$ (fm^2)	0.44	0.34 ± 0.05
$\pi^+ \to e^+\nu_e\gamma$	$h_V(m_\pi^{-1})$	0.027	0.029 ± 0.017
	h_A/h_V	0.46 (input)	0.46 ± 0.08
$K^+ \to e^+\nu_e\gamma$	$(h_A + h_V)(m_\pi^{-1})$	0.038	0.043 ± 0.003
$\pi^+ \to e^+\nu_e e^+e^-$	r_A/h_V	2.6	2.6 ± 0.6
$\gamma\pi^+ \to \gamma\pi^+$	$(\alpha_E + \beta_M)(10^{-4})$ fm^3	0	1.4 ± 3.1
	$\alpha_E(10^{-4}$fm$^3)$	2.8	6.8 ± 1.4[6]
			2.2 ± 1.1[7]
$K \to \pi e^+\nu_e$	$\xi = f_-(0)/f_+(0)$	-0.13	-0.20 ± 0.08
	λ_+ (fm^2)	0.067	0.065 ± 0.005
	λ_0 (fm^2)	0.040	0.050 ± 0.012

Table 2: Comparison between chiral predictions and experimental values for parameters in the Goldstone sector.

with that found at SLAC in $\gamma\gamma \to \pi^+\pi^-$.[7] This is clearly an area which deserves further study and work in this regard is presently underway at DAΦNE, Fermilab and Mainz.

3 Baryons

In the case of pion-nucleon intereactions, chiral perturbative calculations can also be performed. However, things are much less clean for reasons which will become clear. One begins as before with the simplest πN Lagrangian having chiral symmetry

$$\mathcal{L}_{\pi N} = \bar{N}(i \not{D} - m_N + \frac{g_A}{2} \not{u}\gamma_5)N \tag{5}$$

where D_μ is a covariant derivative, g_A is a coupling constant to be determined, and $u_\mu = iu^\dagger \nabla_\mu U u^\dagger$ with $u^2 = U$. Expanding to lowest order we find

$$\mathcal{L}_{\pi N} = \bar{N}(i \not{\partial} - m_N - i g_A \vec{\tau}\cdot \not{\vec{A}}\gamma_5 - \frac{g_A}{F_\pi}\vec{\tau}\cdot \not{\nabla}\vec{\phi} + \ldots)N \tag{6}$$

so that g_A is to be identified with the axial coupling in neutron beta decay. Also we see that chiral symmetry requires the Goldberger-Treiman relation between g_A and the πNN coupling constant[9]

$$F_\pi g_{\pi NN} = m_N g_A \tag{7}$$

$\pi N \to \pi N$	Scattering lengths, $\sigma_{\pi N}$
$\pi N \to \pi\pi N$	LET's, $\pi\pi$ Scattering lengths
$\gamma N \to \gamma N$	Polarizabilities, DHG sum rule
$\gamma N \to \pi N$	LET's
$\gamma^* N \to \pi N$	LET's, $g_A(q^2)$
$\gamma N \to \pi\pi N$	Chiral loop effects in $\pi^0\pi^0$

Table 3: Examples of nucleon reactions which have been examined via χpt.

Using the best present values we have

$$1201\,\text{MeV} = 92.4\,\text{MeV} \times 13.0 \quad \text{vs.} \quad 939\,\text{MeV} \times 1.26 = 1183\,\text{MeV} \tag{8}$$

and the agreement to better than two percent strongs confirms the validity of chiral invariance in the nucleon sector. A second probe in this arena arises from the feature that the nucleon matrix element of the axial current also includes a pion pole contribution, leading to a prediction that in the muon capture process one requires[10]

$$r_P = \frac{g_P(q^2 = -0.9m_\mu^2)}{g_A(q^2 = -0.9m_\mu^2)} = \frac{2m_N m_\mu}{m_\pi^2 + 0.9m_\mu^2} = 7.0 \tag{9}$$

which can be experimentally checked. Present results are

$$r_P = \begin{cases} \mu^- \,\text{capture}\, {}^3\text{He} & 7.4 \pm 2.0^{11} \\ \mu^- \,\text{capture H} & 6.5 \pm 2.4^{12} \\ \text{radiative}\, \mu^- \,\text{capture H} & 10 \pm 1^{13} \end{cases} \tag{10}$$

Obviously the discrepancy in the case of the radiative capture needs to be further explored.

In order to go further one requres loop corrections, just as in the mesonic case. A problem arises here that for the nuclons one has an additional dimensionful paramter—m_N—which is the same size as the chiral scale, which makes the entire renormalization procedure doubtful. This problem can be gotten around, however, by using so-called heavy baryon methods, which are equivalent to the use of a Foldy-Wouthuysen transformation and which make a consistent power counting scheme possible. Of course, renormalization introduces new low energy constants (six, *e.g.* at $\mathcal{O}(p^2)$) but nevertheless this program has been carried out, predominantly by the group of Bernard, Kaiser, and Meissner,[14] and applications are reported in many systems: One area of particular interest here is that of pion photoproduction. In this case for charged

	theory	expt.
$E_{0+}^{\rm th}(\pi^+ n)$	26.3	27.9 ± 0.5 [15]
		28.8 ± 0.7 [16]
$E_{0+}^{\rm th}(\pi^- p)$	-31.3	-31.4 ± 1.3 [15]
		-31.2 ± 1.2 [17]

Table 4: Threshold values of E_{0+} for charged pion photoproduction ($\times 10^{-3}/m_\pi$).

production the feature that the pion derivative in Eq. 6 is covariant leades to the Kroll-Ruderman predictions for the E_{0+} (electric dipole) multipole at threshold

$$E_{0+}^{\rm th} = \begin{cases} \sqrt{2}D(1 - \frac{3}{2}\mu + \mathcal{O}(\mu^2)) & \pi^+ n \\ -\sqrt{2}D(1 - \frac{1}{2}\mu + \mathcal{O}(\mu^2)) & \pi^- p \end{cases} \tag{11}$$

where $D = eg_{\pi NN}/8\pi m_N = 23.9\,(\times 10^{-3}/m_\pi)$ and $\mu \equiv m_\pi/m_N$. In this case theory and experiment are in good agreement, as shown below However, the experimental numbers quoted are from old emulsion measurements and resutls from the recent SAL experiment are eagerly awaited.

In the case of neutral pion photoproduction, things are more interesting. In this case the venerable Low Energy Theorems (LET) for the threshold E_{0+} multipole

$$E_{0+}^{\rm th} = \begin{cases} -D(\mu - \frac{1}{2}\mu^2(3 + \kappa_p) + \mathcal{O}(\mu^3)) = -2.3 & \pi^0 p \\ -D(\frac{1}{2}\mu^2\kappa_n + \mathcal{O}(\mu^3)) = +0.5 & \pi^0 n \end{cases} \tag{12}$$

were shown in 1991 to be incorrect due inappropriate assymptions concerning analyticity.[18] New loop contributions at $\mathcal{O}(\mu^2)$

$$\Delta E_{0+}^{\rm th} = -D\mu^2(\frac{m_N}{4F_\pi})^2 \tag{13}$$

are large and destroyed the agreement which appeared to exist between the original LET predictions and preliminary $\pi^0 p$ experiments at both Mainz and at Saclay. However, there has been a good deal of recent activity. On the theoretical side, Bernard *et al.* have performed an analysis at $\mathcal{O}(p^4)$.[18] In doing so they require the values of two counterterms. Estimating these via Δ, ρ, ω dominance, they predict a value for E_{0+} in good agreement with new experiments performed at both MAMI and SAL This may be somewhat accidental, as the convergence of the series appears slow—$E_{0+} = C(1 - 1.26 + 0.59 + \ldots)$. However, it has been pointed out that the P-wave calculations do not suffer from this slow convergence, yielding what should be a very solid prediction

$$\frac{1}{|\vec{q}|}P_1 \equiv (M_{1+} - M_{1-} + 3E_{1+})$$

	theory	expt.
$E_{0+}(\pi^0 p)(\times 10^{-3}/m_\pi)$	-1.2	-1.31 ± 0.08 [20]
		$-1.32 \pm 0.05 \pm 0.06$ [21]
$P_1/\lvert\vec{q}\rvert(\pi^0 p)(\times \mathrm{GeV}^{-2})$	0.480	0.47 ± 0.01 [20]
		0.41 ± 0.03 [21]

Table 5: Threshold parameters for neutral pion photoproduction.

$$= \frac{D}{M_N}[1 + \kappa_p - \mu(1 + \frac{1}{2}\kappa_p - \frac{g_{\pi NN}^2(10 - 3\pi)}{48\pi}) + \ldots] = 0.480\,\mathrm{GeV}^{-2} \tag{14}$$

There may be a remaining problem in the size of the E_{1+} multipole,[21] but this involves a significant theoretical cancellation.

A second arena of activity is that of Compton scattering. In this case recent precise measurements of both the proton and neutron polarizabilities have been performed yielding values as shown below. As can be observed, the electric polarizability of the neutron is comparable to but slightly larger than its proton counterpart. This is interesting since a valence quark model cannot produce such a result, yielding instead a prediction[22]

$$\alpha_E^p - \alpha_E^n = \frac{\alpha}{3M_N}(< r_p^2 > - < r_n^2 >) \approx 4.6(\times 10^{-4}\,\mathrm{fm}^3) \tag{15}$$

On the other hand a chiral expansion of the polarizability does not have this problem, starting off from equal values for the proton and neutron[23]

$$\alpha_E^p = \alpha_E^n = 10\beta_M^p = 10\beta_M^n = \frac{e^2 g_A^2}{192\pi^3 F_\pi^2 M_N}(\frac{5\pi}{2\mu} + \ldots) = 12.4(\times 10^{-4}\,\mathrm{fm}^3) \tag{16}$$

A full $\mathcal{O}(p^4)$ calculation of both electric and magnetic polarizabilities, with counterterms evaluated via resonance dominance yields very satisfactory results as shown Table 6, although again the convergence of the series for β_M is in question.[24] Similarly a good deal of work has been done in the case of polarized Compton scattering, both experimentally and theoretically, but we do not have the space to discuss this here.

4 Conclusions

Obviously in a short report such as this the discussion above can provide only a brief introduction to the multitude of work which is presently underway. Additional areas of activity include

	theory	expt.
α_E^p	10.5	$11.6 \pm 0.6 \pm 0.6$ [25]
β_M^p	3.5	$2.6 \mp 0.6 \mp 0.6$ [25]
α_E^n	13.4	$12.6 \pm 1.5 \pm 2.0$ [26]
β_M^n	7.8	$3.2 \mp 1.5 \mp 2.0$ [26]

Table 6: Experimental values of electric and magnetic polarizabilities compared to chiral predictions at $\mathcal{O}(p^4)$. (All $\times 10^{-4}\,\text{fm}^3$)

i) $\gamma p \to \pi^0\pi^0 p$ for which the significant near threshold cross section observed via the TAPS group at Mainz is being confronted with the pion loop corrected amplitude. The loop correction is large but it is too early to tell whether it agrees with the experimental findings.

ii) $\gamma^* p \to \pi^0 p$ for which $k^2 \simeq -0.1$ GeV^2 data from both NIKHEF and MAMI seems to contradict at least some of the chiral perturbative predictions. However, this value of momentum transfer is probably above the range where one can expect agreement, and we await the lower k^2 data to be taken at Mainz.

iii) $\pi N \to \pi\pi N$ for which previous analysis in order to extract the $\pi\pi$ scattering lengths has utilized the Olsson-Turner parameterization, which is inconsistent with a modern chiral analysis.

iv) $\vec{\gamma}\vec{p} \to \gamma p$ measurements at CEBAF and elsewhere combined with precise resonance photoproduction data should shed light on the validity of the DHG sum rule.

v) on the theoretical side it is important to include effects of the $\Delta(1240)$ as a baryonic degree of freedom and not just as a heavy state which contributes to a counterterm. Work to this end is underway and should appear soon.

Overall then I hope that I have been able to convey the sense that the area of low energy tests of the standard model via chiral perturbation theory is an active and exciting one, with plenty of work remaining to be done on both the theoretical *and* experimental sides.

Acknowledgements

This research is supported in part by the National Science Foundation. It is a pleasure to acknowledge the warm hospitality of the Oberlin College physics department where this work was performed.

References

1. S. Gasiorowicz and D.A. Geffen, *Rev. Mod. Phys.* **41**, 531 (1969).
2. S. Weinberg, *Phys. Rev. Lett.* **17**, 616 (1966).
3. J. Gasser and H. Leutwyler, *Ann. Phys. (NY)* **158**, 142 (1984); *Nucl. Phys.* **B250**, 465 (1985).
4. S. Weinberg, *Physica* **A96**, 327 (1979).
5. See, *e.g.* B.R. Holstein, *Int. J. Mod. Phys.* **A7**, 7873 (1992).
6. Yu. M. Antipov *et al.*, *Phys Lett.* **B121**, 445 (1983).
7. D. Babusci *et al.*, *Phys. Lett.* **B277**, 158 (1992).
8. J.F. Donoghue and B.R. Holstein, *Phys. Rev.* **D48**, 137 (1993).
9. M.L. Goldberger and S.B. Treiman, *Phys. Rev.* **110**, 1478 (1958).
10. For a modern treatment see V. Bernard, N. Kaiser, and U.G. Meissner, *Phys. Rev.* **D50**, 6899 (1994).
11. L.B. Auerbach *et al.*, *Phys. Rev.* **138**, B127 (1965); D.R. Clay *et al.*, *Phys. Rev.* **140**, B586 (1965).
12. G. Bardin *et al.*, *Phys. Lett.* **B104**, 320 (1981).
13. Preliminary TRIUMF result, M. Blecher (private communciation).
14. For a useful summary, see V. Bernard, N. Kaiser and U.G. Meissner, *Int. J. Mod. Phys.* **E4**, 193 (1995).
15. J.P. Burg, *Ann. De Phys. (Paris)* **10**, 363 (1965).
16. M.J. Adamovitch *et al.*, *Sov. J. Nucl. Phys.* **2**, 95 (1966).
17. E.L. Goldwasser *et al.*, *Proc. XII Int. conf. on High Energy Physics, Dubna, 1964*, ed. Ya.-A.Smorodinsky (Atomizdat, Moscow, 1966).
18. V. Bernard, J. Gasser, N. Kaiser and U.G. Meissner, *Phys. Lett.* **B268**, 291 (1991).
19. V. Bernard, N. Kaiser, and U.G. Meissner, *Z. Phys.* **C70** 403 (1996).
20. M. Fuchs *et al.*, *Phys. Lett.* **B368**, 20 (1996).
21. J.C. Bergstrom *et al.*, *Phys. Rev.* **C53**, R1052 (1996).
22. B.R. Holstein, *Comm. Nucl. Part. Phys.* **20**, 301 (1992).
23. V. Bernard, N. Kaiser and L.G. Meissner, *Phys. Rev. Lett.* **67** 1515 (1991); *Nucl. Phys.* **B373**, 364 (1992).
24. V. Bernard, N. Kaiser, A. Schmidt, and U.G. Meissner, *Phys. Lett.* **B3119**, 269 (1993).
25. F.J. Federspiel *et al.*, *Phys. Rev. Lett.* **67**, 1511 (1991); E.L. Hallin *et al.*, *Phys. Rev.* **C48**, 1497 (1993); A. Zeiger *et al.*, Phys. Lett. **B278**, 34 (1992).
26. J. Schmiedmyer *et al.*, Phys. Rev. Lett. **66**, 1015 (1991).

LOW ENERGY TESTS WHICH PROBE BEYOND THE STANDARD MODEL

L. J. Hall
Department of Physics and Lawrence Berkeley National Laboratory
Berkeley, CA 94720, USA

Low energy precision measurements, such as $(g-2)_\mu$, and rare processes, such as $\mu \to e\gamma$ and $\mu \to e$ conversion, could provide the first sign of superpartners at the weak scale. Rare flavor and CP violating phenomena provide a powerful probe of flavor physics of unified supersymmetric theories.

1 Introduction

1.1 A Conventional Approach

New physics at a scale M, larger than the weak scale, can be probed by the properties of new particles of mass M or via virtual effects amongst the standard model particles. Direct production of the new heavy particles is likely to lead to a more detailed and less ambiguous interpretation of the new physics, however, low energy tests can reveal important information about very high energies. A standard technique for assessing the merits of various low energy tests is to write an effective Hamiltonian for the virtual effects as $H_{eff} = (1/M)^p O_{p+4}$ where O_{p+4} are operators of mass dimension $p+4$ which involve the fields of the Standard Model and which are classified according to the symmetries they break. A given experiment will probe the mass scale M up to some definite value, as shown in Table 1 using current data.

This technique has important advantages: it makes no assumptions about the nature of the new physics as it uses only dimensional analysis and basic properties of quantum field theory. It provides a simple and useful guide for evaluating the power of a given experimental probe. However, it also has shortcomings: it provides no guide as to where new physics lies – it provides no definite targets for experiments to aim at. It also contains hidden assumptions, which is why the numbers in Table 1 may differ from your first estimate.

1.2 My Approach

I will not use the above approach in this talk. Over the last few years a leading candidate for a theory beyond the Standard Model has emerged – supersymmetry – and this theory provides important and attainable targets for rare processes at low energies. If weak scale supersymmetry is correct, then it

Table 1: Mass reach of selected rare processes

	M/GeV	Symmetry
p decay	10^{15}	B
$d_{N,e}$	10^5	CP
$\mu \to e\gamma$ ...	10^5	L_i
m_ν	$10^6, 10^{12}$	L
ϵ_K	10^6	CP
$\bar{\mu}e \to \mu\bar{e}$	10^4	L_i

should be discovered over the next 12 years. This 12 year window is an opportunity for experimentalists working on rare processes to make the discovery first! Nomatter where supersymmetry is discovered, low energy probes give information about the high energy and flavor structure of the theory which cannot be learned by other means. As discussed in 3.2 below, supersymmetric theories have a rich set of flavor and CP violating interactions. In this talk I will briefly mention some of these tests. In the next section I make the case for a supersymmetric extension of the standard model.

2 Beyond The Standard Model: Supersymmetry

2.1 Is there anything Beyond The Standard Model?

The minimalist view of particle physics is, perhaps understandably, becoming pervasive. The Standard Model has held up to years of scrutiny and precision tests and the Higgs mechanism is a very economical origin for mass – a single new particle is the only expense for creating all the observed masses for the quarks, leptons and gauge bosons. Perhaps the Higgs will be found and nothing else: perhaps the Standard Model *is it*. There will be no new physics beyond the Standard Model, except physics at the Planck scale which gives signals too weak to discover.

The Standard Model is indeed economical – and it appears likely to me that a Higgs will be discovered – however there is a very solid reason for believing that it cannot be the whole story of physics at the weak scale. Over the last 25 years we have learnt that the most powerful tool for understanding particle physics is *symmetry*. Symmetries constrain what particles can exist and how they may interact. The Standard Model has only three mass scales, one for each force. The masslessness of the photon is understood from EM gauge invariance, the QCD scale is understood from the asymptotically free

Table 2: Spacetime structure of the Electron

Symmetry Generator	Electron states	Discovered
Momentum	$e(m)$	1897 J.J. Thompson
Angular Momentum	$(e^{\uparrow}, e^{\downarrow})$	1922 Stern-Gerlach
Lorentz Boost	$(e, \bar{e})$	1932 Anderson
Supersymmetry	$(e, \tilde{e})$	??

dynamics of SU(3), but *what is the symmetry understanding of the weak scale?* The Standard Model provides no answer, yet there must be an answer and it must have to do with new physics at the weak scale.

2.2 What is Supersymmetry?

There is a unique tool for constructing a theory which has the economical elementary Higgs description of electroweak symmetry breaking while providing a symmetry description for the weak scale: that tool is an extension of spacetime symmetry called *supersymmetry.* The idea of supersymmetry is introduced in Table 2. As we have learnt more about the nature of spacetime, so the electron (and the other particles) have been found to have more structure. Stern and Gerlach found a doubling of the electron states, and Anderson discovered the Lorentz-partner of the electron. Supersymmetry further enlarges spacetime, and hence also the multiplet structure of the particles. It differs from the previous extensions since it is a broken symmetry: the superpartner of the electron is not degenerate with the electron.

2.3 Supersymmetry and Electroweak Symmetry Breaking.

Since superpartners have not been seen, supersymmetry must be broken. For supersymmetry to provide a symmetry understanding of the weak scale, it must be broken at the weak scale, and supersymmetry breaking must trigger electroweak symmetry breaking. Such a theory possesses many fundamental scalars: the Higgs bosons and all the superpartners of the quarks and leptons. Why do the squarks and sleptons all obtain a positive mass-squared while the Higgs receives a negative mass-squared to induce electroweak symmetry breaking? It is remarkable that a negative mass-squared is induced only for the Higgs, and not for all the other scalars, precisely through the dynamics of the large Yukawa coupling, λ_t, which the Higgs has to the top quark: $\Delta m_h^2 \approx 3/8\pi^2 \lambda_t^2 m_{\tilde{t}}^2$ where $m_{\tilde{t}}$ is the top squark mass. Electroweak symmetry breaking

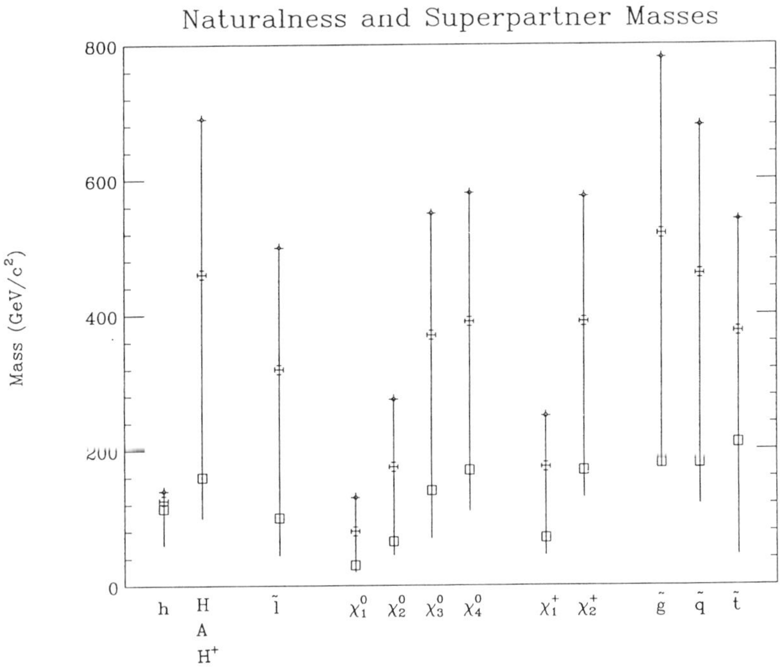

Figure 1: Expectations for the spectrum of superpartners. For each particle, the line extends up to a point where the fine tuning of parameters is 1 part in 10.

is a heavy top quark effect – with a magnitude governed by the superpartner masses. The Higgs potential depends on the parameters of the theory which determine $m_{\tilde{t}}$ and also on other supersymmetry breaking mass parameters. If these parameters are all chosen larger than the weak scale, the Z mass can be kept at 91 GeV only by a fine tuned cancellation. Limiting the size of such cancellations gives a superpartner spectrum shown in Figure 1, supplied by G. Anderson [1], for the case of the minimal supersymmetric standard model.

3 Low Energy Tests of Supersymmetry.

3.1 $(g-2)_\mu$

The supersymmetric contribution to the anomalous magnetic moment of the muon, Δa_μ^{susy} is expected to be of order $(\alpha_2/8\pi^2)(m_\mu/\tilde{m})^2 \tan\beta \approx 1.5 \times 10^{-9}$ $\tan\beta(100\text{GeV}/\tilde{m})^2$ from diagrams with internal superpartners of mass $\tilde{m}$, where the ratio of electroweak vevs is $\tan\beta$. The aim of BNL-E821 is to improve the present sensitivity by a factor of 20 to 0.4×10^{-9}. It is also believed that the standard model hadronic uncertainties will be at the level of $\pm 1 \times 10^{-9}$, so that BNL-E821 will provide a significant test of weak scale supersymmetry.

As usual, there is a caveat: supersymmetric theories have a large parameter space – perhaps several diagrams cancel, or perhaps the relevant supersymmetric mass is larger than expected. A recent paper [2], which contains references to earlier works, shows that for the relevant smuon, Higgsino and wino mass parameters less than 200 GeV, the minimal supersymmetric standard model will give a signal larger than 10×10^{-9} for $\tan\beta$ larger than 30. The absense of a signal would cast a serious doubt on the case of large $\tan\beta$. Smaller values of $\tan\beta$ can also be probed: keeping the Higgsino and wino mass parameters at 200 GeV, but reducing the smuon mass to 100 GeV, gives a signal of 2.5×10^{-9} for $\tan\beta = 3$. The precision measurement of $(g-2)_\mu$ is an exciting experimental challenge. Current thinking in supersymmetric theories suggests that a signal will be present at or above the level of 1×10^{-9}. Furthermore this signal does not rely on the flavor or CP violating structure of the supersymmetry breaking interactions. A word of caution is obligatory: theories with weak scale supersymmetry triggering electroweak symmetry breaking do not necessarily mandate that the smuon (or its sneutrino) is lighter than 1 TeV – such special theories have been constructed and do not give a signal in $(g-2)_\mu$.

3.2 Flavor and CP violation

Flavor and CP violation arises in the standard model when the quark mass matrices are diagonalized – the CKM matrix, V, results as a relative rotation between u_L and d_L type quarks in the 3-dimensional generation space. The W gauge boson couples to $\bar{u}_{Li} V_{ij} d_{Lj}$. If either the three up quarks or the three down quarks were degenerate, V could be rotated to the unit matrix by rotating the basis for the u_L or d_L quarks. Thus the standard model CP and flavor violation is a consequence of the non-trivial flavor structure of the quark mass matrices.

In any supersymmetric extension of the standard model, there are 3×3 mass matrices for the squarks and for the sleptons, in addition to those for the quarks and leptons. If these scalar mass matrices are non-trivial, then the supersymmetric gauge interactions will involve new flavor mixing matrices, W, analogous to the CKM matrix. For example, a relative rotation between the up quarks and the up squarks leads to a matrix, W^u, appearing at the gluino vertex: $\tilde{u}_i^\dagger W_{ij}^u u_j$. Similarly if there is a relative rotation between the charged leptons and charged sleptons necessary to diagonalize the mass matrices, then the photino vertex allows CP and flavor violation via a matrix W^e: $\tilde{e}_i^\dagger W_{ij}^e e_j$. Such additional W mixing matrices are a generic feature of supersymmetric theories. They would be made trivial if the scalar superpartners of each charge were degenerate, but this is not the generic situation. The approximate U(3) flavor symmetry of the standard model, which treats the three generations as a triplet, is badly broken to U(2) by the large top quark mass, while the U(2) symmetry of the lighter generations is broken only weakly by the masses of the second generation fermions. In a supersymmetric theory, this approximate symmetry is expected to apply to the entire theory, ie to squarks and sleptons as well as to quarks and leptons. Hence the diagonalized scalar mass matrices are expected to have the form $(m_1, m_1 + \delta, m_3)$ along the diagonal. There is only a small mass splitting, δ, between the masses of the scalars of the first two generations, while the scalar of the third generation can have a very different mass.

Hence, for theories with superpartners at the weak scale, extra contributions to rare flavor and CP violating processes are to be expected. These include: flavor violation in $\mu \to e\gamma$, $\mu \to e$ conversion, $B^0\bar{B}^0$ mixing and $b \to s\gamma$; and CP violation in the K system, ϵ_K, and in the B system, $\sin 2\alpha, \sin 2\beta$ and $\sin 2\gamma$, as well as in the electric dipole moments of the electron and neutron. It is possible that significant supersymmetric contributions to all these processes could be discovered in the next decade!

A prediction of the sizes of these supersymmetric effects requires a supersymmetric theory of flavor. A simple such theory of flavor based on the approximate flavor group $U(2)$ has recently been constructed [3]. In this particular theory, flavor changing muon interactions and ϵ_K are of great interest, but still the effects depend on the mass splitting $m_1 - m_3$. While this mass difference is expected to be large, giving large effects, it cannot be predicted. An important feature of grand unified theories is that a large contribution to this mass splitting can be *computed*.

4 Low Energy Tests of Supersymmetric Grand Unification

4.1 Supersymmetric Grand Unification

In grand unified theories the strong and electroweak forces are low energy manifestations of a single unified force, based for example on the group SU(5) or SO(10). The quarks and leptons of a generation are similarly different aspects of a unified multiplet. Such concepts are clearly elegant, but can they be tested and proven correct? The spinor representation of SO(10) predicts the quantum numbers of the quarks and leptons under the three low energy gauge groups $SU(3), SU(2)$ and $U(1)$ – this is already a considerable achievement. The unification of the three gauge forces also allows the prediction of the strength of one of these forces. If the theory at the weak scale is supersymmetric, the weak mixing angle is predicted, $\sin^2\theta = 0.233 \pm 0.003$, whereas the experiments at LEP and SLC have measured $\sin^2\theta = 0.2313 \pm 0.0003$. This is the only parameter of the standard model which has been successfully predicted at the percent level of accuracy. Without supersymmetry, the unified prediction for the weak mixing angle is incorrect. This does not prove that supersymmetry or grand unification are correct; but it does suggest that other predictions of unified supersymmetric theories should be studied, both experimentally and theoretically.

4.2 Flavor and CP violation

While there are no known predictions from the gauge sector of unified supersymmetric theories, there are many signals in the flavor sector. Proton decay and neutrino masses are well-known examples, but others may well turn out to be more significant. Theories of the flavor sector have been constructed which make considerable progress in understanding fermion masses and mixings. However, the progress is partial – how can we be sure that these theories really are making inroads on the flavor problem? Eventually, when superpartners have been discovered, it will be possible to learn a great deal from the superpartner spectrum and the flavor changing interactions of the superpartners. In the near term, however, the best hope lies in precision measurements of flavor changing and CP violating phenomena.

Many of the new interactions of the unified theory, like those which violate baryon number, decouple from low energy physics and produce effects which are suppressed by inverse powers of the unification mass. However, this is not the case for the flavor and CP violating interactions – they leave an imprint in the squark and slepton mass matrix, giving rise to new physics directly at the weak scale [4]. In particular

1. Unified theories of flavor lead to predictions for the new flavor mixing matrices W; since quarks and leptons are unified, there is flavor mixing in the lepton sector, $W^e \neq I$. In many theories, the off-diagonal entries of the W matrices have magnitudes similar, but not identical, to the corresponding elements of the CKM matrix.
2. The mass splitting $m_3^2 - m_1^2$ between the scalars of the first and third generations has a contribution proportional to m_t^2 which can be calculated [5]. This heavy top quark effect is not small, ensuring that the flavor and CP violation of the W matrices will not be screened from low energy phenomena.

In this talk I cannot accurately summarize the status of the predictions for the flavor changing and CP violating phenomena. The reader is refered to the several detailed calculations in the literature [6], the results depend on the parameter space of the theory at the weak scale, and on other aspects of the theory, such as the unified gauge group. The lepton flavor and CP violating signals are certainly much larger in SO(10) theories, which fully unify the quarks and leptons, than in SU(5) theories, which have only partial unification of fermions. If the smuon mass is 100 GeV, much of the parameter space of SO(10) theories is already excluded by the present limit on $\mu \to e\gamma$. If the smuon mass is 300 GeV, towards the upper end of expectations as shown in Figure 1, then each factor of three improvement in this branching ratio probes large areas of parameter space. If no signal is seen after a factor 100 improvement, then the theory is restricted to small regions of parameter space. A similar situation applies to $\mu \to e$ conversion. (In the SO(10) theory the conversion rate in titanium is 0.005 of the $\mu \to e\gamma$ branching ratio.)

The electric dipole moment of the electron is a very sensitive probe of the SO(10) theory, but is very small in the minimal SU(5) theory [7]. Since the EDM is an amplitude rather than a rate, it drops off less rapidly as the superpartners are made heavy. In the minimal SO(10) theory $d_e = 1.3 \times 10^{-27} \sin\phi \sqrt{B.R.(\mu \to e\gamma)/10^{-12}}$ e cm, where ϕ is the sum the the CKM phase and another unknown phase. The electric dipole moment of the neutron is also predicted to be at an interesting level in these theories, although the ratio d_N/d_e is typically a factor 30 smaller than in the case where CP violation originates from supersymmetry breaking [8].

5 Conclusions

I have discussed low energy tests only for supersymmetric theories. Precision electroweak data suggests that electroweak symmetry is broken by a Higgs boson which is elementary at the weak scale. In this case, the only known way to describe the weak scale in terms of a symmetry is by introducing weak scale supersymmetry. Furthermore, unification of the 3 gauge couplings provides a successful, percent level, prediction of the weak mixing angle only if the theory contains weak scale superpartners.

Supersymmetric theories provide strong motivation for further precision tests of the standard model, as they suggest that, in a variety of processes, deviations from the standard model are not far off. If the squarks and sleptons are somewhat heavier than the W and Z, then the supersymmetric radiative corrections are expected to be somewhat below the level of the weak corrections. Eventually deviations from the standard model should occur in $(g-2)_\mu, \epsilon_K, \Delta M_B, b \to s\gamma$ and in the CP observables in B decays. Precision experimental and theoretical studies are required to demonstrate a conflict with the standard model.

Supersymmetric theories possess several new flavor mixing matrices, which, in unified theories, generically give large lepton and CP violating phenomena quite unlike anything which the standard model could produce. Hence $\mu \rightarrow e\gamma, \mu \rightarrow e$ conversion, d_e and d_N provide outstanding signals for supersymmetric unified theories.

References

1. G. Anderson and D. Castano, *Phys. Rev.* D **52**, 1693 (1995).
2. T. Moroi, *Phys. Rev.* D **53**, 6565 (1996). This paper cites the earlier literature on $(g-.2)_\mu$ in supersymmetric theories.
3. R. Barbieri, G. Dvali and L.J. Hall, *Phys. Lett.* B **377**, 76 (1996).
4. L. J. Hall, A. Kostelecky and S. Raby, *Nucl. Phys.* B **267**, 415 (1986).
5. R. Barbieri and L.J. Hall, *Phys. Lett.* B **338**, 212 (1995).
6. R. Barbieri, L.J. Hall and A. Strumia, *Nucl. Phys.* B **445**, 219 (1995); J. Hisano, T. Moroi, K.Tobe and M.Yamaguchi, LBL preprint LBL-38653 (May 1996), hep/ph-9605296; P. Ciafalone, A. Romanino and A. Strumia, *Nucl. Phys.* B **458**, 322 (1996); N. Arkani-Hamed, Hsin-Chia Cheng and L.J. Hall, *Phys. Rev.* D **53**, 413 (1996).
7. S. Dimopoulos and L. J. Hall, *Phys. Lett.* B **344**, 185 (1995).
8. R. Barbieri, A. Romanino and A. Strumia, *Phys. Lett.* B **369**, 283 (1996).

PROSPECTS FOR 50 GEV PS AND ITS COMPLEX IN JAPAN

S. NAGAMIYA
Department of Physics, Columbia University
538 W120th St., New York, NY 10027, U.S.A.

In Japan a new plan to construct a 50 GeV PS is in progress. This plan consists of a 3 GeV booster synchrotron at 200 μA and a 50 GeV proton synchrotron at 10 μA. Also, heavy-ion acceleration is planned. The desire is that construction starts in 1997-8 and the first beams in 2002-3. INS at the University of Tokyo and KEK will create a new joint institution for this project. Physics and progress of the project are briefly overviewed.

1. Recent Progress and Overview of the Project

1.1. Recent Progress

Recently a project to construct a 50 GeV PS at the KEK site has been discussed extensively. This new project is an updated revision of the previous project (called JHP = Japan Hadron Project) in which construction of a high-intensity 1 GeV Linac was planned since 1986. Because this JHP had faced several difficulties for smooth construction, a subcommittee was formed in November, 1994, under the Japanese Nuclear Physics Committee, to review the JHP and to recommend the best strategy of the JHP for the 21st century. In February, 1995, this subcommittee presented the first proposal: a) JHP#1 should be a 3 GeV synchrotron, b) JHP#2 must be planned in the vicinity of 50 GeV at beam current higher than that of the AGS, c) acceleration of heavy-ion beams and polarized protons should also be considered, d) a full utilization of the existing KEK-PS infrastructure, and e) the entire facility must be open to the international community.

Subsequently, this proposal was discussed at both the Institute for Nuclear Study (INS), University of Tokyo, and the KEK. In June, 1995, both Directors suggested that the 50 GeV PS could be a central construction project for a new JHP. Around that time, the subcommittee's report[1] was submitted to the Japanese Nuclear Physics Committee. This Committee immediately endorsed this proposal and decided that the 50 GeV PS should be the central nuclear physics project for the community. The proposal was then sent to the Japanese High-Energy Physics Committee in the summer of 1995 and this Committee also decided to support it.

An international expectation of this project grew in 1995, so that the first international workshop on the 50 GeV PS was held in December, 1995.[2] In the spring of 1997, the INS of the University of Tokyo plans to move to Tsukuba to promote this project by forming a new joint organization with the KEK.

1.2. Overview of the Project

The present plan of the project is shown in Fig. 1. Newly constructed elements are: a 200 MeV proton Linac, a 3 GeV booster synchrotron and a 50 GeV main synchrotron. Projected currents are: 200 μA for 3 GeV Booster and 10 μA for 50 GeV PS. The Booster intensity corresponds to 0.6 MW and this Booster will be installed inside the existing 12 GeV PS tunnel. Also, many existing counter halls will be used for physics research with this Booster. On the other hand, at the 50 GeV PS a new counter hall and a neutrino beam line will be constructed.

In addition, the existing PS injectors (ion source, Linac and 500 MeV Booster) will be used for a heavy-ion injector (and, perhaps, for polarized proton beams), since the need for intensity is not high.

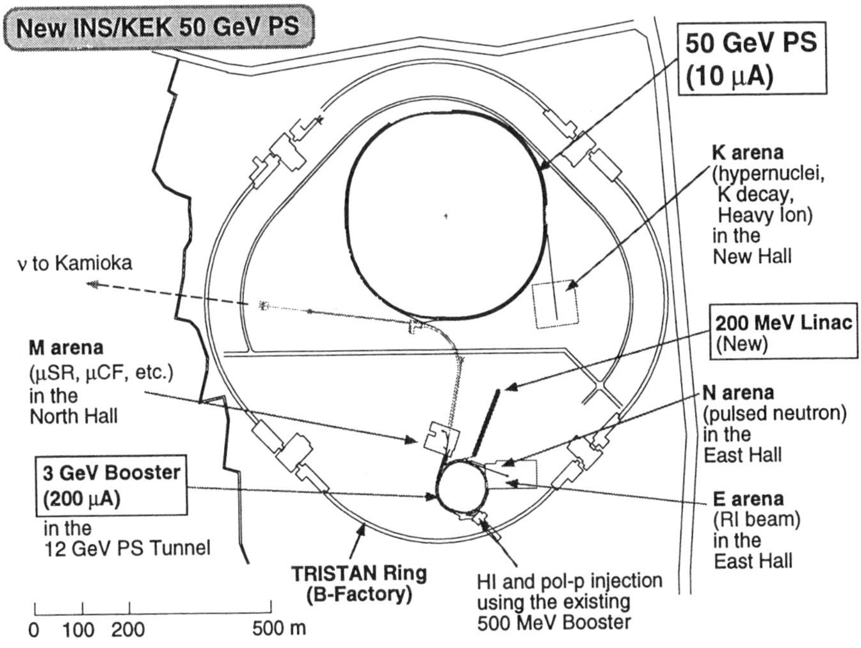

Figure 1: Site plan for a new 50 GeV PS. The outer ring is the existing TRISTAN tunnel in which a new B-factory is under construction. The site of the present KEK 12 GeV PS ring will be used for a new 3 GeV Booster. The 50 GeV PS tunnel will be a brand new one.

The basic guideline for the physics program is to pursue experiments with high-intensity primary and secondary beams, where the primary beams are protons and heavy-ions while the secondary beams are π's, K's, hyperons, neutrons, muons, antiprotons, neutrinos, etc. The physics programs which I describe in this article are only examples. From a wide international community new ideas and proposals will be collected and some of them will be implemented in the project.

2. Nuclear Physics at 50 GeV PS

2.1. Hadron Identity in Nuclear Matter

If one tries to symbolize nuclear physics motivation in a few words, they will be *"From Surface Physics to Matter Physics"*. First, let me describe an important question in nuclear physics. It is well known that nuclei on the Earth have almost a constant density of 0.3 billion tons per cm^3 ($\cong 0.17$ nucleons/fm^3), regardless of nuclear species, except for the low-density surface region. This means that the internucleon distance (d) is almost constant ($d \cong 1.8$ fm). This value of d is only twice the radius of the nucleon ($r_n \cong 0.8$ fm). In addition, the form factor of the free nucleon extends beyond this radius. Therefore, wave functions of neighboring nucleons might significantly overlap in nuclear matter. The question is if the identity of nucleons or, in general, that of hadrons is kept in nuclear matter.

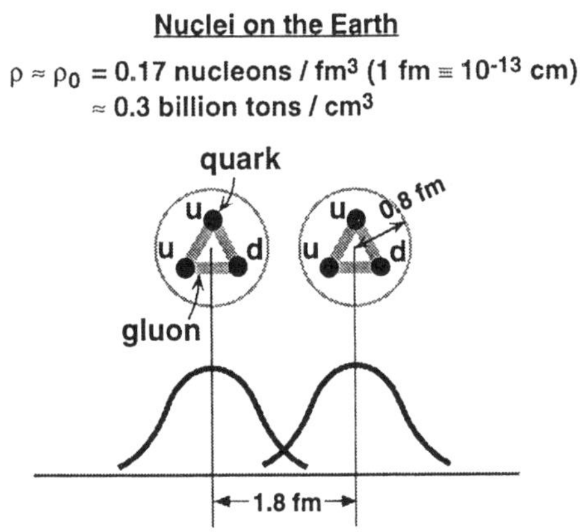

Figure 2: Nucleons inside the normal nuclear matter.

Several years ago, this question was raised explicitly in deep inelastic lepton scatterings. An interesting observation was reported at that time: the quark momentum distribution in nuclear matter is different from that in the free nucleon. This observation, called the EMC effect,[3] suggested a hint that the nucleon in nuclear matter could be different from the nucleon in free space.

On the other hand, it is also well known that the nuclear shell model, in which the nucleon is treated as a fundamental particle, has been very successful in the description of the nucleus. By looking at the data very carefully, however, one finds that we still do not know exactly whether the shell structure exists in the entire nucleus. For example, the available (e,e'p) data[4] do not clearly show that all nucleons obey the shell model, since clean peaks have been observed for nucleons near the Fermi surface alone and not for deeply bound nucleons.

In my opinion, studying the question of hadron identity (or more precisely the distortion of hadron propertieis) in nuclear matter is an important task for the future nuclear physics program. One powerful method for this study is to implant hadrons (other than nucleons) inside the nucleus and to measure properties of these hadrons.

2.2. Probing an Interior of Nuclear Matter at Normal Density

Both baryons and mesons are hadrons. The first method to study the above question is to implant baryons such as Λ, Σ, $\Lambda\Lambda$, etc. These hyperons occupy levels which are close to the ground state and, thus, they tend to probe the interior of the nucleus. In Japan a strong hypernuclear physics community exists. Their recent interest is to study the extremely fine level structure of hypernuclei with an energy resolution down to 2 keV. Note that the past work of hypernuclear spectroscopy has suffered severely from crude energy resolutions on the order of a few MeV. In order to extend the spectroscopy in the future, important tools are a) high energy-resolution detectors and b) high-intensity and high-resolution beams, in particular, high-resolution kaon beams.

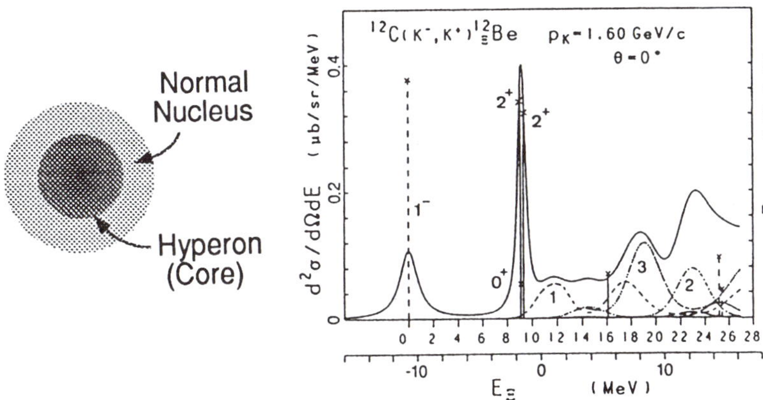

Figure 3: Hyperons as a probe of the interior of the nucleus (left) and a theoretical prediction of a sharp bound state of Ξ^- in ^{12}C (right).[5]

A somewhat much more exotic hypernuclear spectroscopy is the study of $S = -2$ hypernuclei. In (K^-,K^+) reactions both Ξ^- and $\Lambda\Lambda$ could be created inside the nucleus. Theorists expect that Ξ^- would form a sharp and deeply bound state, as shown in Fig. 3.[5] In addition, they expect that, due to a strong mixing between $\Xi^- p$ and $\Lambda\Lambda$, anomalous mixed states of these two states in the nucleus would also create very specific energy levels of the $S = -2$ hypernucleus. This type of spectroscopy will answer the question if baryons keep their free identity in nuclear matter.

In order to understand the behavior of hyperons in nuclear matter, it is also very important to have detailed data on hyperon-nucleon (YN) interactions in free space. So far, very little information on YN interactions is known.[2] High quality, high statistics data are definitely needed. At the 50 GeV PS a proposal was presented to measure YN interactions to the level comparable to the known NN interaction data. These high-quality data have also been anticipated by theorists, in order to establish the chiral effective flavor SU(3) theory, the current most reliable theoretical framework in nuclear physics.

Another method to study the interior of the nucleus is to implant mesons such as ω, ϕ, J/ψ inside the nucleus and to study masses and/or widths of these mesons. As shown in Fig. 4, a theory of chiral condensates predicts that the meson mass will change significantly if the condition of nuclear matter is changed.[6] Theoretical curves are plotted as a function of T (temperature) and ρ (density) of nuclear matter. It must be noted that the meson mass could change by 10-20% if the meson is implanted in nuclear matter at a normal nuclear matter density. Dilepton spectroscopy to measure masses and widths of vector mesons in nuclei is, thus, very interesting. Plans for these measurements are in progress at KEK, GSI, and possibly at the AGS. A high-resolution lepton-pair spectrometer, together with high-intensity meson beams, is needed. For production of the J/ψ inside the nucleus, it was pointed out that high-intensity anti-proton beams are particularly useful.[7]

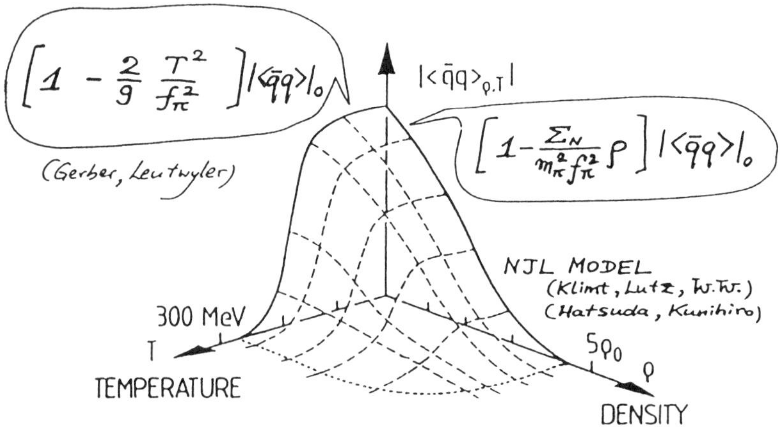

Figure 4: Expected values of the mass term, $<\bar{q}q>$, evaluated by a theory of chiral condenstaes. This picture was created by W. Weise (Ref. 6).

2.3. Probing an Interior of Nuclear Matter at High Density

If one can change the condition of nuclear matter and study the behavior of impurity mesons in such matter, the task of dilepton measurements would be even more exciting and challenging. For this purpose, heavy-ion beams are useful. Already, efforts toward this direction have been initiated at the AGS and SPS,[8] but we must note that heavy-ion programs in these two facilities will be closed soon (within a few years). Also, note that no dilepton measurements have been carried out in heavy-ion programs at the AGS.

Fig. 5 shows an expected density to be reached in heavy-ion collisions. Currently, it is believed that the most efficient creation of high-density matter is expected at beam energies of 20-50 GeV per nucleon. At the 50 GeV PS, the energy for heavy-ion beams will be 20-25 GeV per nucleon, so that this energy is ideal for the creation of high-density matter. As shown in Fig. 4, masses of vector mesons depend strongly on the density of matter but only weakly on the temperature of matter. Studying the meson behavior in high-density matter is an attractive subject for the 50 GeV PS.

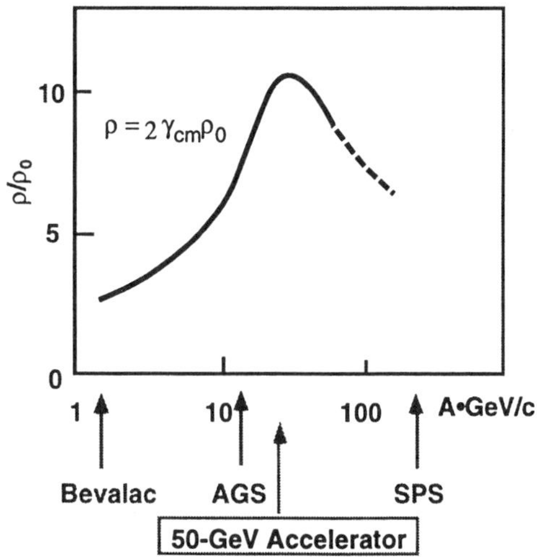

Figure 5: Expected density of nuclear matter to be created in heavy-ion collisions.

The study of bulk properties of high-density matter itself is also an interesting subject. So far, no firm evidence on new phases of high-density nuclear matter has been discovered. A search for multi-strange matter, search for new phases such as kaon condensation, Lee-Wick matter, etc. require extensive work and systematic studies of heavy-ion data. The effort has just begun and it should be pursued somewhere in the world. The 50 GeV PS could offer us this opportunity.

3. Particle Physics at 50 GeV PS

3.1. Rare Decays

There are three approaches in experimental studies of rare decays:

1. To measure branching ratios accurately and determine parameters of CP-violated processes, and compare these values with the values predicted by the Standard Model,
2. To measure T and CP violated processes beyond the Standard Model, and
3. To measure processes like lepton-flavor mixing to explore brand new physics phenomena.

Experiments in Category-1 are, for example, $K^+ \to \pi^+ \nu\bar{\nu}$ or $K^0 \to \pi^0 \nu\bar{\nu}$, to determine the value of V_{td} of the CKM matrix or the value of η in the Wolfenstein parametrization of this matrix, respectively. For the latter process, the branching ratio is given by[9]

$$B(K^0 \to \pi^0 \nu\bar{\nu}) = 8.2 \times 10^{-11}\, A^4 \eta^2\, (m_t/m_W)^{2.3} \approx 4 \times 10^{-11}.$$

The present limit measured at FNAL is $< 5.8 \times 10^{-5}$. At the 50 GeV PS a trial to measure the above quantity down to the level of 10^{-12} is planned, so that the branching ratio can be measured at the accuracy of 5%. Clearly, this type of experiment is not easy to perform. A pencil beam of K_L is required. Pulsed beams will help for the determination of K_L momentum and to reduce the backgrounds from $K^0 \to \pi^0\pi^0$. An active design work is in progress.

Experiments in Category-2 are, for example, the measurement of transverse polarization of μ^+ in $K^+ \to \pi^0 \mu^+ \nu$, the measurements of longitudinal polarization of μ in $K^0 \to \mu^+\mu^-$, etc. At the KEK-PS and at the BNL-AGS these measurements have been started. The Standard Model predicts polarization less than 10^{-3}, while models such as a Higgs-doublet model predict a value larger than 10^{-3}. Measuring muon polarization at the 50 GeV PS with kaon beams is very challenging.

Finally, the third category includes experiments such as $\mu \to e\gamma$, $\mu N \to eN$, $K^+ \to \pi^+\mu^+ e$, $K^0 \to \mu e$, etc. Although some of these measurements can be done more efficiently at the 3 GeV Booster, experiments at the 50 GeV PS would be most interesting. If one tries to trace a historical trend on the values of the experimental upper limit for lepton flavor violation, they are given by $\cong 10^{-[0.2\times(x-1935)]}$ for $x >$ 1940, where x indicates the year when the measurement was carried out.[10] Namely, the present limit in 1995 is 10^{-12} while in the year of 2005 we could expect to reach 10^{-14}. At the level of 10^{-14}, completely new phenomena beyond the Standard Model can appear.[10] The effort to design a new experiment is in progress.

3.2. Neutrino Oscillation

Recently a Kamiokande experiment on atmospheric neutrinos at energies of about

GeV showed a very intriguing result.[11] Let me explain this experiment first. Generally speaking, high-energy neutrinos are generated in the interactions of primary cosmic rays with the air which surrounds the Earth. These cosmic-ray interactions create pions first and, then, these pions decay into muons and electrons by emitting high-energy neutrinos. Basic processes are $\pi^+ \rightarrow \mu^+\nu_\mu$, $\mu^+ \rightarrow e^+\bar{\nu}_\mu\nu_e$ (or, similar processes for a π^- decay). Therefore, in these processes, the expected production rate for ν_μ's is by a factor of 2 larger than that for ν_e's. The observed ratio of ν_μ to ν_e is, however, $\cong$ 1.2 which is smaller than the expected ratio of 2.0.

In addition, the angular distribution of these high-energy neutrinos is such that the ratio is in agreement with the Monte-Carlo result at $\theta = 0$ (neutrinos from overhead) while the ratio at $\theta = \pi$ (neutrinos came up from down below) is by a factor of 2 smaller than the Monte-Carlo result, as illustrated in Fig. 6. A possible interpretation for this result is that a neutrino oscillates ($\nu_\mu \leftrightarrow \nu_\tau$) with an oscillation length of the order of the Earth's diameter. It is, thus, very important to perform neutrino oscillation experiments using accelerator beams.

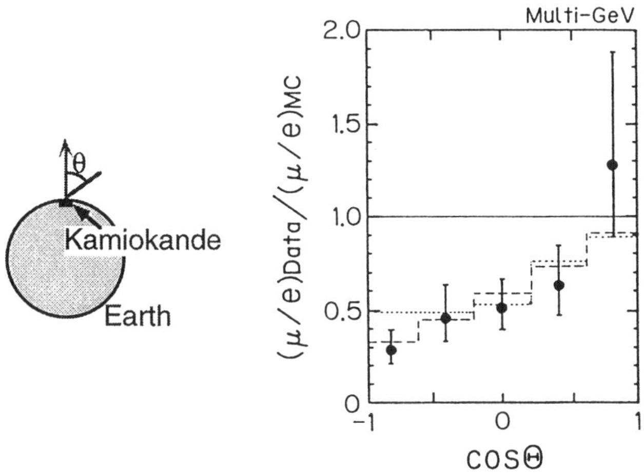

Figure 6: Data of atmospheric neutrinos measured at the Kamiokande. Data from Ref. 11.

In Japan, an experiment to measure the neutrino oscillation is in progress at the SuperKamiokande using the KEK 12-GeV PS as a neutrino source. Data taking will start in 1998. In this experiment the goal is to measure the disappearance of ν_μ, since the energy of ν_μ's from the KEK 12-GeV PS is too low to induce the reaction of $\nu_\tau N \rightarrow \tau N$ at the SuperKamiokande. At the 50 GeV PS, however, an experiment to measure the appearance of ν_τ is possible. The design work on the detection of τ-leptons at the SuperKamiokande is in progress. Clearly, this experiment will compete directly with the MINOS experiment at FNAL. The advantage in Japan is that the Superkamiokande detector exists already at 250 km from the KEK site.

4. Physics at 3 GeV Booster

The 3 GeV Booster is another very strong component of the entire project. Three areas of research are under active consideration and discussion: a) spallation isotope beams to study nuclear physics beyond the nuclear stability line, b) spallation neutron beams for condensed matter physics, magnetism, liquid and amorphous, atomic physics, polymer, biology studies, chemistry, etc., and c) muon beams for μSR studies of magnetism, diffusion, surface phenomena, muonium, muon catalyzed fusion, archeology, medicine, chemistry, etc. It should be noted that this Booster has a power of 0.6 MW proton beams. Each subject is a frontier in basic science. However, since these topics are beyond the subjects to be discussed at this Conference, I will not touch on these topics. An important remark is that scientific programs at this Booster are equally important as the physics programs at the 50 GeV PS.

5. Overall Strategy

For nuclear physics studies, beams of high-intensity kaons, anti-protons, and heavy-ions are very useful. For particle physics studies, beams of high-intensity kaons, muons and neutrinos are vital. In addition, for those subjects which I have not described in this article, other beam species such as polarized protons, pions, etc. might be needed. The basic principle of accelerator construction of this 50 GeV PS Complex is to allow future extensions of the facility into a variety of beams and into a variety of experiments, although the initial scientific programs will be focused on limited subjects.

Presently, the management at both INS and KEK believes that the first beams at the 50 GeV PS will be available in the year of ≅ 2003. Under this assumption, the basic strategy is that on Day-1 both pion and kaon beams will be delivered for studies of strangeness-related nuclear physics, rare decays, and neutrino oscillation. In addition, on Day-1 light-ion acceleration (up to $A = 40$) is planned. On Day-2 (two years after Day-1) truly heavy-ion beams (up to $A = 200$) and anti-proton beams will be delivered.

The facility will be totally open to the international community. Useful advise and cooperation from the international community are particularly sought.

Acknowledgments

I would like to thank many people who have been working on this project. They are: H. Sugawara (Director of KEK), S. Yamada (Director of INS, U. Tokyo), M. Ishihara (Chair of Japanese Nuclear Physics Committee), N. Ikeda (neutron physics), K. Nagamine (muon physics), T. Nomura (ISOL-type fragmentation physics). The talk presented here is a product of the work by the members of the 50 GeV-PS Working Group: Y. Akaishi, J. Chiba, H. Enyo, T. Fukuda, H. Hamagaki, O. Hashimoto, T. Hatsuda, R. Hayano, D. Hertzog, K. Imai, J. Imazato, T. Inagaki, T. Kishimoto, Y. Kuno, L. Littenberg, T. Matsui, Y. Mori, S. Nagamiya, T. Nagae, K. Niwa, K. Nishikawa, M. Oka, J. C. Peng, S. Sugimoto, Y. Sumi, A. Suzuki, Y. Yamazaki, K. Yazaki. I am deeply indebted to these people.

References

1. Y. Akaishi *et al.* (The 50 GeV PS Working Group), "Physics with a 50 GeV PS (in Japanese)", June, 1995.
2. Proceedings of the International Workshop on Physics with 50 GeV PS, HP-Supplement-18, edited by T. Fukuda, H. Hamagaki and S. Nagamiya, April, 1996. (Institute for Nuclear Study Publication)
3. EMC Collaboration, Phys. Lett. **B123**, 275 (1983); A. Bodek *et al.*, Phys. Rev. Lett. **50**, 1431 (1983); R. G. Arnold *et al.*, Phys. Rev. Lett. **52**, 724 (1984).
4. A. E. L. Dieperink and P. K. A. de Witt Huberts, Ann. Rev. Nucl. Part. Sci. **40**, 239 (1990).
5. K. Ikeda *et al.*, Prog. Theor. Phys. **91**, 747 (1994).
6. W. Weise, in Proceedings listed in Ref. 2, p. 19 (1996).
7. K. Seth and D. Hertzog, presentation at the Workshop "AGS 2000", BNL, May, 1996.
8. J. Stachel, Invited talk given at this Conference, PANIC'96 (1996).
9. G. Belanger and C. Q. Gen, Phys. Rev. **D43**, 140 (1991).
10. Y. Kuno, private communication, 1996.
11. Y. Fukuda *et al.*, Phys. Lett. **B335**, 237 (1994).

Probing the Fifth State of Matter with Relativistic Heavy Ions: A Theoretical Overview

Berndt Müller
Department of Physics, Duke University, Durham, NC 27708-0305

Abstract
I review the current status of lattice Monte-Carlo computations of the equation of state of QCD, our current understanding of the thermalization process at collider energies, and two new theoretical developments, one pertaining to the ab-initio calculation of the low-x gluon structure of heavy nuclei, the other to the chaotic behavior of QCD. At the end I give a brief overview of the status of various quark-gluon plasma signatures.

1 Introduction

In the preceding talk, Johanna Stachel [1] surveyed the status of relativistic heavy ion experiments with fixed targets at the Brookhaven AGS and the CERN-SPS. Whereas the observed hadron yields and spectra provide strong evidence that the final state—at the moment of break-up—is approximately thermal, we have little information about when the thermalization is achieved during the course of the nuclear reaction. Since the energy deposition processes at AGS and SPS energies are dominated by soft strong interactions, we currently also lack the theoretical tools to make reliable calculations. Hence we do not know yet, and may never know, whether a thermalized quark-gluon plasma is formed in these reactions.

The situation appears to be much more favorable at the higher energies where the future heavy ion colliders, RHIC and LHC, will operate. Two developments over the past few years have raised the hope that accurate calculations of the initial phase of nuclear reactions at such energies will eventually be possible. Firstly, numerical simulations of lattice QCD are rapidly approaching the stage where quantitatively reliable predictions of the equations of state of strongly interacting matter and its dynamical properties will be feasible. Secondly, the inclusion of color screening into parton cascades and nuclear structure functions opens up the prospect of a seamless description of the reaction dynamics within the framework of perturbative QCD from the initial nuclear structure up to a thermalized quark-gluon plasma. If this expectation comes to fruition, it will be possible to test quantitative theoretical predictions at RHIC and at the LHC.

In this talk I will review the current status of lattice Monte-Carlo computations of the equation of state of QCD, our current understanding of the thermalization process at collider energies, and two recent new theoretical developments, one pertaining to the ab-initio calculation of the low-x gluon structure of heavy nuclei, the other to the chaotic behavior of QCD. At the end I will give a brief overview of the status of various quark-gluon plasma signatures.

2 The QCD Phase Diagram

Rigorous results about the phase diagram of QCD are based on Monte-Carlo simulations of lattice-QCD at finite temperature. For the pure SU(3)-gauge theory, quantitatively reliable results including an extrapolation to the continuum limit have recently become available. Calculations by the Bielefeld group [2] on lattices up to $32^3 \times 8$ yield a critical temperature (in units of the string tension σ) of

$$T_c/\sqrt{\sigma} = 0.629 \pm 0.003, \tag{1}$$

or $T_c = 260$ MeV if the value of σ derived from the Regge slope is used. The phase transition is of first order with a latent heat of slightly less than $2T_c^4$. The simulations provide clear evidence of significant interactions extending far above T_c. Fitting to predictions of thermal perturbation theory one finds, e.g.,

$$g^2(5T_c) \approx 1.5. \tag{2}$$

The simulations also provide information about quantities of dynamical relevance, such as the speed of sound, which is found rising from $c_s^2 < 0.1$ near T_c to $c_s^2 \approx 1/3$ at $T = 5T_c$.

Simulations of full lattice-QCD with light dynamical quarks have not yet reached this advanced stage, but have also made significant progress. State-of-the-art calculations [3] show a steep jump in the quantity ϵ/T^4 at $T_c \approx 150 - 200$ MeV, indicating a rapid unthawing of new degrees of freedom. The uncertainty in the value of T_c is mainly caused by discrepancies between different ways of setting the scale (string tension or hadron masses). Although it remains unclear whether thermodynamic quantities will show a singularity at T_c in the continuum limit and what order phase transition it may be, the simulations clearly demonstrate that the quark condensate $\langle\bar{\psi}\psi\rangle$ drops steeply at exactly the same temperature: color deconfinement and chiral symmetry restoration occur at the same temperature. The results are expected to reach a similar quality as those presently existing for the pure gauge theory when the next generation of parallel computers with close to teraflops performance will become available.

Perturbative approaches, valid far above T_c, have also made significant progress. The equation of state is now known [4,5] up to order g^5, and a scheme yielding a complete perturbative result with only minimal lattice input has been worked out [6]. The "bad news" is that it is now understood that the convergence of the perturbative series (even in the sense of an asymptotic series) requires $g \leq 1$ or $\alpha_s(T) \leq 0.1$, which only holds for temperatures above the electroweak unification scale [7].

Nevertheless, thermal perturbation theory provides important insight into the dynamics of the high-temperature phase of QCD:

1. quarks and gluons develop dynamical masses of order gT.
2. Long-range color forces (except static color magnetism) are dynamically screened on distances of order $(gT)^{-1}$, providing an effective infrared cut-off for many transport coefficients.
3. Colored particles propagating through the plasma constantly exchange their color with the medium, rendering the quark gluon plasma a poor color conductor. As will be discussed below, this "color chaos" facilitates the rapid equilibration of degrees of freedom in the plasma.

3 Initial conditions at RHIC and LHC

Most recent theoretical predictions for the initial conditions at which a thermalized quark-gluon plasma will be produced at heavy ion colliders are based on the concept of perturbative partonic cascades. The parton cascade model [8] starts from a relativistic transport equation of the form

$$p^\mu \frac{\partial}{\partial x^\mu} F_i(x,p) = C_i(x,p|F_k) \tag{3}$$

where $F_i(x,p)$ denote the phase space distributions of QCD quanta. The collision terms C_i are obtained in the framework of perturbative QCD from elementary $2 \to 2$ scattering amplitudes allowing for additional initial- and final state radiation due to scale evolution of the perturbative quanta. To regulate infrared divergences, the parton cascade model requires a cut-off for the $2 \to 2$ scattering (usually $p_T^{\min} = 1.5 - 2$ GeV/c) and a cut-off for time-like branchings ($\mu_0^2 = 0.5 - 1$ GeV$^2/c^2$).

Numerical simulations of such cascades for heavy nuclei provide a scenario where a dense plasma of (predominantly) gluons develops in the central rapidity region between the two colliding nuclei shortly after the impact. Detailed studies [9] show that the momentum spectrum of partons becomes isotropic and exponential, i.e. practically thermal, at a time $\tau \approx 0.7\Delta z$ in the rest frame of a slab of width Δz at central rapidity. To allow for a hydrodynamic description, the width of the slab should exceed the mean free path of a parton. Including color screening effects, one finds that the mean free path of a gluon in a thermalized plasma is $\lambda_f \approx (3\alpha_s T)^{-1}$ where T is the thermal slope of the parton spectrum. For the very high initial values ($T \geq 0.7$ GeV) one concludes that a thermal hydrodynamic picture makes sense after $\tau_i \approx 0.3$ fm/c.

The high density of scattered partons in $A + A$ collisions makes it possible to replace the arbitrary infrared cut-off parameters $p_T^{\rm min}$ and μ_0^2 by dynamically calculated medium-induced cut-offs [10]. The dynamical density-dependent screening of color forces eliminates the need for $p_T^{\rm min}$, and the suppression of radiative processes provided by the Landau-Pomeranchuk-Migdal effect makes the virtuality cut-off μ_0^2 superfluous. Note that the viability of this concept crucially depends on the high parton density: the dynamical cut-off parameters must lie in the range of applicability of perturbative QCD. Since the density of initially scattered partons grows as $(A_1 A_2)^{1/3}(\ln s)^2$, this condition requires both large nuclei and high collision energy. The calculations indicate that this criterion will be met at RHIC and LHC but not at the presently accessible energies of the SPS and AGS. The framework is also not applicable to pp or $p\bar{p}$ collisions at current energies because the parton density remains too low.

The dynamic screening of parton cascades can be implemented as follows [11]. According to the uncertainty principle, a parton-parton collision can be considered as complete after a "formation time" $\tau_f \approx \hbar/p_T$ in the c.m. frame, where p_T is the momentum transfer. Accordingly, harder collisions are completed first. One can then consider the hard collisions as effectively screening the softer ones. The screening mass applicable to parton collisions of scale p_T is then determined as

$$\mu^2(p_T) = -\frac{3}{\pi^2}\alpha_s(p_T^2)\int_{p_T}^{\infty} d^3k|\nabla_k\ f(k)| \tag{4}$$

where $f(k)$ is the momentum density of more violently scattered partons. Feeding $\mu^2(p_T)$ back into the differential cross sections determining the number of scatterings one obtains a differential equation for $f(k)$ or, equivalently, $\mu^2(p_T)$. For Au + Au collisions at RHIC energy, the screening mass saturates at slightly below 1 GeV at small p_T, and at 1.5 GeV for Pb + Pb collisions at the LHC. Both these values are comfortably within the range of applicability of perturbative QCD, demonstrating that there may be no need for an artificial infrared cut-off. The screening of parton scattering by already scattered partons is analogous to the interaction among ladders in the traditional picture of soft hadronic interactions [12]. It would be interesting to rederive these results from this alternative point of view.

The self-consistent parton cascade makes parameter-free (though somewhat model dependent) predictions about the initial conditions achieved at RHIC and LHC. For the heaviest nuclei one expects thermalization to occur at $T = 730$ MeV (RHIC) or $T_c = 1150$ MeV (LHC) and initial energy densities of about 60 GeV/fm^3 (RHIC) or 430 GeV/fm^3 (LHC). At the initial moment ($\tau \approx 0.25$ fm/c) the parton plasma is not chemically equilibrated. It takes another several fm/c to achieve chemical equilibrium. Following the evolution with the hydrodynamical model including self-consistent screening, the quark-gluon plasma is expected to last for about 5 fm/c until the temperature falls to T_c at RHIC and somewhat longer at the LHC. Final multiplicities are predicted to be $dN/dy \approx 1700$ (RHIC) and ≈ 8000 (LHC).

4 New Theoretical Approaches

4.1 Semiclassical parton structure

While representing a major advance in the application of perturbative QCD to relativisitic nuclear collisions, the self-screened parton cascade still relies on the input of experimentally measured parton structure functions of the colliding nuclei. A new approach, due to the Minneapolis group [13] promises to make these structure functions themselves calculable within perturbative QCD. The basic idea underlying this approach is that, as seen by partons with $x \leq 10^{-2}$, the valence quarks in a heavy nucleus constitute a very dense, sheet-like random color source. The large area density $\rho \approx 3A/\pi R^2$ defines a large scale parameter $\mu^2 = \rho$ at which the QCD coupling $\alpha_s(\mu^2)$ is weak. Introducing light-cone coordinates it is then possible to formulate a systematic program for calculating gluon and quark (sea) structure functions at small x as generated by classical random color fields and their quantum fluctuations. At the classical level, corresponding to the Weizsäcker-Williams approximation, the gluon density for not too small p_T is given by

$$\frac{1}{\pi R^2}\frac{dN}{dx dp_T^2} = \frac{\alpha_s \mu^2 (N_c^2 - 1)}{\pi^2 x p_T^2}. \tag{5}$$

Loop corrections will introduce $(\ln x)/x$ contributions, possibly leading to a power-like behavior at small x upon resummation.

The approach can be extended to the collision between two nuclei, viewed as the interaction among two counter-propagating sheets of valence quarks [14]. Since screening effects are already included in this approach, it may provide another description of the initial state produced in a nuclear collision at RHIC or LHC.

5 Chaotic Glue and Thermalization

A completely different approach to the thermalization problem in non-Abelian gauge theories emerges from solutions of the clasical Yang-Mills equations. Numerical calculations have shown that these equations form a strongly chaotic, infinite-dimensional dynamical system [15,16]. The evidence is derived from the study of the (properly defined) distance between two initially neighboring field configurations as a function of time. Neighboring configurations diverge at an exponential rate:

$$||A_1^\mu(t) - A_2^\mu(t)|| \sim e^{\lambda t}. \tag{6}$$

The complete spectrum of Lyapunov exponents λ_i can be obtained from numerical solutions of the Yang-Mills equations on a lattice in Minkowski space. There is evidence that the lattice fields thermalize (in the micro-canonical sense) and the Lyapunov exponents scale as (g^2T).

A fundamental result of nonlinear dynamics is that the sum of all positive Lyapunov exponents (the so-called Kolmogorov-Sinai entropy) measures the rate of entropy growth after coarse graining. The Lyapunov exponents, therefore, provide a measure of the thermal equilibration time. Using the largest Lyapunov exponent for the SU(3) gauge theory [17], one finds a characteristic time of order 0.2 - 0.3 fm/c in the range of temperatures (300-700 MeV) relevant for RHIC. This agrees nicely with the results from parton cascade simulations.

6 Quark-Gluon Plasma Signatures

A wide variety of signatures for the formation of a quark-gluon plasma in heavy-ion collisions has been proposed. Considerable progress has been made in recent years in understanding the background to many of these signals [18,19]. For example, it will be very difficult, if not impossible, to identify direct lepton pairs emitted by the plasma, because of a formidable background of decay leptons from D-mesons [20]. On the positive side, the suppression of charmonium states observed in $p + A$ collisions is becoming understood as due to the absorption of the color octet component in nuclei [21]. This provides a well-defined benchmark against which additional suppression effects can be identified. Indeed, the very high initial temperature now predicted for RHIC and the LHC will allow even charmed quarks to get into chemical equilibrium before hadronization. There is also a chance that even Υ mesons will be significantly suppressed at the LHC. The predicted production of multistrange baryons above thermal yields also remains a promising signature.

It has recently been understood that the energy loss of hard partons in a quark-gluon plasma may be much higher than expected [22]. It may even grow with the length of the traversed medium. Accordingly, the quenching of jet production, or of leading high-p_T hadrons, has attracted increased interest as a probe of the early stage of the evolution of the plasma.

The formation of chirally disorientated domains of the quark condensate in the vacuum is a new, intriguing signature of the chiral phase transition [23]. Such domains correspond to coherent excitations of the pion field; they would decay by producing large fluctuations away from 1/3 in the π^0/π ratio. Detailed numerical studies [24] of the linear sigma model have shown how disorientated domains can grow if the chiral transition occurs rapidly, producing a temporarily unstable state of the chiral order parameter $\langle\bar{\psi}\psi\rangle$. Sophisticated techniques for the identification of domain structures, such as wavelet analysis [25], may help observing these domains if they are produced experimentally.

7 Summary

As RHIC proceeds toward completion, theoretical advances during the last few years are giving us a much clearer view of the physics to be expected in heavy ion collisions at that energy. Lattice-QCD simulations have unambiguously established the rapid "unthawing" of color and the restoration of chiral symmetry at a temperature below 200 MeV. Calculations of parton transport processes based on perturbative QCD predict very high initial temperatures, above 500 MeV at RHIC and 1 GeV at the LHC. New approaches to low-x parton structure and the thermalization problem hold the promise of a model and parameter-independent description of the formation and evolution of a quark-gluon plasma at RHIC energy and beyond. Finally, the various plasma signatures are becoming much better understood, providing valuable guidance for the experimental program at RHIC which will start in 1999.

Acknowledgments

This work was supported in part by a grant from the U.S. Department of Energy, Office of Energy Research (DE-FG02-96ER40945). The author also thanks the organizers of the PANIC-96 conference for support.

References

1. J. Stachel, Exploring the Phase Boundary to the Quark-Gluon Plasma in Collisions Between High Energy Nuclei, these Proceedings.
2. G. Boyd, et al., Thermodynamics of SU(3) Lattice Gauge Theory, preprint BI-TP 96/04, ⟨hep-Lat/9602007⟩.

3. C. Bernard. et al., Two Flavor Staggered Fermion Thermodynamics at $N(T) = 12$, preprint UUHEP-96-3, ⟨hep-lat/9605028⟩.
4. P. Arnold and C.X. Zhai, *Phys. Rev.* **51**, 1906 (1995).
5. C.X. Zhai and B. Kastening, *Phys. Rev.* **D52**, 7232 (1995).
6. E. Braaten and A. Nieto, *Phys. Rev.* **D51**, 6990 (1995).
7. E. Braaten and A. Nieto, *Phys. Rev. Lett.* **76**, 1417 (1996).
8. K. Geiger and B. Müller, *Nucl. Phys.* **B369**, 600 (1992); K. Geiger, *Phys. Rep.* **258**, 237 (1995).
9. K.J. Eskola and X.N. Wang, *Phys. Rev.* **D49**, 1284 (1994).
10. T.S. Biró, et al., *Phys. Rev.* **C48**, 1275 (1993).
11. K.J. Eskola, B. Müller, and X.N. Wang, *Phys. Lett.* **B374**, 20 (1996).
12. S.G. Matinyan and A.G. Sedrakyan, *Sov. J. Nucl. Phys.* **24**, 440 (1976); S.G. Matinyan, M.G. Ryskind and A.G. Sedrakyan, *Sov. J. Nucl. Phys.* **27**, 256 (1978).
13. L. McLerran and R. Venugopalan, *Phys. Rev.* **D49**, 2233 and 3352 (1994).
14. A. Kovner, L. McLerran, and R. Venugopalan, *Phys. Rev.* **D52**, 6231 (1995).
15. B. Müller and A. Trayanov, *Phys. Rev. Lett.* **68**, 3387 (1992).
16. C. Gong, *Phys. Rev.* **D49**, 2642 (1994).
17. C. Gong, *Phys. Lett.* **B298**, 257 (1993).
18. For a recent review see: B. Müller, *Rep. Prog. Phys.* **58**, 611 (1995).
19. J.W. Harris and B. Müller, The Search for the Quark-Gluon Plasma, to appear in *Annu. Rev. Nucl. Part. Sci.* **46**, 71 (1996).
20. S. Gavin, P.L. McGaughey, P.V. Ruuskanen, and R. Vogt, preprint LBL-37981, ⟨hep-ph/9604369⟩.
21. D. Kharzeev and H. Satz, *Phys. Lett.* **B366**, 316 (1996).
22. R. Baier, et al., *Phys. Lett.* **B345**, 277 (1995).
23. K. Rajagopal and F. Wilczek, *Nucl. Phys.* **B404**, 577 (1993).
24. M. Asakawa, Z. Huang, and X.N. Wang, *Phys. Rev. Lett.* **74**, 3126 (1995).
25. Z. Huang, I. Sarcevic, R. Thews, and X.N. Wang, *Phys. Rev. D* (in print), ⟨hep-ph/9511387⟩.

SUPERALLOWED BETA DECAY: A NUCLEAR PROBE OF THE ELECTROWEAK STANDARD MODEL

J.C. HARDY, E. HAGBERG, V.T. KOSLOWSKY, G. SAVARD AND I.S. TOWNER
AECL, Chalk River Laboratories, Chalk River, Ontario, Canada, K0J 1J0

Superallowed $0^+ \rightarrow 0^+$ nuclear beta decay provides a direct measure of the weak vector coupling constant, G_v. The nine most accurately determined transitions, ranging from the decay of ^{10}C to that of ^{54}Co, are surveyed, and the results used to test the conservation of the weak vector current (CVC) and the unitarity of the Cabibbo-Kobayashi-Maskawa matrix.

1. Introduction

Nuclear beta decay between $(J^\pi,T) = (0^+,1)$ analogue states provides a straightforward, yet demanding probe of the weak vector current. Because the axial current cannot contribute between spin-0 states, the experimental *ft* value is directly related to the vector coupling constant, G_v, via [1]

$$ft(1+\delta_R) = \frac{K}{G_V^2(1+\Delta_R^V)\langle M_V\rangle^2} \tag{1}$$

with

$$\langle M_V\rangle^2 = 2(1-\delta_c) \text{ for } T = 1$$

$$K/(\hbar c)^6 = 2\pi^3 \ln 2\hbar/(m_e c^2)^5 = (8120.271 \pm 0.012) \times 10^{-10} \text{ GeV}^{-4} \text{s},$$

where f is the statistical rate function, t is the partial half-life for the transition, and $\langle M_v\rangle$ is the Fermi matrix element. The calculated radiative corrections, δ_R and Δ_R^V, and charge correction δ_c, are each of order 0.01, with δ_R and particularly δ_c being dependent on nuclear structure. These correction terms introduce theoretical uncertainties but they are at least an order of magnitude smaller, so eq. 1 offers an experimental method for determining G_v to about one part in a thousand if the *ft* value can be measured to that level of precision or better.

To date, nine $0^+ \rightarrow 0^+$ "superallowed" transitions have been measured with sufficient precision: the decays of ^{10}C, ^{14}O, ^{26m}Al, ^{34}Cl, ^{38m}K, ^{42}Sc, ^{46}V, ^{50}Mn, and ^{54}Co. In analyzing these results, we find it convenient to define a "corrected" $\mathcal{F}t$ value, which is derived from the conventional *ft* value by the removal of all known effects that depend on nuclear structure:

$$\mathcal{F}t = ft(1+\delta_R)(1-\delta_c). \tag{2}$$

Since, now

$$\Im t = \frac{K}{2G_V^2(1+\Delta_R^V)}, \tag{3}$$

it is evident that the $\Im t$ value must be the same for all transitions if indeed G_v is a constant, as required by the Conserved Vector Current (CVC) hypothesis. The nine measured transitions span a wide mass range, $10 \le A \le 54$, and so constitute a demanding test of CVC.

Their significance also extends to a second fundamental test: the unitarity of the Cabibbo-Kobayashi-Maskawa (CKM) matrix. In the Standard Model, the quark mass eigenstates are related to the weak-interaction eigenstates through the CKM matrix:

$$\begin{bmatrix} d' \\ s' \\ b' \end{bmatrix} = \begin{bmatrix} V_{ud} & V_{us} & V_{ub} \\ V_{cd} & V_{cs} & V_{cb} \\ V_{td} & V_{ts} & V_{tb} \end{bmatrix} \begin{bmatrix} d \\ s \\ b \end{bmatrix} \tag{4}$$

If the model is correct, the nine CKM matrix elements, which can in principal be determined experimentally from weak decays of the relevant quarks[2], should satisfy the unitarity conditions.

The value of V_{ud} can be obtained directly from the best value of G_v, which in turn follows from the experimental $\Im t$ values (as long as these show consistency with CVC):

$$\begin{aligned} V_{ud}^2 &= G_V^2 / G_F^2 \\ &= \frac{K}{2G_F^2(1+\Delta_R^V)\overline{\Im t}}, \end{aligned} \tag{5}$$

where G_F is the Fermi coupling constant derived from the pure-leptonic muon decay and $\overline{\Im t}$ is the average result from the nine superallowed decays. This leads to the most stringent unitarity test of the CKM matrix currently possible, in which

$$|V_{ud}|^2 + |V_{us}|^2 + |V_{ub}|^2 = 1 \tag{6}$$

must be satisfied.

2. Experiments

Three measured quantities are required to determine the $\Im t$ value of a particular β-transition: the decay energy, Q_{EC}, which is used in calculating the statistical rate function, f; the half-life, $t_{1/2}$, of the β-emitter and the branching ratio, R, for the transition of interest, which together with the calculated electron-capture probability yield the partial half-life, t. To ensure that the experimental contribution to the overall uncertainty in $\Im t$ is smaller than that from the theoretical corrections, Q_{EC} measurements must be accurate to better than 100 ppm, and half-lives and branching ratios to better than about 500 ppm. These requirements are at the limits of what is currently possible and have led to experimental techniques highly refined for the purpose.

2.1 ***Decay-energy measurements***. The nine well-measured superallowed β-emitters populate daughters with stable ground states. As a result, reaction Q-value measurements, which lend themselves to experimental precision, can be used to determine decay energies instead of direct measurements of positron end-point energies, which are fraught with precision-limiting difficulties. So far, three different experimental techniques have yielded the required accuracy. If a transition is designated $A(\beta^+)B$, these are: the measurement of the B(p,n)A threshold energy, with the energy of the proton beam related to a one-volt standard (for example, see ref [3]); the measurement of the reaction Q-values for two reactions C(p,γ)A and C(n,γ)B on a common target, with energies determined by comparison with known resonances and γ-ray standards (see ref [4]); and the in-beam comparison with a second superallowed transition, $A'(\beta^+)B'$, via the reactions $B(^3He,t)A$ and $B'(^3He,t)A'$, in which a composite target of B and B' is employed (see ref [5]). This last technique — the one developed at Chalk River — involved the application of a periodic voltage (up to 150 kV) to the reaction target so that two triton groups, one from each reaction, could alternately follow the same path through a magnetic spectrometer. The Q-value difference could thus be accurately established much the same way atomic mass differences are determined from a mass spectrometer.

2.2 ***Half-lives***. Though deceptively simple when only coarse precision is required, half-life measurements do not easily yield high accuracy in spite of the temptation, yielded to by many a researcher, to quote the statistical output of an automated decay-curve analysis. Sample impurity, electronic count-rate dependence, and inappropriate statistical analyses are only a few of the difficult problems faced when uncertainties below 0.1% are sought, particularly for short half-lives. To obviate these problems, we have developed rigorous procedures at Chalk River[6]. Currently, all emitters except ^{14}O and ^{10}C have been measured with an on-line isotope separator. Typically, thousands of multi-scaled spectra are acquired from successive samples observed in a large-solid-angle, high-efficiency, low-background gas proportional counter; the counting electronics have a well-defined, non-extendible dead time. The analysis procedure uses modified Poisson variances and is based on a simultaneous fit of up to 500 decay curves with the same half-life but with individual intensity parameters for each decay curve. This procedure is checked with hypothetical data, computer-generated to simulate closely the experimental counting conditions, and has been proven correct at the 0.01% level.

2.3 ***Branching ratios.*** In all cases but for the decay of ^{10}C, the superallowed transition dominates ($\geq$99%) the decay of the parent nucleus and is best determined via measurement of the other decay branches, their branching ratios being subtracted from 100%. For these other branches, precision is not as important as completeness; therefore we have developed special techniques to observe β-delayed γ rays from branches as weak as a few parts per million[7]. Since the main difficulty in observing these branches results from the intense background generated by the prolific and energetic positrons from the superallowed branch, we use thin plastic scintillators to detect the direction of outgoing positrons, thus rejecting events in which positrons have entered the HPGe γ-ray detector.

For the only case in which the superallowed transition is itself weak — the decay of ^{10}C — a recent experiment at Chalk River[8] has, for the first time, used a large γ-ray array to measure a β-decay branching ratio with high precision: the intensities of β-delayed γ rays in ^{10}B, produced via the reaction $^{10}B(p,n)^{10}C(\beta^+)^{10}B^*$, were compared with the same γ rays produced via $^{10}B(p,p')^{10}B$ in alternating off-line, on-line measurements with the 8π spectrometer.

3. Theoretical Corrections

In a nucleus, the charged particles involved in β decay interact with the electromagnet field from the nucleus, which leads to effects of order $(Z\alpha)^m$, m=1,2..., where Z is the atomic number of the daughter nucleus and α is the fine-structure constant. For the electron (or positron), these effects are accounted for in the calculation of f through the use of exact solutions to the Dirac equation for electrons in the field of a modelled nuclear charge distribution. For the proton, these effects are reflected in the breaking of nuclear isospin symmetry and a resulting correction, δ_c, to the Fermi matrix element.

There are also effects that arise from interactions between the charged particles themselves and their associated bremsstrahlung emissions. These are of order $Z^m\alpha^n$, with m<n, and lead to radiative corrections to the bare β-decay process. By convention these corrections are grouped into two terms, δ_R and Δ_R^V, the former incorporating corrections that are nuclear-structure dependent, the latter, those that are not.

The most controversial of these correction terms is δ_c. Both Coulomb and charge-dependent nuclear forces destroy isospin symmetry between the initial and final states in superallowed β decay. Consequently, there are different degrees of configuration mixing in these two states, and, because the odd proton in the initial state is less bound than the odd neutron in the final state, their radial wave functions differ as well. Calculations accommodate both effects by splitting δ_c

into two components, δ_{c1} and δ_{c2}, to account for configuration mixing and radial mismatch, respectively. Shell-model calculations[7,9,10] have been used to determine δ_{c1} and have been tested experimentally[7] by measurements, in a few of the same nuclei, of weak non-analogue $0^+ \rightarrow 0^+$ transitions, which are also predicted by the calculations and are sensitive to the same effects as δ_{c1}. The δ_{c2} correction has been calculated by full parentage expansions in terms of Woods-Saxon wave functions suitably matched to the experimentally known separation energies[11]; also by a self-consistent Hartree-Fock calculation, in which the central potential was adjusted to match the single-particle eigenvalues with the experimental separation energies[10]; and, most recently, by Hartree-Fock calculations that incorporate charge-symmetry and charge-independence breaking forces in the mean-field potential to take account of isospin impurity of the core[12]. The various calculations for δ_c are displayed in figure 1 where it can be seen that they agree remarkably well with one another, except for ^{34}Cl and ^{38m}K, where the RPA calculations[12] are inadequate to deal with the details of the nuclear structure.

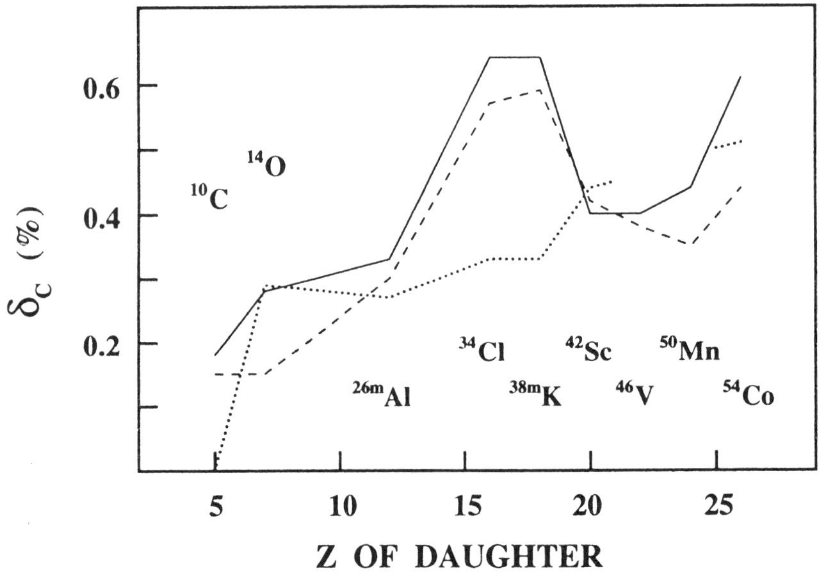

Figure 1 The results of various calculations of δ_c. The solid line joins calculations[7,9,11] that use the shell model for δ_{c2}; the dashed line is for the combined shell model and Hartree-Fock matched to separation energies[10]; and the dotted line is for RPA and Hartree-Fock with core effects included[12].

The radiative corrections, δ_R and Δ_R^v, have been calculated[13] up to order $Z^2\alpha^3$, and recently have been refined to include detailed calculations of order-α axial-vector photonic contributions from both one and two nucleons, the latter contribution being dependent on nuclear structure[14].

Table 1. Experimental results (Q_{EC}, $t_{1/2}$ and branching ratio, R) and calculated corrections (δ_c and δ_R) for $0^+ \rightarrow 0^+$ transitions.

	Q_{EC} (keV)	$t_{1/2}$ (ms)	R (%)	ft (s)	δ_c (%)	δ_R (%)	$\Im t$ (s)
^{10}C	1907.77(9)	19290(12)	1.4638(22)	3040.1(51)	0.16(3)	1.30(4)	3074.4(54)
^{14}O	2830.51(22)	70603(18)	99.336(10)	3038.1(18)	0.22(3)	1.26(5)	3069.7(26)
^{26m}Al	4232.42(35)	6344.9(19)	≥99.97	3035.8(17)	0.31(3)	1.45(2)	3070.0(21)
^{34}Cl	5491.71(22)	1525.76(88)	≥99.988	3048.4(19)	0.61(3)	1.33(3)	3070.1(24)
^{38m}K	6043.76(56)	923.95(64)	≥99.998	3047.9(26)	0.62(3)	1.33(4)	3069.4(31)
^{42}Sc	6425.58(28)	680.72(26)	99.9941(14)	3045.1(14)	0.41(3)	1.47(5)	3077.3(24)
^{46}V	7050.63(69)	422.51(11)	99.9848(13)	3044.6(18)	0.41(3)	1.40(6)	3074.4(27)
^{50}Mn	7632.39(28)	283.25(14)	99.942(3)	3043.7(16)	0.41(3)	1.40(7)	3073.8(27)
^{54}Co	8242.56(28)	193.270(63)	99.9955(6)	3045.8(11)	0.52(3)	1.39(7)	3072.2(27)
					Average $\overline{\Im t}$		3072.3(10)
					$X^2/(N-1)$		1.1

4. Current status of CVC and unitarity tests

Relevant world data were thoroughly surveyed[9] in 1989 and again last year[1]. Table 1 contains all requisite experimental and calculated information for the determination of $\Im t$ values, the results for which appear in the last column together with their weighted average, $\overline{\Im t}$, and scaled uncertainty. The δ_c values shown are the unweighted average of the results of ref[10] and ref[9] updated by ref[7], with a "statistical" uncertainty representing the scatter of the two calculations about their mean. (The results of ref[12] only cover seven of the nine cases and are less reliable for two more, as discussed in section 3; they do, however, corroborate the behaviour of the two complete calculations.)

As demonstrated in table 1 and illustrated in figure 2 there is no statistically significant evidence of inconsistencies in the data, thus verifying the expectation of CVC at the level of 4 x 10^{-4}, the fractional uncertainty quoted on the average $\Im t$ value.

Proceeding next to test CKM unitarity, we must go beyond the "statistical" uncertainties in δ_c and incorporate a "systematic" uncertainty that arises from the small systematic difference between the two independent model calculations of δ_c (ref[10] vs refs[7,9]); our approach is elaborated in ref[9]. This results in $\overline{\Im t} = 3072.3 \pm 2.0$s, which, together with the value of G_F derived from muon decay[2] and the calculated[15] $\Delta_R^V = 2.40 \pm 0.08)$ %, yields via eq (5) the result:

$$V_{ud} = 0.9740 \pm 0.0005 \qquad (7)$$

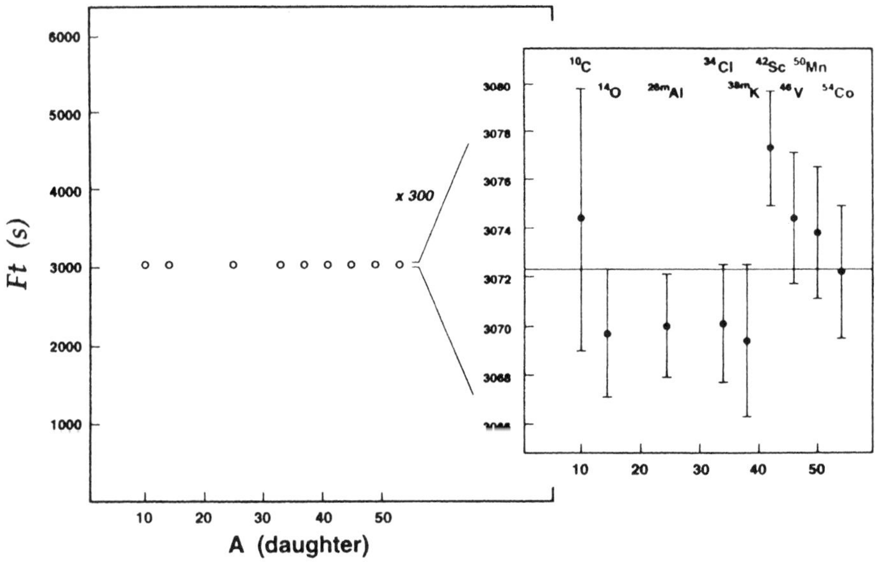

Figure 2: $\Im t$ values for the nine best known superallowed decays and the least-squares one-parameter fit.

The quoted uncertainty is dominated by uncertainties in the theoretical corrections, particularly Δ_R^V; experimental errors on the input data for the $\Im t$ values contribute less than ±0.0002 to the uncertainty in eq (7).

Now, we take the value[2] $|V_{us}| = 0.2205 \pm 0.0018$, which is the average of two results, one from the analysis of K_{e3} decay, the other from hyperon decays. The small value[2] of $|V_{ub}| = 0.0032 \pm 0.0009$, which is derived from the semi-leptonic decay of B mesons, makes it negligible in the unitarity test. The result

$$|V_{ud}|^2 + |V_{us}|^2 + |V_{ub}|^2 = 0.9972 \pm 0.0013 \qquad (8)$$

differs from unitarity at the 98% confidence level.

5. Conclusions

It is too early to assess the significance of the apparent non-unitarity of the top row of the CKM matrix. Taken at face value, it indicates the need for some extension of the three-generation Standard Model — perhaps in the form of right-hand currents or additional gauge bosons. However, it may instead reflect a more "trivial" flaw in the evaluation of V_{us} (as already suggested for other reasons in ref[16]). It seems unlikely, though, that inadequacies in δ_c can explain the failure of the test as has also been suggested[17]: the $\Im t$ data do not support any residual Z-dependence, nor does the most recent δ_c calculation[12] offer any support to there being significant undiscovered effects from the nuclear core.

Finally, it is interesting to note that the world data for neutron decay (as surveyed in ref[1] with the addition of ref[18]) yields a value of $V_{ud} = 0.9767 \pm 0.0026$. This is consistent with the result from superallowed β decay, given in eq (7), but is five times less precise. In the neutron case, the main sources of uncertainty are experimental.

References

1. I.S. Towner and J.C. Hardy in *Symmetries and Fundamental Interactions in Nuclei*, eds. W.C. Haxton and E.M. Henley (World Scientific, Singapore, 1995) pg. 183.
2. Particle Data Group, Phys. Rev. **D50**, 1173 (1994)
3. C.A. Brindhaban and P.H. Barker, Phys. Rev. **C49**, 2401 (1994)
4. S.W. Kikstra, Z. Guo, C. van der Leun, P.M. Endt, S. Raman, T. Walkiewicz, J.W. Starner, E.T. Jurney and I.S. Towner, Nucl. Phys. **A529**, 39 (1991)
5. V.T. Koslowsky, J.C. Hardy, E. Hagberg, R.E. Azuma, G.C. Ball, E.T.H. Clifford, W.G. Davies, H. Schmeing, U.J. Schrewe and K.S. Sharma, Nucl. Phys. **A472**, 419 (1987)
6. E. Hagberg, V.T. Koslowsky, I.S. Towner, J.C. Hardy, J.G. Hykawy, G. Savard, T. Shinozuka, P.P. Unger and H. Schmeing, *Proc. 6th Int. Conf. on Nuclei far from Stability*, Bernkastel-Kues (Inst. of Physics Conf. Series **132**, 1992) pg. 783
7. E. Hagberg, V.T. Koslowsky, J.C. Hardy, I.S. Towner, J.G. Hykawy, G. Savard and T. Shinozuka, Phys. Rev. Lett. **73**, 396 (1994)
8. G. Savard, A. Galindo-Uribarri, E. Hagberg, J.C. Hardy, V.T. Koslowsky, D.C. Radford and I.S. Towner, Phys. Rev. Lett. **74**, 1521 (1995)
9. J.C. Hardy, I.S. Towner, V.T. Koslowsky, E. Hagberg and H. Schmeing (Nucl. Phys. **A509**, 429 (1990)
10. W.E. Ormand and B.A. Brown, Phys. Rev. **C52**, 2163 (1996)
11. I.S. Towner, J.C. Hardy and M. Harvey, Nucl. Phys. **A284**, 269 (1977)
12. H. Sagawa, N.V. Gai and T. Suzuki, Phys. Rev. **C53**, 2163 (1996)
13. A. Sirlin, Phys. Rev. **D35**, 3423 (1987); W. Jaus and G. Rasche, Phys. Rev. **D35**, 3420 (1987)
14. I.S. Towner, Phys. Lett. **B333**, 13 (1994)
15. I.S. Towner, E. Hagberg, J.C. Hardy, V.T. Koslowsky and G. Savard, in *Proc. of Int. Conf. on Exotic Nuclei and Atomic Masses*, eds. M de Saint Simon and O. Sorlin (Editions Frontieres, Saclay, France, 1995) pg. 711
16. A. Garcia, R. Huerta and P. Kielanowski, Phys. Rev. **D45**, 879 (1992)
17. D.H. Wilkinson, Nucl. Phys. **A511**, 301 (1990); Nucl. Inst. and Meth. **335**, 172, 182, 201 (1993); Zeit. Phys. **A348**, 129 (1994)
18. H. Abele, S. Baessler, D. Dubbers, J. Last, C. Metz, T.M. Müller, V. Nesvizhevsky, C. Raven, O. Schärpf and O. Zimmer, these proceedings

NEUTRINO OSCILLATION EXPERIMENTS
Are oscillations really discovered?

F. VANNUCCI

LPNHE, Université de Paris VI et VII, Tour 33 - Rez-de-chaussée, 4 Place Jussieu, BP200, 75252 Paris Cedex 05, France

We have now three indications of neutrino oscillations. This paper presents the results obtained recently at accelerators, compares them with results from astrophysics and discusses the possible compatibilities. Perspectives in the field are sketched.

1 What are oscillations?

Neutrino oscillations were proposed almost thirty years ago.[1] Many experiments have looked for them and several have claimed their discovery. So, where are we today? Oscillations are spontaneous conversions between neutrinos of different flavours; they are a direct consequence of mixing. Classically the phenomenology has been developed on a 2-neutrino basis, arguing that it applies to the channel corresponding to the largest mixing.

It may be time to expand the phenomenology to the 3-neutrino system in order to understand the possible compatibility between some of the present results, and to outline the framework for a general analysis.[2] Here we limit ourselves to the three known neutrinos. To allow for more states would open a Pandora's box.

The 3 weak eigenstates are related to the 3 mass eigenstates through a CKM matrix:

$$\begin{pmatrix} \nu_{\mathrm{e}} \\ \nu_\mu \\ \nu_\tau \end{pmatrix} = \begin{pmatrix} U_{\mathrm{e}1} & U_{\mathrm{e}2} & U_{\mathrm{e}3} \\ U_{\mu 1} & U_{\mu 2} & U_{\mu 3} \\ U_{\tau 1} & U_{\tau 2} & U_{\tau 3} \end{pmatrix} \begin{pmatrix} \nu_1 \\ \nu_2 \\ \nu_3 \end{pmatrix}$$

Let us consider the oscillation between two physical neutrinos, for example ν_{e} and ν_μ. Several terms will participate:

$$P(\vec{\nu_{\mathrm{e}}}\nu_\mu) = (a+b-c)U_{\mathrm{e}1}^2 U_{\mu 1}^2 + (a-b+c)U_{\mathrm{e}2}^2 U_{\mu 2}^2 + (-a+b+c)U_{\mathrm{e}3}^2 U_{\mu 3}^2 \,,$$

where $a = 2\sin^2(1.27\ R/E\ \delta m_{21}^2)$, and b and c have the same oscillatory pattern but with δm_{31}^2 and δm_{32}^2, respectively.

R is the distance between production and detection, and E the energy of the neutrinos.

If one of the three masses is large compared to the other two ($m_3 \gg m_1,\ m_2$) there will be one leading oscillation, and the probability becomes

$$P = 2\ c\ U_{e3}^2 U_{\mu 3}^2\ .$$

Thus there is the possibility of oscillation between ν_e and ν_μ which is an effect of m_3. This corresponds to a kind of indirect oscillation.

As a consequence, if the experiment is carried out at small distances, one can probe the oscillation which corresponds to the large δm^2 even though it may have a small mixing. On the other hand, the same physical channel can show the effect of a large mixing but at large distances when the small δm^2 has time to develop. In practice, this means that the solar neutrino deficit and the LSND result may test the same ν_e to ν_μ channel, one reflecting the small-mixing, large-δm^2 term, while the other (the Sun) reflects the large-mixing, small-δm^2 term.

2 Oscillation experiments

2.1 Disappearance or appearance?

How do we look for oscillations? Oscillation changes a neutrino ν into a different neutrino ν'. There are two complementary ways to detect a signal.

If the experiment is only sensitive to the original flavour present at production, the search consists in finding a decrease of the flux with distance, namely a disappearance of the initial flavour. This can be performed by either calculating the initial flux or using two detectors at two different distances and comparing the spectra. This method is not very precise since it relies on the subtraction of one large number from another. However, it is the method used at reactors, and it also applies to solar neutrinos.

The appearance method consists in looking for the new flavour absent in the original flux. The detector must be able to identify unambiguously the two different flavours. This method can be very sensitive since a single event of the new flavour is, in principle, enough to prove the oscillation. In practice, there are contaminations and backgrounds; however, the appearance method, which is used at accelerators, has proved to be very sensitive to test small mixings. But the distance/energy ratio is small, which explains why the accelerator results test relatively large δm^2.

2.2 Recent negative results

Four new results have been published recently.

- Charm2 [3] looked for the ν_μ into ν_τ channel, extracting quasi-elastic ν_τ events with the τ decaying in the single-pion mode. The signature is clean because of the fine granularity of the detector. No excess is found.

- Karmen[4] at the ISIS spallation source did not find evidence for the ν_μ ν_e oscillation. The beam has a very characteristic time structure which allows a clean separation of initial neutrino flavour.

- CCFR[5] also searched for the ν_μ into ν_τ channel, studying the ratio NC/CC as a function of the total measured energy. An excess of ν_τ would give an increase in NC/CC.

- IHEP-JINR[6] has just published a limit on the disappearance of ν_e in a short neutrino beam enriched with ν_e (5%).

Figure 1 summarizes the negative results obtained so far in the three physical channels. Since the oscillation phenomenon is analysed as a function of two parameters [the mixing $\sin^2(2\Theta)$ and δm^2], the results appear as excluded regions in a 2-dimensional plot. A more general analysis is more difficult to visualize. For maximum mixing, the level of $\delta m^2 \simeq 10^{-2}$ eV^2 has been reached. This is already a good achievement considering that direct measurements of neutrino masses give the much poorer limits[7]:

$$m(\nu_e) \;\leq\; 3.4 \text{ eV}$$

$$m(\nu_\mu) \leq 170 \text{ keV}$$
$$m(\nu_\tau) \leq 24 \text{ MeV}$$

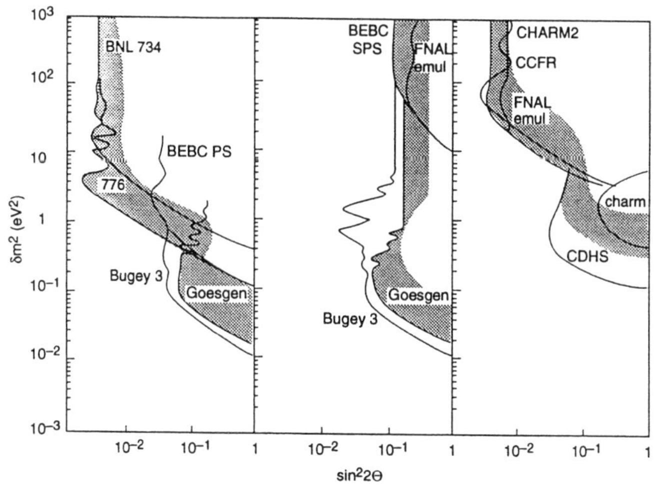

Figure 1: Present excluded regions in oscillation searches for the 3 physical channels.

3 Evidence of oscillations

Now for the more controversial results. We are faced with three different hints of possible oscillation discoveries. In order of seniority, they are:

3.1 Solar neutrinos

This problem has been with us for many years. The Sun produces an abundant flux of ν_e. Four experiments using different techniques have indeed detected ν_e but at a level significantly smaller than the calculated prediction.[8] At present, the most plausible interpretation of this deficit relies on the MSW effect, a resonant oscillation in the Sun interior. The solution gives a parameter $\delta m^2 \simeq 10^{-5}$ eV^2 and a mixing of order 1%.

3.2 Atmospheric neutrinos

From primary cosmic rays interacting in the atmosphere, a ratio of the order of 2 between ν_μ and ν_e is expected. Five underground experiments have published results and three of them, notably the Kamiokande experiment,[9] claim a disappearance of ν_μ. When interpreted by oscillation, the parameter δm^2 comes out to be about 10^{-2} eV^2 and the mixing is large.

3.3 LSND result

In 1995, the LSND experiment published a first result based on 9 events for a background of 2.1, claiming evidence for the oscillation ν_μ into ν_e with a probability[10]

$$P = (0.34 \pm 0.20 \pm 0.07)10^{-2} \ .$$

LSND consists of 167 tonnes of liquid scintillator observed by 1220 photomultipliers. The experiment detects the reaction

$$\nu_e + p \to e^+ + n$$

followed by a delayed 2.2 MeV γ ray from the capture $n + p \to d + \gamma$.

Very recently more data and an improved analysis have been published.[11] The excess is now statistically more meaningful with 22 events detected for a background of 4.6 events. This gives a probability of oscillation of

$$P = (0.31 \pm 0.10 \pm 0.05)10^{-2} \ .$$

3.4 Discussion

These three sets of data give us too many oscillations. Figure 2 shows the three preferred regions together with the excluded regions already displayed. Note that the LSND result is in contradiction with several other experiments except for a band at low δm^2.

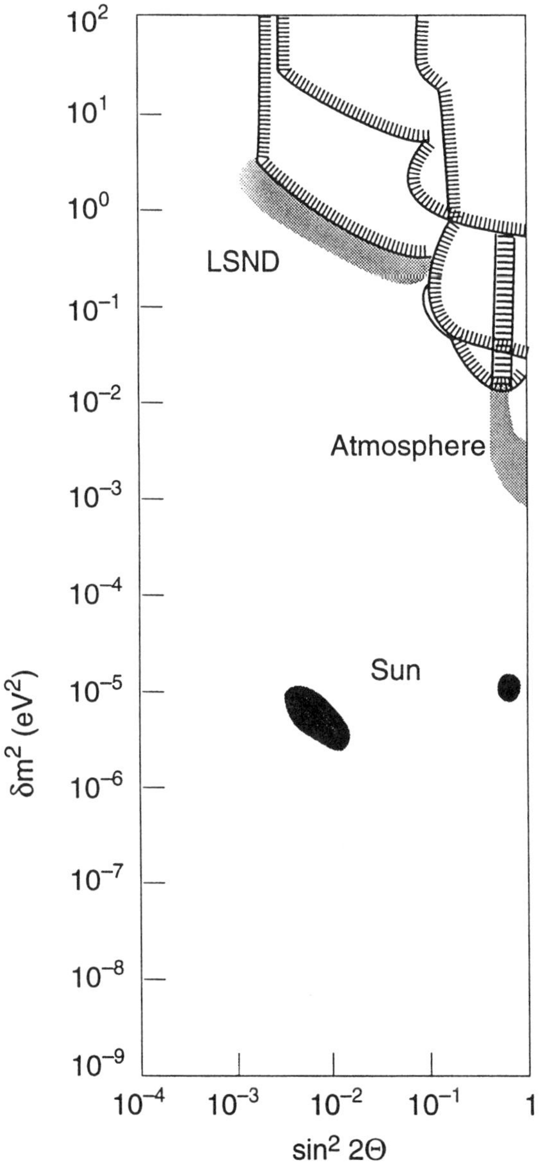

Figure 2: Preferred regions for the LSND, atmospheric and solar results. They do not necessarily apply to the same physical channel.

Clearly, the regions shown in Fig. 2 for the three sources of evidence do not overlap. Can we make sense out of such divergent information?

Assuming only three neutrinos with normal interactions and a hierarchy between the three masses as proposed by the see-saw mechanism, it is obvious that the three sources of evidence are incompatible.

If one trusts the solar neutrino result, the mass of the ν_μ is fixed at $m(\nu_\mu) \simeq 3 \times 10^{-3}$ eV, the mass of the ν_e being much smaller. But we have a choice for the ν_τ: $m(\nu_\tau) \simeq 0.1$ eV if one adopts the atmospheric hint; $m(\nu_\tau) \simeq 1$ eV if one prefers the LSND claim. Note that these two solutions are too small to give the ν_τ a cosmological *raison d'être.*

The experimental situation needs clarification.

4 Current experiments

Two experiments are currently taking data in the CERN high-energy neutrino beam. They are searching for the oscillation ν_μ into ν_τ in the mass region which is relevant for cosmology.

4.1 ν_τ as a dark matter candidate

The Big Bang model predicts the presence of fossil neutrinos in the entire volume of the Universe at a level of about 100 per cm^3 for each flavour. Astrophysicists observe an invisible mass in the Universe which could represent more than 90% of the total mass. It is tempting to relate these two facts. Assuming that the ν_τ is the heaviest of the three neutrinos, a 60-eV ν_τ would close the Universe. Present models of the missing mass prefer a mixture of hot and cold dark matter. This would favour a mass of 10–20 eV for the ν_τ.

Such a mass is extremely small for any direct measurement but is easily accessible by oscillations if the corresponding mixing is not too small. This is the justification for the two CERN experiments which are currently exploring the ν_μ ν_τ channel.

4.2 *The CERN experiments*

The experimental challenge is to identify a ν_τ interaction since the produced τ lepton travels only 1 mm before decaying. Two complementary approaches have been adopted.

- CHORUS uses a target made of emulsion. This is the ideal medium to examine the region around the interaction with high spatial resolution. Figure 3a shows one event with a short-lived particle reconstructed in the experiment. However, the target is passive and requires a sophisticated spectrometer to enrich the sample of data to be scanned.

- NOMAD is a complete spectrometer with a light target inside a magnetic volume, able to measure as precisely as possible and identify particles in order to apply kinematical cuts to extract a signal. Figure 3b shows a reconstructed K^0 mass obtained from the data.

Both experiments hope to reach a limit of a few 10^{-4} in mixing for δm^2 above 30 eV^2.

5 Perspectives

5.1 *Basic rules of oscillation searches*

In order to explore smaller mixings, an experiment must increase its statistics and control its systematical uncertainties. In the case of no background the limit on mixing decreases linearly with the statistics. In going to smaller δm^2, the statistics play little role. The limit on δm^2 depends primarily on the parameter distance $\div$ energy, and small δm^2 are reached at large distances and/or small energies. This explains the present plans with regard to oscillation searches.

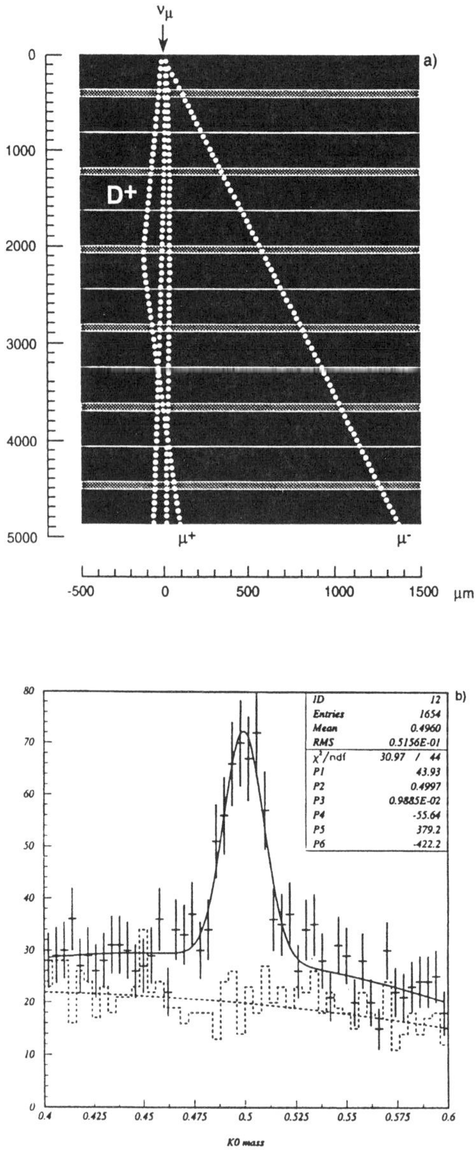

Figure 3: (a) One event with a short-lived particle reconstructed in the emulsion of the CHORUS detector; (b) K^0 signal reconstructed in the NOMAD target.

5.2 Experiments, projects and dreams

LSND will continue to take data for two more years, as will Karmen, which has improved its veto system. CHORUS and NOMAD have runs scheduled until the end of 1997.

In parallel there are two long-distance reactor experiments (Chooz and San Onofre) with detection at 1 km, which will start taking data this year. They should cover the region of the atmospheric anomaly.

In the more distant future long base-line searches are projected. A 1-GeV ν_μ beam from KEK should cross the Superkamiokande detector, 250 km away, in 1999. Around 2002 the Cosmos experiment should improve on the ν_μ ν_τ mixing down to a few 10^{-5}, and at the same time the beam from Fermilab should hit a detector in the Soudan mine 730 km away. The same distance can be explored with the Icarus detector under the Gran Sasso laboratory and a beam from CERN.

In 10 years from now, the excluded regions may extend down to 10^{-5} in $\sin^2 2\Theta$ and 10^{-3} in δm^2 or better, but the *terra incognita* may still look as vast as it does today.

Neutrinos are mysterious particles. Many questions concerning their nature remain unanswered, and experimental puzzles regularly arise. Let us hope that oscillations will be firmly discovered before long.

References

1. B. Pontecorvo, Sov. Phys.-JETP **26**, 984 (1968).
2. C.W. Kim and H. Nishiura, Preprint JHU–HET 8506 (1985);
   K.S. Bahn, et al. *Phys. Lett.* **B359**, 351 (1995);
   S.M. Bilenky et al., *Phys. Lett.* **356**, 273 (1995).
3. M. Gruwe et al., *Phys. Lett.* **B309**, 463 (1993).
4. B. Armbruster et al., *Nucl. Phys.* **B38**, 235 (1995).
5. K.S. McFarland et al., Preprint FNAL–PUB–95/153.
6. A.A. Borisov et al., *Phys. Lett.* **B369**, 39 (1996).
7. See for example J. Bonn in *Proc. 8th Rencontres de Blois: Neutrinos, Dark Matter and the Universe,* Blois, France, 1996 (to be published).

8. GALLEX Collaboration, *Phys. Lett.* **B327**, 377 (1994); SAGE Collaboration, *Phys. Lett.* **B328**, 234 (1994); Homestake Collaboration, *Nucl. Phys.* **B38**, 47 (1995); Kamiokande Collaboration, *Nucl. Phys.* **B38**, 55 (1995).
9. K.S. Hirata et al., *Phys. Lett.* **B280**, 146 (1992).
10. C. Athanassopoulos et al., *Phys. Rev. Lett.* **75**, 2650 (1995).
11. C. Athanassopoulos et al., Preprint LA–UR–96–1326.

CONCLUDING REMARKS

D.H. WILKINSON
TRIUMF, 4004 Wesbrook Mall, Vancouver, B.C. Canada V6T 2A3
and
University of Sussex, Falmer, Brighton, BN1 9QH, England

In 1693, I seem to remember, William & Mary founded the place in which we have been enjoying these past few pleasant and profitable days, seeking after Nature's ways, for the pursuit of "good arts and sciences" together with "good letters and manners". We have indeed enjoyed good science and good manners and are most grateful to our local hosts both of William & Mary and of CEBAF, since the beginning of our conference transmogrified into the Jefferson Laboratory, for their care of our intellectual and material well-being.

It seems, and indeed is, a long time since 1963 when, in Geneva under the chairmanship of Viki Weisskopf, Torleif Ericson, also here today, and Amos de Shalit, I delivered the first lecture at the first of the series of conferences of which this is the 14th. Over the interim things have changed. The first several conferences went under the name ICOHEPANS for International Conference on High Energy Physics and Nuclear Structure, a name that persisted until, after the 9th conference, in Versailles in 1981, it became PANIC, Particles and Nuclei International Conference, on the urging of John Domingo, co-chairman for our present meeting. I mention this not just for historical reminiscence but to emphasize that things have indeed changed and not just in name only although that change is a useful peg upon which to hang my remark. "Structure" has dropped out but "Nuclei" remain. The early meetings, in the ICOHEPANS era, were dominated by nuclei and their structure: the ways in which our vision of that structure could be illuminated by the nuclear interaction of particles of high energy both in the kinetic sense and, for certain particles, in the sense of the energy transmitted as rest energy on their nuclear absorption; reciprocally, the ways in which our understanding of the strong and weak interactions could be enlarged through the use of complex nuclei of known structure. But as time has passed interest that was originally primarily in complex nuclei and their structure has shifted more and more towards the structure not of nuclei but of what we earlier were content to regard as "elementary" particles. Thus at our 7th meeting, in Zürich in 1977, only 10% of the recorded talks did *not* have as their primary objective our vision of the structure of complex nuclei; at the 10th meeting, in Heidelberg in 1984, that proportion had risen to 35% and at the 14th meeting, just concluding, of the plenary talks it has been 85%. What do we make of this drift? Is it a loss of sight of our primary mission, that should be deplored? I think not: it just reflects PANIC's role in a process of readjustment and redefinition across the whole sweep of nuclear and particle physics: the nominally particle physics conferences used to devote a large fraction of their business to *nucleon* structure and before the nucleon had a structure, other than vaguely in the sense of some sort of effective disseminated spatial extension, even to the nucleon-nucleon interaction; now nucleon structure falls squarely in the domain of PANIC while the nucleon-nucleon interaction is even slipping out of PANIC, where it resided for some years, and into the explicitly nuclear structure conferences themselves. But we should at least be aware of this shift: PANIC remains PANIC but we, its clientele, our tastes and interests, are a moveable feast.

But now to the substance of our meeting which has been one of great richness and variety with some 25 plenary talks accompanied by ten times that number in the parallel sessions. It would be absurd to attempt to distill the whole into a formal summary even if that were not forbidden by the great unities of time and space. What I will do is to pick out just a few of what struck me as important novelties and growth points and to highlight some of the great unresolved issues: such is, of course a highly subjective and personal judgement for which I can make no apology.

Firstly, I need to pronounce a kind of blanket disclaimer that relates to the great hopes that we all have for the enlightenment that must flow from the machine in whose dedication we participated a few days ago. Now I have just pointed out that the properties of complex nuclei as such have almost slipped out of PANIC meetings. But that does not mean that we know all that there is to be known about complex nuclei, even in the most general terms and before we start talking about superdeformation and neutron halos; in particular we certainly do not know all there is to be known about what makes a complex nucleus different from the Z protons and N neutrons into which it can be dissociated. That there are important differences, a least in the response of a nucleus to certain types of questioning, we have known for many years. The study of the influence and contribution of mesonic exchange currents, isobars and so on, together with relativistic effects, has been, and remains, central to the PANIC agenda. Now, however, we also know that nucleons themselves are structured objects with their constituent quarks held together by the same forces of gluon exchange that, in the last analysis, hold the nucleons together in the complex nucleus either directly or, through their role in the generation of exchange mesons, indirectly. The ineluctable consequence of this is that a nucleon placed inside a complex nucleus must respond to that context: the structure of bound nucleons must, *to some degree*, differ from that of those in the free state; *at some level*, we shall have to take the quark substructure of nucleons explicitly into account in our description of complex nuclei. Now you may argue that we are there already: the EMC effect has demonstrated it. But that is not what I am talking about. What I am looking for is an unequivocal signal, the famous smoking gun, that tells me that quarks, as such, must, literally, come into the reckoning for our full understanding of the regime of classical nuclear structure physics namely *exclusive processes involving well-defined initial and final quantum states of complex nuclei.* The EMC effect indeed tells us that complex nuclei contain quarks, but that we knew already because their constituent nucleons contain quarks; and it tells us that the quarks in a complex nucleus are not behaving in the same way as are the quarks in Z free protons plus N free neutrons but beyond that there are no specific and unique conclusions; the EMC effect does not tell us how, if at all, other than by being taken off the mass shell, nucleons in the nucleus differ from those in the free state; the only thing that it tells us beyond peradventure is that bound nucleons are not free ...

As of today we know nothing about nuclei in the exclusive regime that I have just indicated that *compels* us to the belief that nucleons in the nucleus differ in any way from those in the free state other than through essentially trivial binding energy effects. Of course, as I have just remarked, there is more to our understanding of complex nuclei than neutrons and protons: explicit reckoning of mesonic contributions is demanded even in low-energy properties of heavy nuclei as E.K. Warburton demonstrated convincingly through his study of rank zero first forbidden beta-decay in the lead region where the effect of mesonic enhancement of the time-like part of the axial current is systematically worth a factor 2 in the transition rate. Sometimes the mesonic contributions are even larger that this: the cross section for ^{3}He(n,γ)^{4}He with thermal neutrons is enhanced by mesonic processes by an order of magnitude over that anticipated for nucleons-alone as shown by J. Carlson and colleagues. But we look to CEBAF to take us onto the next sheet in our understanding of the complex nucleus: to show that in our account of A+B→C+D, where A,B,C and D are all old friends in their eigenstates of known $J^\pi T$, we must talk the quark language and that the extended constituent model of N, Δ, π, ρ ... is not good enough.

Needless to say, "must ... not good enough" will arrive at different times for different persons; the shade of Ptolemy will, for some, take a long time to fade; it was well into the 20th century before such giants as Mach and Ostwald were convinced of the existence of atoms; perhaps Mach never was although he admitted to Einstein in 1913 that atoms were a useful physical hypothesis but even then he did not grant them "real existence". And so it will doubtless be with quarks in nuclei; there will be those for whom a rococo elaboration of an ever-more-extended Ptolemaic constituent model remains preferable to Copernican quarks. But, as good scientists, we must wait to be *compelled* under threat of Occam's razor and not claim that we can run before we have demonstrated our ability to walk; the fact that, in certain regime, $d\sigma(\gamma d \to np)/dt \sim s^{-11}$, as already beautifully demonstrated at CEBAF, and reported to this meeting, does not prove that quarks *must* be brought into play although it may be *consistent with* the expectation of perturbative QCD based on counting rules in which momentum transfer is large to all the constituents of the system and no form factors are involved; the famous strictures of Isgur and Llewellyn Smith, that at one time were even (mis)used by some to oppose the funding of CEBAF, must not be forgotten. It is the prime job of CEBAF, to lend conviction to the role of quarks *in exclusive nuclear processes.*

Within the proper arena of E≫1 GeV perturbative QCD goes from strength to strength in exposing the asymptotic inner workings of the nucleon; next-to-leading-order corrections are becoming critically important and are demanded by the data. However, large theoretical uncertainties remain in extracting the desired quantities from the data, particularly in the context of certain sum rules. Structure function data require extrapolation to x=0 with evolution to fixed Q^2 and for this theory provides as yet no unique guidance although "naive" QCD-based extrapolation looks reasonable while Regge extrapolation is not so attractive.

We should dearly love and, indeed, need to probe the quark structure of the neutron with the precision now afforded in our examination of the proton. The deuteron obviously offers itself but in certain respects ^{3}He may be thought of as a "free neutron" and to be able to bring us more direct news. However, we must not forget that at its centre the charge density of ^{3}He is 60% greater than that at the centre of the lead nucleus and that we cannot simultaneously discuss the modification of nucleons on their incorporation into nuclear matter and claim ^{3}He as a "free neutron".

Today's confidence in our knowledge of the neutron, however, permits us not only to confirm the touchstone of partonic nucleon structure, the Bjorken sum rule:

$$\int_0^1 (g_1^p - g_1^n)dx = \frac{1}{6}\frac{g_A}{g_V} + \text{QCD}$$

but quantitatively to analyze the QCD correction which, through fourth order, gives us a value for α_s. We may here note that various values of α_s, reported at this meeting, derived from scattering experiments and evolved from about 10 GeV to M_Z were $0.111 \pm 0.004 \pm 0.003, 0.113 \pm 0.005$ and 0.113 ± 0.002 to be compared with LEP's "direct" determination at M_Z of 0.123 ± 0.006. While this agreement is certainly a matter for lack of dis-satisfaction we must also note reservations expressed as to its significance after such evolution over such a range.

The seat of the nucleon's spin continues to excite and fascinate. The only thing that we (think we) know for sure is that:

$$J_z = S_z^q + L_z^q + S_z^g + L_z^g = \pm\frac{1}{2}$$

because we know nowhere else where the spin might reside. After much study the contribution of $S_z^{u,d}$ seems to be settling down to about $\frac{1}{3}$ of that $\frac{1}{2}$ with S_z^s

contributing about -0.1 ± 0.03 of that $\frac{1}{2}$. (We note, in passing, that the πN σ-term namely $m<p|s\bar{s}|p>/m<p|u\bar{u}+d\bar{d}+s\bar{s}|p>$ is now taking the value 0.15 to 0.2 while from deep inelastic scattering we have $p_s + p_{\bar{s}} = 4.1 \pm 0.4\%$ so all indications are for a significant (putatively) strange contribution to the make-up of the nucleon although inter-comparison of the different kinds of estimates is not straight-forward. We may also note, in this context, that the lattice-gauge computations of K-F. Liu and his colleagues have suggested a $s\bar{s}$ contribution to the nucleon mass of about 6% – and with a further contribution of similar magnitude from $c\bar{c}$ – hence my "putatively" above. But where is the rest of the nucleon's spin. In S_z^g? In the Ls? We do not know but various possibilities for addressing L are now emerging; we must wait, if impatiently, PANIC by PANIC.

An immensely impressive addition to our armoury since the last PANIC has come in the full flowering of HERA and for this meeting important fruit has been borne. 27 GeV electrons or positrons on 820 GeV protons at $\sqrt{s}$ = 300 GeV bring news from x-values down to 10^{-4} or so i.e. vastly lower than those accessible to fixed-target machines; the structure function data match beautifully, in the region of $x \simeq 10^{-2}$, onto those from fixed targets. Some of the fruit have not so much added fresh understanding as having been of wonderful pedagogical value, for example the measurement of the cross section for $p(e^-, \nu_e)$: previous measurements of the neutrino-induced cross section over the accessible energy range had shown, with considerable accuracy, and as expected, a rise linear with the neutrino energy, a linear rise that could not, however, continue indefinitely owing to the constraints of unitarity. But now the $\sqrt{s}$ = 300 GeV point form HERA indeed falls below the earlier linear rise, reflecting the influence of the finite W-propagator and yielding M_Z = 81.5 ± 6 GeV in conformity with the directly-determined value. That was a sheer delight to behold.

A related inference concerns the possible scale of quark sub-structure where an upper limit of 10^{-16} cm or so is found from HERA. (It is impressively amusing to note that a similar limit on the "size of the electron" comes from the table-top g-2 measurement of H. Dehmelt and his colleagues through comparison with the previsions of QED. Of course, all such size-limits are model-dependent and the relevance of the quark-electron comparison depends on one's attitude to grand unification.)

A most interesting observation from HERA concerns the jet structure resulting from the e^+p collisions. Without any but the most general reference to the reaction mechanism but using for its description just lowest-order Feynman diagrams we expect the mediating exchange objects to be coloured and therefore the primary products of the interaction, the "lepton current" and the "proton current", proceeding rapidly in opposite directions, also to be coloured. But they cannot so emerge into view: complicated and speedy colour-trading is necessary to ensure that the constituents of the hadronic recoil associated with the lepton current and the proton remnant jet in the proton direction are all colourless. This violent trading, necessarily associated with the multi-fragmentation of the rapidly-stretching gluonic flux tubes, implies, in particular, that the proton remnant jet will be a complicated one. But sometimes is isn't; sometimes it simply isn't there: the overall distribution shows a large rapidity gap; sometimes the proton, literally, goes down the tube, by itself or with very little accompanying it. In such events the exchange object of the primary interaction must itself be colourless so that the fore-aft colour trading negotiations following a primary colour exchange are not necessary and the proton merely suffers, in effect, a diffractive nudge and does not come to great harm. What is this colourless exchange object? It is easy to call it a pomeron but that is little better than a tautology because pomerons, colourless objects with the quantum numbers of the vacuum, come in different shapes and sizes depending upon their role. Further analysis suggest that the "object" is about 90% gluonic in its make-up with the rest a (presumably) u,d,s colour singlet.

Another recent development of considerable interest is our ability to address the gluon density of the nucleon through:

$$\frac{\partial \ln F_2}{\partial \ln Q^2} \sim \alpha_s g(x, Q^2)$$

which opens up to us a new dimension of nucleon structure. (As an aside I return to the EMC effect in response to a question as to whether we might now be able similarly to access the gluon density of complex nuclei. The answer is no because the meat of the EMC effect, the reduction below unity of the ratio of the structure function of a complex nucleus to that of deuterium at intermediate values of x, is sandwiched between the Fermi rise towards $x = 1$, which we understand in principle but which is not very interesting in the "fundamental" sense, and the increase in the ratio above unity for smaller values of x ("enhancement") and its subsequent plunge below unity towards $x = 0$ ("shadowing") neither of which we understand well enough to subtract them out to reveal the gluons.)

A final comment from the world of F_2 is a cautionary note: as already remarked, fixed-target and HERA data fit together beautifully and now cover the whole range of x down to 10^{-4} over a wide range of Q^2. Futhermore, all data can be well represented by a QCD parameterization that superficially lends considerable credence to the whole partonic construct upon which it is based. The cautionary note, raised, in particular, by the "molecular" modeling of the Jülich group, is that this successful parameterization does not mean that the physical sense and content of the nucleon are uniquely captured by it: for example it obscures the possible importance of an explicitly mesonic component to the physical nucleon.

Finally, before I leave the nucleon, a glance across the chasm, just hinted at, that separates its structure in repose, its low-energy properties, from its response to a probing well above the QCD characteristic energy: the chasm that separates non-perturbative from perturbative QCD. We are firmly committed to QCD: the many impressive successes of our developing command of next-to-leading-order corrections and their demonstrated correctness and relevance, as reported to this conference, confirm us in that commitment. But the major questions of the real world pose themselves non-perturbatively and we struggle. We struggle with the lattice, which has become a major intellectual industry, through which we hope to build an explicit and reliable bridge across the chasm and we struggle with the QCD sum rules by means of which we hope to leap across the chasm but both these struggles are long and hard. The lattice, despite its multi-gigaflops remains very coarse in its scale and in its approximations; quarks of realistically-small mass and the full relaxation of the quenched approximation are still but fond hopes. The sum rules, despite impressive successes, remain in large measure mysterious, involve mystical commitment and lack a rigorous basis: we leap but cannot know where we shall land.

To solar neutrinos is a far cry from the nucleon but the message they bring may be even "more fundamental", if such a term has any sense, than that from the structure functions. There is no definite conclusion as yet but experimental data are firming up nicely and many new experimental approaches are underway. Of the now-classical data bases: (i) the Homestake (chlorine) results, sensitive chiefly to ^{8}B and other high-energy neutrinos, are updated and stable but have no direct calibration: $\pm$ 10%; (ii) SAGE and GALLEX (gallium), chiefly sensitive to low-energy *pp* and other neutrinos, are now in mutual agreement and are calibrated using ^{51}Cr: $\pm$15%; (iii) Kamiokande (νe scattering) sensitive to neutrinos of energy greater than 7 MeV, chiefly from ^{8}B, is calibrated using 9 MeV gamma-rays: $\pm$10%. The current statement is that all approaches slow substantial deficits with respect to the expectations of the standard solar model and have no significant correlation with the sunspot cycle. In particular, the ^{7}Be neutrinos appear to be strongly depressed. No *reasonable* juggling of nuclear cross sections or solar models seems able to account for all the facts as presented in particular the apparent strong depression of the ^{7}Be flux. Are we then forced to neutrino oscillations, presumably via the MSW effect? But can we not juggle *unreasonably*: the S_{17} problem of the ^{7}Be$(p,\gamma)^8$B rate at the temperature of the centre of the sun is still with us; is there a missing ingredient in our solar models? At the moment we must admit that the jury is still (just) out. We await, impatiently, results from SNO which will be sensitive to neutral currents, will have a high rate for certain processes and should be running in 1997 and from Super-Kamiokande, with many times the fiducial volume of Kamiokande and a threshold lowered to 5 MeV, which has already taken its first data. Several other important initiatives, with differing differential spectral responses, are at various stages of readying and will add critical detail to the picture. (We note, with relief, that the SAGE/GALLEX results, which show roughly the same shortfall in the pp neutrinos as earlier reported for those from ^{8}B, which go as a much higher power of the temperature at the centre of the sun, refute the hypothesis that the sum has gone out: the surface, and so we, not yet having learnt about this because of the very slow radiative transfer from the interior for all except neutrinos.)

Exotics, quark-based hadronic structures that cannot be qqq or $q\bar{q}$ and glueballs, containing no quarks at all, continue to fascinate and to elude; many have come and gone. Of the exotic mesons in the sequence $J^{PC}=0^{--}, 0^{+-}, 1^{-+}, 2^{+-}$... the third now has a promising candidate and of the glueballs, $0^{++}, 2^{++}$..., the scalar $f_0(1500)$ candidate may match the lattice's expectation of 1650 MeV although its decay tells us that it cannot be a pure glueball (what would keep any glueball pure?) and must be mixed with $n\bar{n}$ and $s\bar{s}$ $(n = u, d)$. $\xi(2230)$, seen in J/Ψ decay, puts itself forward as a tensor glueball candidate at about the lattice energy.

So far my eclectic picking of personal high-spots has confirmed the general drift upon which I commented earlier: towards particles and away from nuclei. But nuclei there were too and it was impressive to see the way in which through, particularly, exclusive $A(e, e'p)B$ reactions at NIKHEF and Mainz our confident knowledge of what nucleons are actually *doing* inside nuclei is expanding. The longitudinal interaction, unaffected in first order by meson exchange, tells us how the proton is struck and increasingly-refined optical model analyses, inducing secondary effects, tell us how the struck particle separates, or does not separate, from its parent nucleus. Earlier conclusions to the effect that the occupation probability of well-defined shell-model states is only some 50–70%, depending upon the nucleus and depth below the Fermi surface, of that naively to be expected are confirmed although it appears that use of relativistic structural basis wave-functions may raise that range of probability to 60–80%, but still well below unity.

We must not worry that nucleons in the nucleus are not following the prescription of the zero-order shell model. We do not expect them to: nucleons in the nucleus indeed interact with one another and this must tend to displace them from their nominal zero-order orbitals into higher states. The simplest extrapolation, to below the Fermi surface, of the imaginary part of the optical model potential empirically determined in the continuum would lead us to expect just such a depletion of the bound orbitals as we now know to be the case: there is a lowering of the occupation probability below unity beneath the Fermi surface and a corresponding finite population of states above the Fermi surface with a discontinuity in occupation, long famous in condensed matter physics, at the surface itself. All this is well understood, and in large measure quantitatively so, from the work of C. Mahaux and others but it is beautiful to see it emerging so clearly experimentally.

Something that has been a Holy Grail for nuclear structure experimenters for decades is the putting into evidence of NN correlations in the nucleus. Such correlations there must be, as is obvious from a glance at the NN potential with its hard core and they figure critically, of course, in nuclear matter calculations. But until quite recently our inability to probe the nucleus sufficiently cleanly has meant that initial- and final-state interaction uncertainties have prevented our gaining any unambiguous indication of correlations. Similarly, processes that, of their essence, must reflect correlations, such as $(\pi, 2N)$ reactions, are bedevilled in their quantitative interpretation not only by the same problems as afflict knock-out reactions but also by post-prior uncertainties: were the correlations there before the pion was absorbed or were they a consequence of that absorption? Now, however, with effectively CW accelerators and high-resolution detectors together with enhanced understanding of the final-state interaction we are perhaps at last gaining meaningful exclusive data on the high-momentum tail of the nucleon distribution although doubts are still expressed. It even seems to be emerging that the tail is dominated by, as we should expect, meson exchange and isobar effects at high values of the missing energy (nuclear residual excitation) and by long-range (shell-model-related) correlations at low values of the missing energy.

An important issue upon which I have already touched, namely that of the possible modification of nucleons within complex nuclei, is beginning to be meaningfully illuminated by these $(e, e'p)$ studies: there is no present evidence for any change in the operative form factors of bound nucleons from those of those in the free state. Such studies must be of increasing interest as more realistic methods of assessing theoretically, on a quark basis, the mutually-polarizing effects of neighbouring nucleons in the nucleus are devised. CEBAF indeed has a great responsibility in this quest. An unresolved theoretical problem arises: it seems to be necessary, for present practical purposes of data analysis, to use different potentials either side of the electromagnetic vertex and this violates the Ward-Takahashi identities, the consequences of which are obscure.

And finally: a sigh of relief: by doing the experiment in the right way at the right energy and by looking at the results also in the right way the long-standing shortfall in the Coulomb sum rule has been resolved and the number of protons in the nucleus of carbon has increased to 6. Let us only hope that by the time of PANIC XV in Uppsala in 1999, to which we look forward keenly, that number has not risen to 7 ...

Session 1
Lepton Probes of Hadronic Structure

MEASUREMENT OF THE NEUTRON SPIN STRUCTURE FUNCTION g_1^n WITH POLARIZED ^{3}He AT SLAC

GERASSIMOS G. PETRATOS
Department of Physics and Center for Nuclear Research, Kent State University
Kent, OH 44242, USA

REPRESENTING SLAC E142 COLLABORATION

The final results from the detailed analysis of SLAC experiment E142 are reported. The experiment measured the neutron spin structure function g_1^n using polarized electrons and a polarized ^{3}He target. The results are interpreted in the quark-parton model and are used to test the Bjorken sum rule.

The final results from the detailed analysis[1] of SLAC experiment E142 are reported. The experiment measured the neutron spin structure function g_1^n in the Bjorken x range from 0.03 to 0.6 at an average four-momentum transfer squared $Q^2 = 2$ (GeV/c)^2 using a longitudinally polarized beam from the Stanford Accelerator and a longitudinally/transversely polarized ^{3}He target.

The proton and neutron spin structure functions g_1^p and g_1^n are related by the Bjorken sum rule[2] $\int_0^1 [g_1^p(x) - g_1^n(x)]dx = \frac{1}{6}\,|g_A/g_V|\,[1 - \alpha_s(Q^2)/\pi]$, where g_A and g_V are the weak couplings of the nucleon beta decay and $\alpha_s(Q^2)$ is the QCD coupling constant. This sum rule, first derived from current algebra and based on isospin symmetry, is a rigorous prediction of QCD. A separate sum rule for the neutron was derived by Ellis and Jaffe[3] assuming SU(3) symmetry and that the strange quark sea is unpolarized: $\int_0^1 g_1^n(x)dx = \frac{1}{18}(6F - 4D)\,[1 - \alpha_s(Q^2)/\pi]$, where F and D are SU(3) invariant matrix elements of the axial vector current. Though only first order QCD corrections are shown in these sum rules, higher order corrections have been calculated[4]. The interpretation of the latter integral in the quark-parton model is: $\int_0^1 g_1^n(x)dx = \frac{1}{2}(\frac{1}{9}\Delta u + \frac{4}{9}\Delta d + \frac{1}{9}\Delta s)$, where $\Delta q_i^{\uparrow} = \int_0^1 [q_i^{\uparrow}(x) - q_i^{\downarrow}(x)]dx$ represent the integrals over the momentum distributions of the up, down and strange quark of the nucleon. Therefore, a measurement of the integral, can be used, in conjunction with the neutron beta decay relationship $\Delta u - \Delta d = F + D$ and the SU(3) relationship $\Delta d - \Delta s = F - D$, to disentangle the Δu, Δd and Δs quark spin distributions.

Polarized electrons with energy of 19 to 26 GeV energy and 2 μA intensity were produced by photoemission from an AlGaAs source. The beam polarization was monitored during the experiment by measuring the asymmetry in Møller scattering from polarized electrons in magnetized ferromagnetic foils.

It was found to be stable over the duration of the experiment with an average value of $(36 \pm 1)\%$.

The experiment used a polarized ^{3}He target since a polarized ^{3}He nucleus is to first order a polarized free neutron. The polarized target was based on the technique of ^{3}He polarization by spin exchange with Rb vapor[5]. The Rb atoms were polarized via laser optical pumping. The average ^{3}He nuclear polarization was measured to be $(33 \pm 2)\%$ by means of NMR adiabatic fast passage.

Two magnetic spectrometers[6] centered at 4.5° and 7° respectively, detected scattered electrons with energy from 6 to 20 GeV. Each spectrometer was equipped with a pair of Čerenkov detectors for electron identification, and a pair of scintillator tracking hodoscopes and a lead-glass shower calorimeter for measuring the scattered electron energy and angle.

The neutron asymmetries were extracted from the ^{3}He measured cross section asymmetries assuming that the polarization of the neutron in ^{3}He is $\sim$ 87%. A correction for the polarization of the two protons in ^{3}He ($\sim -3\%$ per proton) was applied. False asymmetries were found to be consistent with zero by comparing data with target spins in opposite directions.

The g_1^n values extracted from the asymmetries are shown in Fig. 1. The integral of g_1^n over the measured x range 0.03 to 0.6 was calculated to be $\int_{0.03}^{0.6} g_1^n(x)dx = -0.028 \pm 0.006$ (stat.) ± 0.006 (syst.) at a constant Q^2 of 2 $(\mathrm{GeV/c})^2$. A correction for the unmeasured x range led to the final result: $\int_0^1 g_1^n(x)dx = -0.031 \pm 0.011$, in agreement with the -0.06 ± 0.03 result of CERN SMC experiment[7] and with the -0.037 ± 0.014 result of SLAC experiment E143[8]. Our result is in disagreement with the Ellis-Jaffe sum rule $\int_0^1 g_1^n(x)dx = -0.011 \pm 0.010$ at a Q^2 of 2 $(\mathrm{GeV/c})^2$.

The $\int_0^1 g_1^n(x)dx$ measurement can be used in the quark-parton model to extract the integrals Δq_i of the three quark spin distributions. Using the value for the hyperon decay constants $F/D = 0.58 \pm 0.06$, our measurement yields $\Delta u = 0.86 \pm 0.04, \Delta d = -0.40 \pm 0.04$ and $\Delta s = -0.06 \pm 0.04$. The sum $\Delta q = \Delta u + \Delta d + \Delta s$ of the three distributions is 0.39 ± 0.11, implying that the quarks partially account for the nucleon spin. The remaining part could be attributed to contributions from orbital angular momentum and the spin of the gluons (see Ref. [9] for a recent theoretical review.)

The new measurement of the neutron spin structure function integral can be combined with previous measurements[8] of the proton spin structure function integral to test the Bjorken sum rule. The result is: $\int_0^1 [g_1^p(x) - g_1^n(x)]\, dx = 0.160 \pm 0.015$ for a Q^2 of 3 $(\mathrm{GeV/c})^2$. The Bjorken sum rule prediction including higher order perturbative QCD corrections[4] gives: $\int_0^1 [g_1^p(x) - g_1^n(x)]\, dx =$

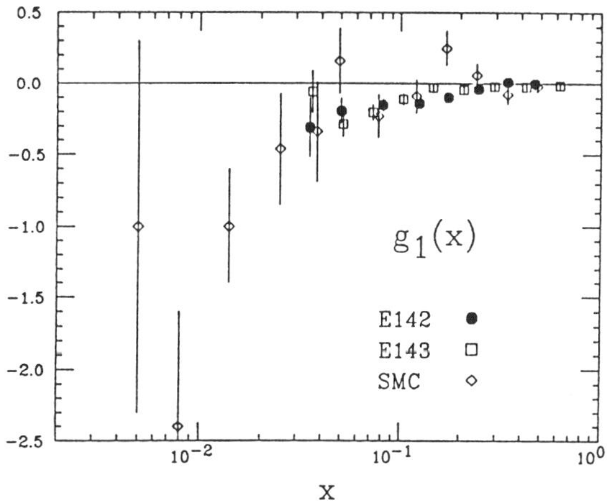

Figure 1: Measurements of the neutron spin structure function g_1^n versus x.

0.176 ± 0.008 (for $\alpha_s = 0.32 \pm 0.05$ at $Q^2 = 3$ $(\text{GeV/c})^2$), which is consistent with the experimental value.

In summary, the final results on g_1^n from the full analysis of SLAC E142 have been presented. The results are consistent with the Bjorken sum rule and suggest that the quarks account for ~0.4 of the nucleon spin. More precise data at higher Q^2 and lower x values will be beneficial for a better test of the Bjorken sum rule and for a clearer interpretation of the spin structure functions in the context of the quark-parton model of the nucleon. A new experiment using an 80% polarized beam of 50 GeV energy, an improved polarized ^{3}He target and new spectrometers, is expected to provide such data[10].

References

1. P. L. Anthony *et al.*, SLAC-PUB-7103 (1996).
2. J. D. Bjorken, *Phys. Rev.* **148** (1966) 1467.
3. J. Ellis and R. Jaffe, *Phys. Rev.* **D9** (1974) 1444.
4. S. A. Larin, Phys. Lett. **B334**, 192 (1994).
5. T. E. Chupp *et al.*, Phys. Rev. **C36**, 2244 (1987); **C45**, 915 (1992).
6. G. G. Petratos *et al.*, SLAC-PUB-5678 (1991) (unpublished).
7. D. Adams *et al.*, Phys. Lett. **B357**, 248 (1994).
8. K. Abe *et al.*, Phys. Rev. Lett. **75**, 25 (1995).
9. J. Ellis and M. Karliner, CERN-TH/95-334 (1995).
10. E. W. Hughes *et al.*, SLAC Proposal E154 (1993).

FIRST RESULTS FROM HERMES: THE $g_1^n(x)$ STRUCTURE FUNCTION

L.H. KRAMER
MIT-Bates Linear Accelerator Center and Laboratory for Nuclear Science, Massachusetts Institute of Technology, Cambridge, Massachusetts 02139

FOR THE HERMES COLLABORATION

The HERMES experiment will study the spin structure of the proton and neutron through inclusive and semi-inclusive deep-inelastic scattering. Data acquisition began in August 1995 with the polarized ^{3}He target serving as an effective neutron target. The first preliminary results for $g_1^n(x)$ and the Ellis-Jaffe sum are presented. An Ellis-Jaffe integral $\Gamma_1^n = -0.032 \pm 0.013_{stat.} \pm 0.017_{syst.}$ at fixed $Q^2 = 3$ $(\mathrm{GeV/c})^2$ is obtained using a subset of the 1995 data in the kinematic range $0.03 < x < 0.6$ and $Q^2 > 1$ $(\mathrm{GeV/c})^2$.

Recent experimental results from deep inelastic lepton scattering experiments indicate that the quarks contribute a relatively small amount to the nucleon's spin. The HERMES experiment will extend these investigations by measuring the deep-inelastic spin-dependent structure functions of the neutron and proton. To extract the structure functions, 27.5 GeV longitudinally-polarized positrons will be scattered from polarized hydrogen, deuterium, and ^{3}He targets. The goal of HERMES is to determine the gluon and flavor-dependent valence and sea quark contributions to the nucleon spin through a comprehensive set of inclusive and semi-inclusive scattering measurements.

HERMES combines two novel experimental techniques to approach the ideal beam-target interaction: a beam of longitudinally polarized positrons in a storage ring and a windowless high-density polarized internal target. The stored positron beam is self-polarized via the Sokolov-Ternov effect. Atomic hydrogen and deuterium sources and an optically-pumped ^{3}He source deliver polarized gas atoms to a storage cell. A magnetic spectrometer detects the scattered positrons with $1 < Q^2 < 20$ $(\mathrm{GeV/c^2})$ and for $0.02 < x < 0.8$. The luminosity is determined by detecting the Bhabha coincidence events with a segmented $\mathrm{NaBi(W0_4)_2}$ calorimeter.

Approximately 5.8 million deep-inelastic scattering events from the polarized ^{3}He target were recorded between August and November 1995. Typical running conditions included beam polarizations of 40-60%, target polarizations of $\sim 50\%$, and luminosities of $2 - 20 \times 10^{31}$ nucleons/cm^2/s.

The first preliminary analysis has been carried out using strict data quality cuts and limiting the kinematic range to $0.03 < x < 0.6$ and $Q^2 > 1$.

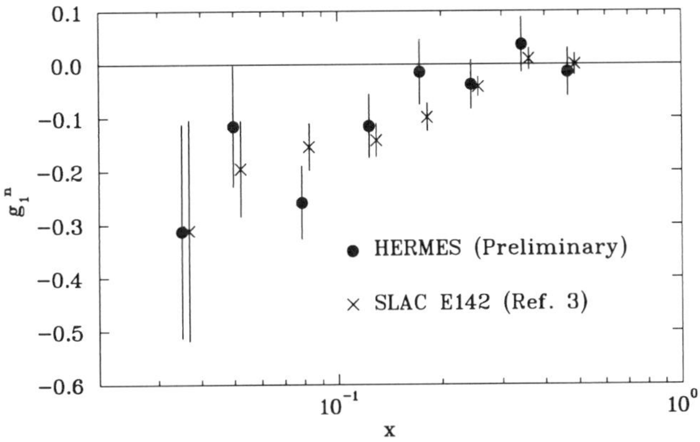

Figure 1: First preliminary result[2] from the HERMES experiment for the spin structure function g_1^n as a function of x. Included are results from SLAC experiment E-142[3]. Only statististical uncertainties are shown.

Radiative corrections were applied according to the prescription of Akushevich and Schumeiko[1]. Detector smearing and nuclear corrections were included in the analysis. Figure 1 shows the preliminary result[2] for the spin structure function $g_1^n(x)$ along with results from SLAC experiment E-142[3]. This result was then used to form the Ellis-Jaffe integral $\Gamma_1^n = \int_0^1 g_1^n(x)dx$ by evolving the data to a constant Q^2 and extrapolating to high and low x. Simple models were used for the extrapolation: $g_1^n(x) = constant$ for $x < 0.03$[4] and $g_1^n(x) \sim (1-x)^3$ for $x > 0.06$[5]. The preliminary result at fixed $Q^2 = 3$ (GeV/c)^2 is

$$\Gamma_1^n = -0.032 \pm 0.013_{stat.} \pm 0.017_{syst.}. \tag{1}$$

The relatively large statistical and systematic uncertainties reflect the current understanding of the experiment. Analysis of the full data set is continuing and the uncertainties are expected to reduce significantly in the final analysis.

References

1. I.V. Akushevich and N.M. Schumeiko, *J. Phys. G: Nucl. Part. Phys.* **20**, 513 (1994).
2. M. Vetterli, *Proceedings of the International Workshop on Deep Inelastic Scattering and Related Phenomena*, Rome, April 1996, to be published.
3. SLAC Publication SLAC-PUB-95-7103, (1995).
4. J. Ellis and M. Karliner, *Phys. Lett.* B **213**, 73 (1988).
5. S.J. Brodsky *et al*, *Nucl. Phys.* B **441**, 197 (1995).

THE SMC MEASUREMENT OF THE NUCLEON SPIN STRUCTURE FUNCTIONS

F. FEINSTEIN
DAPNIA/SPP, CE Saclay, Gif sur Yvette, France, F-91191

On behalf of the Spin Muon collaboration

We have measured the spin-dependent structure function g_1^p and g_1^d of the proton and the deuteron in deep inelastic scattering of 190 GeV polarized muons in the kinematic range $0.003 < x < 0.7$ and $1\,\mathrm{GeV}^2 < Q^2 < 60\,\mathrm{GeV}^2$. The first moment of g_1^p, $\Gamma_1^p = \int_0^1 g_1^p(x)dx$ is smaller than the prediction of the Ellis-Jaffe sum rule by 2.5 standard deviations and is consistent with previous measurements. The first moment of g_1^d $\Gamma_1^d = \int_0^1 g_1^d(x)dx$ is below the Ellis-Jaffe sum rule prediction by 4 standard deviations. Using our determinations of Γ_1^p and Γ_1^d we derive Γ_1^n and we obtain $\Gamma_1^p - \Gamma_1^n = 0.174 \pm 0.034$ which agrees with the Bjorken sum rule prediction 0.187 ± 0.003. The present measurement of g_1^p being performed this year is briefly described.

1 Introduction

Deep inelastic scattering (DIS) of leptons off target nuclei is a powerful technique which has produced detailed information on the internal structure of the nucleons. DIS of polarized muons off polarized protons and deuterons is used by the SMC to get information on the quark contribution to the spin content of the proton and the neutron. Theoretical predictions such as the Bjorken and the Ellis-Jaffe sum rules can be compared with our experimental results.

2 Physics goals

In DIS of a charged lepton off a nucleon the kinematics of the process is entirely described by the quantities ν which is the energy lost by the muon and $Q^2 = -q^2$ where q is the 4-momentum transfer carried by the virtual photon. The fraction of the nucleon momentum carried by the struck quark is $x = Q^2/2M\nu$, M being the nucleon mass. Similar to the unpolarized structure function $F_1(x)$ is the polarized structure function $g_1(x)$ interpreted as $\frac{1}{2}\Sigma e_i^2(q_i^+(x) - q_i^-(x))$ in the quark parton model where $q_i^{(-)+}(x)$ are the quark distributions for which the spin is (anti-)aligned with the nucleon spin. The first moment of g_1 is defined as $\Gamma_1 = \int_0^1 g_1(x)\,dx$.

In 1966 Bjorken [1] derived a sum rule from current algebra and assuming isospin symmetry :

$$\Gamma_1^{\rm p} - \Gamma_1^{\rm n} = \frac{1}{6}\frac{|g_A|}{g_V}$$

where g_A and g_V are extracted from neutron β decay. Presently, QCD corrections to this calculation have been calculated up to the fourth order.

In 1974 Ellis and Jaffe [2] derived independent predictions for $\Gamma_1^{\rm p}$ and $\Gamma_1^{\rm n}$. They assumed $SU(3)_f$ symmetry in hyperons and a strange sea polarization $\Delta s = 0$ in order to derive a_8 and $\Delta\Sigma$, the flavor singlet axial matrix element of the nucleon, from hyperons β decay

$$\Gamma_1^{\rm p(n)} = +(-)\frac{1}{12}\frac{|g_A|}{g_V} + \frac{1}{36}a_8 + \frac{1}{9}\Delta\Sigma.$$

3 Experimental method

We use the CERN SPS 190 GeV μ^+ beam. A DIS event corresponds to a scattered muon detected after a 2 m-thick iron absorber. The muon momentum is analyzed before and after the scattering and the scattering angle is measured. The target consists of two halves polarized in opposite directions. The determination of the vertex of the event allows us to know in which target half the interaction took place. The counting rate asymmetry between the two target halves is independent of the beam flux. In addition to the longitudinal 2.5 T target magnetic field a transverse magnetic field of up to 0.5 T can be applied. A rotation of the field is performed five times a day in order to reverse the target polarisation with respect to the beam momentum thus canceling asymmetries due to different acceptances of the two target halves. The transverse field also allowed to measure transverse asymmetries. The target material consists of 2.5 l of (deuterated) butanol. Dynamic nuclear polarization with frequency modulation of the microwave field is used to obtain target polarizations P_t as high as 0.86 for the protons and 0.50 for the deuterons. The build-up time is of the order of a few hours.

The CERN SPS delivers 450 GeV protons which impinge on a beryllium target producing π^+ and K^+ which decay in flight into $\mu^+\nu_\mu$. The muons are left-handed in the parent meson rest frame due to parity violation. Forward muons which have a high fraction of the meson momentum and a high negative polarization are selected. Two polarimetry methods are used. The shape of the energy spectrum of the positrons from muon decay in flight and the asymmetry in elastic scattering of muons off polarized electrons are measured [3]. The average beam polarization P_μ was -0.78.

The experimental asymmetry is proportional to the virtual photon nucleon asymmetry A_1 : $A_{exp} = A_1 f P_t D P_\mu$ where f accounts for the fraction of polarized target material and D is the depolarization factor of the virtual photon known from the kinematics of each event. A_1 is approximated to g_1/F_1 where $F_1 \approx \frac{F_2}{2x(1+R)}$ is calculated from NMC F_2 and SLAC R parametrizations.

4 Results

The measurements of A_1^p and A_1^d have been performed at $< Q^2 > = 10$ GeV^2 in the x range [0.003, 0.7] [4 5]. They are compatible with EMC [6] and E-143 [7] measurements in the common x range. In order to calculate Γ_1, we chose to evolve our measured points at our average Q^2 [8 9]. We find

$$\Gamma_1^p = 0.131 \pm 0.012\,(\text{stat.}) \pm 0.009\,(\text{syst.})$$

where the contribution to the integral for $x < 0.03$ is significant and

$$\Gamma_1^d = 0.039 \pm 0.006\,(\text{stat.}) \pm 0.004\,(\text{syst.}).$$

Γ_1^p is smaller than the prediction of the Ellis-Jaffe sum rule by 2.5 standard deviations and is consistent with previous measurements. Γ_1^d is below the Ellis-Jaffe sum rule prediction by 4 standard deviations. Using our determinations of Γ_1^p and Γ_1^d we derive Γ_1^n and we obtain

$$\Gamma_1^p - \Gamma_1^n = 0.174 \pm 0.034$$

which agrees with the Bjorken sum rule prediction 0.187 ± 0.003. Using these results in conjunction with the measured parameters of the β decays of the neutron and the hyperons, and assuming $SU(3)_f$ symmetry we obtain the nucleon matrix elements

$$\Delta\Sigma = 0.25 \pm 0.06 \text{ and } \Delta s = -0.11 \pm 0.02$$

In the naive quark-parton model where the axial anomaly is neglected these matrix elements can be interpreted as the contributions of the quarks to the nucleon spin. We have also taken data with the target polarization perpendicular to the beam polarization [10]. Transverse asymmetries have been found compatible with zero. Semi-inclusive asymmetries have been measured using the information from the leading hadron identified in a calorimeter [11]. We have found

$$< \frac{\Delta u_V}{u_V} >= 0.5 \pm 0.1, \quad < \frac{\Delta d_V}{d_V} >= -0.6 \pm 0.2, \quad \Delta\bar{q} \approx 0$$

5 Conclusions and prospects

Since 1992 we have measured $g^{\mathrm{p}}_{1,2}(x)$, $g^{\mathrm{d}}_{1,2}(x)$ and produced data at low x and high Q^2 thanks to CERN high energy muon beam. We are taking data in 1996 using NH3 which dilution factor is 30% higher than that of butanol. This will improve our knowledge of g_1^{p} in the range $0.003 < x < 0.03$ presently accessible only to SMC.

We have confirmed the violation of the Ellis-Jaffe sum rule for the proton and observed it for the deuteron. Altogether this is a $4.7\,\sigma$ effect. Extracting Γ_1^{n} from Γ_1^{d} and Γ_1^{p} we have verified the Bjorken sum rule. Interpreting these data in the naive quark-parton model, we have confirmed that a small fraction of the nucleon spin is carried by the quarks and that the strange sea is negatively polarized. From semi-inclusive data we have measured for the first time the polarization of the valence quarks.

The next step in the description of the spin structure of the nucleon is to measure the gluon contribution. Promising programs are proposed at CERN, looking at open charm in semi-inclusive reactions, at HERA with a polarized proton beam looking at two-jets events as well as the Q^2 evolution of g_1^p, at SLAC looking at the Q^2 evolution of g_1^p and g_1^d and at RHIC looking at selective processes.

References

1. J.D. Bjorken, Phys. Rev. 148 (1966) 1467; Phys. Rev. D1 (1970) 1376.
2. J. Ellis and R.L. Jaffe, Phys. Rev. D9 (1974) 1444; D10 (1974) 1669.
3. S. Eichblatt, these proceedings.
4. SMC, D. Adams et al., Phys. Lett. B**329** (1994) 399, Phys. Lett. B**339** (1994) 332.
5. SMC, D. Adams et al., Phys. Lett. B**357** (1995) 248.
6. EMC, J. Ashman et al., Phys. Lett. B**206** (1988) 364; Nucl. Phys. B**328** (1989) 1.
7. E-143, these proceedings.
8. V.N. Gribov and L.N. Lipatov, Sov. J. Nucl. Phys. 15 (1972) 438 and 675.
9. G. Altarelli and G. Parisi, Nucl. Phys. B**126** (1977) 298.
10. SMC, D. Adams et al., Phys. Lett. B**336** (1994) 125.
11. B. Adeva et al., Phys. Letter B**369**(1996), 93

SEMI-INCLUSIVE PHYSICS WITH HERMES

H. IHSSEN,
ON BEHALF OF THE HERMES COLLABORATION
E-mail: ihssen@hermes.desy.de, DESY/HERMES, Notkestr. 85,
22607 Hamburg, GERMANY

The first semi-inclusive measurement of HERMES and the prospects of future measurements are presented.

1 Introduction

HERMES[1] measures spin-dependent deep inelastic scattering off nucleons at HERA. Positrons with an energy of 27.5 GeV and a longitudinal polarisation of up to 60% are incident on an internal gas target. In 1995 the experiment was commissioned and data were taken on unpolarized hydrogen, deuterium and polarized ^{3}He targets. The ^{3}He nucleons achieved 50% polarisation in a storage cell target. The HERMES spectrometer detects scattered positrons and hadrons with angles between 40 and 220 mrad. A TRD, preshower hodoscope and electromagnetic calorimeter are used to separate positrons from hadrons. Pions above 6.5 GeV are identified using a threshold gas Cerenkov detector.
In semi-inclusive physics the hadronic final states of deep inelastic scattering (DIS) processes are used to extract information about the nucleon.
The key-equation of semi-inclusive physics is the differential cross section

$$\frac{d\sigma^h(z,x)}{dz} = \frac{\sum_f e_f^2 q_f(x)\, D_f^h(z)}{\sum_f e_f^2 q_f(x)} \sigma^T(x), \tag{1}$$

where $\sigma^T(x)$ is the total (inclusive) cross section. The formula is based on the assumption that the quark scattering process and the fragmentation process factorise and that the latter only depends on the fractional energy z. As the fragmentation functions $D_f^h(z)$ are known the measurement of different hadron species will provide us with information on the parton distribution $q_f(x)$. An important aspect of semi-inclusive physics is to separate the hadrons of the current fragmentation region from the hadrons of target fragmentation region. Provided the struck quark has a large enough momentum these two regions separate in the measured variable x_F ("x-Feynman"). At HERMES nearly all detected hadrons have a positive x_F value, indicating they originate mainly from the current fragmentation region and thus allow to extract parton distributions.

2 Unpolarized Semi-Inclusive Physics

From the 95 hydrogen and deuterium data we extracted a preliminary result for the valence quark distribution ratio $d_v(x)/u_v(x)$. Assuming isospin invariance this ratio should be independent of sea quark parton distributions and independent of fragmentation functions. The picture shows the HERMES result in comparison with the results of muon and neutrino experiments and the CTEQ3M [2] fit. Good agreement is observed.

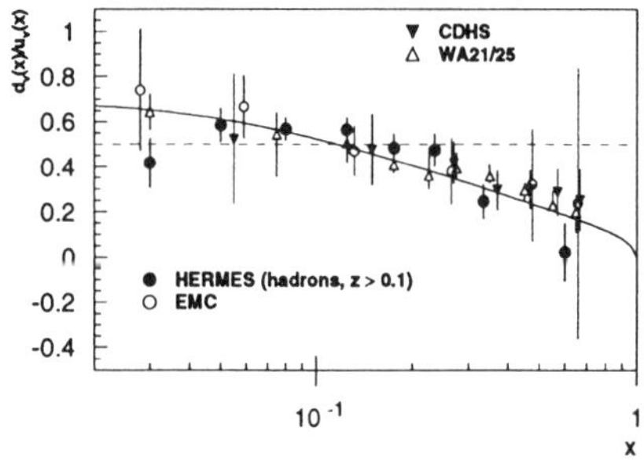

3 Polarized Semi-Inclusive Physics

HERMES will run in 96 with an polarized hydrogen target. These data together with the 95 ^{3}He data will allow to extract the polarised parton distributions $\Delta d_v(x)$ and $\Delta u_v(x)$ from the measurement of the pion asymmetries A_π^P and $A_\pi^{^3He}$:

$$A_\pi(x) = \frac{\left(N_{\uparrow\downarrow}^{\pi^+} - N_{\uparrow\downarrow}^{\pi^-}\right) - \left(N_{\uparrow\uparrow}^{\pi^+} - N_{\uparrow\uparrow}^{\pi^-}\right)}{\left(N_{\uparrow\downarrow}^{\pi^+} - N_{\uparrow\downarrow}^{\pi^-}\right) + \left(N_{\uparrow\uparrow}^{\pi^+} - N_{\uparrow\uparrow}^{\pi^-}\right)}. \tag{2}$$

In addition to pions, K_s, $K^\pm$ and Λ particles were collected by the HERMES experiment. From their asymmetries it might be possible to extract information on the polarised light and strange sea quark distributions.

References

1. , M. Dueren, The HERMES Experiment: From the Design to the First Results, habil. diss. (1995).
2. , H. Lai et al., *Phys. Rev.* D **51**, 4763 (1995).

SPIN STRUCTURE OF THE NUCLEON RESONANCES

P. RAINES[a], S.E. KUHN[b], F. WESSELMANN[b], K.A. GRIFFIOEN[c]
and the E143 Collaboration at SLAC
[a] *University of Pennsylvania, Philadelphia, PA 19104, USA*
[b] *Old Dominion University, Norfolk, VA 23529, USA*
[c] *College of William and Mary, Williamsburg, VA 23187, USA*

We report on the scattering of polarized electrons from polarized hydrogen and deuterium targets in the resonance region. The data, taken as part of E143 at SLAC at the beam energy 9.7 GeV, yield the spin structure function g_1 or a linear combination of A_1 and A_2 for the resonances at $Q^2 = 0.5$ and 1.2 GeV^2.

The spin structure function $g_1(x)$ has a natural interpretation for deep-inelastic scattering in terms of differences in quark distributions with spins parallel and antiparallel to the spin of the nucleon. In the region of the nucleon resonances (invariant mass of the struck nucleon $W < 2$ GeV), however, the more natural spin observables are the photon asymmetries A_1 and A_2. $A_1 = (\sigma_{\frac{1}{2}} - \sigma_{\frac{3}{2}})/(\sigma_{\frac{1}{2}} + \sigma_{\frac{3}{2}})$ measures the difference of final-state spin projections $\frac{1}{2}$ and $\frac{3}{2}$; $A_2 = \sigma_{LT}/\sigma_T$ measures the longitudinal/transverse interference. The observed values of A_1 in the resonance region are a weighted combination of asymmetries for the individual resonances and the inelastic background. Formally, the polarized scattering can be expressed in terms of g_1 and a second structure function g_2 or A_1 and A_2; the choice is a matter of physics emphasis. The goal of measuring the spin asymmetries for the resonance region is to better understand the resonances themselves, and also to determine their influence on the deep-inelastic results, either in terms of radiative corrections or the Q^2-evolution of the integral of g_1 which must formally always include the resonance contribution. The resonant contribution is insignificant at high Q^2, but it dominates the integral below about $Q^2 = 0.5$ GeV^2. In fact, the limits set by the Gerasimov-Drell-Hearn (GDH) sum rule[1] indicate that the integral should change sign somewhere between $Q^2 = 0$ and 1 GeV^2.

The deep-inelastic data from E143 have already been published[2]. For the resonance region, we have extracted both asymmetries and cross section differences for electron and nucleon spins aligned along the beam axis. Both observables yield the same physics with slightly different errors and corrections. Fig. 1 presents two glimpses of the data in the resonance region, expressed in one case in terms of g_1 (corrections due to g_2 are small) and in the other in terms of $A_1 + \eta A_2$ ($\eta \approx 1$ for E143 kinematics). Fig. 1a shows g_1^d for the 4.5-degree spectrometer as a function of the nucleon invariant mass W^2. The data

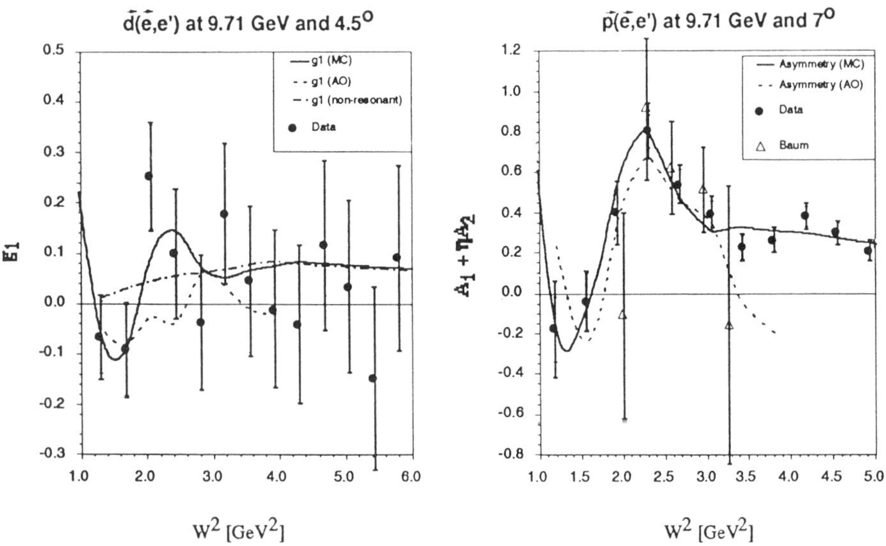

Figure 1: Results for (a) g_1 versus W^2 for ND_3 in the 4.5° spectrometer and (b) $A_1 + \eta A_2$ versus W^2 for NH_3 in the 7° spectrometer.

have been corrected for detector resolution, radiation, dead time, etc. The error bars include both statistical and systematic errors added in quadrature. The solid line is our Monte Carlo model which includes 3 fitted asymmetries for the resonance region, a simple linear interpolation in the regions of overlap, and parameterized deep-inelastic asymmetries. The dot-dashed curve shows the extrapolation of the deep-inelastic measurements into the region of low W^2. The dashed curve is the calculation of Burkert and Li[3] using the code A0. One of the reasons why A0 does not reproduced the data is that it does not include decay channels other than single π emission. Fig. 1b shows the photon asymmetry $A_1 + \eta A_2$ as a function of W^2 as measured in the 7° spectrometer. Plotted as triangles are the older SLAC data of Baum *et al.*[4] taken for similar kinematics. Within errors, the two measurements agree well. Both data sets seem to indicate a negative contribution from the Δ resonance, and a strongly positive contribution just above $W^2 = 2$ GeV^2 where the S_{11} resonance is important. Extractions of the integrals of g_1 at low Q^2 will soon be available using these data and the world's deep-inelastic results.

References

1. S. Gerasimov, *Sov. J. Nucl. Phys.* **2**, 430 (1966); S.D. Drell and A.C. Hearn, *Phys. Rev. Lett.* **16**, 908 (1966).
2. K. Abe, *et al.*, *Phys. Rev. Lett.* **74**, 346 (1995); *Phys. Rev. Lett.* **75**, 25 (1995); *Phys. Lett.* B **364**, 61 (1995); *Phys. Rev. Lett.* **76**, 587 (1996).
3. V. Burkert and Z-J. Li, *Phys. Rev.* D **47**, 46 (1993).
4. G. Baum, *et al.*, *Phys. Rev. Lett.* **45**, 2000 (1980).

A NEW MEASUREMENT OF R AND $\Delta R = R^d - R^p$ IN DEEP INELASTIC MUON SCATTERING

E.-M. KABUSS
Institut für Kernphysik, Universität Mainz, Becherweg 45
D 55099 Mainz, Germany
for the

NEW MUON COLLABORATION (NMC)

Results are presented on deep inelastic muon scattering on the proton and the deuteron at incident energies between 90 and 280 GeV. The x and Q^2 dependence of F_2^p and F_2^d was measured in wide kinematic range as well as R^p and R^d for $0.0045 < x < 0.11$. In an independent analysis σ_d/σ_p was determined with improved statistical and systematic accuracy compared to the structure function analysis and $R^d - R^p$ extracted in the x range between 0.003 and 0.35.

The differential cross section for the one photon exchange can be written in terms of the nucleon structure function $F_2(x, Q^2)$ and of $R(x, Q^2)$ where R is the ratio of longitudinally to transversely polarized virtual photon absorption cross sections. Here, $-Q^2$ is the four-momentum transfer squared and the Bjorken scaling variable x is given by $x = Q^2/2MyE$ with the relative energy transfer y, the incident muon energy E and the proton mass M. At low y a measurement of the cross section is mainly a measurement of F_2, whereas R can be obtained from the small differences in cross section at the same value of x and Q^2 but different values of y, i.e. at different incident energies E.

The New Muon Collaboration (NMC) has performed a simultaneous measurement of deep inelastic muon scattering on hydrogen and deuterium at 90, 120, 200 and 280 GeV incident energy and published recently structure function results using the large angle trigger of the experiment [1]. By the help of a small angle trigger at 200 and 280 GeV the ranges in x and y covered were extended downwards. This has permitted the determination of R as well as F_2. The results for F_2^p and F_2^d cover a kinematic range of $0.002 < x < 0.6$ and $0.5 < Q^2 < 75$ GeV^2. Due to the limited sensitivity of the large x data, R was determined for $x < 0.11$ averaged over Q^2.

Preliminary results for the average R for the proton and the deuteron are shown in fig.1a. The inner error bars represent the statistical errors, the outer ones the full errors. The main contribution to the systematic error stems from a 2% uncertainty in the relative normalisation of the data at different energies and triggers. The results agree well with a prediction from perturbative QCD [2] (solid line) and with the R parametrisation used in previous measurements [4] of F_2 (dashed line).

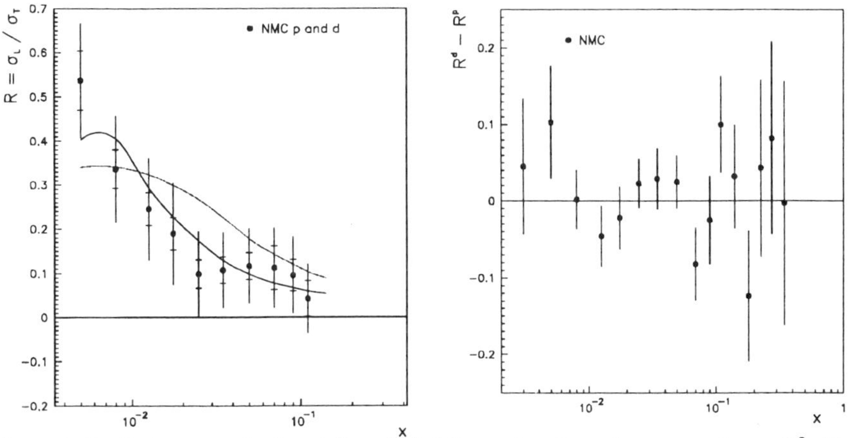

Figure 1: Preliminary results for a) R and b) ΔR as a function of x averaged over Q^2. The curves are discussed in the text.

The difference between R^d and R^p can be investigated most precisely using a direct measurement of the cross section ratio σ_d/σ_p. Due to the complementary NMC target set up cancellation of systematic errors due to spectrometer acceptance and normalisation was obtained allowing measurements in kinematic regions where the detector acceptance is small[3]. Thus, the cross section ratio σ_d/σ_p was obtained in an extended kinematic range of $0.001 < x < 0.8$ and $0.1 < Q^2 < 145$ GeV^2 with small statistical and systematic uncertainties.

To make full use of all measured data including where the data sets do not overlap, a Q^2 averaged value of ΔR was determined in each x bin separately[5]. Preliminary results in the x range between 0.003 and 0.35 are shown in fig.1b with statistical errors bars only. The values of ΔR are small, in particular at small x where R is large. No significant x dependence is observed. Averaging the measurements over x one obtains $\Delta R = 0.004 \pm 0.012(\text{stat})$. [a]

References

1. NMC, M. Arneodo et al., *Phys. Lett.* B **364**, 107 (1995).
2. G. Altarelli, G. Martinelli, *Phys. Lett.* B **76**, 89 (1978).
3. NMC, P. Amaudruz et al., *Nucl. Phys.* B **371**, 3 (1992).
4. L.W. Whitlow et al., *Phys. Lett.* B **350**, 193 (1990).
5. NMC, P. Amaudruz et al., *Phys. Lett.* B **294**, 10 (1992).

[a]This work was supported by the BMBF

Charge Symmetry Breaking from the NA51 Experiment

F. M. Steffens amd A. W. Thomas
Dept. of Physics and Mathematical Physics
University of Adelaide, Adelaide, SA 5005, Australia

We study the effects of charge symmetry breaking between the proton and the neutron in the NA51 experiment and conclude that they may be important.

The NMC experiment [1] has stimulated enormous interest in the flavour structure of the parton distributions [2]. This is because their measurement of the Gottfried sum rule implied that it is not saturated to its expected value of 1/3. In general, it has been assumed that the NMC result reflects a violation of SU(2) flavor symmetry in the sea of the proton. That is, it is concluded that there are more antiquarks of the type d than those of type u in the proton sea. However, the NMC results did not rule out other possibilities such as a large contribution from the unmeasured small x region or charge symmetry breaking between the proton and the neutron.

To discriminate between these possibilities, Ellis and Stirling [3] proposed an experiment based on dilepton production in the Drell-Yan process. Basically, they argued that the $p-n$ cross section asymmetry, defined as $A_{DY}(x) = (\sigma^{pp}(x) - \sigma^{pn}(x))/(\sigma^{pp}(x) + \sigma^{pn}(x))$, can tell whether the sea is symmetric or not. The experiment proposed by Ellis and Stirling was carried out by the NA51 collaboration [4] which measured the Drell-Yan asymmetry A_{DY}. They quoted the following result:

$$A_{DY}(x = 0.18) = -0.09 \pm 0.02(stat) \pm (0.025), \tag{1}$$

from which they derived

$$\lambda_s(x = 0.18) = \frac{\overline{u}(x)}{\overline{d}(x)} = 0.51 \pm 0.04(stat) \pm 0.05(syst). \tag{2}$$

We wish to emphasise that the value quoted on the basis of this experiment is only a lower limit on the asymmetry, $\lambda_s(x)$, of the proton sea – the limit where charge symmetry breaking is exact. If we allow for charge symmetry breaking between the proton and the neutron in the parton distributions, $\overline{u}(x) = \overline{u}^p(x) = \overline{d}^n(x) - \overline{\delta}(x)$, $\overline{d}(x) = \overline{d}^p(x) = \overline{u}^n - \overline{\delta}(x)$, $u_v(x) = u_v^p(x) = d_v^n(x) - \delta(x)$ and $d_v(x) = d_v^p(x) = u_v^n(x) + \delta(x)$, we can write a general expression for the asymmetry:

$$A_{DY}(x) \;=\; \{(4\lambda_v(x) - 1)(\lambda_s(x) - 1) + (4\lambda_s(x) - 1)(\lambda_v(x) - 1)$$

$$
\begin{aligned}
&-(4\lambda_v(x)+1)\overline{\delta}(x)/\overline{d}(x)-(1-4\lambda_s(x))\delta(x)/d_v(x)\\
&-2(4\lambda_s(x)+1)-2(4\lambda_s(x)-1)(\lambda_s(x)-1)\overline{d}(x)/d_v(x)\}\ /\\
&\{(4\lambda_v(x)+1)(\lambda_s(x)+1)+(4\lambda_s(x)+1)(\lambda_v(x)+1)\\
&+(4\lambda_v(x)+1)\overline{\delta}(x)/\overline{d}(x)+(1-4\lambda_s(x))\delta(x)/d_v(x)\\
&+2(4\lambda_s(x)+1)+2(4\lambda_s(x)+1)(\lambda_s(x)+1)\overline{d}(x)/d_v(x)\}\ . \quad (3)
\end{aligned}
$$

To extract any number from Eq. (3) we need to know the value of the sea and valence quark distributions at a given x. As we include isospin breaking terms, we have the problem that there is no standard quark distribution including these corrections. Moreover, the term involving $\overline{\delta}(x)$ is potentially important as it is multiplied by a large factor and divided by a small number (viz. $\overline{d}$). This is true whether the integral over x of $\overline{\delta}(x)$ (and $\delta(x)$) is zero or not. This means that to extract $\lambda_s(x)$ using the measured Drell-Yan asymmetry is at best ambiguous. To estimate the order of magnitude of $\delta(x)$ and of $\overline{\delta}(x)$, we will assume that the quark and antiquark distributions are described by the MRS D'_0 and D'_- parametrizations. For the D'_0 set, the Gottfried sum rule is 0.26 and for the D'_- this value is 0.24, which means that for both sets, $\int_0^1 \overline{\delta}(x)dx \simeq 0$, but this does not mean that $\overline{\delta}(x)=0$ at $x=0.18$. Using the measured asymmetry one finds:

$$
\begin{aligned}
\overline{\delta}(x=0.18) &= 0.169-0.053\delta(x=0.18), \ \ \lambda_s \sim 0.88, \ \ D'_0\\
\overline{\delta}(x=0.18) &= 0.125-0.045\delta(x=0.18). \ \ \lambda_s \sim 0.78, \ \ D'_- \qquad (4)
\end{aligned}
$$

These results indicate that the calculated corrections are of order 30% $\sim$ 40 % of the antiquark distributions themselves. However, we could allow for λ_s to be only 20% bigger then the value quoted by the NA51. This would imply that $\overline{\delta}$ is around 10% of the antiquark distributions. This could well be possible and, from this point of view, the λ_s quoted as being experimentally measured only sets a lower limit corresponding to $\overline{\delta}(x)=\delta(x)=0$.

References

1. NMC, P. Amaudruz et al., Phys. Rev. Lett. 66 (1991) 2712.
2. E. M. Henley and G. A. Miller, Phys. Lett. B 251 (1990) 453; A. I. Signal, A. W. Schreiber and A. W. Thomas, Mod. Phys. Lett. A 6 (1991) 271; S. Kumano and J. T. Londergan, Phys. Rev. D 44 (1991) 717; A. Szczurek and J. Speth, Nucl. Phys. A 555 (1993) 249.
3. S. D. Ellis and W. J. Stirling, Phys. Lett. B 256 (1991) 258.
4. NA51 collaboration, A. Baldit et al., Phys. Lett. B 332 (1994) 244.

DEEP-INELASTIC SCATTERING ON THE PION CLOUD OF THE PROTON

B. POVH
Max-Planck-Institut für Kernphysik, Postfach 103980, 69029 Heidelberg, Germany

B.Z. KOPELIOVICH
Max-Planck-Institut für Kernphysik, Postfach 103980, 69029 Heidelberg, Germany

I.K. POTASHNIKOVA
Joint Institute for Nuclear Research, Dubna, 141980 Moscow Region, Russia

Experiments at HERA looking for deep-inelastic electroproduction of neutrons in the proton fragmentation region are in progress. They are aimed to measure the pion structure function at small Bjorken x. We analyse [1] other sources of the leading neutron, in order to figure out the kinematical region where the one-pion exchange dominates.

The cloud of virtual pions surrounding the nucleon core of the proton is known to be an important source of the sea quarks. One may hope to be able to extract information about the pion structure function, relying upon the dominance of the one-pion-exchange diagram shown in Fig. 1a [2,3,4].

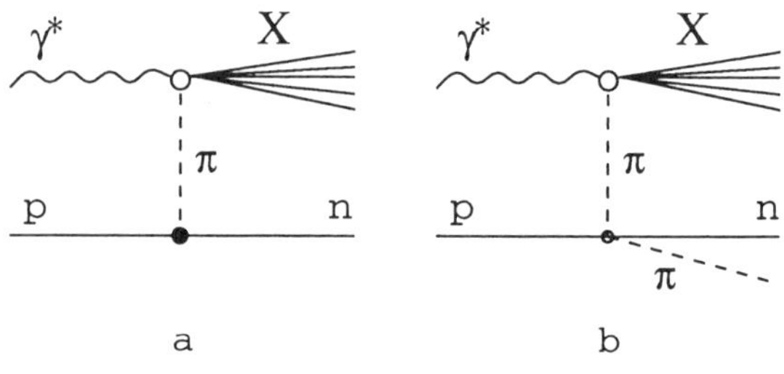

Figure 1:

One has to provide, however, convincing evidence that the sea quarks probed by the virtual photons do really belong to the pion. There are many other sources of sea quarks, which have to be evaluated reliably.

What is specific about the pion, which may help to single out its contribution from other mesons surrounding the proton? First of all, it is the smallness of the pion mass. It allows the pion to be spread around the proton at much longer distances than other mesons. This fact manifests itself as a narrow transverse-momentum distribution of the produced neutrons, $\langle p_T^2 \rangle \sim m_\pi^2$.

The above arguments, however, fail if the pion is far off the mass shell. Such a pion propagates only for a short distance of about $1/\sqrt{|t|}$ from the proton, where $t < 0$ is the pion four-momentum squared.

$$t = -\frac{1}{z}\left[p_T^2 + (1-z)^2 m_N^2\right] \tag{1}$$

z is a portion of the proton light-cone momentum carried by the final-state neutron. In order to make $|t|$ sufficiently small and single out the pion contribution one may select the events with a small value of $1 - z \leq m_\pi/m_N$. However, in this case one faces a new problem, which is related to another pion signature, its spinelessness. Namely, the pion exchange is suppressed by a factor $1-z$ compared to the leading Reggeons, ρ, a_2, having higher intercepts.

We conclude that any attempts to single out the pion-pole contribution relying upon the signatures of the pion, the smallness of its mass and spinelessness, face kinematical restrictions which in some sense exclude each other. In such circumstances one should look for a compromise, i.e. a kinematical region where the pion pole nevertheless dominates[1]

The contribution to the proton structure function corresponding to Fig. 1a reads

$$\left[\frac{dF_2^{p\to n}(x,Q^2)}{dt\ dz}\right]_{1a} = \frac{2g_{\pi NN}^2}{16\pi^2}\,\frac{|t|}{(m_\pi^2-t)^2}\,G_1^2(t)\,(1-z)^{1+2\alpha'_\pi|t|}\,F_2^\pi(x_\pi,Q^2). \tag{2}$$

The Bjorken variable for the pion is

$$x_\pi = \frac{x_p}{1-z-x_p}\ , \tag{3}$$

Function $G_1(t) = \exp[R_1^2(t-m_\pi^2)]$ includes the t-dependence of the formfactor for the pion-nucleon vertex and the virtual pion photoabsorption cross section. The Reggeization of the pion exchange is very important and is included in (1). Numerical values of the parameters are fixed in [1]. Our predictions for z - dependence of the cross section of inclusive electroproduction of neutrons at $p_T = 0$, $Q^2 = 10\ GeV^2$ and $x_p = 5\cdot 10^{-4}$ is depicted in Fig. 2 by the dashed curve. The cross section steeply vanishes towards $z = 1$.

Further contributions are[1]: the pion-pole contribution to the $ep \to e'n\pi X$ (dotted curve in Fig. 2), the Reggeon-exchange contribution (dot-dashed curve in Fig. 2) and the diffractive electroproduction of neutrons amounting to $\sim 6\%$ of events, not shown in Fig. 2.

The contributions of different mechanism of electroproduction of a neutron shown at $p_T = 0$ in Fig. 2, demonstrate that diagram in Fig. 1a dominates by

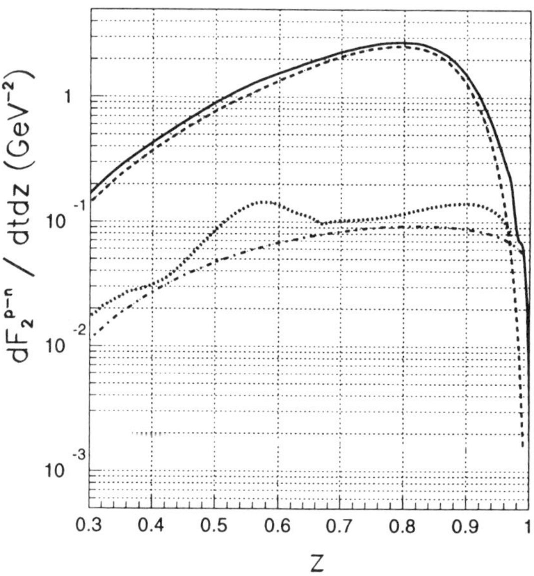

Figure 2: Comparative contributions of the diagrams in Fig. 1a (dashed curve), Fig. 1b (dotted curve) and the Reggeon ($\rho + a_2$) exchange (dot-dashed curve) to the neutron electroproduction cross section at $p_T^2 = 0$. The sum is depicted by the solid line.

about an order of magnitude. It dominates also at small $p_T^2 \leq 0.3\ GeV^2$ as is demonstrated in [1].

However, the left-hand slope of the peak Fig. 2, which corresponds to large $|t|$, is predicted with a large uncertainty [1]. In order to extract reliably the pion structure function one should select the events around the maximum in z-distribution shown in Fig. 2. Dependent on the statistics, one can choose an interval $z \approx 0.7 - 0.9$ and $p_T^2 < 0.2 - 0.3$.

References

1. B.Z. Kopeliovich, B. Povh and I.K. Potashnikova, DESY 96-011, hep-ph/9601291; *Z.Phys.A* in press.
2. J.D Sullivan, *Phys. Rev.* **D5** (1972) 1732
3. G. Levman and K. Furutani, ZEUS note 92-107, 1992; preprint DESY 95-142, 1995
4. H. Holtmann et al., *Phys. Lett.* **B338** (1994) 363

NUCLEON STRANGENESS CONTENT THROUGH VECTOR MESON DOMINANCE

STEPHEN R. COTANCH
Department of Physics, North Carolina State University,
Raleigh, NC 27695-8202

ROBERT A. WILLIAMS
Continuous Electron Beam Accelerator Facility,
Newport News, Virginia 23606

Abstract

It is demonstrated that $N(\pi, e^+e^-)N$ measurements will provide new information about the nucleon time-like form factor in the ϕ meson region including the value of the ϕN coupling $g_{\phi NN}$. Using vector meson dominance, $N(\pi, e^+e^-)N$ calculations reveal a three order of magnitude enhancement from ϕN coupling in the nucleon form factors providing a novel experimental signature for OZI violation and the related strangeness content of the nucleon.

In this paper we report calculations for the $N(\pi, e^+e^-)N$ reaction which document a dramatic, novel effect. Assuming only the validity of vector meson dominance (VMD), we predict three order of magnitude cross section enhancements for $\sqrt{q^2}$ near the low-lying (ρ, ω, ϕ) vector meson masses. The resonances originate from vector mesons coupling in the timelike nucleon form factor which can not be measured for $q^2 < 4M_N^2$ with e^+e^- annihilation reactions. In contrast, $N(\pi, e^+e^-)N$ entails a three-body final state with an effectively unrestricted virtual photon mass $q^2 \geq 4M_e^2 \sim 0$. Measurements will therefore provide significant new information which will permit investigating several important physics issues including OZI violations and the strangeness content of the nucleon.

The model, which is based upon quantum hadrodynamics, is explicitly covariant, gauge invariant and incorporates crossing symmetry and duality constraints. Full theoretical details are contained in ref. [1]. For the baryon form factors we utilize a hybrid VMD formalism [2] which implements Sakurai's universality hypothesis to analyze and describe the baryon octet EM form factors. Our treatment provides a good quantitative description of the nucleon data. In Fig.1 we display the electric proton form factor, $G_E^p(q^2)$, with and without the ϕ contribution. Our ϕN coupling constant relative to ωN is $g^2_{\phi NN}/g^2_{\omega NN} = 0.14$. There is considerable uncertainty in $g_{\phi NN}$ whose magnitude is governed by the currently unknown nucleon strangeness content as well as the small, but much better known, u- and d- quark content of the ϕ. To establish the validity of our reaction model we have successfully analyzed photoproduction data using Born amplitudes (i.e. proton, neutron, pion, and rho graphs) with a phenomenological treatment of the hadronic vertex factors and final state absorption effects. With all model parameters fixed by the photoproduction and form factor data, we employ crossing symmetry to make a consistent prediction for the $p(\pi^-, e^+e^-)n$ reaction, shown in Fig.2. The relatively small ϕ contribution to the nucleon form factors leads to a three order of magnitude enhancement in the cross section. The dashed curve is the same calculation with all ϕN couplings set to zero and hence represents the effect of the $\rho\pi\phi$ coupling due to ρ-exchange between π and N. We conclude that the (π, e^+e^-) process is ideal for investigating the timelike nucleon form factor below $q^2 = 4M_N^2 \sim 3.5\ GeV^2$ and for assessing OZI violation of ϕN coupling.

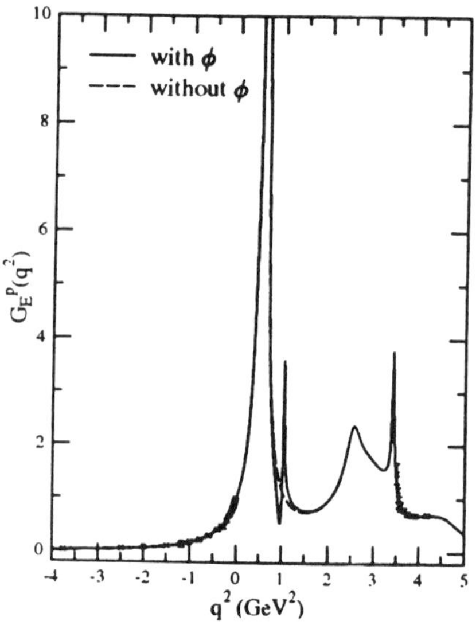

Figure 1.

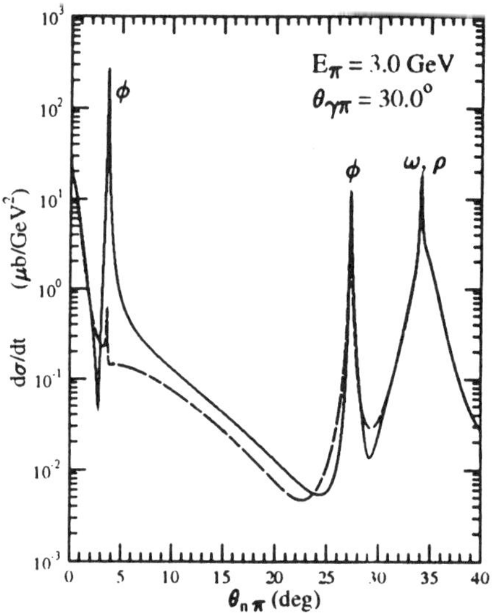

Figure 2.

Acknowledgement

S.R.C. acknowledges support from DOE grant DE-FG05-88ER40461.

References

[1] Robert A. Williams, Chueng-Ryong Ji and Stephen R. Cotanch, Phys. Rev. D41, 1449 (1990); Phys. Rev. C43, 452 (1991); Phys. Rev. C46, 1617 (1992); Phys. Rev. C48, 1318 (1993).

[2] R. A. Williams and C. P. Truman, Phys. Rev. C53, 1580 (1996).

MEASUREMENTS OF PION ELECTROPRODUCTION RESPONSE FUNCTIONS

G. WARREN
Massachusetts Institute of Technology, Cambridge, MA, USA

C. MERTZ
Arizona State University, Tempe, AZ, USA

C. VELLIDIS
University of Athens, Greece

We conducted a set of measurements of the $p\,(e,e'p)\,\pi^0$ and $p\,(e,e'\vec{p})\,\pi^0$ reactions in the $\Delta^+(1232)$ region at $Q^2 = -0.13$ $(\text{GeV/c})^2$ at the M.I.T.-Bates laboratory in 1995. The goal of these measurements is to extract the ratio of the Coulomb quadrupole amplitude to the magnetic dipole amplitude in the $N \to \Delta$ transition.

The $N \to \Delta$ transition provides information on higher orbital angular momentum contributions to the nucleon and Δ wavefunctions. The ratio of the small Coulomb quadrupole amplitude to the dominant magnetic dipole amplitude, the *CMR*, is sensitive to the D state contributions in the nucleon wavefunction. Numerous theoretical investigations and detailed analysis of photoproduction data generally suggest that at low four-momentum transfer the *CMR* is negative and of the order of 5%. Previous electroproduction data yielded contradictory results which, however, were generally characterized by large systematic errors.

We conducted a set of neutral pion electroproduction measurements in order to extract the *CMR* in the $N \to \Delta$ transition. In addition to contributions from the $N \to \Delta$ transition, nucleon and meson pole terms as well as tails of higher resonances contribute to pion electroproduction observables. To extract information about the $N \to \Delta$ transition from these observables, we must understand these "background" contributions. With this objective in mind, we took data at a central Q^2 = -0.13 $(\text{GeV/c})^2$ to extract the R_{LT} response function which is sensitive to the *CMR* and the R^n_{LT} response function which is sensitive to the background contributions.

The R_{LT} response is directly related to the real part of the transverse-longitudinal interference term of the hadron electromagnetic tensor. We took data to extract the R_{LT} response at two central values of the invariant hadron masses, $W = 1170$ and 1232 MeV/c^2, in order to study the mass dependence of the *CMR*. To determine the R_{LT} response, we measured the in-plane coincidence cross section off a liquid Hydrogen target using the magnetic spectrometers MEPS and OHIPS for the detection of electrons and protons respectively. For each invariant hadron mass, we held the electron arm fixed and varied the angle of the proton arm over a range of angles about the central momentum transfer vector. The range of angles between the proton and the momentum transfer in the hadron center-of-mass frame for these measurements was $\Theta_{pq}^{cm} = 0 - 40^{\circ}$ for $W = 1170$ MeV/c^2 and $\Theta_{pq}^{cm} = 0 - 20^{\circ}$ for $W = 1232$ MeV/c^2. We also measured the parallel kinematics cross section at three central values of the invariant hadron mass, $W = 1170$, 1232, and 1293 MeV/c^2.

The R_{LT}^{n} response is related to the imaginary part of the transverse-longitudinal interference term of the hadron electromagnetic tensor. At resonance, the contribution of the pure Δ quadrupole amplitude to this imaginary part is predicted to vanish, but contributions related to the non-resonant processes are predicted to be strong. Therefore, this observable is suitable to constrain model predictions of the background in the extraction of the *CMR*. We will infer the R_{LT}^{n} response function from our measurements of the final state induced proton polarization, P_n, using the Focal Plane Polarimeter (FPP) and of the cross section. Both quantities were measured in parallel kinematics and at a central $W = 1232$ MeV/c^2. Preliminary analysis of P_n yields a large induced polarization, between -0.3 and -0.4 with a 15% error, indicating a significant interference between the resonant terms and the background.

Polarimetry measurements are insensitive to variations in luminosity and are less sensitive to uncertainties in the acceptances of the spectrometers. However, the precession of the spin of the proton through the magnetic spectrometer, when averaged over the acceptance, can effectively reduce the polarization of the proton as measured at the focal plane. Preliminary analysis indicates that the uncertainty of this effect is small compared to the statistical errors for this measurement. In addition, false asymmetries due to chamber misalignment and spatial inefficiencies were determined by measuring polarization observables for elastic $p(e,e'\vec{p})$ scattering. Both the correction due to the false asymmetries and its uncertainty are small compared to the statistical error.

In the near future, the results of this experiment will be expanded by additional pion electroproduction experiments with the Out-Of-Plane-Spectrometers and with the FPP. These results will enable us to improve our understanding of the dynamical ingredients of the pion production.

This paper is presented on behalf of the OOPS and FPP Collaborations.[a] This work has been supported in part by the U. S. Department of Energy.

---

[a]ASU, CSLA, MIT, NSU, ODU, TJNAF, U Athens, U Bonn, U Ga, U Il, U Ma, U Md, U Va, W & M.

THRESHOLD π^0 ELECTRO PRODUCTION AT MAMI

A1-COLLABORATION AT MAMI
Mainz University, Institut für Kernphysik

P. BARTSCH[1], D. BAUMANN[1], A. M. BERNSTEIN[4], W. U. BOEGLIN[6], R. BÖHM[1], K. BOHINC[3], T. CAPRANO[1], N. CLAWITER[1], A. DERBER[1], S. DERBER[1], M. O. DISTLER[1], A. EBBES[1], R. EDELHOFF[1], I. EWALD[1], J. M. FRIEDRICH[1], J. FRIEDRICH[1], R. GEIGES[1], M. HAMDORF[1], P. JENNEWEIN[1], M. KAHRAU[1], M. KORN[1], K. W. KRYGIER[1], V. KUNDE[1], M. KUSS[2], J. LÁC[5], A. P. LIESENFELD[1], H. MERKEL[1], K. MERLE[1], R. MISKIMEN[7], R. NEUHAUSEN[1], M. PAVAN[4], TH. POSPISCHIL[1], M. POTOKAR[3], A. RICHTER[2], G. ROSNER[1], H. SCHMIEDEN[1], G. SCHRIEDER[2], S. ŠIRCA[3], A. WAGNER[1], TH. WALCHER[1], M. WEIS[1], ST. WOLF[1]

[1]*Mainz University,* [2]*TH Darmstadt,* [3]*Ljubljana University,* [4]*M.I.T.,*
[5]*Rutgers University,* [6]*Florida IU,* [7]*UMass*

presented by H. MERKEL
Mainz University, Institut für Kernphysik, Merkel@kph.uni-mainz.de

Pion photo- and electro-production is of fundamental interest for the understanding of photon-hadron interactions and of the structure of the nucleon. This is due to the fact that the pion as lightest meson can be identified with the Goldstone boson of QCD and thus provides an important bridge between the physics of confined quarks and the long range interactions of nucleons.

The interest in these reactions has become rather intense in recent years since calculations on the basis of chiral perturbation theory (ChPT) came available. ChPT is an effective field theory based on the chiral symmetry of QCD, but with the nucleon and pion as effective degrees of freedom.

In the threshold region, i.e. below a center of mass momentum of $\approx 100MeV/c$, a multipole decomposition in s and p waves is sufficient to describe the scattering amplitudes. In this region ChPT applies at low momentum transfers $(-q^2 < 0.1GeV^2/c^2)$ for these multipoles.

Up to now considerable technical problems occured during attempts to access this kinematical range. These situation changed with the new three spectrometer setup of the A1 Collaboration at MAMI (Fig. 1), where an extended program to study these region was started. The high quality continous wave electron beam of MAMI together with a high current liquid Hydrogen target system delivers luminosities of up to $50MHz/\mu barn$ at low background rates. A new target cell was developed to reduced the pathlength in Hydrogen for the recoil proton below $5mm$. With this cell four momentum transfers close to $q^2 = 0.03(GeV/c)^2$ can be achieved.

Last year a first measurement with Rosenbluth separation was performed at an four momentum transfer of $q^2 = -0.1(GeV/c)^2$ [?]. These data will be published soon. Now the kinematical coverage was extended to lower four momentum transfers and higher center of

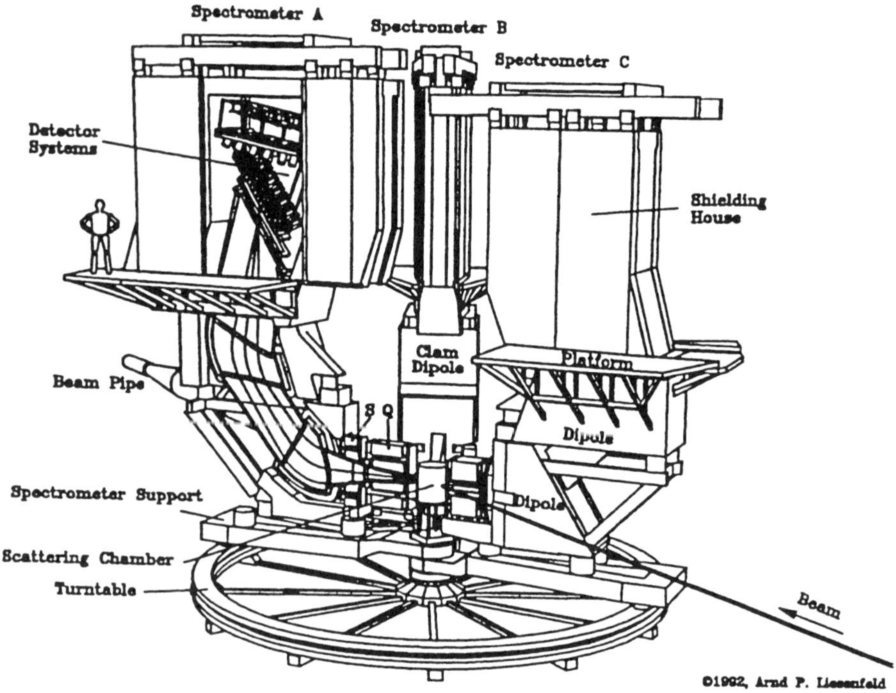

Figure 1: The three spectrometer setup of the A1 Collaboration at MAMI. For this experiment spectrometer A was used for the proton detection, B and C were used for the electron detection

mass energies [3].

Data were taken for the threshold region for $-q^2 = 0.03,\ 0.04,\ 0.06$, and $0.08(GeV/c)^2$. The kinematical coverage for these settings will be up to 4 MeV above threshold, including Rosenbluth separation.

In addition the point at $q^2 = -0.05(GeV/c)^2$ was investigated over an extended momentum range, i.e. up to $p_\pi^{CMS} = 100MeV/c$. These measurements also took place this year. With these data σ_L, σ_T, σ_{TT}, and σ_{TL} can be separated. An out of plane measurement with polarized electrons will be performed in the future to determine $\sigma_{TL'}$.

References

1. For further information see `http://www.kph.uni-mainz.de/A1`
2. M.O.Distler, Doctorate Thesis, in Preparation
3. MAMI PAC proposal A1/1-94.

Unitary Cusp and Isospin Breaking In Electromagnetic Pion Production

Aron M. Bernstein
Physics Department and Laboratory for Nuclear Science, M.I.T., Cambridge Mass., USA

Since the discovery of a large mass difference between the up and down quarks ($m_d/m_u \simeq 1.7$) there has been considerable discussion about the possibility of isospin breaking in the pion-nucleon system[1] . The theoretical consensus is that the magnitude of isospin breaking is not $(m_d - m_u)/m_d + m_u) \simeq 27\%$ but is $\sim (m_d - m_u)/\Lambda_{qcd} \sim 1.5\%$. In this paper it will be shown that the $\gamma p \rightarrow \pi^0 p$ reaction is an excellent candidate to measure dynamic isospin violations due to the up, down quark mass difference at the $\sim 3\%$ level.

There is a static isospin breaking effect in the near threshold $\gamma N \rightarrow \pi N$ reaction due to the threshold difference between the $\gamma p \rightarrow \pi^0 p$ and $\gamma p \rightarrow \pi^+ n$ channels. Of this $\simeq 2/3$ is a Coulomb effect and $\simeq 1/3$ is due to the up,down quark mass difference. In addition the electric dipole amplitude $E_{0+}(\gamma p \rightarrow \pi^+ n) \simeq -20 \cdot E_{0+}(\gamma p \rightarrow \pi^0 p)$. This large magnitude difference leads to the prediction of a unitary cusp in the $\gamma p \rightarrow \pi^0 p$ reaction due to the two step $\gamma p \rightarrow \pi^+ n \rightarrow \pi^0 p$ reaction near the $\pi^+ n$ threshold(see e.g.[2]).

The simplest first principles way to derive the unitary cusp is to us a 3 channel S matrix approach in which unitarity and time reversal invariance are satisfied. The resulting equation is $E_{0+}(\gamma p \rightarrow \pi^0 p) = e^{i\delta_0}[A_0(k_\gamma) + i\beta q_+]$, where the magnitude of the unitary cusp is governed by $\beta = E_{0+}(\gamma p \rightarrow \pi^+ n) \cdot a_{cex}(\pi^+ n \leftrightarrow \pi^0 p)$, δ_0 is the s wave $\pi^0 p$ phase shift(which is negligible in the threshold region), and q_+ is the π^+ center of mass momentum in units of m_{π^+}. The only first principles constraint that we have for $A_0(k_\gamma)$ is that it is a smooth function of k_γ. For photon energies $k_\gamma < k_T(\pi^+ n)$, the $\pi^+ n$ threshold energy, one must analytically continue $q_+ \rightarrow i|q_+|$. This switch of the amplitude from real to imaginary as the secondary threshold opens is the sign of a unitary cusp.

It is important to realize that one can use the $\gamma p \rightarrow \pi^0 p$ reaction to test Weinberg's prediction of dynamical isospin breaking due to the up, down quark mass difference in elastic s wave $\pi^0 N$ scattering and in charge exchange reactions[1] . Due to the connection between the final state scattering and the photopion amplitude (imposed by unitarity) the same predicted isospin violation ($\sim 3\%$) shows up in the unitary cusp in the $\gamma p \rightarrow \pi^0 p$ reaction.

There have been three modern $\gamma p \rightarrow \pi^0 p$ experiments performed with high duty cycle accelerators [3,4,5]. Since the predicted unitary cusp is in the s

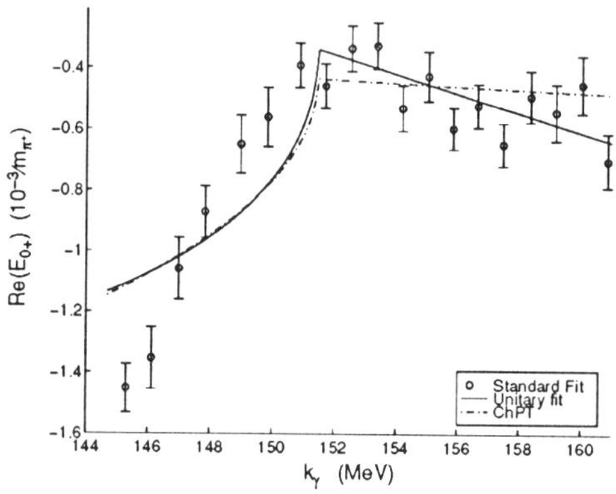

Figure 1: ReE_{0+} vs. photon energy from fits and ChPT

wave, electric dipole, multipole amplitude E_{0+} and since the p wave multipole amplitudes are large, the cusp does not show up e.g. in the total cross section. Rather one must perform a multipole analysis of the data. The results of the recent TAPS/Mainz experiment[4] for E_{0+} versus photon energy is shown in Fig. 1. The curves are a "unitary fit" to the data which were performed using a linear ansatz for $A_0(k_\gamma)$. The other curve shows the chiral perturbation theory fit (there are two low energy constants in the s wave amplitude which have been adjusted to fit the data)[2]. One can see that the predicted unitary cusp has been observed and is in qualitative agreement with the predictions. Although the data from the other experiments[3,5] they are good agreement with the Mainz/TAPS data which are shown in Fig. 1.

However as will be discussed in detail in a future publication[4] the precise value of β has not been determined from these unpolarized experiment. In order to fully measure the unitary cusp and the precise value of β one must perform experiments with polarized beams and targets. In addition to measure the predicted isospin violations one must also measure the $\gamma n \to \pi^0 n$ reaction.

1. S.Weinberg,Transactions of the N.Y.Academy of Science Series II 38 (I.I.Rabi Festschrift),185(1977)
2. V.Bernard, N.Kaiser, and U.G.Meißner, *Nucl. Phys.* B383, 442 (1992),*Z. für Physik* C70, 483 (1996), *Phys. Lett.* in print [hep-ph/9512234].
3. R. Beck et al., *Phys. Rev. Lett.* 65, 1841 (1990).
4. M.Fuchs et al., *Phys. Lett.*, B368, 20 (1996), A.M.Bernstein et al., to be published.
5. J.C. Bergstrom et al., Phys. Rev. C53, R1052 (1996).

Relativistic Model for Exclusive Photoproduction of Eta Mesons on Nuclei[a]

Mohammad Hedayatipoor and H.S. Sherif
Department of Physics, University of Alberta,
Edmonton, Alberta, Canada T6G 2J1

A relativistic model for exclusive quasifree photoproduction of η-mesons on complex nuclei is developed. Starting from an effective Lagrangian we construct an S-matrix which includes contributions from nucleon Born, nucleon resonance and vector meson exchange diagrams. The final state interactions of both the nucleon and eta are taken into account. Relativistic wavefunctions are used. We find that the leading contributions to the reaction come from the nucleon Born and the S_{11} resonance diagrams. Results for the reaction on a number of nuclei and comparison with non-relativistic calculations are discussed.

Eta meson production reactions explore a rich domain of phenomena at the interface of nuclear and particle physics. Until recently the photoproduction of eta mesons has been largely confined to proton targets. The new generation of high duty-cycle electron accelerators has opened up novel possibilities of performing production experiments on nuclei. Some results have been reported recently and there are several experiments already underway[1]. The strong selectivity in the decay of the S_{11} nucleon resonance favoring an eta channel provides an interesting possibility of learning more about this resonance. In particular, the reaction on nuclei can be a means of studying the propagation of the resonance in the nuclear medium.

In this paper we outline a relativistic model for exclusive quasifree eta photoproduction on nuclei. The reaction amplitude is calculated in the framework of a relativistic mean field approach to nuclear dynamics. The elementary interactions are based on the approach of Bennmerrouch *et al.*[2], who develop a covariant and gauge invariant effective Lagrangian for a system of photons, mesons (η, ρ, and, ω), nucleons and nucleon resonances. The diagrams contributing to the reaction are shown in Fig.1. We use the psuedoscalar form for the η coupling to the nucleon states and include contributions from the P_{11} and D_{13} in addition to the S_{11}. The associated S-matrix for the $A(\gamma,\ \eta N)A-1$ reaction is written as

$$S_{fi} \sim \int d^4x \Psi^{(-)\dagger}_{s_f}(x)\Gamma_{total}\Psi_{J_B M_B}(x)\varphi^*_\eta(x)e^{-ik_\gamma\cdot x}, \qquad (1)$$

where $\Psi_{s_f}(x)$ and $\Psi_{J_B M_B}(x)$ describe the outgoing and bound nucleon respectively and are solutions of the Dirac equation with the appropriate potentials[3,4].

[a]Work supported in part by the Natural Sciences and Engineering Research Council of Canada

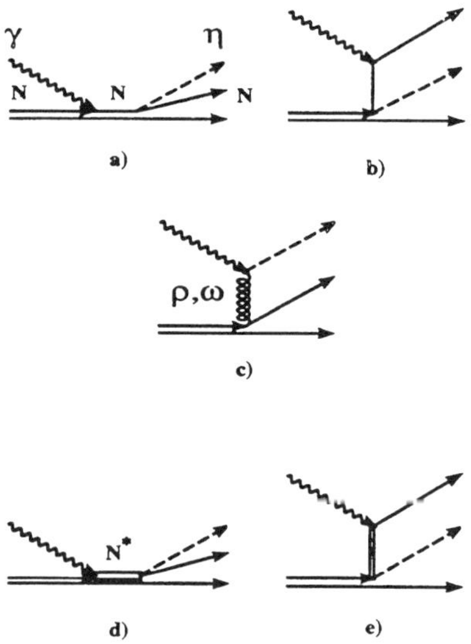

Fig.1

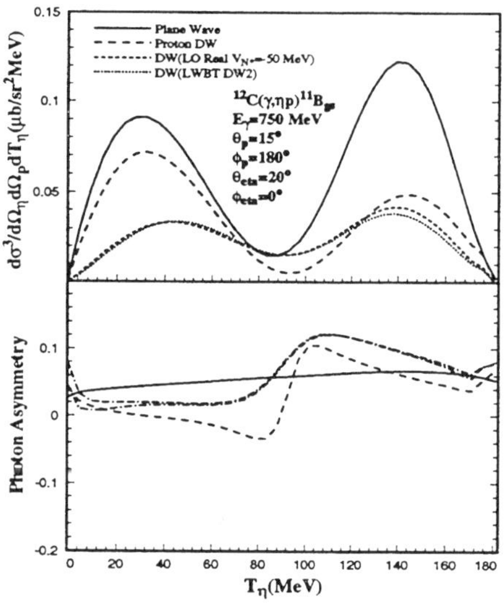

Fig.2

The η is described by the wavefunction φ_η which is a solution of the Klein-Gordon equation with a suitable optical potential[5,6]. The 4×4 operator Γ_{total} is the sum of operators resulting from each of the contributing diagrams shown in Fig.1. For example the contribution from the S_{11} proton resonance is given by

$$\Gamma_R = \frac{\kappa^p_R g_{\eta NR}}{m_P + m_R}\gamma_0 \left(\frac{\not{k}_s + m_R}{k_s^2 - m_R^2}\gamma_5 \not{k}_\gamma \not{\epsilon}_p + \gamma_5 \not{k}_\gamma \not{\epsilon}_p \frac{\not{k}_u + m_R}{k_u^2 - m_R^2} \right) \tag{2}$$

The two terms on the right hand side refer to s- and u-channel contributions, respectively.

Fig.2 shows the results for the reaction on ^{12}C leading to the ground state of ^{11}B. The small dashed and dotted curves show the effect of the eta distortion using optical potentials of references 5 and 6 respectively. Note that final state interactions of the proton and η lead to substantial suppression of the cross section. The predicted cross sections are generally similar in shape but somewhat smaller in magnitude than those predicted by the non-relativistic calculations of Lee *et al.*[6] The associated photon asymmetries, however, are quite different.

We are grateful to J. Ahrens, M. Benmerrouche, C. Bennhold, J.I. Johansson, B. Krusche, F.X. Lee, B. Lopez Alvaredo, E. Oset, M. Roebig-Landau and M. Vineyard for helpful communications concerning their work.

References

1. M. Roebig-Landau *et al.*, *Phys. Lett.* B **373**, 45 (1996); M. Vineyard, praivite communication.
2. M. Benmerrouche *et al.*, *Phys. Rev.* D **51**, 3237 (1995).
3. E.D. Cooper *et al.*, *Phys. Rev.* C **47**, 297 (1993).
4. C.G. Horowitz and B.D. Serot, *Nucl. Phys.* A **368**, 503 (1986).
5. B. Lopez Alvaredo and E. Oset, *Phys. Lett.* B **324**, 125 (1994).
6. F.X. Lee *et al.*, nucl-th /96011001.

PRECISE DETERMINATIONS OF THE NEUTRON MAGNETIC FORM FACTOR

J. Jourdan for the A1–collaboration at MAMI
Department of Physics and Astronomy, University of Basel, Switzerland

The accurate knowledge of the nucleon electromagnetic form factors on the dependence of momentum transfer q^2 is important for a quantitative understanding of the nucleon in terms of QCD. Precise data serve as a sensitive test for model calculations.

The accurate knowledge of the magnetic form factor G_{mn} is of particular importance for the analysis of the various $\vec{D}(\vec{e},e'n)$, $D(\vec{e},e'\vec{n})$, and $^3\vec{H}e(\vec{e},e'n)$ experiments, which will determine the *electric* form factor of the neutron G_{en} as the observable measured in these experiments is proportional to the product $G_{en} \cdot G_{mn}$.

In the past, G_{mn}–data have been determined mostly from quasielastic $D(e,e')$ cross section data. The extraction of G_{mn} requires a L/T separation, and a subtraction of the (dominant) proton magnetic contribution. The difficulty encountered in these experiments is the large uncertainty resulting from the deuteron model and the sensitivity to exchange currents (MEC) and final state interactions (FSI) the effects of which are greatly amplified by the two subsequent subtractions. The accuracy of G_{mn} from these data is no better than ~20%.

To significantly improve upon this knowledge an alternative technique is needed. A measurement of the $D(e,e'n)$–coincidence cross section[1] has been performed, thus avoiding the need to subtract the proton contribution. However, the extraction of G_{mn} still depends on a deuteron model.

In order to minimize the dependence on the deuteron model a number of experiments have measured the cross section ratio $R = D(e,e'n)/D(e,e'p)$ in quasifree kinematics[2,3]. The quantity R is *independent* of the deuteron wave function as it is identical for both proton and neutron. Corrections due to FSI are small and well understood. Restriction of the acceptance to the main quasi-elastic peak minimizes the corrections from MEC which are less well understood.

In all these measurements, the difficulty is shifted to experiment, the need of a neutron detector of well known absolute detection efficiency. The efficiency has been determined, with a bremsstrahlung beam, using either the $D(\gamma,p)n$ or the $H(\gamma,\pi^+)n$ reaction by tagging the outgoing neutron via detection of the outgoing p or π^+. The detector has been placed in this beam of tagged neutrons and its efficiency measured. However, for a number of reasons one has to question the results of such efficiency determinations.

Due the low cross section of the reactions chosen the "tagged" neutron beam has a low intensity. This does not allow for a detailed study of the neutron efficiency which depends on the neutron energy and the spatial distribution over the detector surface. Its precise determination is important due to the different neutron distributions during the efficiency and the $(e,e'n)$–measurement. Corrections of up to 15% can not be neglected for precise measurements.

A major disadvantage of the low cross section are background reactions which become important. For instance a closer look at the recent G_{mn}–determination at Bonn[3] reveals that not the $H(\gamma,\pi^+)n$–reaction but predominantly the $H(e,\pi^+)ne'$–reaction (~90%) was used for the efficiency determination. For this reaction with three particles in the final state an unambiguous reconstruction of the neutron energy and angle from the measurement of the π^+ is not possible. A Monte Carlo simulation suggests that for a significant fraction of the events of order 20%, the "tagged" neutron misses the neutron detector, resulting in an incorrect determination of the neutron efficiency.

Given these deficiencies it does not surprise that the data from the experiments cited[1,2,3] do not significantly improve the knowledge of G_{mn}.

In our approach we employ a novel method which improves upon the weakest point of these experiments, the determination of the neutron efficiency. Precise determinations of the neutron efficiency and a detailed study of the detector response are possible at one of today's high intensity neutron beams. This has been demonstrated in a pilot experiment[4] where we determined G_{mn} with an accuracy of 1.7%.

This determination of the neutron detection efficiency was performed at PSI using a well collimated tagged neutron beam of high intensity and known energy. The tagged neutron beam was produced via $H(n,p)n$–scattering using the 68 MeV neutron beam from PSI incident on a liquid H_2–target. The redundant measurement of the tagged neutron energy and scattering angle was free from contributions of background reactions. With such a beam, the local response of the neutron detector could be studied in detail.

In the present work we have extended our NIKHEF/PSI experiment to higher energies in order to accurately determine the q^2–dependence of G_{mn}. At the 100% duty factor accelerator at MAMI we simultaneously measured the $D(e,e'n)$ and $D(e,e'p)$ cross sections. The simultaneous measurement of the $(e,e'n)$ and $(e,e'p)$ yield makes R independent of the luminosity, dead time effects, and efficiency of the electron arm. The hadron detector, used to detect the coincident nucleon, consisted of two 10 cm thick plastic converters preceeded by 3 thin veto counters to identify the incident nucleon. A lead absorber of thickness, 1–20 mm depending on kinematics, was placed at the entrance window of the detector lead house in order to absorb photons of very low energy.

The neutron detection efficiency was again determined at PSI using the same approach as for the NIKHEF/PSI experiment. For the present experiment we used the high energy neutron beam (100–550 MeV) incident on a liquid H_2–target. The efficiency measurements always bracketed a measurement of R at MAMI in order to check the reliability of the efficiency measurements.

Multiple scattering and absorption effects of the protons, in particular in the lead–absorber, during the $(e, e'p)$–measurements required also an accurate determination of the proton efficiency. These effects were measured at PSI using $H(n,p)$–scattering and placing the detector on the recoil arm. The efficiency determined was as low as 84%, which obviously needs to be corrected for accurately.

Considerable care was taken to obtain accurate energy calibrations. $H(n,p)$ at PSI and $H(e,e'p)$ at MAMI were used to obtain absolute calibrations. In addition, each converter was monitored for gain variations and baseline shifts using two temperature compensated LED's whose light outputs were in turn monitored by very stable PIN-diodes. With these measures the energy scale of the converters was calibrated to an accuracy of 0.4MeV_{ee}.

In the present experiment we determined G_{mn} at 7 kinematical points over a q^2–range from 0.07 GeV/c^2 up to 0.89 GeV/c^2. We achieved a substantial decrease of the uncertainties. This was possible even for the highest q^2 value given the low noise environment at MAMI. The data point at the q^2–value of 0.12 GeV/c^2 in comparison with the NIKHEF/PSI result will provide an ultimate consistency check of the measurements.

Some preliminary results, indicate that accuracies of $< 3\%$ of G_{mn} can be reached over most of the q^2–range measured. The precision of the results achieved so far confirms that it is possible to obtain very precise neutron form factors when using an approach that depends the least on the input of theory, and when performing the needed calibrations using modern high-intensity neutron beams.

References

1. P. Markowitz *et al.*, *Phys. Rev.* C **48**, R5 (1993).
2. P. Stein *et al.*, *Phys. Rev. Lett.* **16**, 592 (1966). W. Bartel *et al.*, *Nucl. Phys.* B **58**, 429 (1973).
3. E.E.W. Bruins *et al.*, *Phys. Rev. Lett.* **75**, 21 (1995).
4. H. Anklin *et al.*, *Phys. Lett.* B **336**, 313 (1994).

MEASUREMENT OF PROTON ELECTROMAGNETIC FORM FACTORS VIA THE SPIN TRANSFER REACTION $^1H(\vec{e},e'\vec{p})$

B.D. Milbrath,[1] D.H. Barkhuff,[1] J.M. Finn,[2] K. Joo,[3] R.W. Lourie,[1] J.I. McIntyre,[2] S.P. Van Verst,[1,3] G.A. Warren,[3] C. Armstrong,[2] W. Bertozzi,[3] D. Dale,[3] K.A. Dow,[3] M.B. Epstein,[4] S. Gilad,[3] M. Jones,[2] P. Markowitz,[5] C. Mertz,[6] J. Mitchell,[1,7] C. Perdrisat,[2] V. Punjabi,[8] A. Sarty,[3] D. Tieger,[3] P.E. Ulmer,[7,9] L.B. Weinstein,[9] R.J. Woo[2]

[1]University of Virginia, Charlottesville, VA 22901
[2]College of William & Mary, Williamsburg, VA 23185
[3]Massachusetts Institute of Technology and Bates Linear Accelerator Center, Cambridge, MA 02139
[4]California State University, Los Angeles, CA 90032
[5]University of Maryland, College Park, MD 20742
[6]Arizona State University, Tempe, AZ 85287
[7]CEBAF, Newport News, VA 23606
[8]Norfolk State University, Norfolk, VA 23504
[9]Old Dominion University, Norfolk, VA 23529

Abstract

A measurement of G_p^E/G_p^M at Q^2 values of 0.38 and 0.50 GeV^2 was performed at the Bates Linear Accelerator Center using a focal plane polarimeter (FPP) to measure the recoil proton polarization following elastic scattering of polarized electrons on a hydrogen target. These are the first measurements of all three recoil polarization components and determination of the proton's form factors via spin transfer rather than Rosenbluth separation. The ratio G_p^E/G_p^M is obtained in a nearly systematic free manner. The measured ratio and polarizations are consistent with the Plane Wave Impulse Approximation (PWIA).

The form factors G_E^p and G_M^p contain information about the finite spatial extent of the proton's charge and magnetization distributions. As such, their accurate measurement has consistently been one of the leading efforts of experimental nuclear physics. In the past this has primarily been done using elastic (e,e$'$) scattering to measure the Rosenbluth cross section:

$$\left[\frac{d\sigma}{d\Omega_e}\right]_{Rosenbluth} = \frac{\alpha^2\cos^2(\theta_e/2)}{4E^2\sin^4(\theta_e/2)}\left(\frac{E'}{E}\right)\left[\frac{G_E^2}{1+\tau} + \tau G_M^2\left(\frac{1}{1+\tau} + 2\tan^2\frac{\theta_e}{2}\right)\right] ,$$

where $\tau = Q^2/4M_p^2$. Extracting the form factors requires performing measurements at fixed Q^2 (but different E and θ_e), making control of systematic errors both difficult and essential. Additionally, at high Q^2, $\tau G_M^2 >> G_E^2$ so that determination of G_E becomes increasingly difficult.

Alternatively, the proton form factors can be determined by measuring polarization transfers in the $(\vec{e},e'\vec{p})$ reaction. Assuming the one-photon exchange, one has the relationships:[1]

$$I_0\, P_t^{tgt} = h\left(-2\sqrt{\tau(1+\tau)}\, G_M^p G_E^p \tan\theta_e/2\right)$$

$$I_0 P_\ell^{tgt} = h\left(2\tau\sqrt{(1+\tau)(1+\tau\sin^2\theta_e/2)}\sec\theta_e/2\tan\theta_e/2\, G_{Mp}^2\right)$$

$$P_n^{tgt} = 0 \quad ,$$

where h is the longitudinal polarization of the electron beam and I_0 is the unpolarized cross section (except for σ_{Mott}):

$$I_0 = G_{Ep}^2 + \tau G_{Mp}^2\ \left[1 + 2(1+\tau)\tan^2\theta_e/2\right] \quad .$$

While the transverse component of the recoil proton's polarization has been measured previously at Mainz,[2] this is the first experiment to measure all three polarization components and determine the ratio G_E^p/G_M^p.

The experiment was performed at the Bates Linear Accelerator Center in early 1995 using the South Hall spectrometers MEPS (electrons) and OHIPS (protons). A focal plane polarimeter (FPP) built by our group was installed on top of OHIPS. The FPP consisted of four multiwire proportional chambers, two on either side of carbon blocks. These were used to establish the proton trajectories before and after scattering in the carbon. The angular distribution of this second scattering:

$$I(\theta,\phi) = I_0(\theta)\left[1 + P_n^{fp} A_y(\theta)\cos\phi + P_t^{fp} A_y(\theta)\sin\phi\right] \quad ,$$

allowed for the extraction of two of the polarization components. Due to precession in the proton spectrometer's vertically-bending dipole and the fact that P_ℓ^{tgt} is helicity-dependent while P_n^{tgt} is not, all three components could be determined.

Figure 1 shows our extracted recoil polarizations while figure 2 shows the ratio G_E/G_M The circles represent the hydrogen data while the diamonds represent deuterium data taken concurrently on the quasi-elastic peak and in parallel kinematics. The ratio in deuterium is consistent with that on a free proton. The equality of the hydrogen and deuterium polarizations at the few percent level is strong confirmation of the validity of the Impulse Approximation via spin observables that are inherently much more sensitive than cross sections to the presence of other amplitudes. The inner error bars are statistical while the outer error bars include systematic errors as well. The data are not final and analysis of systematics is still underway. G_E/G_M data from previous Rosenbluth measurements are shown for comparison.[3]

References

[1] R.G. Arnold, C.E. Carlson and F. Gross, Phys. Rev. **C23**, 363 (1981).

[2] D. Eyl *et al.*, Z. Phys. **A352**, 11 (1995).

[3] T. Janssens *et al.*, Phys. Rev. **142**, 922 (1966); G. Höhler *et al.*, Nucl. Phys. **B114**, 505 (1976); W. Bartel *et al.*, Nucl. Phys. **B58**, 429 (1973).

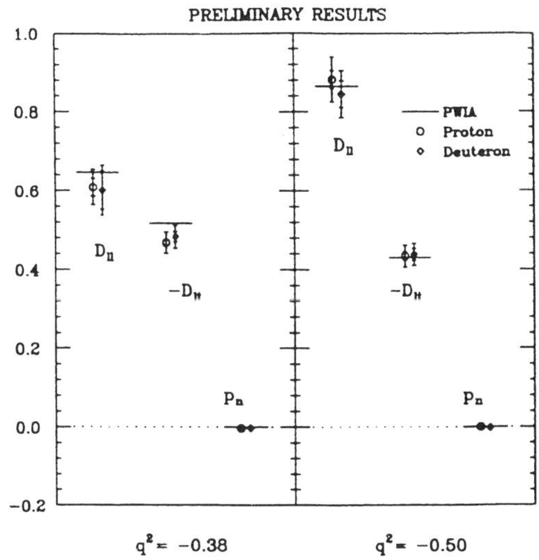

Figure 1. *Preliminary measured recoil polarisations.*

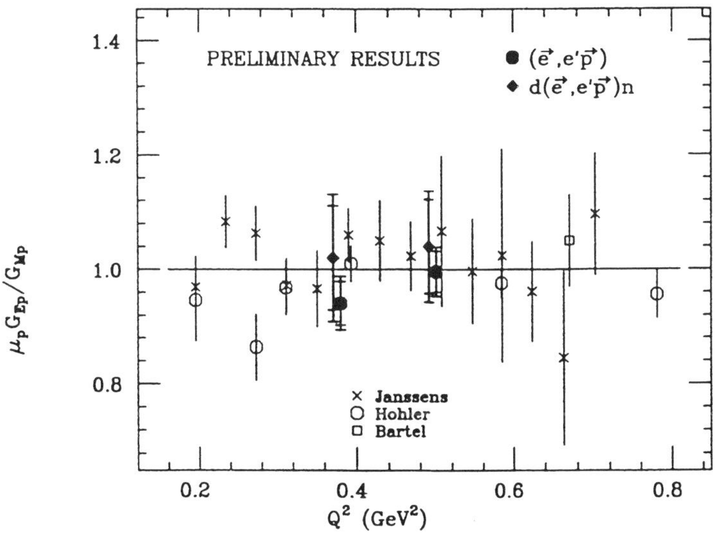

Figure 2. *Preliminary* $\mu_p G_E^p/G_M^p$ *ratio.*

ELECTRON SCATTERING OFF TENSOR POLARIZED DEUTERIUM

M. BOUWHUIS
NIKHEF, P.O. Box 41882, 1009 DB Amsterdam, The Netherlands

for the 91-12 Collaboration [1]

1 Introduction

The electro-magnetic response of the deuterium ground state is described by three form factors (F_M, F_C and F_Q). These are especially sensitive to NN-potentials, but also to meson exchange currents and relativistic corrections. A Rosenbluth separation allows the extraction of F_M and a term containing F_C and F_Q in quadrature. Polarization observables are needed to separate all three. One possibility is to scatter unpolarized electrons off tensor-polarized deuterium, measuring the tensor analyzing power T_{20}.

As part of a long term program to study the deuteron, absolute data on T_{20} have been taken at NIKHEF at two values of the transferred momentum ($Q = 1.5\ \text{fm}^{-1}$ and $2.3\ \text{fm}^{-1}$). Simultaneously the inelastic channel (e,e'p) has been investigated [2].

2 Experimental Setup

The experiment was performed at the internal target facility of NIKHEF with unpolarized electrons stored in the AmPS ring. To reach the high luminosities needed, the tensor-polarized deuterium, produced by an atomic beam source, was fed into a storage cell with two polarimeters monitoring the polarization as described in ref. 3. The sign of P_{zz} was flipped every 10 s. This geometry resulted in a luminosity of 1.6 10^{31} at/s/cm^2 at a flux out of the ABS of 2 10^{16} at/s.

The low background conditions allowed the use of large-acceptance non-magnetic detectors. Electrons were detected in an electro-magnetic calorimeter consisting of 60 CsI(Tl) crystals and two plastic scintillators. Recoiling deuterons as well as knocked-out protons were detected by the Range Telescope, a stack of 16 plastic scintillators. Each detector was preceded by two sets of MWPC's to determine the tracking information. By using the ΔE-E method, Time-of-Flight and the angular kinematical constraints protons, deuterons and electrons were identified.

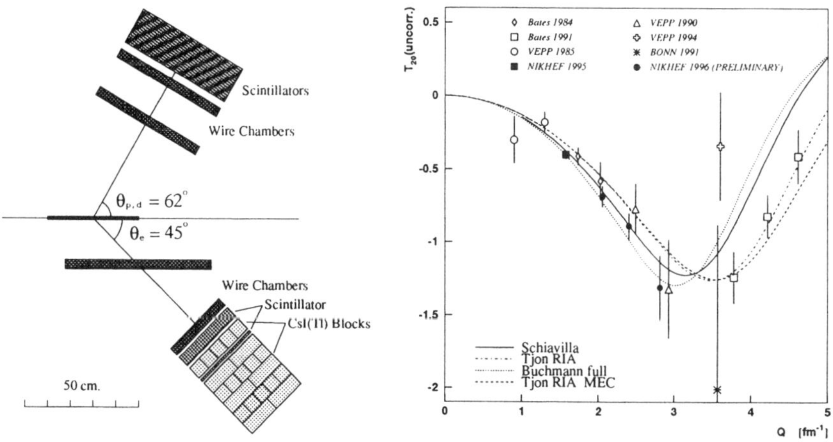

Figure 1: Left: schematic overview of experimental setup, showing Range Telescope and Electron Calorimeter; right: data and theoretical predictions for T_{20}, as a function of momentum transfer (see ref. 4 for references).

3 Preliminary Results

A first measurement has been performed at 565 MeV in January 1995 [2]. In the most recent data run (Jan. 1996) 700 MeV electrons were used to measure T_{20} absolutely with a statistical error of 7 % and a systematical error better than 5 %. In Fig. 1(right) the preliminary data for that run are presented as well as the result from the previous data run [4], the world data set and several theoretical predictions. The non-relativistic impulse approximation calculations appear to be in excellent agreement with the present data.

In the future these measurements will be extended to a transferred momentum of 4.5 fm^{-1}. To this end we will upgrade the ABS (replacing and adding permanent sextupoles), reduce the emittance of the stored beam, change the diameter and length of the target cell to 11 mm and 60 cm, resp., and increase the pumping capacity around the target.

References

1. J.F.J. van den Brand, contrib. to this conference.
2. E. Passchier, Electron scattering off an internal tensor-polarized deuterium target, Ph.D. thesis (1996), University of Utrecht (unpublished).
3. M.M.E. Ferro-Luzzi et al., accepted by Nucl. Instr. Meth.
4. M.M.E. Ferro-Luzzi et al., submitted to Phys. Rev. Let.

Tensor Asymmetries Measured With The $^2\overleftrightarrow{\mathrm{H}}(e, e'p)$ Reaction

Z.-L. Zhou[1], M. Bouwhuis[2], M. Ferro-Luzzi[3], E. Passchier[2], R. Alarcon[4], M. Anghinolfi[5], R. van Bommel[2], T. Botto[2], J.F.J. van den Brand[1], M. Buchholz[1], H.J. Bulten[1], S. Choi[4], J. Comfort[4], S. Dolfini[4], R. Ent[6], C. Gaulard[4], D. Higinbotham[7], C.W. de Jager[2], E.P. van Klaveren[2], E. Konstantinov[8], J. Lang[3], D.J. de Lange[2], M.A. Miller[1], D. Nikolenko[8], G.J. Nooren[2], N. Papadakis[2], I. Passchier[2], H.R. Poolman[2], S.G. Popov[8], I. Rachek[8], M. Ripani[5], E. Six[4], J. Steijger[2], M. Taiuti[5], O. Unal[1], N. Vodinas[2], H. de Vries[2]
[1] *UW-Madison,* [2] *NIKHEF,* [3] *ETH,* [4] *ASU,* [5] *Genova,* [6] *TJNAF,* [7] *UVA,* [8] *Novosibirsk*

Tensor asymmetries for quasi-free $(e, e'p)$ scattering have been measured at NIKHEF using the medium-energy AmPS electron storage/stretcher ring and an internal target [1]. A large acceptance non-magnetic detector system was used to detect electrons in coincidence with protons or deuterons [2]. Data have been taken at a beam energy of 565 MeV, with beam currents up to 125 mA and an average luminosity of 10^{31} atoms/cm^{2}s, and with the target spin directed parallel and perpendicular to the momentum transfer. The target polarization was determined on-line during data taking through ion-extraction polarimetry [3]. Kinematics were selected with four-momentum transfers in the range $0.1 < Q^2 < 0.3$ (GeV/c)^2 and missing momenta up to 200 MeV/c.

The tensor asymmetries are presented as a function of θ_{np} (Fig. 1a and 1b) and ϕ_{np}(Fig. 1c and 1d), the angles of the knocked-out proton with respect to $\vec{q}$, in the cm-frame. Note that the angles ϕ_{np} and θ_{np} are closely related to the missing momentum. A large tensor asymmetry is expected for perpendicular kinematics with $\phi_{np} = 180°$ and $p_m \perp \vec{q}$. If the deuteron spin is perpendicular to the momentum transfer, then the asymmetry is predicted to be negative and its magnitude is expected to increase with θ_{np} (or p_m) due to the D-state contributions to the deuteron ground-state wave function [4]. Fig. 1a shows the tensor asymmetry obtained as a function of θ_{np} for protons in a ϕ_{np}-range between 150° and 210° and with a kinetic energy larger than 40 MeV. The spin was aligned perpendicular to the momentum transfer and oriented in the scattering plane. Data are compared to the results of the PWBA (dashed curve) and the full (solid curve) calculation of Arenhövel *et al.* [4] using a Monte Carlo to average over the kinematic region.

By orienting the spin parallel to the momentum transfer, the asymmetry is expected to change sign. The obtained data shown in Fig. 1b follow this prediction. However, the data deviate from the results of a PWBA calculation. The calculation which includes contributions from FSI, MEC and IC, and relativistic corrections gives a better description of the data. Assuming that in the present kinematics FSI is the dominant contribution, the difference between the results of the two calculations can be related to the spin-dependence of the final-state interaction. The data show that, when the deuteron spin is aligned parallel with respect to the momentum transfer, the struck proton more strongly interacts with the neutron [5].

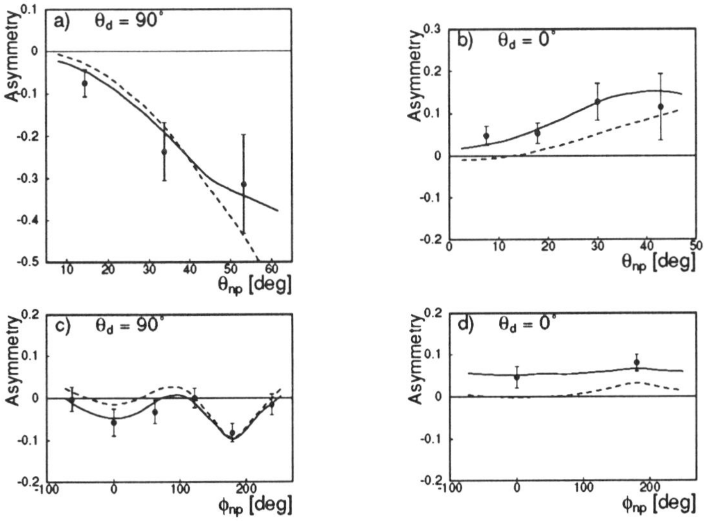

Figure 1: Tensor asymmetries as functions of θ_{np} and ϕ_{np} for the spin parallel and perpendicular to the momentum transfer. Data are compared to the results of a PWBA (dashed curve) and a full (solid curve) calculation of Arenhövel et al. [4].

We plan to pursue measurements with higher luminosities to yield data for missing-momenta around 300 MeV/c [6].

This work was supported in part by the Nederlandse Organisatie voor Wetenschappelijk Onderzoek, the Swiss National Foundation, the National Science Foundation under Grants No. PHY-9316221 (UW), PHY-9200435 (ASU) and HRD-9154080 (Hampton), NATO Grant No. CRG920219.

1. Three contributions to this conference by M. Bouwhuis, J.F.J. van den Brand and M. Ferro-luzzi.
2. E. Passchier *et al.*, to be submitted to *Nucl. Instrum. Methods* A, (1996).
3. Z.-L. Zhou, M. Ferro-Luzzi, J.F.J. van den Brand, H.J. Bulten *et al.*, accepted for publication by *Nucl. Instrum. Methods* A, (May, 1996).
4. H. Arenhövel *et al.*, *Phys. Rev.* C **46**, 455 (1992).
5. S. Jeschonnek *et al.*, contribution to this conference.
6. J.L. Forest, V.R. Pandharipande *et al.*, UIUC-preprint 96-03-022.

D(e,e′np) – THE ENTIRE ANGULAR RANGE MEASURED

H.W.WILLERING, TH.S.BAUER, D.J.BOERSMA, A.MISIEJUK, E.SCHILLINGS
Utrecht University, P.O. Box 80.000, 3508 TA Utrecht, The Netherlands

B. BANTES, U. BONIN, H. BRUNHöBER, D. DOLL, D. DUREK, R.W. GOTHE, H. HAINER, W. HILLERT, A. HUWER, D. JAKOB, G. KREBS, P. MASCHKE, T. REICHELT, B. SCHOCH, D. WACKER, M. WESTERMANN
Universität Bonn, Nußallee 12, 53115 Bonn, Germany

The Deuteron, and the two-nucleon system in general, attracts much attention since it allows to test our understanding of the strong interaction in terms of the underlying degrees of freedom (d.o.f.). Since, at present, one cannot obtain the NN-interaction from the basic d.o.f. – quarks and gluons –, models based on effective d.o.f. (mesons and isobars) are being developed, which need to be gauged against experimental data. Especially interesting is the region of short-range distances, where internal nucleon d.o.f. are expected to play a major role.

We have measured the electro-disintegration of the deuteron, D(e,e′np), at $Q^2 = 0.2$ GeV^2 at the ELSA facility in Bonn. The combination of the ELAN spectrometer (Δp/p = $\pm$12%), the 4 Time-Of-Flight Walls (9 m^2, 0.5 sr), and 12 scintillator telescopes (0.18 sr each) covered a large kinematical range (Figure 1): The entire angular range, $0^o \leq \Theta^*_{np} \leq 180^o$, was sampled while large regions around $\Phi^*_{np} = 0^o$ and around $\Phi^*_{np} = 180^o$ were covered. The acceptance of the ELAN spectrometer enabled us to map out energy transfers ω from Quasi-Free Scattering to beyond the Δ-resonance.

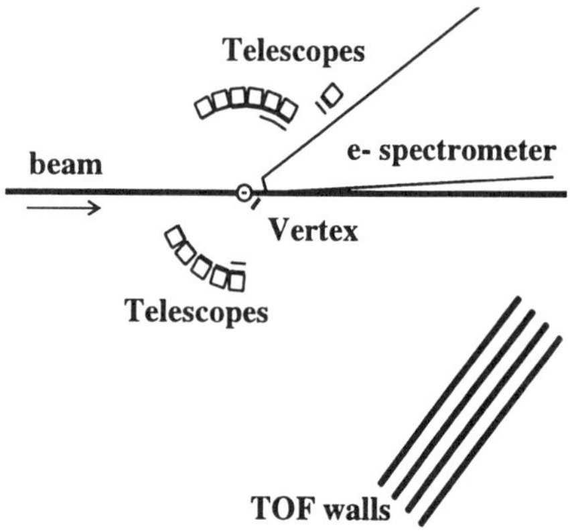

Figure 1: The experimental setup

The analysis of the data taken in January 1996 is in progress. At present, the 12 scintillator telescopes are scanned for proton hits, which are more abundant than neutrons and less affected by background. Figure 2 shows scatter-plots (arranged according to the experimental setup) of pulse-height *vs* time-of-flight for charged particles. One sees a bright pion band – for $\omega \geq 150$ MeV – and at larger flight times a proton band. The TOF-resolution allows to separate the disintegration channel from reactions in which a pion was produced.

With this experiment we have obtained a coherent set of deuteron disintegration data, which serve as a constraint to the available deuteron models.

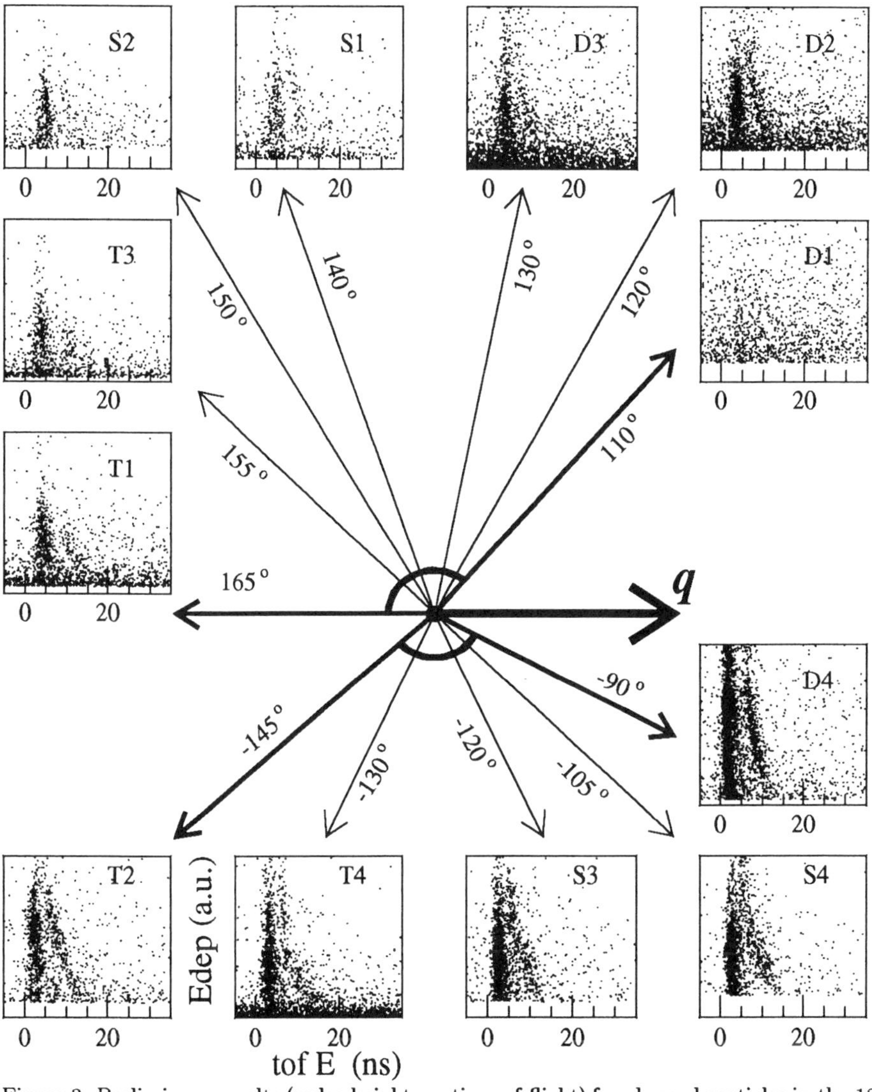

Figure 2: Preliminary results (pulse-height *vs.* time-of-flight) for charged particles in the 12 telescopes. The angles represent Θ^*_{np} for each telescope (for this ω).

RELATIVISTIC EFFECTS IN THE CURRENT OPERATOR AS MEASURED BY THE REACTION ^{2}H(e, e′p)

W-J. KASDORP, E. JANS, L. LAPIKÁS, J.J. VAN LEEUWE, G.J.L. NOOREN,
G. VAN DER STEENHOVEN, J.J.M. STEIJGER, H. DE VRIES
NIKHEF, P.O. Box 41882, 1009 DB Amsterdam, The Netherlands

TH. S. BAUER, J.A. TJON
Rijksuniversiteit Utrecht, P.O. Box 80.000, 3508 TA Utrecht, The Netherlands

W.H.A. HESSELINK, C.J.G. ONDERWATER, A.R. PELLEGRINO,
R. STARINK, J.A. TEMPLON
Vrije Universiteit Amsterdam, de Boelelaan 1081, 1081 HV Amsterdam, The Netherlands

N. KALANTAR-NAYESTANAKI
KVI, Zernikelaan 25, 9747 AA Groningen, The Netherlands

B. MOSCONI
Dipartimento di Fisica, Università di Firenze, I-50125 Firenze, Italy and INFN, Sezione di Firenze, I-50125 Firenze, Italy

The cross-section asymmetry A_ϕ with respect to the direction of momentum transfer $\vec{q}$ in the reaction ^{2}H$(e,e'p)$ has been measured at $Q^2 = 0.2$ (GeV/c)^2 for recoil momenta from 140 to 220 MeV/c at an invariant mass of 1050 MeV. These data are only reasonably well described if a relativistic form of the current operator is used, thus removing ambiguities in the interpretation of pervious asymmetry data at low Q^2 which had considerably larger experimental uncertainties.

It was pointed out by Mosconi [1] that the interference structure function R_{LT} is especially sensitive to details of the current operator, and that relativistic corrections cause significant modifications to R_{LT}. In order to investigate these effects one usually defines an asymmetry A_ϕ, which is less sensitive to systematic uncertainties than R_{LT}:

$$A_\phi = \frac{\sigma^0 - \sigma^\pi}{\sigma^0 + \sigma^\pi} = \frac{v_{LT}R_{LT}}{v_L R_L + v_T R_T + v_{TT} R_{TT}} \tag{1}$$

where $\sigma^0(\sigma^\pi)$ is the cross section in the case of the proton scattered at a forward (backward) angle γ_{pq} with respect to $\vec{q}$, the R_i represent the various structure functions, and v_i their kinematic factors. Quasi-elastic ^{2}H$(e,e'p)$ experiments performed at NIKHEF [2], Bonn [3] and MIT [4] at a relatively low $Q^2 = 0.2$ (GeV/c)^2 showed that both calculations with relativistic corrections

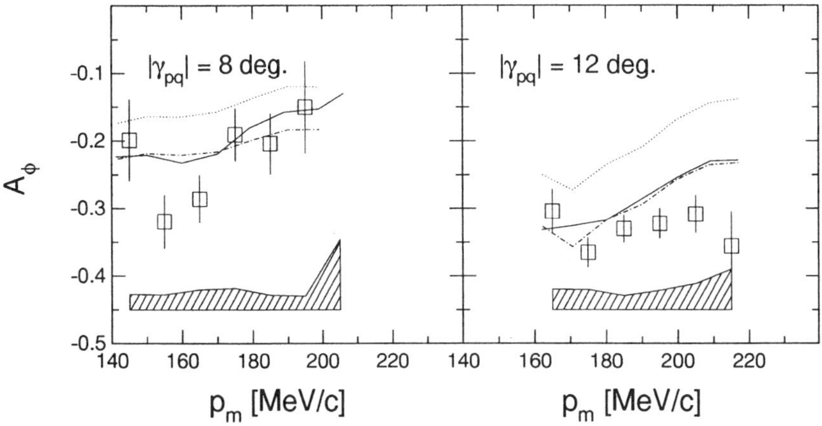

Figure 1: The shaded areas indicates the size of the systematic error. The solid curve: Born+FSI by Tjon; dotted: FULL non-relativistic by Mosconi; dot-dashed: FULL+relativistic corrections by Mosconi

to the current operator [1,5] and fully covariant calculations [6] describe the data better than non-relativistic calculations, although the experimental uncertainties were too large to permit an unambiguous conclusion. However, data from Saclay [7] did not confirm these findings.

The present experiment was performed at NIKHEF with an incident energy of 603.8 ± 0.2 MeV, $Q^2 = 0.2$ (GeV/c)^2 and an invariant mass of 1050 MeV, using a two-spectrometer setup and a newly developed liquid deuterium target. The data are shown in Fig. 1. The right panel shows that both the calculations by Tjon, and those from Mosconi including a relativistic form of the current operator are in fair agreement with the data, whereas the non-relativistic calculations from Mosconi fail do describe the data. The left panel also shows a preference for the relativistic calculations. Hence, we have found unambiguous evidence for the use of a relativistic description of the current operator in the reaction ^{2}H$(e, e'p)$ for relatively low values of Q^2.

1. B. Mosconi and P. Ricci, Nucl. Phys. **A517**, 483 (1990)
2. M. van der Schaar *et. al.*, Phys. Rev. Lett. **68**, 776 (1992)
3. F. Frommberger *et al.*. Phys. Lett. **B339**, 17 (1994)
4. D. Jordan *et al.*, Phys. Rev. Lett. **76**, 1579 (1996)
5. T. Wilbois, G. Beck and H. Arenhövel, Few-Body Systems **15**, 39 (1993)
6. E. Hummel and J. A. Tjon, Phys. Rev. C **49**, 21 (1994)
7. J. E. Ducret *et al.*, Phys. Rev. C **49**, 1783 (1994).

STUDY OF THE EXCLUSIVE *ed*-SCATTERING AT VEPP-3

A.V.Osipov, A.A.Sidorov, V.N.Stibunov
INP, Tomsk, Russia

D.M.Nikolenko, S.G.Popov, I.A.Rachek, D.K.Toporkov
BINP, Novosibirsk, Russia

E.I.Tsentalovich
BINP, present address: MIT Bates, MA, USA

B.B.Wojtsekhowski
BINP, present address: CEBAF, Newport News, VA, USA

We report here the experimental values of the cross section and the target tensor asymmetry in the electro production of the two protons from the tensor polarized deuteron by 2 GeV electrons in a channel with undetected pion.

The data were collected in parallel with the experiments performed by collaboration Novosibirsk, Argonne, Amsterdam, St. Petersburg, Tomsk, which used an internal tensor-polarized deuterium target in the VEPP-3 electron storage ring [1,2]. By using $\vec{d}(e,pp)\pi$ reaction we, for the first time, got the access to the region of large proton momenta in the polarized deuteron.

The data taking was performed at an average electron beam current of 150 mA. The total integral beam charge passed through the target was 386 kC. The target thickness $(3.05 \pm 0.12) \times 10^{11} atom/cm^2$ and the average value of the tensor target polarization $P_{zz} = 0.572 \pm 0.053$ were determined from the electron-deuteron elastic scattering [1].

The particle-detection apparatus [3] consists of two identical two-arm systems to detect the protons in coincidence. Each arm includes 6 drift chamber planes, three thin plastic scintillator counters (4, 10 and 10 mm thick) followed by a 20 cm thick plastic scintillator (top arms) or a 16 cm NaI(Tl) counter (bottom arms). Each proton arm detected particles within the range of angles $\theta = 68° - 82°$ and $\Delta\varphi = 40°$. The central azimuthal angles were 0° and 180° for one of the detection system, while 90° and 270° for the other one. The range of proton kinetic energy was 55 – 180 MeV for the top arms and 46 – 264 MeV for the bottom arms. Two procedures have been applied for calibration of the scintillation hodoscopes:

- measurements of the energy deposition in every layer by cosmic rays;
- data from $d(e,e'p)$ - and $d(e,pn)e'$ - reactions to determine the absolute energy scales.

Separation of protons from pions and deuterons was done by the $(E, dE/dx)$ method. The kinetic energy of the protons is reconstructed combining the values of the deposition in the detector layers. The UNIMOD code [4] was used to calculate the response of the scintillator layers. The resolution for the kinetic energy of the protons was determined to be 8% FWHM for low-energy protons and increased upto 12% FWHM for 260 MeV protons . The effective mass resolution of the *pp*-system was $4.2 - 8.1 MeV/c^2$ according to GEANT simulation. Several cuts were put on the data for background reduction:

- $68^\circ \leq \theta^{lab}_{p_{1,2}} \leq 82^\circ$;
- the proton vertexes in $\pm$ 0.7 *cm* from the beam line;
- the missing mass $M_x \geq m_\pi - \Delta M$ ($\Delta M = 5\ MeV/c^2$).

We selected 876 events for the two detector systems. The background from the walls of the storage cell was retained at the 6% level. For simulation of the $\vec{d}(e, pp)e'\pi^-$ experiment we designed the package NEWGAM. The data analysis has been performed in a nonrelativistic framework. The concept of a virtual photon has been used to relate the production negative pions by an electron scattering to the same proton-proton final state induced by the real photons. The virtual photon spectrum [5], upon integrating over electron scattering angles, has been used for the real photon spectrum, since contributions to that spectrum come predominantly from small scattering angles. One nucleon pion photoproduction operator has been taken from the Metcalf's and Wolker's phenomenological analysis[6]. The result presented in the Fig.1 was calculated with the deuteron wave function obtained by the Paris N-N potential [7].

For generation of events simulating the $d(e, pp)e'\pi^-$ reaction by Monte-Carlo method, the ENIGMA code[8] and the NEWGAM code, based on different theoretical models of the exclusive scattering process, were used. The package ENIGMA uses one nucleon pion photoproduction amplitude of Blomqvist-Laget[9]. By using all these simulating packages (UNIMOD, GEANT, ENIGMA and NEWGAM) we obtained:

- the mass-, momentum-, and angle resolution of the detector system;
- the effective angular acceptance, the efficiency of the detector arms and the counting rates;
- the energy deposition in the scintillation counters;
- all initial kinematic parameters for the accepted events.

The observed effective mass distribution of the *pp*-system is presented in Fig.1 together with the results of simulation by NEWGAM and ENIGMA. One can see a clear disagreement between the experimental data and the theoretical predictions both in shape and in absolute value.

The target asymmetry A_T was calculated from the difference in the count-

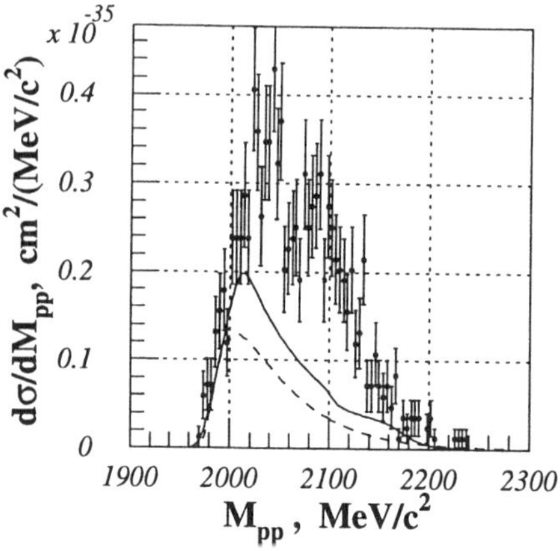

Figure 1: The cross section for $d(e,pp)e'\pi^-$ as function of the pp - system effective mass, integrated within the apparatus acceptance. Solid line – ENIGMA code, dashed line – NEWGAM code.

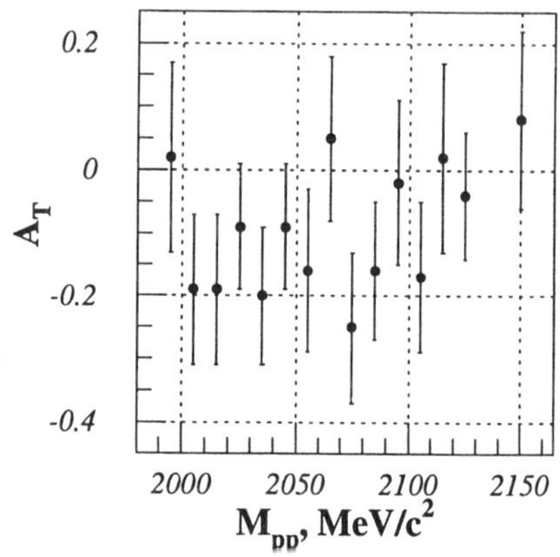

Figure 2: Experimental result for A_T as function of the pp - system effective mass.

ing rates for the opposite signs of P_{zz}. It is represented in Fig.2 as function of the pp - system effective mass. The corresponding region of the "slow" proton momenta is 300 – 550 Mev/c.

This work was supported by the Russian Fund of Fundamental Research under the contract 93-02-03514 and by the International Science Foundation under the contracts RI6000 and RI6300.

References

1. R.Gilman *et al*, *Phys. Rev. Lett.* **65**, 1733 (1990).
2. S.I.Mishnev *et al*, *Phys. Lett.* B **302**, 23 (1993).
3. L.G.Isaeva *et al*, *Nucl. Instrum. Methods* A **325**, 16 (1993).
4. A.D.Bukin *et al*, UNIMOD2. Preprint BINP 94-20, Novosibirsk, (1994).
5. R.H.Dalitz, D.R.Yennie, *Phys. Rev.* v **105**, 1598 (1957).
6. W.J.Metcalf, R.L.Walker, *Nucl. Phys.* B **76**, 253 (1974).
7. M.Lacombe *et al*, *Phys. Rev.* C **21**, 861 (1980).
8. J.L.Visschers, ENIGMA.: An event generator. Proc. of MC93 Int. Conf. on Monte Carlo Simulation in High Energy and Nucl. Phys. Ed.: P. Dragovitsch. World Scientific Publishing, (1994).
9. I.Blomqvist, J.M.Laget, *Nucl. Phys.* A **280**, 405 (1977).

FINAL STATE INTERACTION EFFECTS IN $D(e,e'p)$ AND $\vec{D}(e,e'p)$ REACTIONS

A. BIANCONI
Brescia University, Brescia, Italy, and INFN, Pavia, Italy

S. JESCHONNEK, N.N.NIKOLAEV, B.G. ZAKHAROV
Institut für Kernphysik, KFA Jülich, D-52425 Jülich, Germany

In our analysis [1] of $D(e,e'p)$ scattering in the few-GeV region we have found that final state interaction (FSI) effects completely dominate the observed missing momentum distribution $W(\vec{p}_m)$ at high p_m (over $\sim$1.5 fm^{-1}). They produce a marked angular anisotropy in $W(\vec{p}_m)$, consisting in a dip at $\theta \sim 90^o$ (in perpendicular kinematics) for $p_m \sim$(1-1.5) fm^{-1}, and a prominent peak at the same angle for larger p_m. This anisotropy is mostly due to the FSI distortions of the S-wave contribution. The dominance of FSI is also found for $^4He(e,e'p)$ scattering [2], and leads to a universal pattern in the missing momentum distribution for different target nuclei [3]. The strength of (FSI) between struck proton and spectator neutron in $\vec{D}(e,e'p)$ scattering depends on the alignment of the deuteron [4]. We study the resulting FSI effects in the tensor analyzing power in detail and find substantial FSI effects starting at still low missing momentum $p_m \gtrsim 0.9 fm^{-1}$. At larger $p_m \gtrsim 1.5 fm^{-1}$, FSI completely dominates both missing momentum distribution and tensor analyzing power. We are interested in scattering experiments at CEBAF energies i.e. with momentum transfer over 1 GeV, where Glauber's multiple scattering theory [5] can be applied, which simplifies the evaluation of FSI effects. They are dominated by diffractive incoherent nucleon-nucleon rescattering, and scarcely affected by the details of the nuclear structure. We have studied scattering of unpolarized electrons on unpolarized and polarized deuteron, without detection of the final product polarizations. The reaction is of interest for the study of the short distance behavior of the D-wave component of the deuteron ground state. Due to the non spherical shape of the deuteron, the probability for FSI is different when the deuteron polarization axis is parallel or perpendicular to the path of the ejected proton. This is the origin of a strong FSI contribution to the tensor analyzing power of the deuteron in $(e,e'p)$ at large missing momenta. A relevant finding is that the differences that are present at PWIA level between the two models almost disappear when FSI is included, suggesting small chances of discriminating between different models.

The missing momentum distributions for different polarization states of the target are all heavily affected by FSI. The only quantity that is reminiscent of

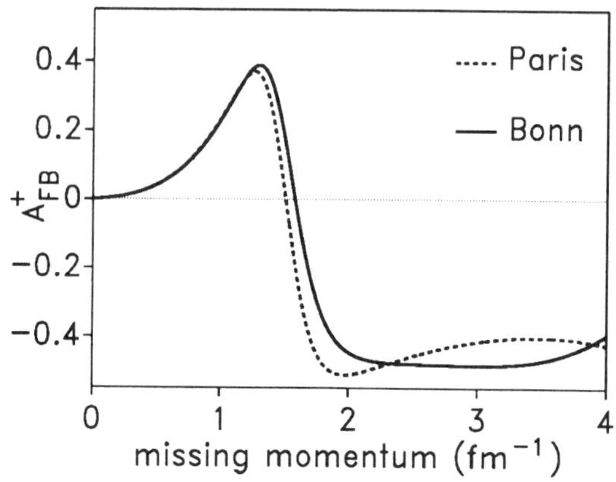

Figure 1: The forward-backward asymmetry A_{FB} vs. p_m for W_+ calculated employing the Bonn[6] wave function (solid line) and the Paris[7] wave function (dashed line).

some PWIA features is the FSI induced forward-backward (F/B) asymmetry. For transversely polarized deuterons the FSI induced F/B asymmetry of the missing momentum distribution is shown to have a node at precisely the same value of p_m as the PWIA (plane wave impulse approximation, i.e. neglecting FSI) missing momentum distribution. The position of this node is not affected by FSI and can be a tool to distinguish experimentally between different models for the deuteron wave function. They are shown in Fig.1.

Summarizing, we show that it would be very difficult to extract information about the large momentum D-wave content of deuteron ground state from $(e, e'p)$ scattering on polarized deuteron. We suggest measurements of the F/B asymmetry as a means to study intrinsic, FSI-independent, nuclear features.

1. A.Bianconi, S.Jeschonnek, N.N.Nikolaev and B.G.Zakharov, *Phys. Lett.* B **343**, 13 (1995).
2. A.Bianconi, S.Jeschonnek, N.N.Nikolaev and B.G.Zakharov, *Phys. Lett.* B **338**, 123 (1994).
3. A.Bianconi, S.Jeschonnek, N.N.Nikolaev and B.G.Zakharov, *Phys. Lett.* B **363**, 217 (1995).
4. A.Bianconi, S.Jeschonnek, N.N.Nikolaev and B.G.Zakharov, *Phys. Rev.* C **53**, 576 (1996).
5. R.J.Glauber and G.Matthiae, *Nucl. Phys.* B **21**, 135 (1970).
6. R.Machleidt, K.Holinde, and C.Elster, *Phys. Rep.* **149**, 1 (1987).
7. M.Lacombe *et al.*, *Phys. Lett.* B **101**, 139 (1981).

Polarization Transfer Measurements of the D($\vec{e}$,e'$\vec{p}$)n Reaction in the Dip Region

Kyungseon Joo
Massachusetts Institute of Technology, Cambridge, MA 02139, USA

The final state proton polarization from deuteron electrodisintegration via longitudinally polarized electrons was measured in the dip region at $Q^2 = 0.31$ GeV^2. The goal of these measurements is to extract all three recoil polarization components and study the reaction mechanism.

In the recent years the theory of the spin structure of the electromagnetic current has been developed. Among others, Donnelly has developed theoretical framework for studying polarization experiments including electron beam helicity, target and recoil polarization, and photon angular correlation measurements. A detailed analysis of the polarization response functions in ($\vec{e}$,e'$\vec{p}$) has been developed by Picklesimer and Van Orden. Their analysis shows that there are 18 independent response functions in the general case, for a polarized electron beam, involving proton polarization and non-coplanar kinematics. In the expression below we display the differential cross section for ($\vec{e}$,e'$\vec{p}$) :

$$\begin{aligned}\frac{d^3\sigma}{d\omega d\Omega_e d\Omega_p} = \frac{mp'}{2(2\pi)^3}\left[\frac{d\sigma}{d\Omega_e}\right]_{Mott} \times \{&v_L(R_L + R_L^n S_n) + v_T(R_T + R_T^n S_n)\\ &+ v_{TT}[(R_{TT} + R_{TT}^n S_n)cos(2\phi) + (R_{TT}^l S_l + R_{TT}^t S_t)sin(2\phi)]\\ &+ v_{LT}[(R_{LT} + R_{LT}^n S_n)cos(\phi) + (R_{LT}^l S_l + R_{LT}^t S_t)sin(\phi)]\\ &+ hv_{LT'}[(R_{LT'} + R_{LT'}^n S_n)sin(\phi) + (R_{LT'}^l S_l + R_{LT'}^t S_t)cos(\phi)]\\ &+ hv_{TT'}(R_{TT'}^l S_l + R_{TT'}^t S_T)\}\end{aligned}$$

- The $v's$ are known kinematics factors.
- ϕ is the angle between the electron and the $(\vec{q},\vec{p})$ reaction plane.
- The $S's$ are the components of the proton spin.
- The $R's$ depend on $\vec{q}$, ω and T_p.

The R's all have differing sensitivities to various effects such as meson exchange currents(MEC), isobar amplitudes(IC) and final state interactions(FSI). In parallel kinematics, each polarization component is proportional to a single response function: $P_t \propto R_{LT'}^t$, $P_l \propto R_{TT'}^l$, $P_n \propto R_{LT}^n$. In these considerations one must remember that the reference directions for n and t polarizations are given as $\vec{q}$x$\vec{p}$ and $\vec{p}$x($\vec{q}$x$\vec{p}$) respectively, in the above expressions. The l component of the polarization is along $\vec{p}$.

At the quasielastic peak and zero recoil momentum, the effects of FSI,

Table 1: Preliminary measured recoil proton polarizations and calculations by Arenhövel. The calculations included FSI, MEC, IC and RC(relativistic corrections).

	P_l	P_t	P_n
Measured	0.53(8)	-0.51(3)	-0.011(8)
Arenhövel	0.60	-0.53	-0.017

MEC, and IC are predicted to be quite small. As one moves from the quasielastic peak to the dip region between the quasielastic peak and delta resonance, one expects the role played by interaction effects to become more important. Studies of other nuclei in such kinematics have indicated a breakdown of the Impulse Approximation. For example, measurements on ^{12}C in the dip region have revealed a nearly uniform distribution of strength in the continuum for the $A-1$ system which cannot be accounted for by simple one-body calculations. Furthermore, the $^3He(\vec{e},e'\vec{p})$ three-body breakup channel falls like the $d(\vec{e},e'\vec{p})$ distribution at high recoil momentum. Since the deuteron is a two-body system and hence is unobscured by many-body effects, it provides a benchmark for nucleon-nucleon interaction effects. Further the relative simplicity of this system allows for complete theoretical calculations to be performed.

The experiment was performed at the Bates Linear Accelerator Center. We scattered longitudinally polarized electron beam off a liquid deuterium target using the magnetic spectrometers MEPS and OHIPS for the detection of electrons and protons respectively. The polarization of the outgoing protons was determined using a focal plane polarimeter (FPP) which was built and installed on the top of OHIPS by our group. The measurement was done at $\vec{q} = 610\ MeV/c$, $\omega = 250\ MeV$ and $\theta_{pq}^{cm} = 0^\circ$ in in-plane kinematics. Table 1 shows our measured recoil polarizations. The data are compared to calculations by Arenhövel. These results are preliminary, but seem to agree well with calculations at this stage of partial analysis.

In the future, the results of this experiment expanded by additional measurements at high recoil momentum and large θ_{pq} using high duty factor electron beam will enable us to improve our understanding of the nucleon-nucleon reaction mechanism.

This paper is presented on behalf of the MIT-Bates FPP Collaboration.[a] This work has been supported in part by the U. S. Department of Energy.

[a]Arizona State U, Cal. State U, MIT, Norfolk State U, Old Dominion U, TJNAF, U Maryland, U Virginia, W & M.

HIGH MISSING-MOMENTUM COMPONENTS IN ^{4}He STUDIED WITH THE (e,e′p) REACTION

J.J. VAN LEEUWE, E. JANS, W.J. KASDORP, L. LAPIKÁS, J.J.M. STEIJGER
NIKHEF, P.O. Box 41882, 1009 DB Amsterdam, The Netherlands

H.P. BLOK, G.E. DODGE, W.H.A. HESSELINK, C.J.G. ONDERWATER, A. PELLEGRINO, C.M. SPALTRO, J.A. TEMPLON
Vrije Universiteit Amsterdam, 1081 HV Amsterdam, The Netherlands

J.F.J. VAN DEN BRAND, H.J. BULTEN, O. UNAL
Department of Physics, University of Wisconsin, Madison, Wisconsin, USA

R. ENT
CEBAF, Newport News, VA. 23606, USA

J.M. LAGET
DAPNIA/SPhN, Saclay, 91191 Gif-sur-Yvette Cedex, France

The energy and momentum distributions of bound nucleons are sensitive to various aspects of the nucleon-nucleon interaction. Model calculations for ^{4}He using realistic NN-interactions indicate that the momentum distribution of the two-body p+t breakup channel at high momenta is mainly sensitive to central short-range correlations, whereas the four-body breakup continuum is also sensitive to tensor correlations.

We have investigated the missing-energy and missing-momentum dependence of the ^{4}He(e,e′p) cross section using the 525 MeV extracted electron beam from the pulse stretcher AmPS at NIKHEF. The scattered electrons were measured in the QDQ magnetic spectrometer of the EMIN endstation, while the knocked out protons were detected in the segmented scintillator detector HADRON-4. It consists of a horizontal and vertical hodoscope, each composed of 20 scintillators of 2 mm thickness having double-sided readout. Behind the hodoscopes 52 scintillators are grouped in 4 layers with a total thickness of 17.6 cm in order to determine the proton energy. The solid angle of the HADRON-4 detector amounts to 550 msr. The virtual-photon characteristics were (q,ω) = (400 MeV/c, 215 MeV). The emitted protons were detected at angles γ_{pq} ranging from 8° to 101° backwards of $\mathbf{q}$.

The missing-momentum dependence of the ^{4}He(e,e′p)^{3}H cross section is shown in figure 1, together with calculations by Laget [1]. In this diagrammatic approach the few-body wave functions that result from variational method calculations [2] having the Urbana potential as input are used to calculate the various overlap functions. The Plane Wave Impulse Approximation (PWIA) prediction for the cross section exhibits a zero around $p_m = 480$ MeV/c.

Clearly two- and three-body effects play an important role in the kinematic domain where the single-nucleon knockout strength, as predicted by PWIA calculations using realistic NN-interactions, becomes very small. Although the full calculation correctly describes the data in the missing-momentum region above 570 MeV/c and between 200 and 300 MeV/c, there is a large discrepancy around the zero point in the $\langle {}^3\mathrm{H}|{}^4\mathrm{He}\rangle$ overlap function, as was observed in a similar experiment performed at Saclay[3]. This discrepancy might be due to an incomplete treatment of Final State Interaction (FSI) and Meson Exchange Current (MEC) effects or to the groundstate wave functions that are used.

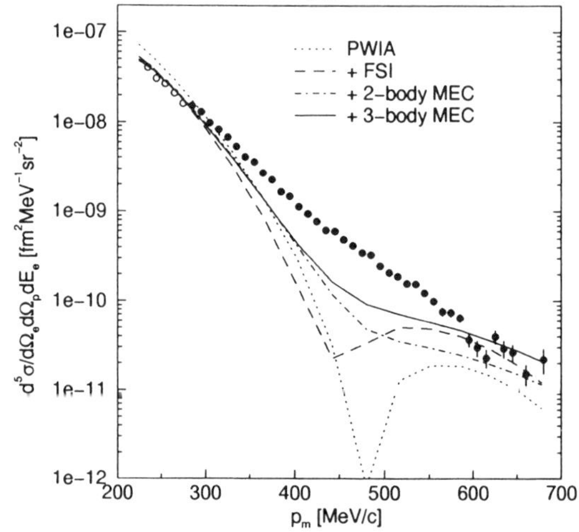

Figure 1: ^{4}He(e,e$'$p)^{3}H cross section as a function of missing momentum.

Calculations for the continuum by Simula[4], in which scattering on a pN-pair is assumed explicitely, and those by Laget, predict that the cross section as a function of the missing energy at fixed values of γ_{pq} exhibits a bump-shaped structure. In the present experiment this has been investigated by measuring the continuum up to a missing energy of 150 MeV. Indeed the missing energy value at which the centroid of the bump is located changes as a function of γ_{pq}, in agreement with the reaction mechanism in which scattering on a nucleon of a correlated pair has taken place. The shape and absolute value of the experimental cross section is well reproduced by the calculations.

References

1. J.M. Laget, *Nucl. Phys.* **A579**, 333 (1994)
2. R. Schiavilla, *et al.*, *Nucl. Phys.* **A449**, 219 (1986)
3. J.M. Le Goff, *et al.*, *Phys. Rev.* **C50**, 2278 (1994)
4. S. Simula, *et al.*, *Phys. Rev.* **C53**, 1689 (1996)

PHOTOABSORPTION AND PHOTODISINTEGRATION PROCESSES ON VERY LIGHT NUCLEI

M.Mac Cormick[a],J.Ahrens[d], G.Audit[a], A.Braghieri[b], N.D'Hose[a], J.Habermann[e], S.Kerhoas[a], L.Murphy[a], P.Pedroni[b], T.Pinelli[b,c], G.Tamas[a]
[a] *CEA DAPNIA-SPhN, C.E. Saclay, Gif sur Yvette, France*
[b] *INFN-Sez. di Pavia, via Bassi 6, Pavia, Italy*
[c] *Dip. Fisica Nucleare e Teorica, Università di Pavia, Italy*
[d] *Institut für Kernphysik, Universität Mainz, Germany*
[e] *Physikalisches Institute, Universität of Tübingen, Germany*

The total photoabsorption and photodisintegration cross sections for ^{1}H, ^{2}H, ^{3}He and ^{4}He have been measured for incident photon energies from 200 to 800 MeV using the large acceptance detector DAPHNE at the MAMI microtron in Mainz.

1 Introduction

Recently, experiments carried out by groups at Mainz and Frascati [1] have provided measurements of σ_{tot} for several medium and heavy nuclei up to 1.2 GeV. The results of these experiments show that the higher $D_{13}(1520)$ and $F_{15}(1680)$ resonances are no longer visible in these nuclei. Furthermore, the total photoabsorption strength per nucleon is reduced. The mechanism behind this apparent "damping" is not as yet well understood.

This paper presents the results of a study that was carried out on ^{1}H, ^{2}H, ^{3}He and ^{4}He with the aim to explore in greater detail the evolution of the cross section in the transition from light to heavy nuclei. The ^{3}He and ^{4}He data are the first for these nuclei. By using the large acceptance detector DAPHNE [2] in conjunction with the tagged photon beam facility of the MAMI accelerator in Mainz [3], results of high precision have been obtained.

2 Data Analysis and Results

The experimental approach was extensively described in [4]. We recall briefly only the general principle of the data analysis method. For photoproduction channels we measure directly ~75% of the total cross section. The remaining 25% is deduced by either, the convolution of the π^0 detection efficiency ($\simeq$ 20% of the unmeasured part), or, by extrapolation of the charged pion angular and momentum distributions from the DAPHNE detector acceptance ($\simeq$ 5% of the unmeasured part). When considering photodisintegration, the unmeasured component of this reaction channels were evaluated using the results obtained

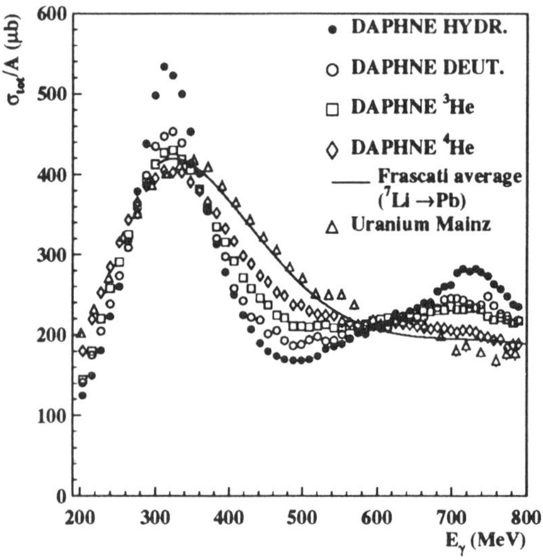

Figure 1: The total photoabsorption cross sections per nucleon for ^1H, ^2H, ^3He and ^4He are compared to previous results.

in the independent studies carried on ^2H [5], ^3He [6] and ^4He. In ^4He, for the first time, the separation between two and three or four body photoabsorption mechanisms was also performed and the so-called "quasi-deuteron" process was found to represent roughly only 50% of the total photodisintegration cross section at 300 Mev. Our results for photoabsorption, depicted in fig.1, show clearly the changes in going from ^1H to ^4H. In particular, the ^4He cross section is very close to the average behaviour of heavy nuclei. These results will provide a strong constraint to the theories that are presently being developed to explain the apparent suppression of higher resonances in nuclei.

References

1. Th. Frommhold *et al.*, *Z. Phys.* A **350**, 249 (1994); N. Bianchi *et al.*, *Phys. Lett.* B **325**, 333 (1994).
2. G. Audit *et al.*, *Nucl. Instrum. Methods* A **301**, 473 (1991).
3. I. Anthony *et al.*, *Nucl. Instrum. Methods* A **301**, 230 (1991).
4. M. MacCormick *et al.*, *Phys. Rev.* C **53**, 41 (1992).
5. R. Crawford *et al.*, *Nucl. Phys.* A, in press.
6. G. Audit *et al.*, submitted to *Nucl. Phys.* A.

QUASI-ELASTIC SCATTERING OF POLARIZED ELECTRONS OFF POLARIZED ^{3}He AND THREE-BODY INTERACTIONS

A. KIEVSKY, M. VIVIANI

INFN, Sezione di Pisa, 56010 S.Piero a Grado, Pisa, Italy

E. PACE

Dipartimento di Fisica, Università di Roma "Tor Vergata", and INFN, Sezione Tor Vergata, Via della Ricerca Scientifica 1, I-00133 Roma, Italy

G. SALME'

INFN, Sezione Sanità, Viale Regina Elena 299, I-00161 Roma, Italy

Quasi-elastic (qe) inclusive experiments involving polarized ^3He are viewed as a possible source of information on the neutron form factors [1], and therefore refined calculations are mandatory in order to obtain a model independent extraction, e.g., of the em form factors. This fact has motivated detailed theoretical investigations, based on the description of polarized ^3He by spin-dependent spectral functions [2,3] obtained from pairwise N-N interaction, i.e. two-body force (2BF). In particular it has been shown that, though the proton contribution in general cannot be disregarded, the ^{3}He asymmetry is very sensitive to neutron properties, if one properly chooses the kinematics [2]. We have further extended the theoretical investigation by considering the effect of 3BF's widely adopted in the literature. In particular we have obtained spin-dependent spectral functions from the ^{3}He wave functions of ref. [4], calculated by the correlated hyperspherical-harmonic expansion, which was proven to be one of the most sophisticated methods for few-nucleon systems, and using the Argonne V14 and different 3BF's. The most important feature we have found is the quite small sensitivity of the asymmetry upon the 3BF's, as shown in Fig.1, where our results (with the Galster nucleon form factors) are compared with experimental data [1]. As a matter of fact, the results corresponding to different 3BF's coincide and therefore only a curve (the dashed one) representing the full calculation (2BF+3BF) is shown. The same independence is obtained in the so called β-kinematics proposed in [2] for minimizing the proton contribution. In summary, since the calculations obtained with different 2BF's and 3BF's are quite similar, the model dependence of the asymmetry upon the interaction in the initial state seems to be under control.

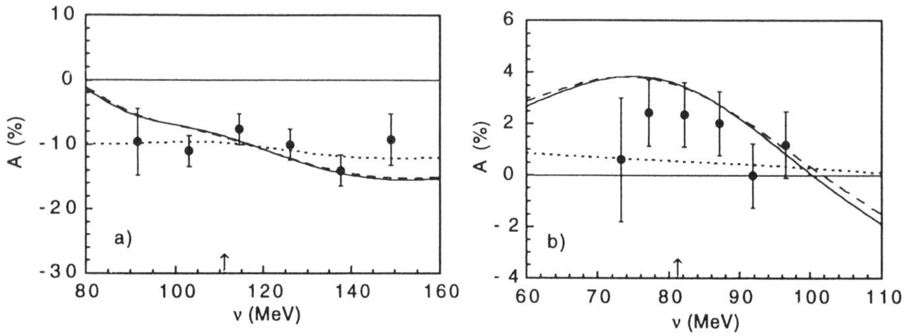

Figure 1: a) The asymmetry vs. the energy transfer ν for $\epsilon_i = 370$ MeV, $\theta_e = -91.4^o$ and a polarization angle $\beta = 42.5^o$; solid line: 2BF only; dashed line: 2BF + 3BF; dotted line: neutron contribution. Data are from [1c]. The arrow indicates the qe peak. b) The same as in a), but for $\theta_e = 70.1^o$. Data are from [1d].

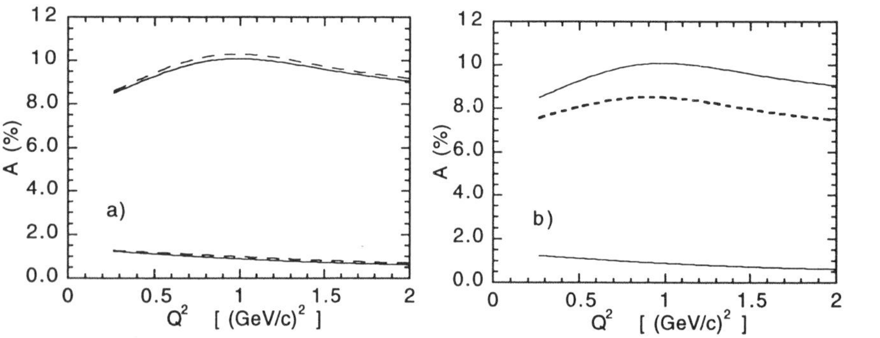

Figure 2: a) The total asymmetry at the top of the qe peak, vs. Q^2, for $\theta_e = 75^o$ and $\beta = 95^o$. The curves corresponding to 2BF only and 2BF+3BF overlap. The curves in the lower part of the figure represent the proton contribution. (cf. [2]). b) The same as in a), but for 2BF; solid line: total asymmetry; dotted line: contribution proportional to $R_{T'}$.

References

1. a) C. E. Jones-Woodward et al., *Phys. Rev.* **C 47** (1993) 110; b) A. K. Thompson et al., *Phys. Rev. Lett.* **68** (1992) 2901; c) H. Gao et al., *Phys. Rev.* **C 50** (1994) R546; d) J.O. Hansen et al., *Phys. Rev. Lett.* **74** (1995) 654.
2. C. Ciofi degli Atti, E. Pace and G. Salmè, *Phys. Rev.* **C 46** (1992) R1591; Phys. Rev. **C 51** (1995) 1108.
3. R.W. Schultze and P.U. Sauer, *Phys. Rev.* **C 48** (1993) 38.
4. A. Kievsky, M. Viviani and S. Rosati, *Nucl. Phys.* **A551** (1993) 241; **A577** (1994) 511.

Data and Discussion on High Momentum in the Deuteron

C. E. Carlson[a], J. Hanlon[c], and K. E. Lassila[b](speaker)
[a]*Physics Department, College of William and Mary, Williamsburg, VA 23187*
[b]*Department of Physics and Astronomy, Iowa State University, Ames, IA 50011*
[c]*Fermi National Accelerator Laboratory, Batavia, IL 60510*

Neutrino-deuteron data from the Fermilab 15 ft bubble chamber is presented. The phase space region studying the neutron-proton wave function is found to have excess spectator protons moving with large momentum against the weak charge current. This excess is consistent with a 1 - 2% component in which the neutron and proton are sufficiently overlapping to form a six-quark state.

We present data on high momentum components in the deuteron (d) wave function from $\nu - d$ reactions, commenting on new data from Mainz.[1] The Mainz electron beam has low energy and limited backward detection so complicated calculations with final state interactions were needed to argue that the largest momentum protons in the deuteron breakup were in this backward region.

High energy c.c. $\nu - d$ DIS collisions cleanly study high momentum in the deuteron. The incident short wavelength probe raises the struck quark charge by one unit, producing a jet. The counting of charged prongs then leads to a statement as to whether this struck quark was in the proton or the neutron. The 2N view of the deuteron sees a loose object with distinct neutron and proton. The unstruck nucleon is left as a spectator; a proton spectator leaves a distinctive, measurable (in the 160 $\sim$ 900 MeV/c range) track in the bubble chamber. Absolute normalization is done from the quasielastic scattering events, $\nu d \rightarrow \mu^- p p_s$, yielding the neutrino energy flux from kinematics. Details of experiment E545 are in our recent preprint[2] giving earlier references.

The unpublished E545 data are the squares in Figs. 1a, 1b, where $d\sigma/dp\, vs\, p$ is given for backward protons p_B of momentum magnitude p in $\nu + d \rightarrow \ell^- + p_B + X$. The circles show the event distribution for CERN WA25[3] normalized to the E545 data at 200 MeV/c. Standard DIS variables (x, y) are used in treating the deuteron on the light front, with p_T the proton transverse momentum. The theory curves shown in the figure have some of these variables integrated over. The low curve in Fig. 1a with the $2N$ label is for the Hulthén model and the lowest curve, $2N - Bonn$, in Fig. 1b, is calculated for the Bonn potential[4], used in Ref. (1). We note the absolute prediction agrees with the normalized data below 0.22 GeV, but at higher momenta, the 2N curve is low. The dotted curve is gotten assuming the individual parton components can not be separated when the neutron and proton overlap, and the charged current sees the nucleons as kneaded into a 6q state. The 6q valence distribution is taken from counting rules, as consistent with that for pions or nucleons. In the

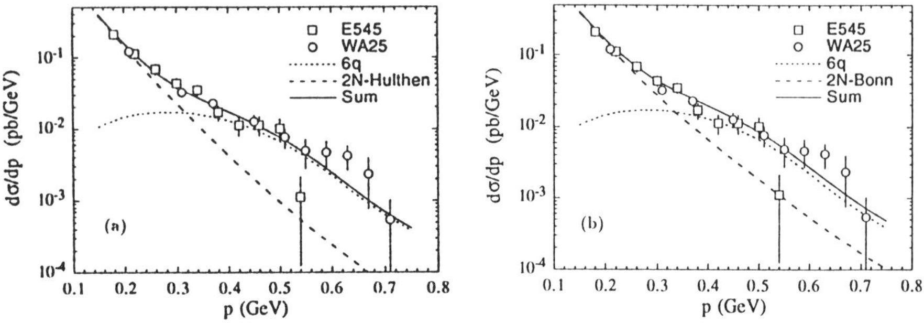

Figure 1: Backward proton cross section for $\nu - d$ breakup vs the magnitude of the three momentum. The data are from Fermilab E545 and BEBC WA25. In (a) the data are compared with predictions of the Hulthén 2N model (dashed). In (b) comparison is made relative to the predictions of the 2N Bonn model. The 6q curve has a fragmentation function with a power-law tailoff transverse part $(1 + p_T^2/\lambda^2)^{-3}$, and a $z^3(1-z)^3$ longitudinal part, normalized to give one proton per 6q cluster breakup. The solid curves include a 2% fraction of 6q state; $z^3(1-z)^3$ leads to a curve of the same shape above 250 MeV/c, and with the same normalization of the 6q fraction is 4%.

x region of valence dominance, the weak current removes a quark and the 5q residuum fragments to produce the backward energetic proton, with fragmentation function going as $(1-z)^3$ at large backward momentum, with z the proton momentum fraction. The solid curves are the sums of these two components, with a 2% fraction of 6q state. The 2N calculation for the Paris potential[2] is higher than that for Bonn, and requires an additional 1% fraction of 6q state, accounting for our estimate of 1 to 2 % of 6q in the deuteron.

Other tests for these high momentum components are of interest: (1) Ratios of cross sections can test the 2N view of the deuteron in plots of the data at fixed x and p bins divided by the corresponding (calculable) cross section on a neutron. In this view the calculated ratio of data to neutron cross section should simply be a horizontal line as a function of x for fixed proton light front momentum and fixed p_T. Deviations from a line would directly signal the presence of more complicated internal deuteron structure. (2) The angular distribution of backward protons from a polarized deuteron target would directly test for characteristic variations of the backward proton angular distribution in the 2N model as a function if backward momentum.

1. Werner Boeglin, paper in these proceedings.
2. C. E. Carlson, J. Hanlon, and K. E. Lassila, Fermilab-pub-96/054 (1996).
3. Quoted in E. Matsinos *et al.*, Z. Phys. C **44**, 79 (1989).
4. R. Machleidt *et al.*, Phys. Rep. **149**, 1 (1987).

Virtual Compton Scattering at MAMI
$\gamma^* p \to \gamma' p'$

D. Lhuillier[a], J. Berthot[b], P.Y. Bertin[b], V. Breton[b], W.U. Boeglin[g], R. Böhm[d], N. D'Hose[a], M. Distler[d], J.E. Ducret[a], R. Edelhoff[d], H. Fonvieille[b], J. Friedrich[d], J.M. Friedrich[d], R. Geiges[d], Th. Gousset[a], P.A.M. Guichon[a], H.Holvoet[c], Ch. Hyde-Wright[e], P. Jennewein[d], M. Kahrau[d], S. Kerhoas[a], M. Korn[d], H. Kramer[d], K.W. Krygier[d], V. Kunde[d], B.Lannoy[c], A. Liesenfeld[d], C. Marchand[a], D. Marchand[a], J. Martino[a], H. Merkel[d], K. Merle[d], G. De Meyer[c], J. Mougey[a], R. Neuhausen[d], E. Offermann[f], Th. Pospischil[d], G. Quemener[b], O. Ravel[b], Y. Roblin[b], J. Roche[a], G. Rosner[d], D. Ryckbosch[c], P. Sauer[d], H. Schmieden[d], S. Schardt[d], G. Tamas[d], M. Vanderhaeghen[a], L. Van Hoorebeke[c], R. Van de Vyver[c], J. Van de Wiele[h], P. Vernin[a], A. Wagner[d], Th. Walcher[d], S. Wolf[d]

[a] *CEA DAPNIA-SPhN, C.E. Saclay, France*
[b] *LPC, Univ. Blaise Pascal, IN2P3 Aubiere, France*
[c] *University of Gent, Belgium*
[d] *Institut für Kernphysik, Universität Mainz, Germany*
[e] *Old Dominium University, Virginia, U.S.A*
[f] *C.E.B.A.F, Virginia, U.S.A*
[g] *Florida International University, Miami,Florida, U.S.A*
[h] *IPN, IN2P3 Orsay, France*

The Virtual Compton Scattering allows us to measure for the first time the generalized polarizabilities of the proton. The experimental method used to extract these new observables is presented as well as some preliminary results obtained at MAMI.

Virtual Compton Scattering is a foundamental exclusive reaction off the proton which provides us with a new insight of the internal structure of the nucleon. Below π^0 production threshold ($M_p < \sqrt{s} < M_p + M_\pi$, but arbitrary Q^2), this experiment measures the generalized polarizabilities of the proton[1]. These are new electromagnetic observables, functions of Q^2 the mass of the incoming virtual photon, which enlarge the concept of electric (α) and magnetic (β) polarizabilities already defined in real compton scattering ($Q^2 = 0$)[2]. At MAMI, we measure for the first time the VCS in order to deduce generalized polarizabilities at $Q^2 = 0.33$ GeV^2.
We can experimentally access VCS by the electroproduction of photons : $ep \to e'p'\gamma$. In this reaction, the final photon can be emitted either by the electron or by the proton. The first process is described by the Bethe-Heitler

[1] P.A.M. Guichon, G. Liu, A.W. Thomas, Nucl. Phys. **A591** (1995) 606
[2] B.E. Mac Gibbon et al., Phys. Rev. **C52** (1995) 2097

(BH) amplitude which is calculable from Quantum Electro-Dynamics (QED). The second process is described by the Virtual Compton Scattering (VCS) amplitude, that can be split in two parts : the Born term containing only the nucleon and anti-nucleon contributions (exactly calculable) and the Non-Born term related to the excited states and parametrized by 10 polarizabilities. The probability $\mathcal{M}^{exp}$ (which is the differential cross section divided by a phase space factor) is therefore a coherent sum of the different amplitudes,

$$\mathcal{M}^{exp} = \frac{1}{4}\sum_{spin}|T^{BH} + T^{VCS}|^2 = \frac{1}{4}\sum_{spin}|T^{BH} + T^{Born} + T^{NonBorn}|^2$$

The low energy theorem from Low[3] says that in an expansion in powers of the outgoing photon energy (q′) in the final photon-proton center of mass system, the two first terms of the probability $\mathcal{M}$, of the order ${q'}^{-2}$ and ${q'}^{-1}$ are only due to the interference of the BH and Born amplitudes and are completely calculable. The effect of the polarizabities appears from the ${q'}^{0}$ term. The experimental method (see Fig.1) is to study the evolution of ${q'}^{2}\mathcal{M}^{exp}$ at small values of q′.
The experiment is performed in the A1 Hall at MAMI. The scattered proton and electron are detected in coincidence in two magnetic spectrometers and the final photon is tagged by the reconstructed missing mass: $M_X^2 = 0$. With the high duty cycle available at MAMI, one is not limited by the accidental counting rates and one can use a typical luminosity of $\mathcal{L} = 2.10^{37}$ cm^{-2}s^{-1}. Thanks to the excellent resolution of the facilities ($\Delta E/E = 10^{-4}$ for the energy of the incident electrons, $\Delta P/P = 10^{-4}$ for the momenta of the two spectrometers and an angular resolution better than 3 mrad) the contribution of the π^0 in the missing mass is very clearly separated from the events characterized by the photon mass (see Fig.2).
Fig.3 shows preliminary results obtained from data acquired in spring 1996. At small q' the expected effect of polarizabilities is very small (see Fig.1). Data points which have still to be corrected for radiative effects, have hence to reproduce the predictions for an on-shell nucleon, that is the contribution from BH and Born terms. The radiative corrections are beeing calculated for the first time at Saclay taking into account all the diagrams of order α^4 in the cross section[4]. Their implementation will provide us very soon with the first test of the low energy theorem for the VCS which represents a severe check of our analysis method and will ensure a reliable extraction of the polarizabilities in the energy domain q′=100 MeV.

---

[3] F.E. Low, Phys. Rev. **96** (1954) 1428

[4] D.Lhuillier, D.Marchand, M.Vanderhaeghen, J.Van de Wiele, in preparation.

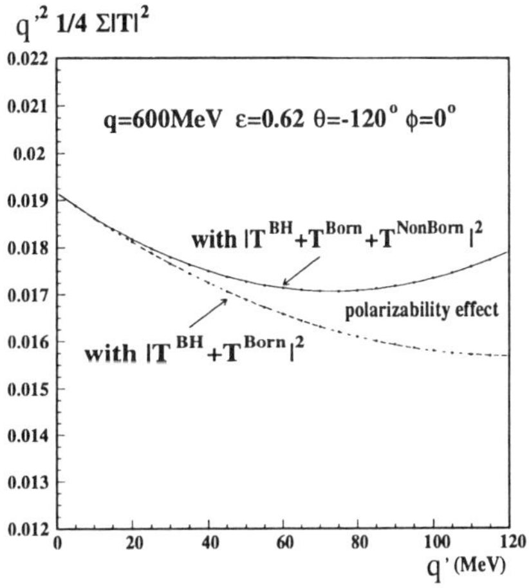

Figure 1: Evolution of $q'^2 \times \mathcal{M}^{exp}$ from calculations of P.A.M. Guichon. et al. The ordinate and slope at $q' = 0$ are due only to the interference between BH and Born, while the polarizability effect appears in the second order of the expansion.

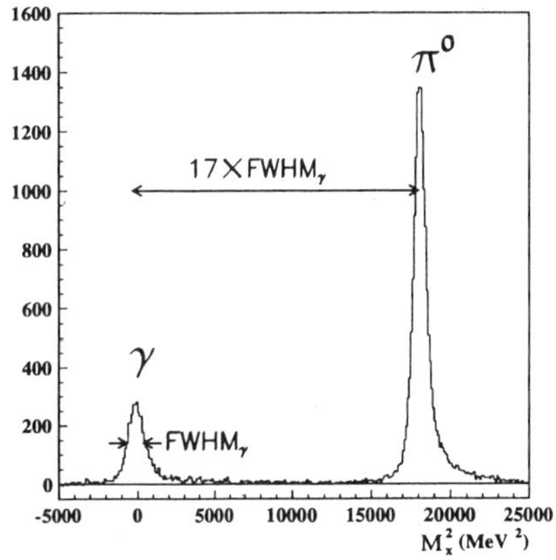

Figure 2: Experimental spectrum of the missing mass in the reaction p(e,e'p)X with a very good separation between Compton and π^0 electroproduction events.

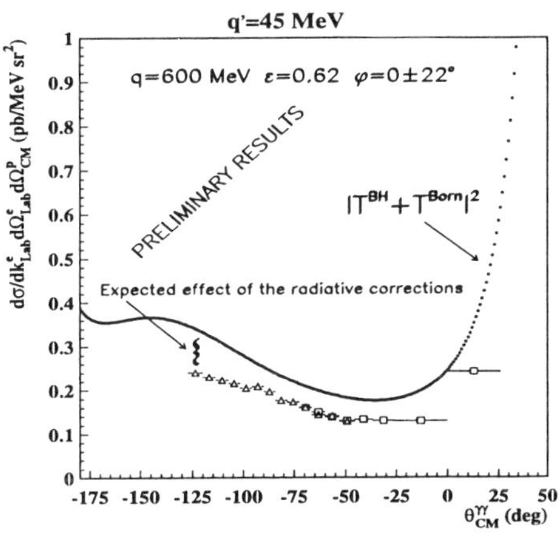

Figure 3: Preliminary results at q'=45 MeV. The two sets of points are obtained with two different positions for the proton spectrometer (but the same electron setting) and are in good agreement in the region of same kinematics. Full radiative corrections are likely to be very soon implemented. At small q' they will provide the first test of the low energy theorem for the VCS. The generalized polarizabilities will be extracted from data at higher q'.

ABSORPTION MECHANISMS FOR THE (γ,*PP*) REACTION

N.R. Kolb, F. Adimi, G. Feldman, E. Hallin, G.V. O'Rielly, R.E. Pywell, D.M. Skopik, D.E. Tiller, J.M. Vogt, *Saskatchewan Accelerator Laboratory*; E. D. Hackett, E. Korkmaz, W.J. McDonald, T. Nakano, A.K. Opper, M. Quraan, N.L. Rodning, F.M. Rozon, *University of Alberta*

The total measured cross section for ^{3}He$(\gamma,pp)n$ as a function of photon energy is displayed in Fig. 1. Software cuts have been placed on the angular

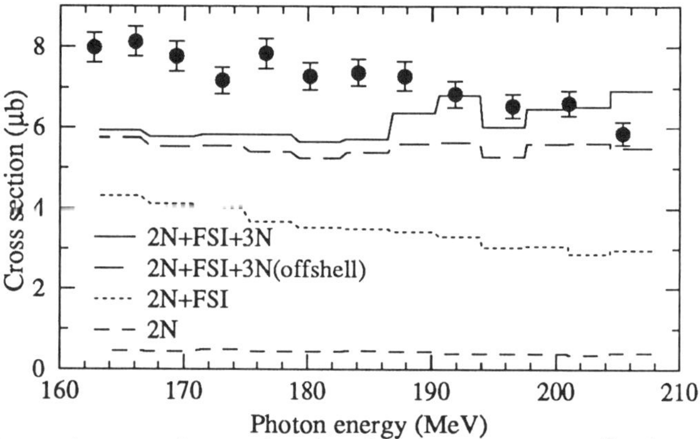

Figure 1: Measured cross section as a function of photon energy for the ^{3}He$(\gamma,pp)n$ reaction.

acceptance ($40° < \theta_p < 140°$) and kinetic energy($T_p > 0.25[E_\gamma - E_{binding}]$). Final state interactions (FSI) have a large effect on the calculated cross section as shown by the difference between the short-dashed and dotted curves in Fig. 1. The strength is only about half that of the measured cross section, clearly additional ingredients are required in the calculation, such as 3N absorption diagrams (solid curve). The long-dashed curve requires that both of the exchanged pions in the 3N diagrams to be offshell. Most of the 3N strength in this photon energy range is from offshell pions and hence is an appropriate place to search for the effect of 3N forces.

The theoretical model of J.M. Laget indicates that a large fraction of the strength in the ^{3}He$(\gamma,pp)n$ reaction is generated by two-nucleon absorption on an *np* pair followed by FSI which results in two energetic protons. One might then expect that the same process would also be important in the ^{4}He(γ,pp) reaction. To test this, it is useful to consider the missing energy and missing momentum for each target. The missing momentum (not shown) is virtually identical for all three targets, strongly resembles phase space, and peaks at a momentum of approximately 200 MeV/c. As the data are taken over a range of photon energy the *reduced* missing energy is defined as $E_{reduced} = \frac{E_\gamma - E_{binding} - T_1 - T_2}{E_\gamma - E_{binding}}$. Fig. 2 displays the reduced missing energy for the (γ, pp) reaction on ^{3}He (dotted), ^{4}He (dashed), and ^{12}C (solid) with the same acceptance as Fig. 1. The histograms have been adjusted to give the

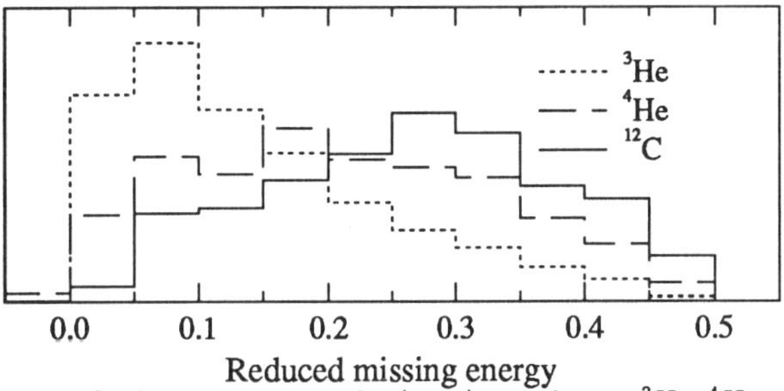

Figure 2: Reduced missing energy in the (γ, pp) reaction on ^{3}He, ^{4}He and ^{12}C.

same area for each of the nuclei. The excess of large missing energy for ^{4}He relative to ^{3}He, coupled with the small missing momentum, implies that more than three nucleons are involved in the absorption process. The distributions for ^{4}He and ^{12}C are nearly identical and strongly resemble 4N phase space. Whatever absorption process is responsible for ^{4}He(γ, pp) seems to dominate the ^{12}C(γ, pp) reaction as well.

If the 'quasi-deuteron' absorption on *np* pairs followed by FSI were to dominate the (γ, pp) reaction in this energy regime then one would expect the (γ, pp) cross section to scale approximately like the 'quasi-deuteron' cross sections (σ_{qd}). Previous measurements of (γ, p) have estimated the ratio of σ_{qd} to the free deuteron cross section to be 1.7, 4.2, and 13 for ^{3}He, ^{4}He, and ^{12}C, respectively. Fig. 3 displays the measured cross sections within the acceptance of the detector. Unfortunately, the acceptance is different from that of Fig. 1 and 2 in that the energy threshold is angle-dependent. The ^{4}He and ^{12}C data are currently being reanalyzed. The ^{4}He(γ, pp) cross section (multiplied by

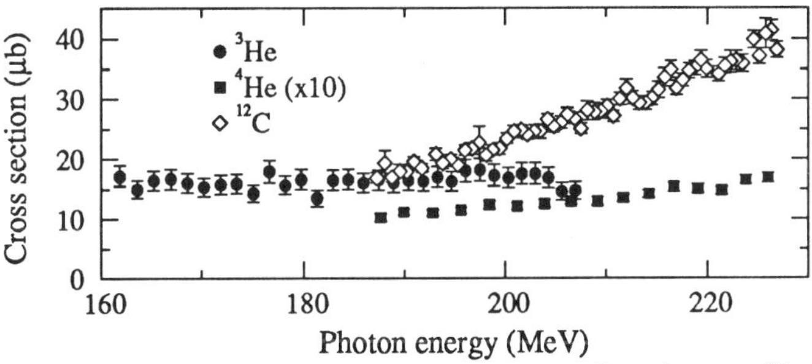

Figure 3: Cross sections in the (γ, pp) reaction on ^{3}He, ^{4}He and ^{12}C.

10 in the figure) is 15 times smaller than the ^{3}He (or ^{12}C) cross section near 200 MeV. This strongly suggests that the absorption mechanism in ^{3}He is suppressed in ^{4}He and ^{12}C.

Mechanisms of the ^{12}C(γ,NN) Reactions at E_γ=114-600 MeV*

P. Grabmayr
Physikalisches Institut, Universität Tübingen, D-72076 Tübingen, Germany

for the PiP/TOF group of the A2 collaboration:
J.R.M. Annand[2], J. Ahrens[4], I. Anthony[2], R. Beck[4], D. Branford[3], G.E. Cross[2], T. Davinson[3], P. Grabmayr[1], S.J. Hall[2], P.D. Harty[2], T. Hehl[1], J.D. Kellie[2], Th. Lamparter[1], J. Leypoldt[1], M. Liang[3], S. MacAllister[2], I.J.D. MacGregor[2], J.A. MacKenzie[3], J.C. McGeorge[2], G.J. Miller[2], R.O. Owens[2], M. Sauer[1], R. Schneider[1], K. Spaeth[1], G.J. Wagner[1], D. Watts[2] and T. Yau[2]
[1] *Physikalisches Institut, Universität Tübingen, D-72076 Tübingen, Germany*
[2] *Department of Physics and Astronomy, University Glasgow, G12 8QQ, UK*
[3] *Physics Department, University of Edinburgh, Edinburgh E49 3JZ, UK*
[4] *Institut für Kernphysik, Universität Mainz, D-55099 Mainz, Germany*

Photon induced two-nucleon emission reactions are considered to be valuable tools for the access to nucleon-nucleon correlations and in particular the short-range part thereof (SRC). However, a profound understanding of the competing one-body and two-body aspects of the photon absorption processes is an indispensible prerequisite. The (γ,NN) experiments were performed at

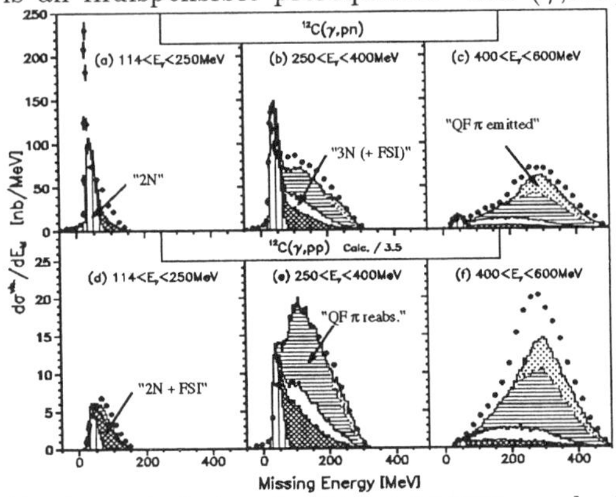

Figure 1: *The missing energy spectra for the $^{12}C(\gamma,np)$ and $^{12}C(\gamma,pp)$ reactions for three ranges of photon energies [4] in comparison to predictions of the Valencia model [6]. The various processes considered are denoted by different hatching. The predictions for (γ,pp) are scaled down by a factor of 3.5.*

the tagged photon facility at MAMI employing the two large acceptance, high resolution detector setups PiP and TOF produced precise data [1-5] over a wide kinematical range with a missing energy resolution of 6 MeV.

* This work was supported by the BMBF, DFG, DAAD, BC, SERC and NATO.

For the ^{12}C(γ,NN) reactions the "visible" cross sections are plotted in Fig. 1 as a function of missing energy E$_m$ for three photon energy ranges. At low photon energies the ^{12}C(γ,np) channel is dominated by a narrow structure close to the 2N separation energy; this has been shown to result due to nucleon removal from $1p$ and $1s$ shells. However, in the (γ,pp) channel this peak is greatly reduced because of the suppression of MEC.At high photon energies the strengths in both reaction channels is located at higher E$_m$ (Fig.1c,f) and both shapes become similar due to the underlying quasifree-π production. The present comparison with model predictions [6] taking into account the experimental constraints reveals a concentration of 2N strength at low E$_m$.

Concentrating on the shell model range (E$_m$< 40 MeV) the relative momentum $\vec{p}_{rel} = \vec{p_p} - \vec{p_n}$ is calculated in the rest frame of the np pair; this removes the effects of Fermi motion (within a PWIA framework). An equal sharing of photon energy and momentum between the two emitted nucleons is assumed. Then $\vec{p}_{rel}$ is also the relative momentum *before* the absorption. Thus we display the yields from Deuterium and Carbon as function of p_{rel}. In Fig.2a the shift between the ridges of two targets originates from the different

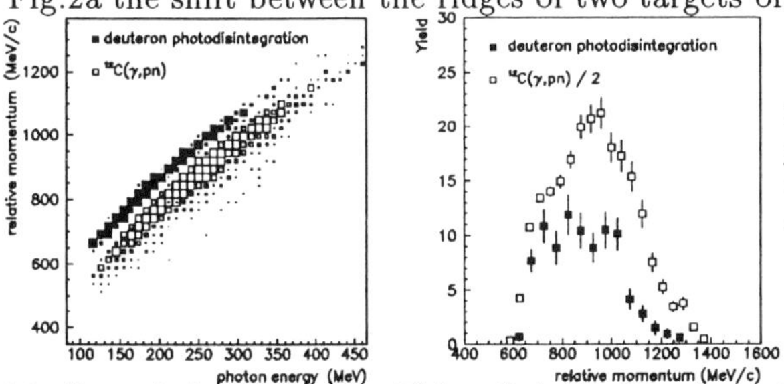

Figure 2: *The (γ,np) yields from Deuterium (full symbols) and ^{12}C (open symbols, E_m < 40 MeV) in the np rest frame.*
a) p_{rel} vs. E_γ
b) projection on p_{rel}

binding of the np pairs. This off-shell effect impedes a direct comparison of photon absorption on a free and a bound np pair because model-dependent corrections will be applied. The two projections (Fig. 2b) are corrected for detector acceptances. Yet, these projections constitute the most direct access to high momentum components ($p_{rel} \approx 1$ GeV/c) and SRCs in free and bound np pairs available to date.

References

[1] G. Cross *et al.*, Nucl. Phys. **A593** (1995) 463
[2] T. Hehl, Prog. Part. Nucl. Phys **34** (1995) 385
[3] P. Grabmayr *et al.*, Phys. Lett **B370** (1996) 17
[4] T. Lamparter *et al.*, Z. Phys. **A355** (1996) 1
[5] P.D. Harty *et al.*, Phys. Lett **B**, in print
[6] R.C. Carrasco *et al.*, Nucl. Phys. **A536** (1992) 445, **A540** (1994) 701

Quasifree π Photo-Production and FSI in the Nuclear Medium at Δ-Resonance Energies

M. Liang[1], D. Branford[1], J. Ahrens[4], J.R.M. Annand[2], R. Beck[4], T. Davinson[1], P. Grabmayr[3], S.J. Hall[2], P.D. Harty[2], T. Hehl[3], D. Johnston[1], J.D. Kellie[2], T. Lamparter[3], J. Leypoldt[3], I.J.D. MacGregor[2], J.A. MacKenzie[1], J.C. McGeorge[2], R.O. Owens[2], M. Sauer[3], R. Schneider[3], A.C. Shotter[1], T. Yau[2]

[1] *Department of Physics and Astronomy, University of Edinburgh, Scotland*

[2] *Department of Physics and Astronomy, University of Glasgow, Scotland*

[3] *Physikalisches Institut der Universität Tübingen, Germany*

[4] *Institut für Kernphysik, Universität Mainz, D-55099 Mainz, Germany*

Differential cross-sections from the first (γ,π^+n) and (γ,π^+p) experiments on ^{12}C are presented together with a microscopic model calculation. The comparison of experimental data with theory confirms that quasifree pion production is the dominant process for the (γ,π^+n) reaction. The comparison of (γ,π^+p) to (γ,π^+n) data shows the importance of final state interactions. The investigation of the (γ,π^+p) reaction has enabled us for the first time to experimentally study the possible mechanisms involved in the final state interaction.

The general features of Quasifree pion (QFπ) production in light nuclei at Δ-resonance energies were first observed from inclusive (γ, $\pi^{\pm}$) and semiexclusive (γ,p) measurements. These measurements also suggest that the reaction products undergo significant final state interaction (FSI). Only exclusive measurements can provide detailed information about the reaction mechanism. There were only two such experiments prior to our measurement, both (γ, π^-p) measurements. The limitation both in the kinematical region and the precision as well as the inadequacy of the theoretical treatment by DWIA calculation require more data with higher precision and better theoretical models to clarify the situation.

To date, the most sophisticated model available for considering medium effects is the Valencia Model(VM), which is based on the concepts of quantum field theory and can treat photon absorption on many-body systems microscopically [1]. According to the VM, the (γ,π^+n) reaction is dominated by QFπ production which proceeds via the Δ-resonance (γ+ p $\rightarrow \Delta^+ \rightarrow \pi^+$ + n), while a fraction of the involved particles may undergo FSI. In the Δ-resonance region, the changes brought about by the FSI to the (γ,π^+n) cross section are relatively small. On the other hand, the VM calculations indicate that the (γ,π^+p) reaction arises almost entirely through processes involving FSI such as charge exchange, rescattering processes, etc. This makes it an ideal channel to investigate the effect of FSI.

The present experiment used the Glasgow tagging facility at the Mainz CW 850 MeV Microtron MAMI. Bremsstrahlung photons were produced in a 4μm thick Ni-foil and were tagged in the range of 150-790 MeV with an overal energy resolution of ~2 MeV. Charged pions were detected on one side of the beam using a Pion-Proton (PiP) scintillator telescope. Particles emitted on the opposite side of the beam and in coincidence with pions were registered in a Time-of-Flight (ToF) scintillator detector array. Both detectors subtended solid angles of ~1 sr. PiP consists of one layer of thin scintillators (2mm thick), which act as a ΔE-detector, and four layers of thick scintillators (thicknesses from 109 to 175 mm), which were used as an E-detector. Positively charged pions were selected using a ΔE–E cut and requiring afterpulse signals from the decays $\pi^+ \to \mu \to e^+$. The ToF detectors were 300×20×5 cm^3 bars placed vertically at distances of 3–6 m from the target. Two layers of small scintillators (2mm thick) close to the target served as discriminators for neutral and charged particles. The pulseheights generated in these scintillators and the ToF detectors provided additional information for charged particle identification. The detection system was calibrated using a 9.4 g/cm^2 CH_2 target which allowed the $\gamma + p \to \pi^+ + n$ reaction to be observed. The missing energy resolution was ~10 MeV for the $(\gamma, \pi^+ N)$ reactions on C.

The results on ^{12}C reported here are part of the first exclusive $(\gamma, \pi^+$N) experiments on nuclei carried out in 1993. The triple differential cross-sections for both ^{12}C$(\gamma, \pi^+$n) and ^{12}C$(\gamma, \pi^+$p) channels are shown in Fig. 1, together with the results calculated with VM. Both the π-energy and n-angular distributions for $(\gamma, \pi^+$n) events are well described by VM. In order to investigate possible reaction mechanisms involved in the $(\gamma, \pi^+$N) reaction, individual processes in the model have been studied by tracing the outgoing particles to determine whether or not they experienced FSI. The analysis shows that most π^+n coincident particles are produced from an initial QFπ process and escape the nuclear environment without FSI. These events are ploted as long-dashed lines for $(\gamma, \pi^+$n) channel in Fig. 1, which is comparable to the calculated total strength (solid line). Results for QFπ events followed by a rescattering or charge exchange reaction, shown as dotted, short-dashed and dash-dotted lines, have negligible effects. In the $(\gamma, \pi^+$p) case, however, discrepancies between experiment and theory show up in both π-energy and p-angular distributions. VM underestimates the magnitude in both distributions, and the calculated shape of the p-angular distribution is completely different from the data. Results for several possible processes included in VM are shown in Fig. 1. The sum of all these processes, presented as a solid line for the $(\gamma, \pi^+$p) channel, overpredicts the cross-section at backward angle but underestimates the forward angle distribution. From the calculation, QFπ production followed by the (n,p) charge exchange reaction (short-dashed line) reproduces the proton backward angle distribution well. QFπ–production followed by pion charge exchange (long-dashed line) and the initial process

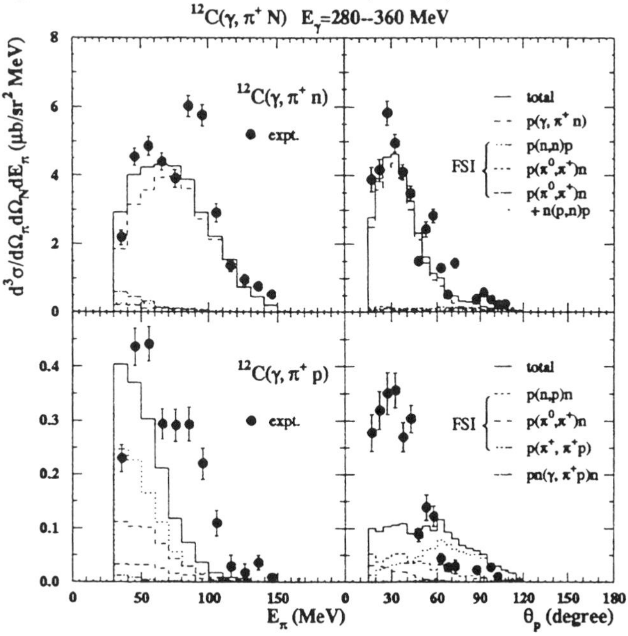

Figure 1: Differential cross sections of ^{12}C$(\gamma,\pi^+$n) and ^{12}C$(\gamma,\pi^+$p). The data are compared to the VM calculations. See text for detail.

in which a photon is absorbed on two nucleons giving rise to a π^+ particle (dash-dotted line) contribute at forward angles but have only minor importance. Obviously, the process of QFπ production followed by π^+ rescattering on the proton (dotted line) is responsible for the extra strength in the calculated results at the backward proton angles. In this case, both detected particles arise from FSI.

The double differential cross-sections versus photon energy are shown in Fig. 2. Since both (γ, π^+n) and (γ, π^+p) reactions were studied simultaneously, the ratio has small systematic uncertainties and gives a measure of the strength of FSI. The experimental results are compared to VM calculations as dashed lines in Fig. 2. The model accounts for the measured differential cross-sections well up to ~ 320 MeV, then progressively underestimates the strengths. One may consider this descrepancy at high photon energy in both channels to be due to the effect of FSI, but the good agreement between the data and the theory for the ratio suggests that the strength of FSI has been treated properly in VM. This agreement of the ratio also indicates that the missing strength of (γ, π^+p) production at high photon energy is caused by the smaller calculated QFπ-production cross-section in the (γ, π^+n) channel. Also the shift of the calculated Δ-resonance peak position in (γ, π^+n) channel compared to the data suggests that the medium effects might not be accounted for properly in VM. In the Δ-resonance region both the Δ-resonance position and the amplitude of the cross section are connected with the medium modification of Δ self-energies, which in turn modifies the Δ-mass (or resonance position) and Δ-width in the nucleus. Although the use of a smaller effective density in the theory would result in both a shift of Δ-resonance position towards higher photon energy and an increase in the magnitude of the cross-section, it is not clear at this stage to what extent is this effect responsible for the observed descrepancies.

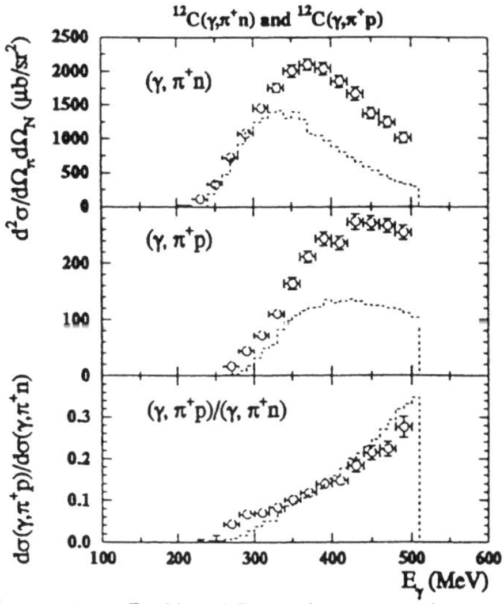

Figure 2: Double differential cross-sections of ${}^{12}C(\gamma, \pi^+ n)$ and ${}^{12}C(\gamma, \pi^+ p)$ and their ratio.

To conclude, the results of first exclusive (γ,π^+n) experiments on nuclei have confirmed that QFπ production is the dominant process at Δ-resonance energy. The study of the (γ,π^+p) channel has enabled the effects of FSI to be investigated experimentally for the first time. The comparision of VM calculations with data shows that the theory has provided a satisfacory description of the reaction mechanism, especially at relative lower photon energies. Our results reveal however that refinements of the theory are needed, in particular the effects related to the Δ medium modification and the details of FSI.

1. R. Carrasco and E. Oset, *Nucl. Phys.* **A536** (1992) 445.

A STUDY OF SHORT-RANGE CORRELATIONS IN NUCLEI USING THE ^{16}O(e, e'pp) REACTION

C.J.G. ONDERWATER, W.H.A. HESSELINK, C. KORMANYOS,
A.R. PELLEGRINO, R. STARINK, M. STEENBAKKERS
Vrije Universiteit Amsterdam, de Boelelaan 1081, 1081 HV Amsterdam, The Netherlands

E.C. ASCHENAUER, D.J. BOERSMA, D.L. GROEP, E. JANS, W-J. KASDORP,
L. LAPIKÁS, J.J. VAN LEEUWE, G. VAN DER STEENHOVEN,
J.J.M. STEIJGER, M.A. VAN UDEN
NIKHEF, P.O. Box 41882, 1009 DB Amsterdam, The Netherlands

E. CISBANI, S. FRULLANI, F. GARIBALDI, M. IODICE, R. PERRINO,
G.M. URCIOLI
Istituto Superiore di Sanitá, Laboratorio di Fisica, INFN, Viale Regina Elena 299, Rome, Italy

Th.S. BAUER, A. MISIEJUK, H.W. WILLERING
Universiteit Utrecht, P.O. Box 80.000, 3508 TA Utrecht, The Netherlands

N. KALANTAR-NAYESTANAKI
KVI, Rijksuniversiteit Groningen, Zernikelaan 25, 9747 AA Groningen, The Netherlands

R. DE LEO
INFN, Via per Arnesano, 73100 Lecce, Italy

L.B. WEINSTEIN
Dept. of Physics, Old Dominion University, Norfolk, VA 23529, USA

The $^{16}O(e,e'pp)$ reaction has been studied in a wide kinematic range. The experimental setup and some first preliminary results are discussed. Evidence for direct knock-out of two protons from the 1p shell in a relative $^1S_0$ state is obtained from the excitation energy spectrum and recoil momentum distributions.

Electron induced two-proton knock-out reactions are promising for studying short-range correlations in nuclei, as was recently shown in a pioneering study of the $^{12}C(e,e'pp)$ reaction [1]. With the high duty factor electron beam extracted from AmPS and two new large solid angle scintillator arrays HADRON-3 and HADRON-4, the triple coincidence $^{16}O(e,e'pp)$ reaction was studied using the waterfall target developed by the Rome group [2].

With a transferred four-momentum of $(\omega,q)=(210$ MeV, 300 MeV/c), data were taken at four kinematic settings covering emission angles of the forward proton between -2° and 40° with respect to $\vec{q}$, and an excitation energy of the residual ^{14}C nucleus up to 70 MeV. Due to the large angular acceptance of the two proton detectors, a substantial part of the angular correlation between the two ejected protons was measured.

The excitation of the 0^+ and 2^+ states in the spectrum shown in fig. 1a is in qualitative agreement with the knock-out of a pp-pair from the 1p shell, of which the relative motion can be described by a $^1S_0$ state [3]. Such a proton pair will have a centre-of-mass angular momentum of $L=0$ or $L=2$.

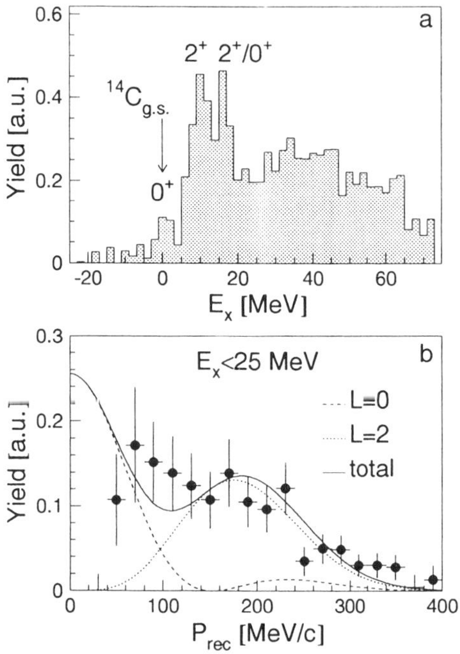

Figure 1: ^{14}C excitation energy spectrum (a) and recoil-momentum distribution (b)

In the PWIA the momentum density of the pair in the initial state is equal to the recoil momentum density. In fig. 1b the preliminary recoil momentum density is shown, together with the prediction of the harmonic oscillator model. The amplitudes of the curves are the result of a fit to the data; the oscillator parameter was derived from a fit to $^{16}O(e,e'p)$ data [4]. The agreement between the data and the calculation confirms that the reaction is dominated by direct knock-out of a pp-pair in a relative $^1S_0$ state. The angular dependence of the cross section will provide further information on the relative motion of the protons, and thus on short-range correlations.

References

1. L.J.H.M. Kester *et al.*, *Phys. Rev. Lett.* **74**, 1712 (1995)
2. F. Garibaldi *et al.* , *Nucl. Instrum. Methods* A **314**, 1 (1992)
3. W.J.W. Geurts *et al.*, accepted for publication in *Phys. Rev.* C
4. M. Leuschner *et al.*, *Phys. Rev.* C **49**, 955 (1994)

2N AND 3N MECHANISMS IN THE CALCULATION OF INCLUSIVE (e,e') PROCESSES IN NUCLEI

A. Gil*, J. Nieves** and E. Oset*

** Departamento de Física Teórica and IFIC Centro Mixto Universidad de Valencia-CSIC Burjasot (Valencia), Spain*

** Departamento de Física Moderna, Universidad de Granada, Granada, Spain*

We present the model used in the calculation of the inclusive (e,e') process in nuclei. We also show an example of our results. Starting for a free (γ^*, π) model and the ph, Δh effective interactions used in the pion-nuclear problem, we develop a systematic many-body expansion in the number of ph excitations in the nucleus. Using this model, we evaluate the cross sections on nuclei in the quasielastic, dip and delta region:

The unpolarized cross section corresponding to the $A(e,e')X$ reaction mediated by one photon exchange is given by

$$\frac{d^2\sigma}{d\Omega_{out}dE_{out}} = \left.\frac{d\sigma}{d\Omega}\right|_{Mott}\left\{\left[\frac{q^2}{|\vec{q}|^2}\right]^2 R_L(\omega,|\vec{q}|)+\left[\frac{q^2}{2|\vec{q}|^2}+\tan^2(\theta/2)\right]R_T(\omega,|\vec{q}|)\right\} \quad (1)$$

in terms of the longitudinal and transverse response functions of the nucleus and where w and $|\vec{q}|$ are the photon energy and momentum.

In order to obtain the response functions, we evaluate them for a Fermi sea of nucleons [1, 2] but we use the local density approximation to determine it in finite nuclei. In considering also that the photon is moving in the z direction, we find

$$R_L = -\frac{1}{\pi e^2}\int d^3r Im\Pi^{00} \quad (2)$$

$$R_T = -\frac{1}{\pi e^2}\int d^3r Im\Pi^{zz} \quad (3)$$

where $\Pi^{\mu\nu}$ is the polarization tensor of the Fermi sea due to ph excitation induced by an external photon.

Since we want to consider an energy transfer range up to 500 MeV, pions can be produced. Therefore we need an elementary model for electropionproduction: using the basic interactions between photons, pions, nucleons and isobars, we get an standard model (the same of Ref.[3] and similar to the one of Ref.[4] with virtual photons in our case) providing a description of the γ^* N$\rightarrow\pi$ N reaction. The Feynman diagrams included in the amplitude are the corresponding to: nucleon pole, crossed nucleon pole, pion pole, delta pole, delta crossed and the Kroll-Ruderman term (which is necessary for gauge invariance).

Once we have the elementary model, we introduce the photon absorption channels in the nucleus:

In order to incorporate the (γ^*,π) channel in our approach, we must take all the diagrams given by the model for the (γ^*,π) T-matrix and fold them to construct a selfenergy diagram with the corresponding (π,γ^*) diagrams. A convenient way to obtain the imaginary part of the self-energy diagrams is the use of the Cutkosky rules. This method has the virtue of showing which of the intermediate particles are put *on shell*, in other words, which is the reaction channel related with the cross section (or the response functions). Our guiding point is that $Im\Pi^{\mu\nu}$ comes from placing the intermediate particles *on shell*.

If in the pionproduction selfenergy diagram we allow the pion to excite a ph component, we introduce a 2ph excitation channel or equivalently a *two-nucleon photon absorption channel*, when the two ph excitations are placed on shell.

Since the delta isobar plays a very important role in this kind of reactions, an important issue in this study is the Δ-selfenergy, which incorporates the medium renormalizations. We take advantage of the study of this piece in nuclear matter [5] and we use the results obtained there for our particular kinematics. *Via* the delta piece we are including contributions to the (γ^*,π),(γ^*,NN) and (γ^*,NNN) channels.

We include also polarization effects (as it is done in Ref.[3]) and the most important contribution (in our energy range) related to the $\gamma^*N\to\pi\pi N$ reaction, which accounts for different multinucleon absorption channels.

We show in the figure an example of our results: the inclusive cross section (full line) and the delta contribution (dotted line) in the dip and delta region. The experimental data are taken from Ref.[6].

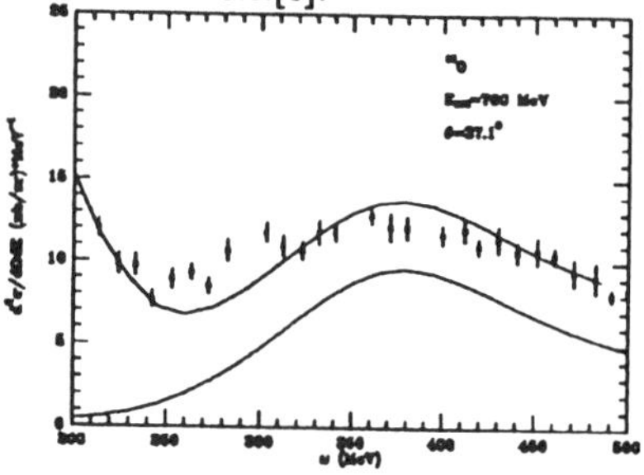

References

[1] S. Nozawa and T.-S. H. Lee, Nucl. Phys. **A513** (1990) 511.

[2] W. Alberico et al. Phys. Rev. **C381** (1988) 1801.

[3] R. C. Carrasco and E. Oset, Nucl. Phys. **A536** (1992) 445.

[4] I. Blomqvist and J.M. Laget, Nucl. Phys. **A280** (1977) 405.

[5] E. Oset and L. L. Salcedo, Nucl. Phys. **A468** (1987) 631.

[6] J.S. O'Connell et al., Phys. Rev. Lett. **53** (1984)

FINAL STATE CHARGE EXCHANGE INTERACTIONS IN THE $^{12}C(e,e'p)$ REACTION

S. KREWALD, S. JESCHONNEK
Institut für Kernphysik, Forschungszentrum, D-52425 Jülich, Germany

A. SZCZUREK
Institute of Nuclear Physics, ul.Razikowskiego 152 PL-31-342 Krakow, Poland

Final state interactions due to charge exchange reactions are important to get a good description of the symmetry properties of the spectral functions derived from a recent $^{12}C(e,e'p)$ experiment performed at Mainz. Implications for quasielastic neutrino scattering and quasielastic neutron knock-out are discussed.

Recently, new data for the reaction $^{12}C(e,e'p)$ have been measured at the electron scattering facility MAMI in Mainz with an energy of the incident electrons of 855.1 MeV [1]. Previously, the same reaction has been measured at NIKHEF in Amsterdam [2] and at Saclay [3]. The data do not agree. Here we want to investigate whether the new data corroborate the importance of final state interactions in the (e,e'p) reaction.

We analyse the data within a model which explicitly takes into account the coupling between the $^{12}C(e,e'p)$ and $^{12}C(e,e'n)$ decay channels and employs a residual interaction derived from a one-boson exchange potential [4]. The results of the calculations are compared with the new Mainz data in Fig.1. Each data point was taken under different kinematical conditions, so the calculations were done under precisely the same conditions, but points characterized by the same average scattering angle were joined by a smooth curve. The spectroscopic factor of Cohen and Kurath produced a good description of the NIKHEF data. An additional factor of 0.76 is necessary to describe the Mainz data. The problem of absolute normalization of the data should be clarified by further experiments. Here we concentrate on the symmetry properties of the data. A standard Distorted Wave Impulse Approximation(dashed line) shows a relatively large asymmetry of the spectral functions for positive and negative missing momenta. The final state interactions reduce the asymmetry, however. In the vicinity of missing momenta of 100 MeV/c the momentum transfer of 280MeV/c is still so small that the coupling of the photon to the neutron via the charge operator is negligible. The charge exchange reaction (e,e'p)(p,n) feeds the (e,e'n) channel at the expense of the (e,e'p) channel. For

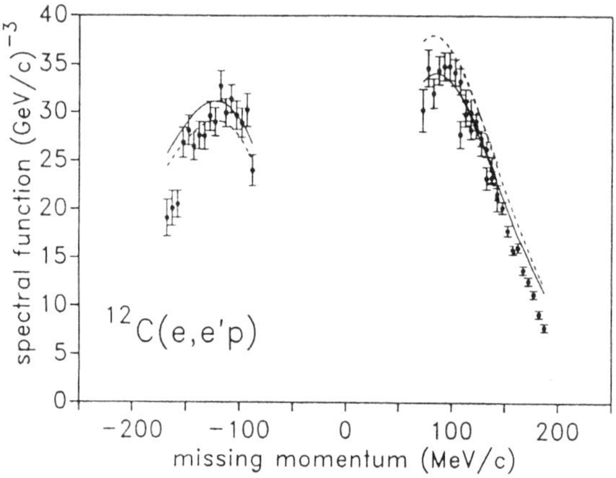

Figure 1: The spectral function as a function of the missing momentum p_m. DWIA-dashed line; Final state interactions included-solid line.

larger momentum transfers, the couling is mainly via the magnetic moments and therefore the effect of the final state interactions is less important.

A good knowledge of charge-exchange final state interactions in the quasielastic region is helpful for the interpretation of an ongoing neutrino scattering experiment at Los Alamos (see Ref. [5] for details) aiming at the determination of the strangeness content of the nucleon by measuring the ratio of neutrino induced quasielastic knock-out of protons and neutrons.

Acknowledgments

We are grateful to J.Friedrich, E.Offermann, J.Speth, and G.van der Steenhoven for stimulating discussions.

References

1. K.I. Blomqvist et al., *Z. Phys.* A **351**, 353 (1995).
2. G. van der Steenhoven et al., *Nucl. Phys.* A **480**, 547 (1988).
3. M. Bernheim et al., *Nucl. Phys.* A **375**, 381 (1982).
4. S. Jeschonnek et al., *Nucl. Phys.* A **570**, 599 (1994).
5. M. Musolf, *Phys. Rep.* **239**, 1 (1994).

Session 2
Mesons and Nucleon Interactions with Nucleons and Nuclei

STUDY OF π^+ ABSORPTION ON ^{12}C WITH THE CHAOS DETECTOR

R. TACIK, E.L. MATHIE
University of Regina, Regina, Saskatchewan, Canada

P. AMAUDRUZ, J.T. BRACK, L. FELAWKA, R. MEIER, D. OTTEWELL, G.R. SMITH
TRIUMF, Vancouver, British Columbia, Canada

F. BONUTTI, P. CAMERINI, N. GRION, R. RUI
University of Trieste and INFN Trieste, Trieste, Italy

T. AMBARDAR, G. HOFMAN, M. KERMANI, K. RAYWOOD, P. TAGLIENTE
University of British Columbia, Vancouver, B.C., Canada

M. SEVIOR
University of Melbourne, Melbourne, Australia

We have simultaneously measured ^{12}C(π^+,pp), ^{12}C(π^+,ppp), and all other reactions with two or more charged particles in the final state following the interaction of π^+'s with ^{12}C, using the CHAOS detector [1] at TRIUMF, at incident pion energies of 280, 240, and 200 MeV. CHAOS is a magnetic spectrometer subtending 360° in the horizontal plane and $\pm7°$ out of this plane, leading to a solid angle coverage of $\sim$ 10 % of 4π sr. A cylindrical plastic scintillator, 6 mm in diameter, was used as an active ^{12}C target. The energy threshold for detected protons was $\sim$ 200 MeV/c, and the momentum resolution was $\Delta p/p \sim 1.5$ %.

In the case of ^{12}C(π^+,pp), two quantities of particular interest are: the sum of kinetic energies of the two detected protons (TSUM = $T_{p_1} + T_{p_2}$), and the missing momentum ($p_m = |\vec{p}_\pi - \vec{p}_{p1} - \vec{p}_{p2}|$). For high TSUM, p_m is consistent with the vector sum of the Fermi momentum of two uncorrelated p-shell nucleons, indicating that only the two detected protons played an active role in the reaction. For lower TSUM, there is more missing momentum, indicating that at least one other undetected particle must have participated in the process. The missing momentum in this region can be compared with p_m calculated from our own data for the ^{12}C(π^+,ppp) reaction, analyzed using only two of the three detected protons. In addition, at the lowest TSUM, we again see events with little missing momentum. We can attribute these to processes in which the pion did not get absorbed, and compare the p_m distributions to those calculated from our data for the ^{12}C(π^+,π^+pp) and ^{12}C(π^+,π^-pp)

reactions, analyzed using only the two detected protons.

If we consider only those events with the highest TSUM (i.e the events which are consistent with two-nucleon absorption), we find angular distributions which are very similar in shape to those for the elementary $\pi d \rightarrow pp$ reaction. By integrating $d\sigma/d\Omega$, and correcting for the number of two-nucleon events with lower TSUM, we can estimate the total two-nucleon absorption cross sections. Our preliminary results for σ obtained in this way are 5.1 mb, 7.9 mb, and 9.0 mb, at T_π = 280, 240, and 200 MeV respectively.

Similarly, in the case of ^{12}C(π^+,ppp), we are able to use missing momentum distributions to distinguish between events in which only the three detected protons played an active role, and those in which additional undetected particles must have participated. A comparison of the measured proton momentum distributions over a range of angles with simple model calculations indicates that a large fraction of the events of the former type can be attributed to a two-step process involving πp quasi-elastic scattering followed by two-nucleon absorption. The momentum, missing momentum, and angular distributions of the latter type of event appear to be reasonably well described by model calculations which include only phase-space for four-nucleon absorption. Addition of distributions obtained from a phase-space model for three-nucleon absorption does not improve the description of the data. Although this does not necessarily exclude a contribution from a direct three-nucleon absorption process, it does appear that any such contribution would have to be small.

In order to obtain the total cross sections for the multi-nucleon absorption processes, we have to estimate what fraction of events have fallen within the CHAOS acceptance, and these estimates will be model-dependent. Using the two models mentioned in the previous paragraph, a preliminary analysis gives σ = 2.8 mb, 3.4 mb, and 3.6 mb for the two-step process at T_π = 280, 240, and 200 MeV. The preliminary cross sections for four-nucleon absorption at the same three energies are 34 mb, 29 mb, and 32 mb. Using the same models, and our normalized (π^+,ppp) data, we can estimate how many of our observed (π^+,pp) events, which we attributed to multi-nucleon absorption on the basis of their large missing momentum can be accounted for. The answer is: all of them. This is surprising, because it implies the cross sections for processes such as (π^+,ppn) and (π^+,$ppnn$) are much smaller than the ones we have mentioned above. Of course, we cannot make any statement about the importance of the (π^+,pnn) and (π^+,$pnnn$) reactions.

References

1. G.R. Smith *et al*, *Nucl. Instrum. Methods* A **362**, 349 (1995).

NEW INSIGHTS INTO PION ABSORPTION ON NUCLEI

A. LEHMANN, G. BACKENSTOSS, J. KÖHLER, M. KROEDEL, H.J. WEYER, M. WILDI
University of Basel, CH-4056 Basel, Switzerland

R. TREZECIAK, H. ULLRICH
University of Karlsruhe, D-7500 Karlsruhe, Germany

P.A.M. GRAM
LAMPF, Los Alamos, NM 87545, USA

H. BREUER
University of Maryland, College Park, MD 20742, USA

N.K. GREGORY, A.O. MATEOS, R.P. REDWINE, D. ROWNTREE, N. ŠIMIČEVIĆ, K.E. WILSON
Massachusetts Institute of Technology, Cambridge, MA 02139, USA

G. KYLE, M. WANG, M.H. WANG
New Mexico State University, Las Cruces, NM 88003, USA

T. DOOLING, A. KLEIN
Old Dominion University, Norfolk, VA 23529, USA

H. DÖBBELING, A. HOFFART, C.H.Q. INGRAM, K. KOCH, B. KOTLIŃSKI, K. MICHAELIAN, U. SENNHAUSER
Paul Scherrer Institute, CH-5232 Villigen PSI, Switzerland

D. ANDROIĆ, D. BOSNAR, M. FURIĆ, T. PETKOVIĆ, M. PLANINIĆ
University of Zagreb, HR-41001 Zagreb, Croatia

The absorption of π^+ on nuclei across the Δ-resonance region is investigated. A better measurement of the total absorption cross section is discussed as well as its breakup into quasi-free two-nucleon absorption and multi-nucleon yields. The differential distributions suggest that two-step processes contribute to the multi-nucleon absorption mode.

Although very important in the energy region of the Δ-resonance the absorption of a pion by nucleons is one of the least understood reactions of the pion-nuclear interaction. We know that the basic mode involves an isoscalar nucleon pair in the absorption process (2NA), but a significant fraction of the total absorption cross section leads to at least three energetic nucleons in the final state (3NA). Since first investigations on the light system ^3He revealed no explicit evidence for multi-step processes (ISI+2NA, 2NA+FSI) being the origin of this mode, it was concluded that new reaction dynamics might exist in pion absorption [1].

The Large Acceptance Detector System (LADS) was built at PSI to study especially the multi-nucleon absorption mode in detail. Its large solid angle coverage ($\approx$98% 4π) and low particle detection threshold ($T_{thr} \approx$ 20 MeV for protons) makes a large fraction of the phase space accessible. More details about the detector may be found elsewhere [2].

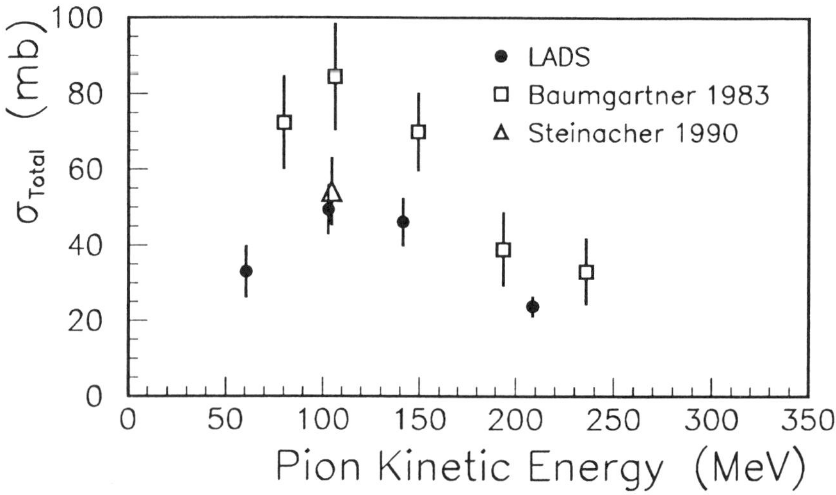

Figure 1: ^4He Total Absorption Cross Section

Total absorption cross sections were measured more accurately than previously for the nuclei ^3He, ^4He, N and Ar for several energies across the Δ-resonance. The excitation function of absorption on ^3He was found to peak at a lower pion energy [3] than earlier measurements suggested, bringing these cross sections more in line with the course of the excitation functions of ^2H and ^4He. The measured total absorption cross section on ^4He (fig.1) is consistent with that measured by Steinacher *et al.* [4] but appears to be systematically lower than the measurements reported by Baumgartner *et al.* [5] across the resonance region. These data remove the previously reported large rise in the absorption cross section from ^3He to ^4He. Also, our total absorption cross sections in the heavier nuclei are roughly 25% lower than previously assumed. This is mainly due to a better exclusion of background from single charge exchange reactions.

For the nuclei ^{3}He and ^{4}He the breakup into 2NA and 3NA was determined. We find that the multi-nucleon yield becomes stronger with increasing incident pion energy and mass number (from about 20% at 70 MeV to roughly 35% at 239 MeV for ^{3}He, and from about 30% at 70 MeV to roughly 55% at 239 MeV for ^{4}He). In contrast to $^3\text{He}(\pi^+, ppp)$ there are several competing multi-nucleon channels in ^{4}He. The 3NA reactions $^4\text{He}(\pi^+, ppn)p$ and $^4\text{He}(\pi^+, ppp)n$ are different in isospin. Therefore their cross section ratio provides information about the importance of possible multi-step processes in 3NA. We find the ratio $\sigma_{ppn}/\sigma_{ppp}$ decreasing from about 3.0 to 1.5 at 70 MeV and 239 MeV, respectively. This finding suggests that ISI+2NA alone cannot be the main contributor to 3NA. The contribution of multi-nucleon strength leading to four fast nucleons (4NA) we find to be less than 10% of the total absorption cross section on ^{4}He.

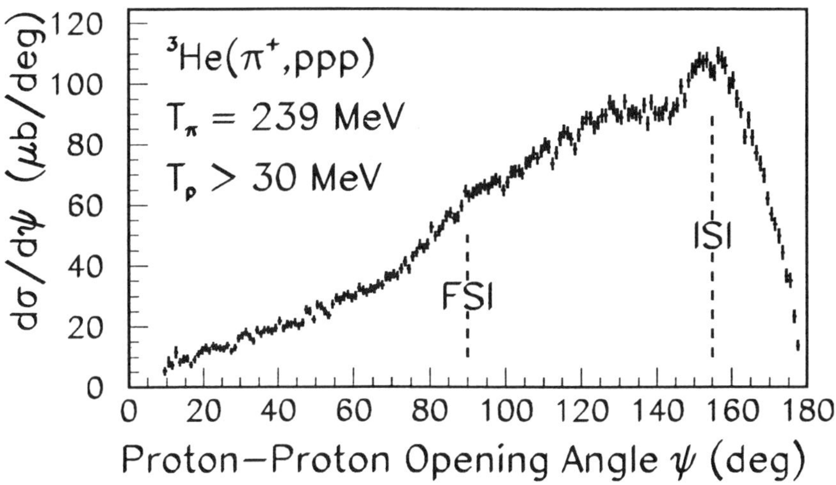

Figure 2: Proton-proton opening angle in the laboratory frame. The positions where ISI and FSI could show up as peak-like structures are indicated.

The kinematic completeness of the LADS detector allows a comprehensive study of the differential distributions of the multi-nucleon mode. From comparisons to simple incoherent simulations of different reaction mechanisms we recently could report the first observation of a signature of a two-step ISI+2NA process in pion absorption [6], which accounts for up to one third of the 3NA yield in ^{3}He. In this process the pion first scatters quasi-elastically from one nucleon before being absorbed on another. One signature (amongst others) of this mechanism is two nucleons emitted almost back-to-back in the laboratory frame. Such a signal is clearly visible in the proton-proton opening angle distribution of fig.2 around 155^0. On the other hand, if a two-step 2NA+FSI process would contribute to 3NA one should find a peak-like structure around 90^0, because one of the two absorption protons is rescattered by another proton. The data in fig.2 show only weak evidence for this mechanism.

From fits to the data we conclude that the importance of initial state interaction (ISI+2NA) in π absorption on ^{3}He increases with the incident pion energy, while final state interactions (2NA+FSI) become weaker. Similar results were obtained for the *ppp* mode in ^{4}He and even in heavier nuclei. Although this seems to be an important step forward in understanding the mechanisms of multi-nucleon pion absorption, a substantial part of the strength remains unexplained.

References

1. G. Backenstoss *et al.*, *Phys. Rev. Lett.* **55**, 2782 (1985).
2. T. Alteholz *et al.*, *Nucl. Instrum. Methods* A **373**, 374 (1996).
3. T. Alteholz *et al.*, *Phys. Rev. Lett.* **73**, 1336 (1994).
4. M. Steinacher *et al.*, *Nucl. Phys.* A **517**, 413 (1990).
5. M. Baumgartner *et al.*, *Nucl. Phys.* A **399**, 451 (1983).
6. G. Backenstoss *et al.*, *Phys. Lett.* B **379**, 60 (1996).

TWO STEP PROCESSES IN PION SINGLE CHARGE EXCHANGE ON ^{2}H AT $T_\pi = 239$ MeV

N.K. Gregory, D. Androić, G. Backenstoss, D. Bosnar, H. Breuer, H. Döbbeling, T. Dooling, M. Furić, P.A.M. Gram, A. Hoffart, C.H.Q. Ingram, A. Klein, K. Koch, J. Köhler, B. Kotliński, M. Kroedel, G. Kyle, A. Lehmann, A.O. Mateos, K. Michaelian, T. Petković, M. Planinic, R.P. Redwine, D. Rowntree, U. Sennhauser, N. Šimičević, R. Trezeciak, H. Ullrich, M. Wang, M.H. Wang, H.J. Weyer, M. Wildi, K.E. Wilson
U. Basel, U. Karlsruhe, LAMPF, U. Maryland, MIT, NMSU, ODU, PSI, U. Zagreb

Results from a measurement of the reaction $^2\mathrm{H}(\pi^+, pp)\pi^0$ at $T_{\pi^+} = 239$ MeV are presented. Because two protons are detected, the measurement is especially sensitive to events which involve both nucleons in a substantial way. Comparisons with simple models are shown.

We have measured the pion single charge exchange reaction $^2\mathrm{H}(\pi^+, pp)\pi^0$ using the Large Acceptance Detector System (LADS) at PSI. The nearly 4π steradian solid angle coverage and low energy detection threshold for protons allowed us to make a kinematically complete measurement without direct detection of the π^0. The charge exchange data were isolated by placing a requirement on the invariant mass of the outgoing proton-proton system. Contamination from other event types such as pion absorption and pion-proton quasi-elastic scattering was minimized using other cuts on the data. Differential quantities of the data were compared with simple Monte Carlo simulations of possible reaction mechanisms. A sample of the comparisons for the kinetic energy and polar angle with respect to the beam axis of the reconstructed π^0 is shown in Figure 1. The comparisons indicate that, as expected for the kinematic acceptance of these data, events where the pion interacted with both nucleons are greatly enhanced with respect to the events where the pion only interacted with one nucleon. We have determined that less than 13% of the data seen by LADS could have come from single scattering events.

In conclusion, we have performed a complete study of a multi-step process in a simple nucleus. The large solid angle of LADS and the measurement method provided access to multiple scattering components of pion interactions with the deuteron. The large solid angle also allowed examination of a number of differential variables. This comprehensive data set should provide stringent constraints on calculations of multi-step pion interactions in light nuclei.

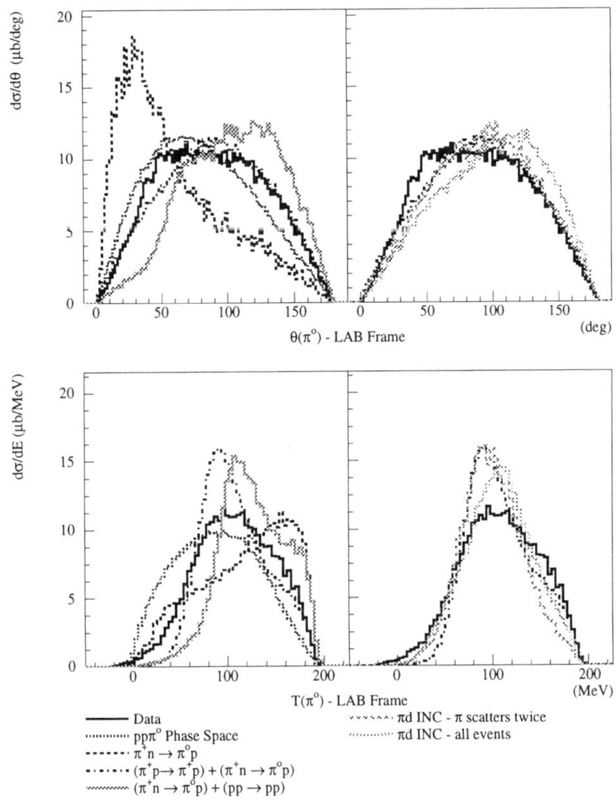

Figure 1: Kinematic quantities of the reconstructed π^0. The upper plot is the polar angle with respect to the beam axis, the lower plot is the kinetic energy of the π^0 in the lab frame. The simulations are normalized to be equal in area to the data.

Inclusive Pion Double Charge Exchange on Light Nuclei above the Delta Resonance Region

B.M. Abramov, S.A. Bulychjov, I.A. Dukhovskoi, A.I. Khanov,
Y.S. Krestnikov, A.P. Krutenkova, V.V. Kulikov, M.A. Matsuk,
I.A. Radkevich, A.N. Starodumov, A.I. Sutormin, E.N. Turdakina -
Institute of Theoretical and Experimental Physics, 117259 Moscow, Russia
M.J.Vicente Vacas - Departamento de Física Teórica and IFIC, Centro mixto
Universidad de Valencia - CSIC, 46100 Burjassot, Spain

The inclusive double charge exchange (DCX) reaction $A(\pi^-,\pi^+)X$ on ^{6}Li and ^{16}O was for the first time measured at incident kinetic energies T_o = 0.6, 0.75 and 1.1 GeV ($<\theta> \approx 5^o$) for outgoing particle momentum region where additional pion production is kinematically forbidden. The experiment was performed at secondary π^- beam of the ITEP proton synchrotron using the 3m magnet spectrometer. The measured double differential cross section shows rather slow decrease with energy. This result is in contradiction with the fast fall of small angle DCX cross section calculated in the standard mechanism of two sequential single charge exchanges (SSCX) of pion on two like nucleons. Therefore, other approaches to DCX (for example, short range NN correlations, meson exchange currents, inelastic Glauber corrections *etc.*) are needed to explain the observed energy dependence.

We studied DCX in the energy region above 0.5 GeV where in the framework of (SSCX) mechanism of DCX the strong decrease of small angle DCX cross section is expected [1]. It was emphasized in [1] that predicted low value of the DCX cross section near 1.3 GeV offers a unique possibility to search for nonconventional mechanisms of DCX reactions. We measured energy spectra of outgoing pions in reaction $A(\pi^-,\pi^+)X$ on ^{6}Li, ^{7}Li, ^{12}C and ^{16}O at $\theta = 0$ - 10° and incident pion kinetic energy $T_o = 600$, 750 and 1120 MeV in the region $\Delta T = T_o - T \leq m_\pi = 140 MeV$ (T is the kinetic energy of the positive pion) where the $(\pi, 2\pi)$ reaction is kinematically forbidden.(See [2] for details and first results.) The experiment was carried out with the help of a 3m magnet spectrometer with spark chambers. The targets were placed in the middle of the magnet and the first part of it was used as a beam spectrometer and the second one as an outgoing particle spectrometer. Pions were separated from protons by TOF and from positrons by Cherenkov counters. Backward pion-proton elastic scattering was used for energy resolution control and for check of absolute cross section measurement.

To study the energy dependence, the measured double differential cross section was integrated over region: $0 \leq \Delta T \leq 140$ MeV. This partially integrated cross section is shown in Fig.1 together with the data [3] for the reactions $^{16}O(\pi^-,\pi^+)X$ and $^6Li(\pi^-,\pi^+)X$.

The results of theoretical calculations for SSCX mechanism of inclusive DCX on ^{16}O in the framework of a cascade style Monte Carlo model [4] are also shown in the Fig.1. The elementary πN cross sections were used and medium effects: Fermi motion, binding energy, Pauli-blocking and absorption, were taken into account. The calculated cross section falls fastly in 0.6 - 1.1 GeV region. Within a SSCX mechanism this fall reflects the fast decrease of a πN single charge exchange amplitude in this energy region.

The measured differential cross sections decrease with incident energy rather slowly. At 0.75 and 1.1 GeV the cross sections are significantly larger than expected in the framework of the usual SSCX mechanism. Energy dependence for ^6Li is rather similar to that of for ^{16}O. Comparison of calculated and measured energy dependence gives clear evidence that contribution of other mechanisms is substantial near 1 GeV. Recently the first calculations of a contribution of meson exchange currents in this energy region have been performed in [5] for exclusive channel $^{16}O(\pi^+,\pi^-)^{16}Ne$. They really demonstrate very weak energy dependence comparable with seen in this experiment for inclusive DCX. But contribution of other effects such as collision broadening and two pion intermediate states must be also estimated for complete understanding of DCX reaction mechanism.

This work was supported in part by the grant INTAS 93-3455 and CICYT Contract No. AEN 93-1205.

1. E. Oset and D. Strottman. Phys. Rev. Lett. **70**, 146 (1993)
2. B.M. Abramov et al. Yad. Fiz. **59**, 399 (1996)
3. S.A. Wood et al. Phys. Rev. **C46**, 1903 (1992). P.A.M. Gram. In: *Pion-Nucleus Physics. Los Alamos, NM 1987* p.75. N.Y. A.I.P 1988.
4. M.J. Vicente Vacas, M.Kh. Khanhkasaev, S.G. Mashnik. Preprint FTUV/94-73 University of Valencia 1994, submitted to Nucl. Phys. A.
5. L.Alvarez-Russo and M.J. Vicente Vacas.J.Phys.G22, L45(1996)

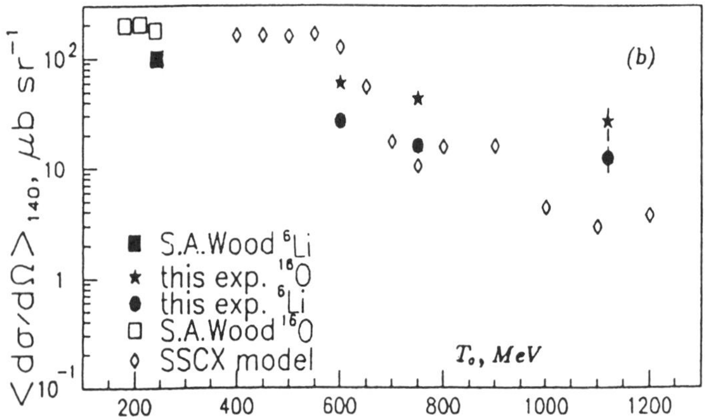

Fig.1. Energy dependence of DCX cross section on ^{16}O and ^6Li.

A SEARCH FOR THE Δ^- WAVE-FUNCTION COMPONENT IN LIGHT NUCLEI

C. L. MORRIS, J. D. ZUMBRO, R. L. BOUDRIE, and P. A. M. GRAM
Los Alamos National Laboratory, Los Alamos, NM 87545, USA

J. L. MATTHEWS, Y. TAN and V. V. ZELEVINSKY
Massachusetts Institute of Technology, Cambridge, MA 02139, USA

E. A. PASYUK
Joint Institute for Nuclear Research, Dubna, Moscow Region, 141980 Russia

B. J. KRISS
University of Colorado, Boulder, CO 80309

G. GLASS
3218C Walnut St., Los Alamos, NM 87544, USA

We have studied the $(\pi^+, \pi^\pm p)$ reactions on ^{3}H, ^{4}He, ^{6}Li, and ^{7}Li at incident energy 500 MeV in quasi-free kinematics. A signature attributable to pre-existing Δ components of the ground state wave function is observed.

Meson models of nuclear binding predict both excess pions and excited nucleons in nuclear matter.[1] Results from a new method of measuring the Δ component of the nuclear wave function using the pion double charge exchange (DCX) reaction (π^+, π^-) are reported in this paper.

If the nucleus contains only nucleons, two like-charge nucleons must participate in the DCX process. If the nuclear wave function contains a Δ^-, the (π^+, π^-) process can occur in a single step. Recent measurements on a range of nuclei have yielded evidence for this effect.[2] In the current work we have extended these studies to ^{3}H, ^{4}He, ^{6}Li, and ^{7}Li.

The experiment was performed using a 500 MeV beam at the Clinton P. Anderson Meson Physics Facility. Pions from $(\pi^+, \pi^+ p)$, NCX, and $(\pi^+, \pi^- p)$, DCX, reactions were observed at an angle of 50° in coincidence with protons at 52° using two magnetic spectrometers, in the kinematics of free πp scattering. The sum of the pion and proton energies was required to exceed 400 MeV.

An estimate of the probability for pre-existing Δ's can be obtained from the data by integrating the cross sections and forming the ratio $R = \sigma_{DCX}/\sigma_{NCX}$:

$$P_\Delta = R \left(\frac{ZN_\Delta}{AN_{\Delta^-}}\right) \left(\frac{\sigma(\pi^+ + p \to \pi^+ + p)}{\sigma(\pi^+ + \Delta^- \to \pi^- + p)}\right) \left(\frac{k_\Delta}{k_F}\right)^2 . \qquad (1)$$

Table 1: Measured cross section ratios and extracted values of P_Δ.

Target	$\sigma_{DCX}/\sigma_{NCX}$ ($\times 10^3$)	P_{Δ^-} (%)	P_Δ (%)
^{3}H	8.9(6)	1.72(12)	3.44(24)
^{4}He	0.90(18)	0.26(5)	1.04(20)
^{6}Li	0.67(13)	0.20(4)	0.73(15)
^{7}Li	5.4(5)	1.58(15)	4.22(40)

where P_Δ is the sum of the probabilities for all Δ charge states, N_{Δ^-}/N_Δ is the fraction of Δ^-'s, Z and A are the proton and nucleon numbers, and k_Δ and k_F characterize the momentum spreads of the Δ's and nucleons. We have used measured branching ratios[3] for N*(1520) $\to \pi$+N and N*(1520) $\to \pi+\Delta$(1232) to estimate the cross-section ratio, and have taken k_F = 200 MeV/c and k_Δ = 400 MeV/c, the peak of the momentum distribution predicted for the calculated pion excess.[1] The ratios of Δ charge states have been obtained by generalizing arguments given in[4] to give

$$\frac{N_{\Delta^-}}{N_\Delta} = \frac{3}{4}\left(1 + \frac{Z}{2(N-1)} + \frac{Z(Z-1)}{N(N-1)}\right)^{-1}, \tag{2}$$

where N is the neutron number of the nucleus. The measured cross section ratios and resulting Δ probabilities are given in Table 1.

The Δ^- probability in ^{3}H is found to agree roughly with theoretical expectations.[5] Moreover, the Δ^- probabilities in the light self-conjugate nuclei are approximately an order of magnitude smaller than those in the T=1/2 nuclei. This is presumably because only core-excited two-particle configurations or $\Delta\Delta$ excitations can contribute to the Δ probabilities for neutron pairs coupled to J=0 within the s-shell.

This work has been supported in part by the U. S. Department of Energy.

References

1. B. L. Friman, V. R. Pandharipande, and R. B. Wiringa, *Phys. Rev. Lett.* **51**, 763 (1983).
2. C. L. Morris, J. M. O'Donnell, and J. D. Zumbro, *Acta Phys. Pol. B* **24**, 1659 (1993).
3. M. Aguilar-Benitez *et al.*, *Phys. Rev.* **D435**, 1 (1992).
4. H. J. Lipkin and T.-S. H. Lee, *Phys. Lett. B* **183**, 22 (1987).
5. W. Strueve *et al.*, *Nucl. Phys. A* **465**, 651 (1987).

DOUBLE DELTA PRODUCTION IN PHOTON-INDUCED REACTIONS OFF THE DEUTERON AROUND 1 GEV

Y. WADA
Meiji College of Pharmacy, Tokyo 154, Japan

J. WIßKIRCHEN, W. NEUERBERG
Physikalisches Institut der Universität Bonn, Bonn, Germany

Representing the SAPHIR collaboration

The double delta production in a photon-induced reaction off the deuteron was investigated experimentally[1] and theoretically[2] around the double delta production threshold. In the experiment[1] a total cross section for double delta production of about 10 μb, in the energy region of 650 MeV, was measured and was found to increase up to 80 μb at 830 MeV. The theoretical calculation[2] is done by using a model developed for the $\pi^+$and π^-photoproduction off the proton and shows good agreement with the experimental results around the double delta threshold. But the calculation includes larger uncertainties as the photon energy increases. Up to now there are no experimental results to discuss the double delta production mechanisms at higher energies.

We measured the d(γ, $\pi^+\pi^-$p)n reaction for an energy range extending from 750 MeV to 1280 MeV. The experiment was done with tagged photons from the stretcher ring ELSA and the large solid angle magnetic spectrometer SAPHIR at Bonn University. We used double arm trigger method, measured the momenta and the time of flight (TOF) of charged particles and, thus, reconstructed the masses of the charged particles. We selected $\pi^+\pi^-$p events and could identify the events γd $\rightarrow$ $\pi^+\pi^-$p n in the channel by these reconstructed masses; furthermore, in addition we performed a kinematical fit which takes into account the energy and momentum conservation. The selected events are kinematically complete. It was determined that events with proton momenta larger than 250 MeV/c come mainly from the quasifree process, γd $\rightarrow$ n(spectator) $\pi^-\Delta^{++}$, but the neutron momentum distribution showed an excess at high momenta.

We calculated the cross section of the quasifree process by using the events with neutron momentum smaller than 250 MeV/c. As shown in Fig.1(a) the cross section is in agreement with the ABBHHM's one[3], which was measured making use of a hydrogen bubble chamber.

We selected the events with proton and neutron momenta higher than 300 MeV/c in order to enhance the two nucleon participated channels. In the

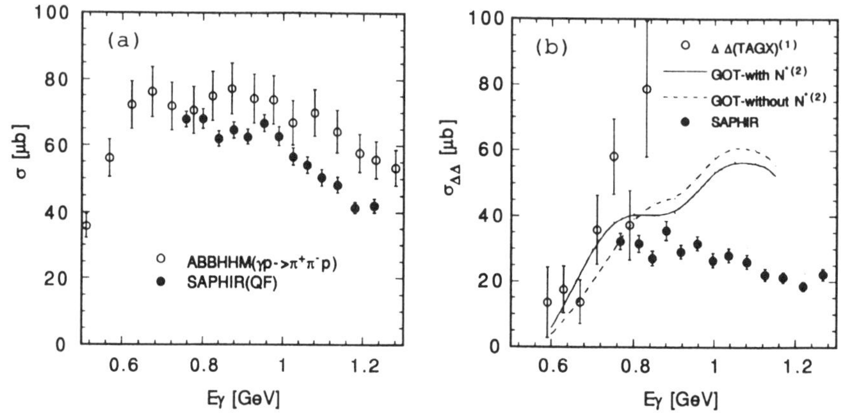

Figure 1: (a) Cross section of the quasifree γ d $\rightarrow$ $\pi^+\pi^-$p n reaction. The solid circle shows this experiment results and the open circle come from the hydrogen bubble chamber experiment of ABBHHM collabolation. (b) Results of the cross sections of the γ d $\rightarrow$ Δ^{++} Δ^- reaction with the previous experiment (open circle) and the theoretical curve (solid and dashed lines)

Dalitz plot of invariant masses for pπ^+versus nπ^-pairs we can see a double delta peak in the corresponding mass region.

The two invariant distributions agree with the simulation of γd $\rightarrow$ $\Delta^{++}\Delta^-$ reaction. The selected events seem to include small contamimations like 4-body phase space. Then, adapting the simulations to the real data and determining their relative strength we calculated the cross section of the γd $\rightarrow$ $\Delta^{++}\Delta^-$ reaction. The result is shown in Fig.1(b) with the previous experiment and the calculation results. But the results are preliminary because the fitting procedure is still not final.

The cross section around 750 MeV agrees with the previous results, but the value decreses slowly as energy increases, while the previously measured cross section increases rapidly. That is not so serious problem because the previous experiment was not statistically significant at 850 MeV. The calculation has large discrepancy in the photon energy greater than 900 MeV. In order to reproduce the data in the calculation, some more higher isober resonances or other mechanisms should be intoduced.

1. M. Asai et al., Z. Phys. A **344** (1993) 335
2. J.A. Gomez Tejedor, E. Oset and H. Toki, Phys. Lett. B **346**(1995) 240
3. ABBHHM Collaboration, Phys. Rev. **175** (1968) 1669

Medium effects from K^+ nuclear total and reaction cross sections

A. Gal (representing AGS Experiment 835*)

Abstract

We present newly determined and largely model independent values of the reaction cross section for K^+ on ^{6}Li, C, Si and Ca at p_L = 488, 531, 656 and 714 MeV/c. These and revised values of total cross sections, derived from the same transmission measurements E835 at the AGS, point out to relatively strong medium effects on the KN interaction. For C, Si and Ca, the excess cross section values with respect to the first-order $t\rho$ optical potential calculation, and relative to ^{6}Li, is found to be nearly energy independent and of magnitude 15-25%.

Total cross sections on ^{6}Li, C, Si and Ca at p_L = 488, 531, 656 and 714 MeV/c were derived from measurements of the attenuation of a K^+ beam as function of solid angle subtended by a detector [1]. In this energy range the KN interaction hardly varies with energy, so given its relative weakness one expects that a first-order $t\rho$ optical potential will reproduce reasonably well the data. However, appropriate calculations [2, 3] have produced disagreement with experiment at the level of 5-15%, leading to various speculations about the role of medium effects in modifying the KN interaction in the nucleus [4, 5, 6].

Here we report on K^+ nuclear reaction cross sections derived for the first time from the same transmission data, and we also provide slightly revised values for the total cross sections with respect to those already published [1] by accounting explicitly for the effect of plural scattering [7] which is particularly important at the lower energies and for the heavier nuclear targets.

Extrapolating the measured attenuation cross sections to zero solid angle $\Omega = 0$, the derived values of the reaction and total cross sections depend on the optical potential assumed in order to generate the necessary corrections. In reanalyzing the transmission experiment [1] we used a first-order $V_{opt} = t\rho$ optical potential

*E. Friedman, A. Gal (The Hebrew University, Jerusalem); R. Weiss, J. Aclander, J. Alster, I. Mardor, Y. Mardor, S. May-Tal Beck, E. Piasetzky, A.I. Yavin (Tel-Aviv University); S. Bart, R.E. Chrien, P.H. Pile, R. Sawafta, R.J. Sutter (Brookhaven National Laboratory); M. Barakat (University of Houston); K. Johnston (Louisiana Tech University); R.A. Krauss (Texas A & M University); H. Seyfarth (KFA, Jülich); R.L. Stearns (Vassar College)

generated by the on-shell KN t matrix. When p waves were singled out to form a gradient term in V_{opt}, the extrapolated values of σ_R and σ_T differed by less than 1% and similarly for the dependence on the choice of matter-density ρ. Details are given in ref. [8]. The derived values of reaction and total cross sections are shown in table I.

Table I

Reaction and total cross sections (in mb) for K^+ interaction with various nuclei

p_L	reaction				total			
(MeV/c)	^{6}Li	C	Si	Ca	^{6}Li	C	Si	Ca
488	65.0	120.4	265.5	349.9	76.6	162.4	366.5	494.6
	± 1.3	± 2.3	± 5.1	± 7.7	± 1.1	± 1.9	± 4.8	± 7.7
531	69.8	129.3	280.4	367.1	78.8	166.6	374.8	500.2
	± 0.8	± 1.4	± 3.4	± 4.5	± 0.7	± 1.3	± 3.3	± 4.4
656	75.6	141.8	306.1	401.1	84.3	174.9	396.1	531.9
	± 1.1	± 1.5	± 3.4	± 5.0	± 0.7	± 0.8	± 2.7	± 4.2
714	79.3	149.3	317.5	412.9	87.0	175.6	396.5	528.4
	(80.0)	(149.2)	(317.7)	(413.4)	(91.2)	(192.1)	(433.9)	(589.6)
	± 1.2	± 1.5	± 3.6	± 5.5	± 0.6	± 0.9	± 2.3	± 2.8

In order to study the model dependence of these derived σ_R and σ_T values we fitted the very recent elastic scattering data [9] for K^+ on ^{6}Li and C at $p_L = 715$ MeV/c by a density dependent optical potential which was then used in the analysis of the transmission experiments at 714 MeV/c. The results of this constrained analysis are shown in parentheses in table I. It is clear that the derived values of σ_R are remarkably model independent, the differences between the values derived by using $t\rho$ and between those derived by fitting V_{opt} to elastic scattering data being considerably smaller than the quoted errors. In contrast, the derived values of σ_T exhibit a substantial model dependence, with differences about 10-20 times larger than the quoted errors.

The total cross sections for K^+ interacting with deuterium were obtained earlier from the differences between CD_2 and C targets and between LiD and Li targets. The effect of plural scattering corrections on these differences was found to be negligibly small. Therefore the published [1] cross sections for K^+D remain unchanged.

It is useful to first note that the experimental values of σ_T/A for ^{6}Li are in agreement with the corresponding values for deuterium [1], as expected when the interaction is dominated by single scattering. This result also suggests that effects due to the nuclear medium are not important in the very light and loosely bound ^{6}Li nucleus. Values of σ_T/A for other nuclei cannot be directly compared to the deuterium values because for heavier nuclei effects due to multiple scattering are not negligible. It is instructive then to study nuclear medium effects on the KN interaction by forming super ratios composed of the ratios $\sigma(exp)/\sigma(t\rho)$ for C, Si and Ca divided by the corresponding ratios for ^{6}Li. Such σ_R super ratios are plotted in fig. 1, showing a 15-25% energy-independent enhancement of the KN interaction relative to the low density limit value of 1. Similar results hold for σ_T super ratios. A comparable degree of enhancement was recently found for the C/^{6}Li forward K^+ angular distribution at 715 MeV/c [9] and also for the quasi free K^+ scattering data on D, C, Ca and Pb at 705 MeV/c [10].

References

1. R. Weiss *et al.*, Phys. Rev. **C49**, 2569 (1994).
2. P.B. Siegel, W.B. Kaufmann and W.R. Gibbs, Phys. Rev. **C31**, 2184 (1985).
3. M.F. Jiang, D.J. Ernst and C.M. Chen, Phys. Rev. **C51**, 857 (1995).
4. G.E. Brown, C.B. Dover, P.B. Siegel and W. Weise, Phys. Rev. Lett. **60**, 2723 (1988).
5. J.C. Caillon and J. Labarsouque, Nucl. Phys. **A589**, 609 (1995); see however J.C. Caillon and J. Labarsouque, Nucl. Phys. **A595**, 189 (1995).
6. M.F. Jiang and D.S. Koltun, Phys. Rev. **C46**, 2462 (1992); C. Garcia-Recio, J. Nieves and E. Oset, Phys. Rev. **C51**, 237 (1995).
7. H.A. Bethe, Phys. Rev. **89**, 1256 (1953); W.T. Scott, Rev. Mod. Phys. **35**, 231 (1963); CERN subroutine PHYS325.
8. E. Friedman, A. Gal, R. Weiss *et al.*, submitted to Phys. Rev. **C**.
9. R. Michael *et al.*, Phys. Lett. B (in press).
10. C. Kormanyos *et al.*, Phys. Rev. Lett. **71**, 2571 (1993); Phys. Rev. **C51**, 669 (1995).

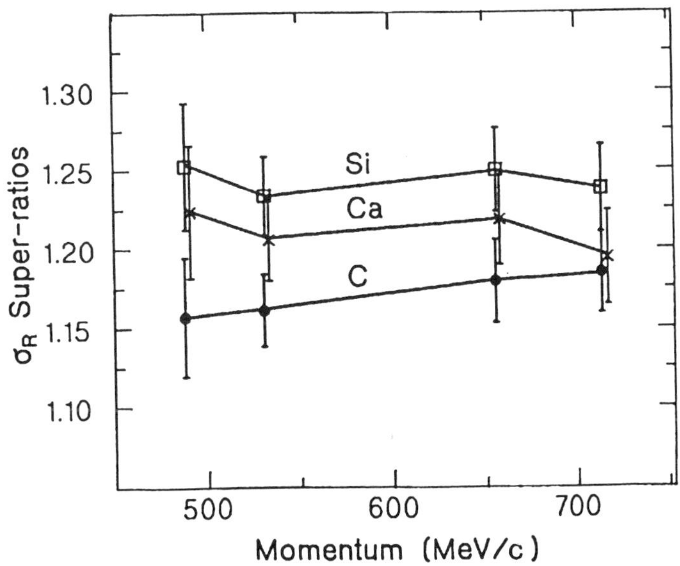

Figure 1. Super ratios: $\sigma_R(exp)/\sigma_R(t\rho)$ divided by the corresponding ratios for ^{6}Li.

η PRODUCTION IN pp AND pA INTERACTIONS AT SATURNE

E. CHIAVASSA, G. DELLACASA, N. DE MARCO,
C. DE OLIVEIRA MARTINS[a], M. GALLIO, P. GUAITA,
A. MUSSO, A. PICCOTTI, E. SCOMPARIN and E. VERCELLIN
Dipartimento di Fisica Sperimentale dell'Università di Torino and
Istituto Nazionale di Fisica Nucleare, Sezione di Torino,
via P.Giuria 1, I-10125 Torino, Italy

An extensive experimental program with the aim of studying η production in p-N and p-nucleus collisions near threshold has been recently carried out at SATURNE. The η-mesons were detected via the $\eta \rightarrow \gamma\gamma$ decay by the two-arm PINOT spectrometer.

The $pp \rightarrow pp\eta$ reaction has been studied from threshold up to 2 GeV, by means of the two-arm spectrometer PINOT. Measurements of the doubly differential cross sections $< d^2\sigma/d\Omega^\star_\eta dT^\star_\eta >$, averaged in different phase space regions, have been performed. From threshold up to T_p=1.4 GeV the whole phase space accessible to the $pp \rightarrow pp\eta$ reaction has been investigated while at higher incident energies we studied only the backward c.m. hemisphere, using the forward-backward symmetry of the reaction in the c.m. frame.

The total cross section has been obtained by integration of the measured $< d^2\sigma/d\Omega^* dT^* >$ over the phase space. The behaviour of σ_{tot} as a function of the incident energy T_p is shown in fig. 1. The data up to T_p=1.5 GeV are already published[1], while at higher T_p they are still preliminary. The slope of the excitation curve from 1.4 up to 2 GeV, where effects due to FSI are no more expected, is well reproduced by a phase space distribution, making the assumption that the η meson is produced via the excitation of the $N^*(1535)$ resonance. The data are compared with the predictions of two versions of the calculation of Laget et al.[2] with constructive and distructive $\pi\rho$ interference and with the Vetter et al.[3] predictions. Even if no calculation is in full agreement with the data at all the incident energies studied, both the predictions of Vetter et al.[3] and of Laget et al.[2] (with distructive $\pi\rho$ interference) are not too far from the experimental points. This seems to indicate that the basic assumptions on the reaction mechanisms made in the calculations are correct and in particular that the crucial role is played by the $N^*(1535)$ resonance excitation.

Concerning η production in p-nucleus collisions the A dependence of the

[a] On leave from Universidade do Estado do Rio de Janeiro, Departamento de Física Nuclear e Altas Energias - 20550, Rio de Janeiro, Brasil

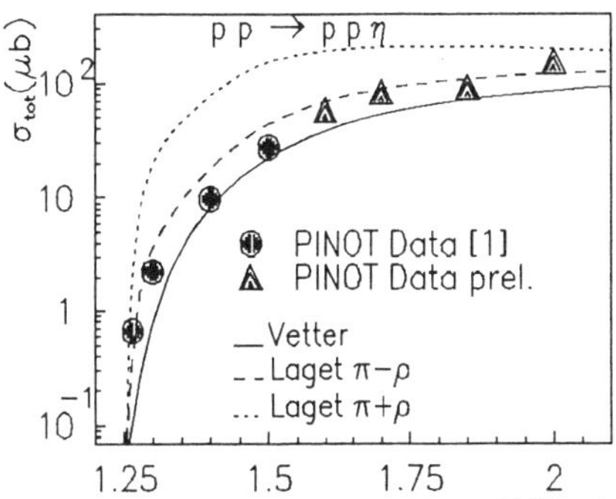

Fig. (1)

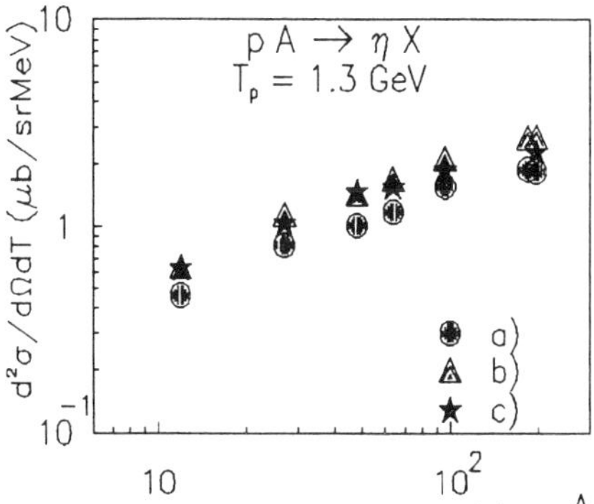

Fig. (2)

doubly differential cross section $d^2\sigma/d\Omega_\eta dT_\eta$ have been measured at T_p=1.3 GeV, just above the pN→pNη threshold. The doubly differential cross sections $d^2\sigma/d\Omega_\eta dT_\eta$ shown in fig. 2 as a function of the target nucleus mass for 3 different η kinetic energy intervals (a) $60 \le T_\eta \le 120$ MeV, b) $120 \le T_\eta \le 220$ MeV and c) $200 \le T_\eta \le 340$ MeV), clearly saturate for large values of A. In order to extract informations on the η-N absorption cross section we have compared our experimental data with the predictions of a folding model[7], which describes η production in nuclei via a one-step process (pN→pNη) and a two-step process (pN→pNπ, followed by πN→ηN). We modified the original code of the theoretical calculation in order to use for the one step process a parametrization of the recent experimental data measured at SATURNE[1,4,5] and CELSIUS[6]. The η reabsorption in nuclei is treated using a semi-classical model[8]. The assumption of a $\sigma_{\eta N}^{abs}$ of the order of 30 mb is consistent with our experimental data for T_η ranging from 60 up to 340 MeV.

References

1. E. Chiavassa *et al.*, *Phys. Lett.* B **322**, 270 (1994).
2. J.M. Laget *et al.*, *Phys. Lett.* B **257**, 254 (1991) and priv. comm..
3. T. Vetter *et al.*, *Phys. Lett.* B **263**, 153 (1991).
4. E. Chiavassa *et al.*, *Phys. Lett.* B **337**, 192 (1994).
5. A.M. Bergolt *et al.*, *Phys. Rev.* D **48**, R2969 (1993).
6. H. Calen *et al.*, *Phys. Lett.* B **366**, 39 (1996) and priv. comm..
7. W.Cassing *et al.*, *Z. Phys.* A **340**, 51 (1991).
8. E. Vercellin *et al.*, *Il Nuovo Cimento* **106 A**, 861 (1993).

Subthreshold K$^+$ Production in Proton-Nucleus Collisions and the Nuclear Spectral Function

M. Dębowski, R. Barth, M. Cieślak, E. Grosse, D. Miśkowiec, E. Schwab, P. Senger
Gesellschaft für Schwerionenforschung, D-64291, Darmstadt, Germany

Y. Le Bornec, M. P. Comets, P. Courtat, R. Gacougnolle, T. Kirchner[a] ,
J. M. Martin, B. Tatischeff, N. Willis
Institut de Physique Nucléaire, IN2P3-CNRS, F-91406 Orsay, France

M. Boivin, R. Wurzinger, J. Yonnet, A. Zghiche
Laboratoire National Saturne, F-91191 Gif-sur-Yvette Cedex, France

Ch. Müntz[b], *Ch. Sturm, A. Wagner*
Technische Hochschule Darmstadt, D-64289 Darmstadt, Germany

W. Waluś
Jagellonian University, PL-30059 Kraków, Poland

In our experiments we studied K$^+$ production on carbon and lead targets. The proton beam kinetic energies were 1.2 and 1.5 GeV. The spectrometer was set at the laboratory angle of $\theta_{lab} = 40^o$ corresponding to particles emitted at midrapidity. The particles were identified by the time-of-flight.

The influence of the internal motion of nucleons on the particle production can be studied quantitatively within the folding model. If we consider collisions of the beam proton with target nucleons with a given momentum distribution we can calculate the resulting cross section as an integral:

$$\sigma_{pA}^{K+} = N \int d^3p \int dE^* \; S(p, E^*) \, \sigma_{pN}^{K+}(\sqrt{s}) \quad , \tag{1}$$

where $N = \sigma_A^{inel}/\sigma_p^{inel}$, p is the internal momentum of a nucleon and E^* the corresponding total energy of this nucleon. $S(p, E^*)$ is the nuclear spectral function giving the distribution of these variables [1]. $\sigma_{pN}^{K+}(\sqrt{s})$ is the free elementary cross section for kaon production. It depends only on the total available energy. A similar approach has been already presented (see i. e. [2]) but up to now nucleons were considered as on-shell particles. The integral (1) takes into account that nucleons in the nucleus are bound and thus are off-shell (they do not obey the formula $E^2 = p^2 + m_N^2$).

---

[a] *Present address:* FZ Rossendorf, D-01314 Dresden, Germany
[b] *Present address:* BNL, Upton, NY 11973, USA

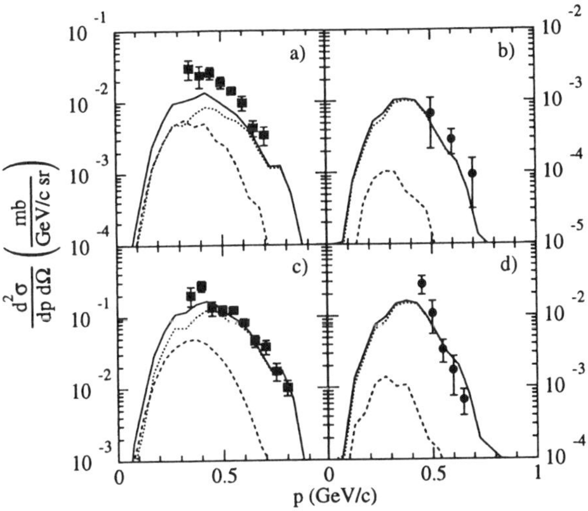

Figure 1: Measured and calculated kaon production cross sections in proton-nucleus collisions. a) – p+C at 1.5 GeV, b) – p+C at 1.2 GeV, c) – p+Pb at 1.5 GeV and d) – p+Pb at 1.2 GeV. The dashed lines are kaons produced in first chance collisions, the dotted lines – kaons from $\pi N \rightarrow KY$ processes and the solid lines are summed both components.

Below the threshold the measured kaon yield cannot be reproduced by first chance collisions even if the Fermi motion of nucleons is taken into account (cf. Fig. 1 – dashed lines). More complex processes like a two-step production mechanisms must be introduced. In the first step a pion is produced in the reaction $NN \rightarrow NN\pi$ and the pion can produce a kaon in the subsequent collision with another nucleon $\pi N \rightarrow K\Lambda(\Sigma)$. This process can also be calculated with the folding model. The distribution of pions produced in the first step is folded with the nuclear momentum distribution as in (1).

The results of the folding model calculation together with our data are shown in Fig. 1. Already at 1.5 GeV beam energy at $\theta_{lab} = 40^o$ about 80% of kaons are produced in two-step processes. At lower bombarding energies the contribution of kaons produced in first-chance collisions is less than 10%.

References

1. I. Sick, S. Fantoni, A. Fabrocini and O. Benhar, Phys. Lett. B **323,** 267 (1994); Nucl. Phys. **A579,** 493 (1994)
2. W. Cassing *et al.*, Z. Phys. A **349,** 77 (1994)

THE REACTION $\vec{d}\, p \to {}^3He\, \pi^0$ NEAR THRESHOLD

V.N. Nikulin,[1,2,3] A. Boudard,[1] M. Clajus,[4] B. Fabbro,[1] M. Garçon,[1]
R.S. Kessler,[4] L.K. Lytkin[2,5] B. Mayer,[1] B.M.K. Nefkens,[4] F. Plouin,[2] J. Poitou,[1]
E. Tomasi-Gustafsson,[2] W.T.H. van Oers,[6] D. White,[4] and C. Wilkin[7]
[1] *Service de Physique Nucléaire, CEA Saclay, F-91191 Gif-sur-Yvette, France*
[2] *Laboratoire National Saturne, F-91191 Gif-sur-Yvette, France*
[3] *St Petersburg Nuclear Physics Institute, 188350 Gatchina, Russia*
[4] *Department of Physics, University of California, Los Angeles, California 90095*
[5] *Joint Institute for Nuclear Research, Dubna, Russia*
[6] *TRIUMF, 4004 Wesbrook Mall, Vancouver, British Columbia, Canada V6T 2A3*
[7] *University College London, London WC1E 6BT, UK*

Angular distributions for the differential cross section and deuteron analyzing powers iT_{11}, T_{20} and T_{22} of the reaction $\vec{d}\, p \to {}^3He\, \pi^0$ have been measured over the whole angular domain at 20 energies close to threshold ($0.03 < T_\pi^{cm} < 10.2$ MeV) [1]. The experiment was carried out at the Laboratoire National SATURNE (LNS), Saclay. The differential cross section and tensor analyzing power T_{20} both show strong variation in energy and angle due to interference between S and P-wave pion production, whereas iT_{11} and T_{22} remain consistent with zero over the whole experimental range.

Six complex amplitudes are required to describe the reaction. The spectator nucleon model of Germond and Wilkin [2] gives estimates for the only amplitudes A and B which are not null in the forward and backward directions at low energies. Such a model immediately explains the striking experimental result that $|B|^2 << |A|^2$ near threshold [3]. The model predicts that the amplitudes A and B should be complex linear functions of $\eta = p_\pi/m_\pi$ at low energies of the form

$$A = A_0 + A_1 \eta \cos\Theta\,,$$
$$B = B_0 + B_1 \eta \cos\Theta\,, \tag{1}$$

though they only considered the cases where $\cos\Theta = \pm 1$.

If one retains only these A and B amplitudes, iT_{11} and T_{22} will remain identically zero at all angles, in accordance with our observations. Both the amplitude-squared, $(p_p/p_\pi)\, d\sigma/d\Omega$, and the tensor analyzing power T_{20} should lie on universal curves in the global variable $\eta\cos\Theta$. Inspection of the combined data sets in Fig. 1 shows this to be a very good approximation.

Since the data are not sensitive to the overall phases of the amplitudes, A_0 and B_0 can be taken to be real and the other terms written as $A_1 = |A_1|\, e^{i\phi}$ and $B_1 = |B_1|\, e^{i\psi}$. This led therefore to fitting the data at all energies and angles in terms of the six free parameters of Eq. (1). The B amplitude is comparatively small and slowly varying and the quality of the fit is not diminished by taking it to be constant, *i.e.* putting $B_1 = 0$.

The best fit to all the present data, yielding a χ^2/degree of freedom of 2.0, is with parameters

$$\begin{aligned} A_0 &= (3.203 \pm 0.004)\,\sqrt{\mu b/sr}\,, \\ |A_1| &= (7.58 \pm 0.04)\,\sqrt{\mu b/sr}\,, \\ \phi &= (0.34 \pm 0.01)\,rad\,, \\ B_0 &= (0.651 \pm 0.005)\,\sqrt{\mu b/sr}\,, \end{aligned} \tag{2}$$

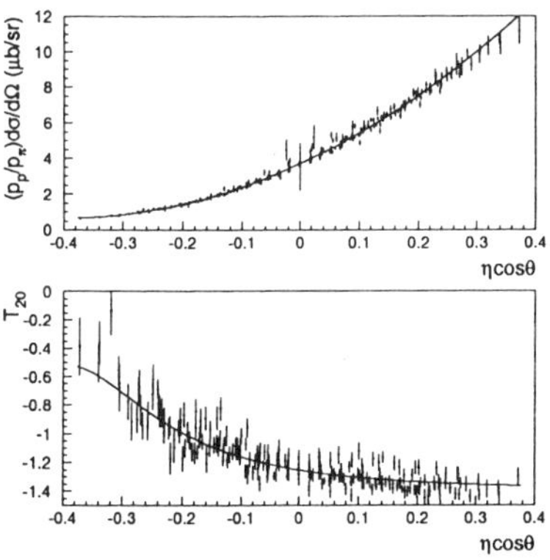

Fig 1: Cross sections and deuteron tensor analyzing power T_{20} at all energies and angles plotted against the universal parameter $\eta \cos \Theta$ and compared with the global fit.

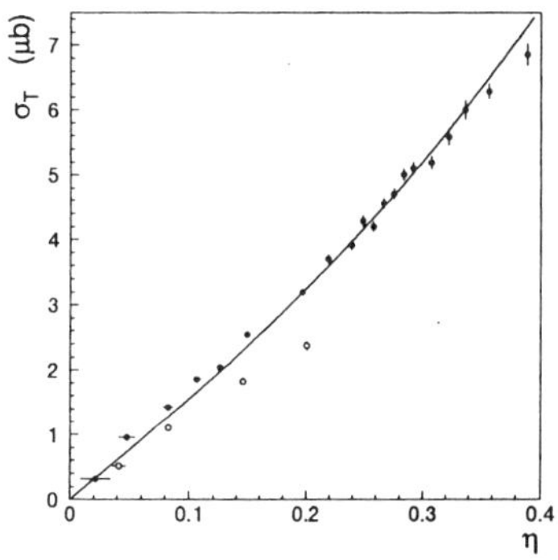

Fig 2: Total cross section σ_T, for the $d\,p \to {}^3He\,\pi^0$ reaction, plotted as function of η. Closed circles are the results of the present experiment whereas the open circles are IUCF data [4]. The curve represents the global fit.

References

1. V. Nikulin et al., report DAPNIA-SPhN-96-15, submitted to Phys. Rev. C for publication.
2. J.-F. Germond and C. Wilkin, J. Phys. G **16**, 381 (1990).
3. A. Boudard et al., Phys. Lett. B **214**, 6 (1988); C. Kerboul et al., Phys. Lett. B **181**, 28 (1986).
4. M.A. Pickar, A.D. Bacher, H.O. Meyer, R.E. Pollock and G.T. Emery, Phys. Rev. C **46**, 397 (1992).

SYSTEMATIC EFFECTS IN THE DETERMINATION OF THE ΠNN COUPLING FROM $\bar{P}P \to \bar{N}N$ DIFFERENTIAL CROSS SECTIONS

T.E.O. ERICSON
Institute of Theoretical Physics, P.O. Box 805,
S-75105 Uppsala, Sweden
and Theory Division, CERN,
CH-1211 Geneva 23, Switzerland
B. LOISEAU
Division de Physique Théorique,
Institut de Physique Nucléaire,
Orsay 91406 and LPTPE
Université Pierre et Marie Curie, Paris 75230, France

We show that the πNN coupling constant extracted model-independently from $\bar{p}p$ charge exchange is subject to a systematic correction, and, more importantly, that the strong absorption in the critical region prevents a determination of the coupling constant to high precision using this process.This removes the possible conflict with the value determined from the np charge exchange cross sections.

It is essential to determine the charged πNN coupling constant convincingly to high precision directly from experiments both in the NN and πN sector. Our Difference Method for the extrapolation to the pion pole using single energy differential np charge exchange cross section data gave with initial data $g^2 = 14.60 \pm 0.30$ [1]. Recent, more accurate Uppsala data give $g^2 = 14.45 \pm 0.15$. Strong objections to our approach have been voiced by the Nijmegen group, which claims a determination of the coupling constant to better than 1% using a global analysis of the entire np data-base with an excellent overall χ^2 [2]. Our detailed examination of their approach has however revealed a number of peculiarities. For example, the coupling constant depends directly on the absolute normalization. This normalization is a free parameter in the Nijmegen analysis! Further, their major constraint comes from the asymmetry parameter A_y, which to leading order does not even contain the quantity claimed to be determined to 1% precision. Finally, an examination of the data base by our experimental colleagues as discussed in the following talk by J. Blomgren reveals that the excellent χ^2 in the Nijmegen analysis has been achieved by rejecting large quantities of experimentally conflicting data. In view of this one cannot accept the

Nijmegen claim that their analysis of np data gives the coupling constant to very high precision. The value they quote is not based on hard facts.

An alternative in the nucleon sector eqivalent to the np charge exchange is the antiproton reaction $\bar{p}p \to \bar{n}n$. A recent experiment from LEAR at 176 MeV provides normalized single energy data of a quality rivalling the best np data. It is therefore potentially a good basis for a determination of the coupling constant. A preliminary analysis using the Chew method demonstrated strikingly that the both the $\bar{p}p$ and np data extrapolate approximately to the same pole residue, but with the very low value $g^2 = 12.80 \pm 0.29 \pm 0.20$, well below other values [3]. However, as for np charge exchange the problem is not one of statistics, but of systematics. The extrapolation is based on the Chew function

$$y(x) = \frac{sx^2}{m_\pi^4 g_R^4} \frac{\mathrm{d}\sigma}{\mathrm{d}\Omega}(x), \tag{1}$$

extrapolated to the pion pole position in the squared momentum transfer. We have repeated their analysis and reproduce their result accurately. Calibration using 10000 equivalent 'pseudoexperiments' generated from the Paris and Nijmegen models with known coupling reveals a systematic underestimate by about 2% (0.3 units), bringing the coupling from 12.8 to 13.1. There are however more fundamental problems in a a precision determination. The dominant pion contributions have a scale of a few pion masses. High momentum transfer data are irrelevant and should be removed. The pion sensitive region experiences a heavy quenching due to the strong absorption by a factor 4-5 compared to the np case. Not only does this eliminate much of the pion information, but the extrapolation function has now a rather deep minimum near the point of vanishing momentum transfer with the extrapolated value four times higher! In spite of these difficulties we have attempted to use the Difference Method comparing with models of known coupling. The method is much less helpful than in the np case largely since there are no strong constraints on the high m0omentum transfer behavior from other data. The method using $q^2 < 5m_\pi^2$ points once more to a value for the coupling of 13.1 to 13.2. The data have however a statistically implausible χ^2 of about 0.45! The statistical and systematic errors in the experiment were added in quadrature. The effect would follow if the systematic errors are highly correlated. The effect on the extrapolation is difficult to evaluate. Consequently, we recommend a value 13.1 ± 1.0 with a systematic error of 0.3 from normalization. It is not possible to claim a significant discrepancy with the value from np charge exchange. the nature of

References

[1] T. E. O. Ericson *et al*, PRL **75**, 1046 (1995).

[2] M.C.M. Rentmeester *et al*, PRL , submitted Comment (1996); V. Stoks *et al*, PRC **47**, 512 (1993).

[3] R. Birsa et al., Phys. Lett. **B339**, 325 (1994); F. Bradamante and A. Martin, *Phys. Lett.* B **343**, 427 (1995).

THE πNN COUPLING CONSTANT AND THE SHAPE AND NORMALIZATION OF np SCATTERING CROSS SECTIONS

J. Blomgren, J. Nilsson and N. Olsson
Department of Neutron Research, Uppsala University, Box 535, S-75121 Uppsala, Sweden

The textbook value of the πNN coupling constant is $g^2_{\pi NN} = 14.3$ to 14.5, deduced mainly from πN data and from forward dispersion relations in pp scattering. In the last few years a Nijmegen group [1] has proposed a smaller value for the charged coupling constant, $g^2_{\pi NN} = 13.58(5)$, on the basis of their analysis of NN scattering data.

The np scattering cross section should be suitable for determination of the charged πNN coupling constant. The CM cross section shows a large maximum in the backward direction, which can be used for extrapolation to the charged pion pole.

We have demonstrated such an approach using the 162 MeV high precision np data from Uppsala [2,3], which cover an angular range of $72° - 180°$. The angular shape differs from older and less accurate data. The main problem is to obtain a reliable normalization N. Absolute differential cross sections were obtained by normalizing to the total np cross section, using the phenomenological SAID solution SM95 in the angular range not covered by the experiment. The total uncertainty in N is estimated to be less than 2%. In addition, the systematic error and the stability of the extrapolation must be controlled precisely. This has been achieved by introducing the novel *difference method* [2], and the result for the coupling constant is $g^2_{\pi NN} = 14.4 \pm 0.2$.

A similar analysis of several previous data sets has revealed a value close to 13.6 [4]. We have scrutinized these data and found that they mainly have their normalization from the cross section of the simultaneously measured $n + p \rightarrow d + \pi^0$ reaction. This cross section is extracted from data on the $p + p \rightarrow d + \pi^+$ reaction, after correction for isospin and Coulomb effects. Renormalizing the published np scattering data, using our method based on the total np cross section, typically yield an increase in the cross section of the order of $10 - 20\%$, which corresponds to an increase of the coupling constant by $5 - 10\%$, resulting in $g^2_{\pi NN} \approx 14.5$.

Our conclusion is that the low value of $g^2_{\pi NN}$ obtained in recent analyses of NN scattering data might, at least partially, be a problem of the normalization of these data. With normalizations compatible with the precisely measured total cross section, values around 14.5 are obtained. Furthermore, the results points to that the normalization technique using pion production is uncertain

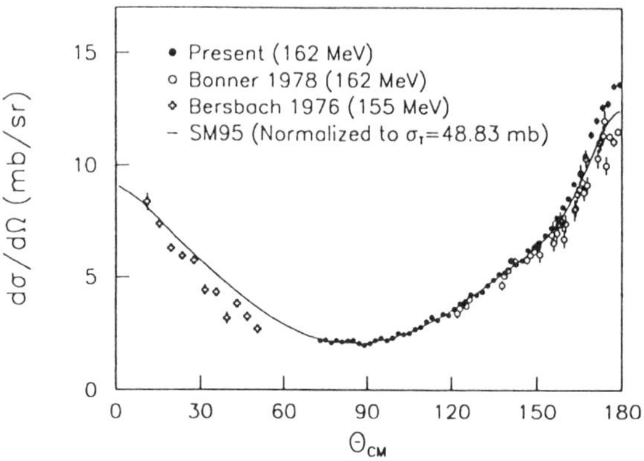

Figure 1: Differential cross sections for elastic np scattering at 102 MeV.

by as much as 10 – 20%. Probably, this is partly due to that this cross section is not known to better than about 10 % above 400 MeV, and is even more uncertain at lower energies.

A second problem with the world np data base is that it is dominated by two large data sets, from Los Alamos[5] and PSI[6], and the shape of the angular distributions differ at backward angles for these. The Los Alamos data - which display a flattish shape, tending to a low coupling constant - have the largest statistical weight, and thus an analysis of the total np data base favours the low value. However, this might soon change, because the PSI measurements have recently been extended [7], and will become the dominant data set when published. Our measurement is in good agreement with the PSI data, and they both favour a value around 14.4.

If this shape discrepancy is of systematic origin, determining which data to believe in cannot be made on statistical grounds. Instead, new precise measurements have to be employed. A new measurement with a different experimental setup is underway in Uppsala, as well as an experiment at IUCF.

1. V. Stoks *et al.*, *Phys. Rev.* C **47**, 512 (1993) and *Phys. Rev.* C **49**, 2950 (1994).
2. T.E.O. Ericson *et al.*, *Phys. Rev. Lett.* **75**, 1046 (1995), and elsewhere in these proceedings.
3. J. Nilsson *et al.*, to be published.
4. R.A. Arndt *et al.*, *Phys. Rev.* C **52**, 2246 (1995).
5. B.E. Bonner *et al.*, *Phys. Rev. Lett.* **41**, 1200 (1978).
6. W. Hürster *et al.*, *Phys. Lett.* B **90**, 367 (1980).
7. H. Schmitt, private communication. J. Franz *et al.*, to be published.

THE PION-NUCLEON COUPLING FROM QCD SUM RULES: FORM FACTOR AND ISOSPIN BREAKING

T.MEISSNER
Department of Physics and Astronomy, University of South Carolina, Columbia, SC 29208, USA

1 The πNN vertex

The matrix element of the pseudoscalar quark current $J_5^0 = \bar{q}(x)i\gamma_5\tau^0 q(x)$ between on shell nucleon states defines the pseudoscalar nucleon formfactor

$$\langle N(p_1)|J_5^0(0)|N(p_2)\rangle = g_P(q^2)\bar{u}_N(p_2)i\gamma_5 u_N(p_1) \tag{1}$$

where $u_N(p)$ denotes a free nucleon spinor in momentum space.

The pion nucleon form factor $g_{NN\pi}(q^2)$ can be determined from $g_P(q^2)$ by choosing the PCAC interpolating field for π and applying the chiral Ward identity

$$m_0\bar{q}(x)i\gamma_5\vec{\tau}q(x) = \partial_\mu\vec{A}^\mu = m_\pi^2 f_\pi\vec{\pi}(x) \tag{2}$$

giving

$$g_P(q^2) = \left(\frac{m_\pi^2 f_\pi}{m_0}\right)\frac{g_{NN\pi}(q^2)}{-q^2+m_\pi^2} \tag{3}$$

This means in fact that the matrix element in Eq. 1 has been saturated with a single pion pole and higher resonances in the pseudoscalar channel (π' ...) have been neglected (pion pole dominance). This assumption is only reasonable if the spacelike q^2 is not too high (less than $\approx 2\text{GeV}^2$).

2 Three point function and OPE

For the QCD sum rule analysis we consider the three point correlator consisting of J_5^0 and two Ioffe nucleon interpolating fields $\eta(x)$ [1]:

$$A(p_1,p_2,q) = \int d^4x_1 d^4x_2 e^{ip_1x_1}e^{-ip_2x_2}\langle 0|\mathcal{T}\eta(x_1)J_5^0(0)\bar{\eta}(x_2)|0\rangle \tag{4}$$

and write down the Borel transformed double dispersion relation with respect to the nucleon momenta ${p_1}^2$ and ${p_2}^2$, respectively. The q^2 is kept fixed at spacelike values. It is convenient to single out the sum rule for the Lorentz structure $\not{q}\gamma_5$, in which only odd dimensional vacuum condensates contribute. On the phenomenological side we saturate with intermediate nucleon states leading to an expression for $g_P(q^2)$ (Eq.1) from which $g_{NN\pi}(q^2)$ can be determined due to Eq. 3.

The contribution of higher nucleon resonances (N^*, ...) in the sum rule can be eliminated by taking the derivative with respect to $\frac{1}{M^2}$ (where M denotes the Borel mass), which generates a new independent sum rule.

3 The πNN form factor $g_{NN\pi}(q^2)$ at spacelike q^2

Following Ref. [1] one can determine the on shell pion nucleon coupling $g_{NN\pi}$ by comparing the residua of the $\frac{1}{q^2}$ pole on both sides of the sum rule. Contributions from higher order poles ($\frac{1}{q^4}$, ...) are small if q^2 is not too small ($\approx 0.5\mathrm{GeV}^2$). The form factor $g_{NN\pi}(q^2)$ is then determined by the polynomial terms in q^2 in an intermediate q^2 window.

The result can ce accurately fitted with the standard monopole form with a monopole mass of $\Lambda_\pi \approx 0.8\mathrm{GeV}$. [2] This value is very close to the one obtained within quenched lattice QCD (0.75GeV)[3] but much smaller than the one commonly used in one boson exchange $N-N$ potentials (≈ 1.5GeV). [4]

4 Isospin breaking

In order to determine the isospin violation in the on shell coupling coming from isospin violation effects in strong interaction one has to take isospin violation in the vacuum condensates on the OPE side into account. In our case the effect is solely due to the difference between u and d quark condensates, $< \bar{u}u >$ and $< \bar{d}d >$, respectively. We obtain a splitting of about $(g_{pp\pi_0} - g_{nn\pi_0})/g_{NN\pi} \approx 0.01$ with an uncertainty of about 40%. [5] This value is in agreement with many other model calculations. It plays an important role in the analysis of the charge symmetry breaking $N-N$ force. The charged pion nucleon coupling is found to be the average of $g_{pp\pi_0}$ and $g_{nn\pi_0}$. It should be mentioned that electromagnetic effects have not been not included.

References

1. L.Reinders, H.Rubinstein and S.Yazaki, *Phys.Rep.* **127**, 1 (1985).
2. T.Meissner, *Phys. Rev.* C **52**, 3386 (1995).
3. K.F.Liu *et al.*, *Phys. Rev. Lett.* **74**, 2121 (1995).
4. R.Machleidt, K.Holinde and C.Elster, *Phys.Rep.* **149**, 1 (1987).
5. E.Henley and T.Meissner, nucl-th/9601030.

DISCOVERY OF DEEPLY BOUND π^- STATES IN THE ^{208}Pb(d,^{3}He) REACTION

A. GILLITZER [a]

Physik-Dept. E12, TU München, James-Franck-Straße, D-85748 Garching

The ^{208}Pb(d,^{3}He) transfer reaction was studied at negative Q values around the π^- production threshold, in order to search for deeply bound pionic states. A narrow peak in the Q value spectrum was observed which we assign to a bound $(2p)_\pi$ state coupled to $(3p_{1/2}, 3p_{3/2})_n^{-1}$ states in ^{207}Pb.

Due to absorption low lying states (e.g 1s or 2p) in pionic atoms with medium heavy and heavy nuclei are not populated in electromagnetic cascades, and these states had so far not been observed experimentally. Toki and Yamazaki [1] realized that nevertheless the deeply bound states should exist even for the heaviest atoms like lead, since the short range s-wave repulsion in the π^--nucleus optical potential reduces the widths of the lowest states to values considerably smaller than the level distance. They proposed [1] to search for their direct population in transfer reactions. Search experiments using this approach indicated some strength in the Q value spectrum in the bound π^- region [2,3], however discrete levels were not resolved.

Following the arguments of ref. [4] we studied the ^{208}Pb(d,^{3}He) reaction at 0° at an incident deuteron energy of 300 MeV/u [5], which due to the low momentum transfer populates dominantly states with small angular momentum. As a forward spectrometer with high resolving power and good background suppression capability we used the GSI Fragment Separator which consists of four bending sections and four focal planes. At the central dispersive focal plane the momentum of the ejectiles is measured by two sets of high-rate drift chambers, whereas the ^{3}He ions are unambiguously identified downstream by the measurement of time of flight and energy loss in three plastic scintillation counters. A 2 mm wide ^{208}Pb strip target with 45 mg/cm^2 thickness was bombarded with up to 10^{11} deuterons/spill at 2.8 s cycle time. The beam momentum spread $\delta p/p \sim 5 \cdot 10^{-4}$ and the energy loss in the target allowed for a Q value resolution of 0.5 MeV. The momentum on the focal plane was calibrated at different $B\rho$ settings of the separator by using the p(d,^{3}He)π^0 reaction, which creates monoenergetic ^{3}He at 1678.7 MeV/c momentum.

---

[a]for GSI-S160: H. Geissel, H. Gilg, A. Gillitzer, R.S. Hayano, K. Itahashi, N. Iwasa, P. Kienle, M. Knülle, M. Münch, G. Münzenberg, K. Oyama, W. Schott, T.Yamazaki

This work was supported by the BMBF (Germany) under contract 06TM365/TP2

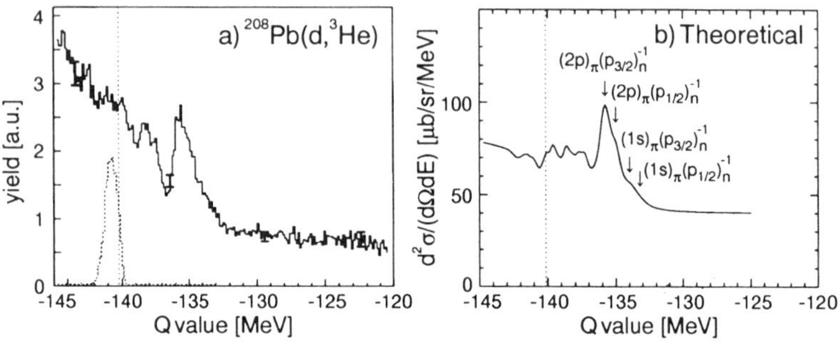

Figure 1: a) Q value sum spectrum taken with three different $B\rho$ settings. The free π^- production threshold is shown by the dotted line. The dotted spectrum is for the $p(d,{}^3He)\pi^0$ reaction in the equivalent Q value scale. b) Theoretical prediction[4] with 0.5 MeV resolution.

In order to exclude instrumental effects as origin of any observed structure in the ^{3}He momentum spectrum, the spectrum was measured for three different $B\rho$ settings. The momentum bite $\Delta p/p \sim \pm 1\%$ corresponds to a 25 MeV interval of the reaction Q value. An acceptance corrected Q value sum spectrum for the three settings for the ^{208}Pb(d,^{3}He) reaction is shown in Fig.1a) in comparison to the the theoretical prediction [4] in Fig.1b). The threshold for free π^- production with ^{207}Pb in its $3p_{1/2}$ ground state corresponds to $Q = -140.147$ MeV. At less negative Q values in the bound π^- region we observe a structure with a pronounced peak at $Q = -135.6$ MeV above a flat continuum. In a comparative run with ^{27}Al target only the rise of the yield for the quasifree π^- production without any structure was seen.

This leads us to the conclusion that deeply bound π^- states in ^{207}Pb have been populated, with the assignment of a $(2p)_\pi(3p_{1/2},3p_{3/2})_n^{-1}$ configuration to the most pronounced peak. The binding energy of the $(2p)_\pi$ orbital was deduced to be $B_\pi(2p) = (5.4\pm0.2)$ MeV with an intrinsic width below 0.8 MeV. These two quantities are directly related to the real and the imaginary part of the pion-nucleus optical potential at short distance, so far not accessible to experimental study. Due to the close contact of the π^- to the nucleus with a kind of halo nature, the deeply bound states may also be interpreted as a new family of narrow nuclear resonances at 130-140 MeV excitation energy.

1. H. Toki and T. Yamazaki, *Phys. Lett.* B **213**, 129 (1988)
2. T. Yamazaki, *Phys. Scrip.* **48**, 169 (1993)
3. N. Matsuoka *et al.*, *Phys. Lett.* B **359**, 39 (1995)
4. S. Hirenzaki, H. Toki and T. Yamazaki, *Phys. Rev.* C **44**, 2472 (1991)
5. T. Yamazaki *et al.*, submitted to Z. Phys. A (1996)

Proton-Proton Elastic Scattering Excitation Functions at Intermediate Energies

H. Rohdjess for the EDDA-Collaboration [1]
ISKP, Universität Bonn, Nussallee 14-16, 53115 Bonn, Germany

The EDDA experiment at the Cooler Synchrotron COSY at Jülich is dedicated to a precision measurement of proton-proton elastic scattering observables in the beam energy range 0.5 – 2.5 GeV. Excitation functions are acquired with an internal target during beam acceleration.

The EDDA experiment has been designed to provide a consistent and precise data set of $\vec{p}$+$\vec{p}$ observables in the COSY energy range. Apart from contributing to the input for a global phase shift analysis, the experiment will be sensitive to the (non-)existence of dibaryonic states coupling to the elastic channel.

The detector consists of a *large acceptance* (85% of 4π), segmented scintillator hodoscope detecting both final state protons in coincidence. An angular resolution of better than 1.5° in the center-of-mass has been obtained.

It is used in conjunction with an *internal* target to acquire excitation functions in small energy steps and over a wide angular range by repeated measurements *during beam acceleration.*

EDDA has just completed the measurement of the unpolarized differential cross-section using a thin CH_2 fiber target. Background from reactions on carbon (mainly quasi-elastic scattering) has been measured with a pure carbon target and subtracted statistically. Relative normalization is accomplished by a concurrent measurement of the δ-electron and secondary electron emission yields from the target. The absolute luminosity is fixed by normalizing to existing data at lower energies[2]. Shifts in the COSY beam position, width and angle during acceleration have been deduced. They are small and are taken into account in the data analysis.

An example of an excitation function at $\theta_{c.m.} = 41°$ is shown in Fig. 1. Statistical error-bars are about the size of the symbols, systematic errors are still under study.

Currently, preparations for measurements of the spin observables A, A_{NN}, A_{SS}, and A_{LS} with a polarized COSY beam and a polarized atomic beam target are under way.

The cooperation of the COSY accelerator team and financial support by the BMBF is acknowledged.

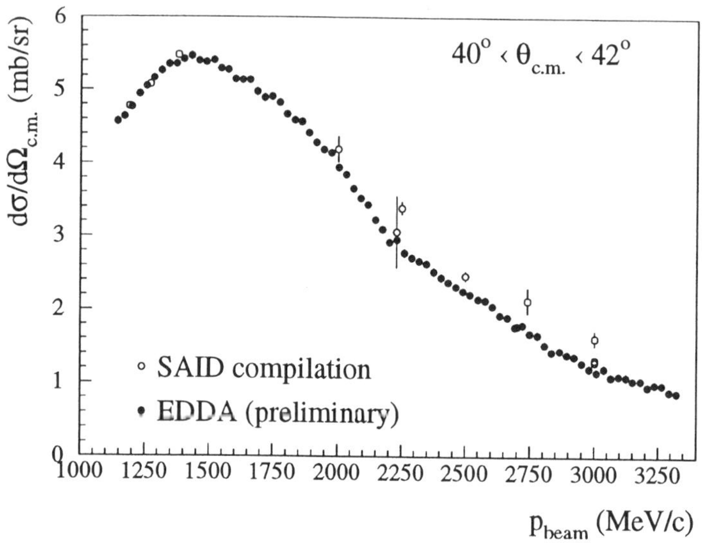

Figure 1: Differential cross section vs. proton beam momentum at $\theta_{c.m.}=41^o \pm 1^o$ measured by EDDA (solid symbols, preliminary). The previous world data set (open symbols) has been taken from the SAID [3] database.

References

1. D. Albers[b], J. Bisplinghoff[a], R. Bollmann[b], K. Büßer[b], P. Cloth[c], R. Daniel[a], O. Diehl[a], F. Dohrmann[b], V. Drüke[c], H. P. Engelhardt[a], J. Ernst[a], P. D. Eversheim[a], D. Filges[c], M. Gasthuber[b], R. Gebel[a], J. Greiff[b], A. Groß[b], R. Groß-Hardt[a], S. Heider[a], A. Heine[a], F. Hinterberger[a], M. Igelbrink[b], R. Jahn[a], M. Jeske[a], I. Koch[b], R. Langkau[b], J. Lindlein[b], R. Maier[c], R. Maschuw[a], T. Mayer-Kuckuk[a], G. Mertler[a], B. Metsch[d], F. Mosel[a], M. Müller[b], M. Münstermann[b], H. Paetz gen. Schieck[e], H. R. Petry[d], D. Prasuhn[c], H. Rohdjeß[a], D. Rosendaal[a], U. Roß[a], P. von Rossen[c], H. Scheid[a], N. Schirm[b], M. Schulz-Rojahn[a], F. Schwandt[a], V. Schwarz[a], W. Scobel[b], B. Steeg[b], S. Steinbeck[b], G. Sterzenbach[c], S. Thomas[a], H. J. Trelle[a], A. Wellinghausen[b], W. Wiedmann[a], K. Woller[b], R. Ziegler[a]
[a] Inst. f. Strahlen- und Kernphysik, Univ. Bonn, [b] I. Inst. f. Experimentalphysik, Univ. Hamburg, [c] Inst. f. Kernphysik, KFA Jülich, [d] Inst. f. Theor. Kernphysik, Univ. Bonn, [e] Inst. f. Kernphysik, Univ. Köln
2. A. J. Simon et al. *Phys. Rev.* **C48** (1993) 662 and **C53** (1996) 30.
3. R. A. Arndt *Phys. Rev.* **D45** (1992) 3995 and interactive program SAID

SPIN OBSERVABLES IN NEUTRON–PROTON ELASTIC SCATTERING AT PSI

A.AHMIDOUCH, Ph.DEMIERRE, N.GOUJON-NAEF, R.HESS †, C.LECHANOINE-LELUC, D.RAPIN, A.TEGLIA, B.VUARIDEL
DPNC Université de Genève, CH-1211 Genève 4, Switzerland

J.ARNOLD, J.FRANZ, H.LACKER, E.ROESSLE, H.SCHMITT, P.SERENI
Fakultät für Physik der Universität Freiburg, D-79104 Freiburg, Fed.Rep.Germany

B. VAN DEN BRANDT, M.DAUM, P.HAUTLE, J.A.KONTER, S.MANGO, P.A.SCHMELZBACH
PSI, Paul Scherrer Institut, CH-5234 Villigen, Switzerland

R.DREVENAK, M.FINGER[c], M.FINGER Jr., A. JANATA, M.SLUNECKA
Joint Institute for Nuclear Research, Dubna, Russia
[c]Faculty of Math. and Phys., Charles University, Prague, Czech Republic

F.LEHAR
DAPNIA, CEN-Saclay, F91191 Gif-sur-Yvette, France

Results for fourteen different two- and three-spin observables [namely A_{oosk}, A_{ookk}, A_{oonn}, A_{ooss}, D_{onon}, D_{osok}, K_{osko}, K_{onno}, K_{osso}, N_{ossn}, N_{oskn} and D_{osos}, N_{osns}, N_{onkk} in half of the angular range] are available at 5 energies between 260 and 540 MeV over the angular range $60^o \leq \Theta_{cm} \leq 160^o$ adding to $\simeq$ 1000 new data points. The two-spin parameters have been measured with an absolute statistical precision of typically 0.03 and the three-spin parameters, measured here for the first time, have a $\simeq$ 0.10 precision.

Two experiments installed at PSI, one behind each other, in the NA2 polarized neutron beam line studied the spin dependence in *np* elastic scattering between 260 and 540 MeV [1]. Both experiments took advantage of the intense polarized neutron beam, created in the charge exchange reaction $C(\vec{p},\vec{n})X$ at 0°. This beam has unique features: (1) continuous energy ; (2) polarization between 15% at 260 MeV to 40% above 450 MeV; (3) intensity $5 \times 10^6\ \vec{n}/s.cm^2$ at 13.7m from the production target using high intensity polarized protons ($\leq 12\mu A$). The beam energy is measured by time-of-flight using the rf-signal from the accelerator.

Experiment I, using a 100 cm^3 polarized target working in frozen spin mode, detected in coincidence both scattered particles, the neutron in a hodoscope made of 11 horizontal plastic scintillation bars and the proton whose

† deceased

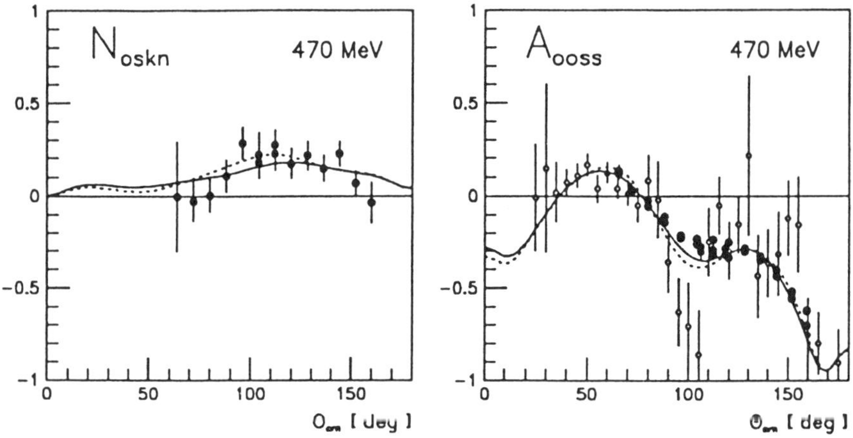

Figure 1: Results for N_{oskn} and A_{ooss} at 470 MeV as a function of cm angle shown as black dots. Open circles are data from LAMPF [2] , full and dashed lines are PSA predictions from SAID-FA95 and Saclay-Geneva respectively.

polarization was measured in a polarimeter made of carbon plates sandwiched by MWPC. Twelve to thirteen different two and three spin-dependent parameters have been measured between 70^o_{cm} and 160^o_{cm} namely A_{oosk}, A_{ookk}, A_{oonn}, A_{ooss}, D_{onon}, D_{osok}, K_{osko}, K_{onno}, K_{osso}, N_{ossn}, N_{oskn} and D_{osos},N_{osns}, N_{onkk} in half of the angular range. Results for N_{oskn} and A_{ooss} are shown in Fig.1 as well as available data [2] and phase shifts (PSA) predictions.

Experiment II measured the $\vec{n}p$ elastic reaction with a liquid hydrogen target between 130^o_{cm} and 180^o_{cm}. A magnetic spectrometer analyzed the momentum of the recoiling protons. Behind it, a polarimeter measured the transverse components of the spin. This experimental configuration allowed to analyze longitudinal components of the spin, since the horizontal components $\hat{s}$ and $\hat{k}$ are rotated in the vertical field of the magnet. Results for K_{osko}, K_{onno}, K_{osso},K_{okko}, K_{okso} have been obtained.

The final goal of our experiments is to perform an I=0 direct amplitude reconstruction using the abundant pp data already measured at PSI [3].

We thank the German BMFT and the Swiss National Funds for support.

References

1. A. Ahmidouch et al., AIP Conf. Proc.339 308(1995).
2. V. Carlson et al, ANL-HEP-PR-95-64
3. A. Aprile et al., *Phys. Rev.* D **34**, 2566 (1986) and 2600(1983).

INTERESTING FEATURES IN N-N INTERACTIONS

J. BALL
on behalf of the Nucleon Nucleon collaboration
Laboratoire National Saturne, CEA DSM and CNRS IN2P3, CE Saclay, 91191 Gif-sur-Yvette Cedex, France

The present report shows several interesting features observed in the energy and angular dependence of nucleon-nucleon (NN) observables. Structures in np and pp total cross section differences and in energy dependence of spin correlation parameters are discussed. They suggest a resonance like behaviour in different isospin and spin states. Throughout this contribution will be used the NN formalism and the four-spin notation[1] of observables.

1 Total Cross Section Differences

The NN total cross section differences $\Delta\sigma_T$ and $\Delta\sigma_L$ together with the spin independent total cross section σ_{0tot} are integral quantities linearly related with three non-vanishing imaginary parts of the NN forward scattering amplitudes via optical theorems. The pp and np observable σ_{0tot} has been measured during the last fifty years from 10^{-2} eV up to thousands of TeV. The measurements of spin-dependent total cross sections are rare due to a lack of polarized beams and targets.

The pp total cross section differences were measured at ANL-ZGS, LAMPF, PSI, SATURNE II and at TRIUMF. Results, at intermediate energies, cover the range from 0.2 to 12 GeV and one point at 200 GeV (FERMILAB). Neutron-proton observables, $\Delta\sigma_T$ and $\Delta\sigma_L$, were measured with free polarized neutrons at JINR-Dubna, LAMPF, PSI, SATURNE II and at lower energies at TUNL-Prague. All existing data smoothly connect in the entire energy region as can be seen in Fig. 1a for the observable $\Delta\sigma_L$. They are compared with the phase shift analysis (PSA) predictions[2]. We observe a dip around 0.6 GeV, followed by a large maximum centered at 0.9 GeV. The JINR results[3] show a fast decrease towards high energies. $\Delta\sigma_L(pn)$ results were also deduced in 1981 from the $\Delta\sigma_L(pd)$ and $\Delta\sigma_L(pp)$ measurement at the ANL-ZGS .

The lowest lying exotic quark configurations in the isospin state $I = 1$ and the spin-singlet state $^1S_0$ with the mass of 2.7 GeV ($T_{kin}(p) \sim 2.0$ GeV) was predicted by Lomon[4] et al.; the authors used the Cloudy Bag Model and an R-matrix connection to long range meson exchange forces. A resonance-like structure had been suggested by the energy dependence of the $\Delta\sigma_L(pp)$ around 2.1 GeV. Using the known $\Delta\sigma_L(pp)$ data one calculates $\Delta\sigma_L(I = 0) = 2\Delta\sigma_L(np) - \Delta\sigma_L(pp)$. The results deduced from existing measurements are

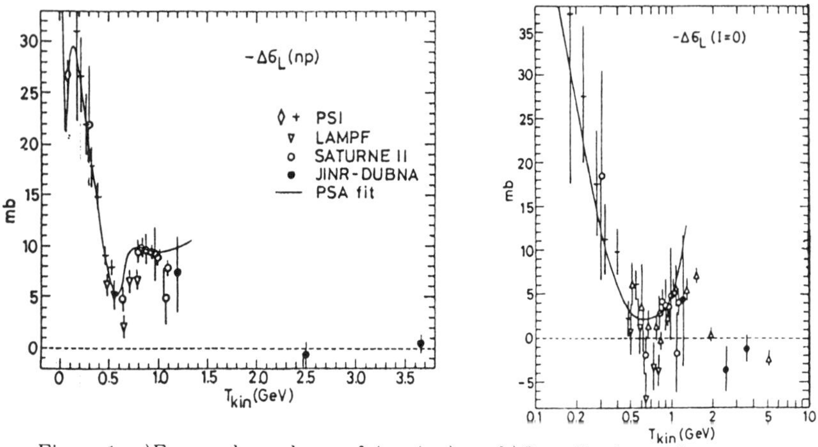

Figure 1: a)Energy dependence of $\Delta\sigma_L(np)$ b)Contribution of $I = 0$.

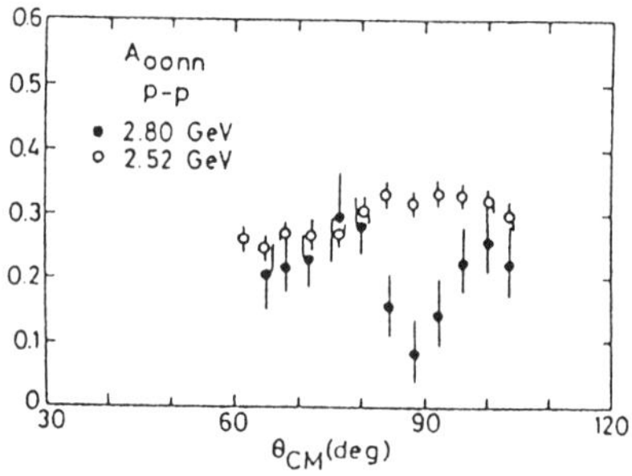

Figure 2: Angular dependence of spin correlation parameter.

plotted in Fig. 1b, together with the PSA predictions [2].

At high energies we observe an unexpected maximum above 1.5 GeV. This behaviour strongly supports the prediction concerning a lowest lying exotic quark configuration for isospin $I = 0$ and the spin-triplet $^3S_1$ around $T_{kin} = 1.8$ GeV. For this observable, the $^3S_1$ wave ($I = 0$) may be predominant.

2 Search for the Singlet-Spin Structure at I=1

The dibaryons, when first proposed, were predicted to be at substantially lower energies using the MIT Bag Model, with an equilibrium radius that would be relevant if the multi-hadron systems were confined and if there were no long range forces. In this lower range of energies, no significant anomaly was never confirmed.

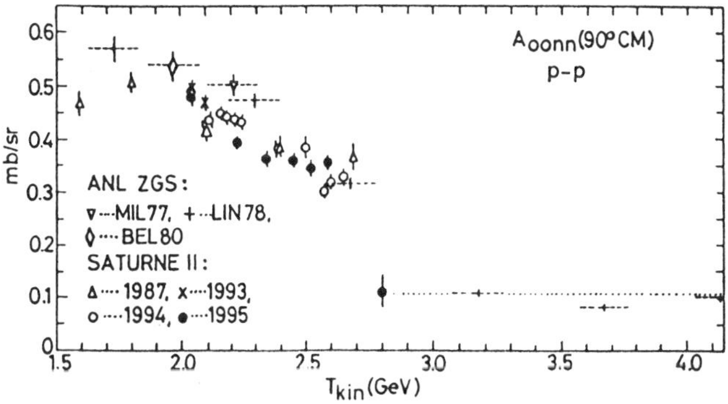

Figure 3: Energy dependence of $A_{oonn}(90^oCM)$.

The spin-singlet structure can be determined by energy dependence of the spin correlation parameter $A_{oonn}(90^oCM)$. Using invariant amplitudes [1], we obtain at this angle $2|b|^2 = (d\sigma/d\Omega)(1-A_{oonn})$. A resonance in the spin-singlet state appears as a shoulder (or a maximum) in the $|b(90^o)|^2$ energy dependence. A fast decrease of $A_{oonn}(90^o)$ in a small energy range would represent a serious indication for this structure.

The shape of the $A_{oonn}(pp)$ over the angular range $60^o \leq \theta_{CM} \leq 108^o$ changes as a function of energy (see Fig. 2). Above 1.8 GeV, A_{oonn} increases with θ_{CM} up to 90^oCM. Above 2.7 GeV, the value at 90° drops down. Around $T_{kin} = 2.1$ GeV, the preliminary results were published [5]. New results, at high energies, are shown in Fig. 3, We observe a decreasing slope above 1.9 GeV, and a very fast decrease between 2.7 and 2.8 GeV. Up to 8 GeV the ANL-ZGS values remain constant.

In order to draw a preliminary conclusion, the energy dependence of the spin-singlet amplitude was calculated. There is a shoulder in $|b(90^o)|^2$ in the vicinity of 2.1 GeV and another one around 2.8 GeV. This behaviour qualitatively agrees with the Lomon predictions for resonances in $^1S_0$ and $^1D_2$ spin-singlet waves.

References

1. J. Bystricky, F.Lehar and P.Winternitz, *J. Physique* **39**, 1 (1978).
2. R.A. Arndt *et al.*, *Phys. Rev.* C **50**, 2731 (1994).
3. E.V. Chernykh *et al.*, *see in these Proceedings.*
4. P. Gonzales, *et al.*, *Phys. Rev.* D **35**, 2142 (1987).
5. J. Ball *et al.*, *Phys. Lett.* B **320**, 206 (1994).

PHASE-SHIFT ANALYSES OF ELASTIC np SCATTERING AT T_L=1.2, 2.5 and 3.6GeV

J. Nagata, H. Yoshino, Y. Yoshino and M. Matsuda
Faculty of Integrated Arts and Sciences, Hiroshima University
Higashi-Hiroshima 739, Japan

We report a preliminary result of phase-shift analyses of elastic np scattering at T_L=1.2, 2.5 and 3.6GeV using the data on $\Delta\sigma_L$ provided by the JINR polarized-spin experiment.

A rich amount of experimental data on spin-correlation parameters of elastic pp scattering in the several GeV region have been provided in 1980's mainly by ANL, CERN and Saclay. Using these data with the pseudo-data obtained by the interpolation of experimental data by means of the cubic spline function method, we have perfomed the phase-shift analyses(PSAs) of elastic pp scattering at P_L=3, 6 and 12GeV/c. The results have shown that the spin-orbit interaction in the range$\lesssim$0.5fm has a strong energy dependence which seems to shrink as the energy of system increases[1]. Only our possible explanation of this behavior may be that a phase transition between quark-gluon phase and hadron phase takes place in the incident energy region T_L=3$\sim$10GeV.

In the np scattering, we expect that a signal of strong energy dependence of the spin-correlated interaction will be also found in the JINR polarized spin-experiment around at T_L=3GeV. In the lower energy region at T_L=0.5$\sim$1.09GeV, we have already perfomed the PSAs of elastic np scattering[2]. For a theoretical support of the future $\Delta\sigma_T^{np}$-experiment at JINR, we perform the preliminary PSAs of np scattering at 1.2, 2.5 and 3.6GeV, respectively, with use of the data base shown in Table 1 containing the data on $\Delta\sigma_L^{np}$ measured at JINR[3]. The predicted values of $\Delta\sigma_T^{np}$ in our preliminary solutions are also given in Fig. 1. The $\Delta\sigma_T^{np}$ calculated by the best-fit solution, which has the minimum value of χ^2 in Table 2, seems to tend to be zero as the energy increases.

It should be noted that all of the absolute values of $\Delta\sigma_{L,T}^{pp,np}$ decrease rapidly above 2GeV, the behaviors of which can not be disregarded to have a possible correlation to the "Spin Crisis." We expect that the future Delta-Sigma experiments of np scattering in several GeV region will significantly contribute to making a study of the physics of nucleon-nucleon system, such as searching the dibaryons and detecting its phase-transition to quark-gluon system.

Table 1: The kinds of observables and their numbers of data used in the present PSA. The ones of pseudo-data are shown to be asterisked.

T_L(GeV)	†Forw.	$d\sigma/d\Omega$	P	A_{ij}	others	total
1.2	7	116(*30)	114(*34)	120	23	385
2.5	7	*31	*34			72
3.6	7	*32	*34			73

†Forw:σ_t, σ_r, $\Delta\sigma_T$, $\Delta\sigma_L$, α, ReF_2, ReF_3

Table 2: The predicted values of $\Delta\sigma_T^{np}$ and χ^2-values of respective solutions.

T_L(GeV)	$\Delta\sigma_T^{np}$(mb)/χ^2		
	Sol.A	Sol.B	Sol.C
1.2	5.9/460	3.6/505	1.4/536
2.5	4.3/ 59	1.0/133	−4.1/433
3.6	3.9/141	0.5/ 80	−4.9/ 82

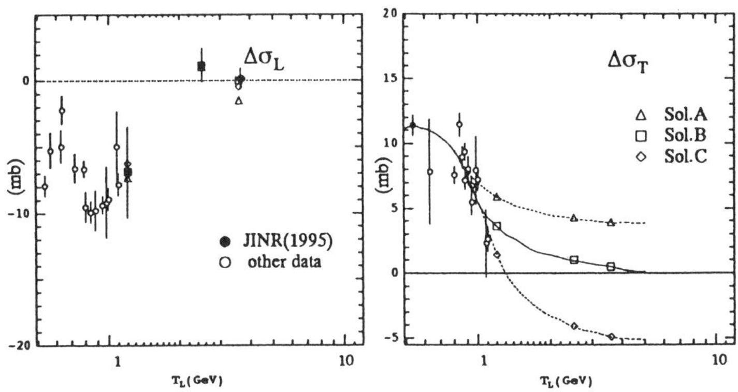

Fig. 1 The experimental data on $\Delta\sigma_L^{np}$ and $\Delta\sigma_T^{np}$ and their predicted values by our preliminary PSAs.

References

1. M. Matsuda *et al.*, Prog. Theor. Phys. **93**, 1059(1995).
2. H. Yoshino *et al.*, Prog. Theor. Phys. **95**, 577(1996).
3. V.I. Sharov *et al.*, Proc. of VIth Workshop on Spin Phenomena in High Energy Physics, September, 18-23, 1995, Protvino.

FIRST MEASUREMENTS OF THE NP TOTAL CROSS SECTION DIFFERENCES FOR PURE HELICITY STATES BETWEEN 1.2 AND 3.7 GEV.

E.V.CHERNYKH, A.A.NOMOFILOV, V.I. SHAROV, L.N. STRUNOV
and S.A. ZAPOROZHETS
Joint Institute for Nuclear Research, Dubna 141980, Moscow reg. Russia

The $\Delta\sigma_L(np)$ – difference of the np total cross sections for antiparallel and parallel longitudinal spin states, was first measured in an energy region of several GeV. The data were taken at the highest energies of free polarized neutrons that can be reached at present. The results can be used both for extending a neutron-proton phase shift analysis up to several GeV and for checking models of NN interaction.

The total cross section differences $\Delta\sigma_L$ and $\Delta\sigma_T$, along with σ_{tot}, are related to a three non-vanishing imaginary parts of the nucleon-nucleon (NN) forward scattering amplitudes via optical theorems. A set of existing data on these observables is used for a direct reconstruction of the amplitudes and for a NN phase shift analysis (PSA)[1,2]. Preceding studies of the spin-isospin observables $\Delta\sigma_{L,T}(pp)$ up to 11 GeV and $\Delta\sigma_{L,T}(np)$ up to 1.1 GeV have been performed during the past two decades[3]. For the first time $\Delta\sigma_L(pn)$ were deduced from $\Delta\sigma_L(pd)$ and $\Delta\sigma_L(pp)$ measurements at Argonne. The energy behaviour of these pn-data is even different from the free np-data and they are not included in the PSA[1].

An advance of spin-isospin structure studies of the basic NN interaction up to highest energy polarized neutrons,provided now only by the LHE JINR accelerator was proposed in ref.[4]. The facility for dedicated to this investigation DELTA-SIGMA experiment was constructed in Dubna at the end of 1994. The experimental setup included both polarized deuteron and polarized free neutron beam lines,two beam line polarimeters (absolute and relative) for measuring and monitoring the deuteron polarization, a polarized proton target, neutron transmission detectors, electronics and data acquisition systems. The results of the $\Delta\sigma_L(\vec{n}\vec{p})$ measurements are $(7.1 \pm 3.7), (-0.85 \pm 1.32)$, and (0.30 ± 0.84) mb at neutron beam kinetic energies of 1.20, 2.50 and 3.66 GeV, respectively. The errors are statistical only. The estimate of systematic uncertainty is $\sim 5\%$ (see refs.[5] for more details).

The measured value of $\Delta\sigma_L$ at 1.2 GeV is close to the nearest existing ones and two others show a fast decrease in contrast to a tendency of the PSA solution[1] . The energy behaviour of $\Delta\sigma_L$ seems unexpected also from the point of view of meson theory[6] giving a reasonable description of $\Delta\sigma_{L,T}(pp)$ and

$\sigma_{tot}(pp)$ but predicting no decrease of $\Delta\sigma_L(np)$ up to 2 GeV. An anomalous behaviour of the energy dependence of $\Delta\sigma_L(np)$ was described rather well up to ~ 1 GeV in ref.[7]. The model predicts a possible manifestation of an exotic $^3S_1$ dibaryon in the energy behaviour of $\Delta\sigma_L(np)$ near 1.8 GeV. Another explanation of the observed dip in the energy dependence of the existing $\Delta\sigma_L(np)$ data near 2π production threshold at 0.6 GeV can be proposed using the contribution to $\Delta\sigma_L(np)$ of the nonperturbative flavour-dependent interaction between quarks induced by instantons estimated in ref.[8]. The second deep is predicted at $2\pi + \eta$ threshold. Precise and detailed $\Delta\sigma_{L,T}(np)$ data in an energy range of 1.2-2.5 GeV will be very useful to verify the predictions of these models.

The results obtained in the highest polarized neutron energy region are surprising for traditional nuclear physics,based on rather low energy NN data,but they agree with QCD motivated predictions. Our new data and upcoming measurements over a new energy range covering the mass of NN exotics will be important to extend the evaluation of the characteristics of NN-interactions to higher energies, and will provide sensitive tests of the spin-isospin structure of the short-range NN-interaction and associated models (e.g.[7]). Future studies of $\Delta\sigma_{L,T}$ are also related to an instanton model of QCD vacuum, which predicts sizable anomalies in the $\Delta\sigma_L(np)$ energy behavior over an energy range of several GeV.

Acknowledgments

This work was in part supported by the ISF and Russian Government Grant No.JHW100 and by the Russian Foundation for Basic Research grants RFBR-93-02-03961,RFBR-96-02-18736-a.

References

1. R.A.Arndt,I.I.Strakovsky,R.L.Workman,Phys.Rev.C50,2731(1994).
2. M.Matsuda et al.,Prog.Theor.Phys.93,1059(1995).
3. M.Beddo et al.,Phys.Rev.D50,N1(1994)104; C.Lechanoine-LeLuc and F.Lehar,Rev.Mod.Phys.65(1993)47 and refs. therein.
4. E.Chernykh et al.Proc.Int.Symp."Deuteron-93",JINR E2-94-95,p.185.
5. V.I.Sharov et al.In:Proc.of SPIN-95,Protvino,1995,to be publ.; F.Lehar et al.DAPNIA/SPP 96-03,to be publ.in Z.Phys.C.,v.71,No.1,1996.
6. T.-S.H.Lee,Phys.Rev. C29,195(1984). Nucl.Phys.A513(1990)442.
7. P.LaFrance and E.L.Lomon.In:Proc.Intern.Conf.'Mesons and Nuclei at Intermediate Energies'.Dubna,3-7 May 1994. World Scientific,1995.
8. N.I.Kochelev,3rd Int.Symp."Dubna Deuteron-95",to be published.

QED-QCD MIXING EFFECTS AND ISOSPIN-VIOLATION IN $NN\ ^1S_0$ SCATTERINGS

Chung-Wen Kao[a,b] and Shin Nan Yang[a,c]
[a] *Department of Physics, National Taiwan University, Taipei, Taiwan 10617*
[b] *Department of Physics & Astronomy, University of Maryland, College Park, Maryland 20742, U.S.A.*
[c] *Nuclear Science Division, Lawrence Berkeley National Laboratory, Berkeley, California 94720, U.S.A.*

The quark effects, which include the up-down quark mass difference, quark-exchange Coulomb interaction and the QED-QCD mixing effect proposed by Stepheson, Maltman and Goldman, on the charge dependence of the NN scattering length are studied in a quark cluster model. It is found that they give rise to a large correction to $a_{pp} - a_{nn}$, about the right size as indicated in the experiments. Its contribution to $a_{pp} - a_{np}$, however, is small.

Most of the theoretical studies on the isospin violation in the nuclear force concentrate on meson-mediated mechanisms. On the quark level, the isospin symmetry is violated due to the up-down quark mass difference and Coulomb interaction between quarks. Both of them have been studied [1,2] and found to give important effects. Recently, Stephenson, Goldman and Maltman (SMG)[3] showed that the interference effect of QED and QCD could also give important contributions to the baryon isomultiplet mass splittings and the binding energy difference between ^3He and ^3H [4]. The mixing effect considered by SMG includes the electromagnetic vertex corrections to the gluon exchange graphs (penguin-like) and corrections involving both the gluon and photon exchanges (box-like). We report here the results of an investigation on the effect of this mechanism on the isospin violation in $NN\ ^1S_0$ scattering length, i.e., the scattering length differences between pp, np and nn scatterings in $^1S_0$.

We treat the quark dynamics with the use of a non-relativistic constituent quark cluster model [1]. The resonating group method is employed to treat the six-quark system and we restrict ourselves to the configurations of two three-quark clusters with nucleon quantum numbers. The $\pi-$ and $\sigma-$mediated quark interactions are included in order to achieve a realistic description of NN phase shifts in S-waves. We also include the $u-d$ quark mass difference Δm_q and the exchange-quark Coulomb interaction, i.e., the Coulomb force between two quarks which are exchanged between the two nucleon clusters, in the calculation.

Using the parameters: Δm_q = 2.46MeV, A = -1.54, B = 6.42, C = 3.12, D = -2.09 (all in units of MeV), which lead to the prediction of $\Delta\Sigma^* =$

$-5.45MeV$ as favored by the SMG's analysis [3] we obtain the following table for the corrections to the NN scattering length (in fm).

	δm_q	EC	Box^E	Box^M	$Penguin^E$	$Penguin^M$	Total
$a_{pp} - a_{nn}$	1.98	0.38	-0.02	-0.67	0.10	0.06	1.81
$a_{pp} - a_{np}$	0.99	0.57	-1.12	-1.14	0.05	0.03	-0.62

One sees that each of the box-like and penguin-like diagrams as suggested by SMG, indeed could give rise to large effects in both charge symmetry breaking and charge independence breaking, consistent with their previous findings in $A = 1$ and 3 systems. It is interesting to see that, for the case of $a_{pp} - a_{nn}$, the total contribution from quark effects, i.e., the sum of the effects of quark mass difference δm_q, quark-exchange Coulomb interaction (EC), box-like and penguin-like (both have induced electric and magnetic effects and denoted as Box^E etc.) graphs is large and almost of the right size as observed in the experiments. However, the quark contribution to $a_{pp} - a_{np}$ is small. A word of caution is probably needed here. The model used by SMG for QED $\times$ QCD mixing is rather crude. The results obtained here should therefore serve as more of a qualitative, rather than as quantitative, estimate of the effects and more studies are definitely needed.

Acknowledgments

We thank Ms Yih Jen Lee for fruitful collaboration who participated in the early stage of this investigation. This work is supported in part by the National Science Council of ROC under grant No. NSC85-2112-M002-007.

References

1. M. Chemtob and Shin Nan Yang, Nucl. Phys. **A420** (1983) 461.
2. M.-Z. Wang, F. Wang, and C.W. Wong, **A483** (1988) 661.
3. G.J. Stephenson Jr., K. Maltman, and T. Goldman, Phys. Rev. **D43** (1991) 860; *ibid.* **C41** (1990) 2764.
4. K. Maltman, T. Goldman, and G.J. Stephenson Jr., Nucl. Phys. **A530** (1991) 539.

Sensitivity of nucleon–nucleus scattering to the off–shell behavior of on–shell equivalent NN potentials

H. F. Arellano[a], F. A. Brieva[a], M. Sander[b] and H. V. von Geramb[b]

[a] *Departamento de Física, Facultad de Ciencias Físicas y Matemáticas Universidad de Chile, Casilla 487–3, Santiago, Chile*

[b] *Theoretische Kernphysik, Universität Hamburg Luruper Chaussee 149, D-22761 Hamburg, Germany*

The sensitivity of nucleon–nucleus elastic scattering to the off–shell behavior of realistic nucleon–nucleon interactions is investigated when on–shell equivalent nucleon–nucleon potentials are used. The study is based on applications of the in–medium full–folding optical model potential [1]

$$\begin{aligned} U(\vec{k}\,',\vec{k};E) &= \frac{4}{(2\pi)^3}\int d\vec{R}\, e^{i\vec{q}\cdot\vec{R}}\rho(R)\frac{1}{\hat{\rho}(R)}\int d\vec{P} \\ &\times\ \Theta[\hat{k}(R)-P]\,\left\langle \tfrac{1}{2}(\vec{K}-\vec{P}-\vec{q})\left|g_{\vec{R}+\vec{P}}(E+\bar{\epsilon};\vec{R})\right|\tfrac{1}{2}(\vec{K}-\vec{P}+\vec{q})\right\rangle_A . \end{aligned}$$

for an explicit treatment of the off–shell behavior of the nucleon–nucleon effective interaction. The optical potential requires the calculation of g matrices off–shell as their relative momenta obey no constraints apart from those imposed by the ground state mixed density of the target. Furthermore, no assumptions are introduced on the nature of the momentum dependence of the optical potential, thus retaining all non localities arising from the genuine momentum dependence of the NN effective interaction and as prescribed by the full–folding integral.

We use the momentum–dependent Paris potential in its parameterization [2]

$$V = V_a + \frac{\hbar^2}{m}p^2 V_b + V_b\frac{\hbar^2}{m}p^2\,,$$

and its local on–shell equivalent as obtained by Gelfand–Levitan and Marchenko inversion for the two nucleon Schrödinger equation [3]. Inversion potentials were also constructed directly from NN phase–shift data (SM94) [4] in the 0–1.3 GeV energy range. For examples of phase shifts and their reproducton by inversion see Fig. 1.

Applications were made at beam energies between 40 and 500 MeV for proton scattering from ^{40}Ca and ^{208}Pb (Fig. 2). Full–folding calculations for nucleon–nucleus scattering using Paris and Paris inversion show only small fluctuations in the corresponding observables. This implies that off–shell features of the NN interaction cannot be unambiguously identified with these processes. The use of phase shift analysis based inversion potentials in proton–nucleus scattering above 200 MeV provide a superior description of the observables relative to those obtained from current realistic NN model potentials.

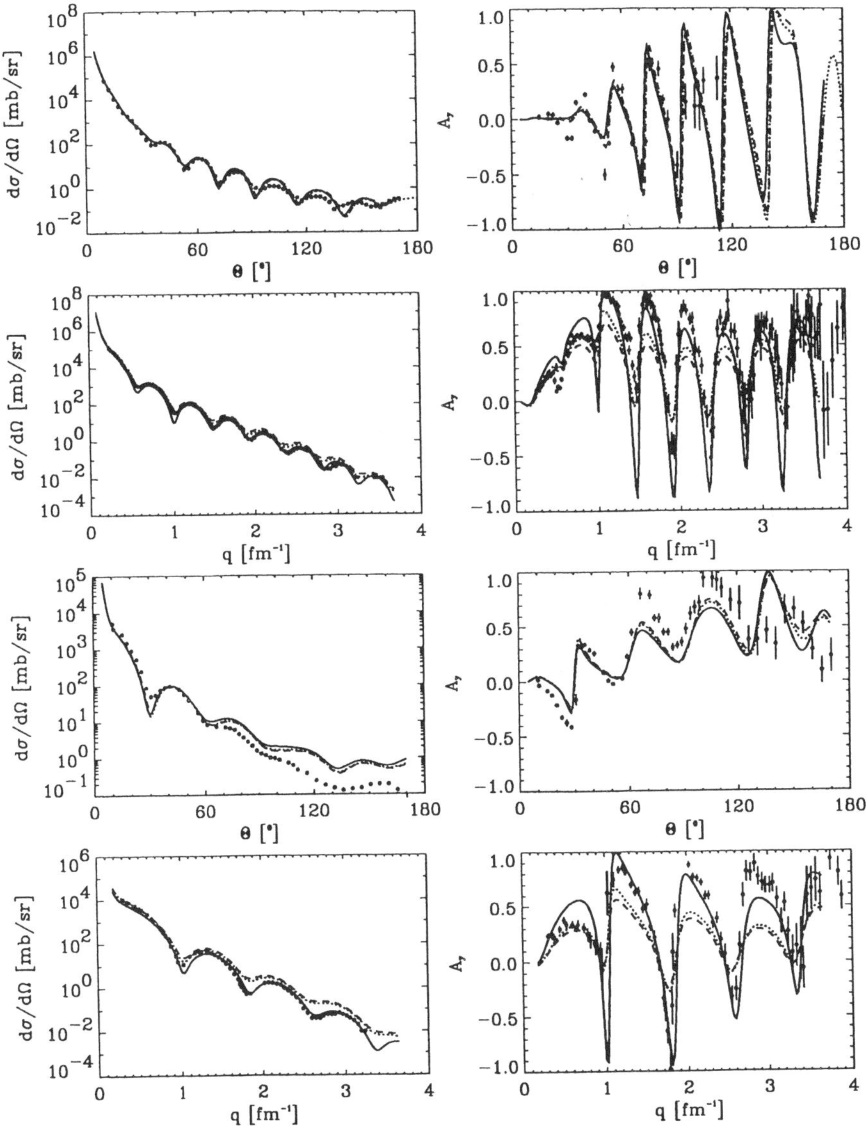

Figure 2: Differential cross sections (left) and analyzing powers (right) for ^{208}Pb(p,p) at 40 MeV (top) and 400 MeV (second row) and ^{40}Ca(p,p) at 40 MeV (third row) and 500 MeV (bottom). We show g–matrix results for SM94 inversion (solid line), the genuine Paris potential (dotted) and Par1.3 inversion (dashed). Experimental data taken from [5].

References

[1] H. F. Arellano, F. A. Brieva and W. G. Love, Phys. Rev. C41, 2188 (1990); *ibid.* C50, 2480 (1994); *ibid.* C52, 301 (1995).

[2] M. Lacombe et al., Phys. Rev. C21, 861 (1980).

[3] H. V. von Geramb and H. Kohlhoff, in *Quantum Inversion Theory and Applications*, edited by H. V. von Geramb, pp 285, (Springer–Verlag, 1994); H. Kohlhoff and H. V. von Geramb, *ibid.* pp 314.

[4] R. A. Arndt, I. I. Strakovsky and R. L. Workman, Phys. Rev. C50, 2731 (1994).

[5] R. H. McCamis et al., Phys. Rev. C33, 1624 (1986); L. N. Blumberg et al., Phys. Rev. 147, 812 (1966); D.A. Hutcheon et al., Nucl. Phys. A483, 429 (1988); W. T. H. van Oers, H. Haw and N. E. Davison, Phys. Rev. C10, 307 (1974).

Supported in part by FONDECYT grants 3940008 and 1931115, and KFA Jülich COSY Collaboration Grant No. 41126865.

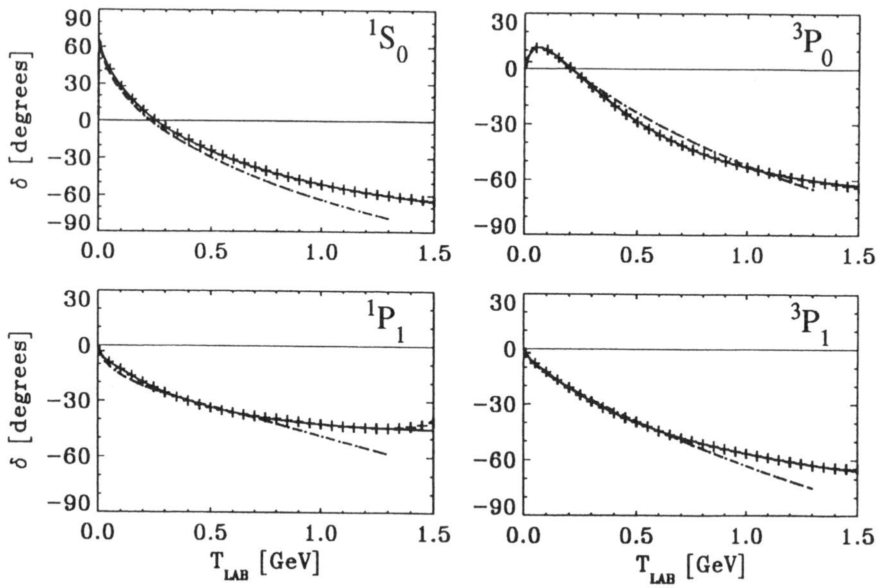

Figure 1: *np* single channel phase shifts for $\ell = 0, 1$ channels. We show the CAPS SM94 phase shift analysis [4] (crosses), the phase function of the SM94 inversion potential (solid line), the genuine Paris phase shifts [2] (dotted) and the phase function of the Par1.3 inversion potential (dashed).

$\pi^+ + d \to p + p$ REACTION BETWEEN 18 AND 44 MeV

E. A. PASYUK, V. YU. ALEXAKHIN, S. I. GOGOLEV, K. O. OGANESYAN
Joint Institute for Nuclear Research, Dubna, 141980 Russia

C. L. MORRIS, M. W. RAWOOL-SULLIVAN
Los Alamos National Laboratory, Los Alamos, NM 82740

M. K. JONES
College of William and Marry, Williamsburg, VA 23187-3542

J. M. O'DONNELL
University of Minnesota, Minneapolis, MN 55455

F. F. GUBER, A. I. RESHETIN
Institute for Nuclear Research, Troitsk, 117312 Russia

I. I. STRAKOVSKY
St. Petersburg Nuclear Physics Institute, Gatchina, 188350 Russia

The experimental data showed a deviation from smooth behavior in the total cross section of the reaction $\pi^+ d \to pp$ close to $T_\pi \sim 30$ MeV.[1] More recent measurements of the partial total cross section for the inverse reaction at SATURNE have verified this effect.[2] In a recent experiments of the Dubna-Gatchina group on pion absorption on carbon[3] and deuteron[4] a dip in the energy dependence of the total cross section has been observed at this energy range. However, these experiments are not accurate enough to reach a conclusion concerning the structure under the discussion. A satisfactory solution fitting the available database has not been found in partial wave analysis, suggesting underlying systematic problems with the data base.

The experiment described here was intended to resolve the discrepancies in existing data and to verify the observed structure in excitation function of pion absorption on the deuteron at low pion energies.

The experiment was performed at the Low Energy Pion channel (LEP) of the Clinton P. Anderson Meson Physics Facility (LAMPF). A large solid angle detector, the LAMPF BGO ball, was used to detect the reaction products in this study. The differential cross sections for six angles and total cross sections have been measured at 15 energies between 18 and 44 MeV with an increment of pion energy of 1 – 2 MeV. Measured differential and total cross sections are presented in Fig. 1 and 2. The number of the experimental data points in this energy range is almost doubled by these results. The data generally follow the trends of previous results and are in a good agreement with recent SM95

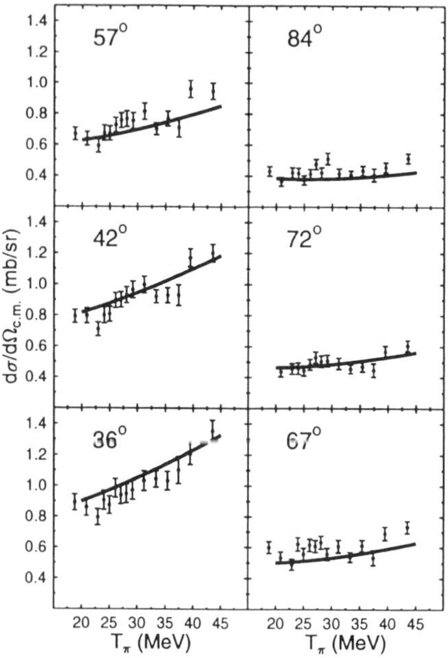

Fig. 1. Measured differential cross sections, solid line - SM95 partial wave analysis.

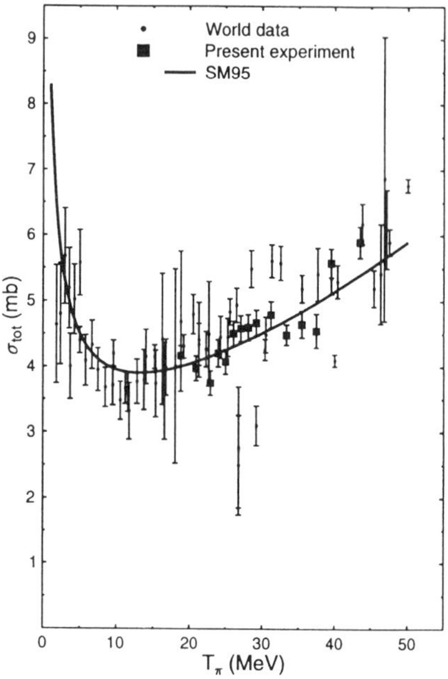

Fig. 2. Total cross section.

solution of VPI partial wave analysis.[5] The previous disagreement in total cross section data is resolved by this experiment. We do not observe any structure either in total cross section or in differential cross section within the accuracy of this experiment. Obtained data are most precise for this energy range.

References

1. I. I. Strakovsky and E. A. Pasyuk, in: *"Experimental Program of the Investigations at Inst. Nucl. Res. Moscow Meson Factory"* (Proc. of 4th All–Union Seminar on Experimental Program of Investigations at INR Moscow Meson Factory, 1985) Moscow, Russia: Institute for Nuclear Research Press, 1986, p. 241.
2. J. Saudinos, *et al.*, in: *Proc. of Int. Workshop on Nucl. Phys. in GeV region, KEK, Nov. 13–15, 1984*, KEK Report 84-20, 1985, p. 228; Nouvelles de Saturne **10**, 25 (1985).
3. Yu. K. Akimov, *et al.*, *Nucl. Phys.* A **451**, 433 (1992).
4. S. I. Gogolev, *et al.*, *Phys. Lett.* B **300**, 24 (1993).
5. R. A. Arndt, *et al.*, *Phys. Rev.* C **48**, 1926 (1993); private communication (1995).

STUDY OF THE REACTION $pp \to d\pi^+$ CLOSE TO THRESHOLD

M. HAWASH, S. IGEL, K. KILIAN, G. LIPPERT, H. MACHNER, R. MAIER, C. NAKE, B. RAZEN, P. VON ROSSEN
Institut für Kernphysik, Forschungszentrum Jülich, Jülich, Germany

M. DROCHNER, G. KEMMERLING, K. ZWOLL
Zentrallabor für Elektronik, Forschungszentrum Jülich, Jülich, Germany

L. JARCZYK, W. KLIMALA, J. SMYRSKI, A. STRZAŁKOWSKI
Institute of Physics, Jagellonian University, Krakow, Poland

L. FREINDL, S. KLICZEWSKI, R. SIUDAK
Institute of Nuclear Physics, Krakow, Poland

T. KUTSAROVA
Institute of Nuclear Physics and Nuclear Energy, Sofia, Bulgaria

D. KOLEV, R. TSENOV
Physics Faculty, University of Sofia, Bulgaria

R. JAHN, J. ERNST, K. SCHO
Institut für Strahlen- und Kernphysik der Universität Bonn, Bonn, Germany

S. FÖRTSCH
National Accelerator Centre, Faure, South Africa

D. FREKERS, W. GARSKE
Institut für Kernphysik, Universität Münster, Münster, Germany

We have measured the total cross section and full angular distributions for the reaction $pp \to d\pi^+$ at five laboratory energies within 8 MeV of threshold. We used a magnetic spectrometer which allowed to detect the recoil deuterons with nearly 4π acceptance in CM system. Comparison to $np \to d\pi^o$ data offers no evidence for charge independence breaking. The s-wave and p-wave strength is deduced.

Recently, the reaction $pp \to d\pi^+$ has again attracted theoretical interest. According to Horowitz[1] the direct term and the rescattering term account for only about one-half of the measured cross sections. In order to explain the missing cross section he introduces heavy meson exchange involving a nucleon-antinucleon pair. In contrast, Niskanen[2] claims that inclusion of the Δ isobar will produce a larger s-wave cross section. However, these findings are based on a rather poor data base. The aim of the present work is, therefore, to measure cross sections for the $pp \rightarrow d\pi^+$ reaction close to threshold over the full angular range, which allow to determine the s-wave and p-wave pion production.

The measurements were done for five beam energies within 8 MeV of threshold equal to 287 MeV. An external proton beam from the COSY (COoler SYnchrotron) in Jülich was focussed onto a cell containing liquid hydrogen. The recoil deuterons which, near threshold, emerge within a narrow cone along the beam direction, were detected in a magnetic spectrometer with an acceptance close to 4π in CM system. The trajectories of deuterons were measured in the focal plane of the spectrometer by two stacks of MWDC's, which allowed to reconstruct the momentum vectors of deuterons at the target point. Differential cross sections were deduced by normalizing the count rate to the target thickness and to the number of incident protons measured by counting scattered protons with calibrated scintillation detectors.

Obtained results for the total cross sections and the anisotropies of the angular distributions of differential cross section are shown in fig. 1 together with all data published after 1967, to the best of our knowledge. All data with two charged particles in the exit channel were corrected for Coulomb effects as proposed by Reitan[4]. The present data seem to be a smooth continuation of the earlier data for the charge particle induced reactions towards threshold. They agree well with measurements of Hutcheon et al.[3] of the isospin related reaction $np \rightarrow d\pi^o$. Thus, no evidence for charge independence breaking was established.

The Coulomb corrected total cross sections were fitted by the Gell-Mann and Watson[5] threshold formula:

$$\sigma_{TOT} = \alpha_0 \eta + \alpha_1 \eta^3 \tag{1}$$

The old Coulomb corrected data for $\eta \leq 0.8$ yield values of α_0=0.22±0.01mb and α_1=0.79±0.01mb. The corresponding cross sections are shown as dashed curve in fig.1. The curve definitely overpredicts the present data as well as the $np \rightarrow d\pi^o$ data. If the present data are added to the data sample the fit yields α_0=0.195±0.003mb and α_1=0.840±0.008mb. However, the χ^2-values obtained indicate poor fits. The first fit yields $\chi^2/nfree$=17.4 and the one including the present data $\chi^2/nfree$=16.2. The dominant contribution to the large value of χ^2 comes from the data with η larger than 0.5. When these data are excluded from the analysis, a value of $\chi^2/nfree$=0.9 is obtained for data including the present ones with coefficients α_0=0.181±0.003mb and α_1=0.933±0.054mb. The corresponding cross sections are shown as solid line in fig. 1. The s-wave strength is in agreement with α_0=0.184±0.005mb obtained by Hutcheon et al.[3], whereas, for the p-wave their value of α_1=0.781±0.079mb differs from ours.

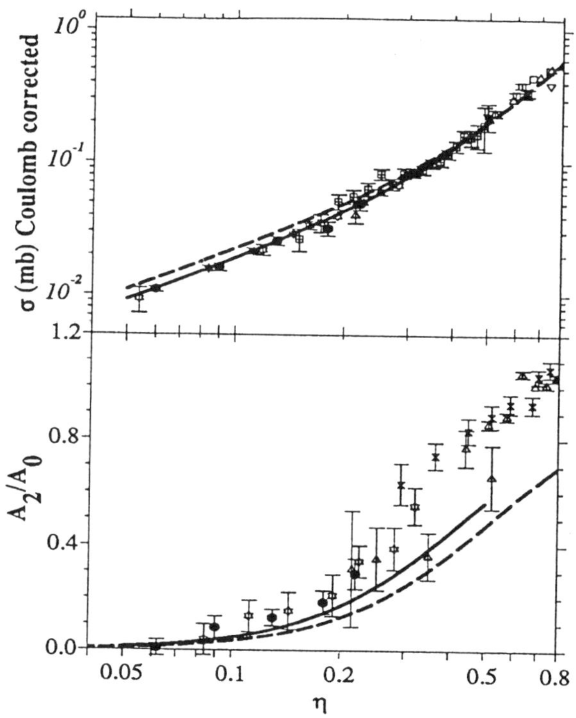

Figure 1: Coulomb corrected total cross sections and anisotropies as a function of the pion momenta. Our data are shown as full symbols and open stars represent the data for $np \to d\pi^0$ of Hutcheon et al.. The curves are predictions of the Gell-Mann and Watson model as described in the text.

The anisotropies obtained from the above fits are shown as corresponding curves in the lower part of fig. 1. They underestimate the experimental data. This could indicate for an interference between the s-wave and d-wave in the pp-channel leading to the p-wave in the πd-system.

1. C. J. Horowitz, Phys. Rev. **C48**(1994)2920
2. J. A. Niskanen, Phys. Rev. **C53**(1996)526
3. D. A. Hutcheon et al., Phys. Rev. Lett. **64**(1990)176
4. A. Reitan, Nucl. Phys. **B11**(1969)170
5. M. Gell-Mann and K. M. Watson, Ann. Rev. Nucl. Sci. **4**(1954

A Signal of a Narrow πNN-Resonance in $pp \to pp\pi^-\pi^+$ [a]

W. Brodowski[1], R. Bilger[1], H. Cálen[2], H. Clement[1], C. Ekström[3], K. Föhl[1], K. Fransson[2], L. Gustafsson[2], S. Häggström[2], B. Höistad[2], A. Johansson[2], T. Johansson[2], K. Kilian[4], S. Kullander[2], A. Kupść[5], G. Kurz[1], P. Marciniewski[5], B. Morosov[6], J. Moehn[2], A. Mörtsell[2], W. Oelert[4], V. Renken[4], R. Ruber[2], M.G. Schepkin[7], U. Siodlaczek[1], J. Stepaniak[5], A. Sukhanov[6(2)], A. Turowiecki[9], G.J. Wagner[1], Z. Wilhelmi[9], J. Zabierowski[8], A. Zernov[6], J. Zlomanczuk[9(2)]

[1]*Physikalisches Institut, Univ. Tübingen, Germany*
[2]*Dept of Radiation Sciences, Univ. Uppsala, Sweden*
[3] *The Svedberg Laboratory, Uppsala, Sweden*
[4]*Institut für Kernphysik, KFA Jülich, Germany*
[5]*Institute for Nuclear Studies, Warsaw, Poland*
[6] *Joint Institute for Nuclear Research, Dubna, Russia*
[7]*Institute for Theoretical and Experimental Physics, Moscow, Russia*
[8]*Institute for Nuclear Studies, Lodz, Poland*
[9]*Institute for Experimental Physics, Warsaw, Poland*

Abstract

We report on the first exclusive measurement of the 2π-production in NN-collisions at $T_p = 750$ MeV using a H_2 cluster jet target and the WASA/PROMICE detector at CELSIUS. In the invariant mass spectrum $M_{pp\pi^-}$ we observe a narrow peak at 2.063(2) GeV with a $> 4\sigma$ deviation relative to the MC-simulations for the conventional 2π-production process. The finding is in agreement with the hypothesis of a narrow πNN-resonance with $I(J^P) =$ even (0^-), called d', postulated recently to understand the peculiar resonance-like behavior of the pionic double charge exchange on nuclei.

1. Introduction - the case of dibaryons

With the realization of quarks being the basic building blocks of hadrons also the idea emerged in the seventies that this substructure should cause nontrivial resonances in the dibaryon system — in addition to the trivial meson-dominated nucleon-nucleon (NN) groundstate, the deuteron. QCD-inspired model calculations [1,2] predicted a large number of 6q states causing a rush for experimental dibaryon searches in the years to follow. Unfortunately, despite a vast number of dedicated experiments no unambiguous evidence for their existence could be found. The bulk of these experiments, however, was dedicated to searches of dibaryons coupled to NN or to NΔ, where the fall-apart decay will cause widths even large to those of baryon resonances. Hence such states should be very difficult to sense in experiments.

2. First evidence for the πNN resonance d' from the pionic double charge exchange

Recently the existence of a narrow, NN-decoupled πNN-resonance, called d', has been proposed [3] to explain the peculiar resonance-like behavior of the pionic double charge exchange (DCX) reaction (π^+, π^-) on nuclei at pion energies below the Δ-resonance. From the analysis of these DCX-data the parameters of d' have been deduced to be $m \approx 2.06$ GeV, $\Gamma_{NN\pi} \approx 0.5$ MeV and $I(J^P) =$ even (0^-). From the small width it was concluded that this resonance is NN-decoupled, hence the isospin should be even. A major critique on this interpretation has been that the DCX-reaction takes place on nuclei, where subtle, hitherto unknown medium effects cannot be excluded unambiguously as a possible alternative reason — though the predicted signatures of d' formation and production in DCX meanwhile have been observed [4,5] even on the lightest nuclei

[a]Supported by BMBF (06 TÜ 669), DFG (Mu 705/3, Graduiertenkolleg), DAAD (313/S-PPP-2/94), the Swedish Institute and the Swedish Natural Science Research Council

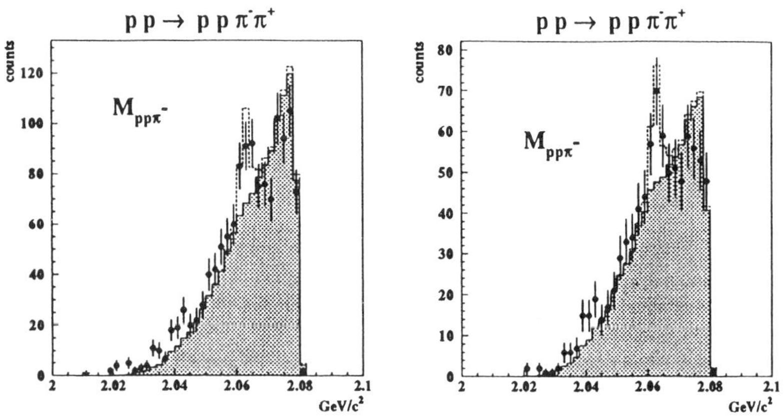

Figure 1: Invariant-mass spectrum $M_{pp\pi^-}$ of the reaction $pp \to pp\pi^-\pi^+$ at $T_p = 750$ MeV with (right) and without (left) constraints on the missing-mass spectrum $MM(pp\pi^+)$. The solid histograms show the MC simulations of the conventional 2π production process, the dashed ones show the result with inclusion of the d' production process.

where this reaction is possible. Unfortunately since neither diproton nor dineutron are stable, the DCX-reaction cannot be carried out on the basic dinucleon system.

3. Signal of d' in the 2π production in NN collisions

Hence it has been proposed [3,6] to look into the 2π-production in pp-collisions according to

$$pp \to d'\pi^+ \to pp\pi^-\pi^+ ,$$

where the NN-decoupled resonance d' could be produced hadronically in a most fundamental way. The d' contribution there has been estimated [6] to be in the order of 10% of the conventional 2π cross section near the d' threshold, which is expected to be around $T_p \approx 710$ MeV.

Here we report on the first exclusive measurement [7] of the reaction $pp \to pp\pi^-\pi^+$ at $T_p = 750$ MeV using a H_2 cluster jet target. All previous measurements of this reaction in this energy region have been inclusive and/or difference measurements on CH_2 and C targets. The experiment has been carried out at the CELSIUS storage ring using the detector setup [8] of the WASA/PROMICE collaboration. For the data presented here only events registered in the forward detector, a segmented scintillation calorimeter covering the angular range $4° \leq \theta \leq 21°$ have been used. Since the present setup contains no magnetic field, the delayed pulse technique has been installed for π^+ identification. The trigger has been set to at least 3-prong events of charged particles in the forward detector, thus suppressing events from single pion production. For the 2π-production events selected we demand two identified protons and one identified π^+ particle. The missing mass spectrum of these identified $pp\pi^+$ events peaks as expected at the mass of the unidentified π^- particle.

Using the momentum and energy information of the identified π^+ from these selected $pp\pi^+$ events together with the corresponding information of the beam, we obtain the $M_{pp\pi^-}$-spectrum as shown on the left of fig. 1.

If in addition we require the missing mass of the registered $pp\pi^+$ event, i.e. the mass of the unobserved π^-, to be in the range of 130 MeV – 150 MeV, then we obtain the $M_{pp\pi^-}$-spectrum as shown on the right of fig. 1. On top of a smooth distribution we observe in both cases a narrow irregularity around 2.063 GeV with a statistical significance of (4 to 5)σ relative to the Monte-Carlo (MC) simulations for the conventional 2π-production. For these simulations a pure phase space distribution of the conventional 2π production process has been assumed. The width of the irregularity is in accordance with the experimental resolution. For details see ref. [7]. Further measurements are in progress.

4. Outlook — renaissance of 6q searches

In conclusion the first exclusive measurement of the reaction $pp \rightarrow pp\pi^-\pi^+$ at $T_p = 750$ MeV exhibits in the $M_{pp\pi^-}$ spectrum a narrow peak sitting upon a continuum arising from the conventional 2π production process. Hence we have reason to believe that d' production has been observed in this reaction in full agreement with predictions based on the d' hypothesis. In consequence that would mean that there are, indeed, more states in the dibaryon system than just the trivial meson-dominated deuteron ground state and the virtual $^1S_0$ level.

A final proof of the existence of d' will have a number of implications. Since it is unlikely that d' would be singular in the dibaryonic world, the search for further resonances could lead to a revival of the hunt for 6q states. The existence of such states will provide a unique opportunity to study for the first time the QCD of baryonic systems with more than three quarks. Among others this could give new insights into the problem of confinement [9]. Another point of interest would be the impact of dibaryonic resonances, in particular their bosonic nature, on the behavior of nuclear matter under extreme conditions.

1. see e.g. P.J.G. Mulders et al., Phys. Rev. **D21**, 2653 (1980)
2. L.A. Kondratyuk et al., Sov. J. Nucl. Phys. **45**, 776 (1987)
3. R. Bilger et al., Z. Phys. **A343**, 491 (1993), Phys. Rev. Lett. **71**, 42 (1993)
4. K. Föhl et al., contribution to this conference
5. R. Meier et al., contribution to this conference
6. M. Schepkin, O. Zaboronsky and H. Clement, Z. Phys. **A345**, 407 (1993)
7. W. Brodowski et al., Z. Phys. **A355**, 5 (1996)
8. H. Calén et al., Nucl. Instr. Meth., in press
9. A.J. Buchmann et al., Prog. Part. Nucl. Phys. **36** 383 (1996)

Pionic Double Charge Exchange and the Signature of the πNN Resonance d'[a]

K. Föhl, D. Schapler, R. Bilger, H. Clement, G.J. Wagner
Physikalisches Institut der Universität Tübingen, D-72076 Tübingen, Germany

A. Denig, W. Kluge, R. Wieser
Institut für Experimentelle Kernphysik, D-76021 Karlsruhe, Germany

M. Schepkin
Institute for Theoretical and Experimental Physics, Moscow 117259, Russia

R. Abela, F. Foroughi, D. Renker
Paul Scherrer Institut, CH-5232 Villigen PSI, Switzerland

Abstract

We report on a systematic study of the pionic charge exchange reaction on nuclei. Its peculiar resonance-like behavior at energies below the Δ resonance has recently been attributed to the signature of a narrow πNN resonance, called d', with $I(J^P)$ = even (0^-) and $m = 2.06$ GeV. The new measurements on closed shell nuclei and on the lightest nuclei possible are in strong support of this hypothesis and render other interpretations very unlikely — in full agreement with the most recent discovery of a d' signal in $pp \rightarrow pp\pi^-\pi^+$, also reported at this conference.

With the realization of quarks being the basic building blocks of hadrons also the idea emerged in the seventies that this substructure should cause nontrivial resonances in the dibaryon system — in addition to the trivial meson-dominated nucleon-nucleon (NN) groundstate, the deuteron. QCD-inspired model calculations [1,2] predicted a large number of 6q states, which caused a rush for experimental dibaryon searches in the years to follow. Unfortunately, despite a vast number of dedicated experiments no unambiguous evidence for their existence could be found. The bulk of these experiments, however, was dedicated to searches for dibaryons coupled to NN or to NΔ, where the fall-apart decay will cause widths even large to those of baryon resonances. Hence such states should be very difficult to sense in experiments.

Recently the existence of a narrow, NN-decoupled πNN resonance, called d', has been proposed [3] to explain the peculiar resonance-like behavior of the pionic double charge exchange (DCX) reaction (π^+,π^-) on nuclei at pion energies below the Δ resonance. From the analysis of these DCX data on energy and angle dependence the parameters of d' have been deduced to be $m \approx$ 2.06 GeV, $\Gamma_{NN\pi} \approx 0.5$ MeV and $I(J^P)$ = even (0^-). From the small width it was concluded that this resonance is NN-decoupled, hence the isospin should be even. A major critique on this interpretation has been that the DCX reaction takes place on nuclei, where subtle, not yet understood medium effects cannot be excluded unambiguously as a possible alternative reason.

Hence, in order to meet the above critique and to test the d' hypothesis on most clear-cut cases, we have carried out new measurements on closed shell nuclei as well as on ^{7}Li, the lightest nucleus where the DCX-process to a discrete, albeit unbound, final state can be observed. In order to keep nuclear structure problems at a minimum and to avoid the additional complexity of an analog route in case of transitions to double analog states (DIAT), we have investigated the non-analog groundstate transitions (GST) on the doubly magic isoscalar nuclei ^{16}O and ^{40}Ca (fig. 1). These necessitate cross shell transitions, which are expected to be particularly weak in case of a conventional reaction process. In the seniority model GSTs on isoscalar nuclei are predicted [4] to have even vanishing cross sections. Also, if the above mentioned subtle medium effects really were responsible for the resonance-like behavior, then the latter would be expected to be much

[a]Supported by BMBF (06 TÜ 669, 06 KA 266), DFG (Mu 705/3, Graduiertenkolleg) and DAAD (313-S-PPP-2/94)

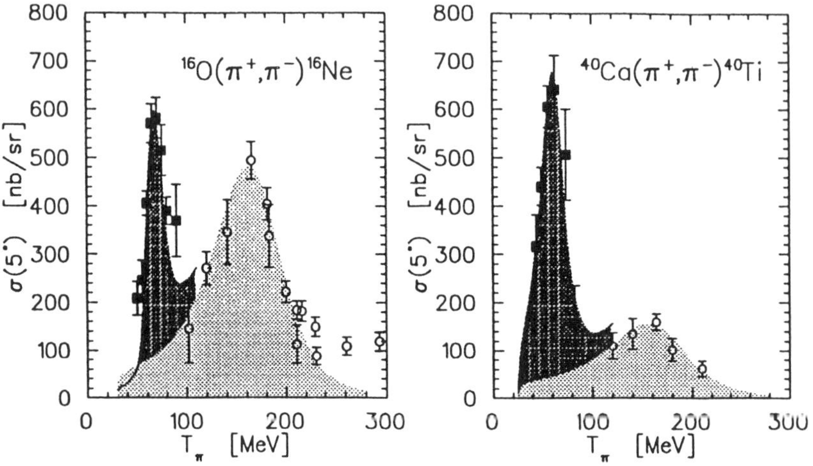

Figure 1: Energy dependence of the DCX groundstate transitions on ^{16}O and ^{40}Ca. Open data points are from work at LAMPF.

less pronounced in the light nucleus ^{16}O than in the much heavier nucleus ^{40}Ca. The cleanest case in this respect appears to be ^{7}Li, which to a good approximation represents the cluster partition $t + \alpha$, and the DCX on ^{7}Li leads to the groundstate of ^{7}B, which is already 2 MeV above the $(3p + \alpha)$-threshold. Inasmuch as the α core can be considered as spectator, this reaction then constitutes already the DCX on a three-nucleon system.

In all cases we again find a pronounced resonance-like behavior in the energy dependence, whereas the angular distributions are smooth, behaving quite regularly. As seen in fig. 1, showing the GSTs in ^{16}O and ^{40}Ca as examples, the energy dependence is characterized by two bumps, a broad one near $T_\pi = 160$ MeV and a pronounced, quite narrow one at lower energies. While the broad structure is understood as being due to the Δ resonance, the narrow one has recently [3] been associated to an additional resonance. Since in the πN system there is no resonance below the Δ, this resonance cannot happen in the πN system. And since the angular distributions do not exhibit any similarity to $P_J^2(\cos\Theta)$, this structure cannot be due to a π-nucleus resonance either. Hence the only possibility left appears to be a resonance in the πNN-system, the basic system of the DCX-process. And indeed, both energy and angular dependences of GSTs and DIATs can be well described [3] by such a resonance, called d', with parameters as given above, and with a spreading width due to the interaction with the nuclear medium. A dominant contribution to the spreading is expected to arise from $d'N \to 3N$ mediated e.g. by virtual pion exchange. The invariant amplitude of this process is given by $(fg/2)\sum_{ikj}\epsilon_{ijk}\bar{u}_i C\gamma_5(\tau_2\vec{\tau})\bar{u}_j \cdot \bar{u}_k\gamma_5\vec{\tau}u \cdot [(P-p_k)^2 - m_\pi^2]^{-1}$, where the πNN decay amplitude is $f \cdot \bar{u}_1 C\gamma_5(i\tau_2\vec{\tau})\bar{u}_2\vec{\pi}$, with $f^2 \sim \Gamma_{\pi NN}$. Here $i,j,k = 1,2,3$ denote outgoing nucleons, $\bar{u}_i$ (u) and p_i (P) are bispinors and 4-momenta of the outgoing (initial) nucleons, respectively, C is the charge conjugation matrix, and g is the pseudoscalar $NN\pi$ coupling constant. The cross section $\sigma(d'N \to 3N)$, multiplied by the velocity of d' and by the nuclear density, gives an estimate for the spreading width of ≈ 7 MeV, which is already quite close to the experimental findings. The experiments, in addition, indicate that the spreading width increases towards heavier nuclei as will be discussed in a forthcoming paper. The darkly shaded areas in fig. 1 show the calculation for d' interfering with the tail of the Δ-resonance process (lightly shaded).

We note that the cross section of the broad bump decreases very much when going from ^{16}O to ^{40}Ca according to the well-known $A^{-4/3}$ dependence of the GSTs in the Δ-region. The narrow structure associated with d' formation on the other hand is of comparable strength in both cases

— in contrast to what would be expected, if medium effects were the reason for its appearance.

Another hint on the nature of this structure at low energies may be obtained from the dependence of its peak energy on the reaction Q-value. If there was no Q-dependence, then the effect could be associated with an initial state interaction. If the peak energy changed just in accordance with the Q-value, then the structure would be attributed to a final state interaction effect. The low-energy data on GSTs and DIATs, however, exhibit roughly a $Q/2$-dependence of the peak energy, which again is in favor of a resonance in the intermediate πNN system.

In conclusion, the assumption of a NN-decoupled d' resonance is supported by all hitherto measured low-energy DCX-data, which range from ^{56}Fe down to ^{7}Li. The reaction on ^{7}Li constitutes already the DCX on a three-nucleon system, if the α core can be considered as a pure spectator. The DCX on the still lighter nuclei ^{3}He and ^{4}He no longer leads to any discrete final state but only to a continuum of identical nucleons. In this case the d' resonance can no longer be formed in the s-channel, however, may be produced in the course of this reaction. This production starting at the d' production threshold could be observed either by a steep rise in the energy dependence of the total cross section [5] or by exclusive mesurements [6,7] of the invariant πNN mass in this reaction. Available data on the DCX on ^{4}He [5-8] again are in support of the d' hypothesis. The most recent discovery [9] of the d' signal in $pp \rightarrow pp\pi^-\pi^+$ will, if confirmed, provide the ultimate proof for the existence of d'.

1. see e.g. P.J.G. Mulders et al., Phys. Rev. **D21**, 2653 (1980)
2. L.A. Kondratyuk et al., Sov. J. Nucl. Phys. **45**, 776 (1987)
3. R. Bilger et al., Z. Phys. **A343**, 491 (1992), Phys. Rev. Lett. **71**, 42 (1993)
4. N. Auerbach et al., Phys. Rev. **C38**, 1277 (1988)
5. H. Clement et al., Phys. Lett. **B337**, 43 (1994)
6. A. Lehmann et al., πN Newsletter **11**, 53 (1995)
7. R. Meier et al., contribution to this conference
8. E. Friedman et al., TRIUMF experiment E 725, priv. communication
9. W. Brodowski et al., Z. Phys. **A355**, 5 (1996) and contribution to this conference

^{4}He(π^+,π^-pp) INVARIANT MASS SEARCH WITH CHAOS

R. MEIER, P. AMAUDRUZ, L. FELAWKA, D. OTTEWELL, G. R. SMITH
TRIUMF, Vancouver B. C., Canada

T. AMBARDAR, G. HOFMAN, M. KERMANI, G. TAGLIENTE
University of British Columbia, Vancouver B. C., Canada

J. CLARK, M. SEVIOR
University of Melbourne, Melbourne, Australia

P. HONG, E. L. MATHIE, R. TACIK
University of Regina, Regina, Saskatchewan, Canada

J. GRÄTER, G. J. WAGNER
University of Tübingen, Tübingen, Germany

F. BONUTTI, P. CAMERINI, N. GRION, R. RUI
University of Trieste and INFN Trieste, Trieste, Italy

This experiment is searching for the hypothetical d′ dibaryon. The d′ was put forward as the explanation for an enhancement in the total cross section of pion double charge exchange in nuclei at low energies [1]. From the analysis of DCX, the mass of the d′ has been deduced to be about 2.06 GeV, with a width of 0.5 MeV. The suggested quantum numbers are $J^P = 0^-$ and $T = 0$. With these quantum numbers, the d′ can not couple to the two nucleon channel, but will decay into a pion and two nucleons. Searches for the d′ are underway in various reaction channels. Recently, the WASA/PROMICE collaboration at CELSIUS reported a four sigma signal at 2063 MeV in the two pion production reaction $pp \rightarrow \pi^+\pi^- pp$ [2].

We are looking for the d′ by investigating the double charge exchange reaction $\pi^+ \, ^4He \rightarrow \pi^- pppp$. If the d′ is there, a large part of the DCX cross section in a region above the d′ production threshold should be due to the reaction $\pi^+ \, ^4He \rightarrow d'pp \rightarrow (\pi^- pp)\, pp$ [3]. The available data for the total DCX cross section on ^{4}He favour the predicted sharp rise of the cross section above the d′ threshold of about $T_\pi = 85$ MeV [3]. Unfortunately, the few data in the critical region from 80 to 120 MeV suffer from large error bars or systematic uncertainties due to extrapolations necessary to account for finite detector acceptances. Therefore, new measurements of the inclusive total cross section were done recently at TRIUMF using the CHAOS detector [4]. Preliminary online results from that experiment indicate a behaviour of the cross section consistent with the d′ hypothesis.

As a direct, model independent test for the existence of the d′, we measured the invariant mass of the d′ decay particles produced in DCX on ^{4}He in the region above the d′ threshold. The experiment employed the CHAOS detector at TRIUMF [5]. The detector was set up to simultaneously detect a π^- and at least two protons from the ^{4}He(π^+,π^-pp)pp reaction. A ^{4}He gas target at standard temperature and pressure was used. From the measured momenta of the three detected particles, the invariant mass of the $\pi^- pp$ system was calculated; the d′ should show up as a peak in the invariant mass spectrum. Data were taken in January/February 1996 at T_{π^+}=115 MeV. Fig. 1 shows the

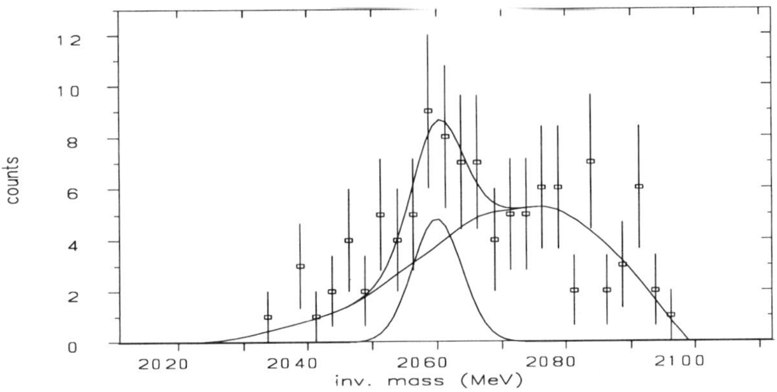

Figure 1: Spectrum of the invariant mass.

invariant mass spectrum for the DCX events identified so far. Clearly, the poor statistics do not allow definitive conclusions. Nevertheless, an enhancement can be seen over the simulated nonresonant DCX background in the region of 2060 MeV. The enhancement has a statistical significance of about two sigma. We'll continue taking data at this energy until the statistics allow a solid statement. Should the enhancement persist, it would be a strong indication for the existence of the d′.

We'd like to thank R. Bilger, J. T. Brack, H. Clement, E. Friedman, and M. Schepkin for useful conversations.

1. R. Bilger et al., Z. Phys. **A343**, 491 (1992); Phys. Rev. Lett. **71**, 42 (1993).
2. W. Brodowski et al., Z. Phys. **A355**, 5 (1996).
3. H. Clement et al., Phys. Lett. **B337**, 43 (1994).
4. E. Friedman et al., TRIUMF Research Proposal E725 (1994).
5. G. R. Smith et al., NIM **A362** 349 (1995).

PRECISION MEASUREMENT OF $\pi^{\pm}p$ DIFFERENTIAL CROSS SECTIONS AT ENERGIES SPANNING THE Δ

MARCELLO M. PAVAN[a]
University of British Columbia/TRIUMF,
4004 Wesbrook Mall, Vancouver, B.C. V6T-2A3, Canada

We present results of a precision measurement of $\pi^{\pm}p$ elastic absolute differential cross-sections at energies spanning the Δ resonance. The data favour the results of recent data sets and V.P.I. partial–wave analyses over older data sets and Karlsruhe analyses.

At TRIUMF, we have completed a precision measurement of $\pi^{\pm}$p absolute differential cross sections at pion kinetic energies from 141 to 267 MeV. Precise and accurate $\pi^{\pm}$p data kind in this energy region are crucial elements of πN partial–wave analysis (PWA). This work was motivated in part to try to resolve long standing discrepancies between data sets in this important energy region, which has implications for determinations of the πNN coupling constant and Σ term. A detailed account of the experiment can be found elsewhere [1].

$\pi^{\pm}$p scattering data were accumulated using six arms of scintillator counter telescopes detecting the scattered pions and recoil protons in coincidence covering CMS angles from $\sim 70°$ to $\sim 160°$. π+p single arm data (detecting only the scattered pions) were obtained for angles from $\sim$25 to $\sim 75°$. Solid (CH2) and liquid targets were used; however, only the liquid target was used for the single arm measurements. The philosophy of the experiment was to measure the cross sections with much redundancy to probe sources of systematic error, and to estimate their uncertainty. This approach proved to be very successful.

For example, results at 141 MeV are shown in figure 1 as ratios to the Karlsruhe–Helsinki KH80 PWA [2] solution. The data sets of Bussey, *et al.* [3], and Brack, *et al.* [4] are plotted as ratios to KH80 at their respective energies. Two V.P.I. analyses, SM95 [5] and SP95 [6], are also shown. (*n.b.* our data have *not* been included in any of these analyses). In general, our data have 1–2% statistical and $\sim$1–1.5% normalization uncertainties, and exhibit excellent self-consistency. The data are consistent with the Brack data near 140 MeV (their highest energy) within the respective normalization uncertainties, and the Sadler, *et el.* [7] data near 263 MeV (their lowest). Below 193 MeV, we clearly disagree with the Bussey, but above the results are compatible, though our uncertainties are smaller. Our data clearly favour the V.P.I. solutions over

[a]Present Address: M.I.T., L.N.S. 26-456, 77 Mass. Ave., Cambridge, MA 02143

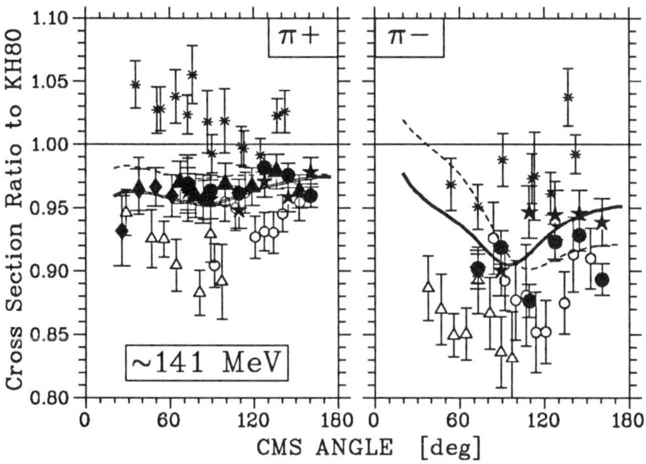

Figure 1: Ratios of absolute cross sections near 141 MeV to the Karlsruhe–Helsinki KH80 [2] PWA result. Our data are the solid circular, triangular (LH_2 πp coincidence), starred (CH_2 coincidence), and diamond–shaped (LH_2 single arm), points. The data of Bussey, *et al.* [3] (stars) and Brack, *et al.* [4] (open symbols) are shown, as are the V.P.I. solutions SM95 [5] (dashed) and SP95 [6] (double line).

KH80, where overall the SP95 solution predict out data better than SM95; interestingly, since in this energy region they these solutions differ only in the P33 partial–wave, where the SP95 solution predicts a Δ mass ≈ 1 MeV larger than SM95 and KH80. Due to P33 dominance, this should have implications for both the πNN coupling constant and Σ term. However, firm conclusions await a new PWA including our results.

References

1. M.M. Pavan, Ph.D. Thesis, University of British Columbia, 1995, unpublished. For a copy, contact: mpavan@bun.mit.edu.
2. R. Koch, E. Pietarinen, NPA **336**, 331 (1980)
3. P.J. Bussey, *et al.*, *Nucl. Phys.* B **58**, 363 (1973).
4. J.T. Brack, *et al.*, *Phys. Rev.* C **34**, 1771 (1986) ; *Phys. Rev.* C **51**, 929 (1995).
5. R.A. Arndt, I.I. Strakovsky, R.L. Workman, M.M. Pavan, *Phys. Rev.* C **52**, 2120 (1995).
6. R.A. Arndt, *et al.*, Unpublished analysis by V.P.I. group.
7. M.E. Sadler, *et al.*, PRD **35**, 2718 (1987).

DETERMINATION OF THE πN s-WAVE SCATTERING LENGTHS IN PIONIC HYDROGEN AND DEUTERIUM

M. Janousch, A. Badertscher, P.F.A. Goudsmit, H.J. Leisi, E. Matsinos, H.-Ch. Schröder, D. Sigg, Z.G. Zhao
Institute for Particle Physics, Eidgenössische Technische Hochschule Zürich, CH-5232 Villigen-PSI, Switzerland

D. Chatellard, J.-P. Egger, E. Jeannet
Institut de Physique de l'Université de Neuchâtel, CH-2000 Neuchâtel, Switzerland

K. Gabathuler, P. Hauser, L.M. Simons
Paul-Scherrer-Institut, CH-5232 Villigen-PSI, Switzerland

A.J. Rusi el Hassani
Ecole Mohammadia d'Ingénieurs, Rabat, Morocco

The 3p-1s x-ray transitions in pionic hydrogen and deuterium were studied with a high-resolution reflecting crystal spectrometer. The measured energy level shifts and the total decay widths of the 1s state, induced by the strong interaction, lead to a determination of the two πN s-wave scattering lengths at threshold.

The pionic hydrogen and deuterium offer a unique possibility to study the pion-nucleon interaction at threshold.[1,2] The hadronic s-wave scattering lengths are related to the strong interaction shift ε_{1s} (defined as the difference of the measured and electromagnetic 1s level energy) and the total decay width of the 1s state Γ_{1s} through Deser type-formulas,[4]

$$\varepsilon_{1s} = -\frac{4B_{1s}}{r_B} a^h_{\pi^- p\to\pi^- p} \left(1+\delta_\varepsilon\right) \tag{1}$$

$$\Gamma_{1s} = \frac{8B_{1s}}{r_B} q\left(1+P^{-1}\right)\left[\left(a^h_{\pi^- p\to\pi^0 n}\right)\left(1+\delta_\Gamma\right)\right]^2 . \tag{2}$$

Here B_{1s} is the electromagnetic binding energy of the π^-p atom in the 1s state, r_B the Bohr radius, P the Panofsky ratio and q the c.m. momentum of the π^0; δ_ε and δ_Γ are small corrections due to electromagnetic and mass-splitting effects.[4] In the case of deuterium the shift and width can still be related with the Deser [2] formula to the πd scattering lengths; the extraction of the πN scattering lengths involves a theoretical treatment to account for multiple scattering and absorption effects.[5]

The experimental setup consisted mainly of three parts: a cryogenic gas target, mounted in the center of a cyclotron trap; six cylindrically bent quartz

(110) crystals on a turn table and an array of four position sensitive charge-coupled devices, mounted on an XY-table for the detection of the x-rays. The energy of the 3p–1s transitions was determined by measuring the Bragg angle *difference* of the pionic K_β line with respect to the precisely known electronic K_{α_1} line of argon. The instrumental resolution was deduced from the measurement of the 4f–3d transition in pionic beryllium which has a negligible natural line width.

Preliminary values for the shift and width of the 1s state are $\varepsilon_{1s}^{\pi^-p} = -7.108 \pm 0.047$ eV and $\Gamma_{1s}^{\pi^-p} = 0.897 \pm 0.082$ eV for hydrogen and $\varepsilon_{1s}^{\pi^-d} = +2.43 \pm 0.10$ eV and $\Gamma_{1s}^{\pi^-d} = 1.02 \pm 0.21$ eV in the case of deuterium. A small correction was applied to the widths due to Doppler broadening after Coulomb deexcitation.[6] Inserting these values into Eq. 1 and 2 results in the following preliminary values for the scattering lengths (hydrogen):[6]

$$a^h_{\pi^-p\to\pi^-p} = b_0 - b_1 = (0.0883 \pm 0.0008)\ \mathrm{m}_\pi^{-1}\ , \tag{3}$$

$$a^h_{\pi^-p\to\pi^0n} = \sqrt{2}b_1 = (-0.1301 \pm 0.0059)\ \mathrm{m}_\pi^{-1}\ , \tag{4}$$

and (deuterium):[3]

$$a^h_{\pi^-d} = [(-0.0259 + i0.0054) \pm 0.0011]\ \mathrm{m}_\pi^{-1}\ . \tag{5}$$

A recent calculation[7] in the framework of heavy baryon chiral perturbation theory beyond lowest order results in a value for the isovector scattering length b_1 that is in excellent agreement with our result (Eq. 4).

Following Thomas and Landau [5] the hadronic scattering length for deuterium can be connected to the isoscalar scattering length b_0. Having three constraints on the isoscalar and isovector scattering length one can test for isospin symmetry in the strong interaction *at threshold*.[8] So far, all three constraints are consistent with the assumption of isospin symmetry at the 1σ-level.

References

1. D. Sigg et al., *Phys. Rev. Lett.* **75**, 3245 (1995).
2. D. Chatellard et al., *Phys. Rev. Lett.* **74**, 4157 (1995).
3. D. Chatellard, PhD. thesis, Université de Neuchâtel, 1995.
4. D. Sigg et al., *Nucl. Phys. A*, submitted for publication.
5. A.W. Thomas and R.H. Landau, *Phys. Rep.* C **58**, 122 (1980).
6. H.Ch. Schröder, PhD. thesis, Swiss Fed. Inst. of Technology, 1996.
7. V. Bernard et al., *Phys. Rev.* C **52**, 2185 (1995).
8. H.J. Leisi et al., in *Chiral Dynamics: Theory and Experiment*, ed. A.M. Bernstein and B.R. Holstein (Springer, New York, 1995), p. 241.

ISOSCALAR SPIN STRENGTH IN ^{90}Zr AND ^{208}Pb

C. DJALALI
Physics Department, University of South Carolina, Columbia, SC 29208, USA and CNRS-IN2P3, I. P. N., 91406-Orsay, France

M. MORLET, J. VAN DE WIELE, A. WILLIS, L. BIMBOT, J. GUILLOT, H. LANGEVIN-JOLIOT, N. MARTY, L. ROSIER
CNRS-IN2P3, I. P. N., 91406-Orsay, France

F.T. BAKER
Physics Department, University of Georgia, Athens, GA 30602, USA

C. GLASHAUSSER
Physics Department, Rutgers University, New Brunswick, NJ 08903, USA

E. TOMASI-GUSTAFSSON
LNS, CEN Saclay, 91191 Gif-sur-Yvette, France

The isoscalar spin strength in nuclei has been studied recently at the Laboratoire National Saturne (France) using the vector and tensor polarized deuteron beam. A clear signature for isoscalar spin-flip strength (S_d^y) has been obtained by measuring spin transfer parameters in the inelastic scattering of polarized deuterons at intermediate energy [1]. This signature has been tested and turned out to be robust even in the presence of distorsion [2]. This has allowed us to carry out the first exploration of isoscalar spin-strength in nuclei. The relative isoscalar spin response (R_s^0) has been extracted and the $\Delta S=1$ and $\Delta S=0$ parts of the cross section have been separated in ^{90}Zr and ^{208}Pb.

In ^{90}Zr, the signature has a maximum around 8 to 10 MeV and then flattens out in the continuum all the way up to 40 MeV of excitation energy. The relative spin response is of the order of 75%. In the spin-flip cross section spectrum a 2 MeV wide structure centered at 9 MeV of excitation energy seems to have a forward peaked angular distribution characteristic of a 1^+ transition. This wide structure would be the first observation of the isoscalar component of the M1 resonance in Zr.

In ^{208}Pb, at low excitation energy, the 4^- and 1^+ states respectively at 3.55 and 5.86 MeV although not visible in the cross-section spectrum, stand out in the spin-flip cross-section. This illustrates once more how powerful the S_d^y signature is in detecting spin transtions in the presence of a large "background" of non spin-flip transitions. Several previously unknown isoscalar spin-flip transitions are observed between 6 and 10 MeV of excitation energy where the isovector 1^+ states have been observed in (e,e') and (p,p') reactions.

In both nuclei the relative spin response in the continuum is of the order of 50 to 60% which is compatible with the Fermi gas response, indicating no collectivity in the isoscalar spin channel. Extensive microscopic DWIA calculations using RPA transition densities have been carried out and compared to the data. The $\Delta S=0$ cross section is overpredicted in the calculation, however the $\Delta S=1$ cross section as well as the signature are reasonably reproduced.

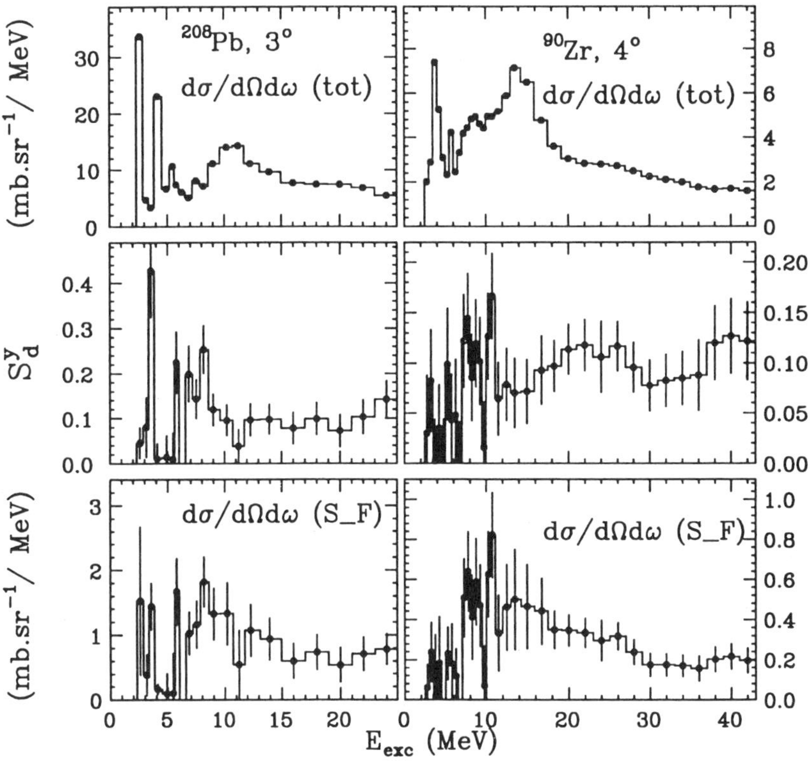

Figure 1: Total cross section, S_d^y signature and spin-flip cross sections measured in inelastic scattering of 400 MeV deuterons.

References

1. M. Morlet et al., Phys.Lett.**B247**(1990)228.
2. B. N. Johnson et al., Phys.Rev.**C51**(1995)1726.

POLARIZATION PHENOMENA IN DEUTERON PROTON SCATTERING: A USEFUL TOOL FOR THE STUDY OF NUCLEON RESONANCES PROPERTIES

M. P. REKALO,[a] E. TOMASI-GUSTAFSSON
LNS, CE Saclay, 91191 Gif-sur-Yvette Cedex, France

Polarization observables for dp reactions in the domain of nucleonic resonances are calculated in the framework of a t-channel ω-exchange model: the results show the sensitivity of the tensor analyzing power to the deuteron form factor and to the ωN vertex .

1 Introduction

The microscopic structure of the nucleon N and its excited states N^* can be determined through the (elastic or inelastic) electromagnetic form factors. Intensive experimental and theoretical studies have been devoted to meson electroproduction processes or elastic $e - d$ scattering. We will show here that hadron induced reactions with isospin zero projetiles, could be an effective method for the study of the nucleon structure, in particular through the measurement of polarization observables. Experimentally the study of the N^*–structure by hadronic probes is favoured by large values of cross sections, advanced technic of high intensity $\vec{d}$, $\vec{p}$ beams, polarized targets and polarimeters, absence of problem of radiative corrections, natural selection of isoscalar N^*-excitation.. Recent data have been obtained at Saclay (Saturne II) and at Dubna [2] on the inclusive process $\vec{d} + p \to d + X$, concerning cross sections and/or tensor analyzing power T_{20}. These data show the presence of at least two mechanisms in the intermediate energy region ($2 \leq E_{kin} \leq 10\ GeV$). One is the coherent excitation of d with pion production which results in a Deck peak [1], in the energy spectrum of the scattered deuterons. The Deck-peak has to decrease when the energy of the colliding particles increases, while the role of a second mechanism, the N^*– excitation, becomes more important.

2 The model

We present a model based on t–channel exchange by mesonic states. In particular the ω-exchange seems to be the best mechanism to describe the N^*-excitation: a spin 1 particle allows to obtain very specific polarization phenomena and to predict the energy dependence of the cross section. Moreover

[a]Permanent address: *National Science Center KFTI, 310108 Kharkov, Ukraine*

the similarity of the spin structure of the $\omega dd-$ and $\gamma dd-$ vertexes suggests that the $t-$ and m^*- dependences of T_{20} for $d+p\to d+X$ can be predicted in terms of electromagnetic deuteron form factors and of the ratio of cross sections of $\gamma_s^* + N \to X$ processes (γ_s^* is isoscalar virtual photon), induced by the longitudinal and transversal polarization of γ^*. For small transversal isoscalar transition in $\gamma_s^* + N \to X$ (in the region of the Roper Resonance) the behaviour of T_{20} for the process $\vec{d}+p\to d+X$ coincides with the tensor polarization t_{20} for $e^- + d \to e^- + \vec{d}$ elastic scattering.

The matrix element $\mathcal{M}^{(\omega)}$, can be written in the following form:

$$\mathcal{M}^{(\omega)} = -\frac{\mathcal{J}_\mu^{(d)}\mathcal{J}_\mu^{(s)}}{(t-m_\omega^2)},\quad t=(p_1-p_3)^2, \tag{1}$$

where the vector (and isoscalar) current $\mathcal{J}_\mu^{(d)}$ corresponds to the ωdd vertex and $\mathcal{J}_\mu^{(s)}$ is the isoscalar current describing the process $\omega + d \to X$. In case of collinear processes five independent transitions are allowed for the process $d+p\to d+N^*$. The vector dominance model suggests a simple relation between the form factors of the ωdd and $\gamma^* dd-$interactions. The spin structure of the other vertex, ωNN^* can be parametrized as a function of the isoscalar form factors of the N^*-electroexcitation. In case of $X = N^*(1440)$, with $\mathcal{J}^P = (1/2)^+$, (the Roper Resonance), the tensor analyzing power T_{20} is expressed as:

$$T_{20} = -\sqrt{2}\frac{V_1^2+(2V_0V_2+V_2^2)r(t)}{4V_1^2+(3V_0^2+V_2^2+2V_0V_2)r(t)}, \tag{2}$$

where the ratio $r(t) = g_e^2(t)/g_m^2(t)$ characterizes the relative role of longitudinal and transversal excitation of N^*, $V_0(t)$, $V_1(t)$ and $V_2(t)$ are the real deuteron form factors which are simply related to the standard electromagnetic deuteron form factors, G_c (electric), G_m (magnetic) and G_q (quadrupole). In Fig. 2 we report the theoretical predictions together with the existing experimental data on T_{20} for $p(\vec{d},d)X$ [2] for different momenta of the incident beam (open symbols) and the tensor polarization t_{20} for the elastic e,d scattering [3] (black stars). All these data show a very similar behaviour: negative values, with a minimum in the region $|q| \simeq 0.3\ fm^{-1}$, an increase toward zero at larger $|q|$ and a scaling as a function of t, with a small dependence on the incident momentum, in the interval 4.5-5.5 GeV/c. The lines are the result of the ω-exchange model for the $d+p\to d+X$ process, Eq. 2, for different values of the ratio r, assumed $q-$constant. The deuteron form factors, have been taken from [4], a calculation based on relativistic impulse approximation, and they reproduce well the $T_{20}-$data for ed elastic scattering [3]. The dashed line corresponds to $r=0$, where our model predicts a constant value of observable in the $d+p \to d+X$, which is not consistent with the data. The dependence on r, in the range 0.01-2, is particularly evident in the region of the minimum, in the point where T_{20} changes sign and in the two points where all theoretical curves cross due to a particular cancellation in the numerator of Eq. 2.

The data seem consistent with the value $r = 0.1$, in the simplified version of the $\omega-$exchange model with a $t-$independent ratio r. It is possible to show that for η-exchange one would obtain $T_{20} = 1/\sqrt{2}$, positive and $t-$independent. For $\sigma-$exchange (spin 0 particle), T_{20} is different from zero only taking into account high order effects (in teraction with derivatives). Other polarization observables like $k_y^{y'}$ show also a large sensitivity to the form factors. The cross section should increase with energy in case of $\omega-$exchange, and become energy independent at large s, when the diffractive regime is reached. It should decrease as the energy increases for η or $\sigma-$exchange.

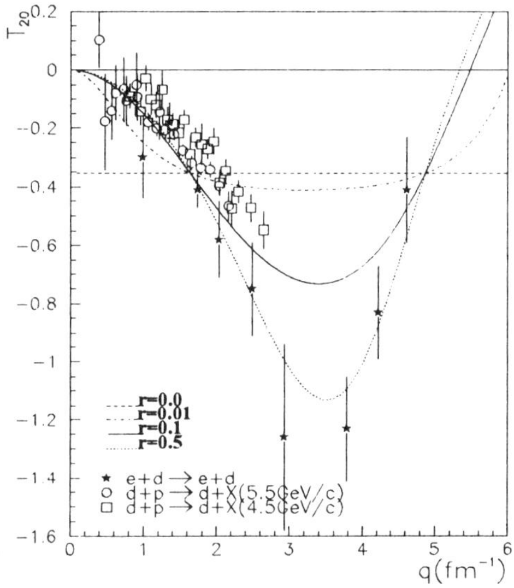

Fig. 1 T_{20} for $(e, \vec{d})$ and $p(d,d)X$ and predictions of the model.

3 Conclusion

Our calculation of cross section and polarization observables for the inclusive $d+p$ reactions in the domain of nucleonic resonances show that it is possible to disentangle the different reaction mechanisms ($\omega-, \sigma-, \eta-$ *exchange*) and bring new information about the electromagnetic form factors of the deuteron as well as of the nucleonic resonances. Existing data on the tensor analyzing power are in agreement with the prediction based on the $\omega-$exchange model.

References

1. L.S. Azghirey et al., Phys. Lett. **B361** (1964) 21.
2. R.T. Deck, Phys. Rev. Lett. **13** (1964) 169.
3. M. Garçon et al., Phys. Rev. **C49** (1994) 2516.
4. P.L. Chung et al. Phys. Rev. **C37** (1988) 2000.

SPIN OBSERVABLES IN $\bar{\Lambda}\Lambda$ PRODUCTION

M.A. ALBERG
Department of Physics, Seattle University, Seattle, WA 98122-4340, USA
Department of Physics, Box 351560, University of Washington, Seattle, WA 98195-1560, USA

E.M. HENLEY and L. WILETS
Department of Physics, Box 351560, University of Washington, Seattle, WA 98195-1560, USA

P.D. KUNZ
Nuclear Physics Laboratory, University of Colorado, Boulder, CO 80309, USA

A quark model which includes both scalar and vector contributions to the reaction mechanism for $\bar{p}p \rightarrow \bar{\Lambda}\Lambda$ is used to calculate the target depolarization D_{nn}, which will be measured at LEAR by PS185.

1 Motivation

Good fits to existing data for $\bar{p}p \rightarrow \bar{\Lambda}\Lambda$ can be obtained for either meson exchange or quark models of the reaction. Haidenbauer et al. [1] have proposed that the measurement of target depolarization D_{nn} might help to determine which model is a better description of the reaction mechanism. It has also been suggested [2] that the measurement of this spin observable in the PS185 experiment [3] may test dynamical mechanisms invoked to explain the proton spin puzzle.

2 Theory

The reaction mechanism in our quark model for $\bar{p}p \rightarrow \bar{\Lambda}\Lambda$ includes both scalar "$^3P_0$" and vector "$^3S_1$" contributions to $\bar{q}q$ annihilation and creation. We use a realistic $\bar{N}N$ potential in a DWBA calculation of total and differential cross-sections, polarizations, and spin correlation coefficients. The free parameters of the calculation, fit to experimental data at laboratory momenta from threshold to 1.695 GeV/c, include the scalar and vector strengths, a quark cluster size parameter, and parameters of the unknown $\bar{\Lambda}\Lambda$ interaction. We find excellent agreement with experiment for a $\bar{\Lambda}\Lambda$ potential which differs from that predicted on the basis of SU(3) arguments. Our results are presented in detail elsewhere[4]. Meson-exchange calculations are also in reasonable agreement with experiment. We have examined the suggestion of Haidenbauer et al. [1] that a measurement of

Table 1: Predictions in Born approximation for D_{nn}

model	reaction mechanism	D_{nn}
meson exchange	"tensor" $\tilde{\sigma}_1 \cdot \hat{r}\, \tilde{\sigma}_2 \cdot \hat{r}$	-1
	S_{12}	-2/3
quark	vector "$^3S_1$"	+2/3
	scalar "$^3P_0$"	≥ 0

the depolarization, D_{nn}, of a polarized proton target may discriminate between the two models for the reaction mechanism.

3 Predictions for D_{nn}

As shown in Table 1, in Born approximation, meson exchange (dominated by tensor terms) and quark model predictions for D_{nn} differ even in sign. We find that these sign differences persist when initial and final state interactions are taken into account [5]. The sign of D_{nn} also distinguishes between two alternative dynamical mechanisms that have been invoked to explain the proton spin puzzle. Models with negatively polarized $\bar{s}s$ pairs in the proton wave fuction predict $D_{nn} < 0$, whereas models with positively polarized gluons predict $D_{nn} > 0$.

4 Acknowledgments

This work has been supported in part by the US Department of Energy under grant DE-FG06-88ER40427.

References

1. J. Haidenbauer, K. Holinde, V. Mull and J. Speth, *Phys. Lett.* B **291**, 223 (1992).
2. M. Alberg, J. Ellis and D. Kharzeev, *Phys. Lett.* B **356**, 113 (1995).
3. B. Bassalleck et al., Proposal P287, CERN/SPSLC 95-13 (accepted as PS185-3).
4. M. Alberg, E.M. Henley, L. Wilets and P.D. Kunz, *Nucl. Phys.* A **560**, 365 (1993).
5. M. Alberg, E.M. Henley, L. Wilets and P.D. Kunz, *Phys. Atom. Nucl.* **57**, 1678 (1994).

EXCLUSIVE $\overline{\Lambda}\Lambda$ AND $\overline{\Lambda}\Sigma^o + cc$ PRODUCTION IN $\overline{p}p$ ANNIHILATIONS NEAR THE REACTION THRESHOLDS

THE PS185 COLLABORATION

K.Röhrich

IKP, Forschungszentrum GmbH, D-52425 Jülich, Germany

The PS185 experiment at LEAR/CERN studies strangeness production in $\overline{p}p$ annihilations in flight. Exclusive final states containing pairs of antihyperons and hyperons have been detected. Angular distributions of cross sections, final state polarizations and spin correlations were determined. The new data at the highest energies available extend the existing data set for the $\overline{\Lambda}\Lambda$ final state. At momenta above 1.65 GeV/c cross sections and polarizations for the annihilation into $\overline{\Lambda}\Sigma^o + cc$ were obtained.

The PS185 collaboration[1] investigates strangeness production in $\overline{p}p$ annihilations at LEAR at CERN. Antiprotons are delivered with momenta up to 2 GeV/c. In our fixed target experiment only $\overline{\Lambda}\Lambda$, $\overline{\Lambda}\Sigma^o + cc$ and $\overline{\Sigma}\Sigma$ pairs can be produced. New data for all three final states have been obtained.

Produced near the reaction threshold the (anti)hyperons emerge in the forward direction in the laboratory. They can proceed a few centimeters before they decay. The experiment makes use of the delayed weak decay of the hyperons into charged final state particles like $\Lambda \rightarrow p\pi^-$. At threshold even the decay baryons are always produced in a forward cone. Employing small CH_2-targets, which are surrounded by veto scintillators, and a larger scintillator hodoscope further downstream of the target, the multiplicity of charges in the final state can be detected and is used in the experiment's trigger. The volume between the sandwich target and the hodoscope is equipped with multi-wire proportional and drift chambers. From the hits in these detectors tracks and vertices are reconstructed. The events are kinematically overdetermined, because both decay vertices are required and thus the final event sample is almost background free. An additional carbon target is used to determine background of $\overline{Y}Y$ production from this nucleus. A solenoid magnet downstream of the hodoscope equipped with three drift chamber planes serves as a particle identifier by reconstruction of the bend of the charges inside the field. The experiment's acceptance is of the order of 50% and its energy resolution better than 1 MeV.

The PS185 has investigated the $\overline{p}p \rightarrow \overline{\Lambda}\Lambda$ reaction for a multitude of momenta between the reaction threshold and the maximum available antiproton momentum, covering an excess energy range from threshold ($\epsilon = \sqrt{s} - m_{\overline{\Lambda}} - m_{\Lambda} = 0$ MeV) up to 200 MeV. The most recent data have been taken at momenta above and around the thresholds for $\overline{\Lambda}\Sigma^o + cc$ and $\overline{\Sigma}\Sigma$ production. The total cross sections rise steeply from the reaction threshold to saturate at values around 80 μb at the highest energies.

Angular distributions of cross sections show a pronounced forward peaking. This is observed already very close to threshold and signals the early onset of higher partial wave contributions. At the largest energies we find contributions from partial waves up to L=4.

The cross sections for $\overline{\Lambda}\Sigma^o + cc$, and even more for $\overline{\Sigma^+}\Sigma^+$, show a stronger forward peaking then $\overline{\Lambda}\Lambda$. This earlier onset of higher partial waves indicates a more complex production mechanism.

The asymmetric weak decay of the (anti)-hyperon into baryon and pion allows to extract their polarizations. They exhibit have very distinct features which are maintained over the whole accessible momentum range. At small reduced momentum transfer $t' = -\frac{1}{2}[(s - 4m_1^2)(s - 4m_3^2)]^{\frac{1}{2}}(1 - \cos\Theta_{\overline{\Lambda}})$, with $\Theta_{\overline{\Lambda}}$ being the production angle, the polarizations are small and positive. They change sign at t' between -0.2 (GeV/c)^2 and -0.08 (GeV/c)^2 from lowest to highest energies and become as large as 50% for intermediate momentum transfers. At t' around -1 (GeV/c)^2 the polarizations are smaller and seem to oscillate from negative to positive values at higher energies. In the case of $\overline{\Lambda}\Sigma^o + cc$ statistics limits the study of the polarizations. The $\Lambda(\overline{\Lambda})$ emerges with similar, but less pronounced polarization characteristics, while the $\Sigma^o(\overline{\Sigma^o})$ polarizations show inverted features: they are negative at small and positive at large reduced momentum transfer.

The polarizations of both final state particles are measured simultaneously and spin correlations can be detemined. The singlet fraction $S_F = \frac{1}{4}(1 + C_{xx} - C_{yy} + C_{zz})$ measures the alignment of the spins in the final state. It is 1, if the spins are antiparallel and drops to zero when the spins are parallel. A statistical mixture of spin singlet and triplet states would have $S_F = 0.25$. For the $\overline{\Lambda}\Lambda$ the singlet fraction is consistent with zero at all energies and angles. In the $\overline{\Lambda}\Sigma^o + cc$ we clearly observe a departure of S_F from zero, but the large statistical errors do not allow for a good determination of its value.

The PS185 experiment has measured a set of observables for the $\overline{p}p \rightarrow \overline{\Lambda}\Lambda$ reaction, which will be completed by a measurement of spin-transfer in autumn 1996. The production of $\overline{\Lambda}\Sigma^o + cc$ has been studied and some few events with charged $\overline{\Sigma}\Sigma$ have been collected. More data will be available in the near future.

1. P.D.Barnes et al., *Observables in High-Statistics Measurements of the Reaction* $\overline{p}p \rightarrow \overline{\Lambda}\Lambda$, CERN-PPE/96-58, submitted to Phys. Rev. C

HIGH PRECISION MEASUREMENT OF THE $\bar{p}p \to \bar{n}n$ DIFFERENTIAL CROSS-SECTION AT LEAR

R. BIRSA, F. BRADAMANTE, A. BRESSAN, S. DALLA TORRE-COLAUTTI, M. GIORGI, M. LAMANNA, A. MARTIN, A. PENZO, P. SCHIAVON, F. TESSAROTTO, A.M. ZANETTI
INFN Trieste and University of Trieste, Trieste, Italy

A. DE FALCO, M.P. MACCIOTTA, A. MASONI, G. PUDDU, S. SERCI
INFN Cagliari and University of Cagliari, Cagliari, Italy

A. AHMIDOUCH, E. HEER, R. HESS†, C. MASCARINI, D. RAPIN
DPNC, University of Geneva, Geneva, Switzerland

J. ARVIEUX, R. BERTINI, J.C. FAIVRE, R.A. KUNNE
LNS, CRNS-IN2P3 and CEA-DSM, CE Saclay, Gif-sur-Yvette, France

M. AGNELLO
INFN Turin and Polytechnic of Turin, Italy

Presented by D. Rapin

Experiment PS206 has measured with good precision and high statistical accuracy the differential $\bar{p}p \to \bar{n}n$ cross-section at the CERN Low Energy Antiproton Ring. The measurements, performed at 601 and 1202 MeV/c incident $\bar{p}$ momenta, show a sharp peak at 0° scattering angle, followed by an energy dependent dip-bump structure. The high precision of the measurement allows the application of the Chew method to extract the charged-pion-nucleon coupling constant f_c^2.

Experiment PS206 has measured at LEAR the charge-exchange $\bar{p}p \to \bar{n}n$ differential cross-section at 601 and 1202 MeV/c. The experimental apparatus consists of three MWPCs, that measure the incoming $\bar{p}$ direction, two small scintillator counters, that define the $\bar{p}$ at the trigger level, a liquid hydrogen target, surrounded by a veto box, that discards reactions with charge particles in the final state as well as most of the $\bar{p}p$ annihilations into neutrals; a C-shaped magnet is used to sweep the unscattered $\bar{p}$ beam. Both the n and the $\bar{n}$ produced in the reaction are detected; the $\bar{n}$'s produced in the c.m. forward hemisphere (i.e. $\theta_{\bar{n}} < 45°$ in the laboratory frame) are detected by means of two identical counters, ANC_1 and ANC_2, while the backward region is measured by the ANC_3 counter, that can detect both n and $\bar{n}$. The two forward detectors are the same as the one already used in Experiment PS199 [1,2]. The $\bar{n}$'s are

† deceased

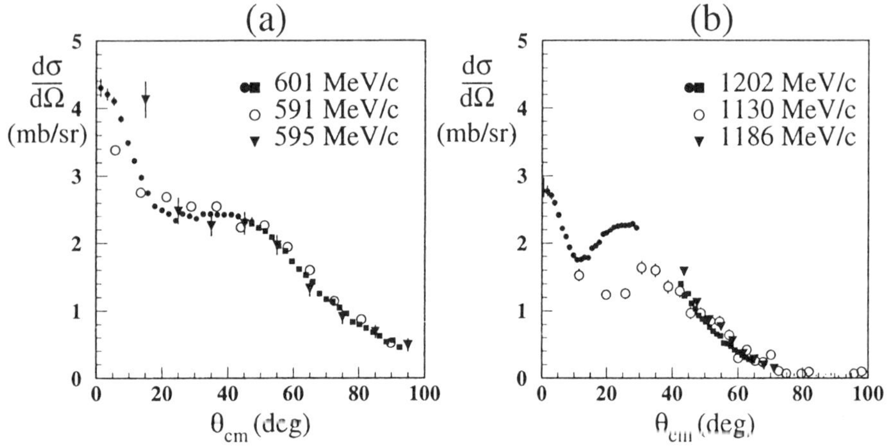

Figure 1: (a) Differential cross-section $d\sigma/d\Omega$ results at 601[5] MeV/c (ANC_1 closed points, ANC_2 closed squares); also shown are the measurements of Nakamura[6] at 591 MeV/c (open points) and Brückner[7] at 595 MeV/c (closed triangles). (b) Differential cross-section at 1202[5] MeV/c as a function of θ_{cm}; also shown are the measurements of Colebourne[8] at 1130 MeV/c (open points) and Ahmidouch[9] at 1186 MeV/c (closed triangles).

identified in ANC_1 and ANC_2 by reconstructing the tracks of the charged particles produced in the annihilation, that give a characteristic 'star' pattern. The backward detector ANC_3 consists of a neutron counter sandwiched by two modules, identical to those of ANC_1 and ANC_2. The neutron counter[3] is a hodoscope of 27 vertical scintillator bars, 8 cm wide, 20 cm thick.

The differential cross-section is measured from the angular distribution of the n̄'s detected by ANC_1 and ANC_2 alone (the measurement in the backward region needs a different analysis, which is in progress). The neutrons are detected only in a small angular region to measure the efficiency of the ANC counters by the associated particle method[4]. The charge-exchange events are identified by the star topology and by the time-of-flight. A cut on the star sphericity is applied to remove false secondary vertexes from pions or photons produced in the n̄ annihilation. Conservative acceptance cuts are also applied to limit edge effect corrections. The measured angular distributions are corrected for small efficiency non uniformity, beam effects, absorption and smearing effects. The final differential cross-section is shown in fig. 1, together with previous measurements, close in energy to our experiment. The estimated overall normalization error, not shown, is just the error on the measured ANCs absolute efficiency, namely 3% at 601 MeV/c and 4% at 1202 MeV/c. The quality of our results is apparent from the figure; the agreement with previous measurements is good at large angles, while the contribution of this experiment shows up in the very forward region. The forward peak is very sharp (the slope at $t=0$ is about 100 $(\mathrm{GeV}/c)^{-2}$), a clear manifestation of the pion pole [10]. At variance with the universality of the forward np $\rightarrow$ pn differential cross-section shape, a strong energy-dependence of the dip-bump structure is seen very clearly going from 601 to 1202 MeV/c.

As suggested almost thirty years ago [11], these data have been used to extract the $\pi^{\pm}$NN coupling-constant f_c^2 by an extrapolation to the pion pole [12]. The procedure, first proposed by G. F. Chew [13], was already successfully applied to the np $\rightarrow$ pn reaction. The result of the fit yields $f_c^2 = 0.071 \pm 0.002$ [12], quite compatible with the estimate ($f_c^2 = 0.0732 \pm 0.0011$) of a recent phase-shift analysis [14], which uses all the available $\bar{\mathrm{N}}$N data, and not only the charge-exchange channel.

References

1. R. Birsa et al., *Nucl. Instrum. Methods* **A 300** 43 (1991)
2. R. Birsa et al., *Nucl. Phys.* **B 403** 25 (1993)
3. A. Ahmidouch et al., *Nucl. Instrum. Methods* **A 326** 538 (1993)
4. A. Martin et al., *Nucl. Instrum. Methods* **A 346** 237 (1994)
5. R. Birsa et al., *Phys. Lett.* **B 339** 325 (1994)
6. K. Nakamura et al., *Phys. Rev. Lett.* **53** 885 (1984)
7. W. Brückner et al., *Phys. Lett.* **B 169** 302 (1986)
8. A. Colebourne et al., *Nucl. Phys.* **B 36** 258 (1962)
9. A. Ahmidouch et al., *Phys. Lett.* **B 369** 116 (1995)
10. F. Bradamante and A.Martin, *Phys. Lett.* **B 343** 427 (1995)
11. R. A. Bryan and R. J. N. Phillips, *Nucl. Phys.* **B 52** 201 (1968)
12. F. Bradamante et al., *Phys. Lett.* **B 343** 431 (1995)
13. G. F. Chew, *Phys. Rev.* **112** 1380 (1958)
14. R. G. E. Timmermans et al., *Phys. Rev.* **C 50** 48 (1994)

COMPARING $\bar{p}p \rightarrow \pi^-\pi^+$ AND $\bar{p}p \rightarrow K^-K^+$

W.M. Kloet
Department of Physics and Astronomy, Rutgers University, NJ

F. Myhrer
Department of Physics and Astronomy, University of South Carolina, SC

A partial wave amplitude analysis of $\bar{p}p \rightarrow \pi^-\pi^+$ and $\bar{p}p \rightarrow K^-K^+$ has been performed for data obtained at LEAR in the range $p_{\rm lab}$ = 360 - 1000 MeV/c. For $\bar{p}p \rightarrow \pi^-\pi^+$ partial wave amplitudes with J = 0, 1, 2, and 3 are required, while for $\bar{p}p \rightarrow K^-K^+$ amplitudes with J = 0, 1, and 2 are sufficient to fit the data in this energy range.

The reactions $\bar{N}N \rightarrow \pi\pi$ and $\bar{N}N \rightarrow KK$ are two of the more basic annihilation and subsequent hadronization reactions. Therefore data on these reactions may reveal details of the underlying mechanisms and clarify the nature of the degrees of freedom necessary to describe these short range hadronic processes. Recent low energy data from the CERN Low Energy Antiproton Ring [LEAR] for both reactions [1] for $d\sigma/d\Omega$ and A_{on} show a rather rich angular dependence which changes rapidly with increasing energy. The recent data agree with earlier data [2,3,4,5] at corresponding momenta, but the LEAR data cover a much wider momentum range and have smaller error bars. Our aim in analyzing these data is to extract information about the annihilation mechanism.

In both reactions there are only two helicity amplitudes. So far two observables $d\sigma/d\Omega$ and A_{on} have been measured. Therefore any analysis has ambiguities unless a third spin-observable like A_{ss} or A_{ls} is measured preferably at the same set of LEAR-energies. This would require polarized beam and target. It is not likely that experiments with polarized antiprotons will be feasible at present. A standard method to avoid ambiguities in such a case is to make use of some theoretical input. We are in a bind here, because the theory of this process is very poorly known. Parity conservation requires that the orbital angular momentum L of the $\bar{p}p$ initial state is different from the orbital angular momentum J of the two-meson final state. This requires the $\bar{p}p$ spin to be S = 1. Since only spin triplet angular momentum states are allowed, the number of parameters in the analysis is restricted considerably. As mentioned above, ambiguities arise because only two observables have been measured. However, it turns out that even from an analysis which uses only the data of two observables, and no theoretical input, very interesting physical

Table 1: Values of χ^2 per degree of freedom for $\pi^-\pi^+$ at momenta below 1 GeV/c.

$p_{\rm lab}$ (MeV/c)	$\chi^2(J_{\rm max}=1)$	$\chi^2(J_{\rm max}=2)$	$\chi^2(J_{\rm max}=3)$	$\chi^2(J_{\rm max}=4)$
360	2.56	1.96	1.77	1.74
404	2.18	1.38	1.12	1.12
467	5.75	1.98	1.31	1.18
497	8.38	3.04	1.50	1.45
523	8.16	2.63	1.45	1.43
585	10.5	1.96	1.51	1.57
679	15.3	2.17	1.50	1.53
783	21.9	2.50	1.49	1.47
886	21.7	3.21	1.23	1.13
988	24.0	4.39	1.85	1.55

requirements can be extracted. In particular the analysis shows which angular momenta are necessary and sufficient. We find that the present data for $\bar{p}p \to \pi^-\pi^+$ starting from $p_{\rm lab} = 360$ MeV/c up to 1 GeV/c, can be fitted with partial wave amplitudes with total angular momentum $J \leq 3$. At the same time the data for the reaction $\bar{p}p \to K^-K^+$ require only angular momentum less or equal to $J = 2$. The corresponding χ^2 per degree of freedom of the fits are shown in Table 1, and Table 2 for all measured momenta in the interval of 360 MeV/c to 988 MeV/c. In the same tables the values of χ^2 are given for smaller and larger sets of partial waves, $J \leq J_{max}$, where J_{max} is the highest partial wave included. From the values of χ^2 for $\bar{p}p \to \pi^-\pi^+$ in Table 1 one concludes that the fits for momenta below 988 MeV/c do not improve beyond $J_{max} = 3$. From Table 2, which shows the values of χ^2 for $\bar{p}p \to K^-K^+$ one concludes that for this reaction the fits for momenta below 886 MeV/c do not improve beyond $J_{max} = 2$. These findings put constraints on theoretical descriptions of these annihilation processes, which present models [6,7,8] do not seem to satisfy.

The fact that very few partial waves are needed in the analysis, indicates that both annihilations are very central processes. The annihilation into K^-K^+ may occur in an even smaller volume than the annihilation into $\pi^-\pi^+$. It should be kept in mind that even at these low energy $\bar{p}$ momenta, the final center of mass momenta of both the $\pi^-\pi^+$ and K^-K^+ channels are very similar. The very few partial wave amplitudes required by data put stringent constraints [9] on possible model descriptions of these basic annihilation reactions.

Table 2: Values of χ^2 per degree of freedom for K^-K^+ at momenta below 1 GeV/c.

$p_{\rm lab}$ (MeV/c)	$\chi^2(J_{max}=1)$	$\chi^2(J_{max}=2)$	$\chi^2(J_{max}=3)$	$\chi^2(J_{max}=4)$
360	2.26	0.93	0.97	0.95
404	2.32	1.23	1.40	1.44
467	1.90	1.00	1.13	0.92
497	3.81	1.40	1.02	1.03
523	4.70	1.03	0.72	0.80
585	3.25	0.95	0.98	0.79
679	3.42	1.44	1.53	1.52
783	6.92	2.45	2.41	2.22
886	5.08	1.59	1.37	1.30
988	4.73	2.80	1.06	1.20

References

1. A. Hasan *et al.*, *Nucl. Phys.* B **378**, 3 (1992).
2. E. Eisenhandler *et al.*, *Nucl. Phys.* B **98**, 109 (1975).
3. A.A. Carter *et al.*, *Phys. Lett.* B **67**, 117 (1977).
4. T. Tanimori *et al.*, *Phys. Rev. Lett.* **55**, 185 (1985); *Phys. Rev.* D **41**, 744 (1990).
5. A.A. Carter, *Phys. Lett.* B **67**, 122 (1977).
6. B. Moussallam, *Nucl. Phys.* A **407**, 413 (1983); *Nucl. Phys.* A **429**, 429 (1984).
7. V. Mull, K. Holinde and J. Speth, *Phys. Lett.* B **275**, 12 (1992). V. Mull and K. Holinde, *Phys. Rev.* C **51**, 2360 (1995).
8. G. Bathas and W.M. Kloet, *Phys. Rev.* C **47**, 2207 (1993).
9. W.M. Kloet and F. Myhrer, *Phys. Rev.* D **53**, 6120 (1996).

BARYON MASSES IN A CHIRAL EXPANSION WITH MESON-BARYON FORM FACTORS

Richard E. Stuckey and Michael C. Birse
Theoretical Physics Group, Department of Physics and Astronomy, University of Manchester, Manchester, M13 9PL, UK

The chiral expansion of the one-loop corrections to baryon masses is examined in a generic meson-cloud model with meson-baryon form factors. For pion loops, the expansion is rapidly convergent and at fourth order in m_π accurately reproduces the full integral. In contrast, the expansion is found to converge very slowly for kaon loops. Despite the importance of high-order terms, relations like that of Gell-Mann and Okubo are well satisfied by the baryon masses.

The cloud of virtual pseudoscalar mesons surrounding a baryon contributes significantly to the properties of that particle. This has long been known, from both chiral or "cloudy" bag models[1] and chiral perturbation theory (ChPT)[2,3]. The cloudy bag model and chiral versions of the colour-dielectric soliton model[4,5,6] are able to give very good descriptions of baryon properties. Even if they cannot be extended to a rigorous chiral expansion, these models may be able to provide useful estimates of the higher-order terms in such an expansion.

In particular these models give rise to form factors at the meson-baryon vertices that regulate the loop integrals involved. Here we examine the chiral expansions of the self-energies of octet and decuplet baryons including one-loop contributions from the octet of pseudoscalar mesons using a generic form factor instead of one calculated using some specific bag or soliton model[7].

At one-loop level in the self-energy of a baryon A is a sum of terms like

$$\Sigma_{AB\nu} = -\frac{3}{4\pi^2}\left(\frac{g_{\pi NN}}{2M_N}\right)^2 \frac{(f^{AB\nu})^2}{25}\int_0^\infty \frac{k^4\mu^2(k)dk}{\omega_\nu(k)\left(\omega_\nu(k)+M_B^{(0)}-M_A^{(0)}\right)}, \quad (1)$$

where the $f^{AB\nu}$ are defined as in Morgan *et al.*[8] and the meson energy is $\omega_\nu(k)=\sqrt{k^2+m_\nu^2}$. For the form factor we have taken a Gaussian with a mass of $M = 660$ MeV, which has been fitted to the colour-dielectric model. In an SU(6) quark model, the bare baryon mass $m_A^{(0)}$ contains a term proportional to the number of strange quarks, but we have also considered more general octet splittings. A bare octet-decuplet splitting δ is also included.

We have examined the expansion of these self-energies in terms of the current quark masses, which are proportional to the squares of the meson masses. The contributions from loops with intermediate octet baryons contain the correct leading nonanalytic terms in this expansion[2], of order $m_\nu^3 \sim m_q^{3/2}$. For pions the expansion shows good convergence to the full integral by order m_π^4. In contrast the expansion for kaons is much slower, needing terms of order at least m_K^7. The small size of the octet-decuplet splitting means that the simple chiral expansion of the loops with intermediate decuplet baryons shows very poor convergence. However, if we make a combined expansion in m_ν and δ, treating both quantities as being of the same order, the convergence is similar to the octet loops[7]. The alternating signs of the terms in the series are similar to what is seen in SU(3) ChPT[9,10].

Despite the importance of higher-order SU(3)-breaking terms, the full masses continue to satisfy relations like the Gell-Mann–Okubo relation and the decuplet equal spacing rules. Both ChPT and $1/N_c$ expansions find similar behaviour[11,12,13], showing that these relations cannot be used as evidence for perturbative breaking of SU(3) in the baryon masses[14].

The contributions of the quark masses to the nucleon mass can be determined by applying the Feynman-Hellmann theorem to the self-energy. For the nonstrange quarks we find $\overline{m}\langle N|\bar{u}u + \bar{d}d|N\rangle_{cloud} \simeq 25$ MeV. This is sufficient to make up the difference between the empirical 45 MeV[15] and the 15 – 20 MeV provided by the valence-quark core. The kaon cloud and intermediate strange baryons lead to a total contribution of $m_s\langle N|\bar{s}s|N\rangle_{cloud} \simeq 25$ MeV to the nucleon mass. Since m_s is an order of magnitude larger than $\overline{m}$, this corresponds to a very small strange-quark density in the nucleon. In contrast to perturbative treatments of SU(3) breaking, meson loops can give rise to a large sigma commutator without a large strangeness content in the nucleon[14].

1. A. W. Thomas, Adv. Nucl. Phys. **13** (1984) 1.
2. J. Gasser and H. Leutwyler, Phys. Reports **87** (1982) 77.
3. V. Bernard *et al.*, Int. J. Mod. Phys. **4** (1995) 193.
4. A. G. Williams and L. R. Dodd, Phys. Rev. **D37** (1988) 1971.
5. J. A. McGovern, Nucl. Phys. **A533** (1991) 553.
6. M.-S. Bae and J. A. McGovern, J. Phys. G **22** (1996) 199.
7. R. E. Stuckey and M. C. Birse, hep-ph/9202312.
8. M. A. Morgan *et al.*, Phys. Rev. **D33** (1986) 817.
9. V. Bernard, N. Kaiser and U.-G. Meissner, Z. Phys. **C60** (1993) 111.
10. B. Borasoy and U.-G. Meissner, Phys. Lett. **B365** (1996) 285.
11. E. Jenkins, Nucl. Phys. **B368** (1992) 190.
12. M. K. Banerjee and J. Milana, Phys. Rev. **D52** (1995) 6451.
13. E. Jenkins and R. F. Lebed, Phys. Rev. **D52** (1995) 282.
14. R. L. Jaffe, Phys. Rev. **D21** (1980) 3215; Nucl. Phys. **A478** (1988) 3c.
15. J. Gasser, H. Leutwyler and M. E. Sainio, Phys. Lett. **B253** (1991) 252.

QUARK-MESON COUPLING MODEL FOR FINITE NUCLEI

P.G. BLUNDEN and G.A. MILLER
Institute for Nuclear Theory and Department of Physics,
University of Washington, Seattle, WA 98195

A Quark-Meson Coupling (QMC) model is extended to finite nuclei in the relativistic mean-field or Hartree approximation. The ultra-relativistic quarks are assumed to be bound in non-overlapping nucleon bags, and the interaction between nucleons arises from a coupling of vector and scalar meson fields to the quarks. We develop a perturbative scheme for treating the spatial nonuniformity of the meson fields over the volume of the nucleon as well as the nucleus. Results of calculations for spherical nuclei are given, based on a fit to the equilibrium properties of nuclear matter.

In a recent series of papers, Saito and Thomas [1] have considered a model for nuclear matter involving mesons and color-singlet clusters of quarks. This is the Quark-Meson Coupling (QMC) model originally proposed by Guichon [2] in which the quarks are assumed to be bound in non-overlapping nucleon bags, and the interaction between nucleons arises from a coupling of scalar and vector meson fields to the quarks. Here we report on the implementation of this model for finite nuclei. Details and further results are given elsewhere.[3]

Relativistic quarks at position $\vec{r}$, in a nucleon centered at position $\vec{X}$, are confined by some confining potential $V_{\rm con}(\vec{r}-\vec{X})$. In addition, there are external vector and scalar mean field potentials, $V_v^q(\vec{r})$ and $V_s^q(\vec{r})$ respectively, arising from meson exchange with quarks bound in other nucleons. Hence the equation of motion for quarks is

$$\left[-i\partial\!\!\!/_r + m_q - V_s^q(\vec{r}) + V_{\rm con}(\vec{r}-\vec{X}) + \gamma^0 V_v^q(\vec{r})\right] \psi_q(t,\vec{r}-\vec{X},\vec{r}) = 0. \tag{1}$$

The mean field vector and scalar potentials obey the equations

$$(-\nabla^2 + m_v^2)\, V_v^q(\vec{r}) = 3(g_v^q)^2 \int d^3X\, \rho_v(\vec{X})\, \rho_v^q(\vec{r}-\vec{X};\vec{X}), \tag{2}$$

$$(-\nabla^2 + m_s^2)\, V_s^q(\vec{r}) = 3(g_s^q)^2 \int d^3X\, \rho_s(\vec{X})\, \rho_s^q(\vec{r}-\vec{X};\vec{X}), \tag{3}$$

with g_v^q and g_s^q the vector and scalar quark-meson coupling constants. These are determined by fitting the properties of nuclear matter. Thus the source term for $V_v^q(\vec{r})$ in Eq. (2) is a convolution of the vector density $\rho_v(\vec{X})$ of the nucleon with the vector density $\rho_v^q(\vec{r}-\vec{X};\vec{X})$ of the quarks in that nucleon. A similar results holds for the scalar quantities.

Because the net quark potential in Eq. 1 is not central in the frame of the nucleon, the ground state will not be spherically symmetric. To avoid this difficulty, we replace $V^q_{v,s}(\vec{r})$ by an "average" potential $U_{v,s}(\vec{X})$. Choosing

$$U_{v,s}(\vec{X}) = \int d^3r\, \rho^q_{v,s}(\vec{r}-\vec{X};\vec{X})\, V^q_{v,s}(\vec{r}) \Big/ \int d^3r\, \rho^q_{v,s}(\vec{r}-\vec{X};\vec{X}) \qquad (4)$$

ensures that the residual interaction $V^q_{v,s}(\vec{r}) - U_{v,s}(\vec{X})$ vanishes in first order perturbation theory.

The feature that $U_{v,s}$ depends only on $\vec{X}$ allows us to simplify the solution of the Dirac equation (1) to the form

$$\left[-\imath\vec{\gamma}\cdot\vec{\nabla}_{\vec{r}} + m^*_q(\vec{X}) + V_{\text{con}}(\vec{r}-\vec{X})\right]\phi(\vec{r}-\vec{X};\vec{X}) = \epsilon^*(\vec{X})\,\gamma^0\phi(\vec{r}-\vec{X};\vec{X}). \quad (5)$$

Equation 5 is of the same form as the equation for quarks in a free nucleon, but with a quark effective mass $m^*_q(\vec{X}) = m_q - U_s(\vec{X})$. Hence the internal structure of a bound nucleon depends only on the scalar potential, and is unaffected by the vector potential.

The nucleon now experiences an effective mass $M^*(X) = 3\epsilon^*(X)$ and a mean field vector potential $V_v(X) = 3U_v(X)$. In analogy with the philosophy of the Walecka model,[4] we solve the Dirac-Hartree equations for the nucleons, and take the meson parameters from a fit to the properties of nuclear matter. There is an additional self-consistency requirement here that the quark and nuclear wave functions must be determined together. To perform the convolutions required in Eqs. 2, 3, and 4, we work in momentum space.

Results have been obtained for several spherical nuclei using both a bag model and a relativistic oscillator model for the quark confining potential.[3] Compared to the Walecka model,[4] the totally binding energy in QMC models is larger, whereas the spin-orbit splitting is somewhat smaller. The charge densities are in reasonable agreement with data. The quantitative prediction of quark wave functions for bound nucleons opens up a number of interesting possibilities for further work, and these are currently being explored.

References

1. K. Saito and A.W. Thomas, *Phys. Lett.* B **327**, 9 (1994); *Phys. Rev.* C **51**, 2757 (1995).
2. P.A.M. Guichon, *Phys. Lett.* B **200**, 235 (1988); *Nucl. Phys.* A **497**, 265c (1989).
3. P.G. Blunden and G.A. Miller, *Phys. Rev.* C **54**, 359 (1996).
4. B.D. Serot and J.D. Walecka, *Adv. Nucl. Phys.* **16**, 1 (1986).

Relativistic and Meson Exchange Current Corrections to the Baryon Currents

K. Dannbom[1], L. Ya. Glozman[2], C. Helminen[1], D.O. Riska[1]

[1] *Department of Physics, University of Helsinki, 00014 Finland*
[2] *Institute for Theoretical Physics, University of Graz, 8010 Austria*

As the constituent quarks are lighter than the nucleon, their electromagnetic and axial current operators have significant relativistic correction terms, if the quarks are treated as Dirac particles [1]. While the magnitude of these relativistic corrections depend on the model for the hyperfine interaction between the quarks [2], their effect is to reduce the magnitude of the predicted values of both the axial coupling constants and the magnetic moments of the baryons. This correction reduces the overprediction of the axial current coupling constants of the nucleons (5/3 vs 1.24) and the hyperons but it worsens the satisfactory predictions for the magnetic moments of the baryons that are obtained with the static quark model. The exchange current correction that is associated with the chiral boson exchange interaction between the quarks [3] can compensate for the relativistic correction in the latter case, while leaving it operative in the case of the axial coupling constants.

The relativistic correction to the axial couplings of the baryons amount to a reduction of $\sim 10 - 20\%$ from the static quark model values if $m_q \sim 340$ MeV and the oscillator frequently ω is in the typical range 160-300 MeV in the harmonic oscillator model. With $g_A^q = 0.88$ [4] this reduction brings the predicted values close to the empirical ones. The corresponding correction to the magnetic moments amount to reductions that are about 2 times larger, thus making the predicted values much too small.

If the hyperfine interaction between the quarks has the pseudoscalar meson exchange form

$$V_S(\vec{r}) = \frac{1}{3} f(r) \vec{\sigma}^1 \cdot \vec{\sigma}^2 \vec{\lambda}^1 \cdot \vec{\lambda}^2, \tag{1}$$

there is an associated exchange magnetic moment

$$\vec{M}_S = -\frac{1}{2} g(r) \{ (\tau^1 \times \tau^2)_3 + \lambda_4^1 \lambda_5^2 - \lambda_5^1 \lambda_4^2 \} \vec{\sigma}^1 \times \vec{\sigma}^2, \tag{2}$$

where the function $g(r)$ is determined by $f(r)$ as

$$g(r) = -\frac{1}{3} \{ 2 \int_r^\infty dr' r' f(r') - \frac{1}{r} \int_r^\infty dr'' r'' f(r'') \}. \tag{3}$$

If the form for $f(r)$ is taken to be that used in ref. [5] the exchange magnetic moments are large, and can compensate the relativistic corrections for the octet baryons. If e.g. $\omega = 160$ MeV as suggested by the nucleon radius, the relativistic corrections to the magnetic moments of the proton and the neutron are -0.54 nm and +0.36 nm respectively. For $\omega = 320$ MeV the corresponding values grow to -0.83 nm and +0.55 nm. The exchange current contributions obtained from (2),(3) are correspondingly $\pm$ 0.58 nm (ω=160 MeV) and $\pm$ 1.36 nm ($\omega = 320$ MeV). These results demonstrate the compensatory aspect of the relativistic corrections and the exchange current corrections.

The exchange magnetic moment operator (2) does not contribute to the magnetic moments of the decuplet baryons nor to those of the Σ^- and the Ξ^-. To compensate the relativistic corrections for these baryons one has to consider the relativistic corrections to the exchange magnetic moment operator, which has a flavor symmetric component.

References

1. F. Close, An Introduction to Quarks and Partons, Academic Press, London (1979)
2. C. Hayne and N. Isgur, Phys. Rev. **D25** (1982) 1944
3. L.Ya. Glozman and D.O. Riska, Phys. Rept. **268** (1996) 268
4. D.A. Dicus et al., Phys. Lett. **B284** (1992) 384
5. L.Ya. Glozman, Z. Papp and W. Plessas, e-print hep-ph 9601353

RELATIVISTIC DEUTERON BREAKUP ON PROTONS

J. PAULO PINTO, A. AMORIM and F. D. SANTOS
Centro de Física Nuclear da Universidade de Lisboa
Av. Gama Pinto 2, 1699 Lisbon, Portugal

Deuteron breakup reactions $A(d,p)X$ in the energy range of 1-9 GeV are an important source of information on the high momentum components of the deuteron wave function. The cross section, tensor analyzing power T_{20} and polarization transfer k have been measured [1-3] in the inclusive breakup reactions where one final proton is detected at 0°. A distinctive feature of the data for various targets ranging from ^1H to Sn is the enhancement of the cross section for a proton momentum q, in the deuteron center of mass, larger than 0.2 GeV/c. Non-relativistic calculations [2], including double scattering, have been unable to reproduce simultaneously this feature of the cross section and the data on T_{20}. Approximate relativistic calculations using light-cone dynamics have been performed for the impulse approximation (IA) and including corrections to the IA [4]. A relativistic calculation of the IA diagram using the deuteron Bethe-Salpeter wave-function (BS) has been reported recently [5].

Here we present a fully covariant calculation of the $^1H(d,p)X$ reaction using BS deuteron amplitudes calculated in the Gross approximation [6]. At 0° the most important process corresponds to the IA with a spectator proton. The resulting covariant cross section is given by

$$\frac{E_p d^2\sigma}{P_p{}^2 dP_p d\Omega_p} = \frac{1}{12(2\pi)^3}\left(\frac{(P_n.P_p)^2 - m_n^2 m_p^2}{(P_d.P_p)^2 - m_d^2 m_p^2}\right)^{\frac{1}{2}} \sigma^{np\to X} \times \sum_{spin}\left(|\Psi^+(q)|^2 + |\Psi^-(q)|^2\right) \quad (1)$$

where $\sigma^{np\to X}$ is the total cross section for the $n+p\to X$ reaction and Ψ^+, Ψ^- are determined by the internal deuteron BS amplitudes and correspond, respectively, to the positive, negative energy component of the off-shell neutron.

The deep minimum in the $T_{20}(0^\circ)$ data is particularly significant since it is a direct manifestation of the vanishing deuteron helicity amplitude $\Phi^d_-(q\,\mathbf{e}_z)$ [7], where the nucleons have opposite helicities. The zero is predicted to occur for $0.25\ GeV/c < q < 0.28\ GeV/c$, using various realistic NN interactions [7] and is related to the dimensions of the toroidal structure of the deuteron. In the IA with a spectator proton $T_{20}(0^\circ) = -\sqrt{2}$ for $\Phi^d_-(q\,\mathbf{e}_z) = 0$ and $T_{20}(0^\circ) = 1/\sqrt{2}$ for $\Phi^d_+(q\,\mathbf{e}_z) = 0$ [7]. Thus the deviations of $T_{20}(0^\circ)$ data from $-\sqrt{2}$ at

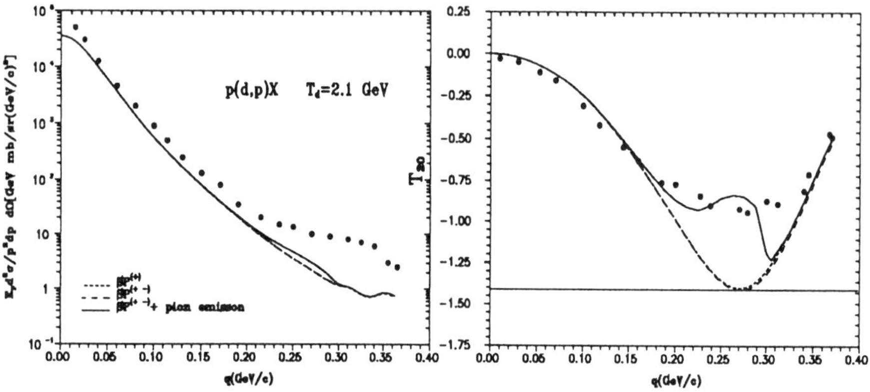

Figure 1: Cross section and tensor polarization T_{20} for the reaction $p(d,p)X$ at $T_d = 2.1$ GeV. The small dashed line gives the contribution of the impulse approximation, considering only the positive component of the deuteron wave function (S and D states), the long dashed line includes also the negative components (P states) and the solid line is the sum of the impulse approximation plus pion production, with a complete deuteron wave function.

the minimum result from other diagrams and also from the deuteron negative energy components. The present calculations shown in Fig. 2 for $T_d = 2.1\ GeV$ indicate that the contribuition of Ψ^- at the minimum is very small.

Covariant calculations of the contribuition from a process where the spectator nucleon emits a π^0 or a π^- have been performed. The effect on the cross section is small at $T_d = 1.25$ and $2.1\ GeV$ and does not explain the enhancement for $q > 0.2\ GeV/c$. However the effect of π emission on $T_{20}(0°)$ is non–negligible for $T_d = 2.1\ GeV$ as shown in Fig. 1. Calculations of other diagrams where the forward detected proton interacts with the target proton and diagrams involving Δ excitation are in progress and will be reported elsewhere.

References

1. C. F. Perdrisat and V. Punjabi, Phys. Rev. C **42**, 1899 (1992).
2. E. Cheung *et al*, Phys. Lett. B **284**, 210 (1992).
3. V. G. Ableev *et al*, Nucl. Phys. A **393**, 491 (1983);
   JINR Rap. Comm. 1 [54], 10 (1990)
4. M. G. Dolidze and G. I. Lykasov, Z. Phys. A **336**, 339 (1990).
5. L. P. Kaptari *et al*, Phys. Lett. B **351**, 400 (1995).
6. W. Buck and F. Gross, Phys. Rev. D **20**, 2361 (1979).
7. F. D. Santos and A. Arriaga, Phys. Lett. B **325**, 267 (1994).

A DWIA Model for Eta Photoproduction from Nuclei

F.X. Lee,[a] C. Bennhold,[b] L.E. Wright,[c] L. Tiator[d]
[a] *TRIUMF, 4004 Wesbrook Mall, Vancouver, BC, V6T 2A3, Canada*
[b] *The George Washington University, Washington, DC 20052*
[c] *Department of Physics, Ohio University, Athens, Ohio 45701*
[d] *Institut für Kernphysik, Universität Mainz, 55099 Mainz, Germany*

First, an elementary eta production operator is constructed in a dynamical, coupled-channels approach. The non-resonant sector contains nucleon Born terms and vector meson contributions. The resonance sector includes the $S_{11}(1535)$, $P_{11}(1440)$ and $D_{13}(1520)$ states. The model uses independent data to constrain all verteces except the ηNN vertex. By comparing with the new Mainz data on the proton (both total and differential cross sections), we find that a pseudoscalar ηNN coupling with a coupling constant of $g^2_{\eta NN}/4\pi$=0.4 is strongly prefered.

Then, we use the operator to study *quasifree* eta photoproduction on nuclei $A(\gamma, \eta N)B$ in the Distorted Wave Impulse Approximation (DWIA) [1], with a focus on the η-nucleus interaction and possible medium modifications.

Two models for the ηN t-matrix, DW1 [2] and DW2 [3], are used to construct a simple ηA optical potential based on a $t\rho$-approximation.

We find that the exclusive $A(\gamma, \eta N)B$ process can be used to study medium modifications of the N^* resonances, particularly if the photon asymmetry can be measured.

The inclusive $A(\gamma, \eta)X$ reaction is compared to new data [4] obtained on ^{12}C, ^{40}Ca in Figs. 1, 2, 3, and is found to provide a clear distinction between different models for the ηN t-matrix.

We would like to acknowledge support from National Sciences and Engineering Research Council of Canada; U.S. DOE under Grant No. DE-FG-02-87-ER40370 and DE-FG-02-95-ER40907; Deutsche Forschungsgemeinschaft (SFB201); NATO Collaborative Research Grant.

1. F.X. Lee, L.E. Wright, C. Bennhold, and L. Tiator, nucl-th/9601001, to be published in Nucl. Phys. A.
2. C. Bennhold, H. Tanabe, Phys. Lett. **B243**, 13 (1990); Nucl. Phys. **A530** (1991) 625.
3. M. Batinic, I. Slaus, A. Svarc and B.M.K. Nefkens, Phys. Rev. **C51**, 2310 (1995).
4. M. Röbig-Landau *et. al.*, Phys. Lett. B (1996), in print.

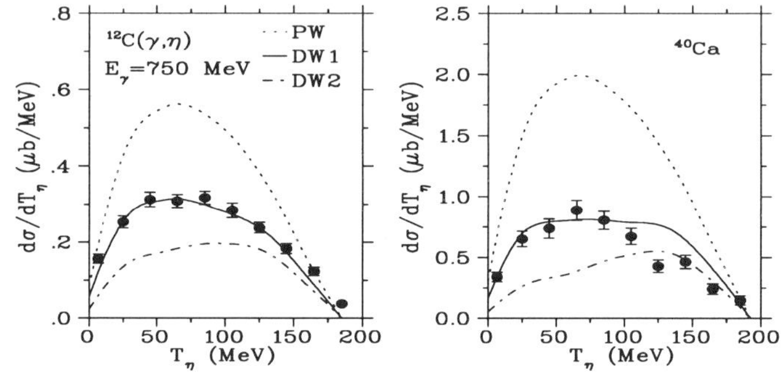

Figure 1: Comparison of our calculations with the Mainz data[4] for the inclusive cross section $d\sigma/dT_\eta$.

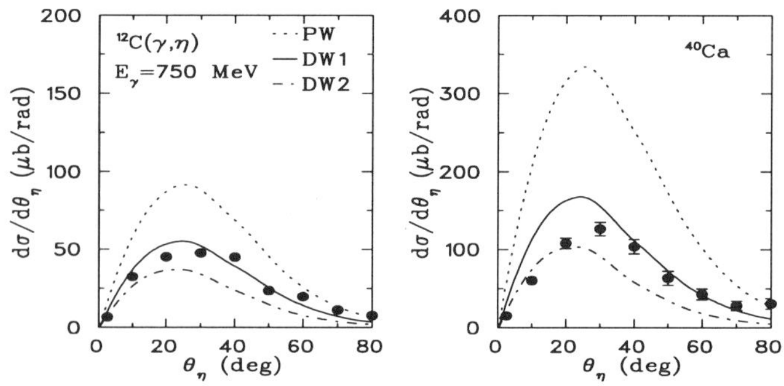

Figure 2: Comparison of our calculations with the Mainz data[4] for the inclusive cross section $d\sigma/d\theta_\eta$

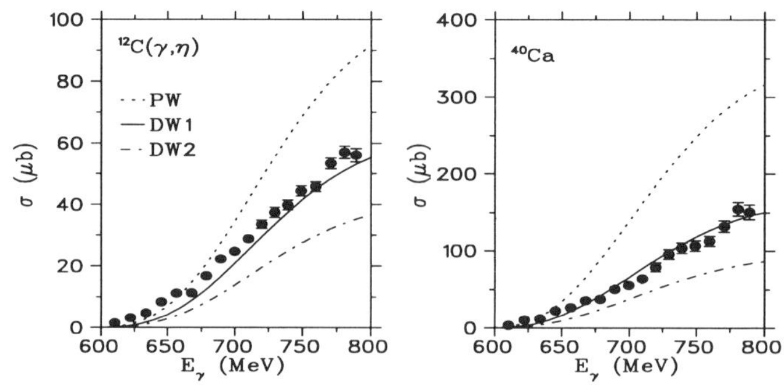

Figure 3: Comparison of our calculations with the Mainz data[4] for the total cross section.

THE $\pi\pi$ FINAL STATE INTERACTION IN THE $K \to \pi\pi$ DECAY

M.P. LOCHER, V.E. MARKUSHIN and H.Q. ZHENG
Paul Scherrer Institute, 5232 Villigen PSI, Switzerland

We have re-examined the final state interaction (FSI) effects in the S-wave $I = 0, 2$ $\pi\pi$ system for $s < 1.2$ GeV^2. Phenomenological models of the $\pi\pi$ interaction were used to describe the scattering amplitudes, and unitarity and analyticity of the form-factors were imposed by using the Omnès-Muskhelishvili (OM) equation.

In order to isolate the role of different mechanisms of the $\pi\pi$ interaction we exploit an interaction model [1] which combines the t-channel exchanges (ρ, f_2) with the explicit s-channel resonance (f_0), Fig. 1a. The s-dependence of the FSI factors for each mechanism is calculated and their relative contribution is analyzed. The ρ exchange in the t-channel, which determines the attractive character of the low energy $\pi\pi$ interaction in the $I = 0$ S-wave, gives the main contribution to the FSI enhancement below the f_0 resonance. The combined effect of the t-channel exchanges in the $K \to \pi\pi$ decay is $F^{I=0}_{t-exch}(m_K^2) = 1.57$.

Next we discuss the role of the $f_0(980)$ resonance. We demonstrate in a simple model that in the case of a coupled channel resonance the resonance pole in the form-factor is accompanied by a nearby zero. The elastic scattering phase $\delta \stackrel{s\to\infty}{\to} \pi$ in that case, and the correct asymptotic behaviour of the form-factor $F(s)$ is automatically preserved by a protective zero in $F(s)$. The Omnès function with a resonance rise in the scattering phase reaching π does not provide this zero in $F(s)$, therefore it must be included in the polynomial appearing in the general solution of the OM equation.

In the case of the f_0 resonance there is direct evidence for its coupled channel nature from the data for the $\pi\pi$ production in the $J = I = 0$ state in the reaction $pp \to pp\pi\pi$ [3]. At energies near the f_0 there is no prominent peak, but only a shoulder and the cross section is almost zero nearby, Fig. 1b. For the $K \to \pi\pi$ decay we therefore expect a nearby zero as well, strongly suppressing the FSI enhancement factor from the f_0 resonance: $F^{I=0}_{f_0}(m_K^2) \approx 1.06$ contrary to some statements in the literature.

The total enhancement factor in the attractive $I = 0$ channel is $F^{I=0}(m_K^2) = 1.62$, Fig. 1c. The suppression of the form-factor in the repulsive $I = 2$ channel in our model is $F^{I=2}(m_K^2) = 0.9$. The $\Delta I = 1/2$ rule is characterized by our net enhancement factor $F^{I=0}(m_K^2)/F^{I=2}(m_K^2) \approx 1.81$ similar to the results found in the literature (see [4,5] and references therein). Combined with the weak matrix elements this result is slightly smaller than required by the data. For details we refer to Ref.[6]

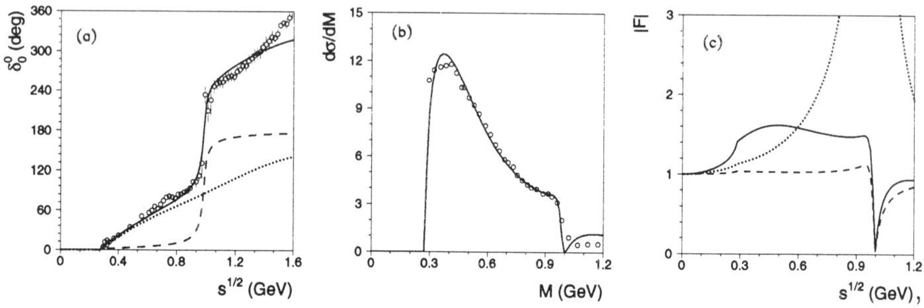

Figure 1. (a) The $\pi - \pi$ scattering phase δ_0^0. The solid line shows the net result for the model combining the ρ and f_2 exchange (dotted line) with the f_0 resonance (dashed line). The data are from Refs.[2]
(b) The effective mass distribution of pion pairs in $pp \rightarrow pp\pi\pi$ vs. $M = \sqrt{s}$. The data are from Ref.[3] and the line represents the coupled channel model.
(c) The pion scalar form-factor $F^{I=0}(s)$ vs. $\sqrt{s}$. The solid line corresponds to ρ and f_2 exchange plus the f_0 resonance including the polynomial (protective zero). The f_0 resonance alone leads to the dotted line (OM equation without a polynomial) and to the dashed line with polynomial.

References

1. B. S. Zou and D. V. Bugg, *Phys. Rev.* **D50**, 91 (1994).
2. G. Grayer *et al.*, *Nucl. Phys.* **B75**, 189 (1974); W. Ochs, Ph.D. thesis, Munich Univ., 1974; L. Rosselet *et al.*, *Phys. Rev.* **D15**, 574 (1977).
3. T. Akesson *et al. Nucl. Phys.* **B264**, 154 (1986).
4. G. E. Brown, J. W. Durso, M. B. Johnson and J. Speth, *Phys. Lett.* **B238**, 20 (1990).
5. N. Isgur, K. Maltman, J. Weinstein and T. Barnes, *Phys. Rev. Lett.* **64**, 161 (1990).
6. M.P. Locher, V.E. Markushin and H.Q. Zheng, Preprint PSI-PR-96-13, 1996.

FIRST EXCLUSIVE MEASUREMENT OF η PRODUCTION IN QUASIFREE PN COLLISIONS

R. BILGER[1], A. BONDAR[2], H. CALÉN[3], H. CLEMENT[1], C. EKSTRÖM[4], K. FRANSSON[3], L. GUSTAFSSON[3], S. HÄGGSTRÖM[3], B. HÖISTAD[3], A. JOHANSSON[3], T. JOHANSSON[3], K. KILIAN[5], S. KULLANDER[3], A. KUPSC[7], A. KUZMIN[2], P. MARCINIEWSKI[7], B. MOROSOV[6], J. MOEHN[3], A. MÖRTSELL[3], W. OELERT[5], A. POVTOREJKO[6] , V. RENKEN[5], R. RUBER[3], U. SCHUBERTH[3], V. SIDOROV[2], B. SHWARTZ[2], J. STEPANIAK[7], A. SUKHANOV[2], A. SUKHANOV[6(3)], G.J. WAGNER[1], M. WATERS[5], Z. WILHELMI[8], J. ZABIEROWSKI[9], A. ZERNOV[6], J. ZLOMANCZUK[3]

[1] *Physicalisches Inst., Tübingen University, Germany;* [2] *BINP, Novosibirsk, Russia;* [3] *ISV, Uppsala University, Sweden;* [4] *TSL, Uppsala, Sweden;* [5] *IKP, KFA, Jülich, Germany;* [6] *JINR, Dubna, Russia;* [7] *INS, Warsaw, Poland;* [8] *Inst. of Exp. Phys., Warsaw University, Poland;* [9] *INS, Lodz, Poland*

Data on the total cross section of the quasifree reactions pp $\to$ ppη, pn $\to$ pnη and pn $\to$ dη for CM excess energies in the range of 15 - 115 MeV are presented.

The cross section for the reaction pp $\to$ ppη has been well measured from the threshold region up to a beam energy of 2 GeV by a number of experiments [1-4]. The η production in pn interactions is not as well known. The existing data indicate a much higher η-production cross section in pn-interactions than in pp-interactions. However these data either have large uncertainties [5] or are from inclusive measurements with a deuteron target [6] where only the η was detected.

In our experiment at the CELSIUS storage ring in Uppsala[7], η-production in pd-collisions has been studied using an internal cluster jet target and a proton beam at 1350 MeV. The detection of both the η, from the $\eta \to \gamma\gamma$ decay, and outgoing particles enables us to separate the different quasifree reaction channels, pp $\to$ ppη, pn $\to$ pnη and pn $\to$ dη. Due to the Fermi momentum of the target nucleons, the CM energy of the system consisting of the beam proton and target nucleon varies. Thus, for each η event, being consistent with one of the channels, the CM energy is reconstructed and the excitation functions can be extracted after correcting for the detector acceptance and the distribution of the Fermi momentum[8].

Our preliminary results on the quasifree cross sections for pp $\to$ ppη, pn $\to$ pnη and pn $\to$ dη are shown in Fig. 1. Phase space distributions have been assumed when calculating the detector acceptance and effects from *e.g.* nuclear shadowing and reabsorbtion of the η-meson have not yet been corrected for. Only the statistical errors are indicated in the figure. The systematical errors are estimated to be of the order 30 %. Our excitation function for pp → ppη is in agreement with the previous measurements [1,2,4] within the systematical errors.

The cross section for pn → pnη is considerably larger than for pp → ppη but the shape is similar over the whole measured energy range (Fig. 2). The excitation function for pn → dη (shown separately in Fig. 3) is more flat and the shape is understood qualitatively by phase space and the excitation of the N*(1535) resonance [9] . The ratio between the pn → pnη and pn → dη cross sections is shown in Fig. 4. The curve is from a model of Fäldt and Wilkin [10] in which the the behaviour at low energy is shown to be dominated by the difference in the s-wave scattering state of the final NN system.

References

1. A. M. Bergdolt et al., *Phys.Rev.* **D48** (1993) R2969
2. E. Chiavassa et al., *Phys.Lett.* **B322** (1994) 270
3. E. Chiavassa et al., *Proc. of the sixth Int. Symposium on Meson-Nucleon Physics and the Structure of the Nucleon*, Blaubeuren/Tübingen, Germany, July 10-14, 1995, πN Newsletter, **10** (1995) 93
4. H. Calén et al., *Phys. Lett.* **B366** (1996) 39
5. E. Chiavassa et al., *Phys.Lett.* **B337** (1994) 192
6. F. Plouin, P. Fleury and C. Wilkin, *Phys. Rev. Lett.* **65** (1990) 690
7. H. Calén et al., *Detector setup for a cooler storage ring with internal target*, to appear in Nucl. Instr. Meth.
8. M. Lacombe et al., *Phys. Lett.* **B101** (1981) 139
9. G. Fäldt and C. Wilkin, private communication
10. G. Fäldt and C. Wilkin, *TSL/ISV-95-0127* December 1995

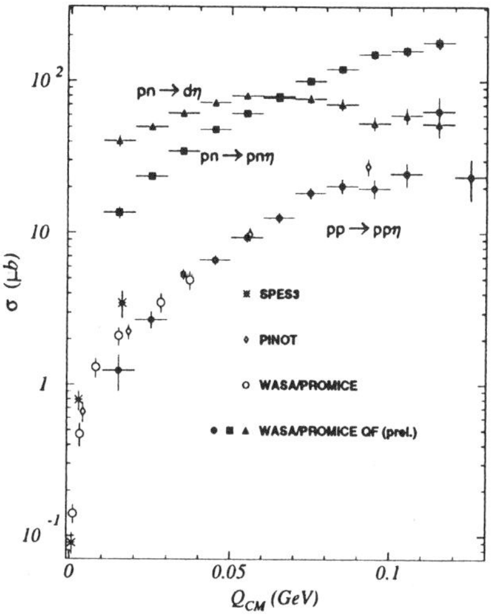

Figure 1: Cross sections for the reactions pp → ppη, pn → pnη and pn → dη as a function of the excess energy in the centre of mass system of the final state.

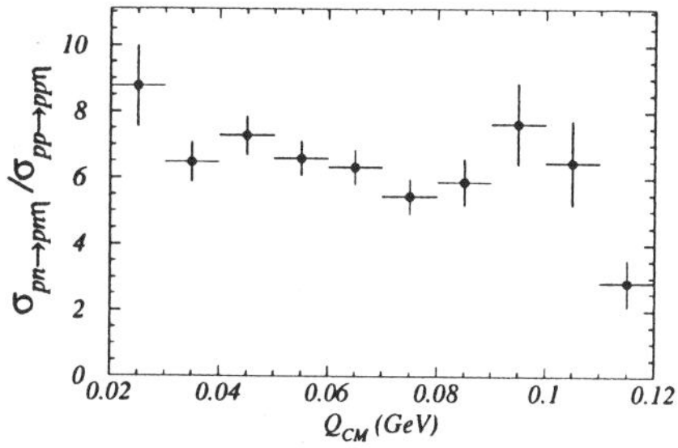

Figure 2: Cross section ratio between the reactions pn → pnη and pp → ppη.

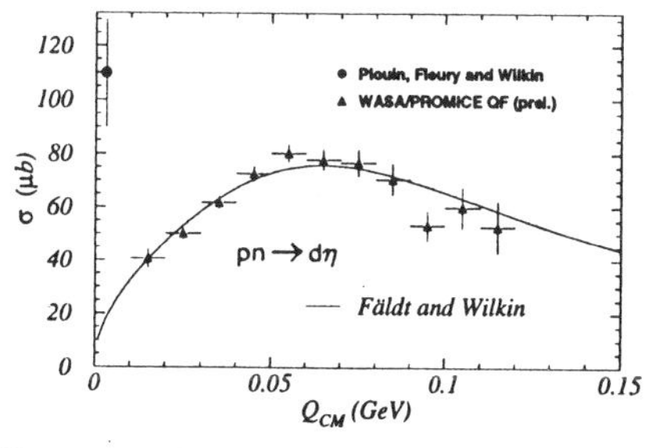

Figure 3: Cross section for the reaction pn → dη.

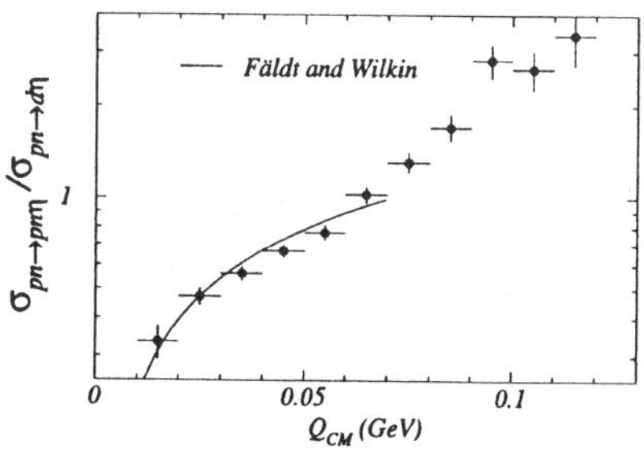

Figure 4: Cross section ratio between the reactions pn → pnη and pn → dη.

THE (γ,p) REACTION AND THE MASS-DEPENDENCE OF MESON-EXCHANGE CURRENTS

E.C. Aschenauer[a,d], C. Van den Abeele[d], J-O. Adler[b], B-E. Andersson[b], L.J. de Bever[a], I. Bobeldijk[a], D. Branford[c], S.A. Bulychjov[b], T. Davinson[c], K. Hansen[b], D.G. Ireland[e], L. Isaksson[b], D. Ivanov[d], D.G. Johnstone[c], A. Khanov[b], L. Lapikás[a], L.J. Lindgren[b], G. de Meyer[d], D. Van Neck[d], B. Nilsson[b], H. Ruijter[b], D. Ryckbosch[d], A. Sandell[b], B. Schröder[b], A.C. Shotter[c], V. Van der Sluys[d], G. van der Steenhoven[a], P. Svensson[b], M.A. van Uden[a] and R.E. Van de Vyver[d]

(a) NIKHEF, P.O. Box 41882, 1009 DB Amsterdam, The Netherlands
(b) University of Lund, Solvegatan 14, 223 62 Lund, Sweden
(c) University of Edinburgh, Edinburgh EH9 3JZ, Scotland, UK
(d) University of Gent, Proeftuinstraat 86, 9000 Gent, Belgium
(e) University of Glasgow, Glasgow G12 8QQ, Scotland, UK

In order to investigate the relative importance of nucleon-nucleon correlations and explicit meson-exchange currents in a heavy nucleus, complementary (e,e'p) and (γ,p) experiments on ^{208}Pb were performed. The results obtained with the reaction ^{208}Pb(e,e'p) demonstrated the importance of long-range and short-range correlations at high recoil momenta [1,2]. The (γ,p) reaction, on the other hand, is expected to be highly sensitive to meson-exchange currents (MEC), because of the purely transverse character of the incident real photon. However, the importance of MEC effects in the (γ,p) reaction has been subject to debate [3,4], and thus requires further study.

The experiment was carried out using the tagged photon beam of the MAX-Lab at the University of Lund. The energy of the tagged-photon beam covered simultaneously the ranges from 41 to 48 MeV and 51 to 57 MeV with a resolution of 230 keV. The target, placed at an angle of 19.8$^o\pm$ 0.58^o with respect to the photon beam, consisted of a self-supporting 99% enriched ^{208}Pb-foil. The knocked-out protons have been detected in two solid state detector telescopes, which subtended a solid angle of 413 $\pm$ 20 msr. Each telescope consists of two Silicon Strip detectors and a Hyperpure Germanium detector which measure the in-plane proton-emission angle and the energy of the proton, respectively.

The obtained energy resolution of 460 keV for the ^{208}Pb(γ,p) data [5] enabled us to make a comparison with the results obtained with the ^{208}Pb(e,e'p) reaction in the same recoil-momentum range. It is thus possible to constrain the direct-knockout part of the (γ,p) reaction (including the effects of NN-correlations) by making use of the ^{208}Pb(e,e'p) data.

The ^{208}Pb(γ,p) cross sections for transitions to the valence states and the continuum are compared to results of various theoretical calculations, like coupled-channel calculations in the random phase approximation (RPA) by Ryckebusch et al.[4,9]. In this model the nuclear wave functions are generated with the effective Skyrme NN-interaction [10], including the coupling to the collective properties of the target and recoil nucleus, multi-step processes and rescattering effects. The MEC-effects are introduced as photon absorption on two-body currents by a minimal substitution in the Skyrme NN-Hamiltonian, preserving gauge invariance. Employing the quasi-particle wave functions and normalization factors as obtained from ^{208}Pb(e,e'p), two different (γ,p) distorted-wave impulse-approximation (DWIA) calculations based on [3,6] and [11] have been performed. The optical model parameters for these (γ,p) calculations have been taken from ref. [7]. In order to estimate the effect of MEC's on the DWIA calculations we have used the Siegert theorem [8] in a method developed by [6]. As an example in fig. 2 the comparison between the cross section of the two valence state doublets for two different average photon energies and the above described calculations is shown. It can be concluded that calculations not including MEC-effects already give a fairly good account of the data. At the same time it must be concluded that the MEC effects seem to be modest, at least when they are properly included at the level of the nucleon current (as has been done in the RPA+MEC approach). The simplified estimate of MEC's through the Siegert theorem is apparently not appropriate for a heavy nucleus such as ^{208}Pb. On average the best agreement with the data is obtained by the RPA+MEC calculations.

It is of interest to note the size of MEC-effects in the (γ,p) reaction is considerably larger for lighter nuclei. This is illustrated in fig. 1, were we have plotted the ratio between the data and an RPA calculation not including MEC effect for various values of A and θ_p. The ratio clearly decreases with A, which is properly accounted for by the RPA+MEC calculations.

1. I. Bobeldijk et al., Phys. Rev. Lett. **73**, 2684 (1994).
2. I. Bobeldijk et al., Phys. Lett. **B353**, 32 (1995).
3. S. Boffi, C. Giusti and F.D. Pacati, Nucl. Phys. **A359**, 91 (1981);
4. J. Ryckebusch et al., Nucl. Phys. **A476**, 237 (1988).
5. I. Bobeldijk et al., Phys. Lett. **B356, 13** (1995).
6. D.G. Ireland and G. van der Steenhoven, Phys. Rev. **C49**, 2182 (1994).
7. R.L. Varner et al., Phys. Rep. **201**, 57 (1991).
8. A.J.F. Siegert, Phys. Rev. **52**, 787 (1937);
   H. Arenhövel, Z. Phys. **A302**, 25 (1981);
9. J. Ryckebusch et al., Phys. Rev. **C46**, R829 (1992).
10. M. Waroquier et al., Phys. Rept. **148**, 249 (1987).
11. V. Van der Sluys et al., to be published

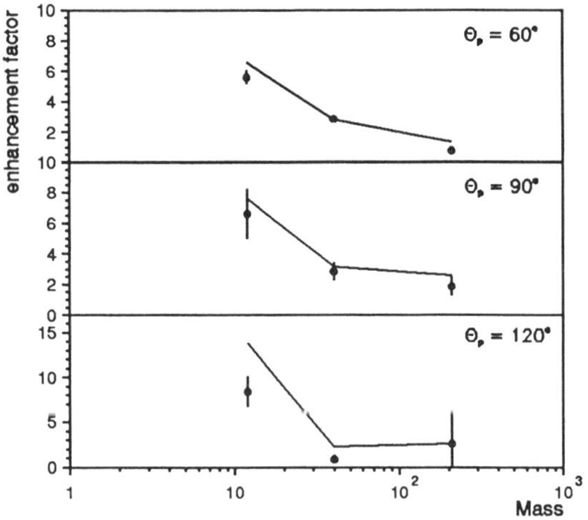

Figure 1: The enhancement of the cross section due to meson exchange currents versus mass for three different proton emission angles at average photon energy of 54MeV. The solid points represent the ratio between the experimental data and the RPA calculations without taking meson exchange currents into account [4,9]. The solid curve represents the ratio between RPA-calculations taking meson exchange currents into account and RPA-calculations without meson exchange currents.

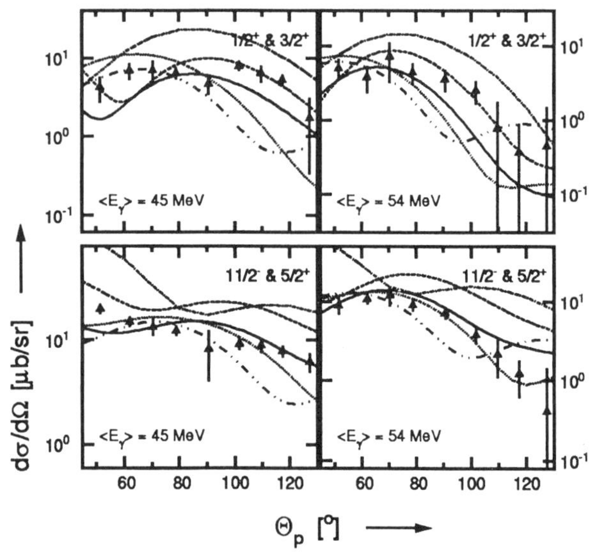

Figure 2: Angular distributions for transitions to the $(\frac{1}{2}^+, \frac{3}{2}^+)$ and $(\frac{11}{2}^-, \frac{5}{2}^+)$ doublets in ^{207}Tl at average photon energies of 45 and 54MeV. The solid, dashed-double-dotted and dotted curves represent RPA [4,9] and DWIA-calculations from ref. [11] and based on [3,6] taking no MEC-effects into account. The dashed and dotted-dashed curves are RPA calculations with MEC-effect and a DWIA-calculations accounting for the MEC-effects via the Siegert approach [8,6].

Session 3
Relativistic Heavy Ions

Extracting the equation of state from a microscopic non-equilibrium model

C. Spieles*, A. Dumitru*, S. A. Bass, M. Bleicher, J. Brachmann, M. Brandstetter, C. Ernst, L. Gerland, J. Konopka, S. Soff, H. Weber, L. A. Winckelmann, J. Maruhn, H. Stöcker, W. Greiner
Institut für Theoretische Physik
Johann Wolfgang Goethe Universität
60054 Frankfurt am Main, Germany

One of the main goals of relativistic heavy ion collisions is the determination of the nuclear equation of state. At high energies, semiclassical cascade models in terms of scattering hadrons have proven to be rather accurate in explaining experimental data. Therefore it is of fundamental interest to extract the equation of state from such a microscopic model, i. e. to investigate the equilibrium limits and bulk properties, which are not an explicit input to the non-equilibrium transport approach with its complicated collision term (unlike e. g. in hydrodynamics). In Fig. 1 we study the thermodynamic properties of infinite nuclear matter with the Ultra-relativistic Quantum Molecular Dynamics 1.0β (URQMD), a semiclassical transport model[1]. The model is based on classical propagation of hadrons and stochastic scattering (s channel excitation of baryonic and mesonic resonances/strings, t channel excitation, deexcitation and decay). For this study the potentials have been switched off. To simulate infinite hadronic systems we construct a box of volumc 250 fm^3 with periodic boundary conditions, and initialize 40 nucleons — that is

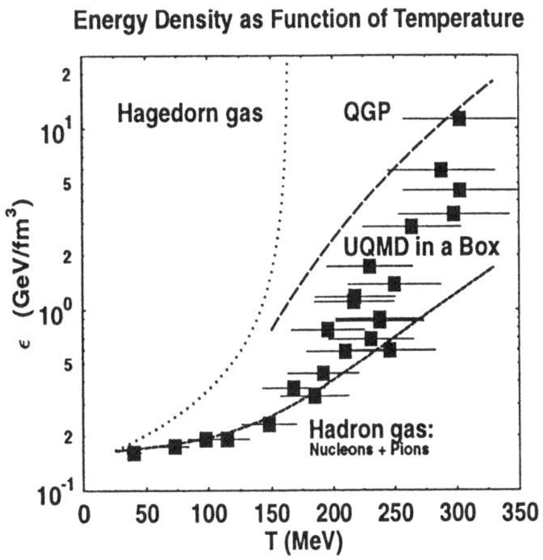

Figure 1: 'Equation of state' of infinite nuclear matter, calculated with URQMD (preliminary). Shown is the energy density as a function of temperature (extracted with a least square fit to the baryon momentum spectrum) at fixed net-baryon density of $\rho = 0.16$ fm^{-3}. The equations of state of a Hagedorn-gas, of a quark-gluon plasma, and of an ideal gas of nucleons and ultrarelativistic pions are also depicted.

*Participant in the PANIC'96 Conference

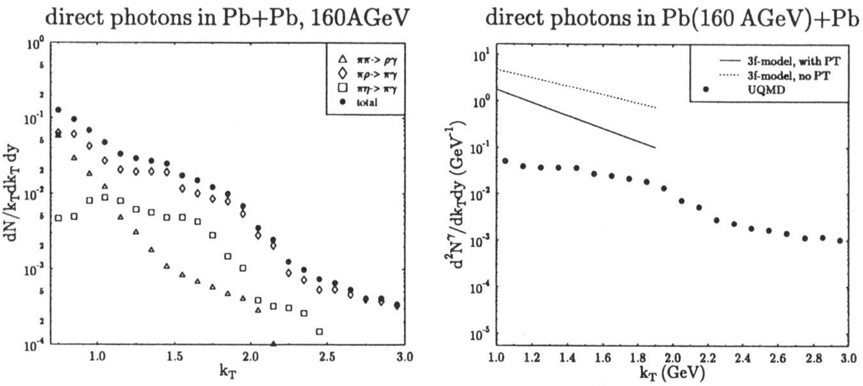

Figure 2: Transverse momentum spectrum of directly produced photons in Pb+Pb collisions at 160 GeV/u calculated with URQMD (left). The contributions of the different processes are shown. The resulting spectrum is compared with hydrodynamical calculations (right). For the photon source, an equation of state with and without a phase transition is assumed.

ground-state nuclear density — randomly in phase space, varying the total energy density. After the system has equilibrated according to the URQMD simulation, the temperature is extracted by fitting the particles' momentum spectra. Alternatively, the temperature can be extracted from the relative abundances of different hadrons, e. g. the Δ/N ratio, which should yield the same temperature. This has been checked in the current simulation. The result of this procedure is plotted in Fig. 1. It appears that the energy density rises faster than T^4 at high temperatures of $T \approx 200 - 300$ MeV. This indicates an increase in the number of degrees of freedom. It may be interpreted as a consequence of the numerous high mass resonances and string excitations, which in a way release constituent quark degrees of freedom (but, of course, no free current quarks as in an ideal QGP). Investigations of equilibration times and relative particle and cluster abundances are in progress. Moreover, the admittedly poor statistics have to be improved.

Experimentally, one can access the equation of state of strongly interacting matter by measuring thermal electromagnetic radiation emitted in heavy ion collisions. We have calculated direct photon production in Pb+Pb collisions at 160 GeV/u within the framework of URQMD 1.0 β. To improve statistics, the total meson-meson cross sections were fixed to 15 mb (independent of

$\sqrt{s}$). For the sake of comparison with earlier hydrodynamical calculations (see [2] and refs. therein) we considered only the processes $\pi\rho \rightarrow \pi\gamma$, $\pi\eta \rightarrow \pi\gamma$ and $\pi\pi \rightarrow \rho\gamma$. The amplitudes are taken from [3]. As can be seen in Fig. 2 (left) the $\pi + \rho \rightarrow \pi + \gamma$ process dominates the spectrum in the transverse momentum range considered. Fig. 2 (right) shows that the direct photon slope from the microscopic calculation equals that from the hydrodynamical calculation without a phase transition in the equation of state of the photon source. In the case of a phase transition the slope becomes significantly steeper due to a lower temperature at given energy density. The lower absolute yields of the direct photons from the URQMD may indicate that the meson abundance at early times is smaller than in the hydrodynamical model of [2], although the final yields are similar (due to higher mass mesons and baryon resonances). Thermal and in particular chemical equilibrium may not be established within URQMD in the early stage of the collision (where the large transverse momentum photons are produced). In contrast, in the hydrodynamical calculation local thermal and chemical equilibrium of the produced particles is assumed from the very beginning. Experimental studies on direct photons, under way at CERN [4], will help to settle these questions in the near future.

Acknowledgments

This work was supported by the GSI, Darmstadt, the BMBF, Bonn, Deutsche Forschungsgemeinschaft, and the Graduiertenkolleg Schwerionenphysik (Frankfurt/Giessen).

References

1. S. A. Bass, M. Bleicher, M. Brandstetter, C. Ernst, L. Gerland, C. Hartnack, J. Konopka, S. Soff, C. Spieles, H. Weber, L. A. Winckelmann, J. Aichelin, N. Amelin, H. Stöcker and W. Greiner: source code and technical documentation, to be published;
   L.A. Winckelmann et al., Proc. of the 12th Int. Conf. on Ultrarelativistic Nucleus-Nucleus Collisions, Quark Matter '96, Heidelberg, May 20-24 (1996)
2. A. Dumitru, U. Katscher, J.A. Maruhn, H. Stöcker, W. Greiner, D.H. Rischke: Phys. Rev. C51 (1995) 2166
3. J. Kapusta, P. Lichard, D. Seibert: Phys. Rev. D44 (1991) 2774
4. T. Peitzmann for the WA98 collab.: Talk given at the 12th Int. Conf. on Ultrarelativistic Nucleus-Nucleus Collisions, Quark Matter '96, Heidelberg, May 20-24 (1996)

RECENT RESULTS OF E866 ON HADRON PRODUCTION IN Au+Au REACTION AT AGS ENERGIES

ZIPING CHEN
Brookhaven National Laboratory, Upton, NY 11973 USA

For the E-802 collaboration

1 Introduction

The aim of experiment E866 is to study relativistic heavy ion reactions by measuring semi-inclusive hadronic particle production in Au+Au reactions at 11.6 A GeV/c under various event characterizations. It was reported by this collaboration that in central Au+Au collisions the rapidity distribution of protons indicates a large degree of stopping, hence high baryon density, in the reaction.

2 Experimental Results

This article first compares low $m_t - m_0$ rise of negative particles for pions and kaons around central rapidity. In Figure 1, the left shows the invariant spectra for π^+, π^-, and protons, as well as the π^-/π^+ ratio in the insert; the right is a similar plot for kaons. Low $m_t - m_0$ rise is clearly seen for π^- over π^+. The K^+ and K^- spectrum shapes are not identical, but the indication for low $m_t - m_0$ rise is not so pronounced for kaons. The left side of Figure 2 shows the centrality dependence of π^-/π^+ ratio at the same rapidity interval. In peripheral reactions, the ratio is flat but larger than one, reflecting there are more neuterons than protons in Au nuclei. When the reaction becomes central, there is not only a rise at low $m_t - m_0$ but a fall at large $m_t - m_0$. The right side of Figure 2 shows the possible existence of radial expansion in the reaction by demonstrating larger increase of inverse slope parameter is oberved for particles of heavier mass as the estimated number of participants increases.

Acknowledgments

The work is a result of the collective effort of E-802 collaboration, supported by the U.S. DOE, NASA, and by the US-Japan High Energy Collaboration Treaty.

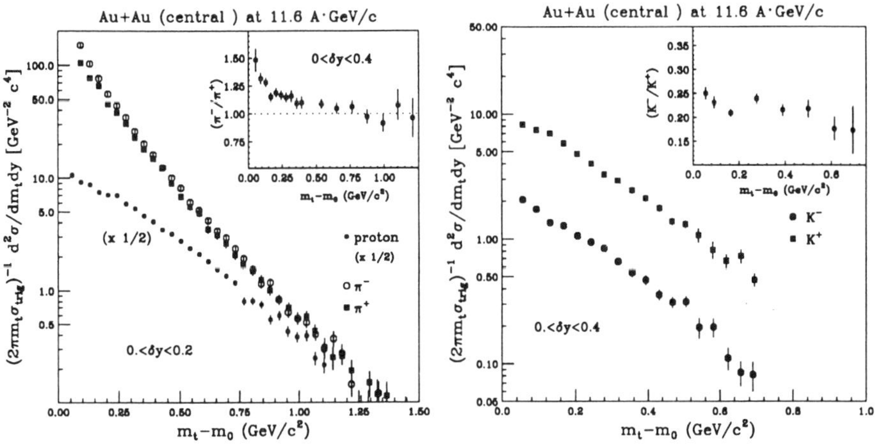

Figure 1: The left side shows the $m_t - m_0$ spectra for π^+, π^-, and protons at central rapidity, and the right side for K^+ and K^-. Low $m_t - m_0$ rise is present for π^+ over π^-, but not so pronounced for K^- over K^+. $\delta y = |y - y_{nn}|$

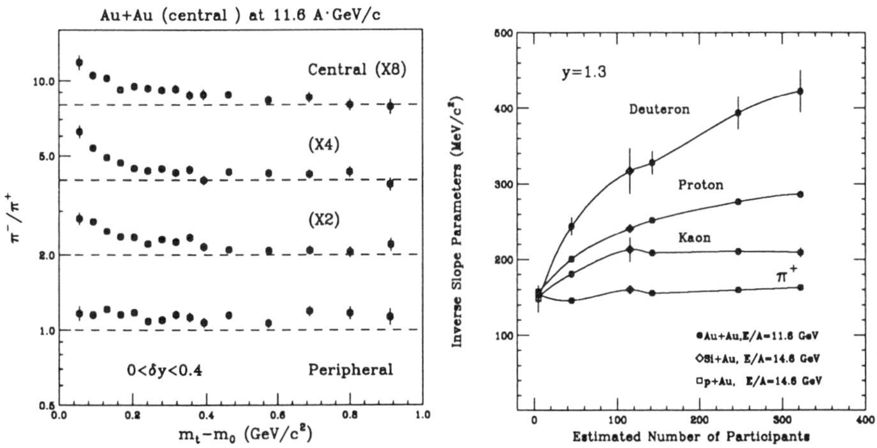

Figure 2: The left shows the centrality dependence of low $m_t - m_0$ rise of ratio π^-/π^+. In peripheral Au+Au, the ratio is flat over $m_t - m_0$, but larger than one (neuteron to proton ratio in the projectile and target). In central reation, it rises at low $m_t - m_0$ and falls at high $m_t - m_0$. The right shows the inverse slope parameter for heavier particles changes larger as the number of participants in the collisions increases, possibly indicating radial expansion.

Λ PRODUCTION IN AU-AU COLLISIONS AT THE AGS

A.C. SAULYS
for the E891 Collaboration
S. Ahmad[d], B.E. Bonner[d], C.S. Chan[b], J.M. Clement[d], S.V. Efremov[d], E. Efstathiadis[b], S.E. Eiseman[a], A. Etkin[a], K.J. Foley[a], R.W. Hackenburg[a], M.A. Kramer[b], S.J. Lindenbaum[a,b], R.S. Longacre[a], W.A. Love[a], J. Marx[c], G.S. Mutchler[d], E.D. Platner[d], A.C. Saulys[a], L.S. Schroeder[c], H.W. Themann[d], K. Zhao[b], Y. Zhu[b]
[a] *Brookhaven National Laboratory, Upton, NY 11973, USA*
[b] *City College of New York, NY 10031, USA*
[c] *Lawrence Berkeley National Laboratory, Berkeley, CA 94720, USA*
[d] *Rice University, Houston, TX 77251, USA*

The results of a measurement of Λ production in $Au-Au$ collisions at 11.6 A GeV/c by Experiment 891 at the Brookhaven AGS are presented. The measurements cover the rapidity region of 2.0 to 3.2 and transverse momenta of 0.0 to 1.4 GeV/c. The results are compared with similar measurements of $Si - Si$ interactions and the predictions of ARC and RQMD models.

1 Experimental method and results

The apparatus for this experiment is similar to that used in an earlier experiment [1]. It is based on charged track reconstruction in TPC. We used the charged track multiplicity for centrality selection. The trigger cross section for this data sample was 270 mb. In fig. 1a we present the Λ transverse mass distributions and in fig. 1b we present the Λ rapidity distributions with the predictions of the ARC [4] and RQMD [5] models for impact parameters 0-4 fermi. We used a global fit to the data points in fig. 1a of the form $A{\cdot}exp((a+b\cdot cosh(y-y_0)){\cdot}m_t+c\cdot(y-y_0)^2)$ where A is an arbitrary constant and a, b, c are independent of rapidity. We scaled the Λ production [2] from $Si+Si$ using the Λ/K_s^0 ratio and the $K^+ + K^-$ data [3] from $Au+Au$.

2 Conclusions

In summary, our measured Λ rapidity distribution for $Au+Au$ central interactions agrees with the scaling of $Si+Si$ measurements [2] and with the predictions of the ARC model. The transverse mass distributions become less steep at mid-rapidity compared to the $Si+Si$ measurements or the predictions of ARC and RQMD models.

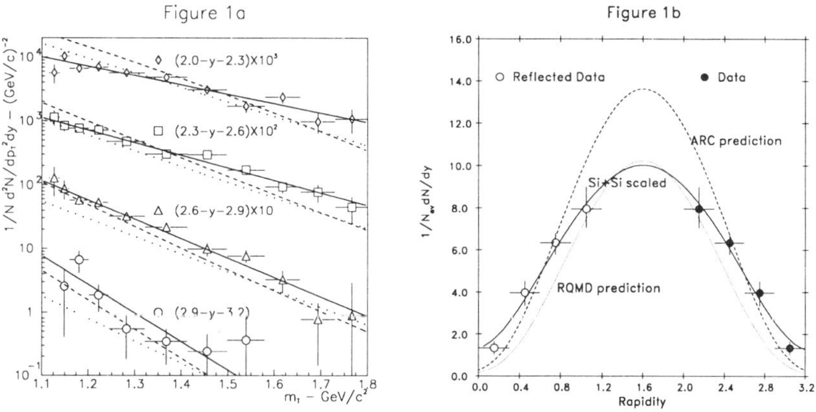

Figure 1: a) Transverse mass (m_t) distributions for Λ's. Each rapidity range has been scaled up by a factor of 10 as shown in the legend on the plot. The solid lines for $2.0 < y(\Lambda) < 2.9$ are fits to the points and for $2.9 < y(\Lambda) < 3.2$ the solid line is the result of the global fit described in the text. Errors shown are statistical only. b)Rapidity distribution for Λ's. The solid curve is the result of scaling $Si + Si$ data as explained in the text. The dashed lines are the ARC predictions and the dotted lines are the RQMD predictions in both figures.

Acknowledgments

This research was supported by the U.S. Department of Energy and the City University of New York PSC–BHE Research Award Program. We wish to thank the members of the AGS department and the MPS staff for their support during this experiment. We thank D. Kahana for providing us with events from the ARC model. We are grateful to H. Sorge for access to the RQMD code and advice on running it.

References

1. S.E. Eiseman, et al., *Phys. Lett.* B **248**, 254 (1990).
2. S.E. Eiseman et al., *Phys. Lett.* B **297**, 44 (1992).
3. F. Vidabaek for E802 collaboration, *Nucl. Phys.* A **590**, 249c (1995).
4. Y. Pang et al., *Phys. Rev. Lett.* **68**, 2743 (1992); S. Kahana et al., Physics at the AGS with a Relativistic Cascade, in: *Heavy-Ion Physics at the AGS* (MIT, Jan. 1993) p. 263.
5. H. Sorge, *Phys. Rev.* C **52**, 3291 (1995); *Nucl. Phys.* A **590**, 571c (1995).

QUANTUM INTERFERENCE EFFECTS IN PRODUCTION OF HEAVY QUARKONIA OFF NUCLEI

B.Z. KOPELIOVICH
Max-Planck-Institut für Kernphysik, Postfach 103980, 69029 Heidelberg, Germany

J. HÜFNER
Institut für Theor. Physik der Universität
Philosophenweg 19, 69120 Heidelberg, Germany

The effective path length of the $c\bar{c}$ wave packet in the nucleus is controlled by the coherence length $l_c = 2E_\Psi/M_\Psi^2$ and grows with the energy. This wave packet turns out to be a specific combination of the J/Ψ and Ψ' states interacting substantially weaker that any of its components. The time evolution of this wave packet is controlled by the formation length, $l_f = 2E_\Psi/(M_{\Psi'}^2 - M_\Psi^2)$. Interplay of the two phenomena results in a nontrivial energy- and x_F-dependence of the nuclear suppression of the charmonium in proton-nucleus and nucleus-nucleus collisions and explains some of the experimentally observed effects.

We study the nuclear suppression factor $S_\Psi^{hA}(E, x_F)$ for charmonium production off nuclei. The experiments have yielded a number of unexpected results, of which we recall the most significant ones:
(i) The value for $S_\Psi^{pA}(E, x_F)$ at small x_F seems to depend on E, the suppression factor being smaller for $E = 800\ GeV$ [1] than for $E = 200\ GeV$ [2].
(ii) Ψ and Ψ' have the same nuclear suppression in pA collisions [1].
(iii) Ψ' is more suppressed than J/Ψ in heavy ion collisions, $S_{\Psi'}^{AB} < S_\Psi^{AB}$ [3].
In the present paper we discuss quantum interference effects, which are not ad hoc mechanisms (and do not need any unknown parameters), but can explain, at least partially, the above effects.

Soft initial/final state interactions, which cannot resolve the $c\bar{c}$ fluctuation, do not cause any shadowing. Charmonia produced on different nucleons add up incoherently, since the longitudinal momentum transfer is large.

An additional contribution comes from a hard diffractive scattering on a bound nucleon producing the $c\bar{c}$ pair on mass shell prior the inelastic interaction. The longitudinal momentum transfer to the target nucleon may be small at high energy, $q_c \approx M_\Psi^2/2E_\Psi \ll 1/R_A$, then different nucleons contribute coherently.

We skip the full expression for $S_\Psi^{\gamma A}$, first derived in [4,5], which simplifies for small $\sigma_{tot}^{\Psi N}\langle T\rangle \ll 1$, where $\langle T\rangle$ is the mean nuclear thickness.

$$S_\Psi^{hA}(E_\Psi) \approx 1 - \frac{1}{2}\,\sigma_{in}^{\Psi}\,\langle T\rangle\left[1 + F_A^2(q_c)\right]\ , \qquad (1)$$

where $F_A^2(q_c) = \int d^2b \left|\int_{-\infty}^{\infty} dz\rho_A(b,z)e^{iq_cz}\right|^2 /A\langle T\rangle$ is the nuclear "longitudinal formfactor". We conclude from (1) that the nuclear shadowing correction at high energy ($F_A^2(q_c) \to 1$) is twice as big as at low energy ($F_A^2(q_c) \to 0$). This result has a natural interpretation: at high energy the lifetime of the $c\bar{c}$ fluctuation of the photon, $t_c = 1/q_c$, called the coherence time, is long, and the mean path of the $c\bar{c}$ pair in the nucleus is twice as long as at low energy.

Glauber approximation (1) is improved by adding off diagonal diffractive interactions of the charmonium. We restrict our consideration to a two-coupled-channel problem. Exact solution of the evolution equation for the charmonium wave packet in nuclear matter is presented in [5]. It is governed by the so called formation length $l_f = 1/q_f = 2E_\Psi/(M_{\Psi'}^2 - M_\Psi^2)$ and depends on the parameters estimated in oscillatory model: $\epsilon = \langle\Psi'|\widehat{f}|\Psi\rangle/\langle\Psi|\widehat{f}|\Psi\rangle \approx -2/3$; $r = \sigma_{tot}^{\Psi' N}/\sigma_{tot}^{\Psi N} \approx 7/3$, where $\widehat{f}$ is the elastic scattering amplitude operator. The relative production rate R^2 of Ψ' to Ψ was evaluated in [5] from the data $|R_{ex}| = 0.48 \pm 0.06$. The produced $c\bar{c}$ wave packet turns out to be nearly an eigenstate of interaction, and has an absorption cross section smaller than any of its components, Ψ or Ψ'. This eigen state does not change its relative Ψ-Ψ' content during propagation through the nucleus, i.e. the relative yields of Ψ' to Ψ has no A-dependence in accordance with the data [1,3]. Our calculations for relative suppression $S_{\Psi'/\Psi}^{pA} = S_{\Psi'}^{pA}/S_{\Psi}^{pA}$ are shown in Fig. 1. Combining the

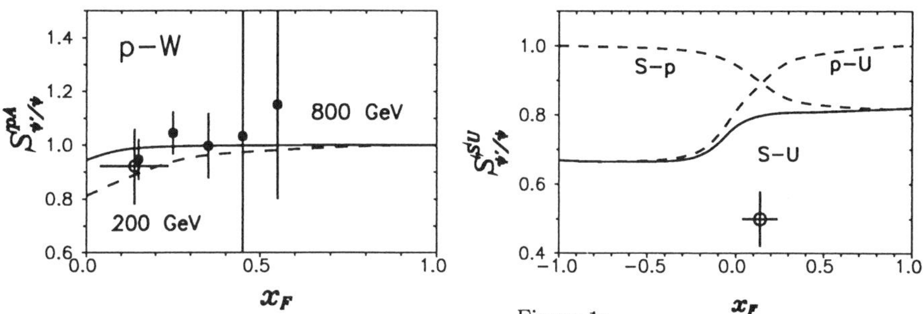

Figure 1:

effects of the the coherence and formation lengths, and assuming smallness of $\sigma_{tot}^{\Psi N}\langle T\rangle \ll 1$ we get (for exact solution see [6])

$$S_\Psi^{hA}(E_\Psi) \approx 1 - \frac{1}{2}\,\sigma_{in}^{\Psi}\,\langle T\rangle\,\left[1 + F_A^2(q_c) + \epsilon\; R\; F_A^2(q_f)\right] \tag{2}$$

$$S_{\Psi'}^{hA}(E_\Psi) \approx 1 - \frac{1}{2}\,\sigma_{in}^{\Psi'}\langle T\rangle\,\left[1 + F_A^2((q_c) + \frac{\epsilon}{r\;R}\;F_A^2(q_f)\right] \tag{3}$$

Interplay of the coherence and formation lengths leads to a nontrivial x_F- and energy-dependence of $S_\Psi^{hA}(E, x_F)$ [6].

We expect the nuclear effects for charmonium production in AA collisions to be quite different from what is known for pA, because the coherence and formation lengths depend on whether we are in the rest frame of the target or of the beam. Due to the inverse kinematics the charmonium wave packet attenuates with different effective cross sections propagating through the two colliding nuclei. The nuclear suppression in A_1A_2 collision is simply related to that in pA_1 and pA_2 interactions,

$$S_{\Psi}^{A_1A_2}(x_F) \approx S_{\Psi}^{pA_2}(x_F)S_{\Psi}^{pA_1}(-x_F) \tag{4}$$

The results of application of (4) to charmonium production in in lead-lead collisions is shown in Fig. 2 at 200 GeV and in the energy range of RHIC-LHC.

The relative nuclear suppression of Ψ'/Ψ in $S-U$ collisions is shown in Fig.1 in comparison with the sole experimental point from [3]. We see that this ratio, which is unity in pA interactions is substantially below one in the case of nuclear collisions. However, this reduction explains only about a half of the effect. This invites one to take into account the interaction of the charmonium with other produced particles, which is the next step to be done.

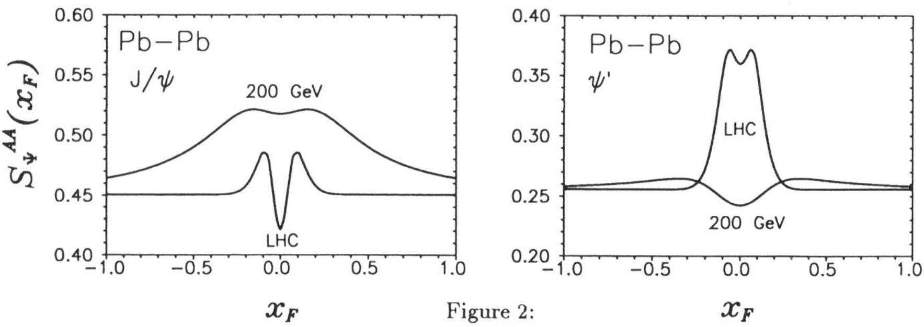

Figure 2:

References

1. D.M. Adle et al., *Phys. Rev. Lett.* **66** (1991) 133
2. The NA3 Collab., J. Badier et al., *Z. Phys.* **C 20** (1983) 101
3. The NA38 Collaboration, C. Baglin et al., *Phys. Lett.* **B345** (1995) 617 and M.C. Abreu et al., presented at QUARK MATTER'95
4. J. Hüfner, B.Z. Kopeliovich and J. Nemchik, DOE/ER/40561-260-INT96-19-03, nucl-th/9605007
5. J. Hüfner and B.Z. Kopeliovich, *Phys. Rev. Lett.* **76** (1996) 192
6. J. Hüfner and B.Z. Kopeliovich, paper in preparation

J/Ψ and Ψ' PRODUCTION IN LEAD-LEAD COLLISIONS AT THE CERN SPS

Presented by E.Vercellin for the NA50 collaboration

M.C. Abreu[6], C. Alexa[2] B. Alessandro[11], J. Astruc[8], C. Baglin[1], A. Baldit[4], F. Bellaiche[12], M. Bedjidian[12], S. Beolè[11] A. Borhani[9], V. Boldea[2], G. Bonazzola[11], P. Bordalo[6], A. Bussière[1], V. Capony[1], J. Castor[4], M. Cerú[3], T. Chambon[4], B. Chaurand[9], I. Chevrot[4], B. Cheynis[12], E. Chiavassa[11], C. Cicalo[3], S. Constantinescu[2], W. Dabrowski[11], A. De Falco[3] G. Dellacasa[11], N. De Marco[11], A. Devaux[4], S. Dita[2], O. Drapier[12], B. Espagnon[4], J. Fargeix[4], F. Fleuret[9], P. Force[4], M. Gallio[11], Y.K. Gavrilov[7], C. Gerschel[8], P. Giubellino[11], M.B. Golubeva[7], M. Gonin[9], P. Gorodetzky[10], J.Y. Grossiord[12], P. Guaita[11], F.F. Guber[7], A. Guichard[12], R. Haroutunian[12], M. Idzik[11], D. Jouan[8], T.L. Karavitcheva[7], R. Kossakowski[1], L. Kluberg[9], A.B. Kurepin[7], Y. Le Bornec[8], G. Landaud[4], C. Lourenço[5], L. Luquin[4], P. Macciotta[3], F. Ohlsson-Malek[12], A. Marzari-Chiesa[11], M. Masera[11], A. Masoni[3] , S. Mourgues[4], A. Musso[11], P. Petiau[9], W.L. Prado da Silva[11a], A. Piccotti[11], J.R. Pizzi[12], G. Puddu[3], C. Racca[10], L. Ramello[11], S. Ramos[6], P. Rato-Mendes[11] L. Riccati[11], A. Romana[9], P. Saturnini[4], E. Scomparin[11], R. Shaoian[5], S. Silva[6], S. Serci[3], P. Sonderegger[5], X. Tarrago[8], P.Temnikov[3], N.S. Topilskaya[7], G. Usai[3], E. Vercellin[11], and N. Willis[8].

[a]On leave from University of Rio de Janeiro

1)LAPP, Annecy-le-Vieux; 2)IAT, Bucharest; 3)University and INFN Cagliari; 4)LPC Clermont-Ferrand; 5)CERN Geneva; 6)LIP Lisbon; 7)INR Moscow; 8)IPN Orsay; 9)Ecole Polytechnique Palaiseau; 10)CRN Strasbourg; 11)University and INFN Torino; 12)IPN Villeurbanne

We present the preliminary results[1] on J/Ψ and Ψ' production in Pb-Pb collisions at 158 GeV per nucleon obtained by the NA50 experiment at the CERN SPS. The detector is an upgrade of the apparatus used by experiment NA38[2]. The $\mu^+\mu^-$ pair from the decay of the J/Ψ and Ψ', produced in the c.m. rapidity region $0 \leq y \leq 1$, is detected by a toroidal magnetic spectrometer located behind a hadron absorber. The event centrality is measured by an electromagnetic calorimeter, a zero-degree calorimeter and a silicon microstrip multiplicity detector, while the incoming Pb-ions are counted by a quartz beam hodoscope.

In the 1995 NA50 run (typical beam intensity $\sim 10^7$ Pb-ions/s, Pb target thickness 7 mm) were recorded about $5{\cdot}10^7$ events, including $5{\cdot}10^4$ J/Ψs.

The yields of the three sources (J/Ψ, Ψ' and Drell-Yan) that account for the $\mu^+\mu^-$ invariant mass spectrum above 2.9 GeV/c^2, have been evaluated by a fit to the spectrum. The cross sections for J/Ψ production $\sigma_{J/\Psi}$, multiplied by the $J/\Psi \rightarrow \mu^+\mu^-$ branching ratio $B_{\mu\mu}$ and divided by the product of the projectile (A) and target (B) mass numbers, is shown in fig. 1. Together with the Pb-Pb point, also the NA38 data, previously obtained with p, Oxygen and Sulphur beams, are plotted as a function of L, the mean length of nuclear matter crossed by the ($c\bar{c}g$) state. The data up to S-U show an exponential behaviour[3] ($\sim exp - (\rho L \sigma_{abs})$, where ρ is the ordinary nuclear density and $\sigma_{abs} = 6.4 \pm 1.3mb$ is the absorption cross section obtained by fitting the data) that is believed to be due to the nuclear absorption of the colour- singlet $c\bar{c}g$ state that is promptly formed in hadron- hadron interaction and that is then converted into a J/Ψ or Ψ' resonance[4]. On the other hand the NA50 preliminary point lies significantly below the extrapolation of the absorption calculation to Pb-Pb.

It is worth to mention that the discontinuity between lighter colliding systems and Pb-Pb is also visible when the data shown in fig. 1 are plotted as a function of $A \cdot B$.

A strong deviation from the absorption picture can also be seen in fig. 2, where the ratio of the J/Ψ to Drell-Yan cross section is plotted vs. L for both p-A and A-B interaction . These last data are divided in different centrality bins, corresponding to different values of L. While the data up to S-U are in agreement with absorption, this is no longer true for the Pb-Pb data where, moving form peripheral to central collisions (i.e. from small to large values of L) the J/Ψ to D.Y. ratio decreases much more than expected .

These facts suggest that, in addition to nuclear absorption, a new suppression mechanism, not present with lighter projectiles, plays a significant role in Pb-Pb. The nature of such a mechanism (onset of deconfinement vs. more "conventional" effects) is not clear at present.

Finally it is interesting to remark that from this preliminary analysis the $\Psi'/J/\Psi$ cross section ratio for Pb-Pb turns out to be very close to the one measured in S-U by NA38. This indicates that in Pb-Pb an extra suppression of the Ψ' also occurs and it is similar of the one observed for the J/Ψ.

References

1. M.Gonin for the NA50 collaboration, *proc. of the Quark Matter 96 conf., May 20-24, Heidelbergh, Germany*, to be published in *Nucl. Phys.* A
2. C. Baglin et al., *Phys. Lett.* B **345**, 617 (1995) and ref. therein
3. C. Gerschel and J. Hufner, *Z. Phys.* C **56**, 71 (1992).
4. D.Kharzeev and H.Satz, *Phys. Lett.* B **334**, 155 (1995); and also *CERN Preprint (1995) TH/95-214r*

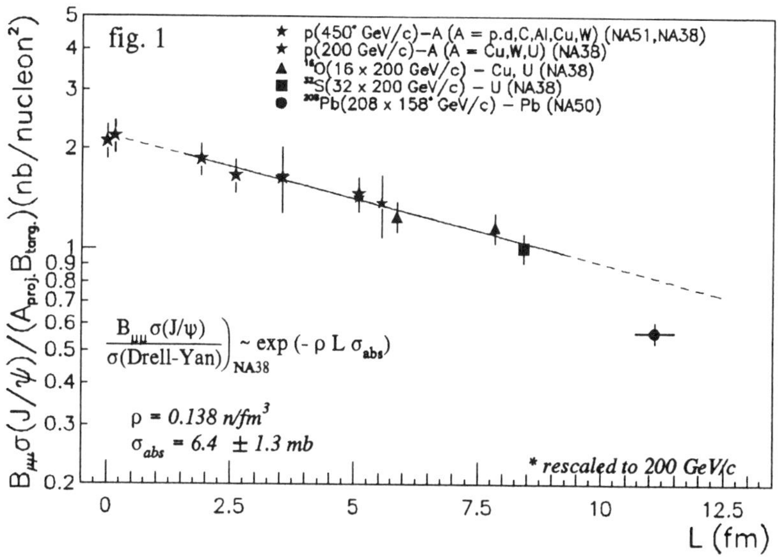

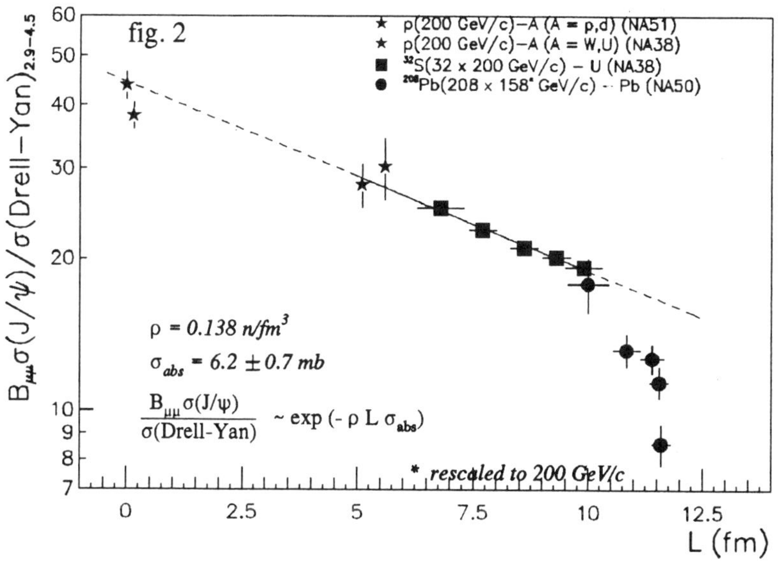

Figure 1: Cross section for J/Ψ production $\sigma_{J/\Psi}$ (multiplied by $B_{\mu\mu}$ and divided by AB) as a function of L; Figure 2: ratio of the cross section for J/Ψ production $\sigma_{J/\Psi}$ (multiplied by $B_{\mu\mu}$) to the Drell-Yan cross section σ_{DY} as a function of L

RHO MESON IN MEDIUM AND DILEPTON PRODUCTION IN HEAVY-ION COLLISIONS

G. Q. LI, C. M. KO
Cyclotron Institute and Physics Department, Texas A&M University, College Station, TX 77843, USA

H. SORGE, G. E. BROWN
Department of Physics, State University of New York at Stony Brook, Stony Brook, NY 11974, USA

Dilepton production in proton- and heavy-ion-induced reaction at SPS energies is studied using the relativistic transport model. It is found that the dilepton spectra in proton-nucleus reaction can be well described by 'conventional' mechanism of Dalitz decay and direct vector meson decay. The decrease of rho meson mass in hot and dense matter provides a quantitative explanation of the enhancement of low-mass dileptons in central heavy-ion collisions.

Dileptons, as well as real photons, have been suggested as useful probes of the initial stage of high energy heavy-ion collisions, as they interact only electromagnetically with the environment. Recent observation of the enhancement of low-mass dileptons in central S+Au and S+W collisions at SPS energies by the CERES [1] and the HELIOS-3 [2] collaboration has generated great deal of interest in heavy-ion community. In this work we carry out a detailed transport model calculation of dilepton production in heavy-ion collisions as well as in proton-nucleus reactions at SPS energies. The dileptons with invariant masses below about 1 GeV come mainly from Dalitz decay and direct vector meson decay. In heavy-ion collisions, vector mesons can be produced from both the initial string dynamics and the later stage thermal processes such as $\pi\pi \to \rho$ and $K\bar{K} \to \phi$. In addition, in heavy-ion collisions, because of the formation of hot and dense matter, hadron properties might be modified. Of particular relevance to low-mass dilepton enhancement is the dropping of rho meson mass in hot dense matter. This is included in our calculation by using the extended Walecka model in which the light quarks of vector mesons are coupled to the scalar field as those in baryons.

Our results for dilepton production proton-nucleus reaction are shown in Fig. 1 [3]. It is seen that both the CERES and HELIOS-3 data can be very well described by the conventional mechanism of Dalitz decay and direct vector meson decay.

The results for dilepton spectra in heavy-ion collisions are shown in Fig. 2. With free rho meson mass, both the CERES and HELIOS-3 data are underestimated by about a factor of 2-4 in the mass region from 0.3 GeV to 0.6 GeV [3]. Similar results have been obtained in other calculations [4–6]. With the in-medium rho meson mass, the agreement with the data is greatly improved. The enhancement of low-mass dilepton thus provides possible evidence for the decrease of vector meson masses in hot and dense matter because of partial restoration of chiral symmetry [7].

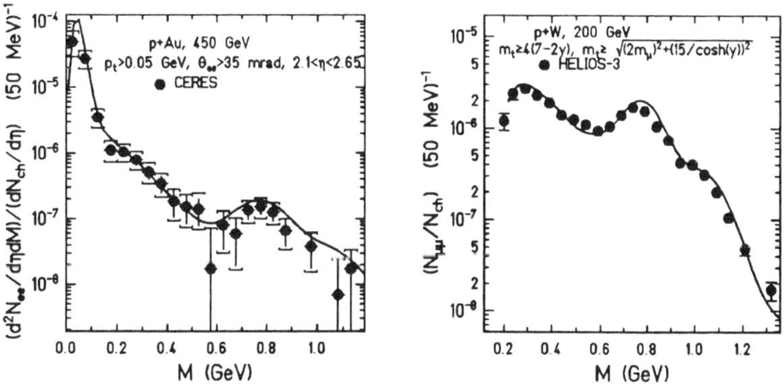

Fig. 1 Dilepton spectra in proton-nucleus reactions.

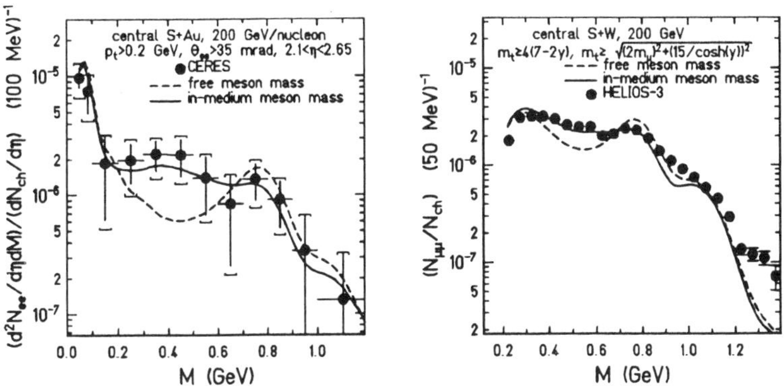

Fig. 2 Dilepton spectra in heavy-ion collisions.

---

[1] G. Agakichiev *et al.*,, Phys. Rev. Lett. 75 (1995) 1272.
[2] M. Masera for the HELIOS-3 collaboration, Nucl. Phys. A590 (1995) 93c.
[3] G. Q. Li, C. M. Ko, H. Sorge, and G. E. Brown, to be published.
[4] G. Q. Li, C. K. Ko, and G. E. Brown, Phys. Rev. Lett. 75 (1995) 4007.
[5] W. Cassing, W. Ehehalt, and C. M. Ko, Phys. Lett. B363 (1995) 35.
[6] D. K. Srivastava, B. Sinha, and C. Gale, Phys. Rev. C53 (1996) R567.
[7] G. E. Brown and M. Rho, Phys. Rev. Lett. 66 (1991) 2720.

SINGLE AND MULTI-PARTICLE SPECTRA FROM NA44

SANJEEV U. PANDEY for the NA44 Collaboration
The Ohio State University, 1302 Kinnear Road, Columbus, OH 43212, USA.

Abstract

Inverse slopes in transverse mass distributions from experiment NA44 for 200 GeV/nucleon S+S and 160 GeV/nucleon Pb+Pb collisions are similar for particles and anti-particles and increase with particle mass. Three dimensional source size parameters measured via two-particle interferometry show a common $1/\sqrt{m_T}$ dependence for all three dimensions. This is consistent with a hydrodynamical model for an expanding source.

1 The NA44 Experiment

The NA44 spectrometer is optimized to measure single and many particle distributions of charged hadrons produced near mid-rapidity for transverse momentum below $\approx$ 1 GeV/c. Excellent particle identification allows measurement of pions, kaons, protons, and deutrons of both charge states with less than 1% contaminations from other particle species. Since it is a focussing spectrometer, the acceptance for pairs of particles with small momentum difference is optimized, allowing small statistical uncertainties in the region of the signal from Bose-Einstein correlations.

2 Results

Fully corrected single particle distributions are well described by exponentials in m_T. The data are fit by single exponentials and inverse slopes for particles and anti-particles are similar. A clear increase in inverse slopes with mass and system size is seen in Figure 1 illustrating the importance of rescattering. Such behavior suggests significant transverse expansion is present and the data can be studied within the context of a thermal model with with global collective expansion[1]. Within this model, a single temperature of $\approx$ 140 MeV for all particle species is extracted as well as a transverse expansion velocity of 0.4c for S+S and 0.6c forPb+Pb collisions. Large transverse flow raises the question of nuclear stopping and indeed π^-/π^+ ratios increase at low p_T as one goes from S+S to Pb+Pb collisions.

Fully corrected two-particle distributions are well fit by gaussians and radius parameters are extracted from the widths of these distributions. Correlations between non-identical particles do not to exist and three-particle correlation results are consistent with those from two particles and serve as a systematic check on the data analysis[2]. In order to accept a hydrodynamical model as a reasonable basis for understanding the data the multi-particle data have to be interpreted within the same framework. Analytical hydrodynamical calculations[3]

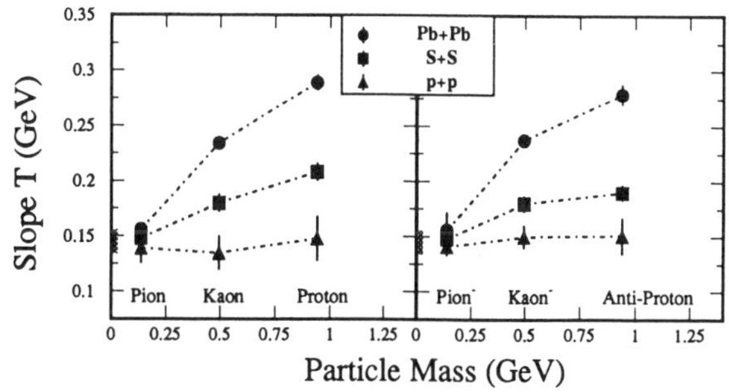

Figure 1: Inverse slopes as a function of particle type.

show that for global collective expansion with superimposed thermal motion, the radius parameters in different directions become more equal and scale as $1/\sqrt{m_T}$. Such behavior is seen in the data and is expected in hydrodynamical expansion where strong correlations exist between the position and momentum of the emitted particles.

3 Conclusions

Single and multi-particle spectra are consistent with the presence of strong transverse expansion in S+S and Pb+Pb heavy ion collisions. Indications of flow suggest a dense and equilibrated intial state which is a necessary but not sufficient condition for a phase transition to a quark gluon plasma.

References

[1] E. Schnedermann et al., Phys. Rev. C **C48** (1992) 2462.

[2] S. U. Pandey et al., QCD and High Energy Hadronic Interactions, Ed. Frontieres. (1994) 659.

[3] T. Csorgo and B. Lorstadt, Lund University preprint LUNFD6/(NFFI-7082) (1994).

Inclusive Particle Spectra and Energy Flow in Pb-Pb Collisions at 158 GeV/nucleon

P. M. Jacobs and the NA49 Collaboration
Lawrence Berkeley National Laboratory, Berkeley, CA 94720

We present preliminary results from the NA49 experiment for Pb-Pb collisions at 158 GeV/nucleon: (i) spectra of negative hadrons, net protons and neutral and charged kaons for central collisions, and (ii) azimuthal correlation of energy flow. For central collisions, negative hadrons show a rapidity density dN/dy≈200 at midrapidity; negative hadron and kaon yields scale with number of participants relative to S+S; and net protons show slightly increased stopping relative to S+S. Azimuthal correlation of energy flow has been observed. This indicates memory of the initial collision geometry in the final state for semi–central collisions.

1 Introduction

The NA49 experiment at the CERN SPS is a large acceptance, fixed target, hadronic spectrometer designed for the study of Pb-Pb collisions at 158 GeV/nucleon. Central collisions of heavy nuclei at these energies produce up to 2000 particles per event. NA49 seeks to track with good momentum resolution and to identify almost all charged particles forward of midrapidity. It is based on four large Time Projection Chambers (TPCs) within and downstream of a magnetic field, with a "Veto" calorimeter at zero degrees for triggering. Particle identification is performed in the TPC via measurement of mean energy loss $< dE/dx >$ in the relativistic rise regime, and via time of flight in four large, highly segmented arrays placed behind the TPCs. A Ring Calorimeter was used in a special run configuration without magnetic field to study energy flow at midrapidity. The physics goals of NA49 include the analysis of inclusive spectra of charged particles and strange mesons and baryons over a large acceptance, two and multi-particle correlations, and energy flow. In addition, the very large acceptance of NA49 and the enormous multiplicity of central Pb-Pb collisions present the opportunity for significant *event-by-event* studies, as are also planned for the STAR detector at RHIC[1]. The construction of NA49 was completed and the first data taken with the full apparatus in November 1995. All data presented here are preliminary.

2 Inclusive Particle Spectra for Central Pb-Pb Collisions

Central collisons are defined as the 5% most central part of the total cross section, as measured by the Veto Calorimeter.

The negative hadron dN/dy$_\pi$ (sum of π^-, K^-, $\bar{p}$) peaks at about 200 at midrapidity and integrates to a total of 680. Comparison with the dN/dy spectrum of the lighter system S+S[2] shows good consistency under scaling with the number of participants (by a factor of 7). This is the scaling expected from the Wounded Nucleon phenomenology extracted from pA data[3], and indicates on an inclusive basis no significant collective contribution to the production of negative hadrons.

The net proton dN/dy$_p$, shown in Figure 1, is measured as the difference between the positive and negative hadron spectra ("plus minus minus"). Corrections for acceptance, efficiency, and the charge imbalance of kaons have been applied, as well as for daughters of Λ decays misidentified as primary particles. The corrections are partially based upon the Venus model[4]. Statistical errors only are shown. The axes have been scaled by y/y_{beam} to facilitate comparison to the lighter S+S system[2] that was measured at 200 GeV/nucleon. The S+S yield is also scaled by the ratio of number of participants. The Pb-Pb spectrum shows somewhat greater proton stopping (rapidity loss) overall. Recalling that energy loss goes exponentially with rapidity loss, this may translate into a greater energy density at midrapidity. Detailed comparisons of the transverse momentum spectra are needed to confirm this.

The dN/dy$_K$ of charged and neutral kaons also exhibit approximate scaling with total participant number relative to the S+S spectrum[5], indicating a saturation of "strangeness enhancement" already in the S+S system. However, systematic errors are at present 30%, restricting the validity of this conclusion.

3 Azimuthal Flow of Energy at Midrapidity

Flow of energy at midrapidity was studied using the NA49 Ring Calorimeter[6], configured to cover the pseudorapidity range $2.1 < \eta < 3.8$. The cylindrical Ring Calorimeter is divided into 240 cells, with 24 azimuthal sectors and 10 radial rings. Azimuthal distribution of energy was studied using the rings $2.1 < \eta < 2.6$ (backward) and $3.3 < \eta < 3.8$ (forward). The gap between these rings reduces the cross talk due to finite shower size effects.

A two-dimensional sphericity tensor is constructed based upon the E_T in each cell. The tensor is diagonalized separately for the backward and forward rings to extract the azimuthal direction of the energy flow in each ring. The difference between these azimuthal angles shows a correlation that is fit well by a function

$$f(\Delta\phi_{fb}) = c * (1 + a_2 * cos(2\Delta\phi_{fb})) \tag{1}$$

Figure 2 shows the coefficient a_2 as a function of E_{veto} (central collisions correspond to low E_{veto}). The correlation peaks at mid-centrality. Also shown are predictions of the Venus [4] and RQMD [7] models, as well as Venus modified with a quadrupole deformation in transverse momentum whose magnitude is fit to reproduce the data [8]. This phenomenon may indicate shadowing by spectator matter or memory in the central region of the initial reaction geometry [9]. Further study, both via calorimetry and via tracking in the TPCs, is in progress.

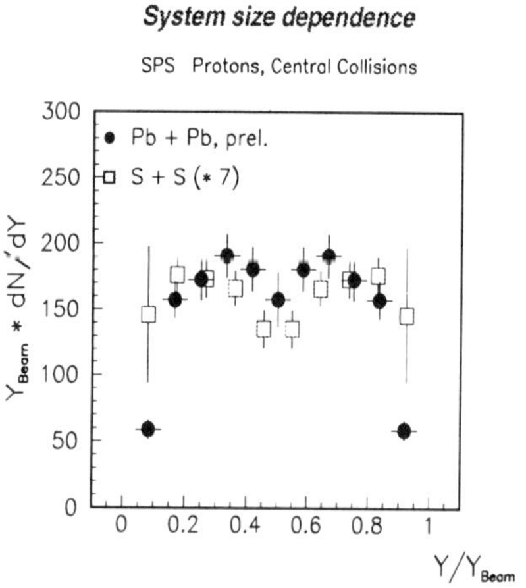

Figure 1: dN/dy of net protons for Pb+Pb and S+S, scaled by ratio of number of participants. Axes described in text.

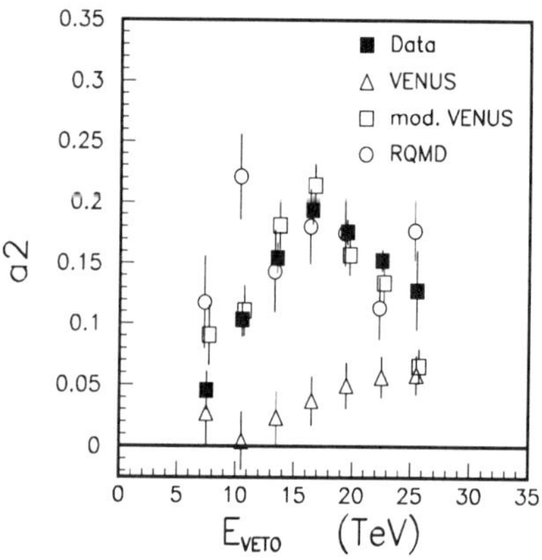

Figure 2: Correlation coefficient a_2 from Eq. 1 as function of E_{veto}. Central collisions correspond to low E_{veto}, peripheral collisions to high E_{veto}.

References

1. STAR Conceptual Design Report, LBL PUB-5347.
2. J. Baechler et al., Phys. Rev. Lett. **72** (1994) 1419.
3. A. Bialas, M. Bleszynski and W. Czyz, Nucl. Phys. **B111** (1976) 461.
4. K. Werner, Phys. Lett. **B208** (1988) 520.
5. T. Alber et al., Z. Physik. **C64** (1994) 195.
6. T. Alber et al., Phys. Rev. Lett. **75** (1995) 3814.
7. H. Sorge et al., Ann. Phys. **192** (1989) 266.
8. T. Wienold and I. Huang, LBL preprint 38506.
9. P. Filip, preprint CU-PH1-96, May 1996.

WAVELET VIEW OF DISORIENTED CHIRAL CONDENSATE[a]

Zheng Huang[a], Ina Sarcevic[a], Robert Thews[a] and Xin-Nian Wang[b]
[a] *Department of Physics, University of Arizona, Tucson, AZ 85721*
[b] *Nuclear Science Division, LBNL, Berkeley, CA 94720*

The possible chiral phase transition in high energy heavy-ion collisions may lead to the formation of a disoriented chiral condensate (DCC). However, the existence of many uncorrelated small domains in the rather large interaction volume and the huge combinatorial background make the experimental search for the DCC signal a rather difficult task. We present a novel method for studying the formation of a DCC in high energy heavy-ion collisions utilizing a discrete wavelet transformation.

In heavy-ion collisions the rapid expansion of the collision debris in the longitudinal direction may lead to supercooling of the interaction region, and as a result, domains of the "unconventionally" oriented vacuum configurations allowed by the chiral symmetry may be formed [1]. Preliminary theoretical investigations on non-equilibrium dynamics using the classical linear σ-model have found some evidence for the growth of long wavelength pion modes, which may indeed lead to domain structure or cluster formation [2,3]. If there are many uncorrelated small domains, the integrated probability distribution of the neutral pion fraction f emitted from a disoriented region, predicted to be $P(f) = 1/2\sqrt{f}$ [1], would become Gaussian. This makes the experimental search for the DCC signal a rather difficult task. We propose a new method which emphasizes not only the behavior of the probability distribution in the full phase space region but also its fluctuation in rapidity η or azimuthal angle ϕ.

The new method[4] is a multi-resolution analysis performed by a discrete wavelet transformation (DWT) which has been found effective in systematically detecting structures on various scales in turbulence, and multiparticle productions [5,6]. For any one-dimensional sample of a point distribution $f(x)$, such as the rapidity distribution of the neutral pion fraction one defines the DWT coefficients

$$\tilde{f}_{jk} = 2^j \int f(x)\psi_{jk}(x)dx, \tag{1}$$

where the basis functions $\psi_{jk}(x)$ have a unique property: they are related to each other by scaling and translation: $\psi_{jk}(x) = \psi(2^j x - k)$. The index j denotes the resolution scale and k the position at each scale. The idea of multiresolution analysis is to find representations of the sample function $f(x)$ at various scales. Roughly speaking, the multiresolution analysis is a journey

[a]Presented by Z. Huang. E-address: huang@physics.arizona.edu

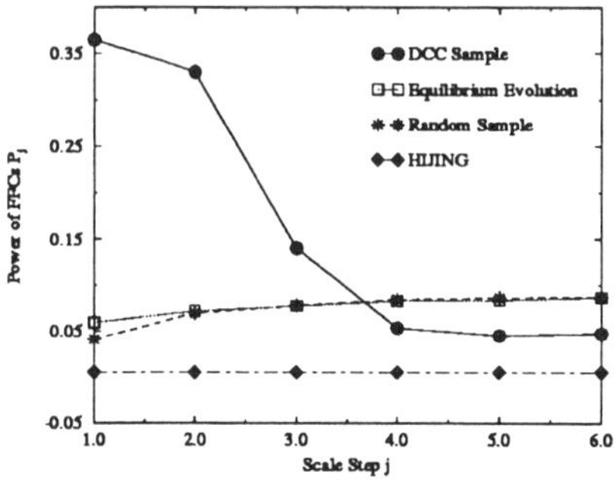

Figure 1: The power spectra of the wavelet coefficients for different dynamical scenarios.

down (or up) a hierarchy of scales and to view the "world" in different "eyes" with magnifying or reducing glasses. We have calculated the wavelet power spectrum for rapidity distribution of the neutral pion fraction in total number of pions simulated using the classical linear σ model[3]. One defines the power spectrum with respect to the wavelet basis at each j as

$$P_j = \frac{1}{2^j} \sum_{k=0}^{2^j-1} |\tilde{f}_{jk}|^2 . \tag{2}$$

P_j is then the power of fluctuations on length scale $L/2^j$. The result is shown in Fig.1. For the random samples and the equilibrium evolution of the classical field where there is no DCC production mechanism, the wavelet power is flat indicating there is no sensitivity on the change of the scale. For the DCC samples, the power starts to build up as j goes down below $j_d \sim 4$ and rises rather quickly, departing distinctively from the equilibrium and the random cases. This suggests the existence of the DCC clustering with a typical size of $\Delta\eta_d = 2\eta_{\max}/2^{j_d} \simeq 0.8$ units in rapidity. Also plotted is the power spectrum from HIJING Monte Carlo data which also features a flat distribution.

1. J. -P. Blaizot and A. Krzywcki, Phys. Rev. D **46**, 246 (1992); J. D. Bjorken *et al.*,SLAC preprint SLAC-PUB-6109.
2. K. Rajagopal and F. Wilczek, Nucl. Phys. **B404**, 577 (1993).
3. Z. Huang and X.-N. Wang, Phys. Rev. D **49**, R4339 (1994).
4. Z. Huang, I. Sarcevic, R. Thews and X.-N. Wang, Phys. Rev. D **54**, July 1 issue (1996).
5. M. Farge, Ann. Rev. Fluid. Mech. **24**, 395 (1992).
6. M. Greiner, P. Lipa and P. Carruthers, Phys. Rev. **E51**, 1948 (1995)

STAR: Present Status and Future Plans

R. Bossingham
Lawrence Berkeley National Laboratory, Berkeley, California 94720

for the STAR Collaboration

STAR (Solenoidal Tracker at RHIC) is now well into its construction phase, with physics data-taking to commence in autumn, 1999. The production and study of the quark-gluon plasma are at the heart of the STAR physics program, with an emphasis on its hadronic signatures—made possible by a large acceptance and an ability to handle high multiplicities. Other topics exploiting RHIC's capabilities are also planned, including measurements of the proton's spin-dependent gluon structure function and meson production in two-photon interactions.

RHIC will collide beams ranging from *p-p* at $\sqrt{s} \leq 500$ GeV and luminosity $\mathcal{L} = 2 \times 10^{32}$ cm^{-2}s^{-1} to Au-Au at $\sqrt{s_{nn}} \leq 200$ GeV and $\mathcal{L} = 2 \times 10^{26}$ cm^{-2}s^{-1} ($\sim$60 central collisions per second). Such flexibility enables the systematic studies needed to establish the existence and properties of the quark-gluon plasma (QGP), and also supports a number of other physics topics, of which two examples will be given.

The STAR[1] QGP strategy focuses on hadronic signatures; the large acceptance of the detector allows many of the global observables (e.g., flavor composition, pion-source size, and "temperature") to be simultaneously measured at the 5–10% level, *event by event*. The correlated variation of such observables with colliding system, impact parameter and incident energy should make a QGP phase transition apparent.

Another STAR physics goal is to use high-luminosity, polarized *p-p* collisions to measure spin-dependent parton distributions in the proton.[2] The gluon structure function can be determined by measuring the final-state photon and jet from gluon-quark scattering; the distribution of sea quarks and anti-quarks can be determined by measuring $W^{\pm}$ production. This information is needed to resolve the "spin crisis."[3–4]

Two-photon scattering processes can also be studied with STAR; neutral meson production in near-collisions of heavy ions scales as Z^4. The relative abundance of the mesons allows one to distinguish conventional $q\bar{q}$ states from exotic states, since photons couple to the electric charge of the quarks. Such measurements will, for example, help to make sense of the apparent over-population of light f_0 states, relative to the expected number of $q\bar{q}$ states.

STAR will be housed in the Wide Angle Hall at RHIC. Its start-up configuration will include a TPC (4-m length and 4-m diameter) in a homogeneous,

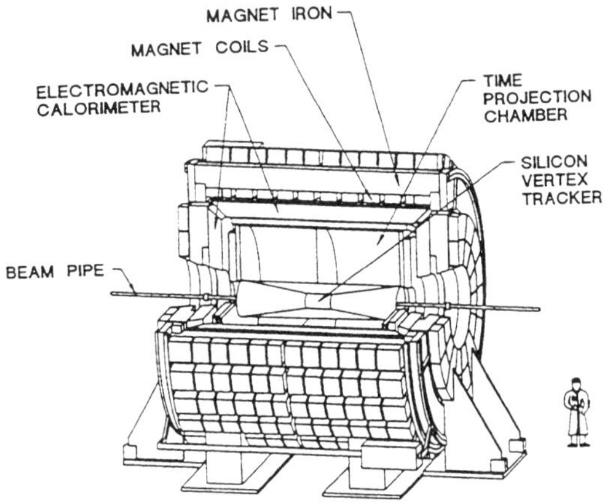

Figure 1: Cutaway view of STAR.

0.5-T magnetic field, together with a scintillator barrel and other trigger counters. An electromagnetic calorimeter (EMC) and time-of-flight (TOF) array will be available in part, to be completed as resources allow. A silicon vertex tracker (SVT) will be inserted after the initial running period, when the beams are thoroughly understood. The layout is shown in Fig. 1.

The EMC is crucial to the spin-physics program, while the TOF array will improve triggering capabilities and extend particle identification to higher momenta. The SVT will, among other things, allow reconstruction of secondary vertices and identification of short-lived particles.

The detector is now well into the construction phase. The TPC sectors and field cage are nearing completion, and cosmic-ray tests are scheduled for early 1997. Roughly half of the magnet iron is on site; installation is scheduled for summer, 1997. Software development is reasonably advanced, with strategies for handling the enormous datasets being developed. Data-taking will begin at RHIC startup in autumn, 1999.

References

1. The STAR Collaboration, LBL PUB-5347 (1992).
2. The STAR EMC Collaboration, LBL PUB-5380 (1994).
3. J. Ashman *et al.*, Phys. Lett. B206, 364 (1988);
   Nucl. Phys. B328, 1 (1989).
4. B. Edeva *et al.*, Phys. Lett. B302, 533 (1993).

AN OVERVIEW OF THE PHENIX EXPERIMENT AT RHIC

J. T. MITCHELL FOR THE PHENIX COLLABORATION
Brookhaven National Laboratory, Physics Department, Building 510C, P.O. Box 5000, Upton, NY 11973-5000

The PHENIX experiment at RHIC will measure electrons, photons, charged hadrons, and muons in 200 GeV/c Au+Au collisions in order to detect and measure a new state of strongly interacting matter, the Quark Gluon Plasma. The PHENIX detector and its physics capabilities are presented.

The primary goal of the PHENIX experiment, currently being constructed at the Relativistic Heavy Ion Collider (RHIC) at Brookhaven National Laboratory, is to detect and measure the properties of a new state of strongly interacting matter - the Quark Gluon Plasma (QGP). The general strategy of PHENIX is to measure as many potential signatures of the QGP as possible to determine if any of the physics observables exhibit a simultaneous anomaly due to its formation. This will be achieved by studying leptonic, hadronic, and photonic signatures of the QGP.

PHENIX consists of a variety of detector subsystems divided into two central spectrometers, or "arms", two muon arms, and a set of inner detectors that measure the start time, vertex, and multiplicity of the interactions. A diagram of the PHENIX detector is shown in Figure 1.

The central arms will detect electrons, photons, and charged hadrons in the pseudorapidity range $-0.35 \leq \eta \leq 0.35$. The central arms consist of a drift chamber providing charged particle tracking in the r-ϕ plane, a set of three pixel pad chambers providing tracking in the r-z plane, a Ring Imaging Cherenkov Counter providing electron identification up to $p \approx 3$ GeV/c, a Time Expansion Chamber providing simultaneous tracking in the r-ϕ plane and electron identification up to $p \approx 2$ GeV/c via dE/dx measurements, an Electromagnetic Calorimeter providing e/γ measurements and further electron identification, and a Time-of-Flight system providing hadron identification and π/K separation up to 2.4 GeV/c.

The muon arms will detect muons with $1.15 \leq \eta \leq 2.44$ and $-2.44 \leq \eta \leq -1.15$. The muon arms consist of 3 sets of multilayer drift chambers followed by a muon identifier consisting of concrete absorber walls interleaved with planes of limited streamer tubes with cathode pad readouts. The muon identifier will distinguish muons from pions and other shower products.

PHENIX will be able to measure many physics variables such as phi mesons in the dielectron, dimuon, and K^+K^- decay channels. Kaon and phi meson production can provide information about the strangeness production in the QGP. Measurements of the phi meson mass and branching ratios $\Gamma(\phi \rightarrow e^+e^-)/\Gamma(\phi \rightarrow K^+K^-)$ can probe the possibility of

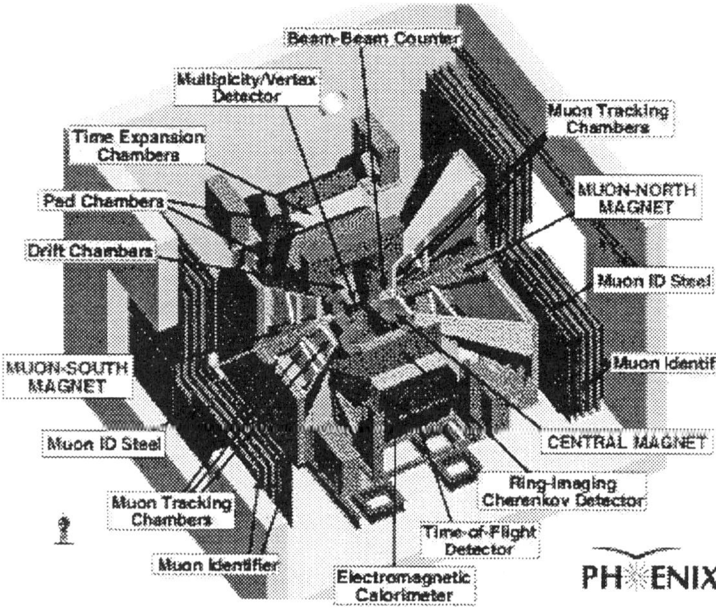

Figure 1: Schematic diagram of the PHENIX detector

a chiral phase transition in relativistic heavy ion collisions. PHENIX will measure the yields of J/ψ and ψ' particles, which are predicted to be suppressed due to Debye screening in the deconfinement transition. PHENIX will also measure p_t spectra for pions, kaons, protons, antiprotons, and photons in order to investigate the basic dynamics of the collision. The properties of the thermal radiation from the collision will be investigated by measurements of prompt photons. Finally, the evolution of the collision will be examined through HBT analyses using pions and kaons.

Rounding out the physics capability is the PHENIX/Spin program that will take advantage of polarized protons with energies up to 250 GeV provided by RHIC in order to study the polarized quark and anti-quark structure functions via the Drell-Yan process and vector boson production. In addition, the Spin program will investigate gluon polarization via heavy quark production (gluon fusion) and gluon compton scattering.

Acknowledgements

Supported in part by the U.S. Department of Energy under Contract no. DE-AC02-76CH00016.

Session 4
The Standard Model

PRECISION TESTS OF THE STANDARD MODEL AT LEP AND RESULTS FROM THE 4-JET FINAL STATE AT LEP 1.5

LEE F. THOMPSON
Department of Physics, University of Sheffield, Sheffield, S3 7RH, England

First a short review is made of current limits on the Standard Model as derived from the electroweak sector concerning the Z boson at LEP. The second part of the paper presents an analysis of the four jet final state at ALEPH, this analysis was performed on LEP data taken at centre of mass energies of 130 and 136 GeV.

1 Tests of the Standard Model at LEP

1.1 Introduction

The results presented here are from approximate totals[1] of 12.3 million hadronic Z decays and 1.3 million leptonic decays recorded by the four LEP experiments during the years 1990 to 1995 inclusive. These statistics permit a precise study of the Standard Model though the coupling of the Z boson to fermion pairs. This reaction is modified by vacuum polarization corrections in the propogator such as t-quark and Higgs boson loops thus allowing a determination of $m_{\rm top}$ and $m_{\rm Higgs}$ to be made via indirect measurements. It should be noted that results for 1993 onwards are regarded as preliminary for all 4 LEP experiments.

1.2 Fitting the Z^0 lineshape

In order to combine electroweak results a common reference point is required. Each of the LEP experiments measures the leptonic and hadronic total cross-sections and the leptonic forward-backward asymmetries from angular distributions in $\cos\theta^*$. These results are fitted[1] according to the Z^0 lineshape to give 9 parameters, namely the Z boson mass and width, $(m_{\rm Z}, \Gamma_{\rm Z})$, the pole hadronic cross-section, $(\sigma^0_{\rm h})$, the ratios of the hadronic to leptonic partial widths, ($R_{\rm e}$, R_μ, R_τ) and the leptonic pole asymmetries $(A^{0,e}_{FB}, A^{0,\mu}_{FB}, A^{0,\tau}_{FB})$. These 9 parameters are then combined[1], along with the individual experimental covariance matices and common systematic errors to give the results in Table 1.

The final two entries in Table 1 represent the values of the partial width ratio and the pole asymmetry under the assumption of lepton universality. This assumption is further explored in Figure 1 where R_ℓ is plotted against $A^{0,\ell}_{FB}$ individually for the 3 lepton species and also under the assumption of lepton universality. The accuracy of these results is remarkable, with the Z boson mass being measured to 24 parts per million.

Parameter	Average Value
m_Z(GeV)	91.1884±0.0022
Γ_Z(GeV)	2.4964±0.0032
$\sigma_h^0(nb)$	41.490±0.078
R_e	20.801±0.058
R_μ	20.795±0.043
R_τ	20.815±0.061
$A_{FB}^{0,e}$	0.0152±0.0027
$A_{FB}^{0,\mu}$	0.0163±0.0015
$A_{FB}^{0,\tau}$	0.0203±0.0022
R_ℓ	20.789±0.032
$A_{FB}^{0,\ell}$	0.0171±0.0011

Table 1: Standard Model parameters from the 5- and 9-parameter fits

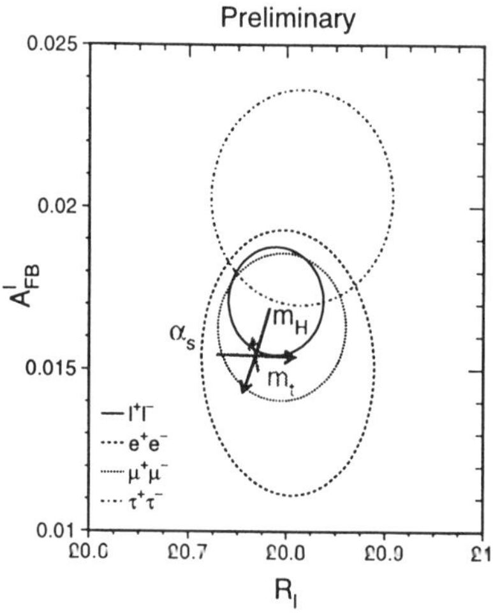

Figure 1: $A_{FB}^{0,\ell}$ versus R_ℓ

1.3 R_b and R_c

In the results above the coupling of the Z to all accessible quark pairs is considered. It is of interest however to investigate the production of individual quark flavours, in particular, pairs of charm and bottom quarks. The ratios

$$R_b = \frac{\Gamma_{Z^0 \to b\bar{b}}}{\Gamma_Z}, R_c = \frac{\Gamma_{Z^0 \to c\bar{c}}}{\Gamma_Z} \tag{1}$$

(where Γ_Z is the total hadronic partial width), are theoretically appealing since the quantum corrections due to m_{top}, m_{Higgs} and QCD effects largely cancel in these ratios. In the case of R_b there are however important vertex corrections due to bWt couplings that result in this ratio being uniquely sensitive to m_{top}.

Experimentally, extracting a pure sample of events with a single flavour content is not easy. Two general techniques are used, so-called "single tag" and "double tag" methods. The "tags" used in these methods rely on characteristic properties of events containing charm and bottom decays and include

- the relatively long lifetime of b-hadrons *"lifetime tagging"*
- b- and c- hadrons have high mass/momentum *"event shape analysis"*
- b- and c- quarks decay semileptonically giving high p and high p_T leptons *"lepton tagging"*

- exclusive decay modes can be reconstructed *"D counting, D* counting"*

In the case of single tag then it is necessary to extract background information and efficiencies from Monte Carlo studies. The double tag method on the other hand involves separating the event into two hemispheres and applying the tagging criteria independently to both hemispheres. This benefits from enabling the efficiency to be evaluated directly from data.

The average LEP values for these quanitities, incorporating the most recent results from all the different techniques are, $R_c = 0.1598 \pm 0.0070$ and $R_b = 0.2202 \pm 0.0016$ compared with Standard Model predictions of 0.172 and 0.2155 respectively. Probability contours in the R_b-R_c plane are presented in Figure 2. The bar to the top left of this figure indicates the Standard Model prediction with a varying top mass as shown. The agreement between measurement and Standard Model is noticably poor.

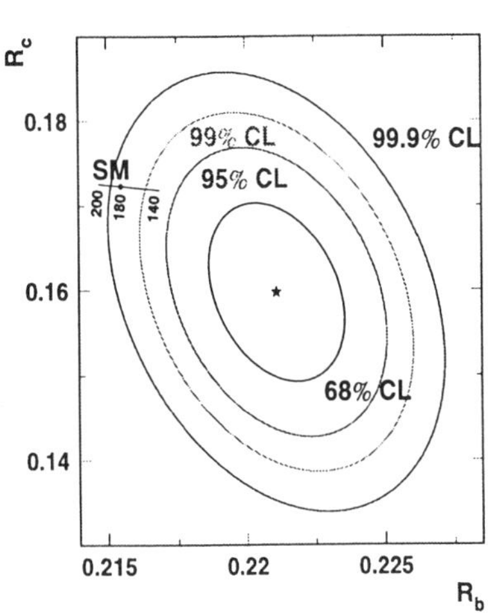

Figure 2: R_b versus R_c along with Standard Model prediction

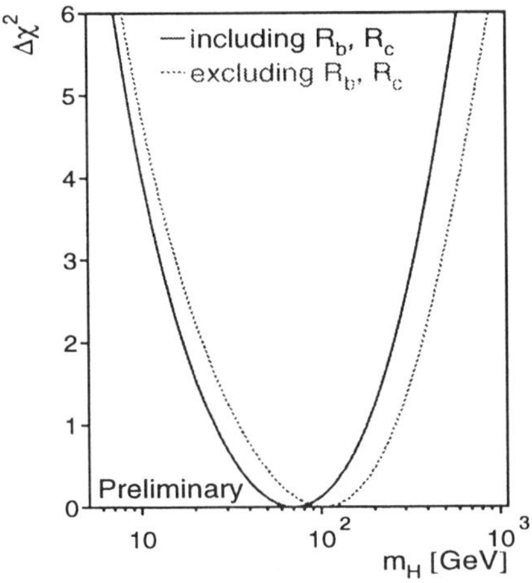

Figure 3: $\Delta\chi^2 = \chi^2 - \chi^2_{min}$ vs $m_{\rm Higgs}$ curves. Continuous line: all data; dashed line: excluding the LEP+SLD measurements of R_b and R_c

1.4 Constraints on the Standard Model

The level of precision on these measurements is such that we can use them to place constraints on the Standard Model parameters α_s, $m_{\rm top}$, $m_{\rm Higgs}$ and hence on m_W, via the radiative electroweak and QCD corrections. The input to this fit includes the results from the lepton universality fit to the Z^0 lineshape discussed above, R_b and R_c as discussed above along with A_e and A_τ from tau polarization measurements[1] and $\sin^2\theta_{eff}^{lept}$ from quark charge asymmetry measurements[1]. This gives a value of m_{top} of

$$m_{top} = 172^{+9+18}_{-10-20}\text{GeV} \tag{2}$$

where the second error corresponds to $m_{Higgs} = 300^{+700}_{-240}$GeV. This is in excellent agreement with the average value of 175±9 GeV from direct observations of top quark production at CDF and D0. The fitted value of α_s is

$$\alpha_s = 0.124 \pm 0.004 \pm 0.002 \tag{3}$$

This leads to a derived value for m_W of

$$m_W = 80.305 \pm 0.053^{+0.016}_{-0.037} \tag{4}$$

in good agreement with the value of 80.33 ± 0.15 GeV from $p\overline{p}$ colliders. In order to constrain significantly the Higgs mass it is important to include the Tevatron measurements on the top mass, otherwise the correlations between m_{top} and m_{Higgs} are strong leading reduced sensitivity to m_{Higgs}. The results from the Standard model fit incorporating the data above along with[1] A_e, R_b, A_b and A_c from SLD, m_W, (1-m_W^2/m_Z^2) from $p\overline{p}$ and νN data and m_{top} from CDF and D0 are presented in Figure 3 in the cases where all the data have been fitted and where the results on R_b and R_c have been excluded from the fit. Here $\Delta\chi^2 = \chi^2 - \chi^2_{min}$. It is of interest to note that lower values of m_{Higgs} are preferred. The Higgs mass is constrained to be lower than 300 GeV at the 95% confidence level (450 GeV excluding results on R_b and R_c).

2 Four Jet Final States in ALEPH at LEP 1.5

2.1 Introduction

Towards the end of 1995 a small amount (5.7pb^{-1}) of data was taken by the LEP experiments at centre of mass energies above the Z mass, namely at 130 and 136 GeV. The analysis, in ALEPH, of events with four jets in the final state[2] is motivated by the search for production of light supersymmetric Higgs particles $e^+e^- \rightarrow hA$ where h and A dominantly decay to $b\overline{b}$. The analysis however also applies to pair-production of charged Higgs or excited quarks.

2.2 Selection Procedure

The main backgrounds to this process are standard $e^+e^- \rightarrow q\overline{q}$ where both quarks radiate a gluon and processes such as $e^+e^- \rightarrow Z\gamma^*$ where $Z \rightarrow q\overline{q}g$

and $\gamma^* \to q\bar{q}$. It is important therefore to develop a selection mechanism that discriminates against these background events whilst retaining a high efficiency for the hA events. The selection procedure is outlined below[2], Table 2 shows the estimated number of events from background processes along with the hA efficiency as a function of the cut applied.

- the event was required to have a final state that is hadronic in appearance, requiring 8 or more good tracks with mometum sum greater than 10% of the centre of mass energy

- events with a high energy initial state photon and thus mediated by a real Z boson ("radiative return" events) were eliminated via a cut on the missing momentum along the beam axis

- the event was clustered into 4 jets using the Durham algorithm with a y_{cut} value of 0.008. Events where the Durham algorithm failed to cluster 4 or more jets were re-clustered with the JADE algorithm ($y_{cut} = 0.0022$). Events with more than 4 jets were forced into 4 jets by combining those jets with smallest invariant mass

- Radiative return events with an ISR photon in the detector were dismissed by placing a requirement that no jet have more than 80% electromagnetic energy and that each jet mass be greater than 1 GeV/c^2

- The events were required to be compatible with a four-body hypothesis by rescaling the jet energies under the condition that the jet velocities were perfectly measured

- Events from the major sources of background typically have two leading jets and two lower energy jets. These events were reduced by requiring a) All di-jet masses to be greater than 25 GeV/c^2; b) The sum of the masses of the lightest two jets to exceed 10 GeV/c^2; c) the sum of the charged particle multiplicity of these two jets to be 10 or more

After the cuts described above are applied, 16 events are observed compared with the expectation from standard processes of 8.6 ± 0.3 events. The Poisson probability for 8.6 events to fluctuate to 16 events is 0.15%.

2.3 The Invariant Mass Distribution

The production of a pair of equal or almost equal particles each decaying to give a four jet final state would be apparent in the distribution of the di-jets masses. In figure 4 this distribution is given for the jet pair combination chosen

Criterion	Data	$q\bar{q}$	ZZ	WW	Total	hA
Hadronic final state	1839	1627	17	0.7	1830*	99.8%
No radiative return	769	780	12	0.5	793	92.4%
At least four jets	153	137	3.3	0.3	141	69.8%
No γ-like jets	37	30.4	1.8	0.25	32.4	67.8%
Four-body compatibility	35	28.2	1.7	0.25	30.1	65.6%
Large di-jet masses	22	14.9	1.2	0.19	16.3	55.0%
Large jet masses	19	11.1	0.50	0.14	11.7	48.6%
Large multiplicities	16	8.3	0.21	0.08	8.6	42.0%

Table 2: Number of events observed in data and expected from various Standard Model processes

to give the smallest di-jet mass difference. Here the bin width has been chosen to be twice the resolution observed from Monte Carlo studies of hA production, namely 3.15 GeV/c^2. As can be seen in the figure, 9 of the 16 events selected populate two bins in the mass sum distribution. The probability for this to happen (independently of that for the excess in the number of events) is 10^{-4}.

2.4 Systematic Checks and Comparison with QCD

A number of systematic checks have been performed to test the hypothesis that the event excess and mass enhancement are some feature of the event selection and/or lack of understanding of the background processes. A full description of these exhaustive checks can be found in ref.2, in conclusion, no systematic bias is seen in any of these investigations to support the theory that the excess seen is due to a detector or Monte Carlo artefact.

2.5 Pair Production Hypothesis

Assuming that the excess comes from particle pair production the observed event excess corresponds to a cross-section of 3.1±1.7 pb, compared to an expected cross-section of 0.49 from hA production. In addition, only 1 event has possible b-quark jets, further dispelling the hA production hypothesis. Further studies of the di-jet mass difference and the (rapidity-weighted) charge separation between the two jet pairings indicate that production of equal mass, zero-width, colour singlets and of electrically neutral particles is also disfavoured.

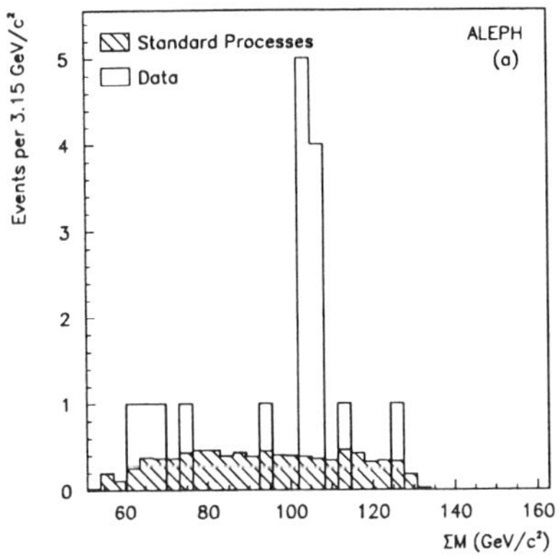

Figure 4: Distribution of the sum of the di-jet masses for the combination with the smallest mass difference

3 Conclusions

The present time is a most exciting one for LEP physics. In the electroweak sector, statistics now allow indirect limits to be placed on the top quark and W boson masses, these are in excellent agreement with those made in direct observation of these particles at $p\overline{p}$ colliders. By incorporating direct top quark mass measurements from the Tevatron then the Higgs mass is constained to be lighter than 300 GeV/c^2 (450/c^2 if the measurements of R_b and R_c are excluded). These limits are expected to become more constraining as the full analysis of the 1993-5 data samples becomes final. Similarly, the observation of an excess of events in data taken at centre of mass energies above the Z pole in 1995 should be verified or otherwise by the high energy run at twice the W mass threshold which is due to commence in the summer of 1996.

Acknowledgments

I would like to that all those people who generously helped with plots and information during the preparation of this presentation. My special thanks go to Mike Hildreth and Bob Clare who were particularly helpful.

References

1. A Combination of Preliminary LEP and SLD Electroweak Measurements and Constraints on the Standard Model, LEP Electroweak Working Group, LEPEWWG/96-01, and references therein
2. Four-jet final states production in e^+e^- collisions at centre-of-mass energies of 130 and 136 GeV, ALEPH Collaboration, CERN-PPE/96-052

e^+e^- PRODUCTION OF Z^0 WITH A POLARIZED ELECTRON BEAM

FRANK E. TAYLOR
Massachusetts Institute of Technology
Cambridge, Massachusetts, 02139, USA

Representing the SLD Collaboration

Precise tests of the Standard Model have been performed with 150,000 $e^+e^- \to Z^0$ events accumulated by the SLD at the Stanford Linear Collider, operating with electron beam polarizations ranging from 63% to 77%, during the period 1993-1995. We report preliminary determinations of the left-right asymmetry of the Z^0 production cross section and the electro-weak interference terms of the Z^0-$b\bar{b}$ and Z^0-$c\bar{c}$ couplings, A_b and A_c respectively. In addition, results are reported on the decay fraction $Z^0 \to b\bar{b}$, R_b. Our preliminary numbers are $A_{LR} = 0.1524 \pm 0.0044$, $A_b = 0.858 \pm 0.054$, $A_c=0.58 \pm 0.10$ and $R_b=0.2176 \pm 0.0037$.

1 Introduction

In the Standard Model (SM) the amplitude of the Z^0-gauge boson coupling to a fermion-antifermion pair has both axial vector (a_α) and vector (v_α) terms which are determined by the fermion weak isospin, electric charge and the electro-weak mixing angle, θ_w. In this note we report our preliminary measurements of the electro-weak interference terms, $A_\alpha = 2v_\alpha a_\alpha/(v_\alpha^2 + a_\alpha^2)$, for electrons and for b and c quarks. These measurements provide precise tests of the SM. A_e is a potent measure of $\sin^2\theta_w^{eff}$ and of Z^0 propagator effects arising from heavy quark loops and higgs scalars while A_b and A_c provide a cross check of parity violation in the b and c quark - Z^0 couplings and are sensitive to right-handed terms. Another test of the SM is provided by our determination of the Z^0 branching fraction into bb quarks defined by $R_b = \Gamma_b/\Gamma_{hadrons}$, probing bottom-top quark vertex corrections.

The SLD is an ideal detector for these measurements. It has a nearly hermetic calorimeter, charged particle tracking, particle identification by means of a ring imaging Cerenkov detector and excellent vertex detection provided by a pixel vertex detector[1]. The Stanford Linear Collider (SLC) is a unique facility with an electron beam polarization of $|P_e| \approx 77\%$ ('94-'95 data) and a very small and stable beams intersection point allowing secondary vertices to be detected with good efficiency.

2 Results

The values of A_e and $\sin^2\theta_w^{eff}$ are determined by the Z^0 production cross section asymmetry for the two electron beam polarization states determined by counting the number of hadronic events accepted with incident left-handed electrons versus the corresponding number with right-handed electrons, the average polarization (using a Compton polarimeter) and small corrections due to backgrounds and experimental asymmetries. Our preliminary values[2] of A_e , corrected to the Z^0 pole value, are '94-'95 data [combined '92-'95 data] = 0.1524 ± 0.0042 (stat) ± 0.0012 (sys) [0.1551±0.0040] and the corresponding values of $\sin^2\theta_w^{eff}$ '94-'95 data [combined '92-'95 data] = 0.23084 ± 0.00054 (stat) ± 0.00015 (sys) [0.23049± 0.00050].

The electro-weak interference term of the final state fermion is measured directly in the SLD by the improved forward-backward asymmetry obtained by selecting forward and backward going fermions produced by left and right handed incident electrons. Heavy quark flavor states are selected by observing displaced vertices from the primary intersection point. Our preliminary value[3], A_b = 0.858 ± 0.054, is an average over a jet charge, lepton and K^+ tag analyses, using the average of the LEP and SLD lepton asymmetries. Several different analyses are used to extract A_c. The experimental challenge was to isolate the $c\bar{c}$ candidates from background requiring the detection of exclusive channels and the simultaneous fit of A_b and A_c. The preliminary value is A_c = 0.58 ± 0.10.

For the R_b analysis the $b\bar{b}$ event sample was selected by a double tag technique[4]. Suppression of charm and light quarks is effected by computing the transverse mass of charged tracks by adding the missing P_t of the tracks associated with the secondary vertex about the line connecting the primary and secondary vertices resulting in a B-purity of 97% and an efficiency of 37%. A total of 71k events passing the standard SLD hadronic event selection results in our preliminary value of R_b = 0.2176 ± 0.0033(stat) ± 0.0017(sys) ± $0.0008R_c$.

No significant deviations from the SM are observed in our data. Our values of A_b, A_c and R_b are within 1.4, 0.9, 0.6 standard deviations, respectively, of the SM. Our value of $\sin^2\theta_w^{eff}$ is within 2.3 standard deviations of the LEP average[5] and the corresponding electro-weak corrections are consistent with the measured values M_w and M_{top}.

References

1. M. Breidenbach, SLAC-PUB-3798 (1985).
2. K. Abe *et al., Phys. Rev. Lett.* **73**, 25 (1994).
3. D. Su, SLAC-PUB-95-700 (1995).
4. E. Etzion, SLAC-PUB-96-7170 (1996).
5. LEP-Electo-weak Working Group, LEPEWWG/95-02.

THE CLEO III DETECTOR UPGRADE

RAINER WANKE
Physics Department, Ohio State University, 174 W 18th Avenue, Columbus, Ohio 43210, USA

(Representing the CLEO collaboration)

Over the next two years, the e^+e^- collider ring CESR at Cornell University will be upgraded to achieve luminosities up to $17 \times 10^{32}\mathrm{cm}^{-2}\mathrm{s}^{-1}$. We describe the improvements of CESR and of the CLEO detector, which is located at CESR. The detector upgrade, called CLEO III, includes a new silicon vertex detector, a new wire drift chamber and a Ring Imaging Cherenkov detector (RICH) for particle identification.

The CLEO detector is located at the symmetric e^+e^- collider CESR at Cornell University, which is operating at the energy of the $\Upsilon(4S)$ resonance (10.6 GeV), which is an unique environment to study b and c quark and τ decays. With a CESR peak luminosity of $3.6 \times 10^{32}\mathrm{cm}^{-2}\mathrm{s}^{-1}$ the CLEO collaboration has already been able to find a series of rare B meson decays like $b \to u$ transitions and penguin decays. However, to perform precision measurements of the involved standard model parameters V_{cb}, V_{ub}, V_{td}/V_{ts}, etc., as well as to study CP violation with decays like $B \to K\pi$, $B \to \pi\pi$, requires at least an order of magnitude higher statistics than collected so far. To achieve this goal, the CESR group and the CLEO collaboration are currently working on upgrades for the so-called CESR phase III and the CLEO III detector.

The main changes being made at CESR are the installation of superconducting RF cavities to increase the current limit to 500 ma/beam and and the installation of new permanent and superconducting IR quadrupole magnets very close to the interaction region (Fig. 1). It is planned to run CESR III with 45 bunches in a 9×5 mode. The peak luminosity is expected to reach $17 \times 10^{32}\mathrm{cm}^{-2}\mathrm{s}^{-1}$. The first superconducting cavity will be installed end of 1996, the second (out of four) in spring 1998. The installation of the new IR quadrupoles is also planned for the beginning of 1998.

This schedule sets the outline for the CLEO III detector upgrade, which we plan for spring 1998. It includes a new silicon vertex detector, a new drift chamber (DR3) and a Ring Imaging Cherenkov detector (RICH) for particle identification.

The innermost part of CLEO III will be a new silicon vertex detector, consisting of 4 cylindrical layers with 447 wafers in total. It will extend to a radius of 13 cm and cover about 93% of 4π. Each layer has double sided

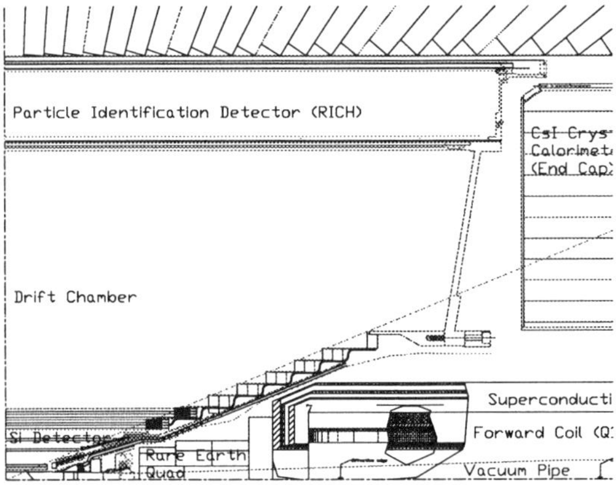

Figure 1: Quarter section of the inner region of the CLEO III detector.

readout for z and $r\phi$ information and the total number of channels is about 125,000. A particular challenge is the mechanical design of the detector, since we want to reduce the amount of support structure as much as possible in order to reduce multiple scattering.

The design of the new drift chamber (DR3) is mainly constrained by the space needed by the new IR quadrupoles, which results is an unconventional shape of DR3 (fig. 1). The inner portion is tapering towards the interaction region, conserving as much efficiency at high $|cos\theta|$ as possible. It consists of 16 axial layers. The outer part of DR3 has conical shaped endplates with 31 stereo layers. The DR3 will be run with a helium based gas mixture with a radiation length > 330 m and a lorentz angle $< 46°$. Though DR3 will have a smaller outer radius than the current drift chamber, we will have a better tracking resolution due to the reduction of scattering material in the inner skin and in the gas.

For identifying rare decays like $B \rightarrow K\pi$, $B \rightarrow \pi\pi$ we need an excellent K/π separation up to 2.8 GeV. This will be achieved by building a Ring Imaging Cherenkov Detector (RICH) between DR3 and the CsI calorimeter. Incoming tracks produce Cherenkov photons in a layer of LiF radiators with an optical index $n = 1.5$. The Cherenkov cone is allowed to expand in a N_2 filled gap, and the Cherenkov photons are finally detected in a photosensitive multi wire proportional chamber, filled with a TEA–methane mixture. The photon positions are read out by the cathode pads of the chamber. A first RICH prototype has been studied for feasibility and resolution last year, a full length prototype module will be tested in November at Fermilab.

Probing the Standard Model at the Tevatron

S. Eno
Department of Physics, University of Maryland, College Park, MD 20742, USA

I report on measurements of the properties of the top quark, the W Boson, and the jet inclusive cross section using data from the Tevatron run I.

1 Introduction

The Tevatron, a $p\bar{p}$ collider operating at the $\sqrt{s}$=1.8 TeV, has just completed a run of approximately 100 pb^{-1} at the world's highest center-of-mass energy. The data includes the world's only top quark sample and the world's largest sample of W bosons. I report on their measured properties and on the jet inclusive cross section. The two major experiments located at the Tevatron are the CDF and the DØ experiments.

2 Inclusive Jet Cross Section

The measurement of the jet inclusive cross section at the Tevatron has been of great interest since the publication of the CDF run Ia data. Their result showed a higher cross section than predicted by the standard model [1] for the highest energy jets, with transverse energies greater than 250 GeV. Such an excess could be the signature of quark substructure, but also could indicate instead a lack of understanding of the parton distribution functions for the proton. Since then, the DØ collaboration has presented preliminary results for the same measurement. [2] The CDF and DØ results were analyzed by Tung and Lai [3] and found to be consistent with each other. In addition, Huston *et al.*, [4] found that by altering the gluon distribution functions in the x region 0.01 to 0.5 they could get good agreement with the new Tevatron results and all previous results which had contributed to the original determination of the parton distribution functions. In addition, CDF has recently presented results on the angular distribution of jets in events with two jets which is in good agreement with the standard model prediction, and in contradiction to the quark compositeness hypothesis.

3 The Top Quark and the W Boson

Top quarks are pair-produced at the Tevatron. The top is expected to decay 100% of the time to a b quark and a W. Both CDF and DØ can identify top quark events where at least one of the W's decays semi-leptonically to electrons or muons. The reported cross-sections for $t\bar{t}$ production are $7.5^{+1.9}_{-1.6}$ pb for CDF [5] and 5.2 ± 1.8 pb for DØ [6]. Both collaborations do a mass reconstruction analysis on the events. The measured masses are 175.6 ±5.7(stat)±7.1(syst) GeV for CDF [5] and 170 ± 15 ± 9.2 for DØ [6].

W's are copiously produced at the Tevatron. They are identified through their semi-leptonic decays to electrons or muons. Using a sample of 5962 $W \rightarrow e\nu$ events, the DØ collaboration reports a new measurement of the mass of the W boson of 80.35 ± 0.140 (stat) ±0.160 (energy scale) ± 0.165 (syst) GeV[7]. The CDF measurement reported last year is 80.41±0.180 GeV[8].

Both CDF and DØ also search for events which contain 2 vector bosons, Wγ, Zγ, and WZ. Theories with anomolous couplings between the bosons, for example if theW and Z are composite objects, predict an excess of events with bosons having large transverse momentum relative to the beam axis. Both collaborations see good agreement with standard model expectations, except maybe in the Zγ channel [9].

References

1. F. Abe *et al.*, preprint FERMILAB-PUB-96/020-E, Jan 96.
2. J. Blazey *et al.*, preprint FERMILAB-CONF-96/132-E, Jan 96.
3. H.L. Lai and W.K. Tung, preprint MSU-HEP-60508, May 96.
4. Huston *et al.*, preprint MSU-HEP-50812, May 96
5. F. Tartarelli, preprint FERMILAB-CONF-96/099-E, Mar 96
6. R. Hall, proceedings of Moriond-Electroweak, March 1996
7. K. Streets, preprint FERMILAB-CONF-96/149-E, Jun 96.
8. F. Abe *et al.*, *Phys. Rev.* D **52**, 1995 (.)
9. D. Neuberger, proceedings of European Physical Society conference, Brussels, 1996

Constrained Minimal Supersymmetry

W. de Boer[1]
Inst. für Experimentelle Kernphysik, Univ. of Karlsruhe
Postfach 6980, D-76128 Karlsruhe, Germany

1 Introduction

The Minimal Supersymmetric Standard Model (MSSM) [1] has become the leading candidate for a low energy theory consistent with the GUT idea. The precisely measured coupling constants at LEP exclude unification in the Standard Model, while in the MSSM unification is possible [2–4].

Within the Constrained Minimal Supersymmetric Model (CMSSM) it is possible to predict the low energy gauge couplings and masses of the 3 generation particles from a few parameters at the GUT scale. Moreover, the CMSSM predicts electroweak symmetry breaking due to large radiative corrections from the Yukawa couplings, thus relating the Z^0 boson mass to the top quark mass via the renormalization group equations (RGE) [5]. In addition, the possible contributions to the R_b anomaly at LEP are discussed [6].

2 The Constrained MSSM parameter space

In this section the various low energy GUT predictions are compared with data. The most restrictive constraints are the coupling constant unification and the requirement that the unification scale has to be above 10^{15} GeV from the proton lifetime limits, assuming decay via s-channel exchange of heavy gauge bosons. They exclude the SM [2–4] as well as many other models [3,7,8]. The only model known to be able to fulfill all constraints simultaneously is the MSSM.

The requirement of bottom-tau Yukawa coupling unification strongly restricts the possible solutions in the m_t versus $\tan\beta$ plane. With the top mass measured by the CDF and D0-Collaborations [9,10] only two regions of $\tan\beta$ give an acceptable χ^2 fit, as shown in the bottom part of fig. 1 for two values of the SUSY scales $m_0, m_{1/2}$, which are optimized for the low and high $\tan\beta$ range, respectively, as will be discussed below. The curves at the top show the solution for m_t as function of $\tan\beta$ in comparison with the experimental value of $m_t = 179 \pm 12$ GeV. The m_t predictions were obtained by imposing gauge coupling unification and electroweak symmetry breaking for each value of $\tan\beta$, which allows a determination of μ, α_{GUT}, and M_{GUT} from the fit for the given choice of $m_0, m_{1/2}$. Note that the results do not depend very much on this choice.

The best χ^2 is obtained for $\tan\beta = 1.7$ and $\tan\beta = 41$, respectively. They correspond to solutions where $Y_t \gg Y_b$ and $Y_t \approx Y_b$, as shown in the middle part of fig. 1. The latter solution is the one typically expected for the SO(10) symmetry, in which the up and down type quarks as well as leptons belong to the same multiplet, while the first solution corresponds to b-tau unification only, as expected for the minimal SU(5) symmetry.

In fig. 2 the total χ^2 distribution is shown as funtion of m_0 and $m_{1/2}$ for the two values of $\tan\beta$ determined above. One observes clear minima at $m_0, m_{1/2}$ around (200,270) and (800,70), as indicated by the stars in the projections. The different shades correspond to $\Delta\chi^2$ steps of 2. The contours show the regions excluded by different constraints used in the analysis: the requirement for the LSP to be the lightest neutralino, constraints from the relic density, $BR(b \to s\gamma)$ and the top quark mass. The main contributions to χ^2 are different for the two $\tan\beta$ regimes: for large $\tan\beta$ only small values of $m_{1/2}$ yield good fits, because of the simultaneous constraints of $BR(b \to s\gamma)$ and the large corrections to the b-quark mass, while at low $\tan\beta$ most of the $m_0, m_{1/2}$ region is eliminated by the requirement $\Omega h_0^2 < 1$. The small effect of the relic density constraint in the case of the large $\tan\beta$ solution can be explained by the larger Higgsino admixture in the neutralino mass eigenstate. This leads to an enhancement of $\tilde{\chi} - \tilde{\chi}$ annihilation via the s-channel Z boson exchange pushing the relic density Ωh_0^2 to lower values. As a result, in the large $\tan\beta$ case the constraint $\Omega h_0^2 < 1$ is almost always satisfied unlike in the case of low $\tan\beta$.

---

[1]Email: wim.de.boer@cern.ch

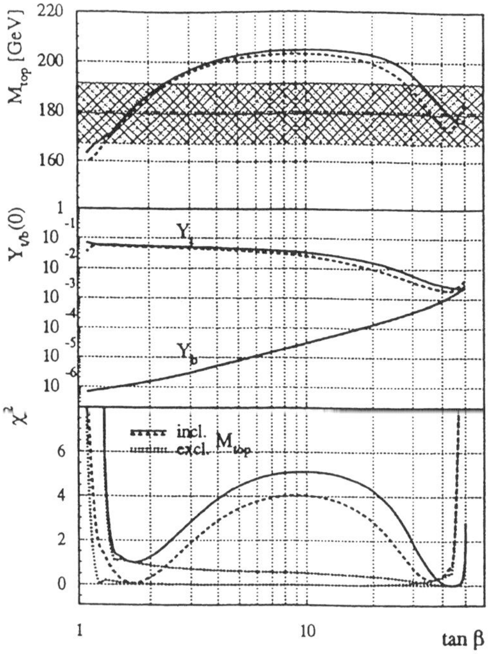

Figure 1: The top quark mass as function of $\tan\beta$ (top) for values of $m_0, m_{1/2}$ optimized for low and high $\tan\beta$, as indicated by the dashed and solid lines, respectively. The dependence shown is mainly determined by the $b-\tau$ Yukawa coupling unification. The middle part shows the corresponding values of the Yukawa coupling at the GUT scale and the lower part the obtained χ^2 values. If the top constraint ($m_t = 179 \pm 12$, horizontal band) is not applied, all values of $\tan\beta$ between 1.2 and 50 are allowed (thin dotted lines at the bottom), but if the top mass is constrained to the experimental value, only the regions $1 < \tan\beta < 3$ and $15 < \tan\beta < 50$ are allowed.

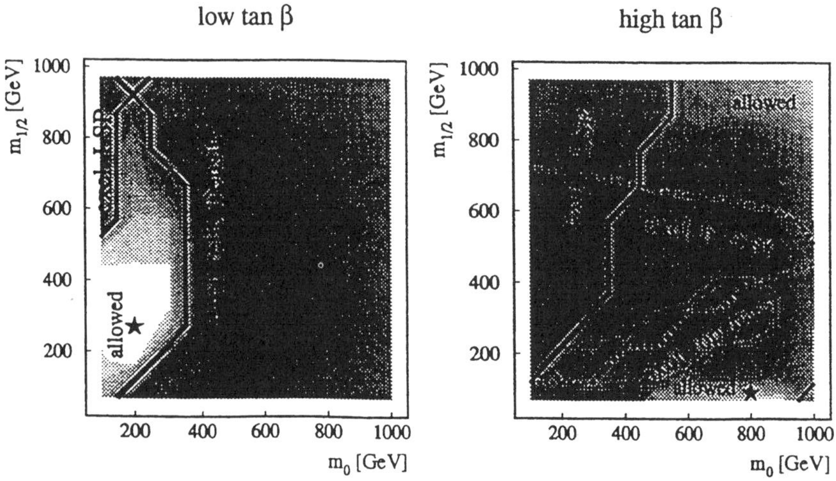

Figure 2: Contours of the χ^2-distribution for the low and high $\tan\beta$ solutions. The different shades indicate steps of $\Delta\chi^2 = 2$, so basically only the light shaded region is allowed.

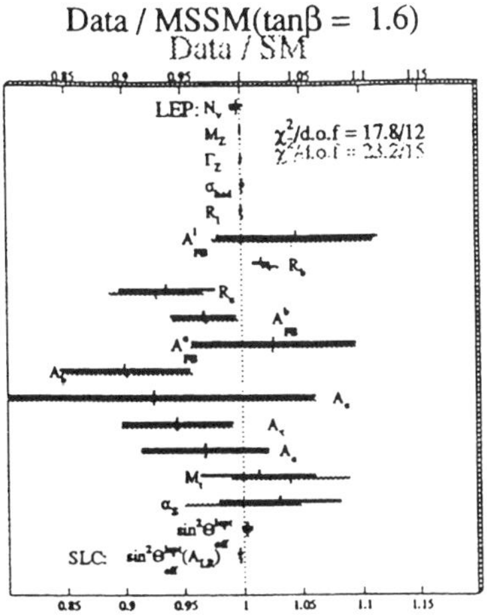

3 R_b anomaly

Global fits of the Standard Model (SM) and the Minimal Supersymmetric Standard Model (MSSM) to electroweak data from LEP, Tevatron, and SLC are performed. For the SM the fit gives a probability below 10%, mainly because of the too high experimental value for the partial width of $Z^0 \rightarrow b\bar{b}$ and the too low value of $Z^0 \rightarrow c\bar{c}$ [11]. In the MSSM the fit yields a considerably better χ^2 value with light charginos and stop or pseudoscalar Higgs masses [6], which contribute positively to the $b\bar{b}$ vertex. This positive contribution to the width is compensated by a decrease in the strong coupling constant: for the SM one finds $\alpha_s = 0.123 \pm 0.005$, while in the MSSM one finds for $\tan\beta = 1.0(1.6)$ $\alpha_s = 0.108(0.118)$, which is in better agreement with the value from deep inelastic scattering $\alpha_s = 0.113 \pm 0.005$. The value $\tan\beta = 1.0$ gives the best χ^2/dof=14.6/12, but yields a too low $BR(b \rightarrow s\gamma)$ rate and is therefore excluded [12]. However, including $BR(b \rightarrow s\gamma)$ and all lower limits on SUSY particles from LEP yields a χ^2/dof of 17.8/12, which is to be compared with the SM value of 23.2/15. The ratio of data and SM (MSSM) predictions is shown in the figure on the left.

References

[1] For references see the review papers:
H.-P. Nilles, Phys. Rep. **110** (1984) 1;
H.E. Haber, G.L. Kane, Phys. Rep. **117** (1985) 75;
A.B. Lahanas and D.V. Nanopoulos, Phys. Rep. **145** (1987) 1;
R. Barbieri, Riv. Nuo. Cim. **11** (1988) 1.

[2] J. Ellis, S. Kelley, and D. V. Nanopoulos. *Nucl. Phys.* **B373** *(1992) 55.*

[3] U. Amaldi, W. de Boer, and H. Fürstenau. *Phys. Lett.* **B260** *(1991) 447.*

[4] P. Langacker and M. Luo. *Phys. Rev.* **D44** *(1991) 817.*

[5] W. de Boer et al. *Preprint Karlsruhe, IEKP-KA/96-04, hep-ph/9603350, to be published in ZP.*

[6] LEP II Working Group on Physics at LEP II, *CERN 96-01, p. 463, hep-ph/9602207, and references therein. .*

[7] U. Amaldi, W. de Boer, P. H. Frampton, H. Fürstenau, and J.T. Liu. *Phys. Lett.* **B281** *(1992) 374.*

[8] H. Murayama and T. Yanagida, *Mod. Phys. Lett.* **A7** *(1992) 147;*
T. G. Rizzo, *Phys. Rev.* **D45** *(1992) 3903;*
T. Moroi, H. Murayama and T. Yanagida, *Phys. Rev.* **D48** *(1993) 2995.*

[9] CDF Collab., F. Abe, et al. *Phys. Rev. Lett.* **74** *(1995) 2626.*

[10] D0 Collab., S. Abachi, et al. *Phys. Rev. Lett.* **74** *(1995) 2632.*

[11] The LEP Collaborations: ALEPH, DELPHI, L3 and OPAL and the LEP electroweak Working Group;. *Phys. Lett.* **276B** *(1992) 247;*. Updates are given in LEPEWWG/96-01, March 1996.

[12] W. Hollik, *talk at the Int. Conf. on High Energy Phys., Brussels (1995), to be published in the Proc.;*
W. de Boer et al.,, *to be published. .*

BROOKHAVEN RARE KAON DECAY EXPERIMENTS

John S. Haggerty
Brookhaven National Laboratory, Upton, New York 11973, USA

There are three large rare kaon decay experiments running at the Brookhaven AGS. These experiments are seeking, and in some cases measuring, strangeness changing neutral currents that occur as higher order weak interactions in the Standard Model. Several of the experiments are searching for lepton number violating processes, processes forbidden in the Standard Model, which would signal new interactions at high mass scales. The current status of these experiments is reviewed, and prospects for the future considered.

Rare kaon decays provide a unique laboratory for exploring the physics of the Standard Model, as well as physics well beyond the Standard Model. Kaon physics repeatedly pointed the way in particle physics, from the discovery of associated production and strangeness to symmetries, to the observation CP violation, to the absence of strangeness changing neutral currents and GIM cancelation. Extremely intense kaon beams are relatively easy to produce, and lower energy experiments are smaller and less expensive to perform. The current program of rare kaon experiments at the Brookhaven AGS with charged and neutral kaons continues to probe the physics forefront.

E787 is primarily a search for the decay $K^+ \to \pi^+\nu\bar{\nu}$, which is a strangeness changing neutral current occurring at second order in the weak interaction in the Standard Model, through loop diagrams involving heavy quarks. This decay is cleanly calculable in the Standard Model, because it is related to the decay $K^+ \to \pi^0 e^+ \nu$ by an isospin rotation. The top quark dominates the loops, and thus the decay is sensitive to the far off-diagonal CKM matrix element V_{td}. The branching ratio is expected to be in the range of 0.6 to 3.0 $\times 10^{-10}$, so it presents a formidable experimental challenge.

The E787 detector[1] is a large solenoid detector which has been optimized for redundant kinematic measurements, excellent π^0 vetoing, and π/μ separation using observation of the decay chain $\pi^+ \to \mu^+\nu$, $\mu^+ \to e^+\bar{\nu}_e\nu_\mu$ using custom made 500 MHz transient digitizers.[2]

The present limit on $K^+ \to \pi^+\nu\bar{\nu}$, set by this experiment from data taken from 1989-1991, is 2.4×10^{-9} at 90% C.L.[3] Data taken in 1995 and 1996 with an upgraded detector should increase the sensitivity by a factor of five or more, bringing the single event sensitivity to 1 or 2 $\times 10^{-10}$. Other physics accessible to this detector include the processes $K^+ \to \pi^+\mu^+\mu^-$ and $K^+ \to \pi^+\gamma\gamma$, which have been observed for the first time in this experiment.

E865 is optimized to search for $K^+ \to \pi^+\mu^+e^-$ using a 6.5 GeV beam.

The previous limit was set by an earlier run of the experiment to be less than 2.1×10^{-10} at 90% C.L.[7] With upgrades to the spectrometer and data acquisition system, E865 should reach a sensitivity of about 4×10^{-12} from data taken in the last two years.

E865 will also improve the measurement of $K^+ \to \pi^+ e^+ e^-$, measured by its predecessor[8] to be $(2.75 \pm 0.23 \pm 0.13) \times 10^{-7}$. E865 should collect several thousand such events, and will also search for some other rare decays, like $K^+ \to \pi^+ \mu^+ \mu^-$ and $K^+ \to \mu^+ \nu e^+ e^-$.

E871 is a large magnetic spectrometer optimized for searching for the decay $K_L^0 \to \mu^\pm e^\mp$. If this lepton number violating decay occurs via a new intermediate gauge boson, a limit on the branching ratio can be related to the decay $K^+ \to \mu^+ \nu$, with the mass of the exchanged boson growing as the fourth root of the branching ratio. For the limit set by E871's predecessor, E791, of 3.3×10^{-11} at 90% C.L.,[4] this leads to an exchanged boson mass of from 1 to 100 TeV, depending on the details of the model.[5] Thus, searching for rare, forbidden decays provides a window to very high mass scales which cannot be probed directly.

This experiment's predecessor[6] measured the branching ratio of the Standard Model decay $K_L^0 \to \mu^+ \mu^-$ to be $(6.86 \pm 0.37) \times 10^{-9}$ with several hundred events; several thousand events are expected from E871.

With two long runs in 1995 and 1996, E871 should reach a sensitivity of about 10^{-12} or less in $K_L^0 \to \mu^\pm e^\mp$, probing a mass scale more than twice as large, despite destruction of some upstream chambers caused by an implosion of a vacuum window at the end of the 1995 run. The straw chambers were rebuilt quickly enough to be ready for the beginning of the 1996 run.

References

1. M.S. Atiya *et al.*, *Nucl. Instrum. Methods* **A321**, 129 (1992).
2. M. Atiya, M. Ito, J. Haggerty, C. Ng, and F.W. Sippach, *Nucl. Instrum. Methods* **A279**, 180 (1989).
3. S. Adler *et al.*, *Phys. Rev. Lett.* **76**, 1421 (1996).
4. K. Arsiaka *et al.*, *Phys. Rev. Lett.* **70**, 1049 (1993).
5. L. Littenberg and G. Valencia, *Ann. Rev. Nucl. Part. Sci.* **43**, 729 (1993).
6. A.P. Heinson *et al.*, *Phys. Rev.* D **51**, 985 (1995).
7. A.M. Lee *et al.*, *Phys. Rev. Lett.* **64**, 165 (1990).
8. C. Alliegro *et al.*, *Phys. Rev. Lett.* **68**, 278 (1992).

SEARCH FOR SPONTANEOUS MUONIUM – ANTIMUONIUM CONVERSION AT PSI

W. BERTL, R. ABELA, D. RENKER, H. K. WALTER
Paul Scherrer Institute, CH-5232 Villigen PSI, Switzerland

D. KAMPMANN, A. KLAAS, G. OTTER, R. SEELIGER
III. Physikalisches Institut B der RWTH Aachen, D-52056 Aachen, Germany

V. BARANOV, V. KARPUCHIN, I. KISEL, A. KORENCHENKO, S. KORENCHENKO, N. KRAVCHUK, N. KUCHINSKY
Joint Institute of Nuclear Research, RU-141980 Dubna, Russia

A. GROSSMANN, K. JUNGMANN, B.E. MATTHIAS, J. MERKEL, V. MEYER, T. PROKSCHA, G. zu PUTLITZ, I. REINHARD, P. V. SCHMIDT, K. TRÄGER, C. WASSER, L. WILLMANN
Physikalisches Institut der Universität Heidelberg, D-69120 Heidelberg, Germany

J. BAGATURIA, D. MZAVIA, T. SAKHELASHVILLI
Tbilisi State University, 380086 Tbilisi, Georgia

V.W. HUGHES
Yale University, New Haven Ct. 06520, USA

R. ENGFER, A. LEUSCHNER, R. MENZ, H.S. PRUYS, W. REICHART, H. P. WIRTZ
Physik-Institut der Universität Zürich, CH-8057 Zürich, Switzerland

At PSI, Switzerland, an experiment searching for the lepton flavour violating conversion of muonium into antimuonium is in progress. After improving the sensitivity by almost two orders of magnitude as compared to previous experiments the process was not observed. The aim is now to improve the sensitivity further by a similar amount to $\approx 1-2\cdot 10^{-10}$. The process would become observable, if its coupling is stronger than $3\cdot 10^{-3}\times G_F$ with G_F being the Fermi coupling constant.

Lepton flavour (LF) conservation as embedded in the standard model of electroweak interactions can be violated in the framework of more general theories proposed to derive some of the *ad hoc* parameters of the model from fundamental principles. Typical for most of these theories is the prediction that lepton flavour violating processes occur in a natural way with model dependent probabilities. The muonium–antimuonium conversion process $\mathrm{M} \equiv (\mu^+ e^-) \rightarrow \overline{\mathrm{M}} \equiv (\mu^- e^+)$ is favoured in various models where LF violation is caused by particles postulated to exist in the TeV mass region [1].

At the Paul Scherrer Institute experiment R 89–06 is searching for $M \rightarrow \overline{M}$ conversion by simultaneously detecting both constituents of antimuonium in its decay. Muonium atoms are formed by stopping a μ^+-beam in a SiO_2 powder target. A fraction of them diffuses into the surrounding vacuum where decays of antimuonium are searched for. The electron from the decay of the muon is measured in an axial magnetic field of 0.1 T by 5 cylindrical concentric multiwire proportional chambers (MWPC) and a scintillator hodoscope. The positron left behind after the muon has decayed is electrostatically accelerated up to 8 keV and guided by a magnetic field to a microchannel plate (MCP), where time-of-flight and impact position is measured. A barrel shaped segmented CsI calorimeter surrounds the MCP to detect photons from the positron's annihilation in the MCP. Muonium decays have been observed to monitor muonium production by reversing the polarity of the accelerating field. After completing a first stage of the experiment without observing any antimuonium decay, an upper limit on the coupling constant of an effective (V-A)×(V-A) four fermion coupling has been derived [2]: $G_{M\overline{M}} \leq 1.8 \cdot 10^{-2} \times G_F$ (90 % C.L.).
The aim of stage 2 of the experiment is to further improve the sensitivity to the conversion process by another two orders of magnitude. In order to reach this sensitivity the acceptance of the spectrometer has been improved considerably, mainly by using a secondary electrons emitting foil in front of the MCP [3] and by replacing 4 MWPC's, all now equipped with cathode strip readout to enhance tracking quality and efficiency. Furthermore, the experiment has been moved to the new beam line $\pi E5$ gaining from its higher rate and better quality. In addition, background rates were reduced significantly. Combining all improvements the process would become observable, if $G_{M\overline{M}} \geq 3 \cdot 10^{-3} \times G_F$.
This experiment is supported by the German BMFT, the Swiss National foundation and the Russian Fund of Fundamental Research.

References

1. P. Herczeg and R.N. Mohapatra, Phys. Rev. Lett. 69, 2475 (1992).
   A. Halprin and A. Masiero, Phys. Rev. D48, 2987 (1993).
   H. Fujii et al., Phys. Rev. D49, 559 (1994).
   G.G. Wong and W.S. Hou, Phys. Rev. D50, R2962 (1994).
2. W. Bertl et al., Weak and Electromagnetic Interactions in Nuclei, Osaka, eds. H.Ejiri et al., World Scientific Publ., Nov. 1995, p117.
3. P.V. Schmidt et al., Preprint Univ. Heidelberg, HD-PY 96/01 and accepted for publication by Nucl. Instr. & Meth.

TEST OF LFC IN $\mu \rightarrow e$ CONVERSION ON TITANIUM

PETER WINTZ
Paul Scherrer Institut, CH-5232 Villigen PSI, Switzerland

Representing the SINDRUM II collaboration
W. Bertl, V. Djordjadze, C. Dohmen, S. Eggli, R. Engfer, C. Findeisen, E. Hermes, J. Hofmann, W. Honecker, J. Kaulard, T. Kozlowski, B. Krause, F. Muheim, C. Niebuhr, G. Otter, S. Playfer, H. Pruys, M. Rutsche, A. v. d. Schaaf, P. Wintz.
III. Physikalisches Institut B der RWTH Aachen, D-52056 Aachen, Germany
Paul Scherrer Institut, CH-5232 Villigen PSI, Switzerland
Physik-Institut der Universität Zürich, CH-8057 Zürich, Switzerland

A new search for the neutrinoless $\mu^- \rightarrow e^-$ conversion in muonic titanium was carried out at PSI with the SINDRUM II spectrometer. No candidate for this lepton flavor nonconserving decay was found. The improved experimental sensitivity resulted in a new upper limit $B_{\mu e} < 8.4 \times 10^{-13}$ (90% CL).

Experimental tests of lepton flavor conservation (LFC) are very sensitive probes of physics beyond the standard model. One of the most stringent tests is the search for $\mu^- \rightarrow e^-$ conversion in muonic atoms as it is carried out at PSI with the SINDRUM II spectrometer [1].

Negative muons are focused on a titanium target, the stopped muons get captured and form muonic atoms which quickly reach their ground state. Assuming LFC-violation, the muons may then convert into an electron, $\mu^- Ti \rightarrow e^- Ti$. The coherent fraction of this hypothetical process is expected to dominate by more than 90% [2] and is of special interest since the signature is a monoenergetic electron emitted at the kinematical endpoint (104.3MeV) of bound muon decay, $\mu^- Ti \rightarrow e^- \nu_\mu \overline{\nu}_e Ti$. The probability of nuclear muon capture, $\mu^- Ti \rightarrow \nu_\mu Sc^*$ is 85.3% (f_{cap}).

The electron energy is measured with a magnetic spectrometer consisting of various cylindrical detectors positioned around the target. Two drift chambers provide the tracking while scintillation and Cerenkov hodoscopes are used for timing and electron identification. The energy resolution of 2% (FWHM) permits to resolve the hypothetical μe-events from the background of bound muon decay down to a level of 10^{-14}. A scintillation counter in the beam helps to recognize prompt background from beam electrons scattering off the target or γ-produced electrons from radiative nuclear capture of beam pions.

After first measurements on Ti and Pb in the years 1989 and 1992 [3] a more sensitive search on Ti was carried out in the year 1993. During an effective measuring time of 60 days 3×10^{13} (N_{stop}) muons stopped in the

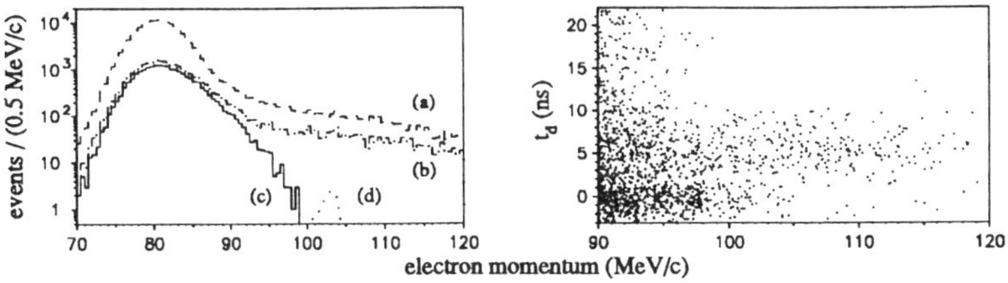

Figure 1: Momentum distribution of all reconstructed events (a) and after applying the cuts of prompt beam (b) and cosmic background (c). (d) shows the simulated μe-distribution assuming a branching ratio of $B_{\mu e} = 4 \times 10^{-12}$. The beam counter time t_d is used to remove background of scattered electrons ($t_d \simeq 0ns$) and radiative pion capture ($t_d \simeq 5ns$).

target. **Fig.1** shows momentum distributions of the reconstructed events. After applying cuts to suppress the prompt beam and cosmic background **no evidence for μe-conversion was found**[4]. A maximum likelihood analysis was used to derive the quantitative results. For a final data sample of 2005 events above 90MeV the probabilities to be of event type μe-conversion, bound muon decay, radiative pion capture, and electron scattering were calculated in a five-dimensional parameter space, three for the electron trajectory and the remaining two for time and deposited energy in the beam counter. The upper limit on the number of μe-events ($n_{\mu e}$) was obtained by integrating the likelihood function. The probability for a μe-event to be selected in the final data was 24% ($\epsilon_{\mu e}$) which accounts for geometric acceptance, trigger and reconstruction efficiency, and losses due to target and cosmic cuts. The resulting new upper limit on the branching ratio improves on the previous one by a factor five.

$$B_{\mu e} = \frac{n_{\mu e}}{N_{stop} \cdot f_{cap} \cdot \epsilon_{\mu e}} < 8.4 \times 10^{-13} \qquad (90\% \ confidence)$$

The experiment was supported in part by the BMFT, Germany, under contract number 06AC651 and by the Swiss National Science Foundation.

References

1. A. van der Schaaf (spokesman), PSI proposal R-87-03 (1987).
2. H. C. Chiang et al., *Nucl. Phys.* A 559, 526 (1993).
3. SINDRUM II Collab., C. Dohmen et al., *Phys. Lett.* B317, 631 (1993). W. Honecker et al., *Phys. Rev. Lett* 76, No. 2, 200 (1996).
4. P. Wintz, *Ph.D. thesis* RWTH Aachen (1995); S. Eggli, *Ph.D. thesis* Uni Zürich (1995); C. Dohmen, *Ph.D. thesis* RWTH Aachen(1996).

STUDIES OF CP, T AND CPT SYMMETRIES FROM THE CPLEAR EXPERIMENT

Raymond Gamet
Dept. of Physics, University of Liverpool, PO Box 147, Liverpool L69 3BX, UK.

for the CPLEAR collaboration
Athens, Basle, Boston, CERN, Coimbra, Delft, Fribourg, Ioannina, Liverpool, Ljubljana, CPPM-Marseille, CSNSM-Orsay, PSI, CEA, DSM/DAPNIA-Saclay, KTH-Stockholm, Thessaloniki, ETH-IPP Zürich

The CPLEAR experiment at CERN studies the CP, T and CPT symmetries in the neutral kaon system by measuring the rate asymmetries between decays of initially pure K^0 and $\overline{K^0}$ states. Results are reported on η_{+-} , Δm and the first direct measurements of T violation and of CPT invariance in K^0 - $\overline{K^0}$ oscillations. Our results are consistent with CPT conservation and lead to an improved limit on the K^0 - $\overline{K^0}$ mass difference

The CPLEAR experiment investigates CP, T and CPT symmetries in the neutral kaon system. Unlike other experiments which use beams of K_L and/or K_S , CPLEAR measures the difference between the decays of initially pure K^0 and $\overline{K^0}$ to a variety of final states, $\pi^+\pi^-$, $\pi e\nu$, $\pi^+\pi^-\pi^0$ and $\pi^0\pi^0$. The relevant parameters are then extracted from time dependent decay rate asymmetries of the form :

$$A_f(\tau) = \frac{R_{\overline{K}^0 \to \overline{f}}(\tau) - \alpha R_{K^0 \to f}(\tau)}{R_{\overline{K}^0 \to \overline{f}}(\tau) + \alpha R_{K^0 \to f}(\tau)}$$

where α is a normalisation factor due to different tagging efficiencies for K^0 and $\overline{K^0}$. The use of such asymmetries has the advantage that all common acceptances to K^0 and $\overline{K^0}$ cancel, thus considerably reducing systematic uncertainties. K^0 and $\overline{K^0}$ mesons are produced in equal amounts in $p\overline{p}$ annihilation at rest, through the channels

$$p\overline{p} \to K^-\pi^+K^0,\ K^+\pi^-\overline{K^0}$$

each one with a branching ratio of $\approx 2\cdot 10^{-3}$. The neutral kaon strangeness at production is tagged by the charge of the accompanying $K^\pm$.

In the $\pi^+\pi^-$ decay channel, a total of 70 million events in the lifetime region $1 - 20\tau_s$ have been selected. The acceptance-corrected decay rates for initial K^0 and $\overline{K^0}$ separately exhibit the expected interference effect due to CP violation. In the sum of these two rates the interference term cancels and a fit

can be made for the K_S lifetime and the level of K_L decays at long lifetimes. This gives a K_S lifetime in agreement with the world average [1] and a level of K_L decays about twice that expected from CP violating $K_L \to \pi^+\pi^-$, showing that some background remains. This is due to a small fraction of semileptonic decays of K_L which are kinematically inseparable from $\pi^+\pi^-$ decays.
After background subtraction, the time-dependent asymmetry

$$\begin{aligned} A_{+-}(\tau) &= \frac{R_{\overline{K}^0\to\pi^+\pi^-}(\tau) - \alpha R_{K^0\to\pi^+\pi^-}(\tau)}{R_{\overline{K}^0\to\pi^+\pi^-}(\tau) + \alpha R_{K^0\to\pi^+\pi^-}(\tau)} \\ &= -2\frac{|\eta_{+-}|e^{\frac{1}{2}(\Gamma_S-\Gamma_L)\tau}\cos(\Delta m\tau - \phi_{+-})}{1+|\eta_{+-}|^2 e^{(\Gamma_S-\Gamma_L)\tau}} \end{aligned}$$

is constructed (fig. 1) from the time-dependent decay rates and fitted for $|\eta_{+-}|$, ϕ_{+-} and α. We have recently published results for $|\eta_{+-}|$ and ϕ_{+-} [2] using data taken to mid 1994. Using all data to the end of 1995, and the value of Δm from [3], we obtain as preliminary results:

$$\begin{aligned} |\eta_{+-}| &= (2.316 \pm 0.025_{stat} \pm 0.028_{syst} \pm 0.010_{ext})\,10^{-3} \\ \phi_{+-} &= 43.5^0 \pm 0.5^0_{stat} \pm 0.5^0_{syst} \pm 0.4^0_{ext}. \end{aligned}$$

The external error on $|\eta_{+-}|$ is dominated by the uncertainty on the value of τ_s [3], and that on ϕ_{+-} by the uncertainty on Δm. The systematic error on ϕ_{+-} is almost entirely due to the uncertainty in the regeneration correction, which is poorly known in the momentum range of this experiment ($<$ 800 MeV/c). Our value of ϕ_{+-} has a statistical precision as good as any other single result published to date[4] and a smaller systematic error, which arises from completely different sources.

In the semileptonic channel, four independent decay rates can be measured, depending on the initial strangeness of the neutral kaon and on the charge of the decay lepton: $R^+(\tau) = R(K^0 \to \pi^- e^+\nu)$, $R^-(\tau) = R(K^0 \to \pi^+ e^-\overline{\nu})$, $\overline{R}^+(\tau) = R(\overline{K^0} \to \pi^- e^+\nu)$ and $\overline{R}^-(\tau) = R(\overline{K^0} \to \pi^+ e^-\overline{\nu})$. Several asymmetries can then be constructed which are sensitive to different parameters. The asymmetry

$$A_{\Delta m}(\tau) = \frac{\overline{R}^-(\tau) + R^+(\tau) - \overline{R}^+(\tau) - R^-(\tau)}{\overline{R}^-(\tau) + R^+(\tau) + \overline{R}^+(\tau) + R^-(\tau)}$$

is sensitive to Δm , important in the determination of ϕ_{+-} . Results from a fit to this asymmetry have already been published [5] and a preliminary result

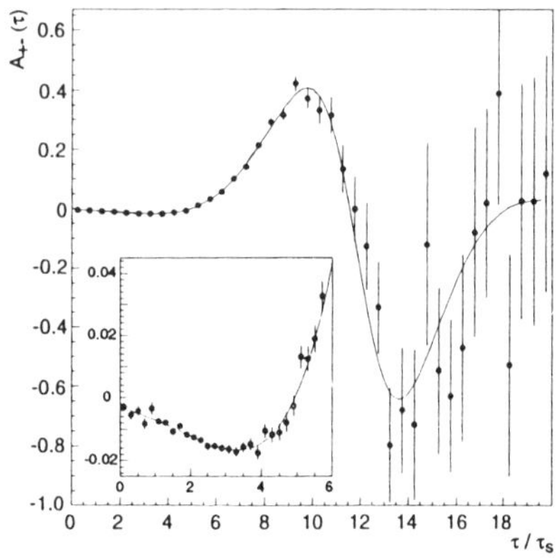

Figure 1: The time-dependent asymmetry A_{+-} for $\pi^+\pi^-$ decays

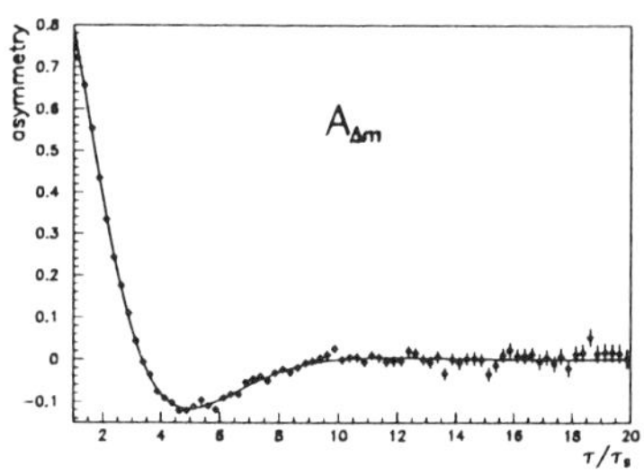

Figure 2: The asymmetry $A_{\Delta m}(\tau)$ for $\pi e\nu$ decays

(fig. 2) using all data to the end of 1995 is:

$$\Delta m = (0.5292 \pm 0.0018_{stat} \pm 0.0005_{syst})\, 10^{10}\, \hbar s^{-1}.$$

Assuming the validity of the $\Delta S = \Delta Q$ rule and no CPT violation in the decay amplitudes, the asymmetry

$$A_T(\tau) \;=\; \frac{\overline{R}^+(\tau) - R^-(\tau)}{\overline{R}^+(\tau) + R^-(\tau)}$$

measures the difference in the rates $R[\overline{\mathrm{K}^0}_{\tau=0} \rightarrow \mathrm{K}^0(\tau)]$ and $R[\mathrm{K}^0{}_{\tau=0} \rightarrow \overline{\mathrm{K}^0}(\tau)]$, and is a direct measure of T violation. A preliminary result (based on 25% of total statistics) for this is:

$$A_T = (6.3 \pm 2.1_{stat} \pm 1.8_{syst})\, 10^{-3}.$$

This is the first direct observation of T violation and the value is in good agreement with $2\delta_l$ [1], and hence with CPT invariance.

The difference between ϕ_{+-} and ϕ_{sw} yields an indirect test of CPT invariance. By assuming CP and CPT conservation in the decay amplitudes, we obtain

$$\eta_{+-} = \varepsilon_T + \delta_{CPT} + i\Delta\phi$$

where ε_T describes CP and T violation in $\mathrm{K}^0 - \overline{\mathrm{K}^0}$ oscillations and has a phase $\phi_{sw} = tan^{-1}(2\Delta m/\Delta\Gamma) = 43.5^0$. The CP and CPT violation parameter δ_{CPT}

is defined as

$$\delta_{CPT} = \frac{i\Delta\Gamma - 2\Delta m}{4\Delta m^2 + \Delta\Gamma^2}\left(M_{\overline{K}^0} - M_{K^0}\right)$$

and is orthogonal to ε_T. A CP violating term arising from the interplay between the $\pi^+\pi^-$ decay amplitudes and $K^0 - \overline{K^0}$ oscillations, $\Delta\phi$, is derived to be

$$\Delta\phi \approx \frac{\Gamma_L}{\Gamma_S}[2BR(K_L \to \pi l\nu)Im(x) + BR(K_L \to 3\pi)Im(\eta_{3\pi} - \eta_{+-})]$$

where we assume $\eta_{+-0} = \eta_{000} = \eta_{3\pi}$ and that the possible violation of the $\Delta S = \Delta Q$ rule (parameterised by x) is identical for μ and e. Using previous measurements of $Im(x)$ and $Im(\eta_{+-0})$ [1] would result in a contribution of $\approx 3^0$ to the error in the difference $\phi_{+-} - \phi_{sw}$. Using the CPLEAR measurements (not reported here) this source of error is reduced by a factor of 10. Thus we can give the limit

$$|\delta_{CPT}| < 4.5 \cdot 10^{-5} \quad (90\% C.L.)$$

and hence

$$\left|M_{\overline{K}^0} - M_{K^0}\right| < 4.5 \cdot 10^{-19} GeV \quad (90\% C.L.)$$

which is the best indirect test of CPT invariance.

Other results from semileptonic, $\pi^+\pi^-\pi^0$ and $\pi^0\pi^0$ decays have been obtained, which space has not permitted us to discuss. Analysis of data already collected and of simulated data will enable many statistical and systematic errors to be further reduced. Dedicated measurement of regeneration parameters in 1996 will help to decrease our systematic error in ϕ_{+-} by more than a factor of 2.

References

1. Particle Data Group, Phys. Rev. D50 (1994) 1175.
2. CPLEAR Coll., Phys. Lett. B363 (1995) 243.
3. CPLEAR Coll., Phys. Lett. B369 (1996) 367.
4. B. Schwingenheuer et al., Phys. Rev. Lett. 74 (1995) 4376.
5. CPLEAR Coll., Phys. Lett. B363 (1995) 237.

A SEARCH FOR T-VIOLATION IN $K^+ \to \pi^0\mu^+\nu$ DECAY

M. Blecher, representing KEK E246

1) Virginia Tech, Blacksburg, VA, USA 24061
E-mail: blecherm@vtinte.phys.vt.edu
2) National Laboratory for High Energy Physics (KEK)
3) University of Tokyo
4) Tsukuba University
5) Institute for Nuclear Study, University of Tokyo
6) Tokyo Institute of Technology
7) Osaka University
8) Yonsei University
9) Institute for Nuclear Research, Russian Academy of Science
10) TRIUMF
11) Univesity of British Columbia
12) University of Montreal
13) University of Saskachwan
14) Korea University
15) National Taiwan University
16) Princeton University

ABSTRACT

A search for a new source of CP or T violation using the decay $K^+ \to \pi^0\mu^+\nu$ is described. A positive signal would be a non-zero measurement of the transverse μ^+ polarization, $P_T = \vec{\sigma}_{\mu^+} \cdot (\vec{P}_{\pi^0} \times \vec{P}_{\mu^+}) / \mid \vec{P}_{\pi^0} \times \vec{P}_{\mu^+} \mid$.

The minimal Standard Model (SM) cannot explain the lack of anti-matter in the universe or the observed baryon to photon ratio. As pointed out by Sakarov in 1967 both C and CP (or T) violation are required. The strength of the CP violation allowed by the SM is too small to explain the present baryon asymmetry. Thus new sources of violation of time-reversal invariance (T) are required. KEK E246 searches for such by seeking to detect, with high precision, the transverse muon polarization (P_T) in $K^+ \to \pi^0\mu^+\nu$ decay at rest. The current limit, $P_T = (-3.1 \pm 5.3) \times 10^{-3}$, is due to a decay-in-flight measurement at the BNL-AGS. That limit when translated to the K^+ rest frame is $(-4.2 \pm 6.2) \times 10^{-3}$. We hope to reduce the limit to $< 10^{-3}$. It should be noted that the SM prediction for P_T is $< 10^{-4}$ (zero in leading order) and that final state interactions can mimic a false P_T at the 10^{-6} level. However, an expanded Higgs sector could produce an effect reachable in this experiment.

The experimental method is briefly described below. A Cherenkov counter selects K^+ from a mixed beam of K^+ and π^+. The K^+ are stopped in a live scintillating fiber target, composed of 256 0.5×0.5 cm^2 fibers, that provides the x,y of the stop point.

Surrounding the target is a fiducial scintillator segmented in 12 sections, which in turn is surrounded by a cylindrical drift chamber. The former, which has an inside shape that matches that of the neighboring fibers, provides a start signal for the μ^+ time-of-flight, while the latter provides position information for momentum fitting. These in turn are surrounded by a CsI(Tl) detector, which is segmented in 768 trapezoidal crystals that view approximately 75% of 4π sr. There are 12 holes in the CsI detector, corressponding to the sectors of a superconducting toroidal magnet. μ^+ pass through these holes and are deflected by the magnetic field into a pure Al stopper. Along the path position and time-of-flight measurements are made by 3 additional planar wire chambers and a scintillator mounted in each sector. In Al the polarization is preserved and there is also a modest fringe field (100 G) to further preserve the spin. In the present data taking runs μ^+ from $K^+ \rightarrow \pi^0\mu^+\nu$ and π^+ from $K^+ \rightarrow \pi^0\pi^+$ are observed. The latter correspond to a unique momentum, P=205 MeV/c, for which our measured resolution is 3% (FWHM), after correction for energy loss in the target.

The energy and position of photons from π^0 decay are measured from pulse heights and times in the Cs(Tl) modules. These positions combined with the stop position as determined by the target rod (x,y) with maximum energy and the fitted μ^+ track (z) determine the π^0 direction and thus its momentum. During the present data runs this calorimeter has been calibrated. The resolution achieved in reconstruction of the π^0 mass is 15% (FWHM) and the timing resolution achieved is 13 ns (FWHM).

The direction of the e^+ from μ^+ decay determines the μ^+ spin direction. These are detected by scintillators mounted on both the clockwise and counter-clockwise side of each sector. Great care has been taken to keep each of the sector geometries and detectors identical. The clockwise and counter-clockwise sectors are summed together for an enhanced P_T signal. Also it should be noted that P_T reverses for forward as compared to backward going π^0 and so by taking the ratio of forward to backward events many systematic errors cancel. For π^0 going perpendicular to the beam direction no P_T is expected, which yields a convenient null calibration of the experiment.

There are many backgrounds, for example $K^+ \rightarrow \pi^0\pi^+$, which can be kept small by appropriate kinematic cuts. However, it should be noted that such backgrounds only dilute the signal. They cannot produce a positive effect.

CPT, STRINGS, AND NEUTRAL MESONS

V.A. KOSTELECKÝ
Physics Department, Indiana University, Bloomington, IN 47405, U.S.A.

A sketch of our results on CPT and strings is given. A mechanism for spontaneous CPT violation in string theory is briefly reviewed, and recent theoretical progress is summarized. Possible CPT-violating contributions to a four-dimensional low-energy effective theory could produce effects at levels testable in neutral-meson systems. Likely experimental signatures and attainable bounds are outlined in candidate experiments with correlated and uncorrelated neutral mesons.

CPT invariance is a fundamental theoretical property of local relativistic field theories of point particles.[1] Moreover, interferometric experiments with neutral kaons have tested it to a high degree of precision.[2] These features make CPT violation an ideal candidate signature for new, non-particle physics.[3]

Currently, string theory is among the most promising frameworks for a unified description of the fundamental forces. Since strings are extended objects, they do not automatically satisfy the axioms leading to CPT invariance. Indeed, an explicit stringy mechanism for CPT breaking is known,[3] connected to spontaneous Lorentz breaking.[4] Briefly, it turns out that couplings in string field theory exist that are impossible in conventional four-dimensional renormalizable gauge theories. When scalars acquire expectation values, these couplings can induce instabilities in effective potentials for Lorentz tensors. Nonzero expectation values for the latter can ensue, leading under suitable circumstances to CPT violation.

Additional support for this scenario is provided by a recent detailed analysis of Lorentz- and CPT-breaking solutions of the equations of motion for the open bosonic string.[5] Dominant terms of the action in a level-truncation scheme are computed analytically, and the equations of motion are solved. For some solutions, over 20,000 nonvanishing terms in the action are used. The associated nonzero expectation values for various tensor fields are found to exhibit predicted features.

If a string theory indeed describes the Universe, then the above mechanism might induce CPT-violating terms in the effective four-dimensional low-energy theory, i.e., in the standard model. Since no zeroth-order CPT violation is observed, it must be suppressed. The natural suppression factor is the ratio $\sim 10^{-17}$ of the low-energy scale to the Planck scale. It turns out that a suppression of this magnitude can produce effects around or just below those detectable in current experiments with neutral-meson systems.[3,6]

The issues of modeling and detecting these effects have been examined[6,7,8] in the K system, the D system, and the two B systems. Candidate experiments include ones with uncorrelated mesons or with correlated mesons (factories). Several interesting features emerge from the analyses.

• In the string scenario, the magnitude of possible CPT violation depends on coupling constants that can vary among different quark flavors as, for example, do the usual Yukawa couplings. This means it would be desirable to test for CPT violation in all the neutral-meson systems, not only the K system.

• The only bounds in the literature to date are those on the components of the K-system CPT-violating parameter δ_K. These may be improved soon, e.g., in ϕ factories, under favorable circumstances possibly to one part in 10^5.

• Testing CPT in the D system is hard because the mixing is so small. Nonetheless, definite tests can be performed at a τ-charm factory. They could include bounding the K-system parameter δ_K and the corresponding quantity δ_D in the D system.

• Testing CPT in the neutral-B_d system is particularly interesting because it involves the b quark, and in some string scenarios CPT violation is largest in this system. Although no bound is currently in the literature, sufficient data already exist to bound the CPT-violating parameter δ_B at about the 10% level. Prospects at the B factories appear interesting.

My thanks to Don Colladay, Rob Potting, Stuart Samuel, and Rick Van Kooten for enjoyable collaborations leading to the results described here.

1. See, e.g., R.F. Streater and A.S. Wightman, *PCT, Spin and Statistics, and All That,* Benjamin Cummings, Reading, 1964.
2. See, e.g., Review of Particle Properties, Phys. Rev. D **50**, 1173 (1994).
3. V.A. Kostelecký and R. Potting, Nucl. Phys. B **359** (1991) 545.
4. V.A. Kostelecký and S. Samuel, Phys. Rev. D **39** (1989) 683; *ibid.,* **40** (1989) 1886; Phys. Rev. Lett. **63** (1989) 224; *ibid.,* **66** (1991) 1811.
5. V.A. Kostelecký and R. Potting, Phys. Lett. B, in press (hep-th/9605088).
6. V.A. Kostelecký and R. Potting, Phys. Rev. D **51**, 3923 (1995); see also V.A. Kostelecký, R. Potting and S. Samuel, in S. Hegarty et al., eds., *Proc. 1991 Joint Intl. Lepton-Photon Symp. and Europhys. Conf. on High Energy Phys.* (World Scientific 1992); V.A. Kostelecký and R. Potting, in D.B. Cline, ed., *Gamma Ray–Neutrino Cosmology and Planck Scale Physics* (World Scientific 1993) (hep-th/9211116).
7. D. Colladay and V.A. Kostelecký, Phys. Lett. B **344**, 259 (1995); Phys. Rev. D **52**, 6224 (1995).
8. V.A. Kostelecký and R. Van Kooten, Phys. Rev. D, in press.

Session 5
Nuclear and Particle Astrophysics

ASTROPHYSICAL FACTOR FOR THE REACTION $^7Be(p,\gamma)^8B$ AND THE HIGH ENERGY SOLAR NEUTRINO PROBLEMS

A.M. MUKHAMEDZHANOV AND R.E. TRIBBLE
Cyclotron Institute, Texas A&M University, College Station Texas, 77843, USA

We present a survey of the latest results for the astrophysical S_{17}-factor for the reaction $^7Be(p,\gamma)^8B$. They definitely favor the value of $S_{17}(0) \approx 17\,eVb$, which is lower than $S_{17}(0) = 22.5\,eVb$ used in the latest calculations of the high energy solar neutrino flux.

Recent developments in the experimental and theoretical analysis of the reaction $^7Be(p,\gamma)^8B$ show that its astrophysical factor $S_{17}(0)$ is significantly smaller than it was assumed in different solar models when evaluating the high energy solar neutrino flux [1,2]. The extrapolation of the latest direct measurements of Filippone *et al*[3] using a new parametrization of the $^7Be(p,\gamma)^8B$ resonance cross section yeilds $S_{17}(0) = 16\,eVb$[4]. The extrapolation of Coulomb breakup of ^8B in the Coulomb field of ^{208}Pb data down to zero energy gives the value of the astrophysical factor in the interval $S_{17}(0) = 15 - 17\,eVb$[5,6]. The simultaneous analysis of both the Filippone *et al* data and the data of Motobayashi *et al* leads to $S_{17}(0) = 16.9^{+}_{-}2.0\,eVb$[4].

We calculated the S_{17}-factor using a simple relation between the asymptotic normalization coefficients of ^8B wave function in the two-body channel $^7Be + p$ and $S_{17}(0)$. Our value is $S_{17}(0) \approx 17.6\,eVb$[7]. The recent calculation by Barker of S_{17} within the framework of the R-matrix approach, using the parameters derived from an analysis of the mirror reaction $^7Li(n,\gamma)^8Li$, gives $S_{17}(0) = 17^{+}_{-}3\,eVb$[8]. Finally, the information about the tail of the ^8B wave function extracted from ^8B breakup gives the interval $S_{17} = 12 - 20\,eVb$[9].

Thus all the latest results for $S_{17}(0)$ published in 1993-1995, with the exception of the microscopic calculations[10], definitely favor the value of $S_{17}(0) \approx 17\,eVb$, which is smaller than the value currently adopted in astrophysical calculations, $S_{17}(0) = 22.5\,eVb$[1,2], by 25%.

Consider now the impact of the new value for $S_{17}(0)$ on the theoretical predictions of the high energy solar neutrino flux and solar neutrino problem.

1. The Chlorine experiment. The ratio of the experimental data ($Y_H = 2.55\,SNU$) to the theoretical predictions is $R_H = 0.27$ for the SSM (with diffusion) and $R_H = 0.40$ for the TCL model. The updated ratios of the experimental data to the theoretical predictions derived for the low value of $S_{17}(0) = 17\,eVb$ are $R_H = 0.34$ for the SSM (with diffusion) and $R_H = 0.49$ for the TCL model. Thus the difference between the chlorine experimental

data and theoretical predictions with the new value of $S_{17}(0)$ is still significant but less severe than it was.

2. Kamiokande experiment. The ratio of the Kamiokande III data to the theoretical predicitions (calculated using $S_{17}(0) = 22.5\,eVb$) is $R_K = 0.43$ for the SSM (with diffusion) and $R_K = 0.64$ for the TCL model. If we use as the theoretical prediction the solar neutrino flux calculated with the lower value of $S_{17}(0) = 17\,eVb$, then the updated ratio is $R_K = 0.57$ for the SSM (with diffusion) and $R_K = 0.85$ for the TCL model. Thus, if the lower value of the S_{17}-factor is true, the discrepancy between the experimental data of the Kamiokande III and the theoretical predictions is on the level of uncertainties for the TCL model.

It follows that even for the small value of S_{17} the main problems for high energy solar neutrinos for now are
1) the apparent contradiction between Kamiokande and Chlorine detector data (so-called second solar neutrino problem),
2) the absence of ^7Be neutrinos,
3) the contradiction between the SSM and TCL model predictions.

Acknowledgments

This work was supported in part by the U.S. Department of Energy under Grant No. DE-FG05-86ER40256.

References

1. J. N. Bahcall and M. H. Pinsonneault, *Rev. Mod. Phys.* **67**, 781 (1995).
2. S. Turck-Chièze *et al*, *Phys. Rep.* **230**, 57 (1993).
3. B.W. Filippone *et al*, *Phys. Rev. Lett.* **50**, 412 (1983); *Phys. Rev.* **C28**, 2222 (1983).
4. Y.E. Kim *et al*, in AIP Conference Proceedings 334, Few-Body Problems in Physics, Williamsburg, VA 1994, ed. F. Gross (AIP Press, 1995), p. 795.
5. T. Motobayashi *et al*, *Phys. Rev. Lett.* **73**, 2680 (1994).
6. C.A. Bertulani, *Nucl. Phys.* **A587**, 318, (1995).
7. H.M. Xu *et al*, *Phys. Rev. Lett.* **73**, 2027 (1994).
8. F.C. Barker, *Nucl. Phys.* **A588**, 693, (1995).
9. K. Riisager and A.S. Hansen, *Phys. Lett.* B **301**, 6 (1993).
10. A. Csòtò *et al*, *Phys. Rev.* **C52**, 1130 (1995).

REACTION STUDIES OF $^{8}Li(\alpha,n)^{11}B$ AND BIG BANG NUCLEOSYNTHESIS

Y. YAMAMOTO, K.-I. KUBO
Department of Physics, Tokyo Metropolitan University,
1-1, Minami-osawa, Hachioji, Tokyo, 192-03, JAPAN

T. KAJINO
Division of Theoretical Astrophysics, National Astronomical Observatory
Mitaka, Tokyo, 181, JAPAN

K. OGAWA
Department of Physics, Chiba University,
1-33, Yayoi-cho, Inage-ku, Chiba-shi, Chiba 263, JAPAN

Astrophysical nuclear reaction, $^{8}Li(\alpha,n)^{11}B$ was investigated theoretically. Calculation was performed in the framework of DWBA as the previous work of Rauscher *et al.*[1]. We included not only stripping reaction mechanism but also the knock-on reaction mechanism. Our S factor is about five times as large as Rauscher's and is in good agreement with the recent experimental observation. Consequently our result leads to the enhancement of A=12 or larger nuclei abundances in the big-bang nucleosynthesis from the previous nucleosynthesis simulation.

Unstable nuclear reaction, $^{8}Li(\alpha,n)^{11}B$, is one of the key reactions for the production of intermediate-to-heavy mass elements in inhomogeneous primordial nucleosynthesis. Two independent experimental groups have ever measured the cross section of this reaction in different methods: Since ^{8}Li is a radioactive nucleus which beta decays to ^{8}Be in 838 msec, the Greek group [2] measured the inverse reaction cross section for $^{11}B(n,\alpha)^{8}Li$ and converted it to the forward (α,n) cross section by the use of the principle of detailed balance. Only the contribution from the ^{8}Li(ground state)-^{11}B(ground state) transition was measured in this method. Independently, the US-Japan collaboration team[3] measured directly the $^{8}Li(\alpha,n)^{11}B$ recation cross section which includes the transitions to the excited bound states of ^{11}B as well as to the ground state. They found that the total reaction cross section for the production of ^{11}B is at least five times that of Paradellis' measurement. It is very challenging to understand why the total reaction cross section is this large.

We tried DWBA to this reaction system in order to understand the reaction mechanism microscopically. Since the energy region of astrophysical interest is below 1 MeV, many different reaction processes are presumed to contribute to the total cross section at these low energies below the Coulomb barrier.

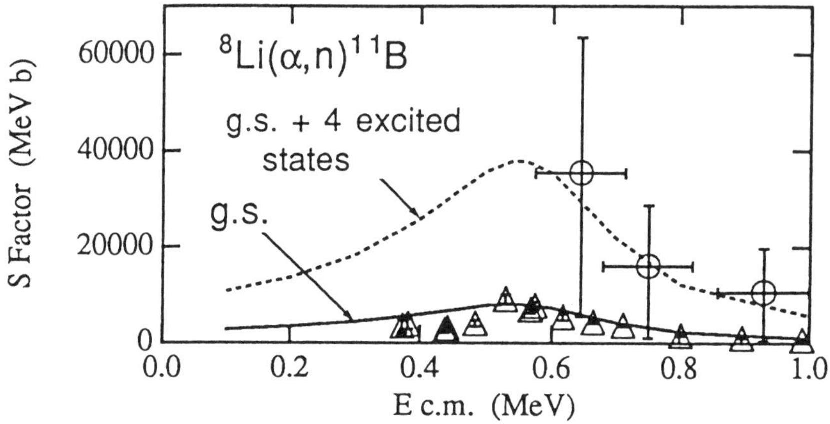

Figure 1: The S factor of $^{8}Li(\alpha,n)^{11}B$. Triangles and open circles are experimental data from refs. 2 and 3 respectively. Our calculations are shown by solid line and dashed line corresponding to these experimental results.

We therefore included all possible processes in the theoretical caluclations, including stripping and knock-on reactions and others. Fig. 1 is shown our calculated result, compared with measured cross sections from two independent groups. Our result is in reasonable agreement with both measured data.

In the previous big-bang nucleosynthesis calculations, the thermonuclear reaction rate from $^{8}Li(\alpha,n)^{11}B_{g.s.}$ alone was used. We have updated the reaction rate by including the contribution from the production of excited bound states of ^{11}B, and recalculated the primordial nucleosynthesis. We have then found that the new rate, which is five times larger than the previous calculated by Rauscher *et al.* [1], leads to an enhancement of the abundance of the A=12 nucleus and the heavier by factor of 2.5. This result means that large amount of heavy elemants could be synthesized in very early stage of our Universe.

References

1. T. Rauscher, K. Grün, H. Krauss and H. Oberhummer, Phys. Rev. **C45**, 1996 (1992).
2. T. Paradellis *et al.*, Z.Phys. **A337**, 211 (1990).
3. X. Gu *et al.*, Phys. Lett. **B343**, 31 (1995); R.N. Boyd *et al.*, Phys. Rev. Lett. **68**, 1283 (1992).

COULOMB BREAKUP OF A LIGHT NUCLEAR PROJECTILE IN THE FIELD OF A HEAVY ION

B.F. IRGAZIEV, H.T. ERGASHBAYEV
Institute for Fundamental Studies, Tashkent, 700000, P.O. Box 4547, Uzbekistan

A.M. MUKHAMEDZHANOV
Cyclotron Institute, Texas A&M University, College Station, Texas, 77843, USA

The expression for the Coulomb breakup amplitude giving by the sum of one-step (direct) and two-step (through the resonance of the light projectile) amplitudes has been derived within the three-body approach. We calculated the cross section of the breakup reaction $^6Li + {}^{208}Pb \to {}^6Li(2.185MeV) + {}^{208}Pb \to \alpha + d + {}^{208}Pb$ at the incident energy of ^6Li $156\,MeV$.

It is well known that breakup of light projectiles in the Coulomb field of heavy targets can be a very useful tool to extract the cross section of the radiative capture reactions important for nuclear astrophysics [1]. We have investigated here the Coulomb two-step breakup (through the excitation of the resonance of light projectile) in the field of a heavy target within the framework of three-body approach. To compare with experimental data the one-step (direct) breakup has also been taken into account.

Consider the breakup reaction (12)+3 $\to$ 1+2+3. The reaction amplitude is given by

$$M = < \Psi_C^{(-)} \mid \Delta V \mid \Phi^{(+)} > + < \Psi_C^{(-)} \mid V_{12}^N G(E+i0)\Delta V \mid \Phi^{(+)} > . \qquad (1)$$

Here $\Phi^{(+)}$ is the initial state wave function, $\Psi_C^{(-)}$ is the three-body Coulomb scattering wave function describing the relative motion of particles 1, 2 and 3 in the final state, $\Delta V = V_{13}^C + V_{23}^C - U^C$ is the transition operator, $V_{\nu\beta}^C$ ($V_{\nu\beta}^N$) is the Coulomb (nuclear) interaction potential between particles ν and β, U^C is the "channel" Coulomb potential between the center of mass of (12) and target 3, $G(E+i0)$ is the three-body Green function of system 1+2+3 with the total energy E. The first term defines the direct breakup. The breakup through the resonance can be extracted from the second term. We found explicit behavior of the resonance term. The so-called post decay Coulomb acceleration effects can shift the location of the resonance, change the resonance shape line and renormalize the resonance cross section.

The differential cross section of the disintegration $^6Li \to \alpha + d$ at the incident energy of ^6Li $E_{Li} = 156\,MeV$ in the Coulomb field of ^{208}Pb has been calculated in the kinematic region of experiment [2]: scattering angle of

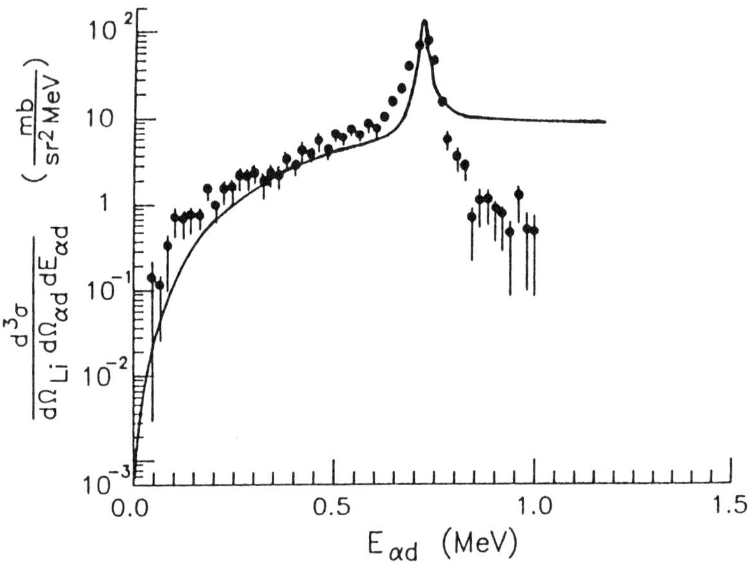

Figure 1: Comparison of the triple differential cross sections of the Coulomb breakup $^{208}Pb(^6Li, \alpha\, d)^{208}Pb$ as the function of the relative kinetic energy of α and d at the incident energy of ^6Li 156 MeV, scattering angle of ^6Li $\theta_{Li} = 3^o$ and relative speed of α-particles and deuterons $v_{\alpha\, d} < 0$: dots are experimental data[2], solid line is our calculation.

^6Li $\theta_{Li} = 3^o$. Calculated cross section reproduces quite well the experimental one at relative kinetic energies of the final fragments α and d in the resonance region ($E_{\alpha d} \approx 710\, keV$) and at $E_{\alpha d} < 710\, keV$ (Fig. 1). The apparent disagreement of theoretical calculations with experimental data at $E_{\alpha d} > 710\, keV$ is due to the neglecting of the nuclear interaction in the final state.

Acknowledgments

This work was supported in part by the U.S. Department of Energy under Grant No. DE-FG05-86ER40256.

References

1. G. Baur *et al*, *Nucl. Phys.* **A458**, 188 (1986).
2. J. Kiener et al, *Phys. Rev.* **C44**, 2195 (1991).

Search for Dark Matters by means of CaF_2 Detector

R. Hazama[a], S. Ajimura[a], H. Ejiri[b], K. Fushimi[c], T. Kishimoto[a], N. Kudomi[b], K. Kume[b], K. Matsuoka[a], H. Miyawaki[a], T. Nitta[a], H. Ohsumi[a], and N. Suzuki[a]
a)Dept. Phys. and Lab. Nucl. Studies, Osaka Univ., Toyonaka, Osaka 560, Japan
b)Research Center for Nucl. Phys., Osaka Univ., Ibaraki, Osaka 567, Japan
c)Dept. Phys., The Univ. of Tokushima, Tokushima 770, Japan

Recently it has become quite interesting to explore the spin coupled dark matter of the universe. Cold dark matters are required to explain the structure of the universe [1]. It is likely that the cold dark matters are unknown particles that interact weakly with matters. Such particles are classified by the interactions of vector and axial vector couplings. So far experiments have far below the necessary sensitivity to exclude the dark matters that have axial vector coupling with matters (spin coupled dark matters), while ultrapure germanium detectors have already ruled out vector coupled one like the heavy dirac neutrino [2]. Among such spin coupled dark matters the lightest neutral super partners (neutralino) are one of the strongest candidates, since the supersymmetry seems to be quite promising for the theory beyond the standard model. These considerations make the search for the spin coupled dark matter quite attractive.

We have been developing a new detector system; ELEGANT VI, which consists of 25 CaF_2(Eu) crystals with dimension of 45 mm cube. They are surrounded by active and passive shields. In order to find the tiny and rare signal of nuclear recoil by WIMPs background reduction is essential. The background is reduced by the design of 4π active shield. The CaF_2(Eu) crystal has light guide of pure CaF_2 crystal which are also sensitive to gamma rays. The signal from the light guide is separated by the roll-off ratio which makes the detector effectively 4π shielded. The CaF_2(Eu) crystals used here has been purified to the level about 0.05ppb for U and 0.24ppb for Th.

The dark matter gives quite low energy signal to the detector. It is thus quite important to obtain detailed information of the detection efficiency and the response to the nuclear recoil in the low energy region. We have determined the efficiency of the low energy signal by using the signal from a light emitting diode (LED). It is found that the detector system is sensitive to 2 keV signal with 40 % efficiency.

The response of the detector to the nuclear recoil was measured by the ^{19}F(n,n') reaction [3]. The experiment was carried out at the PELLETRON accelerator of Tokyo Institute of Technology. Neutrons were obtained by the d(d,n)^{3}He reaction. The ~3.7 MeV neutron gives similar energy to a ^{19}F nucleus as the scattering by the dark matters. The low energy signal from the detector was measured in coincidence with scattered neutrons detected by NE213 liquid scintillator. The obtained recoil response is similar to what has been measured for other detectors [4] although our detector is a little bit better than those.

The engineering run was carried out by using only 9 elements of CaF_2 detector in the overground laboratory [5]. Here outer detectors were used as veto counters for the central detector. Excluded region by the present measurement is shown in Fig. 1. The experimental condition is far below the final setup. However, it already has similar precision as BPRS Collaboration [4]. It is noted that the present result is obtained by the subset of ELEGANT VI at the sea level without radon purge by Nitrogen gas. We expect much better results when the detector is fully operational at the underground laboratory to be constructed at Nara (Oto Cosmo Observatory) [6].

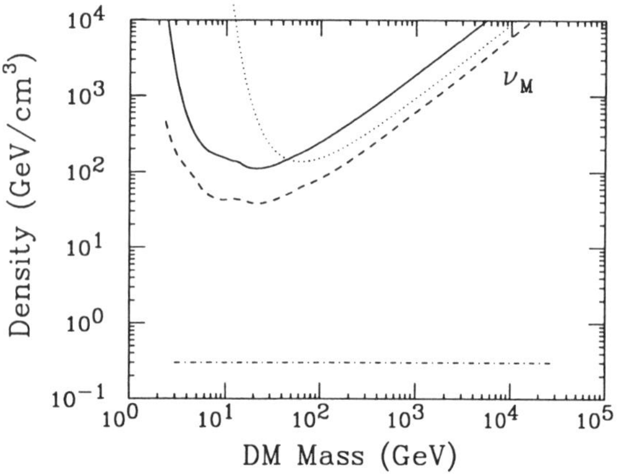

Figure 1
The observed counting rate is converted to the halo density of Majorana neutrino (ν_M). Upper regions from the lines are excluded. Dotted line show the result of BPRS Collaboration [4]. The dashed line is the exclusion plot with 100% detection efficiency. The dash-dotted line indicates the critical density of $0.3 GeV/cm^3$.

1. J. Primack et al., *Phys. Rev. Lett.* **74**, (1995) 2160.
2. H. V. Klapdor, *Proc. IV Int Symp. on Weak and Electromagnetic Inteactions in Nuclei"WEIN'95, June 1995, Osaka, World Scientific* , (1995) 174.
3. R. Hazama, et al., paper in preparation; OULNS annual report, (1995) in press.
4. BPRS Collaboration, *Astroparticle Phys.* **2**, (1994) 117.
5. R. Hazama, et al., *Proc. IV Int Symp. on Weak and Electromagnetic Inteactions in Nuclei"WEIN'95, June 1995, Osaka, World Scientific*, (1995) 635.
6. H. Ohsumi, et al., *Proceeding of FRONTIER'96* (1996) in press.

DARK MATTER-NUCLEUS SCATTERING

YIHARN TZENG[a], T.T.S. KUO[b]

[a] *Institute of Physics, Academia Sinica, Taipei, Taiwan, R.O.C.*
[b] *Department of Physics, SUNY-Stony Brook, New York 11794, USA*

An accurately computed G-matrix is used as the input to construct the BCS states from which the nuclear structure functions for the dark matter-nucleus scattering are calculated.

Increasing amounts of experimental efforts have been made to explore the existence of the dark matter through the detecting of possible dark matter-nucleus interactions. An accurate theory to predict the dark matter-nucleus scattering cross sections is thus more desirable than ever.

The neutralino has been proposed as one of the possible candidates for the dark matter. It was shown [1], based on the impulse approximation, that the neutralino-nucleus scattering cross section is

$$\begin{aligned}
\frac{d\sigma}{dq^2} &= \frac{8G_F^2}{(2J+1)v^2}[S_A(q)+S_s(q)], \\
S_A(q) &= \sum_{odd\ L}(|<J||T_L^{el5}(q)||J>|^2+|<J||L_L^5(q)||J>|^2), \\
S_s(q) &= \sum_{even\ L}|<J||\sum_i c_0 j_L(qr_i)Y_L(\hat{r}_i)||J>|^2, \\
T_L^{el5}(q) &= \frac{1}{2\hat{L}}\sum_i(a_0+a_1\tau_3^i)[-\sqrt{L}M_{L,L+1}(q\mathbf{r_i})+\sqrt{L+1}M_{L,L-1}(q\mathbf{r_i})], \\
L_L^5(q) &= \frac{1}{2\hat{L}}\sum_i(a_0+\frac{a_1\tau_3^i m_\pi^2}{|\mathbf{q}|^2+m_\pi^2})[\sqrt{L+1}M_{L,L+1}(q\mathbf{r_i})+\sqrt{L}M_{L,L-1}(q\mathbf{r_i})].
\end{aligned} \tag{1}$$

$S_A(q)$ and $S_s(q)$ have been expanded in vector spherical harmonics, and $\hat{L}=\sqrt{2L+1}$, $M_{L,L'}(q\mathbf{r_i})=j_{L'}(q\mathbf{r_i})[Y_{L'}(\hat{r}_i)\sigma_i]^L$ and a_0, a_1 are the parameters from which the type of incident neutralino can be determined [1].

In this work, We propose a reliable and accurate way to calculate the nuclear structure functions $S_A(q)$ and $S_s(q)$. The framework used here is the BCS approximation as it is one of the best ways to describe pairing effects in nuclei. The BCS states are constructed with the inputs of the accurately calculated G-matrix elements obtained via the method derived in [2, 3]. For a general model-space P, we define the Brueckner G-matrix by the integral equation [3]

$$G(\omega) = V + VQ_2 \frac{1}{\omega - Q_2TQ_2} Q_2G(\omega), \tag{2}$$

where ω is an energy variable, Q_2 a two-body Pauli exclusion operator, and T is the kinetic energy. The exact solution of this G-matrix is $G = G_F + \Delta G$ [2, 3], where G_F is the "free" G-matrix, and ΔG is the Pauli correction term

$$\Delta G(\omega) = -G_F(\omega)\frac{1}{e}P_2 \frac{1}{P_2[\frac{1}{e} + \frac{1}{e}G_F(\omega)\frac{1}{e}]P_2} P_2\frac{1}{e}G_F(\omega), \tag{3}$$

where $e \equiv \omega - T$, and the projection operator $P_2 = (1 - Q_2)$.

We choose the nucleus of Ge^{73} to illustrate our method. The realistic NN Paris potential is used here to calculate the $G-$ matrix. We have vary the parameters a_0 and a_1 so that different incident neutralinos can be considered. An example of the results with 1 qp state is shown in the following figure, while results with higher qp states will be reported elsewhere shortly.

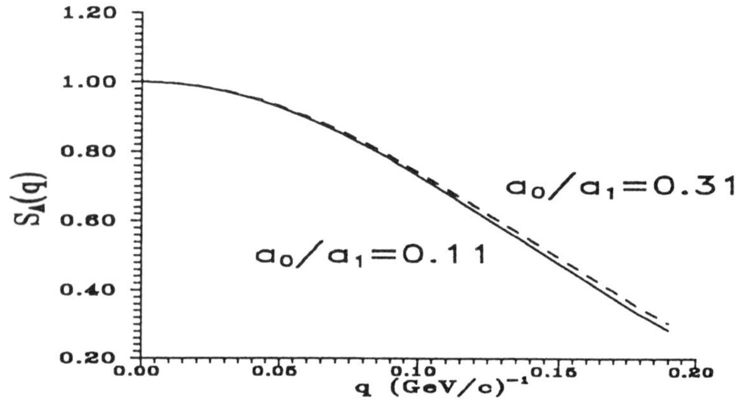

This work is supported in part by the NSC of the R.O.C.(Taiwan) under the grant of NSC85-2112-M-001-009, and the USDOE Grant DE-FG02-88ER40388.

References

[1] J. Engel, S. Pittel and P. Vogel, *Int. J. of Mod. Phys.* **E1**, 1 (1992).

[2] S.F. Tsai and T.T.S. Kuo, *Phys. Lett.* B**39**, 427 (1972).

[3] E.M. Krenciglowa, C.L. Kung, T.T.S. Kuo and E. Osnes, *Ann. Phys.* (N.Y.) **101**, 154 (1976).

NONEQUILIBRIUM WEAK PROCESSES IN KAON CONDENSATION AND DYNAMICAL EVOLUTION OF NEUTRON STARS

T. MUTO[a], T. TANAKA, H. FUJII, T. TATSUMI
Department of Physics, Kyoto University, Sakyo-ku, Kyoto 606-01, Japan

N. IWAMOTO
Department of Physics and Astronomy, The University of Toledo, Toledo, Ohio 43606-3390, USA

We investigate a relaxation process in kaon condensate, where the process proceeds through the nonequilibrium weak reactions which include the kaon-induced Urca and the modified Urca reactions.

1. Introduction

Kaon condensation in dense hadronic matter has been investigated from viewpoints of both astrophysics and heavy-ion physics. [1] In the K^--condensed phase, net strangeness is produced through the weak processes, one of which is a nucleon pair bremsstrahlung process (K), $n + n \to n + p + K^-$. Because time scales for the weak reactions are longer than those for the strong and electromagnetic interactions and for the gravitational collapse($\sim 10^{-3}$sec), it takes some time to reach chemical equilibrium. Here we investigate the relaxation process in kaon condensation, and discuss how normal matter is converted to the K^--condensed phase through the nonequilibrium weak processes.

2. Formulation

The relevant weak reactions are the kaon-induced Urca reaction (KU), $n(p) \to n(p) + e^- + \bar{\nu}_e$, $n(p) + e^- \to n(p) + \nu_e$,[1] and the modified Urca process (MU), $n + n \to n + p + e^- + \bar{\nu}_e$, $n + p + e^- \to n + n + \nu_e$.[2] Dynamical process is controled by the rate equations, which give temporal changes of the number densities of the chemical components. E.g., for the electrons, we get

$$\begin{aligned} dn_e(t)/dt &= \Gamma^{(\mathrm{KU})}(\xi^{(\mathrm{KU})}(t)) - \Gamma^{(\mathrm{KU})}(-\xi^{(\mathrm{KU})}(t)) \\ &+ \Gamma^{(\mathrm{MU})}(\xi^{(\mathrm{MU})}(t)) - \Gamma^{(\mathrm{MU})}(-\xi^{(\mathrm{MU})}(t)) , \end{aligned} \tag{1}$$

where $\Gamma^{(\alpha)}(u)$ and $\Gamma^{(\alpha)}(-u)$ (for α=KU,MU) are the forward and backward reaction rates, respectively, and $\xi^{(\mathrm{KU})}(t) \equiv (\mu_e(t) - \mu_K(t))/T$ and $\xi^{(\mathrm{MU})}(t) \equiv (\mu_p(t) + \mu_e(t) - \mu_n(t))/T$ are the chemical potential differences. Together with Eq.(1), the chiral angle $\theta(t)$, which is an order parameter of the condensate, may change adiabatically from zero to the equilibrium value θ^{eq}.

---

[a]Present address: Department of Physics, Chiba Institute of Technology, Shibazono 2-1-1, Narashino-shi, Chiba 275.

3. Numerical Results and Concluding Remarks

In Fig. 1, we show the time dependence of $\theta(t)$ for the baryon number density, $n_B=3n_0$ with the standard nuclear density $n_0=0.17$ fm^{-3}. The temperature is taken to be $T = 1\times10^{10}$ K, and the KN sigma term, $\Sigma_{KN}=400$ MeV. The reaction rates for the processes (KU) and (MU) are also shown in Fig,2, by dotted lines and dashed lines, respectively.

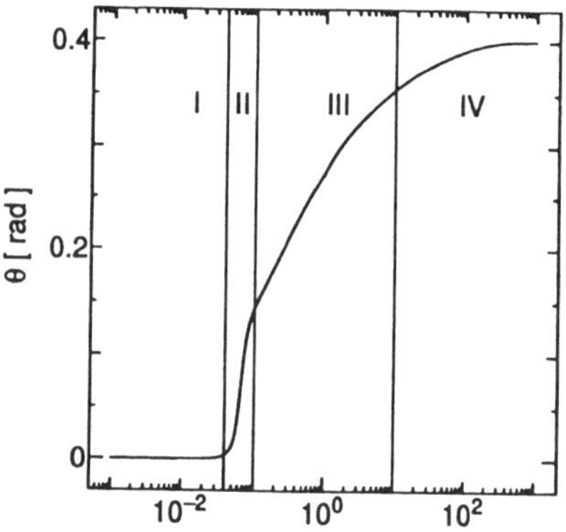

Figure 1

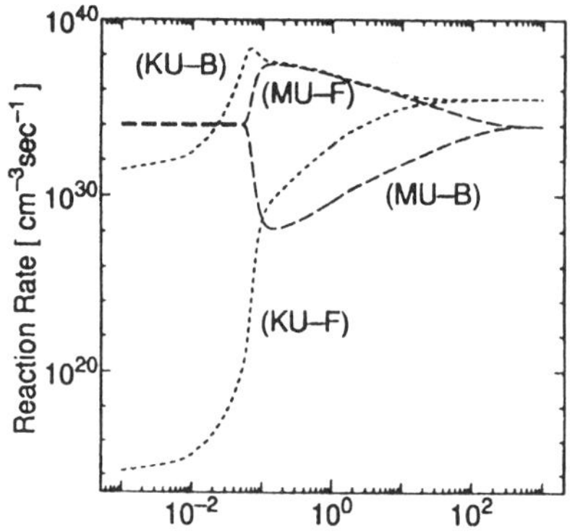

Figure 2

One can see that the dynamical evolution of the K^--condensate in the kinetic regime is divided into the following characteristic stages: (I) A small seed of a condensate grows very slowly to a non-negligible value during a characteristic time, τ_1, due to a small reaction rate for the process (KU). (II) A rapid rise of the condensate, where the backward process for (KU)(denoted by (KU-B)) dominates in the relaxation process. (III) Competition of the backward process for (KU) and the forward process for (MU) (denoted by (MU-F)), where the change in the electron number density due to the reaction (KU) induces the reaction (MU), leading to a monotonous increase in the proton number density. Hereafter, the system gradually approaches chemical equilibrium, and (IV) the asymptotic region, where the system is near equilibrium.

The dynamical behavior of the chiral angle $\theta(t)$ is sensitive to the baryon number density. For example, the "stagnation" time τ_1 is several tens of 10^{-3}sec for $n_B=3n_0$, while, it is shorter by five order of magnitude for $n_B=4n_0$.

The temporal behavior of the K^--condensed equation of state (EOS) can also be obtained. It is shown that the EOS in chemical equilibrium is settled by the saturation of the condensate (the chiral angle θ).

References

1. For a review, T. Tatsumi, Prog. Theor. Phys. Suppl. **120** (1995) 111.
H. Fujii, T. Maruyama, T. Muto and T. Tatsumi, Nucl.Phys.**A597** (1996) 645.
2. B. L. Friman and O. V. Maxwell, Astrophys. J. **232** (1979) 541.

QCD Phase Transition at high Temperature in Cosmology

H. Suganuma[a], H. Ichie[a], H. Monden[a], S. Sasaki[a], M. Orito[b], T. Yamamoto[b] and T. Kajino[c]

a) Research Center for Nuclear Physics (RCNP), Osaka University, Osaka 567, Japan

b) Dept. of Astronomical Science, The Graduate Univ. for Advanced Studies, Tokyo 181, Japan

c) Division of Theoretical Astrophysics, National Astronomical Observatory (NAO), and The Graduate Univ. for Advanced Studies; Dept. of Astronomy, Tokyo Univ., Tokyo 181, Japan

The cosmological QCD phase transition is studied in terms of the color confinement at finite temperature using the dual Higgs theory of QCD. The confinement force is largely reduced at high temperature, which leads to the swelling of hadrons. We derive analytical formulae for the surface tension and the boundary thickness of the mixed phase from the effective potential at T_c. We predict a large reduction of the glueball mass near T_c. We investigate also the process of the hadron-bubble formation in the early Universe.

1. QCD, Color Confinement and Dual Higgs Mechanism

Quantum chromodynamics (QCD) is the fundamental theory of the strong interaction. In spite of the simple form of the QCD lagrangian, it miraculously provides quite various phenomena like color confinement, dynamical chiral-symmetry breaking (DχSB), non-trivial topologies, quantum anomalies and so on. Therefore, QCD can be regarded as *an interesting miniature of the history of the Universe*, where a quite simple Big Bang also created various things including galaxies, stars, lives and thinking reeds. This is the most attractive point of the QCD physics.

As a modern progress in QCD, confinement physics is providing an important current of the hadron physics in '90s, since recent lattice QCD studies shed a light on the confinement mechanism. In the 't Hooft abelian gauge [1], QCD is reduced into an abelian gauge theory with the QCD-monopole, which appears from a hedgehog configuration corresponding to the non-trivial homotopy group $\pi_2(\mathrm{SU}(N_c)/\mathrm{U}(1)^{N_c-1}) = Z_\infty^{N_c-1}$ on the nonabelian manifold. In this gauge, the nonperturbative QCD (NP-QCD) vacuum is described as the dual Higgs phase with QCD-monopole condensation. Due to the dual Meissner effect, color-electric flux is squeezed like a one-dimensional flux-tube, which leads to the linear quark confinement potential [1]. Thus, the origin of color confinement can be recognized as the dual Higgs mechanism by monopole condensation.

Many recent studies based on the lattice QCD [4] show QCD-monopole condensation in the confinement phase and abelian (monopole) dominance for NP-QCD, e.g., linear confinement potential, DχSB and instantons [4]. Abelian (monopole) dominance means that the essence of NP-QCD is described only by abelian (monopole) variables in the abelian gauge [4]. Therefore the condensed monopole in the 't Hooft abelian gauge is nothing but the *relevant collective mode for NP-QCD*, and the NP-QCD vacuum can be identified as the dual-superconductor in a realistic sense.

As a remarkable fact in the duality physics, these are *two "see-saw relations" between the electric and magnetic sectors.*

(1) Due to the Dirac condition $eg = 4\pi$ [1] in QCD between the electric charge e and the magnetic charge g, a strong-coupling system in one sector corresponds to *a weak-coupling system in the other sector.*

(2) The long-range confinement system corresponds to *a short-range interaction system in the other sector.*

As the most attractive point in the dual Higgs theory, the highly-nonlocal confinement system can be described by *a short-range interaction theory with the dual variables.*

2. QCD Phase Transition in Dual Ginzburg-Landau Theory

The dual Ginzburg-Landau (DGL) theory is the QCD effective theory based on the dual Higgs mechanism, and can be derived from the QCD lagrangian and the QCD nature (monopole condensation and abelian dominance). The DGL lagrangian [1-3] for the pure-gauge system is described with the dual gauge field B_μ and the QCD-monopole field χ,

$$\mathcal{L}_{\rm DGL} = {\rm tr}\left\{-\tfrac{1}{2}(\partial_\mu B_\nu - \partial_\nu B_\mu)^2 + [\hat{D}_\mu, \chi]^\dagger [\hat{D}^\mu, \chi] - \lambda(\chi^\dagger\chi - v^2)^2\right\},$$

where $\hat{D}_\mu \equiv \hat{\partial}_\mu + igB_\mu$ is the *dual covariant derivative.*

The dual gauge field $B_\mu = \vec{B}_\mu \cdot \vec{H} = B_\mu^3 T^3 + B_\mu^8 T^8$ is defined on the *dual gauge manifold* $\mathrm{U}(1)_m^3 \times \mathrm{U}(1)_m^8$ [1-3], which is the dual space of the maximal torus subgroup $\mathrm{U}(1)_e^3 \times \mathrm{U}(1)_e^8$ embedded in the original gauge group $\mathrm{SU}(3)_c$. The abelian field strength tensor is written as $F_{\mu\nu} = {}^*(\partial \wedge B)_{\mu\nu}$, so that the role of the electric and magnetic fields are interchanged in comparison with the ordinary A_μ description.

The QCD-monopole field χ is defined as $\chi \equiv \sqrt{2}\sum_{\alpha=1}^{3}\chi_\alpha E_\alpha$ [3]. In the QCD-monopole condensed vacuum with $|\chi_\alpha| = v$, the dual gauge symmetry $\mathrm{U}(1)_m^3 \times \mathrm{U}(1)_m^8$ is spontaneously broken. Through the dual Higgs mechanism, the dual gauge field B_μ acquires its mass $m_B = \sqrt{3}gv$, whose inverse provides the hadron-flux-tube radius. The dual Meissner effect causes the color-electric field excluded from the QCD vacuum, which leads to color confinement. The QCD-monopole fluctuations $\tilde{\chi}_\alpha \equiv \chi_\alpha - v$ (α=1,2,3) also acquire their mass $m_\chi = 2\sqrt{\lambda}v$ in the QCD-monopole condensed vacuum. As a relevant prediction, one QCD-monopole fluctuation $\tilde{\chi} \equiv \sum_{\alpha=1}^{3} \tilde{\chi}_\alpha$ appears as a color-singlet scalar glueball in the confinement phase, although the dual gauge field B_μ and the other two combinations of the QCD-monopole fluctuation are not color-singlet and cannot be observed.

In the DGL theory, the QCD phase transition is characterized by QCD-monopole condensate $\bar{\chi} \equiv |\chi_\alpha|$, which is an order parameter on the confinement strength. For the study of the QCD phase transition, we formulate the effective potential $V_{\rm eff}(\bar{\chi};T)$ at finite temperature as the function of $\bar{\chi}$ using the *quadratic source term* to avoid the imaginary scalar-mass problem [1-3].

We study the relation between $V_{\rm eff}(\bar{\chi};T_c)$ and the surface tension σ [1,3], which characterizes the strength of the first order in the phase transition, and is very important for the shape of the boundary surface in the mixed phase. Using the *sine-Gordon (SG) kink ansatz* [1,3] for the boundary profile, $\bar{\chi}(z) = \bar{\chi}_c \tan^{-1} e^{z/\delta}$, we derive simple formulae for the surface tension $\sigma \simeq \frac{4\sqrt{3}}{\pi}\sqrt{h}\bar{\chi}_c \simeq (112{\rm MeV})^3$, and the phase-boundary thickness $2\delta \simeq \frac{2\sqrt{3}}{\pi}\bar{\chi}_c/\sqrt{h} \simeq 3.4$fm, where h is the "barrier height" between the two minima in $V_{\rm eff}(\bar{\chi};T_c)$.

At high temperature, the hadron string tension $k(T)$ becomes smaller and drops rapidly near T_c [2,3]. Therefore, the slope of the inter-quark potential is reduced and the inter-quark distance inside hadrons increases at high temperature. In addition, the color-electric field spreads according to the decrease of m_B. Thus, *the reduction of the confinement force leads to the swelling of hadrons at high temperature* [3].

We predict also *a large mass reduction on the QCD-monopole (a scalar glueball) near T_c. The violent excitation of QCD-monopoles (scalar glueballs) with a reduced mass would promote the QCD phase transition.*

3. Hadron Bubble Formation in the Early Universe

Finally, we study the hadron bubble formation [3,5] in the early Universe using the DGL theory [3]. As Witten pointed out, if the QCD phase transition is of the first order, the hadron and QGP phases should coexist in the early Universe. During the mixed-phase period, there appears the inhomogeneity on the baryon density distribution [5], which can strongly affects the primordial nucleo-synthesis.

The hadron bubble created in the supercooling QGP phase is described by the SG-kink type profile of the QCD-monopole condensate [3], $\bar{\chi}(r;R) = C \tan^{-1} e^{(R-r)/\delta}$, where R and 2δ correspond to the hadron-bubble radius and the phase-boundary thickness, respectively. The total energy of the hadron bubble with radius R can be estimated using $V_{\rm eff}(\bar{\chi};T)$, $E(R;T) = 4\pi \int_0^\infty dr r^2 \{3(\frac{d\bar{\chi}(r;R)}{dr})^2 + V_{\rm eff}(\bar{\chi};T)\}$, which is roughly estimated as the sum of the positive surface term and the negative volume term [3]. The hadron-bubble energy $E(R;T)$ takes a maximal value at a critical radius R_c. Hence, the hadron bubbles with $R < R_c$ collapse, and only large hadron bubbles with $R > R_c$ grow up with radiating the shock wave [3,5]. On the other hand, the creation probability of large bubbles is strongly suppressed [3] because of the thermodynamical factor $P(T) \equiv \exp\{-\frac{4\pi}{3}R_c(T)^3 h(T)/T\}$, with a large barrier height $h(T)$ in $V_{\rm eff}(\bar{\chi};T)$. Thus, the only small bubbles are created practically, although its radius should be larger than R_c [3].

As the temperature decreases, the smaller hadron bubbles are created, while the bubble formation rate becomes larger [3]. Then, we can imagine how the QCD phase transition happens in the Big Bang scenario: (a) Slightly below T_c, only large hadron bubbles appear, but the creation rate is quite small. (b) As temperature is lowered by the expansion of the Universe, smaller bubbles are created with much formation rate. During this process, the created hadron bubbles expand with radiating shock wave, which reheats the QGP phase around them. (c) Near $T_{\rm low}$, many small hadron bubbles are violently created in the unaffected region free from the shock wave. (d) The QGP phase pressured by the hadron phase is isolated as high-density QGP bubbles [3], which provide the baryon density fluctuation [5].
Thus, the numerical simulation using the DGL theory would tell how the hadron bubbles appear and evolve quantitatively in the early Universe.

References

1. H. Suganuma, S. Sasaki and H. Toki, Nucl. Phys. **B435** (1995) 207.
   H. Suganuma, S. Sasaki, H. Toki and H. Ichie, Prog.Theor.Phys.(Suppl.) **120** (1995) 57.
2. H. Ichie, H. Suganuma and H. Toki, Phys. Rev. **D52** (1995) 2944.
3. H. Ichie, H. Monden, H. Suganuma and H. Toki, Proc. of *Nuclear Reaction Dynamics of Nucleon-Hadron Many Body Dynamics*, Osaka, Dec. 1995, in press.
4. H. Suganuma, A. Tanaka, S. Sasaki & O. Miyamura, Nucl. Phys. **B**(Suppl.)**47** (1996) 302.
5. T. Kajino, M. Orito, T. Yamamoto and H. Suganuma, *Confinement '95*, (World Scientific, 1995) 263.

COSMOLOGICALLY SCREWY LIGHT

BORGE NODLAND[a] and JOHN P. RALSTON
Department of Physics & Astronomy, and Institute for Theoretical and Computational Science, University of Kansas, Lawrence, KS 66044

We present evidence for a new phenomenon in the propagation of electromagnetic radiation across the Universe, a corkscrew rotation of the plane of polarization not accounted for by conventional physics.

Electromagnetic radiation propagating on cosmological distances provides an exceedingly sensitive laboratory for new phenomena. Here we summarize observation of an unexpected systematic rotation of the plane of polarization of radio signals from galaxies with redshift $z > 0.3$. We find that the data are explainable two ways: either invoking unnatural conspiracies among the sources, or proposing new physics.

Radio polarization is produced by synchrotron radiation. However the observed plane of polarization from distant sources does not usually align with the symmetry axis of the source (denoted ψ). For decades this has been studied in terms of Faraday rotation in the intervening medium. We study a residual quantity remaining after the Faraday effect is taken out.

Extracting Faraday rotation does not depend on models, because the Faraday angle of rotation for wavelength λ goes like $\alpha\lambda^2$. Consistent linear dependence on λ^2 is indeed observed. The problem is that the data fit requires more: for each source (i), the fits are given by $\theta_i(\lambda) = \alpha_i\lambda^2 + \chi_i$. The residual polarization angle χ_i does not generally align with the galaxy major axis; statistics on the differences $\chi_i - \psi_i \pm \pi$ have puzzled astronomers for 30 years.[1]

Analysis of the data is challenging, due to several complications. The data set is highly non-uniform in its angular distribution on the sky and in the distances $r(z)$ to the sources. We employed Monte Carlo methods to search for correlations.[2]We make thousands of fake data sets with random polarization and major axis orientations, while keeping the positions of the galaxies the same as the real data. We then calculate the probabilities of linear correlations observed in the data relative to the random sets.

We find[2] an amount of residual rotation beta described by a dipole rule $\beta = r/\Lambda_s\ cos(\gamma)$, where γ is the angle between the propagating wavevector and an axis $\vec{s}$ fit to the data. The dependence on the propagation direction is quite novel, indicating anisotropy. Making a cut on $z > 0.3$, which selects

[a]Present address: Rochester Theory Center for Optical Science and Engineering, University of Rochester, Rochester, NY 14627.

the most-distant half of the data set, we find a striking correlation with probabilities that the observed correlation would be produced by random angular fluctuations less than 10^{-3}; the effect is 3.7σ. A separate study eliminates bias from fitting the s-direction to the data, by fitting the best $\vec{s}$-direction data-set-by-data- set in the Monte Carlo, and calculating the probability to finding the data's correlation relative to the optimized sets so constructed. This gives a probability less than 0.006, corresponding to 2.7σ. The fits to the parameters are $\Lambda_s = 1.1 \times 10^{25} h_0/h\ m$, where h_0/h is the ratio of 100 kms^{-1} Mpc^{-1}to the Hubble constant, while $\vec{s} = (0^0 \pm 20^0$ decination, 21 ± 2 hrs. right ascension). The scale Λ_s is 1/10 the Universe's size; the effect is far too small to affect laboratory measurements. We do not find a significant correlation for $z < 0.2$; in our full data set we also do not find a significant correlation of $\beta = const\ \ r$.

We are not the first to find puzzling statistics in the residual angle[1]; the problem is that our correlation cannot be reconciled with conventional ideas. Explaining the effect via population dependence among the sources requires an unnatural conspiracy across the sky. The role of systematic errors, with the possibility that the angular correlation might arise in a local effect, is crucial. The observers deny this possibility; more importantly, the proposal is ruled out by the fact that the near half of the data does not have the correlation. New physics explanations are available; perhaps the most conventional proposal would be domain walls of an axion–like condensate. The prediction of exactly the dipole rule in a different model,[2] and the search for a limit, led to the present investigation.

Acknowledgments

This research supported by DOE Grant Number DE FG02 85 40214 and the Kansas Institute for Theoretical and Computational Science.

References

1. F.F. Gardner and J.B. Whiteoak, *Nature* **197**, 1161 (1963); *Ann. Rev. Astr. Astrophys.* **4**, 245 (1966); J.N. Clarke *et al.*, *Mon. Not. R. Astron. Soc.* **190**, 205 (1980); P. Birch, *Nature* **298**, 451 (1982); D. Kendall and G.A. Young, *Mon. Not. R. Astron. Soc.* **207**, 637 (1984); M. Bietenholz and P. Kronberg *Astrphys. J.* **L1**, 287 (184).
2. Borge Nodland and John P. Ralston, Kansas preprint (1995); J.P. Ralston, *Phys. Rev.* **D51**, 20181 (1995).

Session 6
Hadron Spectroscopy

Observation of $f_0(1500) \to (\pi\pi)_S(\pi\pi)_S$ in $\bar{p}p$ annihilation at rest

Curtis A. Meyer, Carnegie Mellon University
The Crystal Barrel Collaboration[1]

We report here on the first observation by the Crystal Barrel of the four-pion decay of the glueball candidate $f_0(1500)$. The Crystal Barrel detector has been described elsewhere. [2] We only mention that it is a nearly 4π solid angle device for detecting both charged particles and photons with good spatial and energy resolution. The detector is used at the Low Energy Antiproton Ring at CERN to observe $\bar{p}p$ annihilations at rest. The $f_0(1500)$ has been observed in several *glue-rich* production mechanisms [3], and its decay to two pseudoscalar mesons has been studied in detail by us. [4] The unusual decay pattern into two pseudoscalar mesons has led to an interpretation of the $f_0(1500)$ as the ground state scalar glueball mixed into the normal scalar nonet. [5]

We observe $f_0(1500)$ in $\bar{p}p \to f_0(1500)\pi^\circ \to [\pi^\circ\pi^\circ\pi^\circ\pi^\circ]\pi^\circ$ [8]. The data for this 10γ final state come from 18×10^6 $\bar{p}p$ annihilations to final states with zero charged particles. The data are kinematically fit to a $5\pi^\circ$ final state, while events satisfying other 10γ hypothesis such as $\eta 4\pi^\circ$ are rejected. Additionally, events with more than one acceptable $5\pi^\circ$ combinations are rejected. Finally, events satisfying $\eta\pi^\circ\pi^\circ$ with $\eta \to \pi^\circ\pi^\circ\pi^\circ$ are removed. From the resulting large (on order 100,000) data sample, 25,000 events are used in the spin-parity analysis. 45,000 Monte Carlo events satisfying the same cuts are used for normalization.

A spin parity analysis is performed in the framework of the isobar model. An unbinned maximum likelyhood fit is done over the 8-dimensional phase space to determine what intermediate states are present. In performing the analysis, it is assumed that S-wave initial states dominate the annihilations. Additionally, C-parity prevents the $^3S_1$ initial state of the $\bar{p}p$ atom from decaying to $5\pi^\circ$. Only initial $^1S_0$ states are considered. From known branching ratios, the annihilation to $5\pi^\circ$ via $f_2(1270)\pi^\circ$ and the $a_1(1260)(\pi^\circ\pi^\circ)$ should play a very small role. This is confirmed by the fit, and they are neglected. Additionally, the $\rho(770)$ does not decay to $2\pi^\circ$.

This leaves as input several spin zero mesons. in the following decays We consider up to three f_0 states: $f_0(1300)$, $f_0(1500)$ and $f_0(1750)$. (1) $f_0 \to (\pi\pi)_S(\pi\pi)_S$ where $(\pi\pi)_S$ is the isopin zero, $J^{PC} = 0^{++}$ two pion scattering amplitude. [6] (2) $f_0 \to \pi^*(1300)\pi$, $\pi^* \to (\pi\pi)_S\pi$, where π^* is the $\pi(1300)$ of the PDG. [7] (3) $\bar{p}p \to \pi^*(1300)(\pi\pi)_S$. As all spins and angular momenta are 0, we anticipate having difficulty distinguishing a broad $f_0(1300)$ from (1) and the π^* decay from (3). This is indeed what the fit finds.

The one very stable component of the fit is the $f_0(1500) \to 4\pi^\circ$. The mass and width agree quite well with measured values. Independent of the admixture of (1), (2), and (3), the $f_0(1500)$ represents $(12.7 \pm 1.6)\%$ of the $5\pi^\circ$ final state. Combining this with other Crystal Barrel analyses, we find:

$$r = \frac{B(f_0(1500) \to 4\pi)}{B(f_0(1500) \to 2\pi)} = 3.4 \pm 0.8,$$

where 4π means decays not proceeding via $\rho\rho$. The fit is unable to unabigiously separate decays (1) and (2) above; both are present but the exact admixture can not be determined. Additionally, we find two equally good solutions for the remainder of the decay channels. These will not be discussed here.Several projections of the data compared to the best fit are shown in Fig. 1. One can see that the fit does an excellent job of describing the data.

In order to understand this final state and resolve the ambiguities of this analysis, a new analysis of $\pi^+\pi^-3\pi^\circ$ [9], and an analysis of $\bar{p}d \to \pi^+2\pi^-2\pi^\circ p$ are being performed. The $\pi^+\pi^-3\pi^\circ$ data currently confirm the 4π decay of the $f_0(1500)$.

References

[1] A. Abele, B. Adomeit, C. Amsler, D.S. Armstrong, C.A. Baker, B.M. Barnett, C.J. Batty, M. Benayoun, A. Berdoz, K. Beuchert, P. Blüm, K. Braune, J. Brose, D.V. Bugg, T. Case, A. Cooper, O. Cramer, K.M. Crowe, T. Degener, H.P. Dietz, N. Djaoshvili, S. v. Dombrowski,

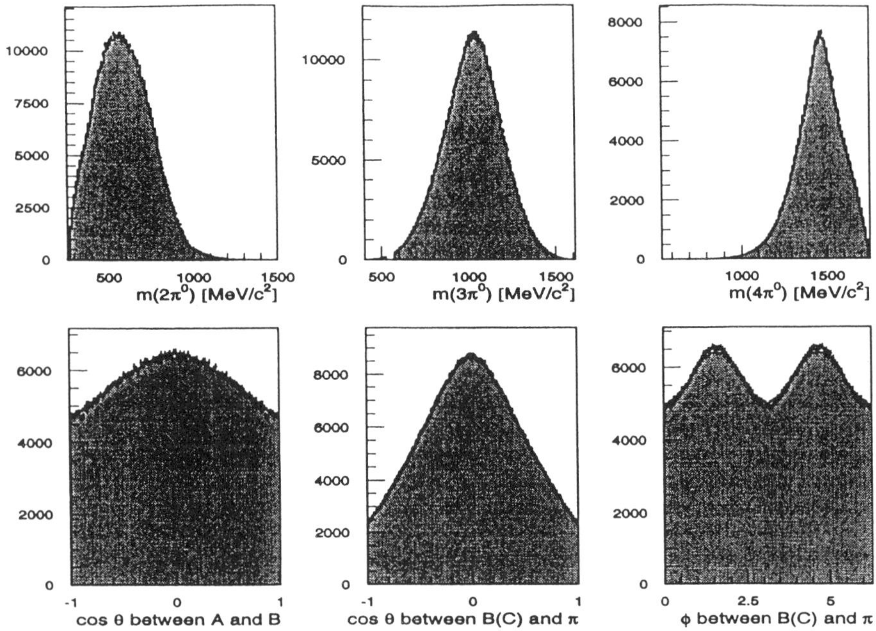

Figure 1: Best fit compared to invariant mass and angular distributions.

M. Doser, W. Dünnweber, D. Engelhardt, M.A. Faessler, P. Giarritta, R. Hackmann, R.P. Haddock, F.H. Heinsius, M. Herz, N.P. Hessey, P. Hidas, D. Jamnik, H. Kalinowsky, B. Kalteyer, B. Kämmle, P. Kammel, T. Kiel, J. Kisiel, E. Klempt, H. Koch, C. Kolo, M. Kunze, M. Lakata, R. Landua, J. Lüdemann, H. Matthaey, R. McCrady, J.Meier, C.A. Meyer, L. Montanet, A. Noble, R. Ouared, F. Ould-Saada, K. Peters, C.N. Pinder, G. Pinter, C. Regenfus, J. Reißmann, S. Resag, W. Roethel, P. Schmidt, I. Scott, R. Seibert, S. Spanier, H. Stöck, C. Straßburger, U. Strohbusch, M. Suffert, U. Thoma, M. Tischhaeuser, D. Urner, C. Völcker, F. Walter, D. Walther, U. Wiedner, B. Zou, Č. Zupančič.

[2] E. Aker *et al.*, Nucl. Instrum. Methods **A321**, 69, (1992).

[3] D. V. Bugg, I. Scott, B. S. Zou, V. V. Anisovich, A. V. Sarantsev, T. H. Burnett, S. Sutlief, Phys. Lett. **B353**, 378, (1995). F. Antinori *et al.*, Phys. Lett. **B353**, 589, (1995).

[4] C. Amsler, *et al.*, Phys. Lett. **B355**, 425, (1995). Also see S. Spanier, in these proceedings.

[5] C. Amsler, F. E. Close, Phys. Lett. **B353**, 385, (1995). C. Amsler, F. E. Close, Phys. Rev. **D53**, 295, (1996).

[6] K. L. Au, D. Morgan and M. R. Pennington, Phys. Rev. **D35**, 1633, (1987).

[7] K. Hikasa, *et al.*, Phys. Rev. **D50 II**, (1994).

[8] A. Abele, *et al.*, A Study of $f_0(1500)$ ecays into $4\pi^0$ in $\bar{p}p \to 5\pi^0$ at Rest. Accepted for publication in Phys. Lett. B., (1996).

[9] C. Amsler, *et al.*, Phys. Lett. **B322**, 431, (1994).

Observation of $f_0(1500)$ decay into $K_L K_L$

Stefan Spanier
University of Zuerich, Switzerland and CERN-PPE

for the Crystal Barrel Collaboration [a]

Today the $f_0(1500)$ is considered as a good candidate for the glueball ground-state [1]. The Crystal Barrel Collaboration studies the decays of this object and scalar mesons in proton-antiproton annihilation. An essential information to dis-antangle the nature of mesonic resonances appearing in the final state interaction of annihilation products is the coupling to kaons. Here the analysis of the final state $K_L K_L \pi^0$ is presented which contributes to that knowledge [2].

The Crystal Barrel collaboration performs light meson spectroscopy using proton-antiproton annihilation. This production mechanism is expected to deliver conventional $\bar{q}q$ mesons as well as exotic hadrons like hybrids and glueballs. The latter are bound states of gluons only as predicted by QCD allowing the self-interaction of gluons. The understanding of scalar mesons is based on an extensive analysis of different decay modes. The $f_0(1500)$ is seen in its decay into $\pi^0\pi^0$, $\eta\eta$, $\eta\eta'$, coupled described in [3], and $4\pi^0$ [4]. The $K_L K_L \pi^0$ final state will deliver first results of $\bar{K}K$ couplings of I=0 and I=1 resonances.

The momentum and energy of the π^0 and the direction of the interacting K_L are sufficient to reconstruct the kinematics of an event. Starting from 13.2 Mio all events which were triggered for the absence of charged paticles a sample of 37400 events could be identified as $K_L K_L \pi^0$ candidates and used for the partial wave analysis. The trigger was based on the two layers of multiwire proportional chambers of the Crystal Barrel experiment covering 99% and 97% of solid angle 4π, respectively, and surrounding the liquid hydrogen target. The two photons of the π^0 decay and the hadronic interaction of one of the K_L were detected in the CsI(Tl) calorimeter which consists of 1380 individual crystals each covering 6° in polar and 6° in azimuthal angle (except for the three sets of layers closest to the beam axis where the acceptance in azimuthal angle is increased to 12°) and all together 95% of 4π solid angle. One K_L was considered missing. For a momentum of 400 MeV/c 96% of all K_L reach the calorimeter and interact hadronically with a probability of about 57%. A cluster of crystals with energy deposit pattern similar to a photon is initiated in 28% of all cases. In general the K_L does not deposit all its energy leaving its direction as infomation to calculate the kinematics of the event. The π^0 can be fully reconstructed. The 2-photon invariant mass spectrum for events with 3

[a] our WWW pages: http://afuz01.cern.ch/cb.html

distinct electromagnetic showers in the crystals and no charged track is shown in fig.1 (left) . The width of the π^0 amounts $\sigma = 7.5$ MeV. A mass window was set around the π^0 and background subtracted using its neighbouring regions which were kinematically extrapolated under the signal. The π^0 momentum was required to be smaller than its maximal value of 666 MeV/c. The determination of the decay angle of the interacting K_L with respect to the π^0 in the rest frame of $K_L K_L$ yields ambiguities in the region $m(K_L K_L) < 1125$ MeV. Therefore, the reconstruction efficiency close to the $\bar{K}K$ threshold is reduced but apart from that varies slowly over the whole phase space. Both $K_L\pi^0$ subsystems were corrected relatively for the K_L interaction probability in dependence on the K_L momentum. The residual background in the channel is $(3.4 \pm 0.5)\%$ and is distrubuted uniformly.

In the acceptance corrected Dalitz plot (fig. 1 (right)) the dominant resonance features are marked by arrows. The $K^*(892)$ shows an angular distribution $\sim cos^2\theta$ along its mass position and interferes strongly with other resonances. In the $\bar{K}K$ system one sees activity due to I=0 and I=1 scalar and tensor resonances. The data were analyzed in the helicity formalism [5] in term of the isobar model using K-matrix formalism in the P-vector approach [6]. The $K\pi$ S-wave was introduced by fitting the scattering amplitude, which was written in terms of the K-matrix, to the phase shift from LASS [7]. It turned out, that the parametrization of the scalar intensity in the $\bar{K}K$ system with only one Breit-Wigner resonance at about 1400 MeV was not satisfactory. A good description of the data could only be achieved by splitting it into two objects which were fit compatible in mass and width to $f_0(1370)$ and $f_0(1500)$. Since one also expects the I=1 resonance $a_0(1450)$, which cannot definitely distinguished from the I=0 scalars, it was introduced additionally with an intensity of 7.7% of the final state as expected from the $\pi^0\pi^0\eta$ analysis [8] and SU(3). This did not affect the quality of the fit and the parameters of the $f_0(1500)$ (the $f_0(1370)$ mass moved slightly downwards).
The $f_2'(1525)$ is observed for the first time in $\bar{p}p$-annihilation at rest. Assuming no strange quark in the nucleon and quark line rule one obtains the tensor mixing angle Θ_2 from the $f_2(1270)$ and $f_2'(1525)$ production rates: $\Theta_2 = (25.3 \pm 1.1)^o$ (the linear mass formula yields 26°).

With the previous assumption for $a_0(1450)$ one obtains the following relative ratios (after correction for phase space):
$f_0(1370)$: $\bar{K}K/\eta\eta = 0.85 \pm 0.31$
$f_0(1500)$: $\bar{K}K/\eta\eta = 0.76 \pm 0.27$
Independent ratios will arise from the study of the $K^0K^{\pm}\pi^{\mp}$ final state which is under investigation. The ratios make $f_0(1370)$ and $f_0(1500)$ rather similar and both more likely being the $\bar{u}u + \bar{d}d$ member of the SU(3) nonet. The

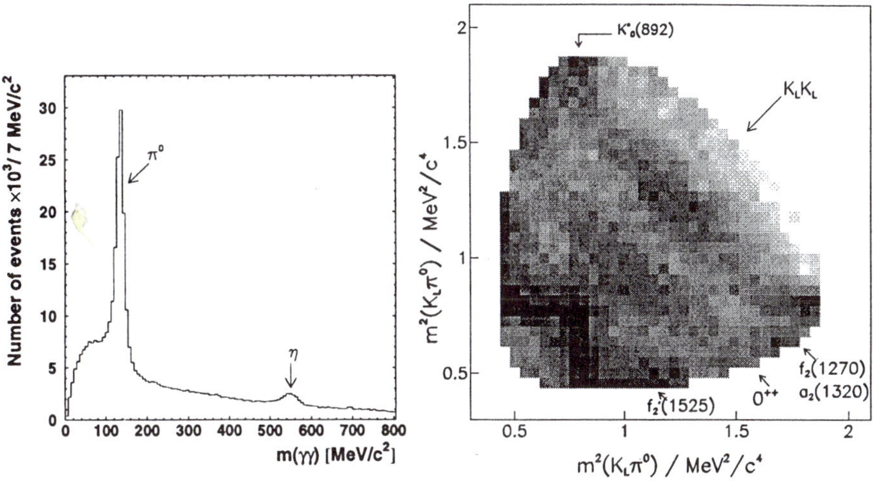

Figure 1: left: The $\gamma\gamma$ invariant mass. right: The $K_L K_L \pi^0$ Dalitz plot.

$\bar{K}K/\pi\pi$ ratios of both resonances are also close and support that hypothesis. Its small width of about 100 MeV classifies the $f_0(1500)$ as an outsider compared to the other scalar states and its mass matches the scale calculated for glueballs on the lattice [9].

References

1. C.Amsler and F.E.Close, *Phys. Rev.* D **53**, 295 (1996).
2. A.Abele et al. (Crystal Barrel), submitted to *Phys. Lett.* B, Ph.D. thesis of Sven v. Dombrowski, University of Zuerich, 1996.
3. C.Amsler et al. (Crystal Barrel), *Phys. Lett.* B **355**, 425 (1995).
4. A.Abele et al. (Crystal Barrel), submitted to *Phys. Lett.* B.
5. C.Amsler and J.C.Bizot, Comp.Phys.Comm. **30**, 21 (1983).
6. S.U.Chung et al., Ann.Phys. **4**, 404 (1995).
7. D.Aston et al., *Nucl. Phys.* B **296**, 491 (1988).
8. C.Amsler et al. (Crystal Barrel), *Phys. Lett.* B **333**, 277 (1994).
9. Lattice'94 conference, *Nucl. Phys.* B **42**, 1 (1995).

THE CRYSTAL BARREL PROGRAM ON $\bar{p}N$ ANNIHILATION INTO KAONIC FINAL STATES USING A NEW K_S TRIGGER

THE CRYSTAL BARREL COLLABORATION
represented by Peter Kammel
University of California and LBNL, Berkeley, CA 94720, USA

The Crystal Barrel experiment at LEAR has been upgraded with a new K_S trigger to obtain high statistics/low background data on $\bar{p}N$ annihilation into kaonic final states. The primary goal of this program is the comprehensive investigation of the $K\bar{K}$ and $K\pi$ interaction and the interpretation of new scalar states, observed recently at Crystal Barrel. Data taking at rest in liquid hydrogen and deuterium has been completed successfully. Further plans include studies of $\bar{p}N$ annihilation at rest in gaseous hydrogen and in flight in liquid hydrogen.

The basic goal of the new K_S program with the Crystal Barrel (CB) detector is the systematic investigation of the $K\bar{K}$ and $K\pi$ interaction in $\bar{p}N$ annihilation at a precision comparable to our recent results on the $\pi\pi$, $\eta\eta$ and $\pi\eta$ channels. In these experiments three new scalar resonances $f_0(1370)$, $a_0(1450)$ and $f_0(1500)$ have been observed [1]. This suggests a major reclassification of the scalar nonet. Moreover, the $f_0(1500)$ is considered a promising candidate for the ground state glueball which is predicted in the 1.5–1.8 GeV region by lattice QCD. Precision information on the $K\bar{K}$ coupling of scalar mesons is urgently needed to clarify the flavor-gluon content of these resonances and to further constrain their interpretation. Specific questions of high priority are the nature of $a_0/f_0(980)$, the coupling of the new states to $K\bar{K}$, the search for the expected $s\bar{s}$ scalar ground state and the investigation of the structure of $f_J(1710)$. Key information to these topics will come from the 3-body final states $\bar{p}p \to K_S K_S \pi^0, K_S K^{\pm}\pi^{\mp}$ and $\bar{p}n \to K_S K_S \pi^-, K_S K^- \pi^0$.

Other interesting physics topics will be investigated, like 3-particle resonances in $K\bar{K}\pi\pi$ and $K\bar{K}\pi\pi\pi$ final states as well as hyperon production in deuterium by Pontecorvo reactions.

At present experimental information on $\bar{p}N$ annihilations into kaonic final states is very limited. We plan to increase the world statistics on final states including K_S by one to two orders of magnitude. For this purpose the CB detector [2] was upgraded with a central vertex detector (SVX). The K_S trigger is based on the decay $K_S \to \pi^+\pi^-$ outside the SVX which leads to a charged multiplicity increase between the SVX and the surrounding jet drift chamber (JDC).

The SVX consists of 15 individual detector elements (size $74mm \times 8.4mm \times 0.37mm$) located close to the target at an average radius of 12.5 mm to reduce the trigger bias. For every element, both the 128 strips ($50\mu m$ pitch) as well as the common backplane are read out. Both signals are included in the hierarchical CB trigger. At the first stage allowed combinations of the SVX backplane multiplicity and the JDC multiplicity are selected by programmable logic. Further event selection in the next stage is based on the strip information. Hardware improvements (new target design and beam counters) and pile-up rejection were necessary to suppress the dominant pionic annihilation channels by typically a factor of 250.

Thus far, data taking at rest in LH_2 and LD_2 with 10.5×10^6 and 6.8×10^6 triggered events, respectively, has been completed successfully. The new trigger system performed extremely well. Up to 3 triggers were run simultaneously to select different event topologies. Depending on the event class one or more K_S can be reconstructed in $\sim$ 5–18% of the triggered events. The resulting K_S mass peaks have good resolution and very low background. For the main $K\bar{K}\pi$ final states (2–5)$\times 10^4$ events are expected in the final Dalitz Plots, compared to 500–3000 events in existing bubble chamber data. The data is presently being analyzed at Berkeley and other collaborating institutions.

Further planned experiments in 1996 include $\bar{p}p$ annihilation at rest in a 14 bar gas target to enhance P-wave annihilation as well as measurements in LH_2 with $\bar{p}$ momenta of 600-900 MeV/c to extend the available phase space.

Combining this data with our complementary efforts[3] on kaonic channels with K_L we will perform a comprehensive analysis on the $K\bar{K}$ S-wave and its implication on the scalar meson sector.

Acknowledgments

This work is supported by the British Particle Physics and Astronomy Research Council, the German Bundesministerium für Bildung, Wissenschaft, Forschung und Technologie, the U.S. Department of Energy and the National Science Research Fund Committee of Hungary.

References

1. C. Amsler et al., (CB–Collab.), *Phys. Lett.* B355 (1995) 425;
   C.A. Meyer et al., contribution 6–61 to this conference
2. E. Aker et al., (CB–Collab.), *Nucl. Instrum. Methods* A321 (1992) 69
3. S. Spanier et al., contribution 6–77 to this conference

GLUONS VERSUS PSEUDOSCALAR MESONS IN BARYON SPECTROSCOPY

Z. DZIEMBOWSKI
Department of Physics, Temple University, Philadelphia, PA 19122, USA

M. FABRE de la RIPELLE
Institut de Physique Nucleaire, Universite de Paris-Sud, 94106 Orsay, France

G.A. MILLER
Department of Physics, University of Washington, Seattle, WA 98195-1560

We include the effects of pseudoscalar meson exchange and one gluon exchange potentials between quarks in calculating baryon spectra.

The recent work of Glozman and Riska[1] has revitalized interest in the theory of baryon spectroscopy. These authors state that "the spectra of the nucleons, Δ resonances and the strange hyperons are well described by the constituent quark model, if in addition to the harmonic confinement potential the quarks are assumed to interact by exchange of the $SU(3)_F$ octet of pseudoscalar mesons". The idea of Glozman and Riska contradicts the long-standing belief that one-gluon exchange is a basic element of quantum chromodynamics QCD and the success of that interaction in baryon spectroscopy.

The present note describes our recent efforts [2] to include both effects in calculating the baryon spectra using a non-perturbative technique. We believe that both kinds of effects are required. Both are strong interactions so that a non-perturbative, all-orders treatment is needed to assess whether or not either of those two elements can be ignored. We employ the hyperspherical methods of Fabre de la Ripelle *et al.* [3] to compute the energies of the baryons.

Our constituent quark model Hamiltonian includes the effects of one gluon exchange (OGE) and the exchange of pseudoscalar mesons mandated by broken chiral symmetry, V_χ, in addition to the kinetic energy T and confinement terms. Thus $H = T + V_{con} + V_{OGE} + V_\chi$. We limit ourselves to light quarks in this first calculation, but work on including strange baryons is in progress.

The published calculations assume a linear confining interaction $V_{conf} = \sum_{i<j} A \mid \vec{r}_i - \vec{r}_j \mid$, with A to be determined phenomenologically. The one gluon exchange interaction between different quarks is given the standard form [4,5] with the replacement of the usual delta function form by a Yukawa of range r_0. This is intended to include the effects of the finite sized nature of the constituent quarks and exchanged bosons. The strong coupling constant α_s

is a parameter to be determined phenomenologically. We ignored the spin-orbit and tensor terms because our first calculation was intended as a broad comparison of the non-perturbative effects of gluon and meson exchange.

The effects of pseudoscalar meson octet exchange are described by the interaction $V_\chi = \sum_{i<j} \alpha_{q\pi} \frac{\vec{\sigma}_i \cdot \vec{\sigma}_j}{3} \frac{\vec{\lambda}_i^F \cdot \vec{\lambda}_j^F}{4\, m^2} [\mu^2 \frac{e^{-\mu\, r_{ij}}}{r_{ij}} - \frac{e^{-r_{ij}/r_0}}{r_0^2\, r_{ij}}]$, with the quark mass m taken as 336 MeV to represent the non-perturbative effects which influence the properties of a single confined quark. We allow the strength of the meson exchange potential, $\alpha_{q\pi}$, to vary away from the expected [1] value of 0.67. This is in the spirit of the work of Glozman and Riska [1] who fit a very few matrix elements of V_χ to a few mass differences and predict the remainder of the spectrum.

We calculate the masses of the ground state and first radial excitation for S, P and D waves in the nucleon and Δ channels and compare our predictions with all of these non-strange four and three star baryon resonances with masses below 1800 MeV found in the particle data tables.

We summarize the effects of one gluon and one pseudoscalar meson exchange interactions. If we include only the effects of one gluon exchange, we do not get a good description of the even and odd parity resonant states. This description is improved if we include only the effects of meson exchange. However, both gluonic and pseudoscalar meson exchange are expected from the underlying theory. Thus although we verify several of the statements of Glozman and Riska, a theory which includes both gluon and meson exchange seems somewhat more plausible. Indeed, our most general model is defined by using a value of $\alpha_{q\pi}$ equal to that provided by the pion-nucleon coupling constant and a value of $\alpha_s = 0.35$ about equal to that provided by perturbation theory. With this model, non-relativistic calculations including confinement, one gluon and pseudoscalar meson exchange can describe the light-quark baryon spectrum reasonably well.

References

1. L. Ya. Glozman and D. O. Riska, Phys. Rep. **268**, 263 (1996). See, also these proceedings.
2. Z. Dziembowski, M. Fabre de la Rippele, and G. A. Miller, Phys. Rev. **C53**, R2038 (1996).
3. M. Fabre de la Ripelle, H. Fiedeldey and S.A. Sofianos, Few-Body Systems, Suppl. 2, 493 (1987).
4. A. DeRujula, H. Georgi and S. L. Glashow, Phys. Rev. **D12** (1975) 147.
5. N. Isgur and G. Karl, Phys. Rev. **D18** (1978) 4187; **D19** (1979) 2653.

MANY-BODY HAMILTONIAN APPROACH TO QCD: APPLICATIONS

A. SZCZEPANIAK, E. S. SWANSON, C.-R. JI, S. R. COTANCH
Department of Physics, North Carolina State University
Raleigh, North Carolina 27695-8202 USA

Abstract

A comprehensive, relativistic many-body approach to hadron structure is presented. An effective Hamiltonian is derived from the Coulomb gauge Hamiltonian of QCD using perturbative similarity transformation. Standard many-body techniques, which render the approximations amenable to systematic improvement, are used to approximately diagonalize the effective Hamiltonian.

1 Introduction

The Constituent Quark Model (CQM) has been widely accepted as the phenomenological framework to describe low energy properties of hadrons. Valence, effective, massive constituents appear to saturate hadronic wave functions, while excitations of the nonvalence degrees of freedom, including the gluonic ones, are expected to be important above 1 GeV. One of the most challenging questions is to be able to relate the constituent picture of hadrons to the underlying theory of QCD.

2 From QCD to the Constituent Quark Model

The natural framework which could result in a Fock space based description of hadrons is that of Hamiltonian QCD. Because of freedom in gauge fixing there is, in principle, an infinite number of choices for the canonical, bare Hamiltonian. When deciding which canonical Hamiltonian to choose as a starting point it should be remembered that the issue is not whether an approach will effectively maintain gauge invariance but rather which choice of gauge has the best chance in resulting in a simple constituent picture. For reasons outlined in [1] we have chosen to work in the Coulomb gauge. The procedure used to construct the effective Hamiltonian, which is also explained in more details in Ref.[1], can be summarized as follows. The Coulomb gauge QCD Hamiltonian is canonically quantized in the quark/gluon-parton basis. Short-distance cut-off dependence of the bare Hamiltonian is controlled through a perturbative similarity transformation[2]. Finally a minimal subtraction renormalization scheme

Table 1: Coulomb-gauge QCD Hamiltonian with linear confinement glueball spectrum

J^{PC}	0^{++}	0^{-+}	2^{++}	2^{-+}
m_{gg}[GeV]	1.66	2.2	2.1	2.8

is adopted to define the counter-terms. This yields an effective Hamiltonian which couples free parton states whose energies do not differ by more than Λ, a typical hadronic scale. Finally as explained in Ref. [1], the non-abelian Coulomb gauge potential is replaced by a linear confining potential.

The distinctive feature of the many-body approach is the ability to study hadronic structure beyond valence approximation and on top of a nonperturbative vacuum. The later is obtained through a canonical, Bogoliubov transformation, which introduces two-particle correlations into the perturbative vacuum. Fermionic correlations break chiral symmetry and lead to a reasonable value for the quark condensate. In both fermion and boson sectors these correlations confine color non-singlets and result in typical constituent energies for two, or more single particle excitations in a color singlet state. In particular kinetic energies of each one of the two transverse gluons forming a color singlet glueball obey a dispersion relation,

$$E_g(\mathbf{k}) = \sqrt{m_g(\mathbf{k})^2 + \mathbf{k}^2} \tag{1}$$

with $m_g(0) \sim 800$ MeV. The spectrum of the glueballs have been calculated in the Tamm-Dancoff, (TD) gg approximation and is shown in Tab. 1 [3]. Because of the large kinetic energy of the constituent gluons, the random phase approximation (RPA) results in only a few percent correction, thus confirming the constituent picture of a glueball. Light meson spectroscopy is currently under investigation. An important question, which this framework permits to address, is where the π-ρ mass splitting arises. The Goldstone-boson pion, which in this many-body approach manifests itself through an RPA solution, may be compared with the CQM, $\bar{q}q$ pion obtained as a TD solution. The effects of the RPA components of the pion wave functions are also being studied in the context of hadronic decays.

References

1. A. Szczepaniak and E. S. Swanson these proceedings.
2. S. D. Glazek and K. G. Wilson, Phys. Rev. D**49**, 4214 (1994).
3. A. Szczepaniak, E. S. Swanson, C.-R. Ji and S. R. Cotanch, Phys. Rev. Lett. **76**, 2011 (1996).

HAVE GLUONIC EXCITATIONS BEEN FOUND?

P.R. PAGE
Department of Physics and Astronomy, University of Manchester, Manchester M13 9PL, UK

New experimental information on the non-exotic $J^{PC} = 0^{-+}$ isovector seen at 1.8 GeV by VES yields convincing evidence of its excited gluonic (hybrid) nature when a critical study of alternative quarkonium assignments is made in the context of $^3P_0$ decay by flux-tube breaking. Production of this gluonic excitation via meson exchange is promising, although its two photon production vanishes.

Based on the phenomenological success[1] of the $^3P_0$ hadronic decay model, the decay modes of $Q\bar{Q}$ systems with an explicit gluonic excitation (hybrids) have been predicted[2] in a non-relativistic flux-tube model. Hybrids are predicted to have mass 1.8 – 1.9 GeV[1], exactly in the region where a $J^{PC} = 0^{-+}$ isovector resonance has recently been seen[3]. The mass of this state also makes it a candidate for radial 3^1S_0 $Q\bar{Q}$ (π_{RR}). The decay of hybrids to "S+S"–wave mesons are expected[3] to vanish for identical mesons, and to be suppressed proportional to the difference of their "sizes" [2] for non–identical mesons. The dominant decay channel is hence to "P+S"–wave mesons.

VES[3] (and BNL[3]) detect a prominent resonance at $\sim$ 1.8 GeV with width $\sim$ 200 MeV in the "P+S" channels $\pi f_0(980)$, $\pi f_0(1300)$, $\eta a_0(980)$ and $(K\bar{K}\pi)_S$. On the other hand, the resonance is absent[3] in the "S+S" channels $\pi\rho$ and $\bar{K}K^*$. There is also possible[4] evidence for the (weak) mode $\pi f_0(1500)$ where the gluonic excitation de–exites to the gluonium candidate $f_0(1500)$. The foregoing clearly supports a hybrid interpretation. The predicted widths for a hybrid π_H at $\sim$ 1.8 GeV are[2] (in MeV)

$$\pi f_0(1300) \sim 170;\ \pi f_2 \sim 5;\ \pi\rho \sim 30;\ \bar{K}K^* \sim 5;\ \pi\rho_R \sim 30$$
$$K^*\bar{K}^* \sim 0;\ \rho\omega \sim 0;\ \eta a_0 \sim 120;\ \pi f_0 \sim 160 \qquad (1)$$

where the last two modes assume that a_0, f_0 are $^3P_0$ $Q\bar{Q}$.

The widths expected for π_{RR} are often distinctively *different*[2] from those of hybrids. *(i)* $\pi f_0(1300)$ is very much *suppressed* ($<$ 10 MeV over parameter space) relative to the prediction for π_H (Eq. 1) and either $\pi_{RR} \rightarrow \pi\rho, \pi f_2$ or $\bar{K}K^*$ whereas the data show that it is much larger than all of these. *(ii)* The same is true of $\bar{K}K_0^*(1430)$, which is threshold forbidden and manifested as $(K\bar{K}\pi)_S$. For π_H at 2 GeV $\bar{K}K_0^*$ is substantial at 200 MeV, consistent with the data, while for π_{RR} it is *suppressed* at 0 – 20 MeV due to a node in the amplitude. The strong affinity of $K\bar{K} \rightarrow f_0(980)$ is probably responsible[2] for the observed strong[3] coupling to $\pi f_0(980)$. *(iii)* For π_{RR} the $\rho\omega$ channel is expected to be *prominent*[2] at $0-120$ MeV. In contrast, $\rho\omega$ vanishes for π_H (Eq. 1) independent of the wave functions assumed in the flux–tube model[2]. The $\rho\omega$ signal[4] builds up significantly below 1.8 GeV with width $\sim$ 300 MeV, although a resonant signal has not yet been established. It is tempting to suggest that this indicates the detection of a seperate state, the π_{RR}, different from the 1.8 GeV π_H with width $\sim$ 200 MeV. *(iv)* The possible existence of a seperate state is corroborated by the πf_2 channel which may also be distinctive. For π_H πf_2 is small (Eq. 1) whereas it is possibly *larger*[2] ($0-30$ MeV) for π_{RR}. The data[3] show a small πf_2 peak at 1.7 GeV, certainly below the 1.8 MeV region, though further analysis and data are required.

π_H and π_{RR} have in *common* that $\pi\rho$ is suppressed ($0-30$ MeV for π_{RR} due to a node) consistent with the data[3] which show no signal in the 1.7 to 2 GeV mass region. We suggest searching for coupling to the $\pi\rho$ channel, and further determinations of the mass and width of the state seen in $f_2\pi$ and $\rho\omega$.

At both VES and BNL the 0^{-+} was produced in $\pi^- N \to 0^{-+} N$ at high energy via either diffractive or ρ exchange. In the case of ρ exchange the width corresponding to the $\pi\rho$ vertex of π_H is bounded above[2] by 150 MeV, and is expected to be $\gtrsim 20\%$ of this value (see Eq. 1) since the ρ is off–shell and hence of potentially very different "size" than the on–shell π. This may lead to significant production of π_H in photoproduction on nuclei through π exchange, with the photon coupling to ρ (with upper bound 270 keV[2]); and would be especially significant at low energy facilities like an upgraded CEBAF where π exchange would be dominant.

An unfortunate corollary of the lack of coupling of π_H to $\rho\omega$ mentioned before, is that when the ρ and ω couple to photons, the two photon width and production of π_H vanish. In addition, the photoproduction of π_H via ρ or ω exchange vanishes[2]. Photon coupling via intermediate vector mesons is currently the only way of effecting flux–tube model photonic couplings for π_H.

References

1. E.S. Swanson *et al.*, *Phys. Rev.* D **52**, 5242 (1995); *Phys. Rev.* D **50**, 6855 (1994); ORNL-CTP-96-03.
2. P.R. Page *et al.*, *Nucl. Phys.* B **443**, 233 (1995); *Phys. Rev.* D **52**, 1706 (1995); OUTP-95-47P.
3. P.R. Page, D.I. Ryabchikov, D.P. Weigand, *Proc. of HADRON95*, ed. M.C. Birse *et al.*; D.V. Amelin *et al.*, *Phys. Lett.* B **356**, 595 (1995); *Jad. Fiz.* **59**, 1021 (1996).
4. Private communication, Y.A. Khokhlov, D.I. Ryabchikov, A.M. Zaitsev.

Microscopic Models of Strong Decays

E.S. Ackleh[1], T. Barnes[1], and E.S. Swanson[2a]
[1] *Computational and Theoretical Physics Group, Oak Ridge National Laboratory, Oak Ridge, TN 37831-6373*
[2] *Department of Physics, North Carolina State University, Raleigh, NC 27695-8202.*

Open-flavor strong decays are mediated by $q\bar{q}$ pair production, which is known to occur dominantly with $^3P_0$ quantum numbers. The relation of the phenomenological $^3P_0$ model of these decays to "microscopic" QCD decay mechanisms has never been clearly established. We investigate $q\bar{q}$ meson decay amplitudes assuming pair production from the scalar confining interaction and from one gluon exchange. Scalar pair production predicts decay amplitudes of approximately the correct magnitude and D/S amplitude ratios in $b_1 \to \omega\pi$ and $a_1 \to \rho\pi$ which are close to experiment. The one gluon exchange decay amplitude is found to be subdominant in most cases, a notable exception being $^3\mathrm{P}_0 \to {}^1\mathrm{S}_0 + {}^1\mathrm{S}_0$. The full scalar plus gluon amplitudes differ significantly from $^3P_0$ model predictions in some channels and can be distinguished experimentally, for example through an accurate comparison of the D/S amplitude ratios in $b_1 \to \omega\pi$ and $a_1 \to \rho\pi$.

Developing an understanding of hadronic decays is important for several reasons. At the most direct level, there are strong recouplings in some sectors of the meson spectrum (notably the 0^{++} sector) which must be sorted out. Similarly, final state interactions bedevil the analysis of many processes. Both of these need to be understood to assist in the identification of possible exotic states. Ultimately, a successful model of strong decays will assist in comprehending QCD. For example, both decays (which are order $1/m_q$) and heavy meson splittings (order $1/m_q^2$) probe the Dirac structure of the effective $q\bar{q}$ interaction of QCD.

Previous models of strong decays include the Cornell model, the $^3S_1$ model, the $^3P_0$ model, and the flux tube model. Except for the Cornell model none of these are microscopic in the sense of having no free parameters. Unfortunately the Cornell model does not create a quark antiquark pair with $^3P_0$ quantum numbers – and these are known to reproduce the data more effectively than $^3S_1$ pair production. We have therefore attempted to determine if the naive quark model may be extended in a simple way to describe strong decays [1].

Quark model lore requires the central potential to be a Coulomb plus linear combination. Heavy mesons splittings may be successfully modelled by assuming that the Coulomb interaction arises from one gluon exchange (OGE) and that the confinement is due to the interaction of scalar quark currents.

[a]Presented at PANIC 96, May 23–28, 1996, Williamsburg, VA

Thus we have modelled strong decays with a very simple extension of the quark model embodied in the following interaction Hamiltonian

$$H_{int} = \frac{1}{2}\int \bar{\Psi}T^A\Psi(x)\frac{3}{4}b|\mathbf{x}-\mathbf{y}|\bar{\Psi}T^A\Psi(y) + \int \bar{\Psi}T^A\gamma_\mu\Psi(x)\frac{\alpha_S}{|\mathbf{x}-\mathbf{y}|}\bar{\Psi}T^A\gamma^\mu\Psi(y)$$

Zero parameter predictions for decays may now be made by fixing the string tension, the strong coupling, and the quark masses to their standard quark model values (0.18 GeV^2, 0.6, and 0.33 GeV, respectively). We have calculated the contributions of each of these terms (in the nonrelativistic limit) to several experimentally well determined decays. The scalar confinement contribution dominates both the Coulomb and transverse gluon portions of the one gluon exchange kernel (except in a few cases such as $f_0(1300) \to \pi\pi$). Furthermore, the magnitude of the scalar piece is roughly correct – within a factor of two for all amplitudes tested. The Coulomb piece is always negligible. This may be seen in the limit where the interaction is constant in configuration space since then the Coulomb quark pair creation operator is proportional to the quark number operator squared and, hence, vanishes. Furthermore, we have seen that OGE cannot describe the D to S amplitude ratios in a_1 and b_1 decays. In contrast, the scalar potential works quite well in this regard. The ratio of D to S amplitude ratios is wavefunction independent for all decay operators except transverse gluon exchange (due to a spin flip in the spectator line). Thus an accurate measurement of this ratio will help determine the extent of transverse gluon exchange contributions to hadronic decays.

It appears that both heavy meson splittings and strong decays require a scalar confinement interaction. The standard linear confinement potential serves both purposes admirably. For decays it leads to a $^3P_0$-type vertex of strength $\sim b/(m_q\beta)$ where β is the wavefunction width parameter. This is roughly a factor of two too strong – a problem which may be alleviated by working relativistically or by employing more accurate wavefunctions. Hints of the OGE contribution to pair production may be seen in the ratio of D to S amplitude ratios in $a_1 \to \rho\pi$ and $b_1 \to \omega\pi$ and in the width of the $f_0(1300)$.

References

1. E. Ackleh, T. Barnes, and E.S. Swanson, submitted to PRD; hep-ph/9604355.

PARTIAL WAVE ANALYSES OF THE $\eta\pi^-$ AND $\eta'\pi^-$ SYSTEMS FOR REACTIONS $\pi^-p \to \eta\pi^-p$ AND $\pi^-p \to \eta'\pi^-p$ AT 18 GEV/C

S.A. TAEGAR, D.L. STIENIKE, AND D.R. THOMPSON FOR THE E852 COLLABORATION
University of Notre Dame, Notre Dame, IN 46556, USA

Preliminary results for the partial wave analyses of the $\eta\pi^-$ and $\eta'\pi^-$ systems produced in π^-p interactions at 18 GeV/c are presented. In both reactions, the $J^{PC} = 2^{++}$ and 1^{-+} partial waves dominate.

This paper reports preliminary partial wave amplitude analysis results on the reactions $\pi^-p \to \eta\pi^-p$ and $\pi^-p \to \eta'\pi^-p$ at 18 GeV/c. The data come from Brookhaven experiment E852 carried out at the BNL MPS configured for E852 with a downstream drift chamber and lead glass hodoscope[1]; a cylindrical drift chamber[2] and cesium iodide barrel[3] surrounding a 30-cm hydrogen target; and a new DAQ. E852 had a data run in '94 on which the present analysis is based and a run in '95 which will increase the data by about a factor of five.

The $\eta\pi^-$ and $\eta'\pi^-$ systems are of particular interest because either state in a relative P-wave has quantum numbers $J^{PC} = 1^{-+}$. These values are not allowed for a $q\bar{q}$ bound state. Thus observation of a P-wave resonance would constitute unambiguous evidence of an exotic meson. Previous experimental results[5,6,7] have suggested the presence of significant P-wave in these systems.

The $\eta\pi^-$ analysis is based on a sample of 213,000 events which have been kinematically constrained by the fitting program SQUAW [4] to the $\eta\pi^-p$ final state and have satisfied quality cuts including topological and fiducial volume cuts. Shown in Fig. 1 is the $\eta\pi^-$ effective mass distribution for these events which is clearly dominated by $a_2(1320)$ production. The lower dashed curve is the non-η background estimated from side bands in the two-photon mass distribution (not shown). Shown in Fig. 2 is the acceptance-corrected forward-backward decay asymmetry (in the G-J) frame versus the $\eta\pi$ mass. There is a strong asymmetry which can only arise from the interference of even and odd partial waves and is an indicator of the presence of P-wave in the data. Preliminary results for the amplitude analysis of this data are shown in Fig. 3. Shown are the D-wave intensity (dominated by the $a_2(1320)$), the P-wave intensity, and the relative P-D phase. We are studying the results of this analysis to determine whether or not the P-wave system is resonant.

Shown in Fig. 4 is the $\eta\pi^-\pi^+$ effective mass distribution for the events from the reaction $\pi^-p \to \eta\pi^-\pi^-\pi^+p$ showing a strong η' signal (as well as an $f_1(1285)$ peak). In Fig. 5 is shown the $\eta'\pi^-$ mass distribution for the subset of events (1800 total) which are consistent with the reaction $\pi^- p \rightarrow \eta' \pi^- p$ and satisfy other quality cuts. The preliminary results of a partial wave analysis of the data (not shown) indicate strong P- and D-wave production consistent with earlier VES [7] results.

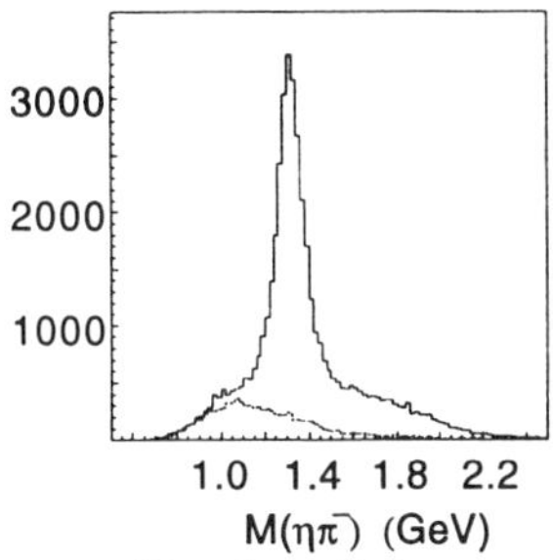

Figure 1 - $\eta\pi$ effective mass distribution in 20 MeV bins.

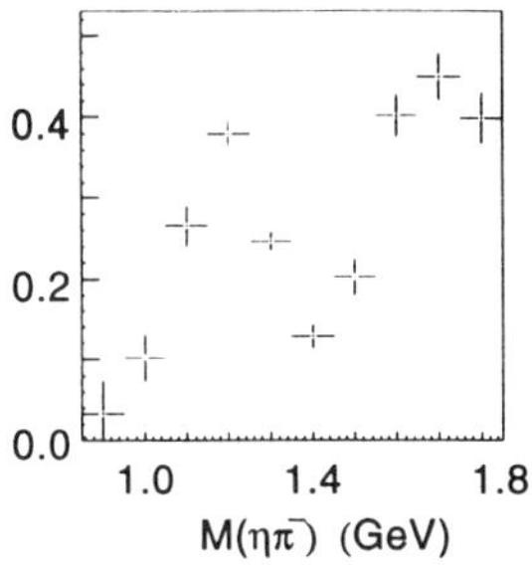

Figure 2 - $\eta\pi$ decay asymmetry F-B/F+B.

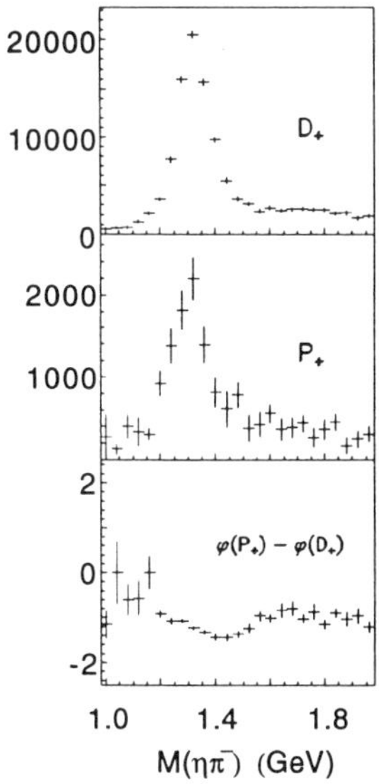

Figure 3 - $\eta\pi$ D-wave and P-wave intensities and the P-D phase difference in 40 MeV bins.

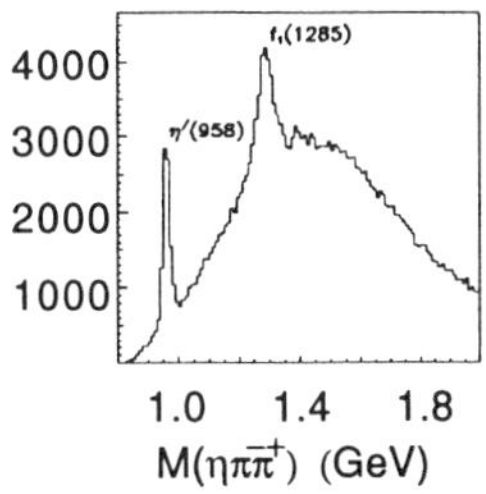

Figure 4 - $\eta\pi\pi$ effective mass distribution in 10 MeV bins.

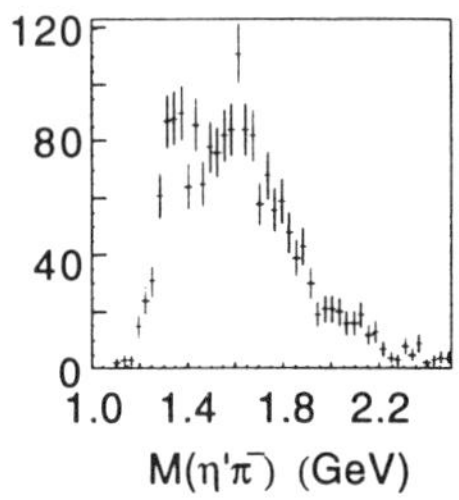

Figure 5 - $\eta'\pi$ effective mass distribution in 30 MeV bins.

This work is supported in part by the U.S. DOE and the NSF.

References

1. R.R. Crittenden et al, "A 3000 element lead glass electromagnetic calorimeter", *Nucl. Instrum. Methods* A, submitted for publication.
2. Z. Bar-Yam et al, "A cylindrical drift chamber with azimuthal and axial readout", *Nucl. Instrum. Methods* A, to be submitted for publication.
3. T. Adams et al, *Nucl. Instrum. Methods* A **368**, 617 (1996).
4. T. Day, "SQUAW a kinematic fitting program", University of Maryland Technical Report 649.
5. D. Alde et al, *Phys. Lett.* B **205**, 397 (1988).
6. H. Aoyagi et al, *Phys. Lett.* B **314**, 246 (1993).
7. E.B. Berdnikov et al, *Nuovo Cimento* **107**, 1941 (1994).

PARTIAL WAVE ANALYSIS OF THE EXCLUSIVE REACTION $\pi^- p \to \eta\pi^-\pi^+ n$ AT 18 GEV/C

JOSEPH MANAK FOR THE E852 COLLABORATION
University of Notre Dame, Notre Dame, IN 46556, USA

Preliminary results for the partial wave analyses of the $\eta\pi^+\pi^-$ system produced in $\pi^- p$ interactions at 18 GeV/c are presented. The data can be characterized by a strong 1^{--} $\rho\eta$ wave consistent with $\rho(1700)$ production. A significant amount of $a_0\pi$ is also observed in both 1^{++} and 0^{-+} waves, consistent with $f_1(1285)$, $\eta(1295)$ and $\eta(1440)$ production.

1 Introduction

This note presents preliminary results of the partial wave analysis being performed at the University of Notre Dame on the $\eta\pi^-\pi^+$ data set. This analysis is based on about 48M triggers from the summer 1994 run of Brookhaven experiment E852, representing 20% of the total data set.

The following event selection criteria were applied to obtain the final $\eta\pi^-\pi^+$ sample: (1) Each selected event had a SQUAW[1] kinematic fit to $\pi^- p \to \eta\pi^-\pi^+ n$ with a confidence level greater than .07; (2) a total energy deposit of less than 20 MeV in the CsI target-region veto[2] was required for each event to insure an exclusive sample of neutrons in the final state; and (3) the vertex of each event was required to be within the hydrogen target. The observed $\eta\pi^-\pi^+$ mass distribution is shown in Fig. 1. The signal to background ratio is estimated to be about 10:1.

A partial wave analysis, using the maximum likelihood method was performed on the $\eta\pi^-\pi^+$ system in the 1.2-1.6 GeV mass range. The intermediate isobars σ and ρ were used for the $\pi^-\pi^+$ system and the $a_0(980)$ was used for the $\eta\pi$ system. In the PWA the σ used was a parameterization of the $\pi\pi$ S wave, and the standard width and mass of the ρ were used. The analysis was done in 30 MeV $\eta\pi^-\pi^+$ mass bins. A solution was obtained which included of 13 waves (IJPxyLME): 00-$a_0(980)\pi$00+, 01+$a_0(980)\pi$10+, 10-$a_0(980)\pi$00+, 11+$a_0(980)\pi$10+, 11-$\rho\eta$10-, 11+$\rho\eta$00+, 11+$\rho\eta$01+, 11+$\rho\eta$01-, 11-$\rho\eta$11+, 00-$\sigma\pi$00+, 01+$\sigma\pi$00+, 01+$\sigma\pi$01+, 01+$\sigma\pi$01+, plus a non-interfering background wave.

The dominant waves in the low mass region were the 0^{-+} and the 1^{++} $a_0\pi$ waves. The 0^{-+} wave is shown in Fig. 2 and exibits two structures, one at 1.28 GeV and another at 1.42 GeV. The low mass peak is associated with the $\eta(1295)$ and the high mass peak with the $\eta(1440)$. The 1^{++} wave shown in Fig. 3 displays clear evidence of an $f_1(1285)$ signal and an absence of any structure in the 1.4 GeV region. The f_1 was fit with a Breit Wigner with an experimentally determined mass of 1.29 GeV and a width of 53 MeV. The results of this analysis are consistent with the results from KEK E179[3], showing only one $a_0\pi$ state in the 1.4 GeV mass region.

In order to investigate the properties of the $\rho(1700)$ a fit was done in the 1.4-2.4 GeV mass region. The following waves were used to characterize the data in this region in addition to the waves used in the low mass fit (IJPxyLME): 12-$\rho\eta$10+, 12-$\rho\eta$11+, 12-$\rho\eta$11-, 12+$\rho\eta$20-, 12+$\rho\eta$21-, 12+$\rho\eta$21+, 13-$\rho\eta$30-, 13-$\rho\eta$31-, 13-$\rho\eta$31+, 12-2π00+, 12-2π01+ 12-2π01-. The dominant wave in this region is the 1^{--} $\rho\eta$ wave shown in Fig. 4. A fit with a single Breit Wigner yielded a mass of 1.68 GeV and a width is 287 MeV.

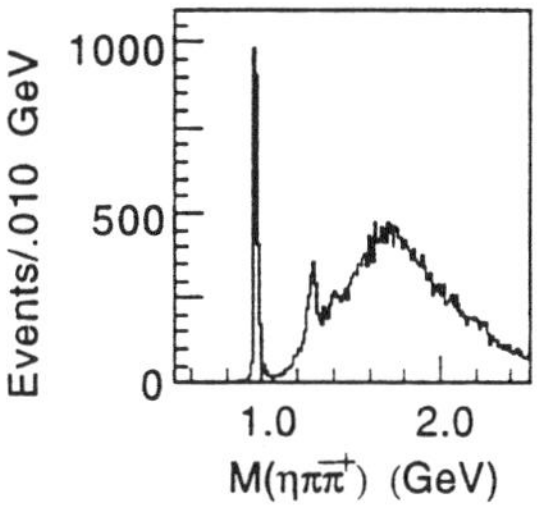

Figure 1 - ηππ effective mass distribution in 10 MeV bins.

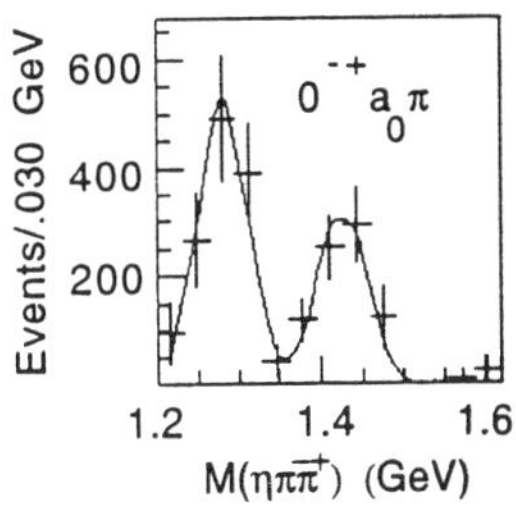

Figure 2

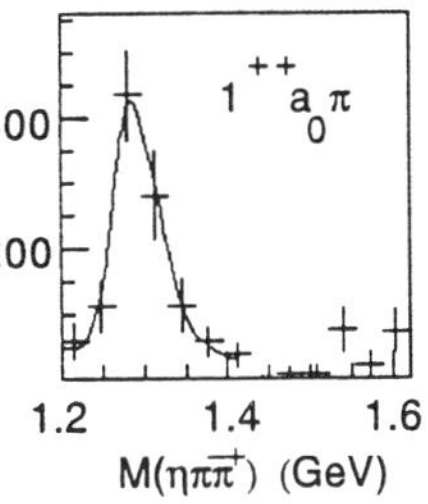

Figure 3

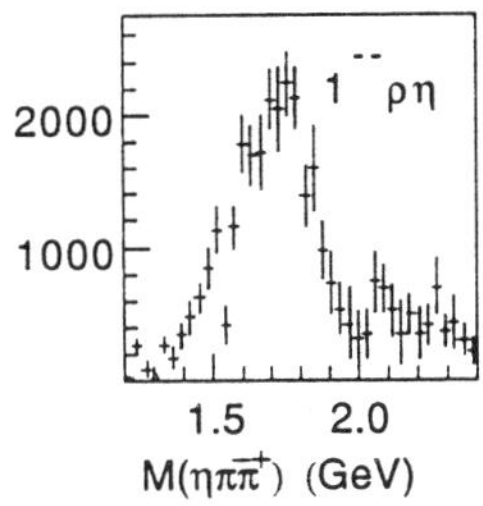

Figure 4

This work is supported in part by the U.S. DOE and the NSF.

References

1. T. Day, "SQUAW a kinematic fitting program", University of Maryland Technical Report 649.
2. T. Adams et al, *Nucl. Instrum. Methods* A **368**, 617 (1996).
3. S. Fukui et al, *Phys. Lett.* B **267**, 293 (1991).

Meson Spectroscopy Results from OBELIX

presented for the OBELIX Collaboration by
A. ROTONDI
Dipartimento di Fisica Nucleare e Teorica and
Istituto Nazionale di Fisica Nucleare, Sezione di Pavia
Via Bassi, 6 -27100 Pavia ITALY

The OBELIX spectrometer, devoted to the study of the $\bar{N}N$ annihilation at the CERN Low Energy Antiproton Ring (LEAR), is briefly described and the results of the spin-parity analysis on some of the numerous channels under investigation are outlined.

1 Introduction

The OBELIX spectrometer, working at the CERN Low Energy Antiproton Ring (LEAR), was conceived as a general purpose, large acceptance and high resolution detector, aimed to the systematic study of the antinucleon annihilation [1]. A unique characteristic of the apparatus is the possibility of changing the probe and/or the target without modifying the experimental set-up, to study the $\bar{N}N$ annihilation from different mixtures of initial states.

Data with $\bar{p}$ at very low momentum ($\approx 50 MeV/c$) have been taken, corresponding to annihilations from $^1S_0$ and $^3S_1$ states, with a P-wave component less than 4% [2].

In annihilations at rest, the partition among the possible $^{2S+1}L_J\,(J^{PC})$ quantum numbers of the protonium six initial states:

$$^1S_0\,(0^{-+}),\quad ^3S_1\,(1^{--}),\quad ^1P_1\,(1^{+-}),\quad ^3P_0\,(0^{++}),\quad ^3P_1\,(1^{++}),\quad ^3P_2\,(2^{++}),$$

depends on the target density. In a LH_2 target the Stark-mixing effect is believed to work strongly in favour of S-wave annihilation from the $^1S_0$ and $^3S_1$ states; on the other hand, in a low density gaseous H_2 target, P-wave annihilations should become the dominant process. A large amount of data has been collected with GH_2 targets with a selectable density (just varying the gas pressure from a few bar down to 1 mbar) and with LH_2 target, to change the relative weights of the initial state quantum numbers of the detected annihilation channels, leaving the detector practically in the same conditions [3].

Finally, with the $\bar{n}$ beam (momentum of 100-400 MeV/c), produced through the $\bar{p}p \rightarrow \bar{n}n$ charge exchange reaction in the OBELIX conversion target placed on the primary $\bar{p}$ beam, one can fix the isospin of the initial state to I = 1, as for the $\bar{p}n$ state in deuterium, but without any complication due to the spectator proton.

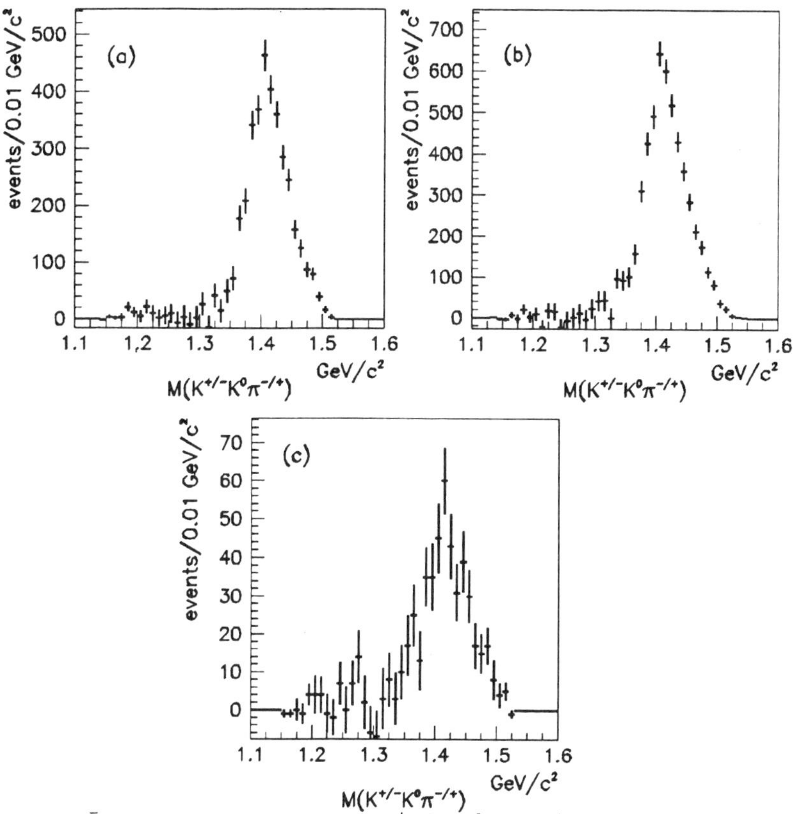

Figure 1: $K\bar{K}\pi$ invariant mass from $\bar{p}p \to K^{\pm}\pi^{\mp}(K^0)_{miss}\pi^+\pi^-$ events with (a) LH2, (b) NTP, and (c) 5 mbar pressure targets. The combinatorial background has been subtracted.

2 Results on meson spectroscopy

Here we briefly list the main results of the spin-parity analysis of the $\bar{n}p \to \pi^+\pi^+\pi^-$, $\bar{p}p \to \pi^+\pi^-\pi^0$, $\bar{p}p \to 2\pi^+2\pi^-$ and $\bar{p}p \to (E/\iota)\pi\pi \to (KK\pi)\pi\pi$ reactions taken with different target and beam arrangements.

The results on the $\bar{p}p \to \pi^+\pi^-\pi^0$ annihilation at rest in liquid, NTP and 5 mbar pressure hydrogen targets and on the $\bar{n}p \to \pi^+\pi^+\pi^-$ annihilation in flight show, beside the production of the known $\rho(770)$ and $f_2(1270)$ mesons, also the presence of the new $f_0(1500)$ and $f_2(1520)$ mesons. The $\bar{p}p$ data at rest, where the neutral ρ production is allowed, also require the significant contribution from a $\hat{\rho}$ signal with a mass of about $1200 MeV/c^2$ and large width ($> 500 Mev/c^2$) not yet well determined. This last set of data has been analysed with a global fit on the samples taken at the three different target densities, having as free parameters, for each density, the weights of the initial states and, within each state, the resonance production rates common to all the densities.

A decay amplitude containing the $\pi\pi$ σ interaction from the $\bar{p}p \to 3\pi$ data, the $\hat{\rho}(1200)$ and the well-known $\rho(770)$, $f_2(1270), a_1(1260)$, and $a_2(1320)$ mesons well describes the $\bar{p}p \to 2\pi^+2\pi^-$ annihilation in flight at low momentum, which occurs from singlet and triplet S waves. The data analysis on the same reaction taken at rest with liquid and gaseous hydrogen targets is in progress.

In the E/ι region good fits for the $\bar{p}p \to (KK\pi)\pi\pi$ reaction at rest in liquid hydrogen are obtained with two $\eta(1416)$ and $\eta(1460)$ resonances, having width of 50 and $102 MeV/c^2$, respectively [5]. The possible presence of an axial vector meson, reported by other experiments, being in this case produced from P-wave initial states, is actively investigated in the data taken at NTP and 5 mbar pressure hydrogen targets. The collected statistics (about 10 000 events in the E/ι region) exceeds of an order of magnitude the previous data coming from the annihilation at rest (fig. 1).

3 Conclusion and perspectives

The spin-parity analysis of the data shows clear signals of the $f_0(1500)$ (0^{++}), $f_2(1520)$ (2^{++}), $\eta(1416)$ and $\eta(1460)$ (0^{-+}) mesons.

The assessment of all these states in the 2^1S_0 (0^{-+}), 2^3P_0 (0^{++}), 2^3P_2 (2^{++}) radially excited nonets seems problematic.

The nature of the $\hat{\rho}$ signal, which appears in interference with the $\rho(770)$, is at present investigated by extending the K-matrix formalism (which preserves unitarity in the case of overlapping resonances) also to the $\pi\pi$ (1^{--}) wave.

The analysis of the other data collected by the Collaboration, as the $\bar{p}p \to K^+K^-\pi^0$, $\bar{p}p \to \eta\pi^+\pi^-$, $\bar{p}p \to K^\pm K^0_S\pi^\mp$ annihilations and the $\bar{p}p \to E/\iota\pi^+\pi^-$ annihilation with $E/\iota \to K^0_s K^0_s\pi^0$, $K^0_S K^\pm\pi^\mp$, $\eta\pi^+\pi^-$, is in progress. These forthcoming results will add new information on the existence and on the possible exotic nature of these states.

References

1. A. Adamo et al., OBELIX Collaboration, *Sov. J. Nucl. Phys.* 55 (1992) 1732.
2. A. Agnello et al., OBELIX Collaboration. *Phys. Lett.* B369(1996)77
3. A. Adamo et al., OBELIX Collaboration, *Phys. Lett.* B285(1992)15
4. M. Agnello et al., OBELIX Collaboration, *Phys. Lett.* B 287(1992)368
5. A. Agnello et al., OBELIX Collaboration, *Phys. Lett.* B36(1995)187

Scalar mesons and the $K\overline{K}$ thresholds

O. Krehl, J. Haidenbauer, K. Holinde, G. Janssen and J. Speth
Institut für Kernphysik, Forschungszentrum Jülich GmbH, D-52425 Jülich, Germany

B. Pearce
Dept. of Physics and Math. Physics, University of Adelaide, Adelaide 5005, Australia

We investigated the structure of the scalar mesons $f_0(980)$ and $a_0(980)$ in the framework of a meson exchange model for $\pi\pi$ and $\pi\eta$ scattering [1]. The latter was obtained by a consistent extension of the Jülich model for the $\pi\pi$ interaction [2]. Our solution for the $\pi\pi$ scattering amplitude is compatible with the available experimental data sets (cf. Fig. 1). In particular, we are also able to describe data on $J/\psi \rightarrow \phi\pi\pi/\phi K\overline{K}$ decay which has been considered to be an important testing criterion for models of the $f_0(980)$. By exploring the pole positions of the $\pi\pi$ T-matrix we found the $f_0(980)$ to be a narrow $K\overline{K}$ bound state of relatively large mass (m_{f_0}=1015 MeV). In addition we located a very broad $\sigma(400)$ pole on the [bt] sheet (cf. Ref. [3] for sheet structure and notation) which is responsible for the large $\pi\pi$ S-wave phase shifts below 1.0 GeV. This pole is generated by the ρ-exchange in the t-channel and remains essentially unchanged if one removes all s-channel pole diagrams ($q\overline{q}$-mesons) from the model [1]. For high energies, the pole of a heavy scalar particle with $q\overline{q}$ structure is observed on the [bb] sheet which is interpreted as a mixture of two scalar mesons, such as $f_0(1400)$ and $f_0(1590)$.

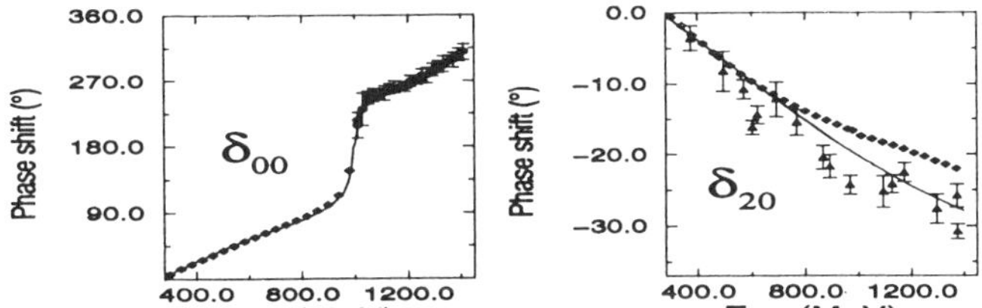

Figure 1: Phase shifts δ_{IJ} of the $\pi\pi$ interaction.

A consistent extension of our model to the $\pi\eta$ channel generates a pole on the [bt] sheet relatively close to $K\overline{K}$ threshold. Looking at the corresponding $\pi\eta$ cross section we observed a peak structure in reasonable agreement with the experimental information available for the parameters of the $a_0(980)$. However the mass and width derived from the pole position show the $a_0(980)$ to be rather broad. We investigated the role of the direct $K\overline{K}$ interaction for our results and found it to be not essential for the generation of a pole (though it improves the agreement with experimental data). Rather it is the transition potential $\pi\eta \rightarrow K\overline{K}$ which is responsible for the appearance of the [bt] pole assigned to the $a_0(980)$.

Presently we are re-analyzing our results in the particle basis taking into account the proper mass splitting between $\pi^{\pm}$ and π^0, and K^+ and K^0, respectively. Due to this mass splitting the thresholds of the K^+K^- and $K^0\overline{K^0}$ channels are now well separated leading to a more complex sheet structure.

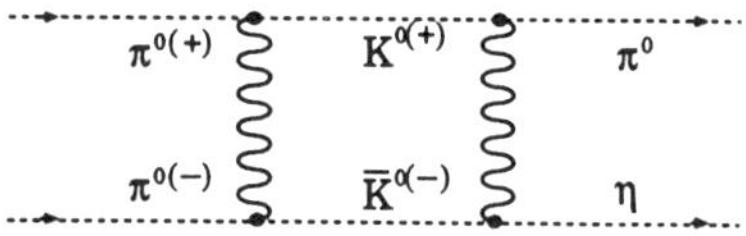

Figure 2: $f_0(980)$ and $a_0(980)$ mixing via $K\overline{K}$ intermediate state.

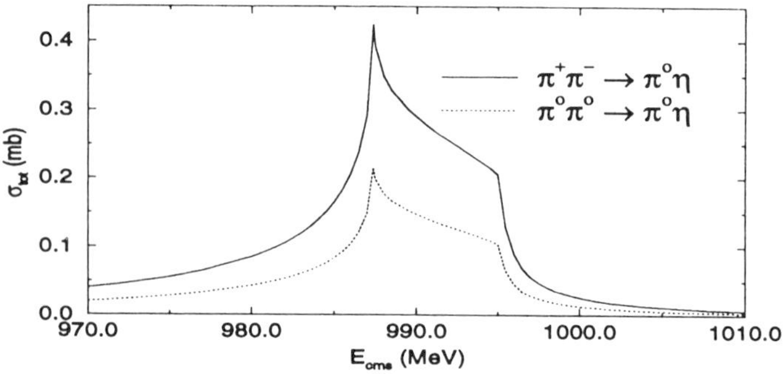

Figure 3: Predicted S-wave cross sections for $\pi\pi \rightarrow \pi\eta$.

We confirm our findings about the nature of the $f_0(980)$ and $a_0(980)$ mesons. An important new feature of this extended model is the fact that it gives rise to a coupling between the I=0 $\pi\pi$-channel and the I=1 $\pi\eta$-channel via $K\overline{K}$ intermediate states (Fig. 2). The isospin violation is due to the mass splitting and the $K\overline{K}$-state allow the mixing of the G-parity. Predictions for the $\pi\pi \rightarrow \pi\eta$ transition cross sections are shown in Fig. 3.

References

1. G. Janssen et al., *Phys. Rev.* D **52**, 2690 (1995).
2. D. Lohse et al., *Nucl. Phys.* A **516**, 513 (1990).
3. B. Pearce and B. F. Gibson, *Phys. Rev.* C **40**, 902 (1989).

$K\bar{K}$ THRESHOLD PHENOMENA, $f_0 - a_0$ INTERFERENCE AND KAONIUM PRODUCTION

B.Kerbikov
Institute of Theoretical and Experimental Physics
Moscow 117259, Russia
S.V.Bashinsky
M.I.T., Cambridge, MA,USA

We develop a new formalism to study $K\bar{K}$ threshold phenomena with resonances f_0, a_0 and K^+K^- atom (kaonium) included. The approach provides two possible scenarios for $f_0 - a_0$ mixing. Drastic interference patterns can be observed in K^+K^-electroproduction at CEBAF, $pd \to {}^3HeX$ reaction, and $N\bar{N}$ annihilation.

It is well known that there is an interesting sometimes controversial physics bearing upon the $K\bar{K}$ threshold phenomena and the nature of f_0, and a_0 resonances. We try to get a different handle of these issues by constructing an effective Hamiltonian which describes $\bar{K}K$ threshold phenomena, f_0 and a_0 resonances, and kaonium [1,2].

Consider reaction of the type $a + b \to X + M \to \pi^+\pi^- + M$ with invariant mass of X being in the vicinity of the $K\bar{K}$ threshold(s). This might be, e.g., $pd \to {}^3He\pi^+\pi^-$, or $\bar{p}p \to \pi^+\pi^-\eta$, or electroproduction reaction. Since f_0 and a_0 mesons play an important role in $K\bar{K}$ threshold region, we write the amplitude in the following form

$$T = < \hat{V}|\hat{G}|\hat{A} > + T^0, \tag{1}$$

where T^0 is the background (not via f_0) $\pi\pi$ production amplitude,

$$\hat{V} = (V_f, V_a), \hat{A} = \begin{pmatrix} A_f \\ A_a \end{pmatrix}, \hat{G} = \begin{pmatrix} G_{ff} & G_{fa} \\ G_{af} & G_{aa} \end{pmatrix} = (m_X - \hat{H} - \Delta\hat{H})^{-1}. \tag{2}$$

Here A_f and A_a are the f_0, a_0 production amplitudes, the propagator G includes their mixing (see below), V_f and V_a couple f_0 and a_0 to other states (note that $a_0 \to \pi\pi$ is forbidden). Finally

$$\hat{H} = \begin{pmatrix} E_f^{(0)} & 0 \\ 0 & E_a^{(0)} \end{pmatrix}, \Delta\hat{H} = \sum_\alpha \hat{V}^+ \frac{|\alpha><\alpha|}{m_X - m_\alpha + i0} \hat{V}, \tag{3}$$

where $|\alpha>$ indicates any of the states to which f_0 and a_0 are coupled.

Now we are furnished to provide two possible $f_0 - a_0$ mixing scenarios. One of them has its origin in the $K^0\bar{K}^0$ and K^+K^- threshold splitting of 8 MeV. Taking $\Delta\hat{H}$ as a sum of two terms corresponding to K^+K^- and $K^0\bar{K}^0$ channels we get

$$\hat{H}+\Delta\hat{H}=\begin{pmatrix} E_f-i\frac{\Gamma_f}{2} & 0 \\ 0 & E_a-i\frac{\Gamma_a}{2}\end{pmatrix}+$$

$$+D\begin{pmatrix} 1 & \zeta \\ \bar{\zeta} & |\zeta|^2\end{pmatrix}(-i)\sqrt{\frac{m_X-2m_{K^+}}{m_K}+i0}\ +$$

$$+D\begin{pmatrix} 1 & -\zeta \\ -\bar{\zeta} & |\zeta|^2\end{pmatrix}(-i)\sqrt{\frac{m_X-2m_{K^0}}{m_K}+i0}, \tag{4}$$

$$D=\frac{m_K^2}{4\pi}|<K^+K^-|\hat{V}|f>|^2,\quad \zeta=\frac{<K^+K^-|\hat{V}|a>}{<K^+K^-|\hat{V}|f>}. \tag{5}$$

Second mixing scenario is due to kaonium production. Kaonium with its Bohr radius of 109.6 fm and binding energy of 6.57 keV is formed by electromagnetic interaction and its wave function has two isospin components $I=0,1$ thus providing f_0-a_0 mixing. As shown in [1,2] mixing parameter depends on whether these mesons are $K\bar{K}$ molecules or genuine quark states.

The two mechanisms of mixing have very different energy scales - of few MeV and few keV correspondingly. Typical plots are shown in Figs. 1,2.

Figure Captions

Fig. 1. The typical plots of the matrix element squared for f_0-a_0 mixing due to $\bar{K}K$ mass splitting.

Fig.2. Same as in Fig. 1. with mixing due to kaonium.

References

[1] B.Kerbikov, Z.Phys. **353**, 113, 1995.

[2] S.V.Bashinsky and B.Kerbikov, Preprint DAPNIA/SPhN 95 35, Saclay 1995, to appear in Yad. Fiz. 1996.

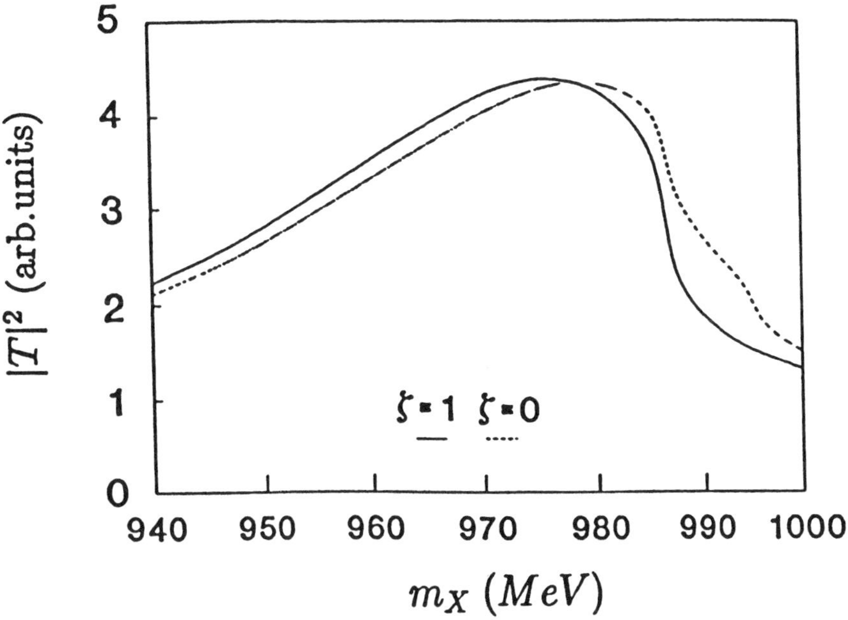

5
4
3
2
1
0
940
950
960
970
980
990
1000
$|T|^2$ (arb.units)
$\zeta = 1$ $\zeta = 0$
m_X (MeV)

Fig. 1

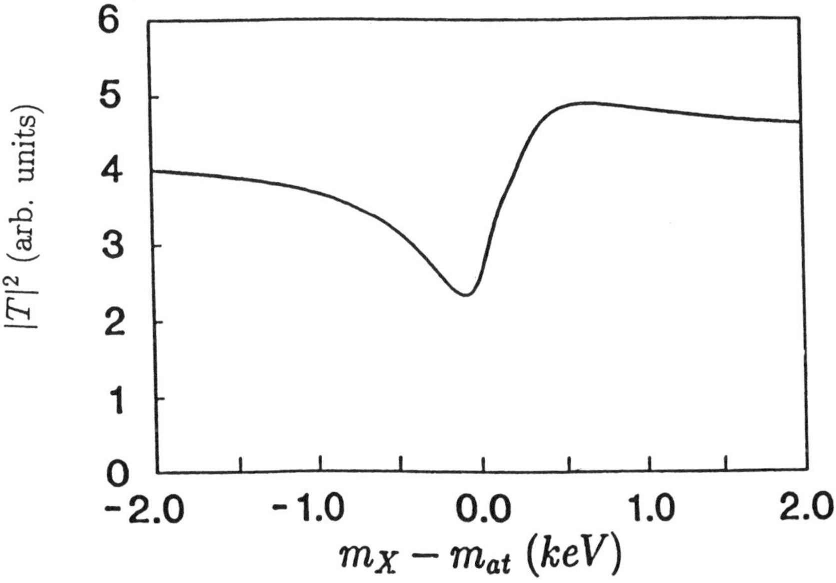

6
5
4
3
2
1
0
-2.0
-1.0
0.0
1.0
2.0
$|T|^2$ (arb. units)
$m_X - m_{at}$ (keV)

Fig. 2

Baryon spectroscopy with π–mesons at PNPI

I. V. LOPATIN, V. V. ABAEV, V. S. BEKRENEV,
E. A. FILIMONOV, A. B. GRIDNEV, M. R. KAN, N. G. KOZLENKO,
S. P. KRUGLOV, L. V. LAPOCHKINA, A. Yu. MAJOROV,
D. V. NOVINSKY, A. B. STAROSTIN, V. V. SUMACHEV
*Petersburg Nuclear Physics Institute, Gatchina,
Leningrad district, 188350 Russia*

B. M. K. NEFKENS, J. W. PRICE, D. B. WHITE, R.M.CLAJUS
*University of California at Los Angeles, Los Angeles,
California, 90024 USA*

M. E. SADLER, L. D. ISENHOWER, S. E. GARNER,
J. R. PHILLIPS, J. A. REDMON
Abilene Christian University, Abilene, Texas, 79699 USA

The main aim of πN research program is to solve problems of the baryon spectroscopy by means of obtaining new precise experimental data and performing a new phase shift analysis.

During last years, differential cross sections and polarization parameters P, R, A for $\pi^- p$ and $\pi^+ p$ elastic scattering and $\pi^- p$ charge exchange scattering were measured at PNPI [1] and LAMPF [2] for many energies. It is important that all experiments at PNPI were carried out using one and the same pion channel; the similar situation was at LAMPF. The new phase shift analysis PNPI-94 was performed in the energy range 160 to 600 MeV [3].

In PSA PNPI-94 we obtained new very precise partial πN amplitudes and tested the charge symmetry breaking in $\pi^+ p$ and $\pi^- p$ elastic scattering. Charge splitting of the P_{33} phase shift is observed. The difference $(\delta^{++} - \delta^0)$ depends on energy. It varies from + 2 degrees at $\sqrt{s} = 1.25$ GeV to – 2 degrees at $\sqrt{s} = 1.43$ GeV (changing sign at 1.35 GeV) and vanishes at $\sqrt{s} = 1.50$ GeV. The parametrization of the P_{33} phase shifts proved to be model-dependent for the resonance and background terms. Our calculations are performed with the relativistic Breit–Wigner resonance term and with the background $\delta_b(q) = q^3 \prod(q - q_i)$ where q is the pion momentum in the c.m.s. and q_i is the momentum corresponding to i-th resonance. In resonance $q = q_i$, hence the background is equal to zero; in particular, the background becomes zero when $q = q_1$ (the first P_{33}–resonance). After fitting we have obtained the following parameters of the P_{33}–resonance:

$$M^{++} = (1230.5 \pm 0.2) \text{ MeV} \quad M^0 - M^{++} = (2.6 \pm 0.4) \text{ MeV}$$
$$M^0 = (1233.1 \pm 0.3) \text{ MeV} \quad \Gamma^0 - \Gamma^{++} = (5.1 \pm 1.0) \text{ MeV}$$

In 1995 we started at PNPI new measurements of $\pi^- p \to \pi^0 n$ differential cross sections in order to decrease the errors of phase shifts for $\pi^- p$ scattering.

The most recent experimental results were obtained by the PNPI group in collaboration with the group of Prof. V. Kanavets from ITEP (Moscow) for the spin rotation parameters A and R in π^+p elastic scattering at pion beam momentum 1.43 GeV/c [4]. The main parts of the apparatus are the polarized proton target with horizontal orientation and the carbon-plate polarimeter using optical spark chambers with television read-out. Results are presented in Fig.1 in comparison with the predictions of the phase shift analyses KH [5], CMU-LBL [6] and VPI [7]. New results do not confirm at 1.43 GeV/c the predictions of PSAs KH and CMU-LBL. At the same time, the analysis VPI describes satisfactorily the results obtained in this experiment.

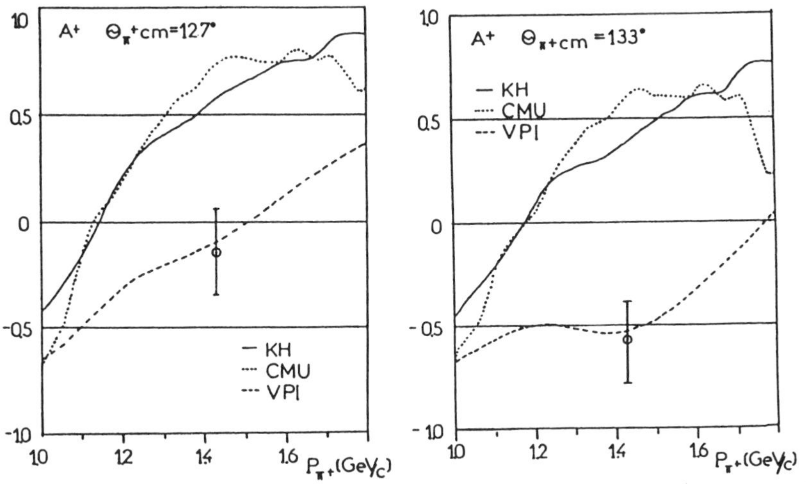

Figure 1: Results of measurement of the spin rotation parameter A^+ in comparison with the predictiions of different PWAs.

It is important to emphasize that the production of η-mesons in π^-p reactions is a very useful new tool for baryon spectroscopy. One of the advantages is the isospin selectivity due to the fact that the η-meson has isospin $I = 0$. The reaction $\pi^-p \rightarrow \eta n$ selects the $I = 1/2$ resonances. The PNPI 1000 MeV synchrocyclotron (pion beams up to 800 MeV/c) was used for studying the reaction $\pi^-p \rightarrow \eta n$ in the region of the $S_{11}(1535)$-resonance.

Yields of the reaction $\pi^-p \rightarrow \eta n$ have been measured in the near-threshold region, at momenta of incident pions 670, 680, 690, 695, 700, and 710 MeV/c. The experiment was carried out by means of detecting neutrons produced in this reaction and measuring their energies by TOF-technique [8]. Measured momentum dependences of the reaction yield for nine angles were compared with results of Monte Carlo simulation (code GEANT) performed for η mass M = 547.45 MeV [9] – see Fig.2. The results of GEANT calculations describe satisfactorily the experimental data for all angles.

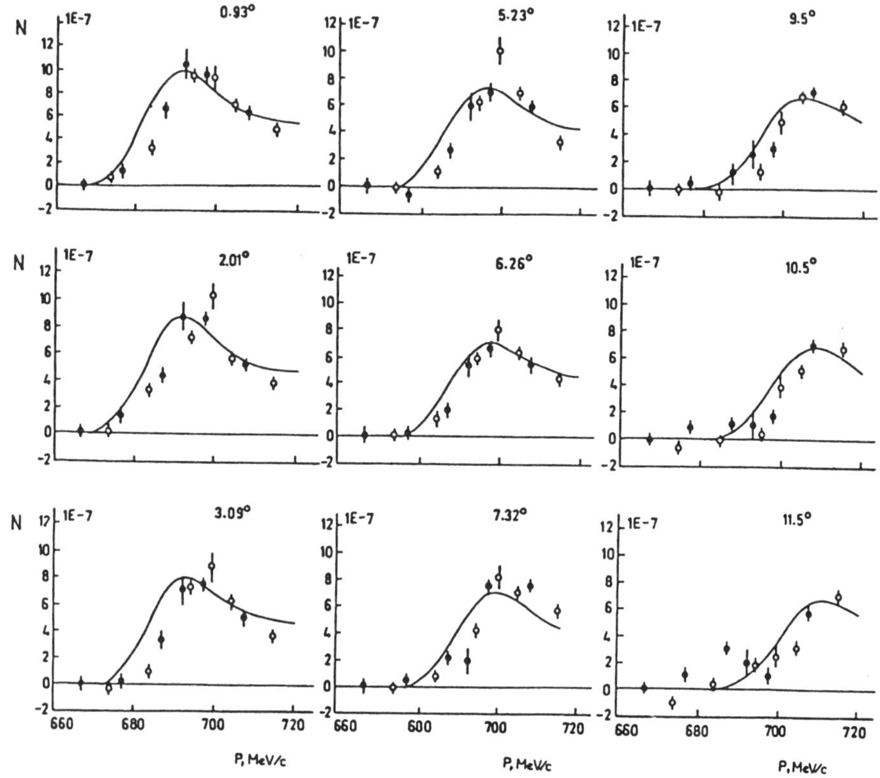

Figure 2: Yields of the reaction $\pi^- p \to \eta n$ *vs* momentum of incident pions for different neutron angles (in the lab. system). Shown by curves are results of Monte Carlo simulation normalized in peaks to the experimental data.

This work is supported by the Russian Foundation for Basic Research.

References

1. *S.P.Kruglov et al.*, πN Newsletter, No 8 (1993) 42.
2. *M.E.Sadler*, Proc. Int. Conf. on Mesons and Nuclei at Intermediate Energies, Dubna, Russia, 3 – 7 May 1994, World Scientific, p.3.
3. *V.V.Abaev and S.P.Kruglov*, Z.Phys., A352 (1995) 85.
4. *I.G.Alekseev et al.*, Phys.Lett., B351 (1995) 585.
5. *G.Höhler et al.*, Handbook of Pion-Nucleon Scattering. Physics Data, No 12–1 (Fachinformationzentrum, Karlsruhe, 1979).
6. *R.E.Cutkosky et al.*, Phys.Rev., D20 (1979) 2839.
7. *R.A.Arndt et al.*, Phys.Rev., D43 (1991) 2131.
8. *I.V.lopatin et al.*, Few Body Systems Suppl., 9 (1995) 241.
9. *Particle Data Group*, Phys.Rev., D50 (1994) 1196.

RECENT RESULTS ON SPECTROSCOPY FROM ARGUS

Henning Schröder
DESY-F15, Notkestr.85, D-22603 Hamburg, Germany

Recent evidences from ARGUS are presented on the production of the $\Lambda_c(2593)^+$ in c quark fragmentation and the $f_0(980)$ in the decay $D^0 \to K^+K^-\overline{K}^0$.

1 Evidence for $\Lambda_c(2593)^+$ production

The first observation of an excited charmed baryon was made by ARGUS in 1993[1] with establishing the existence of the $\Lambda_c(2625)^+$ baryon. CLEO and later E687 reported the observation of another state decaying into $\Lambda_c\pi^+\pi^-$ at a mass of ≈ 2593 MeV/c^2. This state is also seen by ARGUS in the c quark fragmentation in e^+e^- annihilation at $\sqrt{s} \approx 10 GeV$ with 23.1 ± 6.0 entries (Figure 1). The Λ_c^+ is reconstructed in the decays $\Lambda_c^+ \to pK^-\pi^+, pK_s^0, pK_s^0\pi^+\pi^-, \Lambda\pi^+$, $\Lambda\pi^+\pi^0$ and $\Lambda\pi^+\pi^+\pi^-$. The fit results are shown in Table 1.

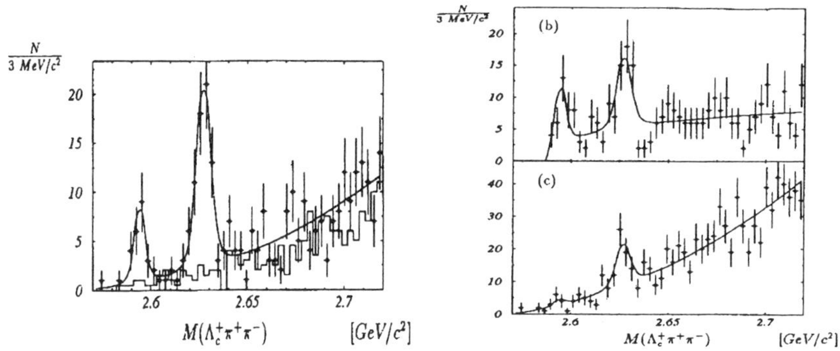

Figure 1: $\Lambda_c^+\pi^+\pi^-$ mass spectra.

Table 1: Fit parameters of the excited charmed baryons.

Baryon state	Mass, MeV/c^2	Γ, MeV/c^2	$\sigma \cdot BR$, pb
$\Lambda_c(2593)^+$	$2594.2 \pm 1.7 \pm 0.8$	$3.4 \pm 2.9 \pm 1.5$	$5.5 \pm 1.4 \pm 1.7$
$\Lambda_c(2625)^+$	$2627.0 \pm 1.1 \pm 0.6$	$3.3 \pm 2.4 \pm 1.4$	$12.5 \pm 2.1 \pm 3.1$

Figure 1 (b) shows the $\Lambda_c^+\pi^+\pi^-$ mass spectrum for $\Lambda_c^+\pi^+$ mass combinations compatible with the production of Σ_c baryons, Figure 1 (c) displays the complement. The analysis of the $\Lambda_c^+\pi^+$ mass spectra yields resonant substructure for the decays of both excited charmed baryons: $\frac{BR(\Lambda_c(2593)^+\to\Sigma_c\pi^\pm)}{BR(\Lambda_c^{*+}\to\Lambda_c\pi^+\pi^-)} = (0.72\pm0.23\pm0.23)$ and $\frac{BR(\Lambda_c(2625)^+\to\Sigma_c\pi^\pm)}{BR(\Lambda_c^{*+}\to\Lambda_c\pi^+\pi^-)} = (0.29\pm0.09\pm0.07)$.

We find good agreement with previous results for masses, widths and production cross section, only the fraction of resonant decays is larger in the case of the $\Lambda_c(2625)^+$ compared to results from CLEO and E687.

2 Evidence for the decay $D^0 \to f_0(980)K^0$ and search for $D_s^+ \to f_0(980)\pi^+$

The decay $D^0 \to K^+K^-\overline{K}^0$ together with its subset $D^0 \to \phi\overline{K}^0, \phi \to K^+K^-$ has been first measured by ARGUS in 1985 [2]. The K^+K^- spectrum in the decay $D^0 \to K^+K^-\overline{K}^0$ has been decomposed by a partial wave analysis allowing at low masses for $L = 0$ or $L = 1$ angular momentum states (Figure 2).

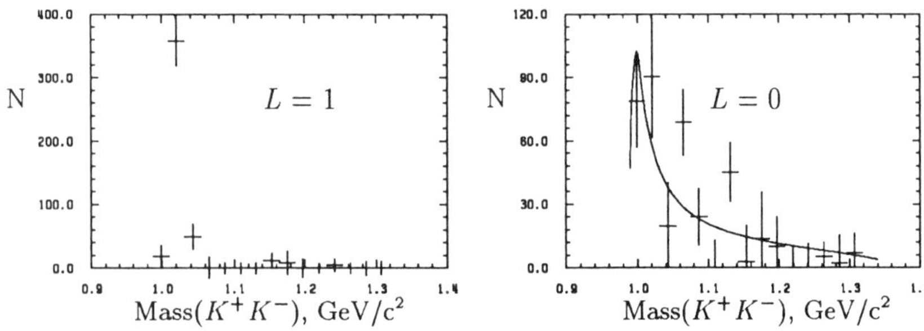

Figure 2: K^+K^- mass spectra for $L = 1$ and $L = 0$.

The $L = 1$ component is completely given by the ϕ meson whereas for $L = 0$ one observes a strong threshold enhancement which is in shape described by a Flatte distribution for the $f_0(980)$. Interpreting the $L = 0$ component as completely due to the presence of an $f_0(980)$ we find: $\frac{\Gamma(D^0\to f_0\overline{K}^0)\cdot Br(f_0\to K^+K^-)}{\Gamma(D^0\to\phi\overline{K}^0)} = (0.48\pm0.07)$. A similar analysis for the decay $D_s^+ \to K^+K^-\pi^+$ yield no signal for an $f_0(980)$ with $\frac{\Gamma(D_s^+\to f_0\pi^+)\cdot Br(f_0\to K^+K^-)}{\Gamma(D_s^+\to\phi\pi^+)} < 0.1$.

References

1. H.Albrecht *et al.* (ARGUS), *Phys. Lett.* B **317**, 227 (1993).
2. H.Albrecht *et al.* (ARGUS), *Phys. Lett.* B **158**, 525 (1985).

AMPLITUDES AND AMBIGUITIES IN THE SCATTERING AND PHOTOPRODUCTION OF PSEUDOSCALAR MESONS

G. KEATON AND R. WORKMAN
Department of Physics, Virginia Tech, Blacksburg, VA 24061

Ambiguities in pion-nucleon scattering amplitudes have analogs in pseudoscalar-meson photoproduction. This uncertainty has implications for the problem of resonance identification in partial-wave analyses. We compare our results with an earlier study by Omelaenko and with make connections with the nodal trajectory method of Tabakin and co-workers. We also mention how these results are connected to some recent spin-rotation measurements from ITEP-PNPI.

1 Introduction

There is a close analogy between the ambiguities found in the photoproduction and the elastic scattering of pseudoscalar mesons. This is particularly evident if one adopts the method of Dean and Lee[1]. In a previous work[2] we considered the problems encountered in constructing a complete experiment. As double-polarization data is scarce, it is more relevant to consider the case in which the cross sections and single-polarization observables are known and double-polarization observables are 'predicted'. This case was considered by Omelaenko[3] and is also relevant to the ITEP-PNPI measurements[4].

2 Fitting Angular Distributions

Omelaenko applied the method of Gersten[5] to photoproduction in order to study the ambiguities associated with a fit to observables measured over an angular range. Each transversity amplitude was written as a polynomial in $\tan\theta/2$ and ambiguities were expressed in terms of the complex conjugation of associated roots[6]. Restricting the polynomial to a fixed and finite order[7] eliminates some of the ambiguities we employed[2] in constructing a 'complete experiment' (at a single energy and angle). However, the case in which *all* roots are conjugated is related to one we[8] have considered.

The above technique is similar to Barrelet's method[9], which has been used in meson-nucleon scattering. In the Barrelet formalism, the operation $w_i \to 1/w_i^*$ (Barrelet conjunction) applied to a root preserves the cross section and polarization. This ambiguity has recently been proposed[10] to explain the large discrepancy between recent spin-rotation measurements[4] and some of the older analyses (which were performed without the existence of this data in the resonance region).

3 Nodes in the Observables

The nodal trajectory method[11] is concerned with the nodes found in photoproduction observables. Observables are split into 'Legendre classes' having similar nodal structure. However, this grouping is not respected by the ambiguities found[2,8] in a generalization of Dean and Lee's formalism[1]. By applying the nodal trajectory method near threshold, using a restricted number of multipoles and Watson's theorem (where valid), this problem can be circumvented. At higher energies, however, further assumptions are required to remove the ambiguities. The ambiguities found by Omelaenko[3] are particularly important, as they exist within the context of a truncated partial-wave expansion.

Acknowledgments

This work was supported in part by the U.S. Department of Energy Grants DE-FG05-88ER40454 and DE-FG05-95ER40709A.

References

1. N.W. Dean and P. Lee, Phys. Rev. D **5**, 2741 (1972).
2. G. Keaton and R. Workman, Phys. Rev. C **53**, 1434 (1996).
3. A.S. Omelaenko, Sov. J. Nucl. Phys. **34**, 406 (1981) [Yad. Fiz. **34**, 730 (1981)]; V.F. Grushin, E.M. Leikin and A. Ya. Rotvain, Short Communications in Physics [Kr. soobshch. po fizike] **11**, 20 (1978).
4. I.G. Alekseev *et al.*, Phys. Lett. B **351**, 585 (1995).
5. A. Gersten, Nucl. Phys. B **12**, 537 (1969).
6. Omelaenko tested the effect of root conjugation in the delta resonance region. However, as this transformation violates Watson's theorem, it is more relevant at higher energies.
7. This is a severe restriction. A finite (but not fixed) order results in further ambiguities.
8. G. Keaton and R. Workman, submitted to Phys. Rev. C.
9. E. Barrelet, Nuovo Cimento **8 A**, 331 (1972).
10. S.P. Kruglov, contribution to this meeting.
11. B. Saghai and F. Tabakin, nucl-th/9509027 ; C.G. Fasano, F. Tabakin, and B. Saghai, Phys. Rev. C **46**, 2430 (1992).

PHOTOPRODUCTION OF ETA MESONS ON DEUTERIUM*

P. Hoffmann-Rothe[1], E. Hourany[1], M. Breuer[1], J.-P. Didelez[1†], M. Rigney[1], J.Ajaka[1], G.Anton[2], J.Arends[2], G.Berrier-Ronsin[1], W.Beulertz[2], G.Blanpied[4], A. Bock[2], G. v. Edel[3], R. Frascaria[1], K. Helbing[2], J. Hey[2], M. Krebeck[3], R. Maass[3], G. Nöldeke[2], B. Preedom[4], L. Rosier[1], B. Saghai[5], M. Schumacher[3], F. Smend[3], S. Whisnant[4], B. Zucht[1]

[1] IN2P3, Institut de Physique Nucléaire, 91406 Orsay, France.
[2] Physikalisches Institut, Nussallee 12, 53117 Bonn, Germany.
[3] Zweites Physikalisches Institut, Bunsenstr. 7-9, Göttingen, Germany.
[4] University of South Carolina, Physics Dept, Columbia, SC 29208, USA.
[5] Laboratoire DAPNIA/SPN, 91191 Gif-sur-Yvette, France.

Measurements of the total and differential cross sections for η-meson photoproduction on ^2D liquid target from threshold to 1.2 GeV, have been taken using the tagged Bremsstrahlung photon beam produced by the electrons extracted from the ELSA storage ring at Bonn. The reaction was identified by detecting the η decay products in the neutral meson spectrometer SPES0-2π, while the recoil baryons (proton, neutron, or deuteron) were detected by a variety of large angle scintillator detectors. We succeeded to identify completely the final states corresponding to the production of an η meson on a Quasi-Free (QF) proton, a QF neutron, and the coherent deuteron. The differential cross sections corresponding to the production of a coherent deuteron in the final state have been measured, from threshold to 800 MeV; they are 10 times smaller than the only previous measurement reported by Anderson and Prepost in 1969. This is consistent with an Isoscalar part of the Amplitude much smaller than the Isovector one. The differential cross sections are in good agreement with the theoretical prediction based on the impulse approximation mechanism; indicating in particular, fairly small contributions from rescattering terms. A direct measurement of the neutron to proton cross section ratios has been obtained by integrating the counting rates on the corresponding QF peaks and is 0.8±0.1, from 700 MeV to 900 Mev, with a small angular dependance. These two results, by comparison to the measured "free proton" data should allow to reconstruct the "free neutron" cross sections in a rather model-independent way. The coherent differential cross sections are presented on Fig.1, and an example of the QF neutron/proton cross section ratios are displayed on Fig.2.

*Supported in part by BMFT, Germany, HC&M, Brussels, and the NSF, USA
†*E-mail address:* didelez@ipncls.in2p3.fr

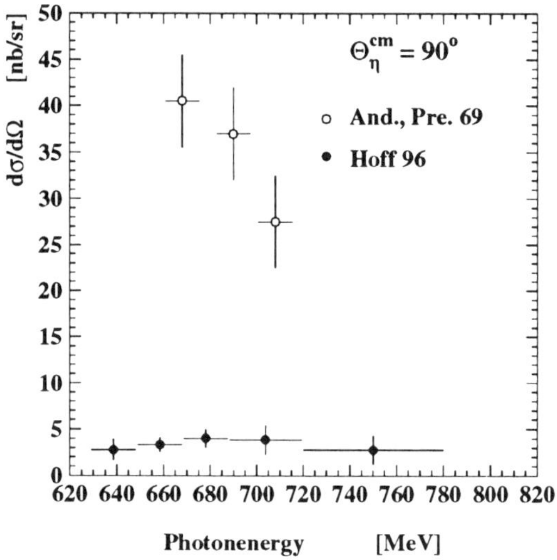

Figure 1. Differential cross sections for the reaction $\gamma\, d \rightarrow \eta\, d$ compared to the former results of Anderson and Prepost.
(*Ref.: Anderson and Prepost, Phys.Rev.Lett. 23(1969) 46*)

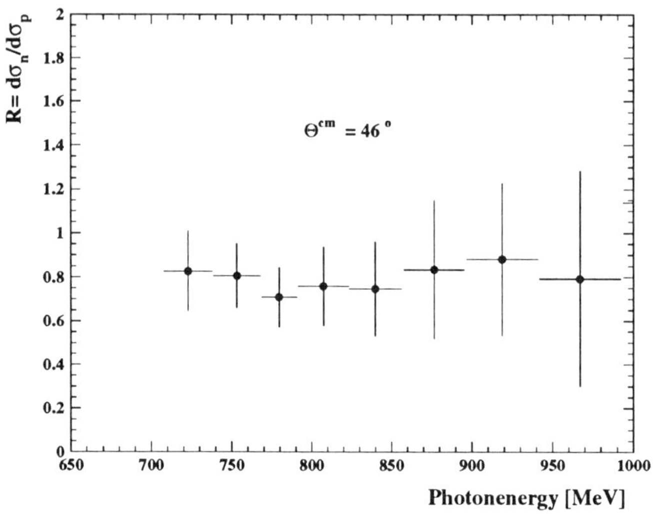

Figure 2. Ratios of $d(\gamma, \eta n)p/d(\gamma, \eta p)n$ 'Quasi Free' reactions, as a function of incident photon energies.

NOVEL THEORETICAL IMPLICATIONS OF THE PHOTOPRODUCTION OF ETA PRIME MESONS

NIMAI C. MUKHOPADHYAY, JI-FANG ZHANG
Physics Department, Rensselaer Polytechnic Institute, Troy, NY 12180-3590

M. BENMERROUCHE
Saskatchewan Accelerator Laboratory, University of Saskatchewan, Saskatoon, SK S7N 5C6

1 Introduction

The structure of eta prime (η') meson has attracted a lot of recent attention. The reasons for this renewed interest [1,2,3,4,5,6] range from the chiral U(1) anomaly to the facility with which it can probe the structure of baryon resonances. It is the last topic that concerns us here. We study this via the photoproduction reaction

$$\gamma + p \rightarrow \eta'(958) + p, \tag{1}$$

with the threshold energy for the photon in the lab of 1447 MeV (E_γ^{th}), corresponding to a cm energy W = 1896MeV. Only experimental data for (1) comes from the 1968 experiment by the ABBHHM collaboration [7]. There are ambitious proposals to explore the reaction(1) at CEBAF and GRAAL. The electroproduction studies are proposed at the HERMES facility using the ^{3}He target.

2 Tree level structure

Our paper [6] has already reviewed the tree-level physics of the reaction(1): it involves the usual nuclear s- and u- channel Born terms involving the nucleon exchange, the t-channel vector meson (ρ^0, ω) exchange and the s- and u-channel N* exchanges. The quark model studies by Capstick and Roberts [5] allow us to identify the important N* exchanges suggested by this approach. This turns out to be a very valueble guideline for our study.

3 The eta-prime-nucleon coupling

For the effective Lagrangian $L = g_{\eta'}[i\bar{N}\gamma N\eta']$, the coupling constant $g_{\eta'}$ to the nucleon is not known. One can use the quark model to determine $g_1/g_8 \approx \sqrt{2}$, where g_1 and g_8 are the singlet and octet η-nucleon coupling, ignoring the strangeness of the nucleon. Using the singlet and octet mixing angle of 20^0, and using the information on g_η, we can obtain $2 \leq g_{\eta'} \leq 6$.

4 Vector meson exchanges

Using the η-photoproduction analysis, radiative decays and quark model, we can constrain the elctrostrong couplings of the vector meson exchanges. In the standard notation [8], we have $\lambda'_\rho g_v^\rho + \lambda'_\omega g_v^\omega \approx 4.1$, $\lambda'_\rho g_t^\rho + \lambda'_\omega g_t^\omega \approx 12.1$.

5 The N* exchanges

Guided by the quark model, we find the role of the N*(2080) excitation to be most prominent. Its treatment in the effective Lagrangian approach is more subtle than that of a spin-1/2 object. It requires fixing of the so-called "off-shell" parameters and the dealing with the s- and u-channel form factors, without which the cross-section explodes. Extraordinarily, the N*(2080) contributes via the E_{o^+} and M_{1^-} multipoles.

6 Interpretation of the ABBHHM data

The role of the N*(2080) excitation in explaining the feature of the ABBHHM data [7] is reasonable. Without this resonance, hardly much cross-section is left. Dominant multipoles are E_{0^+} and M_{1^-}.

7 Angular distributions

Fig.1a shows the angular distribution to be highly asymmetric around $E_\gamma \approx$ 2GeV. Polarization observables show some peculiar variations around backward angles (Fig.1b-d).

8 Conclusions

The eta prime photoproduction emerges as a surprising spectroscopic tool to probe the structure of the N*(2080), a little known resonance in the famous 2GeV region of "missing resonances".

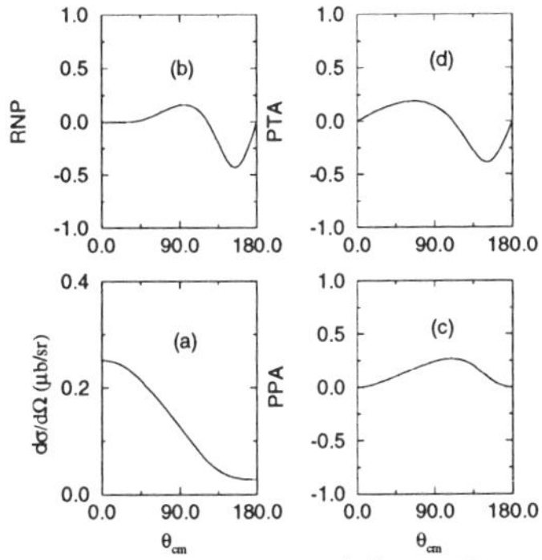

Figure 1: (a) Angular distribution of eta prime, (b) recoil nucleon polarization (RNP), (c) polarized photon asymmetry (PPA), (d) polarized target asymmetry (PTA) as functions of cm angle θ_{cm} at $E_\gamma = 1836.4$MeV.

Acknowledgments

This work is supported by DOE (NCM and JFZ) and by the Natural Sciences and Engineering Research Council of Canada (MB).

1. See various invited talks on the η' structure in the Proceedings of the 1996 Las Cruces Workshop on Eta Prime, to be published. Also S. Weinberg, Phys. Rev. D11, 3583 (1975); G.'t. Hooft, Phys. Rev. Lett. 37, 8 (1976). For a brief review, see S. Coleman, *Aspects of Symmetry* (Cambridge University Press, Cambridge, 1988), p. 307.
2. R. Baltrusaitis *et al.*, Phys. Rev. D32, 2823 (1985).
3. F. Lenz, Nucl. Phys. B279, 119 (1987).
4. F. Iachello, N.C. Mukhopadhyay and L. Zhang, Phys. Rev. D44, 898 (1991).
5. S. Capstick and W. Roberts, Phys. Rev. D49, 4570 (1994).
6. J.-F. Zhang, N.C. Mukhopadhyay and M. Benmerrouche, Phys. Rev. C52, 1134 (1995).
7. R. Erbe *et al.*, Phys. Rev. 175, 1669 (1968).
8. M. Benmerrouche, N.C. Mukhopadhyay and J.-F. Zhang, Phys. Rev. D51, 3237 (1995); M. Benmerrouche, N.C. Mukhopadhyay, Phys. Rev. Lett. 67, 1070 (1991); N.C. Mukhopadhyay, J.-F. Zhang and M. Benmerrouche, Phys. Rev. Lett. 75, 3022 (1995); and Phys. Lett. B 364, 1-7 (1995).

Missing Resonances *via* pseudoscalar Meson Photoproduction

Bijan Saghai
Service de Physique Nucléaire, CEA/DSM/DAPNIA,
Centre d'Etudes de Saclay, F-91191 Gif-sur-Yvette

Frank Tabakin
Department of Physics and Astronomy, University of Pittsburgh

Baryonic resonances predicted by quark-based studies[1] are typically weakly coupled to the πN channel, but should appear in other meson-nucleon systems, such as ηN.

Selected η photoproduction spin observables are potentially useful in detecting some of the low-mass missing nucleonic resonances. Spin observables for pseudoscalar meson photoproduction can be described [2] in the following general form: $\mathcal{O} = \sin^M \theta \sum_{m\geq 0}^{n} a_m \cos^m \theta$, with $M = 0$ for observables $(\sigma, E, C_{z'}, L_{z'})$; $M = 1$ for $(P, H, C_{x'}, L_{x'}, T, F, O_{x'}, T_{z'})$; and $M = 2$ for $(\Sigma; G; O_{z'}; T_{x'})$. The polynomial coefficients, a_m, can be expanded in terms of electric and magnetic multipole amplitudes $E_\ell^{\pm}, M_\ell^{\pm}$; see Ref. 3 for angular momentum of $\ell \leq 2$ expressions. It is assumed that resonances dominate, feed into the multipoles, and contribute to the polynomial coefficients as shown in Table 1. This general dependence of the expansion coefficients on underlying driving resonances (Table 1) has been applied to pion, kaon and eta meson photoproduction in our recent preprint where a more complete table for polynomials with $n \leq 5$ is given.

For eta mesons, starting from the assumption that the process is dominated by only $S\&D$ resonances, one sees from Table 1, for example, that for $T\&F, a_0 = 0$ while $a_1 \neq 0$; therefore, 90° nodes are predicted. Other observables can be similarly analyzed. As additional resonance scenarios are added in, including one of the "missing" resonances [1], the P_{13} at 1.88 GeV, we find [2] that the $M = 2$ single (Σ) and $M = 1$ single and double polarization P, T, H, F asymmetries, are the most likely observables for this missing resonance to play a role in the energy dependence and the nodal structure.

1. R. Koniuk and N. Isgur, *Phys. Rev.* D**21**, 1868 (1980); S. Capstick and W. Roberts, *ibid* D**49**, 4570 (1994).
2. C.G. Fasano, F. Tabakin and B. Saghai, *Phys. Rev.* C**46**, 2430 (1992); B. Saghai and F. Tabakin, *ibid* **53**, 66 (1996) and preprint June, 1996.
3. P. Girard, B. Saghai and F. Tabakin, Saclay Report, DAPNIA-SPhN-96-01 (1996).

	a_0	a_1	a_2
$\mathcal{L}_0$			
$d\sigma$ & $\mathbf{E}$	$\boxed{S} \oplus P_{2Ii} \oplus D_{2Ij}$ $\oplus \boxed{SD_{2Ij}} \oplus PP'$ $\oplus DD'$	$\boxed{SP_{2Ii}} \oplus P_{2Ii}D_{2Ij}$	$P' \oplus D_{2Ij}$ $\boxed{SD_{2Ij}} \oplus PP'$ $\oplus DD'$
$\mathbf{C_{Z'}}$ & $\mathbf{L_{Z'}}$	$\boxed{SP_{2Ii}} \oplus P_{2Ii}D_{2Ij}$	$\boxed{S} \oplus P_{2Ii} \oplus D_{2Ij}$ $\oplus \boxed{SD_{2Ij}} \oplus PP'$ $\oplus DD'$	$\boxed{SP'} \oplus P_{2Ii}D_{2Ij}$
$\mathcal{L}_{1a}$			
$\mathbf{P}$ & $\mathbf{H}$	$\boxed{SP_{2Ii}} \oplus P_{2Ii}D_{2Ij}$	$\boxed{SD_{2Ij}} \oplus PP'$ $\oplus DD'$	$PD' \oplus P'D_{2Ij}$
$\mathbf{C_{X'}}$ & $\mathbf{L_{X'}}$	$\boxed{S} \oplus P_{2Ii} \oplus D_{2Ij}$ $\oplus \boxed{SD_{2Ij}} \oplus PP'$ $\oplus DD'$	$\boxed{SP'} \oplus PD_{2Ij}$ $\oplus P'D'$	$P' \oplus D_{2Ij} \oplus$ $\boxed{SD'} \oplus DD'$
$\mathcal{L}_{1b}$			
$\mathbf{T}$ & $\mathbf{F}$	$\boxed{SP'} \oplus P_{2Ii}D_{2Ij}$	$P' \oplus D_{2Ij}$ $\boxed{SD_{2Ij}} \oplus PP'$ $\oplus DD'$	$PD' \oplus P'D_{2Ij}$
$\mathbf{O_{X'}}$ & $\mathbf{T_{Z'}}$	$P' \oplus D_{2Ij}$ $\oplus \boxed{SD_{2Ij}} \oplus PP'$ $\oplus DD'$	$\boxed{SP'} \oplus P_{2Ii}D_{2Ij}$	$P' \oplus D_{2Ij}$ $\oplus \boxed{SD'} \oplus DD'$
$\mathcal{L}_2$			
$\mathbf{\Sigma}$ & $\mathbf{G}$	$P' \oplus D_{2Ij}$ $\oplus \boxed{SD_{2Ij}} \oplus PP'$ $\oplus DD'$	$PD' \oplus P'D_{2Ij}$	D' $\oplus DD'$
$\mathbf{O_{Z'}}$ & $\mathbf{T_{X'}}$	$\boxed{SP'} \oplus P_{2Ii}D_{2Ij}$	$P' \oplus D_{2Ij}$ $\oplus \boxed{SD'} \oplus DD'$	$P'D'$

Table 1: Multipole dependence of the polynomial coefficients for all single and double polarization observables in pseudoscalar meson production. Here $i = 1$ and $i = 3$ denote the $J = 1/2$ and $3/2$ P-waves ($P \equiv P_{2I1}$ and $P' \equiv P_{2I3}$, respectively); while $j = 3$ and $j = 5$ are the $J = 3/2$ and $5/2$ D-waves ($D \equiv D_{2I3}$ and $D' \equiv D_{2I5}$, respectively). Single letters refer to terms of the type $|E_0^+|^2 \equiv S$; such terms are listed in the first row for each set of observables, when appropriate. In the following rows the interference terms are given in the notation $SD_{2Ij}, \cdots$. The term $P_{2Ii}D_{2Ij}$ is short for $PD \oplus PD' \oplus P'D \oplus P'D'$. The boxed letters show how the strong S-wave contributes to the a_m coefficients for each observable.

SEARCH FOR EXOTIC BARYONS IN THE EXPERIMENTS ON PROTON BEAM OF IHEP ACCELERATOR WITH ENERGY $E = 70\,\mathrm{GeV}$
SPHINX Collaboration (IHEP–ITEP)

V.A. BEZZUBOV, S.V. GOLOVKIN, A.G. KHOLODENKO, A.P. KOZHEVNIKOV, V.P. KUBAROVSKY, A.I. KULYAVTZEV, V.F. KURSHETSOV, L.G. LANDSBERG, M.YU. MATVEEV, V.V. MOLCHANOV, V.A. MUKHIN, V.I. SOLYANIK, D.V. VAVILOV, V.A. VICTOROV
Institute for High Energy Physics, Protvino, Russia

G.K. KLIGER, V.Z. KOLGANOV, G.S. LOMKATZI, A.F. NILOV, V.T. SMOLYANKIN
Institute of Theoretical and Experimental Physics, Moscow, Russia

1. The extensive research program of studying the diffractive production reactions on the proton beam with $E_p = 70\,\mathrm{GeV}$ is carried out in the experiments on the SPHINX setup[1]. The main aims of this program are the searches for cryptoexotic baryons with hidden strangeness $|uuds\bar{s}\rangle$. Only part of these experiments is discussed here.

2. In our previous measurements[2] the coherent diffractive production reactions $p + C \to [\Sigma(1385)^0 K^+] + C$ and $p + C \to [\Sigma^0 K^+] + C$ were investigated. The evidences for new baryon states were obtained in the study of hyperon-kaon effective mass spectra in these two reactions: $X(2050) \to \Sigma(1385)^0 K^+$ with mass $M = (2052 \pm 6)\,\mathrm{MeV}$ and width $\Gamma = (35^{+22}_{-35})\,\mathrm{MeV}$ and $X(2000) \to \Sigma^0 K^+$ with $M = (1999 \pm 6)\,\mathrm{MeV}$ and $\Gamma = (91 \pm 17)\,\mathrm{MeV}$. The unusual dynamical features of these massive states (relatively narrow decay widths, anomalously large branching ratios for decays with strange particle emission) make them very serious candidates for cryptoexotic pentaquark baryons with hidden strangeness. In spite of significant statistical confidence of the observation of new states we considered these results as preliminary, which must be confirmed in further measurements with increased statistics.

3. The first results on new test run with the partly modified SPHINX setup (with a new γ-spectrometer and with better possibilities for $\Lambda \to p\pi^-$ identification) for the coherent reaction $p + C \to [\Sigma^0 K^+] + C$ are present now[3]. These results are in a very good agreement with our previous data for this reaction. In the combined effective mass spectrum $M(\Sigma^0 K^+)$ —

see Fig. 1a, besides some threshold structure with $M \sim 1800\,\text{MeV}$, a strong peak $X(2000) \to \Sigma^0 K^+$ is dominated. The average mass and width of this peak are $M = 1996 \pm 7\,\text{MeV}$ and $\Gamma = 99 \pm 21\,\text{MeV}$. Because of the significant changing in the equipment, the good agreement between previous and new data gave us additional confidence for the real existence of $X(2000)$ anomalous baryon state — the candidate for the baryon exotics $|uuds\bar{s}\rangle$.

4. A narrow threshold structure with $M \sim 1810\,\text{MeV}$ and $\Gamma \sim 60\,\text{MeV}$ was observed in the $M(\Sigma^0 K^+)$ mass spectrum, as it was stated before. It was found that this structure is produced practically only in the region of very small $P_T^2 (\leq 0.01 \div 0.02\,\text{GeV}^2)$ where it is seen quite clearly (see for example Fig. 1b).

5. First data in nonperipheral domain (at $P_T^2 > 0.3\,\text{GeV}^2$) were obtained in the reactions $p + N \to [\Sigma^0 K^+] + N$ and $p + N \to [p\eta] + N$. In both reactions there are evidences for a narrow enchantment with $M \sim 2350\,\text{MeV}$ and $\Gamma \leq 60\,\text{MeV}$ [3]

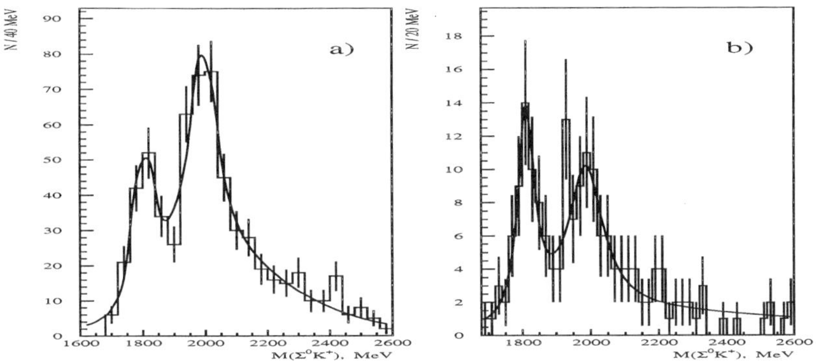

Figure 1: Study of the mass spectra $M(\Sigma^0 K^+)$ in the coherent diffractive production reaction $p + C \to \Sigma^0 K^+ + C$ at $E_p = 70\,\text{GeV}$: a) The combined spectrum $M(\Sigma^0 K^+)$ from the old and new data ($P_T^2 < 0.1\,\text{GeV}^2$). The threshold structure with $M \sim 1810\,\text{MeV}$ and the strong $X(2000)$ peak are clearly seen. b) The mass spectrum from the new data with the region of very small $P_T^2 (< 0.01\,\text{GeV}^2)$. The parameters of this structure are $M = 1812 \pm 7\,\text{MeV}$ and $\Gamma = 56 \pm 16\,\text{MeV}$. This structure is produced practically only in the region of $P_T^2 < 0.01 \div 0.02\,\text{GeV}^2$.

References

1. D.V. Vavilov at al., *Yad. Fiz.* **57**, 241 (1994).
2. S.V. Golovkin et al., *Z. Phys.* C **68**, 585 (1995).
3. V.A. Bezzubov et al., *Yad. Fiz.* **59**, (1996, in press)

OBSERVATION OF A RESONANCE-LIKE STRUCTURE IN THE ENERGY SPECTRUM OF PHOTONS FROM THE REACTION $pp \to pp\gamma\gamma$ AT PROTON ENERGY OF 200 MeV.

A.S. KHRYKIN
and
V.M. ABAZOV, Yu.K. AKIMOV, V.F. BOREIKO, S.N. ERSHOV,
S.B. GERASIMOV, N.V. KHOMUTOV, N.A. KUTCHINSKY, A.B. LAZAREV,
A.G. MOLOKANOV, S.N. SHILOV, Yu.G. SOBOLEV., V.P. ZORIN
Laboratory of Nuclear Problems, Joint Institute for Nuclear Research
141980 Dubna, Moscow region, Russia

A search for exotic dibaryons has been carried out using a new method based on the measurement of the energy spectrum for γ-rays from the $pp \to pp\gamma\gamma$ reaction at 200 MeV. The obtained spectrum clearly exhibits a narrow peak at around 42.0 MeV with a significance of greater than 8σ. Distribution of the reconstructed pp invariant mass showed a narrow peak with a mass of 1923.5 $\pm$ 4.5 MeV and a width(FWHM) of 31.3 $\pm$ 5.0 MeV.

In this paper we report further results [1] of searching for exotic, narrow dibaryon resonances in the reaction $pp \to pp\gamma\gamma$ [2,3]. We believe that among all possible mechanisms of this reaction there is the channel $pp \to \gamma^2B \to pp\gamma\gamma$ leading to formation of dibaryons(^2B's) with such the quantum numbers for which the elastic decay $^2B \to NN$ is either strictly forbidden by the Pauli principle or suppressed to the order $O(\alpha^2)$ by the isospin selection rules and thus the radiative decay $^2B \to NN\gamma$ dominates. The method of searching for these states is to measure the photon energy distribution in the $pp \to pp\gamma\gamma$ reaction, detecting both photons in coincidence. The narrow dibaryons, if they exist, should be seen as sharp γ-lines against a smooth curve due to the photons, coming from the radiative, three-body resonance decay and double pp-bremsstrahlung with an expected good signal-to-background ratio. The experiment was performed by using the proton beam of the JINR phasotron with an energy of 198 MeV and average intensity of $5.5 \cdot 10^8$ p/s. A liquid hydrogen target of a 0.3 g/cm^2 thick has been used. Two photons of the reaction explored were detected in coincidence by an array of seven CsI(Tl) hexagonal crystals, each with an outer diameter 10 cm and length 15 cm and a cylindrical NaI(Tl) crystal of diameter 15 cm and length 10 cm. To reduce a large background of protons scattered by the target, both the γ-detectors were set on either side of the beam to detect backward emitted photons at angles of 111^0 and 240^0, respectively. Prompt particles from the target were attenuated by 20 cm polyethylene bars placed in between the target and each detector. The en-

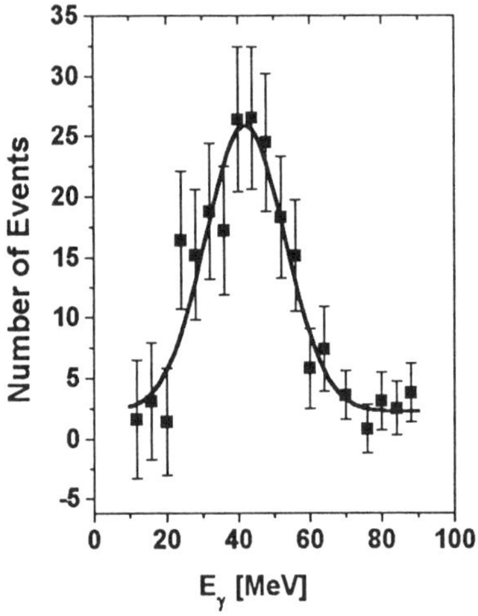

Figure 1: The photon energy spectrum for the S_1 γ-ray detector after subtracting the empty target contribution. The solid line is the result of a Gaussian fit to the data points

ergy spectra of γ-rays were measured in the energy range from 10 to 100 MeV. Data for the empty target and the full one were taken in two successive runs for about 10 hours and 20 hours, respectively. The γ-ray energy spectra for the γ-detector S_1 after subtraction of the empty target contribution is shown in Fig.1. A structure at energy near 42 MeV is clearly seen in this spectrum. A fit to a Gaussian gave the energy of this peak 42.0$\pm$4.5 MeV with width 26.3$\pm$4.0 MeV (χ^2_{14}= 6.7). An analogous peak at the energy about 42 MeV is present in the spectrum for the S_2 γ-detector. For the energy and the width of this peak the fit to a Gaussian gave the values 42.5$\pm$ 6.5 MeV and 31.3 $\pm$ 7.0 MeV (χ^2_{13}= 5.6). If one assumes that a narrow dibaryon resonance is responsible for the peak observed one can reconstruct a corresponding dibaryon mass distribution. It has a peak at $M_B \sim$ 1923 MeV. This distribution when fitted with a Gaussian, gave a fitted mass 1923.5$\pm$4.5 MeV and a width 31.3$\pm$5.0 MeV (χ^2_{14}= 5.0). The significance of the effect exceeds 8σ.

1 Acknowledgements

This work was supported by grants 94-02-04796, 94-02-05702 and 96-02-19147-a from the Russian Foundation of Fundamental Research.

1. A.S.Khrykin, in πN *Newsletter*, **No.10** ,ed. by D.Drechsel, G.Höhler, W.Kluge and B.M.K.Nefkens, October, 1995, p.67
2. S.B.Gerasimov and A.S. Khrykin, Mod. Phys. Lett. **A8**,(1993)2457.
3. S.N.Ershov, S.B. Gerasimov and A.S. Khrykin, Yad. Fiz. **58** (1995)911.

Determination of Baryon Resonance parameters from the data of πN and γN Reactions

T. VRANA, S.A. DYTMAN
Department of Physics and Astronomy, University of Pittsburgh

T.-S.H. LEE
Physics Division, Argonne National Laboratory

The Unitary Coupled-channel Isobar model of Cutkosky et.al.[1] will be reported. Work on extending the model to include photoproduction and electroproducton, aiming at the determination of the $\gamma N \to N^*$ transition matrix element from the wealth of data expected from CEBAF in the next few years, is in progress.

1 The Cutkosky Model

The model used in this paper was one first used by the Carnegie Mellon Berkeley Collaboration in 1979 to extract resonance parameters from πN elastic partial wave T-matrix elements.[1]

This model was chosen because it incorporates many desirable physical constraints such as unitarity, proper threshold dependence, analyticity, and time reversal invariance. It is a coupled channels model where all channels are taken to be two body channels. $\pi\pi$N channels are treated as quasi-two-body channels such as ρN and $\pi\Delta$. Other channels can be included such as ηN, KΛ, ωN,etc...

1.1 T Matrix Elements

The form of the T-matrix elements is:

$$T_{ab} = \sum_{i,j=1}^{N} f_a(s)\sqrt{\rho_a(s)}\gamma_{ai}G_{ij}(s)\gamma_{jb}\sqrt{\rho_b(s)}f_b(s) \tag{1}$$

Written in this way, equation 1 has a unitary normalization:

$$Im(T_{ab}) = \sum_{c=1}^{M} T^*_{ac}T_{cb} \tag{2}$$

The T-matrix is related to the S-matrix via S=1-2iT, where $S^\dagger S$=1.

In equation 1, i,j refer to intermediate states in a scattering process. Two of these intermediate states have their masses below πN threshold and approximate the nucleon pole and other background processes. The indices a, b refer to

Table 1: Extracted Resonance Parameters

	$S_{11}(1535)$	$S_{11}(1650)$	$D_{15}(1675)$
Mass (MeV)	1541 ± 13	1671 ± 10	1682 ± 5
Width (MeV)	144 ± 32	199 ± 44	139 ± 4
$X_{\pi N}(\%)$	37 ± 8	72 ± 7	37 ± 2
$X_{\eta N}(\%)$	51 ± 9	14 ± 8	0 ± 0
$X_{\pi\pi N}(\%)$	12 ± 6	14 ± 8	63 ± 2

initial and final channels such as πN and ηN. G_{ij} is a dress propagator matrix for the intermediate states. G_{ij} is defined in terms of intermediate state masses and channel self energies. ρ is a phase space factor and $f_{a,b}$ is a threshold form factors which give the amplitudes the proper threshold dependence.

The above formalism is valid for two body final states. Three body final states are described by quasi two body states such as ρN and $\pi\Delta$. For these channels the phase space factors and self energies are defined by an averaging procedure which gives the proper analytic form for two body channels where one or both particles have a width.

More details can be found in the original article by Cutkosky et al. [1]

2 Resonance Parameters for the $S_{11}(1535)$, $S_{11}(1650)$ and, $D_{15}(1675)$ Resonances

Extracted Resonance parameters for resonances in the S_{11} and D_{15} partial waves are given in Table 2.

The S_{11} and D_{15} partial waves were focused on because of the contrasting difficulty levels in extracting resonance parameters. The $D_{15}(1675)$ is strongly coupled to the πN channel and has no nearby resonances or inelastic channel openings. The $S_{11}(1535)$ is coupled strongly to both the πN and ηN channels. The ηN channel opens at 1486 MeV, which causes a "cusp" in the πN elastic S_{11} T-matrix element. It also has interference effects with the nearby $S_{11}(1650)$ resonance. These effects are well handled within the Cutkosky Model.

References

1. R.E. Cutkosky,C.P. Forsyth, R.E. Hendrick,R.L. Kelly, *Phys. Rev.* D **20**, 2839 (1979).

ELECTROMAGNETIC COUPLINGS OF NUCLEON RESONANCES[a]

T. FEUSTER, U. MOSEL
Institut für Theoretische Physik, Heinrich-Buff-Ring 16, 35392 Giessen, Germany

An effective Lagrangian calculation of pion photoproduction including all nucleon resonances up to $\sqrt{s} = 1.7$ GeV is presented. We compare our results to recent calculations and show the influence of different width parametrizations and offshell cutoffs on the photoproduction multipoles. We determine the electromagnetic couplings of the resonances from a new fit to the multipole data[1].

Effective Lagrangian models have been used to calculate pion and eta photoproduction on the nucleon; also Compton scattering has been investigated succesfully in this framework. For the pion these models were mainly applied to the $\Delta(1232)$ region and it was found that proper treatment of the πN rescattering is important [2,3]. Compton scattering was also calculated in this energy region only[4]. In the case of eta production the couplings of the $N(1535)$ resonance were extracted from the E^p_{0+} pion multipole and the $\pi N \to \eta N$ data[5,6]. So far no model for all channels exists that includes the nucleon resonances of the first and second resonance region and treats rescattering properly.

In our model we follow an ansatz used by Garcilazo et al. [7] that calculates the resonance amplitudes from the corresponding Feynman diagrams and, at the first stage, neglects rescattering. Once these contributions are under control we then will proceede to fulfill unitarity in a coupled-channel $(\pi, \eta, 2\pi, \gamma)$ K-matrix calculation.

Currently we study the sensitivity of the extracted electromagnetic couplings to the different cutoffs that were used in other calculations. Our results show that a u-channel cutoff is neccessary and that the couplings do not depend on the exact shape of the width, as long as it decreases rapidly enough with energy. We also checked the sensitivity of the fits on the values of the resonance masses used. A better description of the data can be obtained for the pole positions instead of the mass values.

Furthermore we have calculated the offshell contributions of the spin-$\frac{3}{2}$ resonances $\Delta(1232)$ (P_{33}) and $N(1520)$ (D_{13}) to the E^p_{0+} pion multipole. Comparison with other calculations shows that these contributions are comparable to the rescattering effects. Therefore it is not sufficient to extract the $N(1535)$ (S_{11}) couplings from calculations using *only* S_{11} resonances. To illustrate this the E^p_{0+} multipole is shown in Fig. 1. Without the offshell contributions we could not find such a good description of the data.

The total η photoproduction cross section can be reproduced quite well including the Born terms, vector mesons and the $N(1535)$ resonance. For a description of the differential data however, it is neccessary to allow for $N(1440)$ and $N(1520)$ contributions as well.

In the case of Compton scattering a good description of the differential cross section is found using the set of parameters extracted by fitting the pion multipoles. The agreement at higher energies is achived without the use on an artifical form factor for the Born terms that was needed in calculations based on simple Breit-Wigner parametrizations of the resonance amplitudes[8].

[a]Work support by GSI Darmstadt and BMBF

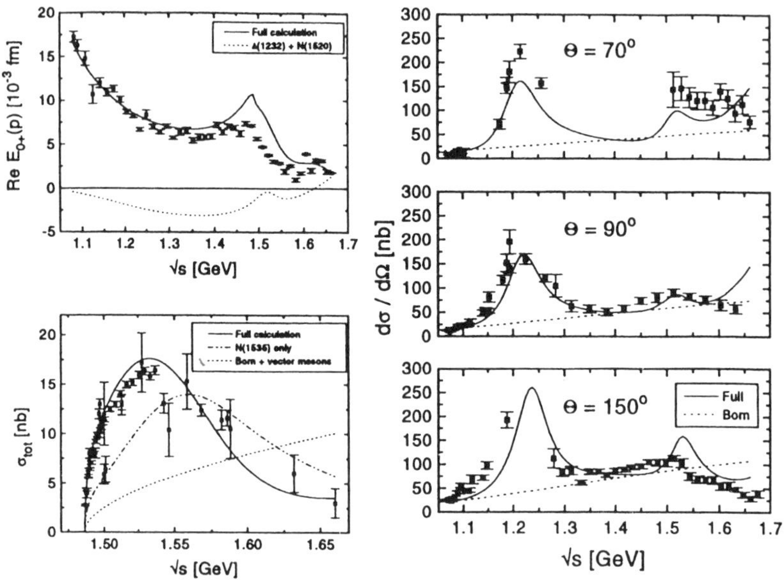

Figure 1: Counterclockwise from left: Real part of E^p_{0+} pion photoproduction multipole, total $\gamma N \rightarrow \eta N$ cross section, differential Compton cross section.

From our calculation it is clear that a full coupled-channel calculation is needed to explain the multipole data in detail and to extract reliable coupling parameter estimates.

References

1. T. Feuster and U. Mosel, submitted to *Nucl. Phys.* A preprint #9604026 available from *nucl-th@xxx.lanl.gov*
2. S. Nozawa, B. Blankleider and T. - S. H. Lee, *Nucl. Phys.* A **513**, 459 (1990)
3. R. Davidson, N. C. Mukhopadhyay and R. Wittman, *Phys. Rev.* D **43**, 71 (1991)
4. V. Pascalutsa and O. Scholten, *Nucl. Phys.* A **591**, 658 (1995)
5. C. Sauermann, B. L. Friman and W. Nörenberg, "Proceedings of the Sixth International Symposium on Meson-Nucleon Physics and the Structure of the nucleon, Blaubeuren, Germany July 1995"
6. M. Benmerrouche, N. C. Mukhopadhyay and J. F. Zhang, *Phys. Rev.* D **51**, 3237 (1995)
7. H. Garcilazo and E. Moya de Guerra, *Nucl. Phys.* A **562**, 521 (1993)
8. Y. Wada *et al*, *Nucl. Phys.* B **247**, 313 (1984)

BARYONS IN A QUARK MODEL WITH CHIRAL DYNAMICS

L. Ya. GLOZMAN[1], Z. PAPP[2], W. PLESSAS[1]

[1] *Inst. for Theor. Physics, University of Graz, A-8010 Graz, Austria*
[2] *Inst. of Nucl. Res., Hung. Acad. of Sci., H-4001 Debrecen, Hungary*

Beyond the scale of spontaneous breaking of chiral symmetry light and strange baryons can be viewed as systems of three constituent quarks which interact by the exchange of Goldstone bosons (pseudoscalar mesons) and are subject to confinement [1]. We report results for the light-baryon spectra obtained from rigorous three-body Faddeev calculations [2]. Besides the confinement potential of linear form, the Goldstone-boson-exchange (GBE) interaction between the constituent quarks is taken into account to all orders. Our results further support the adequacy of the GBE for baryon spectroscopy. In particular, we are able to describe the energies of all 14 lowest states in the N and Δ spectra in close agreement with phenomenology.

For the GBE the spin-spin component of the interaction between the constituent quarks i and j consists of the octet and singlet exchange parts:

$$V_\chi^{octet}(\vec{r}_{ij}) = \left\{\sum_{a=1}^{3} V_\pi(\vec{r}_{ij})\lambda_i^a\lambda_j^a + \sum_{a=4}^{7} V_K(\vec{r}_{ij})\lambda_i^a\lambda_j^a + V_\eta(\vec{r}_{ij})\lambda_i^8\lambda_j^8\right\}\vec{\sigma}_i\cdot\vec{\sigma}_j, \quad (1)$$

$$V_\chi^{singlet}(\vec{r}_{ij}) = \frac{2}{3}\vec{\sigma}_i\cdot\vec{\sigma}_j V_{\eta'}(\vec{r}_{ij}). \quad (2)$$

Here $\vec{\sigma}$ and $\vec{\lambda}$ are the quark spin and flavor matrices, respectively. In the simplest case, when the boson field satisfies the linear Klein-Gordon equation, one has the following spatial dependence for the meson-exchange potentials in (1) and (2):

$$V_\gamma(\vec{r}_{ij}) = \frac{g_\gamma^2}{4\pi}\frac{1}{3}\frac{1}{4m_i m_j}\{\mu_\gamma^2\frac{e^{-\mu_\gamma r_{ij}}}{r_{ij}} - 4\pi\delta(\vec{r}_{ij})\}, \quad (\gamma = \pi, K, \eta, \eta'). \quad (3)$$

Here m_i are the light quark masses whose values we take to be 340 MeV, as it is suggested from nucleon magnetic moments, and μ_γ are the meson masses. Within the pseudoscalar meson octet we work with a coupling constant $\frac{g_8^2}{4\pi} = 0.67$ whose value is extracted from the phenomenological pion-nucleon coupling [1]. In case of the flavor singlet η', which is decoupled from the octet due to the $U(1)_A$ anomaly, a slightly different coupling is employed.

The δ-function term in eq. (3) is strictly valid only for pointlike particles. It must be smeared out as the constituent quarks and pseudoscalar mesons have finite sizes and because the boson fields in a chiral Lagrangian should in fact satisfy a nonlinear equation. Consequently we replace the δ-function by a finite-width two-parameter representation. With the total potential between the constituent quarks, $V_\chi + V_{conf}$, involving only 4 free parameters [2], we succeed in describing all of the 14 lowest states of the N and Δ spectra, including the absolute value of the nucleon (ground state), see the figure below. In the most unfavourable cases deviations from the experimental values do not exceed 3%! In addition all level orderings are correct. In particular, the positive-parity state $N(1440)$ (Roper resonance) lies *below* the pair of negative-parity states $N(1535)$ - $N(1520)$. The same is true in the Δ spectrum with $\Delta(1600)$ and the pair $\Delta(1620)$ - $\Delta(1700)$. Thus a long-standing problem of baryon spectroscopy now appears definitely resolved. The reason for this achievement evidently lies in the flavor-spin dependence of the GBE [1].

References

[1] L.Ya. Glozman and D.O. Riska, Phys. Rep. **268** (1996) 263.

[2] L.Ya. Glozman, Z. Papp, W. Plessas, hep-ph 9601353, Phys. Lett. B, in print.

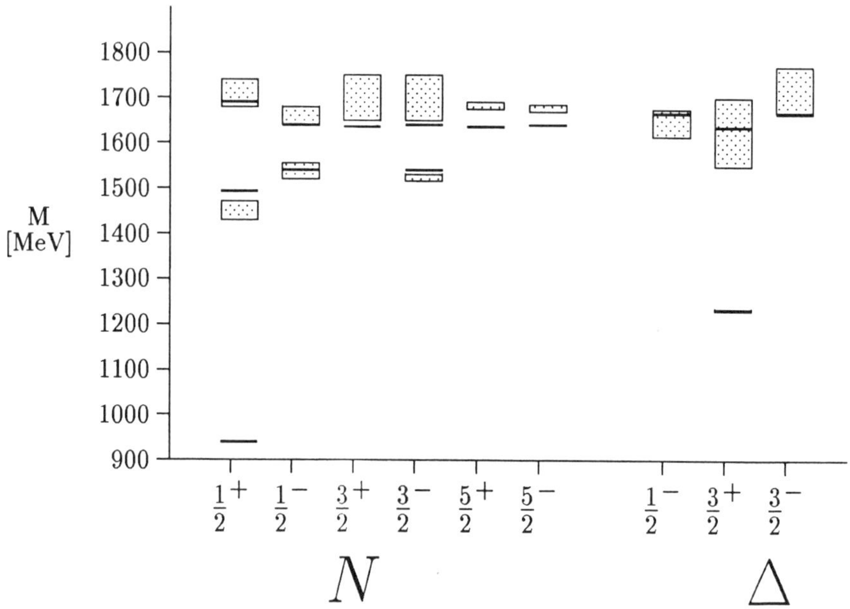

RESONANCES IN $\gamma\gamma \to \pi^+\pi^-\pi^0$

Henning Schröder (ARGUS)
DESY-F15, Notkestr.85, D-22603 Hamburg, Germany

The two-photon reaction $\gamma\gamma \to \pi^+\pi^-\pi^0$ was studied by ARGUS. The reaction is dominated by the formation of the $a_2(1320)$ meson with $\Gamma_{\gamma\gamma}(a_2(1320)) = (0.96 \pm 0.03 \pm 0.13)$ keV. The $\gamma\gamma$ helicity 0 fraction of $\Gamma_{\gamma\gamma}$ is $(6.7 \pm 2.2)\%$. Upper limits for the $\pi(1300)$ and the $\pi_2(1670)$ are presented.

Using the ARGUS detector at the DORIS II e^+e^- storage ring at DESY the reaction $\gamma\gamma \to \pi^+\pi^-\pi^0$ has been investigated [1]. From an integrated luminosity of $\int Ldt = 456pb^{-1}$ collected at $\sqrt{s} \approx 10.4$ in total 3567 event candidates of the reaction $\gamma\gamma \to \pi^+\pi^-\pi^0$ have been selected. This event sample which contains less than 10% of well understood background events allows for a partial wave analysis (PWA) using a maximum likelihood method with the amplitudes (J^P, J_z)

$$\begin{array}{llllll} (2^+,\pm 2) & \to & \rho^\pm\pi^\mp & (0^-,0) & \to & \rho^\pm\pi^\mp \\ (2^+,0) & \to & \rho^\pm\pi^\mp & (0^-,0) & \to & f_2\pi^0 \\ (2^-,0) & \to & \rho^\pm\pi^\mp & (0^-,0) & \to & f_0(980)\pi^0 \\ (2^-,0) & \to & f_2\pi^0 & \text{isotropic} & \to & \pi^+\pi^-\pi^0. \end{array}$$

At low masses ($M_{3\pi} < 1.45$ GeV/c^2) the cross section is dominated by the formation of the $a_2(1320)$ resonance in its helicity 2 state, to which only the first three amplitudes contribute. A partial wave analysis with these amplitudes plus the $(0^-,0) \to \rho^\pm\pi^\mp$ and the isotropic one results in event distributions for these amplitudes (Figure 1). The analysis yields:

$$\Gamma_{\gamma\gamma}(a_2(1320)) = (0.96 \pm 0.03 \pm 0.13) \text{ keV}$$

and the $\gamma\gamma$ helicity 0 fraction of $\Gamma_{\gamma\gamma}$ is determined to be $(6.7 \pm 2.2)\%$. The precision of these results equals or exceeds the precision of the world average values. The ARGUS results confirm the predicted dominance of $J_z = 2$ in the two–photon formation of the a_2 meson. The $J_z = 0$ fraction of about 6% is best accounted for by a relativistic $q\bar{q}$ bound state model [2]

No clear indication is observed for the formation of the $\pi(1300)$ (Figure 1,left bottom) yielding:

$$\Gamma_{\gamma\gamma}(\pi(1300)) \cdot BR(\pi(1300) \to \rho^\pm\pi^\mp) < 0.54 \text{ keV} \quad (90\%\text{C.L.})$$

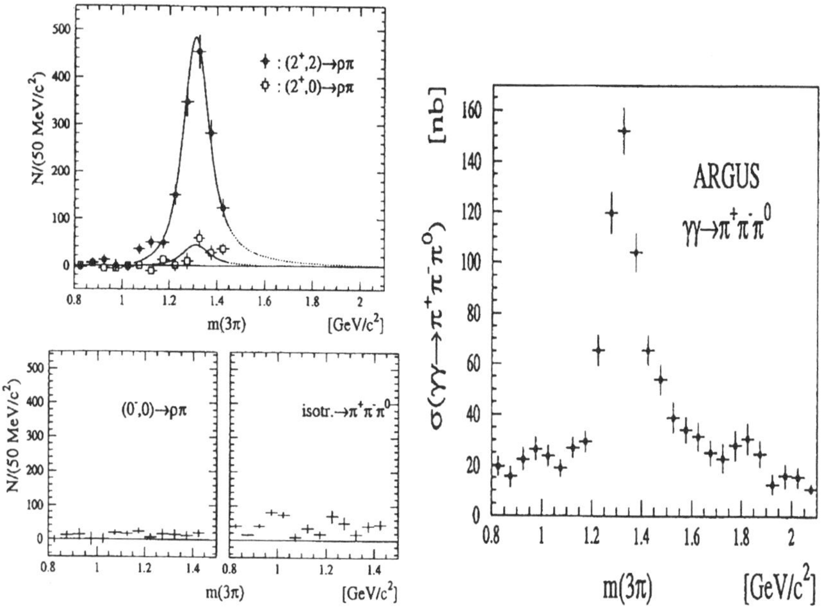

Figure 1: Event distribution for the displayed partial waves (left), the total cross section $\gamma\gamma \to \pi^+\pi^-\pi^0$ obtained by summing up all the partial wave contributions (right).

Figure 1 (right) shows the total $\gamma\gamma \to \pi^+\pi^0\pi^-$ cross section obtained by summing up all the partial wave contributions. It is interesting that in the region above 1.45 GeV most of the cross section belongs to the $J^P = 2^+$ waves and that the $J^P = 2^-$ $\rho\pi$ and $f_2\pi$ wave contributions equal zero within the statistical errors. This is in contradiction to an observation of $\pi_2(1670)$ with a two–photon partial width of $(1.41 \pm 0.23 \pm 0.28)$ keV and $(1.3 \pm 0.3 \pm 0.2)$ keV reported by the Crystal Ball [3] and CELLO [4] collaborations, respectively. ARGUS can set only an upper limit of $\Gamma_{\gamma\gamma}(\pi_2) < 0.19$ keV (90% C.L.) in gross disagreement with the other two experiments.

References

1. T.Podobnik, (ARGUS), *Resonances Observed in the Reaction* $\gamma\gamma \to \pi^+\pi^-\pi^0$, Doctor Thesis, University of Ljubljana, Slovenia (1995).
2. Z.P.Li, F.E.Close, T.Barnes, *Phys. Rev.* D **43**, 2161 (1991)
3. D.Antreasyan *et al.* (Crystal Ball) *Z. Phys.* C **48**, 561 (1991).
4. H.-J.Behrend *et al.* (CELLO) *Z. Phys.* C **46**, 583 (1990).

SPIN PARITY MEASUREMENT OF CENTRALLY PRODUCED $(\pi^+\pi^-)$ IN PROTON-PROTON COLLISIONS AT 800 GEV/C*

K. MARKIANOS, M. C. BERISSO, E. P. HARTOUNI, M. N. KREISLER, S. LEE, M. WANG, D. WESSON
U. of Massachusetts , Amherst, Massachusetts , USA

A. GARA, B. C. KNAPP
Columbia U., Nevis Labsoratory New York, USA

E. GOTTSCHALK
Carnegie Mellon U., Pittsburg, Pennsylvania, USA

D. C. CHRISTIAN, G. GUTIERREZ, A. WEHMANN
Fermilab, Batavia, Illinois, USA

J. FELIX, G. MORENO, M. REYES, M. SOSA
U. of Guanajuato, Leon, Guanajuato, Mexico

Experiment E690* at Fermilab recorded 5.5 billion $p+p \rightarrow p+X$ events using an 800 GeV/c proton beam and a liquid hydrogen target, during the 1991 fixed target run. We use a .5 billion subset of this sample, to perform amplitude analysis for the reaction $p+p \rightarrow p_s(\pi^+\pi^-)p_f$ for dipion invariant mass between threshold and 2.3 GeV. The amplitude analysis provides significant information for phenomenological S-wave production models, and sheds light on the nature of glueball candidate $f_0(1710)$.

We used an 800 GeV/c proton beam on a 2% interaction length liquid hydrogen target. We reconstructed the central $(\pi^+\pi^-)$ system in the multiparticle spectrometer (MS) (geometrical acceptance 700 mrad and momentum acceptance between 0.15 and 20 GeV/c), a Cherenkov counter (with a pion threshold of 2.55 GeV/c) and a time of flight system. The fast proton was reconstructed in the Beam Spectrometer (BS) which had very good momentum resolution between 650 and 800 GeV/c. The trigger required at least one track in the MS and at least one incoming/outgoing track in the BS. We collected 5×10^9 events on tape and we present here results from 10% of this sample.

Selection of the final state $p_s(\pi^+\pi^-)p_f$ we required two π-compatible tracks in the Cherenkov counter and rapidity gap between the p_s and any one of the two pions greater than 1.8. We restrict the dipion system to $-0.1<x_F<0$ and the transverse momentum for both protons $Pt^2<0.1$ (GeV/c)^2. Since the p_s is not directly observed a missing mass cut was applied as shown in Fig. 1a. Figures 1b and 1c present the invariant mass distributions after all cuts. We observe two dramatic drops in the invariant mass distribution at 1 and 1.5 GeV/c^2, and no significant signal in the $f_0(1710)$ region. We perform an amplitude analysis in the Gotfried-Jackson frame allowing only waves with l=0,2 and m=0,1. In Fig. 2 we show the uniquely determined solution dominated by S-wave below 1 GeV/c^2

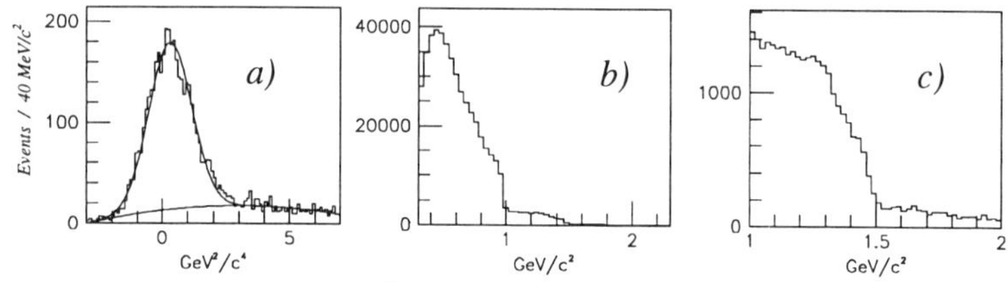

Fig.1 (a) (missing mass)^2, cut is -1,0.5 sigma, (b) Invariant mass, (c) Invariant mass, detail above 1 GeV/c^2. Not acceptance corrected.

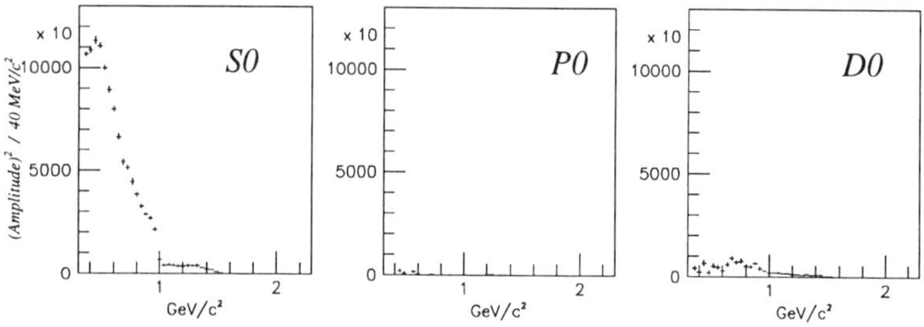

Fig.2 Acceptance corrected (amplitude)^2 solutions for l=0,2 and m=0,1. The reflectivity basis is used

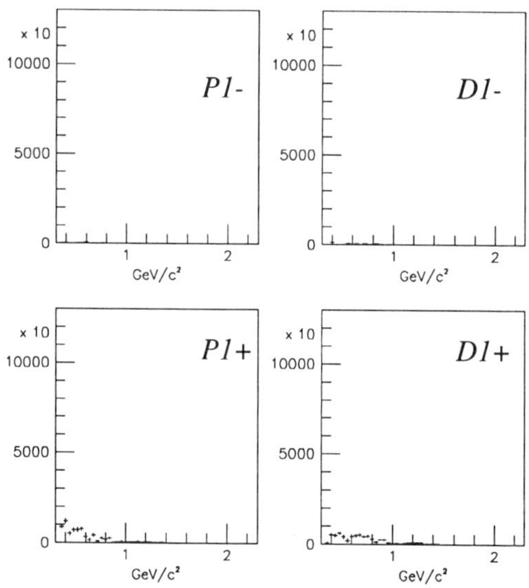

Acknowledgments

*Supported in part by the U.S. National Science Foundation and the U.S. Department of Energy.

MEASUREMENT OF SPIN ROTATION PARAMETERS AT ITEP ACCELERATOR AND ANALYSIS OF THE BARYON SPECTRUM NEAR $\sqrt{S} = 1.9$ GEV.

ITEP-PNPI Collaboration
V.P. KANAVETS
ITEP, B. Cheremushkinskaya 25, 117259 Moscow, Russia

We present the data on spin rotation parameters in the elastic $\pi^+ p_\uparrow$ scattering measured for the first time in the second resonance region and the results of the baryon spectrum analysis near $\sqrt{s} = 1.9$ GeV. The experiment was performed at the positive pion beam momentum 1.43 GeV/c for scattering in the backward hemisphere.

Up to now a large body of data on total and differential crossections and on the polarization parameter P in elastic pion-nucleon scattering has been accumulated. However the information on these observables is not sufficient for the direct reconstruction of the scattering amplitudes which leads to the discrete ambiguities of the partial-wave analysis (PWA). The measurements of the spin rotation parameters A and R allows to exclude these ambiguities and determine the characteristics of the baryon resonances reliably. But up to this experiment the data on spin rotation parameters in the resonance region was restricted by the pions momentum 0.75 GeV/c. The predictions of the three global PWAs, **CMB** [1], **KH** [2] and **VPI**[3], are significantly different in the question of the spectra, characteristics and status of the resonances.

The choice of the kinematic region was motivated by the fact of existence of the interesting resonance structure near $\sqrt{s} = 1.9$ GeV ($P_{lab} = 1.43$ GeV/c), which included 7 nearly degenerated resonances with $I = 3/2$, and the contradiction between the different PWAs predictions of spin rotation parameters in the angular region $120^o < \theta_{CM} < 140^o$ (fig. 1a). The spectra and properties of these resonances, given by various authors are essentially different also.

The experimental setup and data processing can be found in [4]. The parameter A^+ measured in this experiment is shown in fig. 1a in comparison with the predictions of **CMB-80**, **KH-80** and **SM-90** (**VPI**) PWAs. The experiment confirms the angular dependence of A^+ given by **SM-90**. The difference between PWAs in the region $\theta^{cm} = 120 - 136^o$ appears in a wide momentum range from 1.1 to 1.8 GeV/c. So the amplitudes found in **KH-80** and **CMB-80** need a correction.

The main instrument we used to analyze experimental data was transverse amplitudes formalism. Measurement of the differential crossections and polarization allows to extract the modulus of the amplitudes. The relative phase of these amplitudes can be found from the measurements of spin rotation parameters. Barrelet[5] showed that the scattering could be described by one polinom $F(w)$ of the variable $w = e^{i\theta}$ defined in complex plane. The important property of the amplitude $F(w)$ is the invariance of the differential crossection and polarization to the transition of the roots $w_i \rightarrow \frac{1}{w_i^*}$ called Barrelet conjunction. Using of the Barrelet conjunction provided a way of PWA amplitudes correction to account for the spin rotation experimental data. The prediction of PWA **CMB** for the parameter A corrected by the Barrelet conjunction is shown in fig. 1a. It demonstrates a good agreement with the experimental data. In fig. 1b Argand plot for partial amplitude F_{37} from original PWAs and PWAs corrected by Barrelet conjunction of the transverse amplitude zero trajectory at $\theta^{cm} = 120 - 140^o$ is shown. The agreement between different PWAs is better after correction. The are small changes for the resonances with 4-star rating, there is large influence on the 3-star rating resonances (elasticity decreases). The signal of resonance $D_{33}(1940)$ disappears.

1. R.E. Cutcosky et al., *Phys. Rev.* **D20** (1979) 2839
2. G. Höhler, Handbook of Pion-Nucleon Scattering., Physics Data. No 12-1 (Fachinformationzentrum, Karlsruhe, 1979)
3. R.A. Arndt et al., *Phys. Rev.* **D43** (1991) 2131
4. I.G. Alekseev et al., *Phys. Lett.* **B351** (1995) 585
5. E. Barrelet, *Nuovo Cim.* **A8** (1972) 31

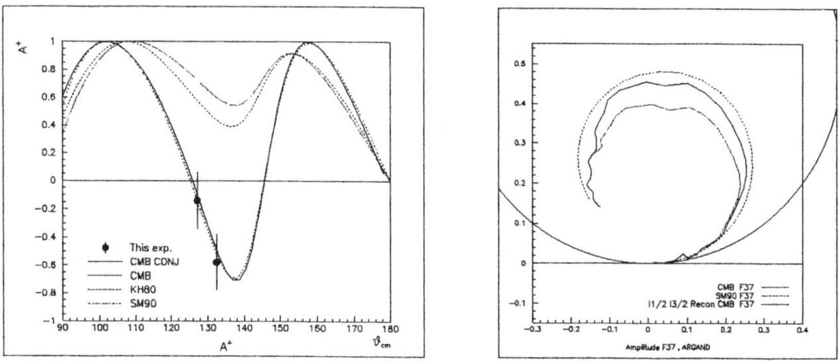

a b

Figure 1: Parameter A^+ for $\pi^+ p$ elastic scattering as function of θ_{cm} (a) and aragand plot for F_{37} (b)

STUDY OF THE π^-, K^- AND Σ^--ATOMS WITH A HIGH ENERGY PROTON BEAM

A.S.Denisov, O.L.Fedin, M.P.Gur'ev, Yu.M.Ivanov, L.P.Lapina, P.M.Levchenko, A.A.Petrunin, Yu.P.Platonov, A.G.Sergeev, A.I.Smirnov, V.M.Suvorov and A.V.Zhelamkov

Petersburg Nuclear Physics Institute, 188350 Gatchina, Russia

A crystal-diffraction method for study of π^-, K^- and Σ^--atoms with high energy proton beam has been developed. The obtained results are presented and prospects are discussed.

Recently we have developed a crystal diffraction method for investigations of the π^-, K^- and Σ^--atomic X-rays with a high energy proton beam. The observed X-ray intensities, the high metrological accuracy and the resolution of the crystal diffraction method allowed to carry out improved measurements of K^- and Σ^--masses.

The experiments was fulfilled at IHEP(Protvino) [1,2,3]. The measurement scheme is shown in Fig.1. As a source of π^-, K^- and Σ^--atomic X-rays we used a target bombarded by a slowly extracted 70 GeV protons. Slow π^-, K^- and Σ^--particles produced in target stopped in it forming exotic atoms. The X-rays emitted by these atoms were detected with the help of the focusing crystal diffraction spectrometer of Cauchois type.

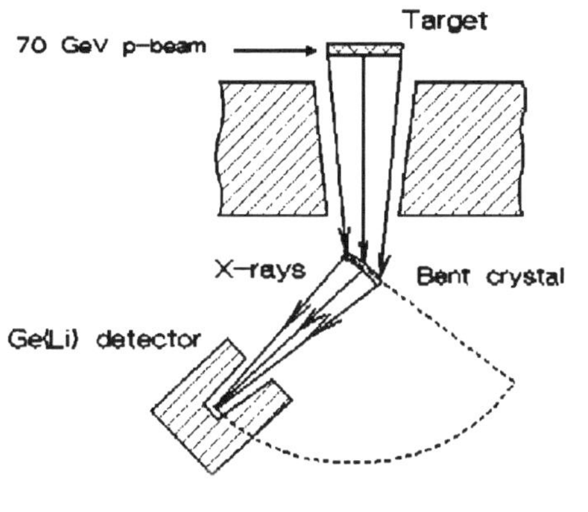

Fig.1. The layout of the experiment at IHEP.

The measured π^-, K^- and Σ^--carbon X-ray lines are shown in Fig.2,3,4. The luminosity of the spectrometer in these measurements was about 2×10^{-9}, resolution 6-7 eV. As reference lines we used $AgK_{\alpha1}$, $AgK_{\alpha2}$ and $SnK_{\alpha2}$ lines. The energies of noncircular 4d-2p $\pi^- - C$ and circular 4f-3d $K^- - C$ and 5g-4f $\Sigma^- - C$ transitions was found to be $(24\,828.36 \pm 0.15 \pm 0.15)$ eV, $(22\,105.61 \pm 0.26 \pm 0.14)$ eV and $(23\,420.47 \pm 0.49 \pm 0.62)$ eV, respectively [1,2,4]. The first error here is a statistical error, and the

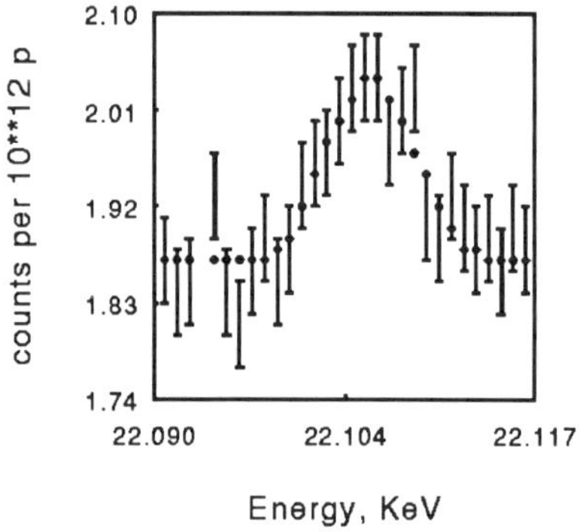

Fig.2. The 4f-3d $K^- - C$ transition[1].

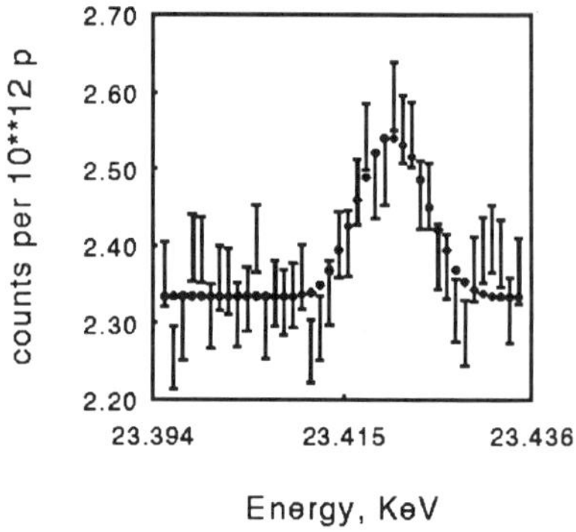

Fig.3. The 5g-4f $\Sigma^- - C$ transition[2].

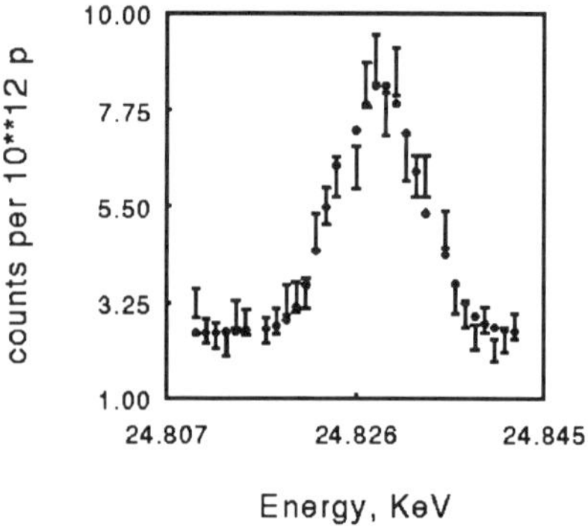

Fig.4. The 4d-2p $\pi^- - C$ transition[1].

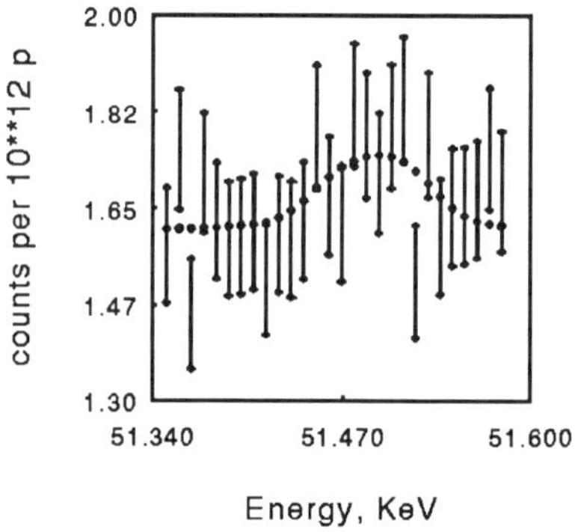

Fig.5. The 3d-2p π^--Ne transition[3].

second is a systematic error. The K^- and Σ^--masses were determined by fitting the calculated transition energies to the experimental values and found to be (493.696 ± 0.007) and $(1\,197.417 \pm 0.040)$ MeV/c^2 [1,2,4]. The obtained values have relative errors 14 ppm and 34 ppm, respectively, and these values are more precise than results of previous experiments [5]. In case of 4d-2p $\pi^- - C$ transition the calculation error of energy levels is about 15 ppm, therefore this transition was used only for testing method, but not for π^--mass determination.

The counting rates for the circular K^- and Σ^--atomic lines in these experiments were about 1-2 counts per 10^{13} incident protons on the target of 0.3-0.7 nuclear interaction length. With such rate, for example, the measurement of the K^--meson mass with an accuracy of 12 ppm has required less than 6 days (reference line measurements included) to collect about 10^4 counts.

The counting rates for circular π^--atomic lines were by two orders of magnitude higher. This fact stimulated us to search a direct production of pionic atoms in proton-nucleus collisions through decays of hadronic resonances near residual nuclei. We tried to find the 3d-2p transition of the π^--Ne atom with the energy 51519 eV from Mg target. The result is shown in Fig.5. The possible bump has (0.36 ± 0.15) count per 10^{12} incident protons, its position corresponds to (51521 ± 14) eV and its width does not exceed 100 eV. From this measurement the upper limit on the cross section for the production of an excited π^--Ne atom in p-Mg collision can be estimated as 0.5 mb[3].

In conclusion, our experiments proved the feasibility and the efficiency of study of exotic atoms with a high energy proton beam using crystal spectrometers in Gatchina target configuration[6]. We believe, the upgrades of FNAL and BNL accelerators will open new possibilities for precise study of pionic, kaonic, hyperonic and antiprotonic atoms with crystal spectrometers. The data in Table confirm this conclusion (estimations were made for installation described above).

Table. The running time for 1 ppm statistics on circular transitions of π^-, K^- and Σ^--carbon atoms at PNPI, IHEP and FNAL Main Injector.

accelerator	GeV	intensity	π^-, days	K^-, days	Σ^-, days
PNPI	1	3×10^{12} / 1s	10		
IHEP	70	3×10^{12} / 9s	1.4	900	900
FNAL MI	120	3×10^{13} / 3s	0.05	20	20

1. A.S.Denisov et al., JETP Lett., **54**, 558, 1991.
2. M.P.Gur'ev et al., JETP Lett., **57**, 400, 1993.
3. M.P.Gur'ev et al., JETP Lett., **58**, 71, 1993.
4. Yu.M.Ivanov, Thesis, Petersburg Nuclear Physics Institute, 1992.
5. Particle Data Group, Phys.Rev.D, **50**, 1173, 1994.
6. V.I.Marushenko et al., JETP Lett., **23**, 72, 1976.

New Results for the H Dibaryon in the Quark Cluster Model

Carl E. Wolfe
Dept. of Physics and Astronomy, York University,
Toronto, Canada, M3J 1P3

Kim Maltman
Dept. of Math and Stats, York University,
Toronto, Canada, M3J 1P3

The H dibaryon channel, $(I = 0, J = 0, S = -2)$, is revisited in the non-relativistic quark cluster model. The $\Lambda\Lambda$, $\Sigma\Sigma$, NΞ basis of previous studies is extended to incorporate an explicit 6q state, which allows access to certain quark configurations not available in the basis of ground state baryon pairs. We find that binding predicted using the two-baryon basis alone is significantly deepened by the addition of the new 6q state. The NRQCM thus appears, contrary to earlier findings, to be incompatible with the experimental information available for this channel.

The quark model of hadrons suggests the possible existence of six-quark bound states. The best known of these is the doubly strange *uuddss* $I = J = 0$ state known as the H dibaryon. It was shown to be bound by about 80 MeV in the $SU(3)_F$ limit if the MIT bag model with respect to the $\Lambda\Lambda$ 'fall-apart' threshold of 2231 MeV [1]. As an exotic state the H has the potential to be a highly non-trivial testing ground for hadronic models in the multiquark sector.

The observation of sequential Λ decay of double Λ hypernuclei [2] rules out the possibility of an H dibaryon bound by more than about 30 MeV, and thus creates a problem for many QCD-inspired quark models which predict more deeply bound states. Among this class of models, the quark cluster model has had considerable success in describing not only the baryon sector but also the qualitative features of short-distance NN and YN interactions. According to previous treatments of the H channel in the model, only rather weak binding, of order 10-20 MeV, is predicted [3,4], compatible with the experimental double Λ hypernuclear decay results.

These treatments of the H, however, which use the Resonating Group Method (RGM), include only the three two-baryon states $\Lambda\Lambda$, $\Sigma\Sigma$ and $N\Xi$ in the RGM ansatz for the H. Since one finds, e.g., that the diagonal $N\Xi$ potential is quite attractive, the $N\Xi$ configuration could well provide access for coupling to states more easily represented as symmetric spatial six-quark configurations (like the bag model H) than as two-baryon states. The analogous case of the I=0, J=3, S=0 channel is instructive. Including only the lowest-lying two-baryon state, $\Delta\Delta$, one finds a weak attractive potential and binding of a few

MeV[5,6] while, after the addition of a localized (hidden color) configuration, the binding is shifted to 260 MeV[6]. The presence of a weak attraction, rather than the strong hyperfine-induced repulsion present in most two-baryon channels, allows access to the hidden-color sector, which strongly alters the predicted binding.

We have explored the effect of a similar expansion of the basis in the H channel by adding to the three-two-baryon-state basis a state, h, modeled on the bag model H, (i.e. the h is taken to be spin-flavor-color antisymmetric and spatially symmetric, with the baryon-cluster and six-quark-cluster radii related as in the bag model). By no means does this state exhaust the space of available configurations, but it is a natural starting point. We then perform a bound-state RGM calculation, first incorporating only the usual cluster model quark Hamiltonian, and then a modified version involving effective πqq, Kqq, and σqq couplings (the latter incorporates the intermediate range attraction, as in Ref.[4]).

We find the following results: (1) with the quark Hamiltonian alone and no h, there is no bound state, as found in earlier quark cluster model treatments; (2) with π's, K's, and the σ, but no h, a bound state with weak binding of order 30 MeV is found, again compatible with earlier treatments; (3) with the h but no π, K, or σ, either no binding or only very weak binding results, depending in detail on the precise values of the model parameters; (4) with both the h and the π's, K's, and σ, a large binding, of order 70 MeV, is obtained. While we have noted considerable sensitivity to some of the model parameters, notably the σ meson mass, the addition of h always leads to an enhancement of binding.

Thus, when one already has a bound state, and hence significant attraction, the inclusion of the additional available configurations in the model (represented here by the h) results in a significant shift in binding. As such, it appears that, contrary to earlier conclusions based on the use of a more restrictive space of basis states, the cluster model fails to treat at least those portions of the six-quark sector where attractive entry channels exist in a manner consistent with experiment.

1. R.L. Jaffe, Phys. Rev. Let. **38** (1977) 195.
2. S. Aoki *et al.*, Prog. Th. Phys. **85** (1991) 1287.
3. M. Oka, K. Shimizu and K. Yazaki, Nucl. Phys. **A464** (1987) 700.
4. U. Straub *et al.*, Phys. Lett. **B200** (1988) 241; Z. Zhang *et al.*, Nucl. Phys. **A578** (1994) 573.
5. M. Oka and K. Yazaki, Phys. Lett. **B90** (1980) 41.
6. K. Maltman, Nucl. Phys. **A438** (1985) 669.

Session 7
Neutrino Physics and Weak Interactions

CONSTRAINTS ON EXTENSIONS OF THE STANDARD MODEL FROM POLARIZATION-ASYMMETRY CORRELATION MEASUREMENTS

P. QUIN
Department of Physics, University of Wisconsin-Madison, USA

J. DEUTSCH, F. GIMENO-NOGUES, J. GOVAERTS, T. OTTO, I. PEPE, R. PRIEELS, E. THOMAS
Institut de Physique Nucléaire, Université Catholique de Louvain, BELGIUM

J. CAMPS, P. DE MOOR, P. SCHUURMANS, N. SEVERIJNS, A. VAN GEERT, L. VANNESTE
Instituut voor Kern- en Stralingsfysika, Katholieke Universiteit Leuven, BELGIUM

M. ALLET, K. BODEK, M. FERRO-LUZZI, J. LANG, R. MÜLLER, S. NAVERT, O. NAVILIAT-CUNCIC, J. SROMICKI, E. STEPHAN, J. ZEJMA
Institut für Teilchenphysik, Eidgenössische Technische Hochschule Zürich, SWITZERLAND

B.R. HOLSTEIN
Department of Physics and Astronomy, University of Massachusetts, USA

B.A. BROWN
National Superconducting Cyclotron Laboratory and Department of Physics and Astronomy, Michigan State University, USA

R. KIRCHNER
Gesellschaft für Schwerionenforschung, Darmstadt, GERMANY

The Standard Model proposes a maximally parity violating left-handed interaction, mediated by charged gauge bosons $W_L^\pm$, to describe nuclear β-decay. The β-decay lepton-quark helicity amplitude[1] $a_{LL}^V \equiv 1$, with all other $a_{\ell q} = 0$ [ℓ or q = L or R]. In extensions of the standard model with right-handed bosons $W_R^\pm$ or several types of leptoquarks [$\mathcal{L}$], the helicity amplitude a_{RL}^V is small for small W_L-W_R or leptoquark mixing, while a_{RR}^V is given by

$$a_{RR}^V = \frac{g_R^2 V_{ud}^R m_W^2}{g_L^2 V_{ud}^L m_R^2} \quad \text{or} \quad \frac{g_{\mathcal{L}}^2 m_W^2}{g_L^2 V_{ud}^L m_{\mathcal{L}}^2}, \tag{1}$$

with m_W, m_R and $m_{\mathcal{L}}$ and g_L, g_R and $g_{\mathcal{L}}$ the masses and coupling constants of $W_L^\pm$, $W_R^\pm$ and $\mathcal{L}$, respectively. V_{ud} is the u-d quark-mixing matrix element. The present results set the most precise limits[1] for $|a_{RR}^V|$ with $|a_{RL}^V| \ll |a_{RR}^V|$.

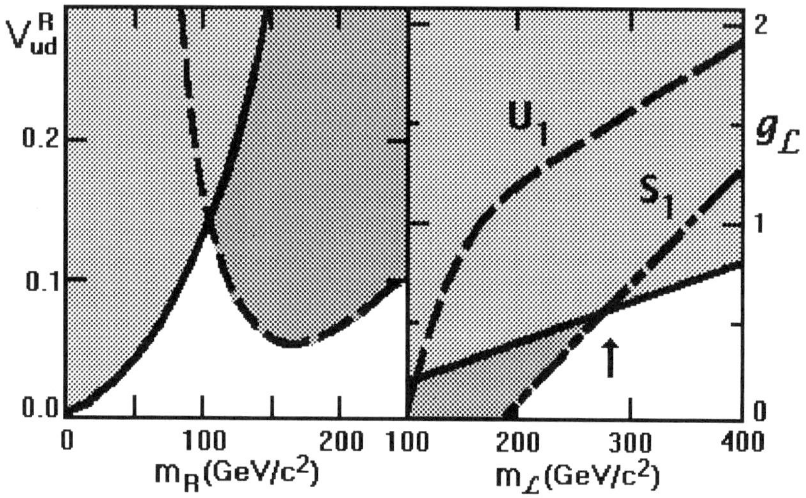

Figure 1: The present 90%CL exclusion limits on right-handed W_R, with $g_R = g_L$, are shown by the solid line and lightly shaded region in the left hand panel. The additional CDF limits[4], insensitive for $m_R \leq m_W$, are shaded darker. The present and ZEUS[5] 95%CL exclusion limits on S_1 scalar and U_1 vector leptoquarks are similarly shown in the right hand panel, with the HERA kinematic limit indicated by the arrow.

We have completed three measurements[2] of the relative positron polarization in the decay of polarized ^{107}In and ^{12}N. These polarization-asymmetry correlation experiments[3], which have enhanced sensitivity to a_{RR}^V and immunity from systematic uncertainties, give $|a_{RR}^V|^2 = 0.0010(27)$. Exclusion limits using Eq. (1) are shown in Fig. 1, along with results from CDF[4] and ZEUS[5]. A new ^{107}In experiment[6] will reduce the β-decay error on $|a_{RR}^V|^2$ to ≈ 0.0017.

This work[2] was supported in part by the U.S. National Science Foundation, the Belgian National Fund for Scientific Research (FNRS/NFWO), and the Swiss National Science Foundation.

1. J. Deutsch and P. Quin in *Precision Tests of the Standard Electroweak Model*, ed. P. Langacker (World Scientific, Singapore, 1995).
2. N. Severijns *et al.*, *Phys. Rev. Lett.* **70**, 4047 (1993); (E) **73**, 611 (1994): M. Allet *et al.*, *Phys. Lett.*, accepted: E. Thomas *et al.*, to be published.
3. P.A. Quin and T.A. Girard, *Phys. Lett.* **B229**, 29 (1989).
4. F. Abe *et al.*, *Phys. Rev. Lett.* **74**, 2900 (1995).
5. S.B. Silverstein, Ph.D. dissertation, University of Wisconsin (1996).
6. J. Camps *et al.*

Beta-Asymmetry Measurement in the Decay of Polarized Free Neutrons

H. Abele[a], S. Baeßler[a], D. Dubbers[a], U. Kania[a], J. Last[b], U. Mayerhofer[a], C. Metz[b] T. Müller[a], V. Neshvizhevsky[b], C. Raven[b], O.Schärpf[b], C. Schmidt[a], O. Zimmer[b]

We have remeasured the angular correlation coefficient A in the decay of free neutrons which describes the correlation between neutron spin and electron momentum. From this quantity we deduce the ratio λ of the coupling strengths of vector and axialvector current. λ is found to be $-1.275(4)$. This was the first measurement with our new spectrometer PERKEO II.

Introduction

In the last years several contradictory values of the vector and axialvector coupling constants g_V and g_A in neutron beta decay have stimulated discussions about the existence of right-handed currents [1] and a possible non-unitarity of the Kobayashi-Maskawa-Matrix. To confirm or solve these problems from the experimental point of view we developed the spectrometer PERKEO II, that is designed to measure angular correlations in neutron decay.

In our experiment we measured this coefficient by observing the difference in electron count rates as a function of energy for two opposite neutron polarization states. The angular distribution of the electrons is given by

$$dw \propto 1 + A\beta \cos\Theta \tag{1}$$

β is the electron velocity in units of the light velocity, Θ the angle between electron momentum and neutron spin, and A the electron-neutron spin-correlation coefficient. A is related to the ratio of the coupling constants $\lambda = g_A/g_V$ by

$$A = -\frac{2\lambda(\lambda+1)}{1+3\lambda^2} \tag{2}$$

Description of the experiment

PERKEO II was installed at the cold neutron beam position PF1 at the Institut Laue-Langevin in Grenoble. The neutrons were polarized by double-sided supermirrors. We were able to reverse the polarization by a current sheet spinflipper, modified for the use in wide beams. The efficiency of the polarizer (P) and the spinflipper (f) was determined with a new method and we obtained $P = 98.50(74)\%$ and $f = 99.18(17)\%$. The polarized beam passed a collimator and entered the decay volume in the spectrometer. A strong magnetic field perpendicular to the beam separated the decay electrons according to their emission angle. The field was nearly homogeneous in order to suppress the magnetic mirror effect.

a Physikalisches Institut, Philosophenweg 12, 69120 Heidelberg, Germany
b Institut Laue Langevin, av. de Martyrs, 38000 Grenoble, France

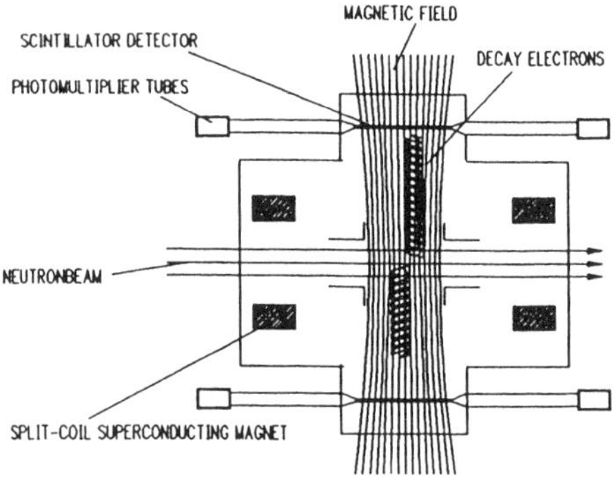

Fig 1: Schematic view of our apparatus

The electrons were gyrating along the field lines to one of the plastic scintillation detectors at both sides of the neutron beam. Thus, the acceptance angle of the spectrometer was truly $2 \times 2\pi$. Backscattering of electrons, which is usually a source of error in low-energy electron spectroscopy, was effectively suppressed by the magnetic field. We installed the detectors far away from the beam to decrease the amount of beam-related background significantly. The energy calibration was performed with conversion electron sources which covered the required energy range.

Result and discussion

If we combine the count rates of the different neutron spinflipper states

$$A_{exp} = \frac{N_{\text{off}}(E) - N_{\text{on}}(E)}{N_{\text{off}}(E) + N_{\text{on}}(E)} \doteq \frac{1}{2} A\beta(E) \quad , \tag{3}$$

many sources of error cancel out and, taking into account the finite resolution of the spectrometer, we can extract A in an one-parameter fit. After correcting for polarization and spin flip efficiency, beam-related background, weak magnetism, proton recoil, and other small effects we obtain $A = -0.1194(13)$ and $\lambda = 1.275(4)$. Together with $\tau_n = 885.9(16)\,\text{s}$ we get $g_V = 1.410(3) \times 10^{-62}\,\text{Jm}^3$ and $g_A = -1.798(2) \times 10^{-62}\,\text{Jm}^3$. Corrections due to experimental imperfections are minimized in this work, as shown in the following table:

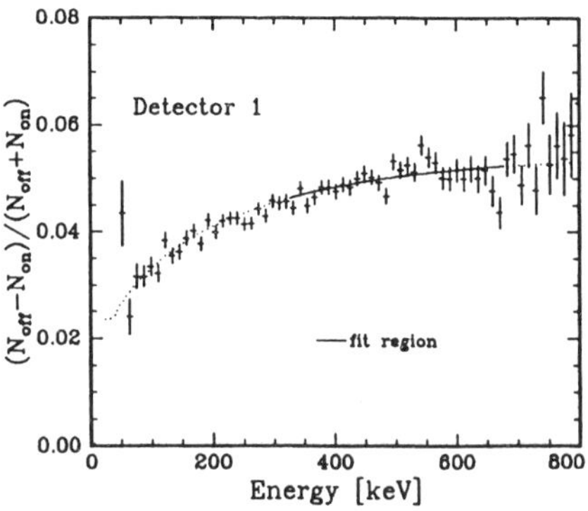

Fig 2: Experimental asymmetries of the first detector. The second one is similar.

Experiment	PERKEO [2]	Gatchina [3]	Schreckenbach [4]	PERKEO II
mirror effect	13%	-	-	0.15%
polarization	2.6%	27%	1.9%	1.5%
$\angle(\vec{\sigma_n}, \vec{p_e})$	small	3%	15%	0.2%
background	$\sim 3\%$			2%
A	−0.1146(19)	−0.1116(14)	−0.1160(15)	−0.1194(13)

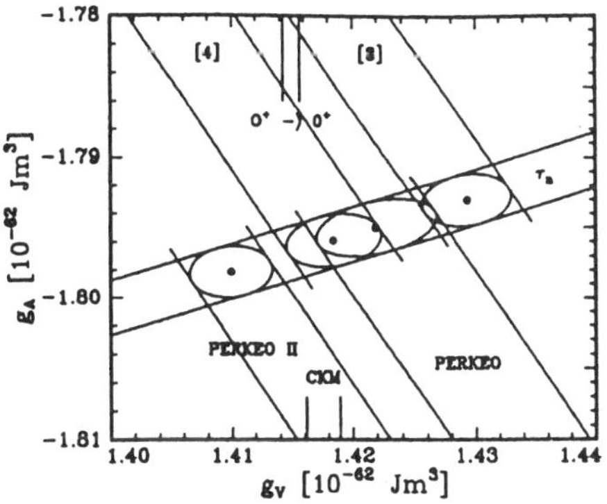

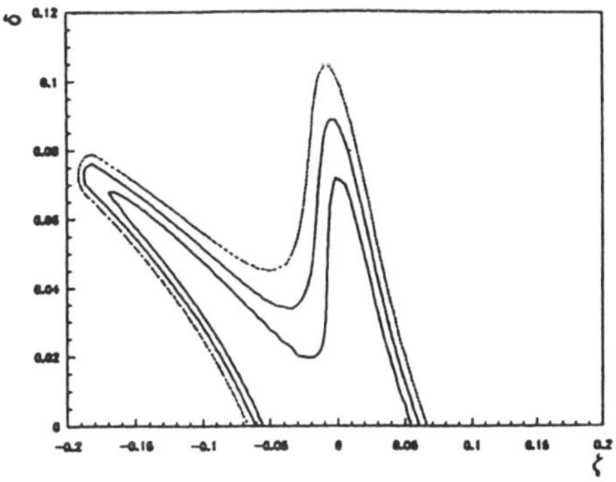

Fig 3: Evaluations of g_A and g_V with different methods: shown are the 1σ -contour lines of β-asymmetry measurements ([2–4], this work), and of the neutron lifetime. g_V is evaluated from muon decay, assuming the unitarity of the Kobayashi-Maskawa-Matrix, and from $0^+ \to 0^+$ nuclear transitions [5].

Fig 4: New exclusion plot for right-handed currents: $\delta = (m(W_1)/m(W_2))^2$, ζ is the mixing angle. The 70%, 90%, and 95% contour lines are shown. We extract $m(W_2) > 270\,\mathrm{GeV}$ (90% confidence level).

In the past, deviations between different methods for determining λ were interpreted in the context of left-right-symmetric extensions of the standard model [6]. Here, two mass eigenstates exist which are linear combinations of the left- and right-handed states:

$$W_1 = W_L \cos\zeta - W_R \sin\zeta$$
$$W_2 = W_L \sin\zeta + W_R \cos\zeta$$

The standard model predicts zero for δ and ζ. Our data support the standard model assumptions of no right-handed currents and the unitarity of the Kobayashi Maskawa Matrix.

This work was supported by the BMBF under contract number 06HD562I.

[1] Yu.V. Gaponov, N.B. Shulgina, P.E. Spivak, Phys. Lett. **B 253**, 283 (1991)

[2] E. Klemt, P. Bopp, L. Hornig, J. Last, S.J. Freedman, D. Dubbers, O. Schärpf, Z. Phys. **C37**, 179 (1988)

[3] B.G. Erozolimskii, I.A. Kuznetsov, I.V. Stepanenko, I.A. Kuida, Yu. A. Mostovoi, Phys. Lett. **B 263**, 33 (1991)

[4] K. Schreckenbach, P. Liaud, R. Kossakowski, H. Nastoll, A. Bussière, J.P. Guillaud, Phys. Lett. **B 349**, 427 (1995)

[5] D.H. Wilkinson, Z. Phys. **A348**, 129 (1994)

[6] M.A.B. Beg, A. Sirlin, Phys. Rep. **88**, 1 (1982)

RELATIVE STRENGTHS OF THE PARITY-NON CONSERVING PROTON-NUCLEUS AND PROTON-PROTON FORCES

B.DESPLANQUES
Université Joseph Fourier, Institut des Sciences Nucleaires, CNRS-IN2P3, UJF, F-38026 Grenoble Cedex

It is shown that phenomenology suggests that the contribution of the parity-non-conserving (pnc) proton-proton force to the pnc proton-nucleus force is too low with respect to what is expected from theoretical models of the pnc nucleon-nucleon force. Various ways to solve this problem are considered

A decade ago, an analysis of pnc effects in various nuclear processes[1] showed that they were quite consistent with DDH expectations[2]. Since then, the absence of pnc effect for the transition 0^- (1.08 MeV) $\rightarrow 1^+$ (g.s) in ^{18}F, as well as more accurate measurements of the longitudinal asymmetry in pp scattering at low energy, lower than previous ones by a factor 1.5, have changed the situation. First, the pnc πNN coupling is found to be significantly lower than the DDH best value. Apart for the fact that the DDH updated estimate for this coupling is now lower[3], it may be noticed that neglecting the strange quark content of the nucleon, as assumed in many works,may provide such a small value. This one makes more difficult to account for the size of some pnc effects, hence a second question.

Using notations of ref. [4], the present situation may be summarized as follows. The strength of the pnc proton-proton force, X_{pp}, is determined by the pnc asymmetry measured in pp scattering at low energy :

$$X_{pp} = 0.9 \times 10^{-6}. \qquad (1)$$

The strength of the pnc proton-nucleus force, $X_N{}^p$, is determined by pnc effects in odd-proton systems (p-α scattering and γ-transitions in ^{19}F,^{41}K,^{175}Lu and ^{181}Ta):

$$X_N{}^p = 3.4 \times 10^{-6}. \qquad (2)$$

The strengths X_{pp} and $X_N{}^p$ involve a lot of physics, especially about the short range weak and strong NN interactions. The relation between them is given by :

$$X_N{}^p\ (\cong 3.4 \times 10^{-6}) = 5.5\ h_\pi{}^1\ (\leq 0.7 \times 10^{-6}) + X_{pp}\ (\cong 0.9 \times 10^{-6}) + X_{pn}{}^+, \qquad (3)$$

where $h_\pi{}^1$,the πNN pnc coupling, has an upper limit of 1.3×10^{-7} (from ^{18}F).

The question arises to know whether relation (3) can be fulfilled in current models of the weak NN interaction. It concerns in particular the size of $X_{pn}{}^+$ (the short range pnc proton-neutron force contribution to X_N^P), once possible uncertainties about some parameters in the model (coupling constants) are raised by fitting X_{pp}.

Assuming the dominance of an isoscalar ρ-exchange and ignoring tensor correlations in the strong interaction leads to $X_{pn}^+ = 1.35 \times 10^{-6}$ ($\cong (0.5 + 1.5\,\mu_v/(\mu_v+2)).X_{pp}$). The r.h.s. of (3) is therefore equal at most to 2.95×10^{-6}, slightly below the value of the l.h.s., 3.4×10^{-6}. The discrepancy is not to be a source of concern, provided h_π^1 is closed to its maximum value (with the right sign). Including various corrections (isotensor and isovector pnc forces and tensor correlations in the strong interaction) leads to a strong decrease of X_{pn}^+ (0.3×10^{-6} instead of 1.35×10^{-6}). With such a value, the r.h.s. of (3) is definitively lower than the l.h.s. This result indicates that either some contribution is missing in the pnc NN interaction or the strength of the pnc proton-nucleus force is enhanced in the nuclear medium. Various plausible candidates may be considered in both cases.

Concerning the pnc NN interaction, one may wonder about the role of multi-meson exchange. In the present case, it would be desirable they lead to a relative enhancement of the $^3S_1$-$^1P_1$ transition with respect to the isoscalar $^1S_0$-$^3P_0$ one.

Concerning medium effects, it should be noticed that the value of X_N^P (eq. 2) relies on calculations which incorporate some nuclear correlations such as pairing and deformation effects. Some check is available in ^{18}F and ^{19}F from β decay of ^{18}Ne and ^{19}Ne, but this only concerns the isovector part of the pnc NN force. Nuclear correlations that would have a different effect on the isoscalar and isovector parts of the pnc nucleon-nucleus force could have been missed. Such an effect may originate from the exchange part of the tensor force for the largest part. It is related to the position of 0^- isovector and isoscalar states in nuclei such as ^{4}He, ^{16}O,... The size of the effect, which tends to enhance the isoscalar part and decrease the isovector one, is difficult to precise. The contribution of the π-exchange alone would lead to an instability in the isoscalar 0^- channel[5]. Different corrections (cut-off by the ρ-exchange force, effective mass, repulsive contribution of the scalar part of the force, ...) tend to reduce the effect. This one, partly supported by an analysis of the position of the isoscalar 0^- states, may provide an enhancement of the isoscalar pnc nucleon-nucleus force in the nuclear medium between 1 and 2.

In conclusion, various effects may explain the too low contribution of the pnc pp force to the p-nucleus one. Determining which one is at work requires further experiments, especially in light systems.

References

1. E.G. Adelberger and W.C. Haxton, Ann. Rev. Nucl. Part. Sci. **35**, 501 (1985)
2. B. Desplanques, J.F. Donoghue and B.R. Holstein, Ann. of Phys. **124**, 449, (1980)
3. B. Desplanques, submitted to Nucl. Phys. A.
4. B. Desplanques and J. Missimer, Nucl. Phys. **A300**, 286 (1978)
5. S.O. Bäckman, O.J. Sjöberg and A.D. Jackson,Nucl. Phys. **A321**, 10 (1979)

UPPER LIMIT OF THE MUON-NEUTRINO MASS AND CHARGED-PION MASS FROM THE MOMENTUM ANALYSIS OF A SURFACE MUON BEAM

M. DAUM, R. FROSCH, P. GHENO, R. HORISBERGER, P.-R. KETTLE
Paul Scherrer Institut, CH-5232 Villigen-PSI, Switzerland

Ch. BRÖNNIMANN, M. JANOUSCH, Th. SPIRIG, C. WIGGER
Paul Scherrer Institut, CH-5232 Villigen-PSI, Switzerland.
Physik-Institut der Universität Zürich, CH-8057 Zürich, Switzerland.

H. FORRER
Paul Scherrer Institut, CH-5232 Villigen-PSI, Switzerland.
Institut für Teilchenphysik, Eidgenössische Technische Hochschule, CH-5232 Villigen-PSI, Switzerland.

K. ASSAMAGAN
Institute of Nuclear and Particle Physics, University of Virginia, Charlottesville, VA22901, USA.

Using a surface muon beam and a magnetic spectrometer equipped with a position-sensitive detector, we have measured the muon momentum from pion decay at rest $\pi^+ \to \mu^+\nu_\mu$, to be $p_{\mu^+} = (29.79200 \pm 0.00011) MeV/c$. This value together with the muon mass and the favoured pion mass leads to an upper limit of 0.17 MeV (90% CL) for the muon-neutrino mass.

We have undertaken a series of measurements of the muon momentum p_{μ^+} in stopped pion decay $\pi^+ \to \mu^+\nu_\mu$. A new method employing a surface muon beam (a beam of muons originating from the decay of π^+-mesons at the surface of a pion production target) was chosen. This gave significant advantages compared to our previous method using a π^+-beam stopped in a scintillator [1]. The increase in the density of muons in phase space, by four orders of magnitude compared to the old method, allowed the momentum resolution of the spectrometer to be substantially improved while maintaining a usable event-rate.

From a total of 44 muon momentum spectra (distribution of hits in the position-sensitive detector at a given magnetic field setting) taken at six different magnetic field settings a muon momentum of:

$$p_{\mu^+} = (29.79200 \pm 0.00011)\ \mathrm{MeV/c} \tag{1}$$

is obtained [2]. This value together with the μ^+-mass [3] and each of the two possible solutions (A&B) of the π^--mass from a recent re-analysis [4] of the pionic X-ray measurements of Jeckelmann et al. leads to two solutions for the squared muon-neutrino mass. This assumes the validity of the CPT-Theorem ($m_{\pi^-} = m_{\pi^+}$) and energy and momentum conservation in π^+-decay. The two solutions are:

$$m^2_{\nu_\mu}(A) = (-0.143 \pm 0.024)\ \mathrm{MeV}^2 \tag{2}$$

$$m^2_{\nu_\mu}(B) = (-0.016 \pm 0.023)\ \mathrm{MeV}^2 \tag{3}$$

where the (A) and (B) correspond to the respective value of m_{π^-} chosen. Solution $m^2_{\nu_\mu}(A)$ is negative by 6 standard deviations and implies an unphysical result. Where as solution $m^2_{\nu_\mu}(B)$ is compatible with zero and shows $m_{\pi^-}(B)$ is consistent with our p_{μ^+}-result (1).

Fig.1 shows the constrained region within the $(m^2_{\nu_\mu}, m_{\pi^\pm})$-plane. From an analysis of the probability density function, which is set to zero for negative values of $m^2_{\nu_\mu}$ (unphysical), one obtains:

$$m_{\nu_\mu} \leq 0.17\ \mathrm{MeV} \quad (90\%\mathrm{CL}) \tag{4}$$

and a combined charged pion mass $m_{\pi^\pm}$ of Jeckelmann 94(B) and Assamagan 96 of:

$$m_{\pi^\pm} = (139.57037 \pm 0.00021)\ \mathrm{MeV}. \tag{5}$$

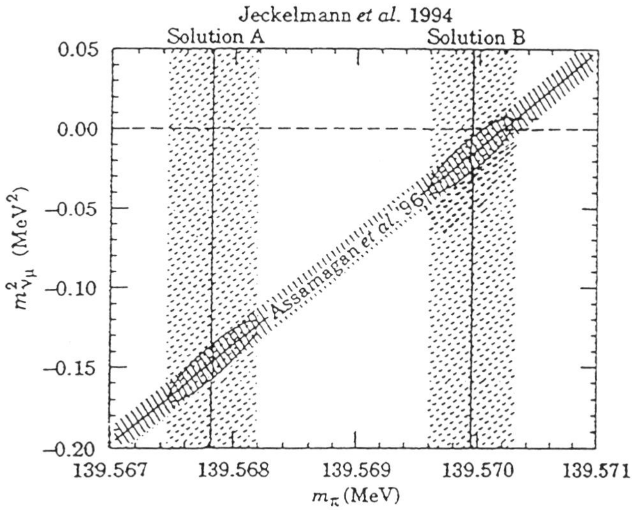

Fig 1 Allowed regions of the $(m^2_{\nu_\mu}, m_\pi)$ plane; shaded bands correspond to $\pm$ 1σ.

References

1. M. Daum et al., *Phys. Lett.* **B 265** (1991) 425.
2. K. Assamagan et al., *Phys. Rev.* **D 53** (1996) 6065.
3. Particle Data Group, *Phys. Rev.* **D 50** (1994) 1175.
4. B. Jeckelmann et al., *Phys. Lett.* **B 335** (1994) 326.

SPIN DEPENDENT μ^- CAPTURE IN LASER POLARIZED MUONIC ^{3}He

P. A. SOUDER, R. HOLMES, J. MCCRACKEN, X. WANG
Syracuse University, Syracuse, NY 13244, USA

P.L. BOGORAD, G. D. CATES, K.S. KUMAR, H. MIDDLETON, D. SIEGEL, A. YOUNG
Princeton University, Princeton, NJ 08544, USA

E.J BRASH, A. GORELOV, O. HAUSSER, P. MICHAUX
Simon Fraser University, Burnaby, BC V5A 1S6, Canada

M.D. HASINOFF, J.C. HUANG, E. SAETTLER
University of British Columbia, Vancouver, BC V6T 1Z1, Canada

W.J. CUMMINGS
TRIUMF, Vancouver, BC V6T 2A3, Canada

K. HICKS, B. LARSON
Ohio University, Athens, OH 45701, USA

W. LORENZON
University of Pennsylvania, Philadelphia, PA 19104, USA

D. TUPA
Los Alamos National Lab, Los Alamos, NM 08754, USA

We report a preliminary result for the first measurement of the spin-dependence of muon capture in ^{3}He.

1 Motivation

One of the most important goals in the field of muon capture is the measurement of g_P, the pseudoscalar form factor of the nucleon. The nucleon is a collection of three light quarks bound by the spontaneoulsy broken chirally symmetric forces of QCD, and this implies that $g_P \approx 2M_P m_\mu g_A/(Q^2 + m_\pi^2)$. Unfortunately, this prediciton is difficult to test precisely because the few observable quantities sentitive to g_P are difficult to measure experimentally.

One of the classic ideas for determining g_P is to measure the spin dependence in the capture of muons on ^{3}He. In particular, the reaction $\mu^- + {}^3\text{He} \to \nu + {}^3\text{H}$ is observed, where the rate can be written $d\Gamma/d\cos\theta \approx \Gamma_0(1 +$

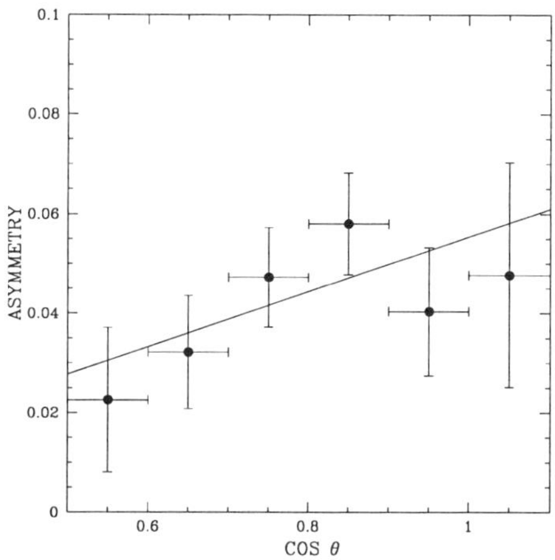

Figure 1: Observed triton asymmetry $A_v P_v$ as a function of $\cos\theta$.

$A_v P_v \cos\theta)/2$. Here Γ_0 is the spin averaged capture rate, θ is the angle between the muon polarization and the triton recoil, P_v is the polarization of the muonic ^{3}He, and A_v is the spin correlation coefficient. A_v is much more sensitive to g_P than is Γ_0. Unfortunately, methods did not exist until recently for producing large values of P_v.

2 Experimental Details

New laser techniques have made it possible to produce highly polarized muonic helium by collisions with optically pumped Rb vapor[1]. In addition, we have developed an ion chamber that detects the recoil tritons inside the polarizing volume. The triton pulses from the chamber are digitized, and a detailed analysis of the pulse shapes identifies the events as tritons and measures their direction relative to the direction of the polarizing laser. With this apparatus, we have measured A_v at TRIUMF (see Fig. 1) and report the preliminary value $A_v = 0.59 \pm 0.09(\text{stat}) \pm 0.10(\text{syst})$. Our result is consistent with the theoretical prediction $A_v = 0.52$ and represents the first significant measurement of this quantity. By using more powerful lasers and a segmented detector, we expect to be able to make significant improvements in the precision of the measurement.

1. A. Barton *et al.*, Phys. Rev. Lett. **70**, 758, (1993)

Muon Capture on ^{23}Na and the Weak Couplings g_a and g_p.

T.P. Gorringe, B.L. Johnson, J. Bauer, M.A. Kovash, R. Porter.
Dept. of Physics and Astronomy, University of Kentucky, Lexington, KY, 40506, USA.

M.D. Hasinoff, D.F. Measday, B.A. Moftah,
Dept. of Physics, University of British Columbia, Vancouver, BC, V6T 1Z1, Canada.

D.S. Armstrong.
Dept. of Physics, College of William and Mary, Williamsburg, VA, 23187, USA.

D.H. Wright.
TRIUMF, 4004 Wesbrook Mall, Vancouver, BC, V6T 2A3, Canada.

We report measurements and calculations of capture rates and hyperfine dependences for six $^{23}\mathrm{Na} \rightarrow {}^{23}\mathrm{Ne}$ muon capture transitions. The measurements were performed at the TRIUMF cyclotron and the calculations were performed with the 1s–0d shell model and USD empirical effective interaction. From fits of the calculated and measured capture rates we obtained an effective coupling $\tilde{g_a} = -1.01 \pm 0.07$, and from fits of the calculated and measured hyperfine dependences we obtained an effective coupling ratio $\tilde{g_p}/\tilde{g_a} = 6.5 \pm 2.4$. The value of $\tilde{g_a}$ is consistent with values of $\tilde{g_a}$ extracted from β^+/β^-–decay and (p,n)/(n,p) charge exchange data, and the value of $\tilde{g_p}/\tilde{g_a}$ is consistent with the predictions of PCAC and pion–pole dominance.

The dynamics of the nucleon's weak axial current A^μ are encoded in the axial and pseudoscalar form factors $F_A(q^2)$ and $F_P(q^2)$ and coupling constants $g_a = F_A(0)$ and $g_p = m_\mu F_P(+0.9m_\mu^2)$. PCAC, grounded in the approximate $SU(2)_L \times SU(2)_R$ chiral symmetry of QCD and the pion's relation to its explicit and spontaneous symmetry breaking, and pion–pole dominance, grounded in the dominance of one pion over heavy meson and multi–meson exchange, yield the prediction $g_p/g_a = 6.7$ [1].

Due to the non–conservation of A^μ, the coupling constants g_a and g_p may be renormalized in nuclei. A number of authors have explored the effects of the nuclear medium on g_a and g_p including Delorme et al. [2] who predicted a small quenching of g_a and large quenching of g_p due to delta excitations. Currently there is a lively debate as to whether recent measurements of radiative muon capture on nuclei do or do not indicate a large renormalization of g_p.

The hyperfine effect in muon capture, i.e. the difference in μ^- capture rates from the $F_+ = J_i+1/2$ and $F_- = J_i-1/2$ hyperfine states in $J_i \neq 0$ nuclei, can be sensitive to g_p/g_a. The experimental exploration of the hyperfine effect

Table 1: Comparison of measured and calculated $^{23}Na \rightarrow {}^{23}Ne$ capture rates and hyperfine dependences. The calculations are using the "best fit" values $\tilde{g_a} = -1.01$ and $\tilde{g_p}/\tilde{g_a} = 6.5$.

E_f (keV)	J_f^π	Λ ($\times 10^3 s^{-1}$) expt.	Λ ($\times 10^3 s^{-1}$) calc.	Λ_+/Λ_- expt.	Λ_+/Λ_- calc.
1017	$1/2^+$	4.9±1.4	5.26	0.18±0.03	0.14
1702	$7/2^+$	1.4±0.3	0.64	≤0.50	0.18
1823	$3/2^+$	4.1±0.9	6.28	0.23±0.04	0.20
2315	$5/2^+$	2.1±0.5	0.71	0.98±0.36	1.66
3432	$3/2^+$	4.1±0.8	2.40	0.87±0.48	0.31
3458	$1/2^+$	10.4±2.2	8.97	≤0.19	0.10

was pioneered by Winston et al. [3] in inclusive μ^- capture and Deutsch et al. [4] in exclusive μ^- capture. We have measured capture rates and hyperfine dependences for six transitions in $^{23}Na(\mu^-,\nu)^{23}Ne^*$ and extracted "best fit" values of g_a and g_p/g_a using the 1s–0d shell model and the USD effective empirical interaction.

We obtained $g_a = -1.01 \pm 0.07$ and $g_p/g_a = 6.5 \pm 2.4$. The value of g_a is consistent with the values of g_a extracted from β^+/β^-–decay and (n,p)/(p,n) charge exchange in 1s–0d shell nuclei. The value of g_p/g_a is consistent with PCAC and pion-pole dominance and does not suggest a large renormalization of g_p/g_a in ^{23}Na. Table 1 shows the calculated and measured capture rates and hyperfine dependences for the six $^{23}Na \rightarrow {}^{23}Ne$ transitions are in general agreement. Note the capture rates correspond to the experimental mix of F_- and F_+ hyperfine states, $\Lambda = 0.97\Lambda_- + 0.03\Lambda_+$.

In general the $^{23}Na \rightarrow {}^{23}Ne$ transitions are dominated by the Gamow-Teller (GT) matrix elements. Comparison of the calculated and measured GT matrix elements (obtained via the (n,p) charge exchange reaction on ^{23}Na [5]) show generally good agreement and build confidence in the nuclear model and the extracted values of $\tilde{g_a}$ and $\tilde{g_p}/\tilde{g_a}$.

1. M.L. Goldberger and S.B. Treiman, Phys. Rev. **110**,1178(1958).
2. J. Delorme, M. Ericson, A. Figureau, and C. Thévenet, Ann. Phys. (NY) **102**, 273 (1976).
3. R.Winston, Phys. Rev. **129**, 2766, (1963).
4. J.P.Deutsch, L.Grenacs, J.Lehmann, P.Lipnik and P.C.Macq, Phys. Lett. **28B**, 178 (1968).
5. B.Siebels, T.P.Gorringe, J.Bauer, J.Evans, S.El–Kateb, K.P.Jackson, A.Trudel and S.Yen, Phys. Rev. **C52**, 1488 (1995).

MEASURING THE INDUCED PSEUDOSCALAR COUPLING IN NUCLEI: the cases of ^{11}B and ^{28}Si

R. PRIEELS
Université catholique de Louvain, 2, Chemin du Cyclotron,
B-1348 Louvain-la-Neuve, Belgium

Results of two experiments are presented. The first one, performed at PSI, measured the ratio of the muon capture rates from the two muonic hyperfine levels in ^{11}B. The second experiment, completed at Dubna, measured the neutrino-gamma angular correlation subsequent to the nuclear capture of polarised muons. The aim of both experiments is a determination of g_P/g_A in the nuclear medium.

The charged weakly interacting hadronic current carries two contributions of different space-time properties: a vector and an axial component. The conservation of the vector current implies that its nuleonic matrix elements are essentialy not modified in a nuclear medium of moderate density and temperature. The situation might be different for the non conserved axial current. The PCAC prediction[1] $g_P/g_A = 6.8$ for the nucleon axial and induced pseudoscalar form factors could possibly be affected by "in medium" renormalisation effects.

Two precision measurements aiming at the measurement of g_P/g_A by the observation of specific muon capture channels are reported. The first experiment was performed at PSI by a Louvain-la-Neuve, JINR, Rez, PSI and University of Zürich collaboration[2]. Among other channels, muons captured on ^{11}B produce ^{11}Be in its excited bound state at 320 keV. Since the latter is the only bound excited state of ^{11}Be, no upper feeding is expected. The muon capture rates λ^+ and λ^- from the two hyperfine states ($F = 2$ and $F = 1$) of muonic ^{11}B are different functions of g_P/g_A. Since there is a substantial transition probability R between the two hyperfine states, the combined capture rate is time dependent. The time distribution of the 320 keV gamma rays, resulting from the de-excitation of ^{11}Be*, reveals this capture rate behaviour. Stopping a 49 MeV/c muon beam in a 99.999% purity natural boron powder target sealed in a PVC box under nitrogen, 2.10^5 gammas of 320 keV were recorded between 0 to 10 μs after the muon stop, with a signal to background ratio of about 3. The background determined from various energy windows on both sides of the 320 keV peak, and the peak were globally fitted in time. This leads to the following, still preliminary result, requiring critical assesment of possible systematic effects:

$$\lambda^+/\lambda^- = 0.015 \pm 0.022 \qquad\qquad R = (175.9 \pm 16)\, 10^3\, s^{-1}$$

The capture rate ratio is related to g_P/g_A through theoretical models [3,4] from which a rather small value for g_P/g_A in comparison to the PCAC prediction is inferred. However the nuclear model calculations [3,4] do not yet account for the neutron halo effects in ^{11}Be, an aspect which Drs. V. Kuzmin and T. Suzuki are presently looking into. No conclusion with regards to g_P/g_A can be drawn from the experiment before this theoretical issue is resolved.

The second experiment undertaken at Dubna was a Dubna, Louvain-la-Neuve, Orsay collaboration. It makes use of the population distribution of the sublevels of the 1^+ excited state at 2201.5 keV of ^{28}Al following capture of polarised muons on ^{28}Si. These sublevel populations are different functions of g_P/g_A, and can be extracted from the neutrino-gamma correlation between the capture neutrino and the subsequent gamma de-exciting the 1^+ level. Due to the short lifetime of the 2201.5 keV level, the neutrino direction can be revealed via the Doppler broadening of the 1229 keV and 2171 keV lines. A shape analysis of these lines is the clue of the method. The parameters are the lifetime of the 2201.5 keV level and an asymmetry parameter α function of the sublevel populations only. The residual polarisation of the muon is measured by μSR techniques. Two different experiments were performed, one by putting the Ge detectors at two positions differently sensitive to the muon spin [5], the second by keeping the Ge detectors fixed but by precessing the muon spin [6]. An important background was subtracted from the 1229 keV peak after careful investigation of his various components. A Triumf experiment [7] on the same nucleus but requiring coincidence with a subsequent gamma, was able to remove this background.Both results agree within errors:

$$\text{Dubna (exp 1+2)} : \alpha = 0.285(26), \qquad \text{and Triumf} : \alpha = 0.360(59)$$

These values must be related to g_P/g_A through a theoretical calculation. However, further work is mandatory to include configuration mixing and to understand any model dependence, before any definite conclusion can be reached.

References

1. V. Bernard, N. Kaiser, *Phys. Rev.* D **50**, 6899 (1994) .
2. V. Wiaux *et al.*, *PSI Annual Report 1995 / annex 1*, p
3. J. Bernabeu, *Il Nuovo Cim.* **A4**, 715 (1971) .
4. V. A. Kuz'min et al., *Phys. At. Nucl.* **57**, 1881 (1993) .
5. V. Brudanin *et al.*, *Nucl. Phys.* A **587**, 577 (1995).
6. V. Brudanin *et al.*, in *WEIN 1995*, ed. H. Ejiri,T. Kishimoto, T. Sato. (World Scientific, Singapore, 1995), p390.
7. B.A. Moftah, *Thesis of University of British Columbia* December 1995.

Muon Transfer in Mixtures of Light Atoms: Experimental Investigation via Detection of Muonic X Rays.

B.Gartner[1], B.Lauss[1], S.Tresch[5], P.Ackerbauer[1], W.H.Breunlich[1], M.Cargnelli[1], D.Chatellard[3], H.Daniel[4], J.-P.Egger[3], A.Fischer[1], F.J.Hartmann[4], R.Jacot-Guillarmod[5], E.Jeannet[3], M.Jeitler[1], P.Kammel[1*], R.C.King[1], G.Kminek[1], A.Kosak[4], J.Marton[1], M.Mühlbauer[4], F.Mulhauser[5], C.Petitjean[2], C.Piller[5], W.Prymas[1], L.A.Schaller[5], L.Schellenberg[5], H.Schneuwly[5], E.Steininger[1], Y.A.Thalmann[5], T.von Egidy[4], A.Werthmüller[5] and J.Zmeskal[1].

[1] Institute for Medium Energy Physics, Austrian Academy of Sciences, A-1090 Wien, Austria
[2] Paul Scherrer Institut, CH-5232 Villigen, Switzerland
[3] Institut de Physique de l'Université, CH-2000 Neuchâtel, Switzerland
[4] Physik Department, TU München, D-85747 Garching, Germany
[5] Institut de Physique de l'Université, Pérolles, CH-1700 Fribourg, Switzerland
[*] present address: University of California and LBL, Berkeley CA, USA

The systematic study of muonic X rays in light atoms (hydrogen and helium isotopes) allowed us for the first time

- to observe directly and investigate systematically excited state muon transfer in protium/deuterium(H/D)-mixtures
- to determine the rates of ground state muon transfer from hydrogen to helium isotopes for several isotopic mixtures and to detect the resulting intermediate muonic molecular ion with high statistics.

Profiting from the unique detection technique of charge coupled devices (CCDs) and its enormous background rejection capability [1], muonic hydrogen X rays were observed with unprecedented accuracy. Additionally, in H/D-mixtures it was possible to separate the different isotopic K-lines of muonic hydrogen in the energy region of 1.9 - 2.5 keV [2]. This allowed the first direct measurement of q_{1s}, the probability that a μp-atom, which is initially formed in an excited state, reaches the 1s state. The determination of q_{1s} solves a longstanding problem in muon catalyzed fusion [3]: q_{1s} determines the initial populations in mixtures of hydrogen isotopes and therefore influences to a large extent the rate of the subsequent process leading to muon catalyzed fusion.

The experimental idea was to compare the intensities of the K-lines of the two isotopes. This is possible because at the investigated densities non-radiative deexcitation to the 1s ground state is negligible [4]. The higher binding energy of the heavier isotope - the energy difference ΔE for muonic protium and deuterium depends on the principal quantum number n: $\Delta E = (135/n^2)$ [eV] - favours the transfer of the muon from H to D, even in excited states. This enhances the intensities of the muonic deuterium K-lines. In a first analysis considering only K_α-intensities this approach yields [2]:

$$q_{1s} \approx \frac{K_\alpha^{p\mu}}{c_p}, \qquad q_{1s} \approx \frac{1 - K_\alpha^{d\mu}}{c_p}$$

$K_\alpha^{p\mu}$ ... Intensity of muonic protium 2 - 1 transition

$K_\alpha^{d\mu}$ ... Intensity of muonic protium 2 - 1 transition

c_p ... Protium concentration

Preliminary results of the measurements at densities of LHD and 1 % of LHD (Liquid Hydrogen Density LHD = $4.25 \cdot 10^{22}$ atoms/cm^3) are displayed in fig.1a. Possible molecular effects have not been observed at a significant level (Fig.1b).

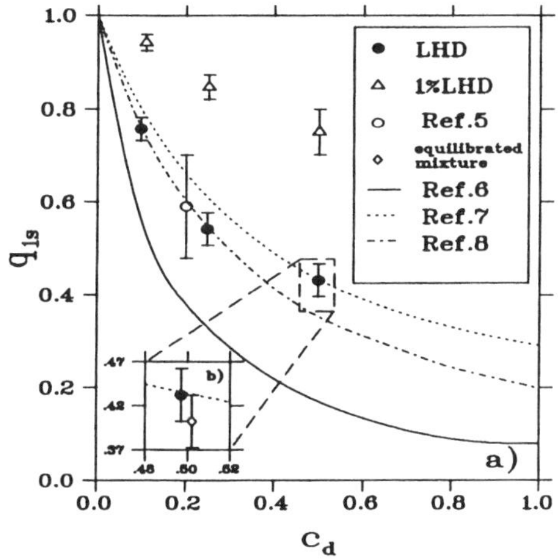

Fig.1a): The dependence of q_{1s} on the deuterium concentration. Full circles display the preliminary results for liquid hydrogen density measurements, the point at $c_d = 0.2$ (open circle) is taken from our test experiment [5]. Triangles show the results at 1 % of LHD. The theoretical expectations for q_{1s} at liquid hydrogen density are taken from [6] (solid line), [7] calculated for a collision energy $\varepsilon_\mu = 6$ eV (dotted line), and [8] (dashed-dotted line).
b) Molecular effects on q_{1s} at c_d = 0.5. Equilibrated H_2/HD/D_2 mixture (dot) and non-equilibrated H_2/D_2 mixture (diamond) at LHD = 1.0.

In mixtures of hydrogen and helium isotopes, a unique charge exchange process via the formation and dissociation of a metastable $(x\mu He)^*$ (x = proton p or deuteron d; He = ^{3}He or ^{4}He) molecular ion is predicted and facilitates ground state muon transfer at an unexpectedly high rate [9]. This excited molecular ion rapidly decays either by particle decay, where the excess energy is transferred to the kinetic energy of the decay products [10], by Auger electron emission, or by X-ray emission (X-ray energy ~ 7 keV) [9]. We have already observed these molecular X rays for three isotopic mixtures [11,12]. Their time distributions allowed us to extract the formation rates of the respective muonic molecules. Fig. 2 displays the observed X-ray energy spectra for mixtures of p-^{4}He (a), d-^{3}He (b) and d-^{4}He (c). The determination of the transfer rates from p to ^{3}He and ^{4}He was also done via the triple mixture method, using a small admixture of neon as a third component [12]. The measurements of the transfer rates depend crucially on the determination of the helium concentrations in the target mixtures. For the triple mixture method, the precise knowledge of the neon concentration is also crucial. Preliminary results are given in table 1. Further evaluation of our CCD-measurements - method described in [11] - will also allow us to determine the branching ratios (radiative/non-radiative decay of the muonic molecules) for all investigated mixtures and to test

isotopic mixture	$\lambda_{He} \cdot 10^8$ [s^{-1}]
p - ^{3}He	0.20 ± 0.03 stat. ± 0.16 syst.
p - ^{4}He	0.46 ± 0.03 stat. ± 0.06 syst.
d - ^{3}He	1.7 ± 0.2 stat. ± 0.5 syst.
d - ^{4}He	9.5 ± 0.3 stat. ± 2.0 syst.

Tab.1: Preliminary results from the measurement of the ground state transfer rates λ_{He} in hydrogen - helium mixtures. The final analysis of our mass spectrometry measurements will considerably improve the systematic error.

the strong isotopic effect predicted for the decay probabilities of the (dμHe)* mesomolecules [10].

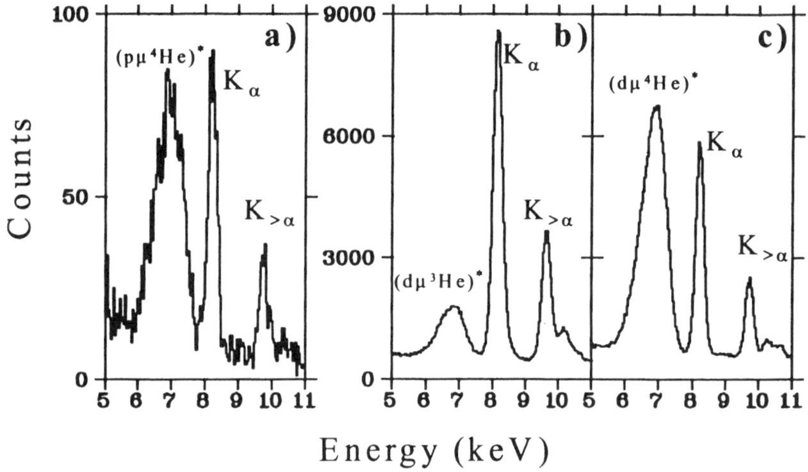

Fig.2: Energy spectra measured with a Ge-diode in a mixture of **a)** p-^{4}He, **b)** d-^{3}He, **c)** d-^{4}He. In addition to the decay line of the (xμHe)* mesomolecule at ~ 7 keV, the K-lines of the muonic cascade in helium ($K_{>\alpha}$ denotes all K-Lines with energies higher than K_α) can be seen.

Acknowledgments:
Financial support by the Austrian Science Foundation, the Austrian Academy of Sciences, the Swiss Academy of Sciences, the Swiss National Science Foundation and the German Federal Ministry of Research and Technology is gratefully acknowledged.

References:
[1] J.-P.Egger et al., Part. World Vol. 3 (1993) 139.
[2] B.Lauss et al., Proc. of the Inter. Symposium on Muon Cat. Fusion, μCF'95, Dubna, Russia, Hyp. Int., in press; B.Lauss et al., Phys.Rev.Lett., accepted for public.
[3] W.H.Breunlich et al., Annu. Rev. Nucl. Part. Sci. 39 (1989) 311.
[4] V.E.Markushin, Phys. Rev. A50, 2 (1994) 1137.
[5] B.Lauss et al., in: Proceedings of the Inter. Workshop on Low Energy Muon Science LEMS'93, Los Alamos publication LA-12698-C, (1994) p.300.
[6] V.Markushin, Abstracts of the Inter. Workshop on Low Energy Muon Science LEMS'93, Los Alamos publication (1994).
[7] W.Czaplinski et al., Phys.Rev.A 50,1 (1994) 525.
[8] E.C.Aschenauer and V.E.Markushin, Proc. of the Inter. Symposium on Muon Cat. Fusion, μCF'95, Dubna, Russia, Hyp. Int., in press.
[9] Yu.A.Aristov et al., [Yad.Fiz.33, (1981) 1066], Sov.J.Nucl.Phys., 33/4 (1981) 564.
[10] Y.Kino and M.Kamimura, Hyp.Int., 82 (1993) 195.
[11] B.Gartner et al., Proc. of the Inter. Symposium on Muon Cat. Fusion, μCF'95, Dubna, Russia, Hyp. Int., in press.
[12] S.Tresch et al., ibid.

THE GOTTHARD EXPERIMENT ON ^{136}Xe $\beta\beta$ DECAY

A. Paić (on behalf of the Caltech-Neuchâtel-PSI collaboration)
Institut de Physique, Université de Neuchâtel, Rue A.L. Breguet - 1
CH 2000 Neuchâtel, Switzerland

The improved Gotthard experiment is now running with a lower background version of the 180 litres Xenon TPC. Alpha spectroscopy of $\alpha\alpha$ and $\beta\alpha$ coincidences provides an estimation of the radioactive contamination of the Xe gas. $\sim 7 \cdot 10^{-12}$ g/g The new 90% c.l. limit on neutrinoless $\beta\beta$ decay is $T^{0\nu}_{1/2} > 4.4 \times 10^{23}$ yr. The 90% c.l. limit on the majoron mode is $T^{\chi^0}_{1/2} > 1.4 \times 10^{22}$ yr.

1 Introduction

Double beta decay [1] is a weak nuclear process in which a nucleus (A, Z) spontaneously transforms into another with the same mass and 2 more (or less) units of charge $(A, Z \pm 2)$. It can occur in nuclei in which the simple beta decay is forbidden for energy or angular momentum reasons.

The neutrinoless (0ν) mode violates lepton number conservation by two units and is possible only if the neutrino is a massive Majorana particle and thus gives a handle on the neutrino mass. As no energy is lost to the neutrinos, the two-electron spectrum should show a peak at the Q value of the reaction. The majoron (χ^0) mode supposes the emission of a hypothetical scalar Goldstone boson in addition to the two electrons.

2 The experiment

The experiment [2] consists of a pressurized xenon time projection chamber (TPC). The Xe fill gas is the $\beta\beta$ candidate (it is 62.5% enriched in ^{136}Xe) and the drift gas.

The signal to noise ratio is enhanced by the tracking capability that allows to select events with two electron tracks emerging from the same point [3], and reject with high efficiency those background components like internal β and α radioactivity, Compton electrons, and cosmic rays.

An upgrade in 1994 consisted in the B_4C shield, new lower activity resistors for the field-shaping cage, a new thinner xy-pad and a new anode plane with crimped wires to avoid the activity from the solder. The effect of this upgrade is a reduction roughly by a factor of 2 of the 2-electron background rate.

Alpha spectroscopy of "short" $\alpha\alpha$ and $\beta\alpha$ coincidences in the natural radioctivity has been performed. The following coincidences were observed:

$$^{220}\mathrm{Rn} - ^{216}\mathrm{Po} \quad (0.15s, \alpha\alpha) \quad \text{and}$$
$$^{212}\mathrm{Bi} - ^{212}\mathrm{Po} \quad (0.3\mu s, \beta\alpha)$$

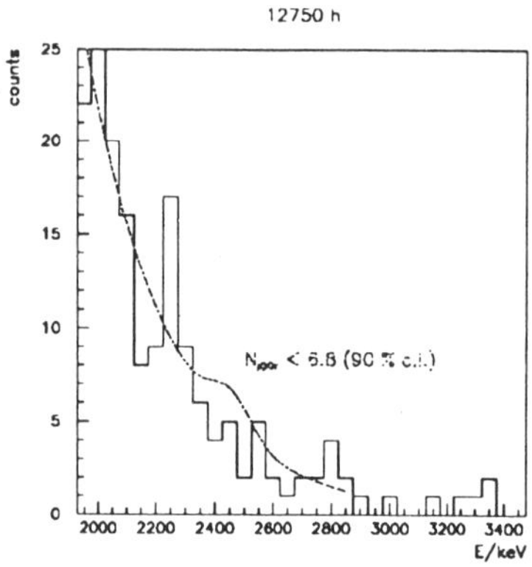

Figure 1: 2-electron spectrum after 1.57 yrs of live time with the 90% confidence limit exclusion curve.

from ^{232}Th,

$$^{214}\text{Bi} - ^{214}\text{Po} \quad (162\mu s, \beta\alpha)$$

from ^{238}U and

$$^{219}\text{Rn} - ^{215}\text{Po} \quad (1.8ms, \alpha\alpha)$$

from ^{235}U.

The observed counting rates provide us with an estimation of the radioactive contamination of the Xe gas, amounting to $\sim 1 \cdot 10^{-11}$ g/g of ^{232}Th, $\sim 7 \cdot 10^{-12}$ g/g of ^{238}U, and $\sim 4 \cdot 10^{-13}$ g/g of ^{235}U.

3 Results

The experiment accumulated a total of 1.57 yrs of live time, of which 0.69 yrs after the background reduction. The latter sample is already superior for the estimation of the 2ν and χ^0 modes, whereas for the 0ν mode the whole data sample gives a better limit. For the latter mode a maximum likelihood 90% confidence level exclusion curve (exponential background plus hypothetical Gaussian peak at the Q value of 2.48 MeV) is plotted in fig. 1.

The preliminary limits attained are:

$$T_{1/2}^{0\nu} \geq 4.2 \times 10^{23} yr$$
$$T_{1/2}^{2\nu} \geq 5.5 \times 10^{20} yr$$
$$T_{1/2}^{\chi^0} \geq 1.4 \times 10^{22} yr.$$

From the 0ν limit, using the matrix elements of refs. 4-6 we extract a mass limit of:

$$m_{\nu_e} < 2.4 - 2.8\text{eV}\ ^4$$
$$m_{\nu_e} < 2.3\text{eV}\ ^5$$
$$m_{\nu_e} < 1.8\text{eV}\ ^6.$$

From the χ^0 mode limit, we extract a limit on the neutrino-majoron coupling of $\langle g_{\nu\chi^0} \rangle < 1.5 \cdot 10^{-4}$.

4 References

1. For an introduction, see: F. Boehm and P. Vogel *Physics of massive neutrinos* Cambridge Univ. Press 2nd ed. (1992).
2. J.C. Vuilleumier *et al.* *Phys. Rev.* **D48** (1993) 1009–20.
3. H.T. Wong *et al.* *Nucl. Inst. Meth.* **A 329,** 163, (1993)
4. J. Engel *et al. Phys. Rev.* **C37** (1988) 731–46.
5. A. Staudt *et al. europhys. Lett.* **13** (1990) 31–36.
6. T. Tomoda *Rep. Prog. Phys.* **54** (1991) 507–46.

Limits on Neutrino-less Double Beta Decay of ^{100}Mo

H. Ejiri[a], K. Fushimi[b], K. Hayashi[a], R. Hazama[b], T. Kishimoto[b], M. Komori[b], N. Kudomi[a], K. Kume[a], K. Nagata[b], T. Nitta[b], H. Ohsumi[b], K. Okada[b], T. Shima[b] and J. Tanaka[b]

a)Research Center for Nucl. Phys., Osaka Univ., Ibaraki, Osaka 567, Japan

b)Dept. Phys. and Lab. Nucl. Studies, Osaka Univ., Toyonaka, Osaka 560, Japan

Double beta decays are of current interest from astroparticle and nuclear physics view points [1]. The neutrino-less double beta decays($0\nu\beta\beta$) give evidences for relevant theories beyond the standard $SU(2)_L \times U(1)$ theory. Previously, we studied the $\beta\beta$ decays of ^{100}Mo and ^{116}Cd[2,3]. In the present work, the most stringent limits on the $0\nu\beta\beta$ and $0\nu\beta\beta$B processes for ^{100}Mo were obtained by the exclusive low background measurement with ELEGANT V[4].

Unique features of the present work are, 1) energy and angular correlations of two β rays are measured by ELEGANT V. Thus limits on the $0\nu\beta\beta$ processes for individual terms of $\langle m_\nu \rangle$, $\langle \lambda \rangle$, $\langle \eta \rangle$ and others, and on the $0\nu\beta\beta$B process can be obtained, 2) ^{100}Mo has large phase space factors because of the large Q value, and 3) ELEGANT V has been improved so that the background rate has been reduced to the level of 0.1counts/day/kg at the $0\nu\beta\beta$ region.

ELEGANT V consists of three drift chambers for measuring two β trajectories, plastic scintillator arrays(PL) for β ray energies and arrival times, and NaI scintillator arrays for γ and X rays[4].

The $\beta\beta$ decays of ^{100}Mo were measured for 7333hrs at the Kamioka underground laboratory with 2700m w.e.(water equivalent) depth. Following selections were required in the off-line analysis to select possible candidates for the true $\beta\beta$ decay events of the present interest. 1. There are two PL signals. 2. There are such signals from DC's that indicate two β-ray tracks with the vertex point on the source plane. 3. There are no γ-ray signals from NaI detectors for $\beta\beta$ decay to the 0^+ ground state. 4. The β-ray energy threshold was set at 150keV. 5. Rejection of β events with $|180° - \theta| < 30°$ (θ-cut), where θ is the angle between the two β-ray tracks. 6. Rejection of such two β events that the time difference between two PL's is not consistent with the flight time difference for the two β rays estimated from the β-ray trajectories in DC's (TOF-cut). 7. Rejection of electron events followed by delayed α events. This is made to avoid the β decay of ^{214}Bi signals, which are followed by α decay of ^{214}Po with the half-life of 164μsec.

The sum energy($E_1 + E_2$) spectrum was obtained for the two electron events selected by the seven conditions given above. The major part of the

Table 1: Limits with 68%CL on the half-lives for the $0\nu\beta\beta$ decay of ^{100}Mo.

mode		E window (MeV)	Yield (counts)	$2\nu\beta\beta$+BG (counts)	eff.	$T_{1/2}$ (10^{21}y)
$0^+ \to 0^+$	$\langle m_\nu \rangle$ term	2.8-3.3	3	4.1	0.19	>52
	$\langle \lambda \rangle$ term	2.8-3.3	3	4.1	0.11	>31
		2.8-3.3	0*	0.6*	0.075*	>39*
	$\langle \eta \rangle$ term	2.8-3.3	3	4.1	0.18	>51
$0^+ \to 2^+$	$\langle \lambda \rangle$ term	2.1-2.8	0	0.23	0.048	>14
Majoron	emission	2.5-3.0	26	28.	0.07	>5.4

* Values derived by additional energy and angular windows for the $\langle \lambda \rangle$ term $0\nu\beta\beta$ process.

spectrum is the $2\nu\beta\beta$ component with the half-life of $T_{1/2}^{2\nu} = 1.15 \times 10^{19}$y [2].

Background contributions from the natural radioactive contaminations were estimated. The ^{214}Bi with $Q_\beta = 3.28$MeV and ^{208}Tl with $Q_\beta = 4.99$MeV are two major isotopes, which may give background events in the $0\nu\beta\beta$ energy window of $E_1 + E_2 \sim 3$MeV.

There are no excess $0\nu\beta\beta$ counts beyond the statistical fluctuations of the $2\nu\beta\beta$+BG events. Since the $0^+ \to 0^+$ $0\nu\beta\beta$ process due to the $\langle \lambda \rangle$ term has unique energy and angular correlation patterns, additional windows of $E_1 < 0.8$MeV and $E_2 > 2.0$MeV and $\theta_{12} < 110°$ were set as λ-selection for preferential selections of the $\langle \lambda \rangle$ term contribution. Then the results give more stringent limit due to the $\langle \lambda \rangle$ term. The most stringent limits with the 68%CL on the half-lives are obtained as in Table 1. Using the nuclear matrix elements given in ref. [5], the half-life limits for the $0^+ \to 0^+$ $0\nu\beta\beta$ and $0^+ \to 0^+$ $0\nu\beta\beta$B processes leads to the upper limits of $\langle m_\nu \rangle < 2.2$eV, $\langle \lambda \rangle < 3.7 \times 10^{-6}$, $\langle \eta \rangle < 2.5 \times 10^{-8}$ and $\langle g_B \rangle < 7.3 \times 10^{-5}$.

1. W.C. Haxton and G.J. Stephenson, *Jr. Prog. in Part. Nucl. Phys.* **12**, (1984) 409; M. Doi, T. Kotani and E. Takasugi, *Prog. Theor. Phys. Suppl.* **83**, (1985) 1.
2. H. Ejiri, et al., *Phys. Lett.* **B 258**, (1991) 17; H. Ejiri, et al., *J. Phys. G. Nucl. Part. Phys.* **17**, (1991) S155.
3. H. Ejiri and H. Toki, *Joun. of Phys. Soc. of Japan,* **65**, (1996) 7.
4. H. Ejiri, et al., *Nucl. Instr. and Meth.* **A 302**, (1991) 304. K. Nagata, et al. *Nucl. Instr. and Meth.* **A 362**, (1995) 261.
5. T. Tomoda, *Rep. Prog. Part. Phys.* **54**, (1991) 53.

SHELL MODEL MONTE CARLO FOR TWO-NEUTRINO DOUBLE BETA DECAY

P. B. RADHA, S. E. KOONIN, K. LANGANKE
W. K. Kellogg Radiation Laboratory, 106-38
California Institute of Technology Pasadena, California 91125 USA

P. VOGEL
Division of Physics, Mathematics and Astronomy
California Institute of Technology, Pasadena, California 91125 USA

We present a method to calculate two-neutrino double beta decay matrix elements using the Shell Model Monte Carlo and obtain a result for the decay of ^{76}Ge that is in reasonable agreement with experiment.

Shell Model Monte Carlo (SMMC) methods (see [1] and references therein) scale more gently than direct-diagonalization methods with the number of valence nucleons and single particle orbits and so allow larger and hence more realistic calculations. In these methods observables are calculated as expectation values in the canonical ensemble of nuclear states. In what follows, the recent application of the SMMC to two-neutrino (2ν) double beta ($\beta\beta$) decay [2] is briefly described.

The 2ν mode of $\beta\beta$ decay is an important element in the interpretation of limits on the neutrinoless mode. The nuclear matrix elements for these decays are highly suppressed and thus are a challenging nuclear structure problem. The $2\nu\beta\beta$ matrix element between the 0^+ ground states of the initial and final even-even nuclei is given by [3],

$$M^{2\nu} = \sum_m \frac{\langle 0_f^+|\mathbf{G}|1_m^+\rangle \cdot \langle 1_m^+|\mathbf{G}|0_i^+\rangle}{E_m - (E_i^0 + E_f^0)/2}. \quad (1)$$

Here, $|0_i^+\rangle$ ($|0_f^+\rangle$) is the ground state of the initial (final) nucleus with energy E_i^0 (E_f^0); $|1_m^+\rangle$ are the 1^+ states of the intermediate odd-odd nucleus with energies E_m; and $\mathbf{G}$ is the GT operator, $\sum_l \boldsymbol{\sigma}_l \tau_l^-$, where $\boldsymbol{\sigma}_l$ is the Pauli spin operator for nucleon l and τ_l^- is the isospin lowering operator that changes a neutron into a proton. Previous shell model calculations for nuclei heavier than ^{48}Ca (see for instance, Ref [3]) have invoked the so-called closure approximation, where the matrix element is written as $M^{2\nu} = M_c/\bar{E}$, with $M_c = \langle 0_f^+|\mathbf{G}\cdot\mathbf{G}|0_i^+\rangle$ and $\bar{E}$ an average energy denominator.

To calculate the exact $2\nu\beta\beta$ matrix element, Eq. (1), we consider the function [2]

$$\phi(\tau,\tau') = \frac{\mathrm{Tr}[e^{-(\beta-\tau-\tau')H}\mathbf{G}^\dagger \cdot \mathbf{G}^\dagger e^{-\tau H}\mathbf{G}e^{-\tau' H}\cdot \mathbf{G}]}{\mathrm{Tr}[e^{-\beta H}]}, \tag{2}$$

where H is the many-body Hamiltonian and the trace is over all states of the initial nucleus. Large values of the parameters $(\beta-\tau-\tau')$ and τ guarantee cooling to the parent and daughter ground states. In these limits, we note that $\phi(\tau,\tau'=0)$ approaches $e^{-\tau Q}|M_c|^2$, where $Q = E_i^0 - E_f^0$ is the energy release, so that a calculation of $\phi(\tau,0)$ leads directly to the closure matrix element. If we then define

$$\eta(T,\tau) \equiv \int_0^T d\tau' \phi(\tau,\tau')e^{-\tau' Q/2}, \tag{3}$$

and

$$M^{2\nu}(T,\tau) \equiv \frac{\eta(T,\tau)M_c^*}{\phi(\tau,0)}, \tag{4}$$

in the limit of large τ, $(\beta-\tau-\tau')$, and T, $M^{2\nu}(T,\tau)$ becomes independent of these parameters and is equal to the matrix element in Eq. (1).

Using these methods, we obtained a value of 0.15 ± 0.07 MeV^{-1} for the exact 2ν matrix element of ^{48}Ca whereas the calculation of Caurier *et al.* [4] gives 0.08 MeV^{-1}. There is thus agreement within the uncertainty.

We have then applied the SMMC method to ^{76}Ge, where direct diagonalization is not possible. Our result for the matrix element is 0.12 ± 0.07 MeV^{-1} while the experimental value of this matrix element is 0.14 ± 0.01 MeV^{-1} [5] (using $g_A = 1.26$) or 0.22 ± 0.01 MeV^{-1} (using $g_A = 1.0$).

Details of these calculations can be found in Ref [2].

This work was carried out under grants from the NSF and the DOE. Computational cycles were provided by the Concurrent Supercomputing Consortium on the Intel Touchstone Delta and the Intel Paragon and on the IBM SP2 at the Maui High Performance Computing Center.

References

1. S. E. Koonin, D. J. Dean, and K. Langanke, Phys. Rep. (1995).
2. P. B. Radha *et al.*, Phys. Rev. Lett. **76**, 2642 (1996).
3. W. C. Haxton and G. J. Stephenson, Prog. in Part. and Nucl. Phys. **12**, 409 (1984).
4. E. Caurier, A. Poves, and A. Zuker, Phys. Lett. B **252**, 13 (1990).
5. A. Balysh *et al.*, Phys. Lett. B **322**, 176 (1994).

STATUS OF THE CHORUS EXPERIMENT

The CHORUS collaboration
Presented by P.F. Loverre - Università di Roma "La Sapienza" and INFN, Italy

The CHORUS experiment[1] searches for $\nu_\mu \rightarrow \nu_\tau$ oscillations. It aims at the detection of $\nu_\tau N \rightarrow \tau X$ interactions in a beam of muon neutrinos. The beam is the CERN SPS Wide Band neutrino Beam (WBB), with an average neutrino energy of the order of 27 GeV. Neutrino interactions take place in a nuclear emulsion target of about 800 Kg (the largest ever used in a high energy physics experiment). The exceptional spatial resolution of the emulsions ($< 1\mu m$) will allow to observe the decay of the τ lepton produced in the charged current ν_τ interactions. The decay path of the τ is in fact of the order of $1mm$.

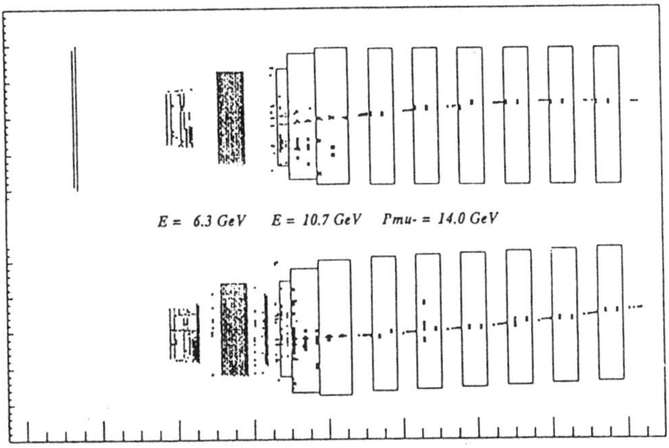

Figure 1: A neutrino interaction in the detector

The Chorus collaboration has built a sophisticated detector which follows the emulsion target and has two purposes: a)-high precision reconstruction of tracks to allow to predict the position of each interaction in the emulsions and b)-full reconstruction of the kinematics of the events. The high resolution tracking is performed by an array of scintillating fibers of $500\mu m$ diameter.The downstream apparatus includes an air-core magnet (0.12 T over 75 cm), followed by a scintillating fiber array, a high resolution "spaghetti calorimeter" and an upgraded version of the CDHS muon spectrometer.

The detector is fully operative since 1994. As an example we show in fig. 1 a charged current (CC) ν_μ induced event.

In the years 1994 and 1995 CHORUS has collected 120000 + 200000 CC ν_μ interactions. The emulsion target was then removed and developed, while a new one was installed in spring 1996. The data taking in the years 1996 and 1997 should allow to collect 400000 more CC events.

The sensitivity to neutrino oscillations is easily related to the number N_{CC} of ν_μ CC interactions collected. In case a number N_τ of candidates for ν_τ interactions is observed, the oscillation probability is given by $P = (N_\tau/\epsilon_\tau)/N_{CC}$. For Chorus, the efficiency ϵ_τ for detecting ν_τ CC interactions is of the order of 0.05. This leads to the expected sensitivity shown in fig. 2 as a function of the relevant parameters.

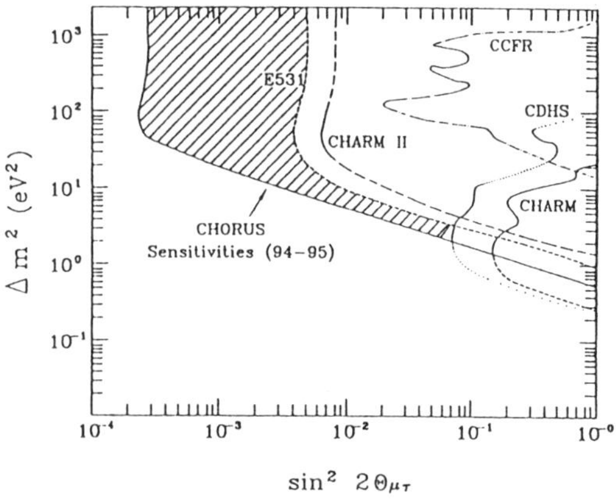

Figure 2: Exclusion plot for $\nu_\mu \to \nu_\tau$

Currently the detector performances meet the expectations and the main effort is devoted to the analysis of the data. The exceptional spatial resolution of the emulsions has its drawback in the very difficult scanning by microscope. Kinematical selections exploiting the momentum unbalance due to neutrinos produced in the τ decay, will allow to reduce by almost one order of magnitude the events to be scanned. Yet the analysis would be impossible without the implementation of the automatic scanning techniques developed mainly in Nagoya. Automatic scanning should allow full analysis of one event in times of the order of 15 minutes. The techniques are now being tuned with data from 1/4 target removed and developed at the end of the 1994 run. Preliminary results from these data will come soon. In one year from now about 20 automatic systems will be operational in Europe, Korea and Japan allowing to reach the discussed sensitivity.

[1] M. De Jong et al., CHORUS Collaboration, CERN-PPE/93-131.

Evidence for Neutrino Oscillations from Muon Decay at Rest

Rex Tayloe
Los Alamos National Laboratory, Los Alamos, NM 87545,
representing the LSND collaboration

The LSND collaboration searches for $\bar{\nu}_\mu \rightarrow \bar{\nu}_e$ oscillations using $\bar{\nu}_\mu$ from μ^+ decays at rest. The analysis of the current LSND data set (1993-95 runs) using strict cuts to identify oscillation candidate events with positron energy in the range 36-60 MeV along with a correlated γ from the neutron capture reaction yields 22 events with only 4.6 ± 0.6 events expected from conventional processes.

1 Introduction

The LSND experiment situated at the LANSCE experimental facility at Los Alamos searches for $\bar{\nu}_\mu \rightarrow \bar{\nu}_e$ oscillations by looking for $\bar{\nu}_e$ events in excess of that expected from conventional processes. The $\bar{\nu}_\mu$ are produced predominantly from the decay chain that starts with π^+ produced by 800 MeV protons incident on water followed by $\pi^+ \rightarrow \mu^+ \nu_\mu$, $\mu^+ \rightarrow e^+ \nu_e \bar{\nu}_\mu$. The $\bar{\nu}_e$ produced from the π^- decay chain are suppressed by approximately 7.5×10^{-4} due to a lower π^- production cross section, as well as π^- absorption and μ^- capture in the beam stop.

The $\bar{\nu}_e$ are detected in the Liquid Scintillator Neutrino Detector (LSND)[1] via the reaction $\bar{\nu}_e\, p \rightarrow e^+\, n$ followed by a correlated 2.2 MeV γ from the neutron capture reaction, $np \rightarrow d\gamma$. This two-fold detection process provides a low-background event signature.

2 Analysis

The data analysis consists of selecting events consistent with the two-fold detection reaction.[2,3] Events with positrons are selected on the basis of an ID parameter that quantifies the agreement of the space, time, and Čerenkov cone reconstruction to that expected for a positron. Additional cuts are imposed to define a fiducial volume and to reduce various cosmic and beam-related backgrounds.

Events with correlated photons are identified using a ratio of correlated to accidental likelihoods, R. These likelihoods are formed for each photon candidate using the expected correlated and accidental photon distributions in positron-to-photon time and distance along with number of photon PMT hits.

3 Results

After selecting events that pass the positron ID cuts with $E_{e+} > 36$ MeV and have a correlated photon with $R > 30$, 22 events remain with 4.6 ± 0.6 expected from backgrounds. The e^+ (beam-off subtracted) energy distribution for these events is shown in Fig. 1. The probability that this signal is due to a statistical fluctuation of these backgrounds is 4.1×10^{-8}. These results may be interpreted as evidence for $\bar{\nu}_\mu \to \bar{\nu}_e$ oscillations.[3]

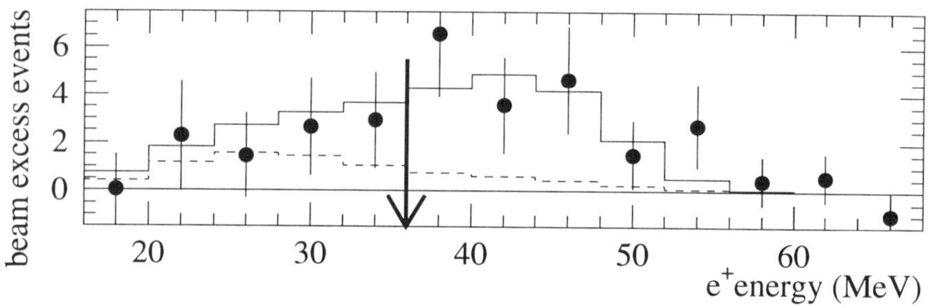

Figure 1: The energy distribution of the beam excess data along with the estimated ν background (dashed) and the expected distribution for ν background plus a large-Δm^2 signal (solid). The arrow indicates the e^+ energy cut.

References

1. C. Athanassopoulos et al., LA-UR-96-1327, submitted to Nucl. Instrum. Methods.
2. C. Athanassopoulos et al., Phys. Rev. Lett. **68**, 274 (1995).
3. C. Athanassopoulos et al., LA-UR-96-1326, submitted to Phys. Rev. C. and LA-UR-96-1582, submitted to Phys. Rev. Lett.

Inclusive and Exclusive ν_eC and ν_μC Cross Section Measurements from LSND

H.J.KIM [a]

Dept. of Physics and Ast., Lousiana State Univ., Baton Rouge, La70803 , USA

ν_eC and ν_μC scattering have been studied at LAMPF with the Liquid Scintillator Neutrino Detector. The LAMPF neutrino source produces ν_μ from π^+ decay in flight and ν_e from μ^+ decay at rest. The measured ν_e ^{12}C $\rightarrow$ e$^-$ ^{12}N$_{gs}$ and ν_e ^{12}C $\rightarrow$ e$^-$X cross sections agree well with previous measurements and theoretical predictions. The ν_μ ^{12}C $\rightarrow$ μ^- ^{12}N$_{g.s.}$ cross section is reported for the first time and is also consistent with theory. However, the measured ν_μ ^{12}C $\rightarrow$ μ^-X inclusive cross section is about half the FGM and CRPA model predictions.

Results are reported here on ν_eC and ν_μC scattering [2] using the Liquid Scintillator Neutrino Detector (LSND) described in reference [1]. Low energy neutrino-nucleus scattering is interesting because it can give improved insight into nuclear structure. Also inelastic neutrino scattering cross sections are necessary for various astrophysical calculations. Five different neutrino-nucleus scattering processes have been studied and preliminary cross sections were obtained.

The reaction $\nu_e +^{12} \mathrm{C} \rightarrow^{12} \mathrm{N}_{g.s.} + \mathrm{e}^-$ is identified by the detection of the e^- followed by the positron from the β decay of the ^{12}N. We measured 515.2±23.4 events and obtained a flux averaged cross section of $(9.1 \pm 0.4 \pm 0.9) \times 10^{-42} cm^2$ in excellent agreement with other experiments [8,9] as well as theoretical predictions [3,4,5,6] as shown in Fig. 1-a). Events from $\nu e \rightarrow \nu e$ scattering are identified by strong forward peaking of the electron. The measured cross section is $(4.2 \pm 0.6 \pm 0.4) \times 10^{-43} cm^2$ compared to the expect value of $3.97 \times 10^{-43} cm^2$. The $^{12}\mathrm{C}(\nu_e, \mathrm{e}^-)^{12}\mathrm{N}^*$ cross section was measured to be $(6.0 \pm 0.6 \pm 0.8) \times 10^{-42} cm^2$ which agrees well with the KARMEN experiment [9] and CRPA calculation [6]. The measured angular distributions of the electron for these processes also agree with expectations. The reaction $\nu_\mu +^{12} \mathrm{C} \rightarrow^{12} \mathrm{N}_{g.s.} + \mu^-$ is identified by the triple coincidence of a $\mu-$, $e+$ from a decay of $^{12}\mathrm{N}_{g.s.}$ and e^- from μ^- decay. The cross section obtained from 62.2± 8.0 events is $(6.4 \pm 1.0 \pm 1.0) \times 10^{-42} cm^2$.This agrees well with a recent calculation [7] as shown in Fig.1-b). On the other hand, the inclusive cross section measurement of $^{12}\mathrm{C}(\nu_\mu, \mu^-)$ is $(11.3 \pm 0.3 \pm 1.8) \times 10^{-40} cm^2$, about half of the CRPA and FGM predictions [6].

The good agreement of the two $^{12}\mathrm{N}_{g.s.}$ measurements shown in Fig. 1

[a] Representing the LSND Collaboration.

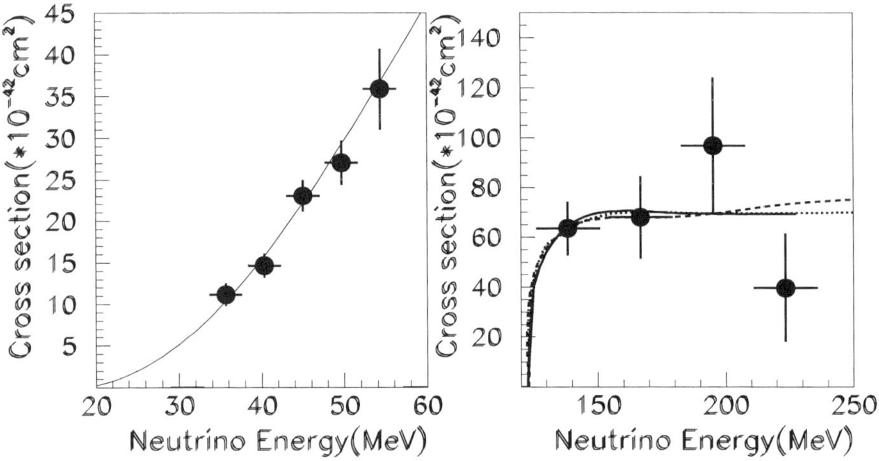

Figure 1: Differential cross section of; a) $^{12}C(\nu_e, e^-)^{12}N_{g.s.}$, b) $^{12}C(\nu_\mu, \mu^-)^{12}N_{g.s.}$ with respect to E_ν. Lines are theoretical predictions from Ref. [3,7].

confirm our understanding of the detector and flux calculation.

References

1. C. Athanassopoulos *et. al.* , LA-UR-96-1327, submitted to Phys. Rev. C
2. C. A. Athanassopoulos *et. al.*, to be submitted to Phys. Rev. C.
3. M.Fukugita et al., *Phys. Lett.* B **212**, 139 (1988).
4. T. W. Donnelly and R. D. Peccei, Phys. Rep. **50**, 1(1979).
5. S. L. Mintz and M. Pourkaviani, S. L. Mintz, J. Phys. **G16**, 569(1990)
6. E. Kolbe, K. Langanke, S. Krewald, Phys. Rev. **C49**, 1122(1994).
7. J. Engel, E. Kolbe, K. Langanke, P. Vogel, nucl-th/9606031, 1996.
8. D. A. Krakauer *et. al.* Phys. Rev. **C45**, 2450(1992).
9. G. Drexlin, KARMEN Collaboration, data presented at PANIC96, to be published in the Proceedings.

NEUTRINO CAPTURE ON ^{12}C: GROUND-STATE TO GROUND-SATE TRANSITIONS

JONATHAN ENGEL
Dept. Physics and Astronomy, Univ. North Carolina, Chapel Hill, NC 27599, USA

E. KOLBE
Physikalishes Institut, Universität Basel, Switzerland

K. LANGANKE
W.K. Kellogg Rad. Lab., California Inst. of Technology, Pasadena, CA 91125, USA

P.VOGEL
Physics Dept., California Institute of Technology, Pasadena, CA 91125, USA

We show that for currently available neutrino energies, $E_\nu \leq 300$ MeV, calculated exclusive cross sections $^{12}C_{gs}(\nu,l)^{12}N_{gs}$ are essentially model independent, provided the calculations simultaneously describe the rates of several other reactions involving the same states or their analogs. The calculations agree well with the measured cross sections, which can therefore be used to check neutrino-flux normalization and detector efficiency.

It is important to understand neutrino scattering from ^{12}C, an ingredient in liquid scintillator detectors that are used to measure neutrino oscillations. Although calculated cross sections for the inclusive reaction $^{12}C(\nu_e,e)^{12}N^*$ (with neutrinos from muon-decay at rest) agree well with measurements,[1] the same is not true [2,3] for the inclusive reaction $^{12}C(\nu_\mu,\mu)^{12}N^*$, measured with higher energy neutrinos (up to about 300 MeV) from pion-decay in flight at LAMPF.[4] The disagreement is disturbing in light of the apparent simplicity of the reaction; here we see whether similar problems affect our understanding of the exclusive reactions [1,4] $^{12}C(\nu_e,e)^{12}N_{gs}$ and $^{12}C(\nu_\mu,\mu)^{12}N_{gs}$.

We have used several models to calculate rates for these reactions, adjusting parameters to simultaneously reproduce the following similar processes: β^+ decay from ^{12}N, muon capture from ^{12}C to the ground state of ^{12}B, and electroexcitation of (and γ-decay from) the 1^+ state at 15.11 MeV in ^{12}C. One set of calculations employs the shell model with only the p-shell active, a framework in which four one-body densities and an oscillator parameter completely determine all of the above reactions, and which has already been used to compute some of them by several people.[5–7]. Here we have adjusted the four densities to reproduce ***all*** the reactions above as well as possible. To include physics from beyond the p-shell we have used as an alternative the Random Phase

Approximation (RPA),[8] which incorporates correlations from higher shells but simplifies the dynamics and requires some ***ad hoc*** renormalization.

We show the calculated and measured exclusive cross sections in Table 1. In both cases the the different shell model calculations and the RPA concur with one another and also agree well with experiment. We conclude that the related reactions mentioned above tightly constrain the neutrino-induced transitions, and that the measured rates for these exclusive processes are therefore very likely correct. This means, among other things, that they can be used to check the normalization of the neutrino flux from the beam and the efficiency of the detector. The discrepancy between calculations and experiment in the inclusive ν_μ-induced reaction, however, remains unexplained.

Table 1: Calculated and measured cross sections for neutrino-induced transitions $^{12}C_{gs} \rightarrow ^{12}N_{gs}$, in units of $10^{-42}cm^2$, averaged over corresponding neutrino spectra.

	$^{12}C(\nu_e, e^-)^{12}N_{gs}$ decay at rest	$^{12}C(\nu_\mu, \mu^-)^{12}N_{gs}$ decay in flight
experiment	8.9±0.6 ± 0.75 [1]	58 ±9 ± 9 [4]
Shell model, this work	9.1 ± 0.1	63.5 ±5
Shell model [5]	9.42	65.4
Shell model [6]	8.8	60.4
Shell model [7]	9.4	62.3
RPA	8.9	63

References

1. B. Zeitnitz, KARMEN Collab., *Prog. Part. Nucl. Phys.* **32**, 351 (1994).
2. E. Kolbe, K. Langanke, F.K. Thielemann, and P. Vogel, *Phys. Rev.* C **52**, 3437 (1995).
3. T. S. Kosmas and E. Oset, *Phys. Rev.* C **53**, 1409 (1996).
4. M. Albert et al., *Phys. Rev.* C **51**, 1065 (1995).
5. J. S. O'Connell, T. W. Donnelly, and J. D. Walecka, *Phys. Rev.* C **6**, 719 (1972).
6. W. C. Haxton, *Phys. Lett.* B **76**, 165 (1978).
7. F. P. Brady et al., *Phys. Rev.* C **43**, 2284 (1991).
8. T. Suzuki, F. Osterfeld, and J. Speth, *Phys. Lett.* B **100**, 443 (1981).

INCLUSIVE AND EXCLUSIVE NEUTRINO REACTIONS IN ^{12}C AND PRESENT EXPERIMENTS

S.L. MINTZ, M. POURKAVIANI
Physics Department, Florida International University
Miami, Florida 33199, USA

We obtain cross sections for the exclusive reaction $\nu_\mu + {}^{12}C \longrightarrow \mu^- + {}^{12}N_{(g.s.)}$ averaged over the LAMPF decay in flight neutrino spectrum. We also obtain cross sections for the inclusive reactions, $\nu_e + {}^{12}C \longrightarrow e^- + X$, $\nu_\mu + {}^{12}C \longrightarrow \mu^- + X$ over appropriate neutrino spectra. We discuss the importance of these reactions and our models for calculating these cross sections.

1 The Inclusive Case

Two experimental results one by the KARMEN collaboration[1] and one by a group at LAMPF[2] exist for the inclusive neutrino reaction $\nu_e + {}^{12}C \longrightarrow e^- + X$ and one experimental result exists for the reaction[3] $\nu_\mu + {}^{12}C \longrightarrow \mu^- + X$. The former reactions are over the Michel spectrum whereas the latter is over a decay in flight spectrum with energies up to 300 MeV. We present a phenomenologically based calculation for these processes. We start from a standard[4] first order weak calculation. The transition matrix element squared for this process may be written as $|M_{ki}|^2 = \frac{G^2 \cos\theta_C}{2m_\mu m_\nu} L^{\sigma\lambda} < k|J_\sigma(0)|^{12}C > < k|J_\lambda(0)|^{12}C >^*$. Here $L^{\sigma\lambda}$ is the well known lepton tensor. We can sum over the final states put in average quantities for the exitation (and hence the momentum transfer) which will depend on the neutrino energy. We can then write the hadronic part of Eq.(1) as a tensor. Arguments[4] simplify the tensor to $Q^{\mu\nu} = \alpha g^{\mu\nu} + \frac{\beta}{M^2} P_i^\mu P_i^\nu$ and lead to a cross section given by:

$$\sigma_c = \frac{G^2 \cos^2(\theta_C)}{4\pi} \frac{< |\vec{p_\mu}| >< E_\mu > D}{M(M + E_\nu)} \tag{1}$$

with $D = \beta - 2\alpha$. The quantity, D, can be determined from inclusive muon capture results as well as from electron scattering and pion electroproduction results[4].

We are now able to calculate results for the reaction $\nu_e + {}^{12}C \longrightarrow e^- + X$ averaged over the Michel spectrum. We quote only those for exitation above the nitrogen ground state, $< \sigma_c >= 8.55 \times 10^{-42}\ cm^2 \pm 25\%$ which is in reasonable agreement with the KARMEN result but not with LAMPF.

Recently LAMPF has performed the reaction, $\nu_\mu + {}^{12}C \longrightarrow \mu^- + X$ over a spectrum of decay in flight produced muon neutrinos. Their result was $<\sigma_c>= (8.3 \pm 0.7\ stat \pm\ 1.6sys) \times 10^{-40}\ cm^2$ and is in reasonable agreement with our calculated result of $<\sigma_c>= 13.1 \times 10^{-40}\ cm^2$ with an error in the 25 to 30 percent range. Two other calculations give results above $20 \times 10^{-40}\ cm^2$. We are improving our calculation for the average excitation energy. This should reduce our results for all these processes by approximately 10 to 15 percent.

2 The Exclusive Reaction

When the LAMPF results were announced for the reaction $\nu_\mu + {}^{12}C \longrightarrow \mu^- + X$, it was suggested that they should consider measuring the reaction, $\nu_\mu + {}^{12}C \longrightarrow \mu^- + {}^{12}N_{(g.s.)}$ as a test both of their experiment and as a test of their spectrum. We have therefore undertaken an elementary particle model calculation for this process[5]. The leptonic part of the first order weak transition matrix element is well known and we will not give it here. The hadronic part has been previously determined[5]. The physics of the nuclear weak current matrix element is contained in the form factors, $F_A(q^2), F_M(q^2), F_E(q^2), and F_P(q^2)$. These have been determined[5]. We note that:

$$F_P(q^2) = \frac{m_\pi^2 F_A(q^2)}{m_\pi^2 - q^2}[1 + \xi(q^2)] \tag{2}$$

where ξ runs according to estimates from 1 to -1.

We therefore obtain cross sections averaged over this spectrum of: $\xi = -1, <\sigma>= .694 \times 10^{-40}\ cm^2; \xi = -.5, <\sigma>= .643 \times 10^{-40}\ cm^2; \xi = 0$, $<\sigma>= .600 \times 10^{-40}\ cm^2; \xi = 1, <\sigma>= .537 \times 10^{-40}\ cm^2$ where these results have a 15 percent error attached. These results are in good agreement with the LAMPF results of around $.6 \times 10^{-40} cm^2$ with about a 30 percent error. Thus this result lends weight to the inclusive LAMPF result. However because the actual cross sections for the exclusive case are exceedingly flat, the spectrum averaged cross sections except near threshold are almost independent of the actual spectrum,.i.e. $<\sigma>\simeq \sigma$. Thus nothing is learned of the LAMPF in flight spectrum for neutrinos. Clearly much remains to be done.

References

1 B.Zeitnitz et al.,Prog. Part. Nucl. Phys.**13**,445(1995).
2 R.C. Allen et al.,Phys. Rev. Lett.**64**,1871(1990).
3 M. Albert,Reaction ${}^{12}C(\nu_\mu, \mu^-)X$ near threshold,LAMPF preprint (1994).
4 S.L. Mintz and M. Pourkaviani,Nucl. Phys.**A594**,346(1995).
5 S.L. Mintz and M. Pourkaviani,The Reaction ${}^{12}C(\nu_\mu, \mu^-)^{12}N_{(g.s.)}$.

Charged current neutrino-nucleus reaction cross sections at intermediate energies

T. S. KOSMAS* , and E. OSET**

* Division of Theoretical Physics, University of Ioannina, GR-45110 Ioannina, Greece
** Departamento de Física Teórica and IFIC, Centro Mixto
Universidad de Valencia-CSIC, 46100 Burjassot (Valencia), Spain.

Inclusive and semi-inclusive neutrino-nucleus reactions cross sections at intermediate energies ($20\,MeV \leq E_\nu \leq 500\,MeV$) are calculated throughout the periodic table for the most interesting nuclei from an experimental point of view. The method used had previously proved to be very accurate in calculating the inclusive reaction cross section for light nuclei (^{12}C and ^{16}O) and in the study of other similar processes and has been further improved to deal with low energy neutrinos. The electron neutrino (ν_e) and muon neutrino (ν_μ) cross sections weighted by their energy distributions are also calculated and discussed in conjunction with the existing experimental results at LAMPF and the KARMEN collaboration.

The method used evaluates accurately the reaction probability in infinite nuclear matter and then makes a mapping into finite nuclei, via the local density approximation which turns out to be a very reliable tool given the short range of the weak interactions.

One takes into account the renormalization of the weak currents in nuclei as done in [1, 2]. Cross section for inclusive (ν, e) and $(\bar{\nu}, e^+)$ in ^{12}C, ^{16}O, ^{37}Cl, ^{40}Ar, ^{71}Ga, ^{81}Br, ^{48}Mo, ^{115}In, ^{127}I and ^{205}Tl, are given as a function of the neutrino energy and the same is done for μ neutrinos. Flux averaged cross sections with the Michel distributions or the ν_μ energy distributions are also evaluated to allow comparison with experiment. Also cross sections where only excited states of the daughter nucleus below the nucleon emission threshold are excited are evaluated to allow comparison with radiochemical experiments.

As an example, in Table I we give averaged cross sections for electron neutrinos on several nuclei, obtained by folding the cross section with the Michel distribution. At the same time we show the results which we obtain for the radiochemical cross sections. The agrement with the data in the case were measurements exit is good.

On the other hand we can compare our results for the $^{127}I(\nu, e^-)^{127}Xe$ with those of [6] which get $\sigma = 2.1-3.1\ 10^{-40}cm^2$, for the radiochemical cross section and those of [7], which get $\sigma = 3.0-6.4\,10^{-40}cm^2$ for the inclusive cross section (the colum of Total in Table I). Another data of interest is the $^{12}C(\nu_\mu, \mu^-)^{12}N$ cross section of the LSND experiment (reported in this session by H.J. Kim) of $\sigma = 11.3 \pm 0.3 \pm 1.8\ 10^{-40}cm^2$.

We get $\sigma = 19\ \ 10^{-40}cm^2$ for this cross section in agreement with the results of [8, 9] which obtain 20 and 19.3 $10^{-40}cm^2$ respectively. The discrepancies with the data of about a factor two are somehow worrisome and require further thoughts by theoretitians and experimentalists. Detailed cross sections for all the nuclei quoted here can be seen in [10].

TABLE I

	Average cross section $\bar{\sigma}(10^{-40}cm^2)$			
(A,Z)	Reaction	Total	Radio-chemical	expt.
$^{12}_{6}C$	$^{12}_{6}C(\nu_e,e^-)^{12}_{7}N$	0.14	-	0.15 ± 0.03 Ref.[3] 0.14 ± 0.03 Ref.[4]
$^{37}_{17}Cl$	$^{37}_{17}Cl(\nu_e,e^-)^{37}_{18}Ar$	1.8	1.4	
$^{40}_{18}Ar$	$^{40}_{18}Ar(\nu_e,e^-)^{40}_{19}K$	1.9	1.3	
$^{71}_{31}Ga$	$^{71}_{31}Ga(\nu_e,e^-)^{71}_{32}Ge$	4.0	2.7	
$^{81}_{35}Br$	$^{81}_{35}Br(\nu_e,e^-)^{81}_{36}Kr$	4.5	3.2	
$^{98}_{42}Mo$	$^{98}_{42}Mo(\nu_e,e^-)^{98}_{43}Tc$	5.3	2.7	
$^{115}_{49}In$	$^{115}_{49}In(\nu_e,e^-)^{115}_{50}Sn$	7.2	4.7	
$^{127}_{53}I$	$^{127}_{53}I(\nu_e,e^-)^{127}_{54}Xe$	7.3	4.3	6.2 ± 2.5 Ref.[5][a]
$^{205}_{81}Tl$	$^{205}_{81}Tl(\nu_e,e^-)^{205}_{82}Pb$	14.0	6.3	

a) This experiment is of a radiochemical type

References

[1] H. C. Chiang, E. Oset and P. Fernández de Córdoba, Nucl. Phys. A510 (1990)591.

[2] S.K. Singh and E. Oset, Phys. Rev. C48 (1993) 1246.

[3] B. Zeitnitz, Prog. Part. Nucl. Phys. 32 (1994) 351.

[4] D.A. Krakauer et al. Phys. Rev. C45 (1992) 2450.

[5] B. T. Cleveland et al, Proc. 23 rd Int. Cosmic Ray Conference, Edts. D.A. Leahy et al., Vol. 2 pag. 865.

[6] J. Engel, S. Pittel and P. Vogel, Phys. Rev. C50 (1994) 1702.

[7] S.L. Mintz and M. Pourkaviani, Nucl. Phys. A 584 (1995) 665.

[8] E. Kolbe, K. Langanke and S. Krewald, Phys. Rev. C51 (1995) 1122.

[9] E. Kolbe, K. Langanke, F.K. Thielemann and P. Vogel, Phys. Rev. C52 (1995) 3437.

[10] T. S. Kosmas and E. Oset, Phys. Rev. C53 (1996) 1409.

Session 8
Strange Matter

RECENT PROGRESS IN THE DECAY OF HYPERNUCLEI

A. RAMOS, A. PARREÑO
Dept. ECM, Facultat de Física, Universitat de Barcelona, 08028 Barcelona, Spain

C. BENNHOLD
Department of Physics, The George Washington University Washington DC 20052, USA

E. OSET, M.J. VICENTE-VACAS
Departamento de Física Teórica and IFIC, Universidad de Valencia, 46100 Burjassot, Valencia, Spain

Recent progress in the nonmesonic decay of Λ-hypernuclei through the $\Lambda N \to NN$ and $\Lambda NN \to NNN$ channels is discussed.

1 Decay Rates and Proton Asymmetry

The weak decay of Λ hypernuclei through the nonmesonic channel $\Lambda N \to NN$ is studied in a shell model framework [1]. A complete weak transition potential, based on meson exchange, is constructed by including the pseudoscalar mesons π, K, η and the vector mesons ρ, ω, K^*. Our model includes realistic ΛN short-range correlations and final state NN interactions. Our results are shown in the Table, where we see a strong influence of the K meson on the decay rate, $\Gamma^{(1)}_{nm}$, and the neutron to proton ratio, Γ_n/Γ_p, which is later compensated by the effect of the vector meson K^*. The results in brackets correspond to a recent estimation of the kaon weak couplings obtained from next-to-leading order in

Table 1: Weak decay observables for ${}^{12}_{\Lambda}$C.

	$\Gamma^{(1)}_{nm}/\Gamma_\Lambda$	Γ_n/Γ_p	a_Λ
π	0.89	0.104	−0.24
$+\eta$	0.83	0.153	−0.28
$+K$	0.46 (0.55)	0.072 (0.10)	−0.33 (−0.31)
$+\rho$	0.47 (0.57)	0.078 (0.11)	−0.20 (−0.20)
$+\omega$	0.52 (0.60)	0.071 (0.10)	−0.30 (−0.30)
$+K^*$	0.81 (0.91)	0.103 (0.12)	−0.43 (−0.42)

χPT.[3] The total rate and asymmetry a_Λ, which represents a measure of the parity-violating to parity-conserving amplitudes, are in agreement with recent experimental data.[2] The ratio Γ_n/Γ_p is very small in the OPE mechanism and is not increased when the other mesons are included.

2 Nucleon spectra and the n/p Ratio

We have calculated the spectra of nucleons from the decay of Λ hypernuclei for both the one-nucleon ($\Lambda N \to NN$) and two-nucleon ($\Lambda NN \to NNN$) induced mechanisms. The calculations start from the momentum distributions of the primary decay events and follow the final state interaction of the nucleons on their way out of the nucleus via a Monte Carlo simulation. The resulting spectra, assuming a value $\Gamma_n/\Gamma_p = 1$, are shown in the Figure.

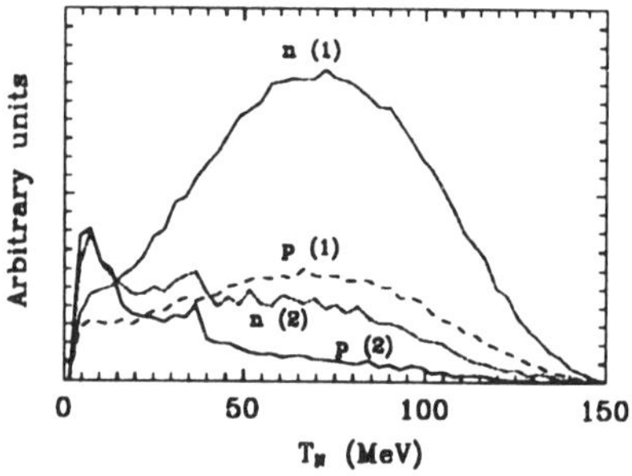

Figure 1: Spectrum of neutrons and protons from the weak decay of ${}^{12}_{\Lambda}$C.

Taking Γ_n/Γ_p as a free parameter, we calculate, from the spectra, the corresponding number of neutrons and protons per decay event. A comparison with the measured values facilitates the extraction of Γ_n/Γ_p. We extend the observations of Ref. [4] to account for the experimental energy cut-offs, and find that the consideration of the 2N induced channel enlarges the band of allowed values for Γ_n/Γ_p to the range 0.15—1.65, a result that is even compatible with the OPE theoretical predictions.

References

1. C. Bennhold, A. Parreño and A. Ramos, Few-Body Systems Suppl. **9**, 465 (1995).
2. J.J. Szymanski *et al*, Phys. Rev. C **43**, 849 (1991); H. Noumi *et al*, Phys. Rev. C **52**, 2936 (1995); S. Ajimura *et al*, Phys. Lett. **B282**, 293 (1992)
3. M.J. Savage and R.P. Springer, Phys. Rev. C **53**, 441 (1996).
4. A. Ramos, E. Oset and L.L. Salcedo, Phys. Rev. C **50**, 2314 (1994).

MESONIC AND NON-MESONIC DECAY WIDTHS OF ${}^{4}_{\Lambda}$He AND ${}^{4}_{\Lambda}$H

H. Outa, M. Aoki, T. Yamazaki
Institute for Nuclear Study, Univ. of Tokyo, 3-2-1 Midori, Tanashi 188, Japan

R. S. Hayano, T. Ishikawa. M. Iwasaki, H. Tamura, E. Takada
University of Tokyo, Bunkyo, Tokyo 113, Japan

Weak decay widths are measured precisely for ${}^{4}_{\Lambda}$H and ${}^{4}_{\Lambda}$He mirror-hypernuclei formed via ^{4}He(K^{-}_{stopped}, π) reaction. The lifetimes of ${}^{4}_{\Lambda}$H and ${}^{4}_{\Lambda}$He and π^0, π^- and $\Lambda p \to np$ decay branching ratios of ${}^{4}_{\Lambda}$He were determined and they are compared with recent theoretical estimations.

The precise determination of the π-mesonic weak decay widths of very light(A=4) hypernuclei has much significance not only because they are sensitive to the ΛN interaction[1,2] but also because quantitative study of non-mesonic weak decay widths of light Λ hypernuclei requires these strengths as inputs[3]. In recent theoretical works, the existence of a central repulsion in the Λ-nucleus potential has been discussed by several authors. In this case, Λ is pushed outwards and overlap of the wave function of Λ and the residual nucleus becomes smaller. We can detect such an effect from precise measurement of the mesonic decay widths. Kumagai*et al.*[2] calculated the mesonic decay widths of ${}^{4}_{\Lambda}$He and ${}^{4}_{\Lambda}$H with two different assumptions on the shape of Λ-nucleus potential: one is the single-range gaussian(SRG) and the other is the Isle potential. The Isle potential is characterized by existence of inner repulsive core. Their results are listed in Table 1. One can distinguish the difference of the potential by the comparison of the ratio of the π^0 and π^- decay widths of ${}^{4}_{\Lambda}$He.

Here results will be presented on measurements of the π-mesonic and non-mesonic weak decay processes of ${}^{4}_{\Lambda}$He hypernucleus, which were obtained from a precise measurement of π^- momentum in the hypernuclear formation from K^- stopped in liquid helium with π^0, π^- and proton coincidence measurements. Since the details of the lifetime measurement of ${}^{4}_{\Lambda}$H are already reported[4], we only summarize the results in Table 1.

Thanks to the good energy resolution of 2.0MeV(FWHM), we can clearly identify ${}^{4}_{\Lambda}$He formation by referring a peak at p_{π} =255.3 MeV/c in the inclusive ^{4}He(K^{-}_{stopped},π^-) spectrum[4]. In order to identify π^0,π^- and proton emission from ${}^{4}_{\Lambda}$He decay, our helium target was surrounded by 128 modules of NaI(Tl) counters covered with thin dE/dX counters[5]. The lifetime of ${}^{4}_{\Lambda}$He (256±27 ps) were obtained from direct measurement of the time difference

Table 1: Various weak decay widths of ${}^4_\Lambda$He and ${}^4_\Lambda$H in unit of the decay width of free $\Lambda(\Gamma_\Lambda)$ with comparison to theoretical estimates(Ref. 2). *Other experimental results are utilized.

Decay widths of ${}^4_\Lambda$He				Decay widths of ${}^4_\Lambda$H			
Decay	Results	SRG	Isle	Decay	Results	SRG	Isle
$\Gamma_{total}/\Gamma_\Lambda$	$1.03^{+0.12}_{-0.10}$			$\Gamma_{total}/\Gamma_\Lambda$	$1.36^{+0.21}_{-0.15}$		
$\Gamma_{\pi^0}/\Gamma_\Lambda$	0.53±0.07	0.49	0.51	$\Gamma_{\pi^0}/\Gamma_\Lambda$		0.16	0.19
$\Gamma_{\pi^-}/\Gamma_\Lambda$	0.33±0.05	0.25	0.31	$\Gamma_{\pi^-}/\Gamma_\Lambda$	$1.00^{+0.18}_{-0.15}$*	0.93	0.88
$\Gamma_{\pi^0}/\Gamma_{\pi^-}$	1.59±0.20	2.02	1.59	$\Gamma_{\pi^- {}^4He}/\Gamma_\Lambda$	$0.69^{+0.12}_{-0.10}$*	0.71	0.61
Γ_p/Γ_Λ	0.16±0.02			$\Gamma_{\pi^- {}^4He}/\Gamma_{\pi^-}$	0.69±0.02*	0.76	0.69
Γ_n/Γ_Λ	0.01±0.05						
$\Gamma_{nm}/\Gamma_\Lambda$	0.17±0.05			$\Gamma_{nm}/\Gamma_\Lambda$	0.17±0.11*		
$\Gamma_{nm}/\Gamma_{\pi^-}$	0.51±0.16						
Γ_n/Γ_p	0.06±0.40						

between K^- injection and proton emission from the $\Lambda p \to np$ non-mesonic decay process. We utilized K^- two-nucleon absorption events (K^- + "np" $\to$ $\Sigma^- p$,$\Sigma^- \to n\pi^-$) for the prompt calibration process for TOF analysis. The π^0-mesonic, π^--mesonic and $\Lambda p \to np$ non-mesonic decay branching ratios of ${}^4_\Lambda$He (0.515±0.035, 0.322±0.034 and 0.160±0.020 respectively) were precisely determined from the analysis of the decay counter system. From these results, we deduced the decay rates of these processes as listed in the Table. 1.

We found that all of our mesonic decay widths are well reproduced only when one assumes the Λ-nucleus potential with inner repulsion. Concerning the non-mesonic decay, it is found that the decay of ${}^4_\Lambda$He predominantly takes place via $\Lambda p \to np$ process($\Gamma_p >> \Gamma_n$), which indicates strong initial Λn spin dependence of the $\Lambda n \to nn$ decay process (little decay width from spin singlet state). We found that the non-mesonic decay widths of ${}^4_\Lambda$H and ${}^4_\Lambda$He are almost the same when we replace the assumed mesonic decay widths of ${}^4_\Lambda$H of Bloch *et al.*[3]'s by our values.

References

1. T. Motoba *et al.*, *Nucl. Phys.* A **534**, 597 (1991); **A 547**123c (1992).
2. I. Kumagai-Fuse *et al.*, *Phys. Lett.* B **345**, 386 (1995).
3. M. M. Bloch *et al.*, *Proc. of Int. Conf. on Hypernuclear Phys.* 63 (1964).
4. H. Outa *et al.*, *Nucl. Phys.* A **547**, 109c (1992); **A 585**109c. (1995).
5. A. Sakaguchi *et al.*, *Phys. Rev.* C **43**, 73 (1991).

HYPERNUCLEAR PHYSICS AT DAΦNE

M. BERTANI
Laboratori Nazionali di Frascati dell'INFN, Via E. Fermi 40, 00044 Frascati, Italy.
For the FINUDA Collaboration[a]

Hypernuclear spectroscopy and decay modes will be studied by the FINUDA experiment at $DA\Phi NE$, the Frascati ϕ-factory.

The $DA\Phi NE$[1] e^+e^- collider is currently under construction at INFN Laboratori Nazionali di Frascati. In 1997 it will run at the center of mass energy of the ϕ meson with an initial luminosity $\mathcal{L}_{in} = 1.3 \times 10^{32} cm^{-2}s^{-1}$ and a target luminosity $\mathcal{L}_{fin} = 5.3 \times 10^{32} cm^{-2}s^{-1}$. The production rate will be $R_\phi = (440 \to 2200)s^{-1}$ from initial to target luminosity.

The main ϕ decay at rest into K^+K^- (49%) will provide the FINUDA experiment with a unique K^- 'beam': monochromatic (16 MeV), slow (127 MeV/c), tagged by the collinear K^+ and background free. Very thin nuclear target ($0.5\, g\, cm^{-2}$) can be employed to stop K^- with minimal straggling very near to the target surface and produce a Y-hypernucleus ($Y = \Lambda, \Sigma, ...$) via the process: $K^-_{stop} + {}^AZ \to {}^A_YZ + \pi^-$. The outgoing prompt pions, emitted isotropically, do not cross any significant amount of the target and therefore do not undergo any momentum degradation. This is an important feature since the spectrum of the π^- is directly related to the energies and excitation strength of the hypernuclear levels. This feature provides unprecedented momentum resolution, as long as very transparent detectors are employed before and after the target.

In the case of $Y = \Lambda$, the following weak-interaction 'decays' are strongly favoured in medium-heavy nuclei: $\Lambda + n \to n + n$, $\Lambda + p \to n + p$, which are interesting for studying the validity of the $\Delta I = 1/2$ rule.

---

[a]M. Agnello[b], P. Amaudruz[c], G. Beer[c], L. Benussi[d], M. Bertani[d], S. Bianco[d], E. Botta[e], T. Bressani[e], L. Busso[f], D. Calvo[e], P. Camerini[g], C. Cattaneo[h], L. Celano[i], N. Colonna[i], G. D'Erasmo[i], A. Donzella[l], F.L. Fabbri[d], A. Feliciello[e], A. Filippi[m], V. Filippini[h], E.M. Fiore[i], L. Fiore[i], P. Gianotti[d], M. Giardoni[d], D.R. Gill[c], N. Grion[g], C. Guaraldo[d], F. Iazzi[b], A. Lanaro[d], L. Lee[c], V. Lucherini[d], S. Marcello[e], G. Marciano[h], B. Minetti[b], N. Mirfakhrai[n], P. Montagna[h], A.A. Nozdrin[c], A. Olin[c], A. Pantaleo[i], V. Paticchio[i], E. Rossetto[e], A. Rotondi[h], R. Rui[g], P. Salvini[h], S. Sarwar[d], A. Sharma[e], G. Smith[c], L. Venturelli[l], S. Yen[c] and A. Zenoni[o]. [b] Politecnico di Torino and INFN, [c] TRIUMF, [d] Laboratori Nazionali di Frascati dell'INFN, [e] Dip. di Fisica Sperimentale Univ. di Torino and INFN, [f] Ist. di Fisica Generale Univ. di Torino and INFN, [g] Univ. di Trieste and INFN, [h] Univ. di Pavia and INFN, [i] Univ. di Bari and INFN, [l] Univ. di Brescia and INFN Sez. di Torino, [m] Univ. di Pavia and INFN Sez. di Torino, [n] Shahid Beheshty Tehran Univ., [o] Univ. di Brescia and INFN Sez. di Pavia.

FINUDA is a high resolution magnetic spectrometer with cylindrical geometry, characterized by a large solid-angle typical of collider experiments ($\sim 70\%$ of 4π), good triggering capabilities, state-of-the-art tracking, particle identification and neutron detection. Hypernuclear formation and decay products will be detected within the same apparatus. The detector [2] consists of an interaction/target region, an external tracker, an outer scintillator array and a superconducting solenoid, providing a maximum magnetic field of 1.1 T. The geometry of the spectrometer, the position of the detectors and the value of the magnetic field have been optimized for maximizing the momentum resolution and acceptance for the prompt π^- from the Λ-hypernucleus formation. For such π^- (250-280 MeV/c) a momentum resolution less then 0.3%(FWHM) is obtained, corresponding to a resolution of 700 KeV in the hypernucleus energy levels. For the $\sim$ 80 MeV protons from non mesonic decays the energy resolution will be $\sim$ 1.3 Mev, while neutrons will be detected by the external scintillator barrel with $\sim$ 15% efficiency and 10 MeV energy resolution. For the typical capture rates of $10^{-3}/K^-_{stop}$ for the hypernuclear ground state, expected trigger rates are of the order of $8 \times 10^{-2}\,Hz$, providing about *80 events/hour* for high resolution hypernuclear spectroscopy.

FINUDA will perform a large program of hypernuclear spectroscopy on all nuclides that can be machined in solid targets, thus enlarging the chart of known hypernuclei; from the non-mesonic decay measurements the Λ-hypernucleus lifetime will be measured with 5% statistical precision and the $\Delta I = 1/2$ rule [3] will be tested in a reasonable time. The completion of this program could allow a step forward in the understanding of the main hypernuclear physics problems: the Λ-N interaction mechanism, the shell model of nuclei, the dynamical aspect of the strangeness degree of freedom in nuclear systems, the short range correlations as well as weak interactions in nuclei.

The FINUDA detector components are in advanced stage of construction and the apparatus will be ready to be installed in the experimental pit at $DA\Phi NE$ commissioning, scheduled for mid 1997.

References

1. S. Bianco in *Nuclear and Particle Physics with Meson Beams in the 1 GeV/c Region*, ed. S.Sugimoto, O.Hashimoto (Tokyo, 1995).
2. For details on the FINUDA detector see: The FINUDA Collaboration (M. Agnello *et al.*), *FINUDA Technical Report*, LNF-95/024(IR), 1995; *The Second DAΦNEPhysics Handbook*, ed. L. Maiani *et al.* (SIS-LNF, Frascati, 1995).
3. T.Bressani, *Il Nuovo Cimento* A **108**, 649 (1995).

Role of Spin-7/2 Resonances in Strangeness Electromagnetic Production

Pascal Girard [1,2], Bijan Saghai [1], and Frank Tabakin [3]
1) Service de Physique Nucléaire, CEA/DSM/DAPNIA, Centre d'Etudes de Saclay, F-91191 Gif-sur-Yvette
2) Institut de Physique Nucléaire, F-91406 Orsay
3) Department of Physics and Astronomy, University of Pittsburgh, Pittsburgh, Pennsylvania 15260

In hypernuclear spectroscopy [1], electromagnetic probes provide information complementary to that obtained from purely hadronic reactions. As a first step toward understanding strangeness production on nuclei, one needs to establish a sufficiently realistic description of the elementary process on a nucleon. In recent years there has been an important effort to established the elementary process operator (see Ref. [2] and earlier works quoted therein). The main difficulty in this topic comes from the fact that *a priori* about thirty resonances can contribute to the reaction mechanism. The most developed theoretical frame, based on effective Lagrangian approaches, does not allow a unique determination of the reaction mechanism. It is hence, highly desirable to find criteria which could offer high enough selectivity to shed a light on the relevant exchanged particles required by the available or forthcoming data.

In this purpose, we apply a *model independent* approach [3] which has already provided [4] some interesting information on the strong points, and shortcomings, of various phenomenological formalisms as well as on the sensitivity of polarization observables to resonances in *s*-, *u*-, and *t*-channels.

Using helicity amplitudes, we have considered the general structure [3] of the full set of 15 single and double polarization spin observables for the reaction $\gamma p \to K^+\Lambda$ [4]. Through this method, we deduced general rules about the angular structure of these observables, especially concerning how nodes arise for each observable from the underlying reaction mechanism. These rules are obtained either in a total angular momentum (J) or a multipole truncated basis. Expressions for the observables as a function of multipoles are reported for $l \leq 1$ and $l \leq 2$ in Refs. [3] and [5], respectively. In Table 1, we summarize the maximum number of allowed nodes for $J \leq 7/2$ and $l \leq 2$. The last column of this Table gives the number of nodes for each observable predicted by the most comprehensive available phenomenological model developed by the Saclay-Lyon group [2] and called SL. This model includes *s*-channel nucleonic resonances with spins up to 5/2, as well as *u*-channel hyperonic resonances ($J - 1/2$) and *t*-channel K^* and $K1$ resonances.

The number of nodes for 14 of the 15 spin observables obey the general rules of Ref. [3] concerning the maximum number of nodes permitted within a helicity basis truncated at $J \leq 5/2$. Only the single polarization observable, Σ, with incident linearly polarized photons deviates from these rules and shows the need for a significant spin-7/2 resonances (the special sensitivity of Σ to higher spin amplitudes due to *t*-channel exchange was

Table 1: Maximum number of nodes for single and double polarization observables.

Class	Obs	$\mathcal{H}$ $J \leq 1/2$	$\mathcal{H}$ $J \leq 3/2$	$\mathcal{H}$ $J \leq 5/2$	$\mathcal{H}$ $J \leq 7/2$	*Mult* $l \leq 1$	*Mult* $l \leq 2$	SL
$\mathcal{L}_0$								
Beam-Target	$\hat{E}$	0	3	5	7	2	4	2
Beam-Recoil	$\hat{C}_{z'}$	1	3	5	7	3	5	3
Target-Recoil	$\hat{L}_{z'}$	1	3	5	7	3	5	3
$\mathcal{L}_{1a}$								
Recoil	$\hat{P}$	0	2	4	6	1	3	1
Beam-Target	$\hat{H}$	0	2	4	6	1	3	1
Beam-Recoil	$\hat{C}_{x'}$	0	2	4	6	2	4	3
Target-Recoil	$\hat{L}_{x'}$	0	2	4	6	2	4	3
$\mathcal{L}_{1b}$								
Target	$\hat{T}$	0	2	4	6	1	3	1
Beam-Target	$\hat{F}$	0	2	4	6	1	3	3
Beam-Recoil	$\hat{O}_{z'}$	0	2	4	6	2	4	2
Target-Recoil	$\hat{T}_{z'}$	0	2	4	6	2	4	4
$\mathcal{L}_2$								
Beam	$\hat{\Sigma}$	0	1	3	5	0	2	4
Beam-Target	$\hat{G}$	0	1	3	5	0	2	1
Beam-Recoil	$\hat{O}_{x'}$	0	1	3	5	1	3	3
Target-Recoil	$\hat{T}_{x'}$	0	1	3	5	1	3	3

emphasized in Ref. [4]). The nodal structure for single polarization observables, whitin the SL model is depicted in Fig. 1. To illustrate further this point, the multipole decomposition of the Σ observable is given in the next page. The expression has a fourth order polynomial structure in $X \equiv cos\theta_{cm}$ and can produce the number of nodes obtained within the SL model. As already discussed [4] in the case of simpler phenomenological models, in fitting data the lack of explicit spin-7/2 resonances in the Saclay-Lyon phenomenological model is compensated by having enhanced t-channel exchanges *via* duality hypothesis.

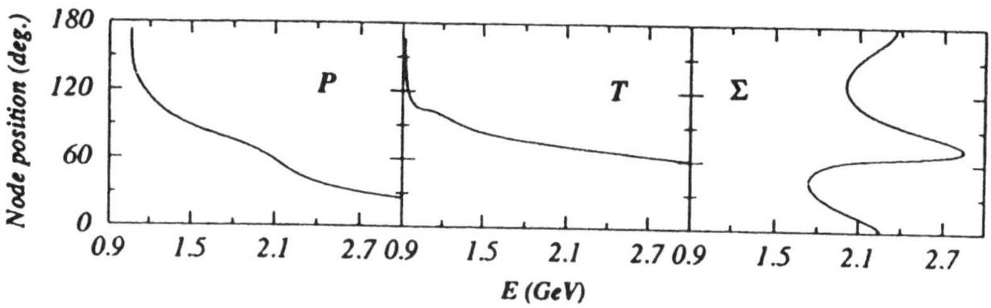

Fig. 1. Nodal structure of the single polarization observables as a function of incident photon lab. energy: recoil hyperon (P), target (T), and beam (Σ).

Our results show clearly that the existing data imply contributions from spin-7/2 nucleonic resonances. We look forward to upcoming data to learn if such resonances are

$$
\begin{aligned}
\hat{\Sigma} \equiv \quad \Sigma\,\mathcal{I} = \tfrac{-\sin^2\theta}{2} Re\Big(& 3\,|E_2^-|^2 + 9\,|E_3^-|^2 + 9\,|E_1^+|^2 + 18\,|E_2^+|^2 + \tfrac{225}{4}|E_3^+|^2 - 9\,|M_2^-|^2 \\
& -18\,|M_3^-|^2 - 3\,|M_1^+|^2 - 9\,|M_2^+|^2 - \tfrac{135}{4}|M_3^+|^2 + 6\,E_0^+\,[M_2^+ - E_2^- - E_2^+ - M_2^-]^* \\
& +E_1^+[18E_3^- - 45E_3^+ + 6M_1^- - 9\,M_3^- - 6\,M_3^+]^* + 3E_2^+[3M_2^- + 7E_2^- - 3M_2^+]^* \\
& +3\,M_1^-[2M_3^- - 5E_3^+ - 2E_3^- - 2M_1^+ + 5M_3^+]^* + 9E_3^-[M_3^+ - M_3^- - 5E_3^+]^* \\
& +6\,E_2^-[M_2^+ - M_2^-]^* + 18\,M_2^{+*}\,M_2^- + 3M_1^+[7M_3^- - 2E_3^- + 5E_3^+ + \tfrac{7}{2}M_3^+]^* \\
& +\tfrac{45}{2}E_3^{-*}[M_3^- - M_3^+]^* - \tfrac{117}{2}M_3^{-*}M_3^+ + \mathrm{X} \quad (\, 30E_0^+[M_3^+ - E_3^- - E_3^+ - M_3^-]^* \\
& +18E_1^+[3E_2^+ - 2E_2^-]^* + 189\,E_2^{+*}\,E_3^- - 6M_1^+[6M_2^- + 5E_2^+ + 4M_2^+]^* \\
& +30M_1^-[E_2^+ - M_2^+]^* + 6E_2^-[5M_3^+ - 4E_3^- + 20E_3^+ - 5M_3^-]^* + 189\,M_3^{-*}\,M_2^+ \\
& +9M_2^-[20M_3^+ + 6M_3^-]^*\ ) + \mathrm{X}^2\ (\ 45\,|E_3^-|^2 + 90\,|E_2^+|^2 - \tfrac{225}{2}|E_3^+|^2 - 90\,|M_3^-|^2 \\
& -45|M_2^+|^2 + \tfrac{135}{2}|M_3^+|^2 + 45E_1^+[M_3^+ - 4E_3^- + 3E_3^+ - M_3^-]^* + 105M_1^-[E_3^+ - M_3^+]^* \\
& -15M_1^+[9M_3^- + 7E_3^+ + 5M_3^+]^* - 45M_3^-[E_3^- - 25M_3^+]^* - 180\,M_2^{-*}M_2^+ \\
& +5E_2^+[9M_2^- - 27E_2^- - 9M_2^+]^* + 45\,M_3^{+*}\,E_3^+ + 45E_3^+[M_3^+ - M_3^- - 21E_3^-]^*\ ) \\
& + \mathrm{X}^3\ (\ -420\,E_3^{+*}E_2^- - 15E_2^+[45E_3^- - 32E_3^+]^* + 210E_3^+[M_2^- - M_2^+]^* \\
& -45M_2^+[15M_3^- - 6M_3^+]^* - 630\,M_2^{-*}\,M_3^+\ ) + \mathrm{X}^4\ (\ \tfrac{2625}{4}|E_3^+|^2 \\
& -\tfrac{1575}{4}|M_3^+|^2 - E_3^+[2100E_3^- - \tfrac{525}{2}M_3^- + \tfrac{525}{2}M_3^+]^* - \tfrac{4725}{2}M_3^{-*}M_3^+\ )\ \Big)
\end{aligned}
$$

indeed required by the reaction mechanism or, as discussed in Ref. [6], the manifestations of such high spin resonances in this energy range are induced by inconsistencies within the present data base.

We wish to thank J.C. David, C. Fayard, and G.H. Lamot for helpful exchanges.

References

[1] *See, e.g.,* H. Bando, T. Motaba, and J. Zofka, *Strangeness nuclear physics, in* Perspectives of Meson Science, Eds. T. Yamazaki *et al.*, North Holland Pub. (1992).

[2] J.C. David, C. Fayard, G.H. Lamot, and B. Saghai, *to appear in* Phys. Rev. C.

[3] C.G. Fasano, F. Tabakin and B. Saghai, Phys. Rev. C **46**, 2430 (1992).

[4] B. Saghai and F. Tabakin, Phys. Rev. C **53**, 66 (1996).

[5] P. Girard, B. Saghai and F. Tabakin, Saclay Report, DAPNIA-SPhN 96-01, 1996.

[6] R. A. Adelseck and B. Saghai, Phys. Rev. C **42**, 108 (1990).

SEARCH FOR THE H DIBARYON AT THE BROOKHAVEN AGS – E813/836

W.D. RAMSAY
University of Manitoba, Department of Physics,
Winnipeg, Manitoba R3T 2N2, Canada
for: the E813/836 collaboration

E813 and E836 are companion experiments designed specifically to search for the H dibaryon. Both experiments look directly for the signature of H formation and are insensitive to decay modes and lifetime. E813 looks for H formation through Ξ^- capture on the deuteron and is most sensitive to a lightly bound H. E836 looks for H formation via the (K^-, K^+) reaction on ^3He and is most sensitive to a tightly bound H. Preliminary results from E836 and the results from analysis of about one-half the E813 data are presented.

1 Introduction

Two similar experiments, E813 and E836, have been performed at the Brookhaven AGS to search for the (uuddss) H dibaryon first predicted by Jaffe in 1977 [1]. The H will be bound against strong decay if it is lighter than the $\Lambda\Lambda$ mass of 2231 MeV. E813 searches the region down to about 100 MeV below the $\Lambda\Lambda$ mass; E836 searches from about 70 MeV to 350 MeV below the $\Lambda\Lambda$ mass. Both experiments have the advantage that they look for the direct signature of H production rather than its decay.

2 E813

E813 uses the two step process, $K^- + p \to K^+ + \Xi^-$, $\Xi^- + d \to (\Xi^- d)_{atom} \to H + n$. The 1.8 GeV/c K^- beam enters the lower, liquid hydrogen, part of the target where Ξ^-s are produced. The outgoing K^+ enter a magnetic spectrometer while the Ξ^-s pass through an energy degrader and a Silicon detector into the upper, liquid deuterium, part of the target where they stop, forming Ξ^-–deuteron atoms. A $(\Xi^-, d)_{atom}$ is predicted [2] to have a substantial probability of forming an H particle and a neutron. Ξ^-s in the right angular range and having the right energy to stop in the liquid deuterium are "tagged". The signature of the production of H particles is the detection of monoenergetic neutrons in coincidence with the tagged Ξ^-s. The main data taking runs for E813 were in 1993 and 1995.

The 1995 data have not been analyzed yet, but the 1993 data show a peak in the neutron spectrum at $\beta^{-1} = 3.04$ which is consistent production of an

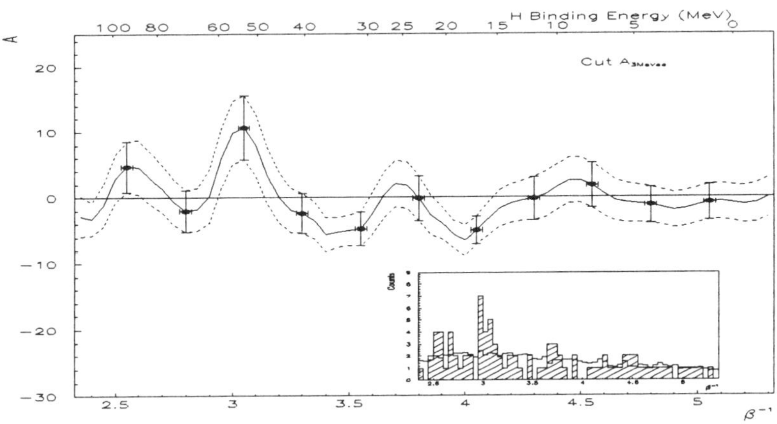

Figure 1: Results from analysis of approximately half the E813 data. The inset shows the neutron β^{-1} spectrum for stopping Ξ^{-}s superimposed on the background. The main figure shows the result of fitting the neutron spectrum with the background plus a series of gaussian peaks of increasing β^{-1}. The solid line is the best fit number of neutrons in the peak, and the dotted lines give the one sigma limits.

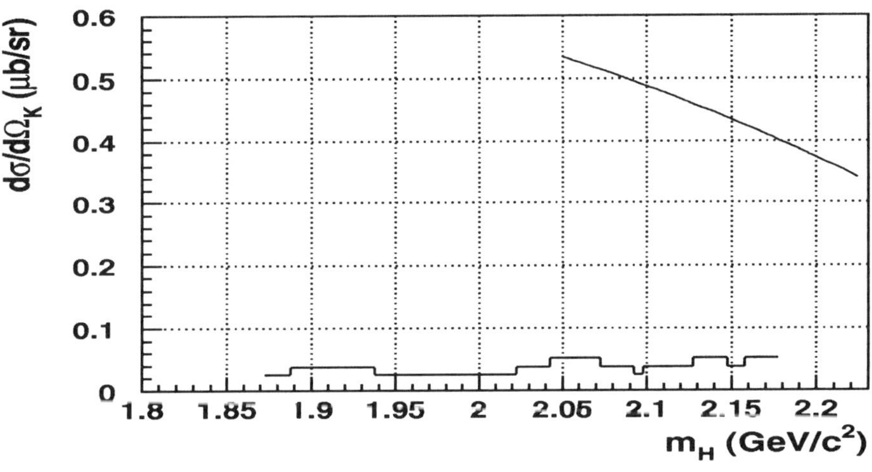

Figure 2: Cross sections for $^3He(K^-, K^+)nH$. The lower curve shows the 90% C.L. upper limit determined by E836 and the upper curve shows the Aerts and Dover prediction.

H particle of mass 2177 MeV (54 MeV bound) (see figure 1). The probability that statistical fluctuations in the background could cause such a peak at 2177 MeV is 0.006, while the probability that statistical fluctuations would cause such a peak anywhere in the neutron spectrum searched is 0.08.

The 1995 data are now being analyzed and are expected to double the total statistics.

3 E836

E836 uses a single target of liquid ^3He. The experiment looks for evidence of H particle production by means of the reaction $^3He(K^-, K^+)nH$ [3,4]. A missing mass spectrum is formed assuming a diproton target and the signal of H production is a peak in this spectrum below the mass of the broad peak from quasi-free Ξ^- production. Although E836 achieves a missing mass resolution of 15 MeV/c^2 FWHM, the expected peak from H particle production is broadened to approximately 35 MeV/c^2 by the presence of the spectator neutron.

E836 took data in 1994. Analysis of the data is still underway, but preliminary results appear to set a 90% C.L. upper limit for H production in the tightly bound region of almost an order of magnitude less than theoretical predictions. Figure 2 shows the E836 results compared to the predictions of Aerts and Dover [3].

References

1. R.L. Jaffe, *Phys. Rev. Lett.* **38**, 195 (1977).
2. A.T.M. Aerts and C.B. Dover, *Phys. Rev.* D **29**, 433 (1984).
3. A.T.M. Aerts and C.B. Dover, *Phys. Rev.* D **28**, 450 (1983).
4. N. Aizawa et al., *Z. Phys. A Hadrons and Nuclei* **343**, 103 (1992)

EFFECTIVE KAON MASS IN SYMMETRIC NUCLEAR MATTER

M. C. RUIVO AND C. A. DE SOUSA

Departamento de Física da Universidade de Coimbra, P3000 Coimbra

We discuss the behaviour of kaon mass in symmetric nuclear matter in the context of a generalized three–flavor Nambu–Jona–Lasinio model. It is shown that two branches with K^- quantum numbers appear: one starting at $\rho = 0$ and a lower one starting around $1.3\,\rho_0$. This second branch is situated below a continuum band of subthreshold resonances of $s\bar{u}$ states and decreases smoothly with density. However the vacuum does not show signs of instability leading to kaon condensation.

In recent years, the possibility of kaon condensation in dense baryonic matter, suggested by Kaplan and Nelson [1], has lead to active studies on this problem. Up to now there is no evidence against or in favour of kaon condensation, so the question remains open. Perturbative treatments lead to a splitting of kaon masses, $m_{K^+}(m_{K^-})$ inceasing (decreasing) linearly with density.

In the present work we are going to adress this problem in the context of a three flavor Nambu-Jona-Lasinio (NJL) model [2] using a non-perturbative approach. The framework we are going to use, extensively developed in our previous works, is based on the Time Dependent Hartree–Fock (TDHF) method [3].

We use a generalized $SU(3)$ NJL model whose Hamiltonian includes two-body scalar-pseudoscalar and the 't Hooft interactions:

$$H = \overline{\Psi}(m - i\gamma.\nabla)\Psi - g_S \sum_{a=0}^{8} [(\overline{\Psi}\lambda^a\Psi)^2 + (\overline{\Psi}i\gamma_5\lambda^a\Psi)^2]$$

$$-g_D\{\det[\overline{\Psi}(1+\gamma_5)\Psi] + \det[\overline{\Psi}(1-\gamma_5)\Psi]\}, \qquad (1)$$

where λ_a are the Gell-Mann flavor matrices, γ_5, γ are Dirac matrices and the other symbols have the usual meaning.

The results for the kaon masses in symmetric nuclear matter ($\rho_s = 0$, $\rho_u = \rho_d$) are plotted in Fig. 1. For reference, we also plot the curve (dashed-dotted) which represents the degenerated masses (which occur for $\rho_s = \rho_u = \rho_d$). Concerning the K^+, there is nothing particularly new. Its mass increases non linearly with density, in a way similar to the perturbative result. The situation is quite different for the K^-. It is well known that K^+ and K^- differ substantially in their interactions with the nucleus. The K^- ($s\bar{u}$) is a quark-antiquark collective mode with strangeness $S = -1$. In symmetric nuclear matter the $s\bar{u}$ states form a continuum band bounded respectively by $\omega_{up} = M_s - M_u$ and $\omega_{low} = (M_s^2+\lambda_f^2)^{1/2} - (M_u^2+\lambda_f^2)^{1/2}$, where λ_f is the Fermi momentum of quark matter (see fig. 1). The presence of these collective states produces repulsive (attractive) contributions above ω_{up} (below ω_{low}). These remarkable effects are reflected on the solutions of the dispersion relations for K^-. So, the novel feature shown in this work is that two branches appear: one starting at $\rho = 0$ ($m_{K^-} = 475.5\,MeV$) and the other starting at $\rho = 1.3\,\rho_0$ ($m_{K^-} \simeq 220\,MeV$). The second collective bound state, situated below ω_{low}, is induced by the attractive $s\bar{u}$ interactions and decreases smoothly with density. It is a collective K^- bound state which resembles the π_s^+ Migdal's mode in the pion sector [4]. The upper bound state starts increasing with density at $\rho \simeq 0.9\,\rho_0$ and comes to the threshold $M_s + M_u$ around $\rho = 2.3\,\rho_0$. The attraction responsible for the lower mode is not enough to produce a K^- condensate.

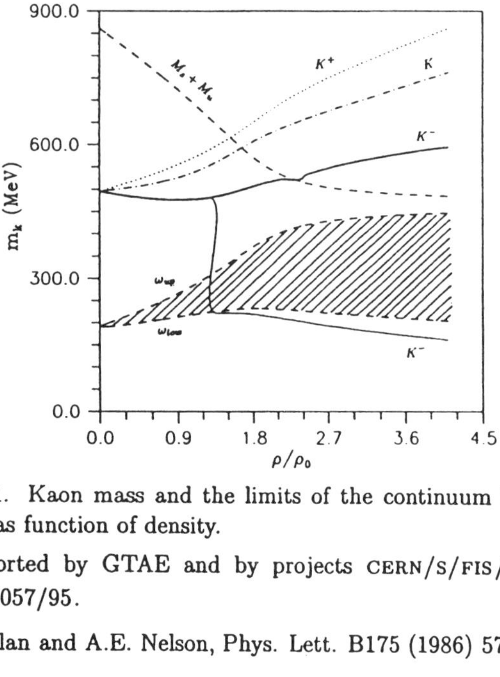

Fig. 1. Kaon mass and the limits of the continuum band (ω_{up}, ω_{low}) as function of density.

Work supported by GTAE and by projects CERN/S/FIS/1034/95 and ESO PESO/S/PRO/1057/95.

[1] D.B. Kaplan and A.E. Nelson, Phys. Lett. B175 (1986) 57; B192 (1987) 193.

[2] Y. Nambu and G. Jona-Lasinio, Phys. Rev. 124 (1961) 246.

[3] J. da Providência, M.C. Ruivo and C.A. de Sousa, Phys. Rev. D36 (1987) 1882.

[4] H. Yabu, S. Nakamura, F. Myhrer and K. Kubodera, Phys. Lett. B315 (1993) 17.

OBSERVATION OF KAONIC HYDROGEN ATOM X-RAYS

M. IWASAKI[a], K. BARTLETT[b], G.A. BEER[b], D.R. GILL[c], R.S. HAYANO[a], T.M. ITO[a], L. LEE[d], G. MASON[b], S.N. NAKAMURA[a], A. OLIN[c], H. OUTA[e], M. SALOMON[c], R. SEKI[g], T.P. TERADA[a], G. TRAYLING[b], Y. YAMASHITA[f] and S. YEN[c]

[a]*Meson Science Laboratory and Department of Physics, University of Tokyo, Bunkyo, Tokyo 113, Japan*

[b]*Department of Physics and Astronomy, University of Victoria, P.O. Box 3055, Victoria, Canada V8W3P6*

[c]*TRIUMF, 4004 Wesbrook Mall, Vancouver, B.C., Canada V6T2A3*

[d]*Department of Physics, University of Manitoba, Winnipeg, MB. Canada R3T2N2*

[e]*Institute for Nuclear Study, University of Tokyo, 3-2-1 Midori, Tanashi, Tokyo 188, Japan*

[f]*KEK, 1-1 Oho, Tsukuba, Ibaraki 305, Japan*

[g]*Department of Physics and Astronomy, California State University, Northridge, CA 91330, and W. K. Kellogg Radiation Laboratory, California Institute of Technology, Pasadena, CA 91123, USA*

The determination of the strong-interaction energy level shifts and widths of the kaonic hydrogen X-rays is one of the most important subjects for the understanding of the $\overline{\text{K}}-\text{N}$ interaction. This interaction is very important not only for the understanding on the low-energy aspect of the strong interaction but also for the study of the structure of the $\Lambda(1405)$ resonance. The present status of the study on this interaction is quite puzzling due to the inconsistency between the sign of the scattering length obtained by the analysis of the low energy scattering data and that of the previous X-ray measurements[1,2,3]. This inconsistency is known to be very difficult to understand in the conventional framework of the various theoretical approaches.

We have performed a new experiment to study this interaction, and have succeeded for the first time in observing with good statistical significance distinct spectral lines in the energy spectrum due to the kaonic hydrogen X-rays. Our result shows a four order of magnitude improvement over the previous experiments in the signal-to-noise ratio due mainly to the use of a gaseous hydrogen target and a new experimental technique involving two pion tagging. Figuer 1 shows the spectrum obtained from our preliminary analysis.

The experiment was performed at the low-momentum kaon-separated beam line (K3) at KEK-PS. Negative kaons of momentum 600 MeV/c were stopped in a gaseous hydrogen target of pressure 3 atm cooled to 100 K. The use of a gaseous target reduced the loss of radiative transitions through Stark mixing of the kaon-atomic orbital, which drastically reduce the kaonic hydrogen x-ray yield. An array of 60 Si(Li) detectors was used to detect the very low yield kaonic hydrogen X-rays. To increase the solid angle intercepted by the Si(Li)'s and to avoid absorption of the X-rays in a beryllium window, the Si(Li) crystals were located right in the hydrogen gas, around the stopping volume for the kaons.

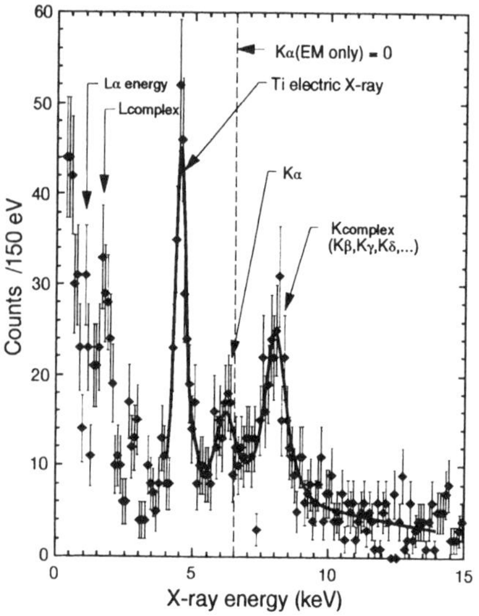

Fig. 1: Preliminary X-ray spectrum from K⁻p atom.

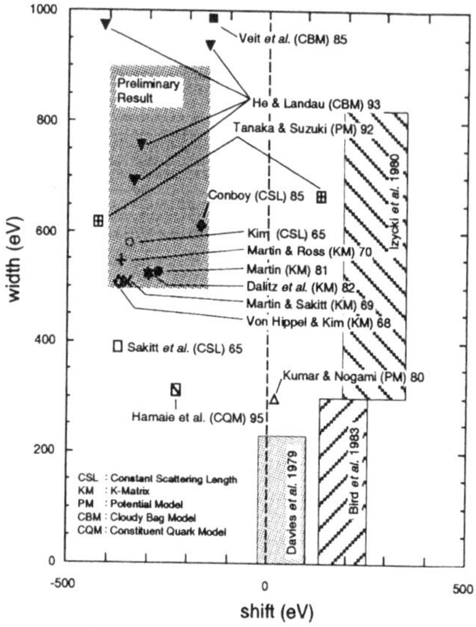

Fig. 2: Preliminary result.

When a kaon stops in the hydrogen target, it undergoes a cascade of atomic transitions, emitting kaonic hydrogen X-rays, and then suffers strong absorption by the proton. 50% of these absorptions eventually emit two charged pions. We applied two charged pion tagging because those events cannot produce energetic γ-rays which are the major background source of the previous experiment. Furthermore, by applying pion vertex point reconstruction and time-of-flight analysis of the kaon in the target, we were able to select only those events in which the kaon stops in the hydrogen gas, and not in the vessel walls nor the kaon in-flight decay. More detailed description of the experiment could be found in ref[4].

As shown in Fig.2, we have obtained, as a tentative result, the shift and width of the kaonic hydrogen 1s state to be -400 $\sim$-150 eV (repulsive) and 500$\sim$900eV, respectively. The result is consistent with the analysis of the low energy scattering data both in shift and width and the sign of the shift is also consistent with the existence of the $\Lambda(1405)$ resonance below K^-p threshold. We are now on the final analysis stage to improve the analysis efficiency by a factor of two and to reduce the systematic error by further understanding of the background which still remains in the final spectrum. The final result will be published in near future.

References

1. J.D. Davies *et al.*, *Phys. Lett.* B **83**, 55 (1979).
2. M. Izycki *et al.*, *Z. Phys.* A **297**, 11 (1980).
3. P.M. Bird *et al.*, *Nucl. Phys.* A **404**, 482 (1983).
4. M. Iwasaki *et al.*, *Nucl. Phys.* A **585**, 239c (1995).

TESTING THE STRANGENESS CONTENT OF THE NUCLEON IN POLARIZATION EFFECTS FOR NUCLEON NUCLEON COLLISIONS NEAR THRESHOLD

M. P. REKALO,[a] J. ARVIEUX and E. TOMASI-GUSTAFSSON
LNS, CE Saclay, 91191 Gif-sur-Yvette Cedex, France

We derived the matrix element for the reactions $NN \to NNV^0$, $NN \to dV^0$, and $\bar{p}p \to \phi\pi$ where N is a nucleon (n or p) and V^0 a vector meson (ϕ, ω, ρ) at threshold and we show that these reactions can give useful information on the presence and the sign of the strangeness content of the nucleon.

1 Introduction

Polarized deep inelastic scattering indicates that the proton wave function may contain strange quarks [1]. Large violation of the OZI rule, observed for ϕ-production through $\bar{p}p$-annihilation at rest [2], as well as the negative Λ-polarization in $\bar{\nu} + N \to \mu^+\Lambda + X$ [3] can be interpreted in the framework of a nucleon model with nonperturbative polarized $\bar{s}s$–component. The near threshold production of the ϕ-meson in NN–collisions may be sensitive to these properties, however it is necessary to disentangle the kinematical and dynamical aspects of the involved processes. The threshold region seems well adapted to the study of the OZI-violation, due to the simplification of the amplitude spin structure. We analyze here the polarization effects in the processes $N + N \to N + N + V$, $N + N \to d + V$ near threshold, where the final particles are produced in S-state and discuss the ϕ mesons production in $\bar{p}p$-annihilation.

2 $N + N \to N + N + V$

The spin structure of the threshold amplitude of the process $p + p \to p + p + V^0$ is defined by a single spin configuration, corresponding to the triplet state of the initial protons:

$$\mathcal{M} = f_1(\tilde{\chi}_2\, \sigma_2\vec{\sigma} \cdot \vec{k} \times \vec{U}^*\chi_1)\,(\chi_4^\dagger\sigma_2\, \tilde{\chi}_3^\dagger), \tag{1}$$

where χ_1 and χ_2 (χ_3 and χ_4) are the two-component spinors of the two incoming (outgoing) protons; $\vec{k}$ is the unit vector along the 3-momentum of the initial proton; $\vec{U}$ is the 3-vector polarization of the produced vector meson;

[a]Permanent address: *National Science Center KFTI, 310108 Kharkov, Ukraine*

f_1 is the complex amplitude corresponding to the triplet interaction of the colliding particles. The polarization observables do not depend on f_1, (Eq. 1), therefore they do not contain information about the dynamics of the process, in particular about the role of the $\bar{s}s$–component of the nucleon.

The spin structure of the matrix element for $n+p \to n+p+\omega(\phi)$ is defined by singlet f_0 and triplet f_1 amplitudes :

$$\mathcal{R} = \frac{\sigma(p+p \to p+p+V^0)}{\sigma(n+p \to n+p+V^0)} = \frac{4|f_1|^2}{|f_1|^2+|f_0|^2}. \tag{2}$$

The ratio $\mathcal{R}$ characterizes both possible correlations of the initial nucleon polarizations in $\vec{n}+\vec{p} \to n+p+\omega(\phi)$:

$$\sigma^{np}(\vec{P}_1,\ \vec{P}_2) = \sigma_0^{np}(1\ +\ \mathcal{A}_1\vec{P}_1\cdot\vec{P}_2\ +\ \mathcal{A}_2\vec{k}\cdot\vec{P}_1\vec{k}\cdot\vec{P}_2), \tag{3}$$

$$\mathcal{A}_1 = -1+\tfrac{\mathcal{R}}{4},\quad \mathcal{A}_2 = \tfrac{\mathcal{R}}{4},\quad \mathcal{A}_2 = 1+\mathcal{A}_1. \tag{4}$$

The elements of the V^0-meson density matrix are independent from the mechanism of the process $N+N \to N+N+V^0$: $\rho_{xx} = \rho_{yy} = \frac{1}{2}$, $\rho_{zz} = 0$. Therefore the $(1+cos^2\theta)$-distribution of the decay products for the $V^0 \to \ell^+\ell^-$ is a direct consequence of the P-invariance of the strong interaction.

The ratio of cross section for $p+p \to p+p+\phi$ and $p+p \to p+p+\omega$ should be very large, near threshold, due to the possible presence of $s\bar{s}$-component in the nucleons, as in the case of the triplet $\bar{p}p$-annihilation at rest:

$$\frac{\sigma(pp \to pp\phi)}{\sigma(pp \to pp\omega)} \approx \frac{\sigma(\bar{p}p \to \phi\pi)}{\sigma(\bar{p}p \to \omega\pi)} \approx 100\cdot 10^{-3}.$$

The P-wave contribution changes essentially the behaviour of the polarization observables in $NN \to NNV$-reactions.

3 $N+N \to d+V$

The processes $p+p \to d+\rho^+$, and $n+p \to d+V^0$, with $V = \rho,\ \omega\ or\ \phi$, are the simplest two-particle reactions of vector meson production in nucleon-nucleon collisions. The matrix element for the process $n+p \to d+\omega(\phi)$ at threshold is defined by a singlet amplitude only:

$$\mathcal{M} = g(\tilde{\chi}_2\ \sigma_2\chi_1)\vec{k}\times\vec{D}^*\cdot\vec{V}^*, \tag{5}$$

where: $\vec{D}(\vec{V})$ is the spin wave function of d (V^0); $\vec{D}$ is an axial vector (as the P-parity of the deuteron is positive), $\vec{V}$ is a polar vector; g is a partial amplitude corresponding to the S-production of the V^0-meson.

The presence of a single amplitude in Eq. 6 allows to predict quantitatively the polarization observables: all non-zero polarization observables have their maximum values. The analyzing powers are zero, but the tensor polarization is different from zero: $T_{20} = 1/\sqrt{2}$, for $\omega-$ and ϕ production at any energy in the threshold region. Therefore the polarization effects near threshold for $n + p \to d + \omega(\phi)$ do not contain any special information on the dynamics of the reaction.

In the presence of a polarized $s\bar{s}$-component in the nucleon one may expect that the ϕ-production is suppressed, with respect to the ω production, due to the singlet initial state near threshold. The fact that the S-wave production is negligible may enhance the P-wave contributions near threshold, which, for $n + p \to d + \omega(\phi)$ are of triplet nature. All these effects must appear earlier in $n + p \to d + \phi$ than in $n + p \to d + \omega$.

A similar analysis applied to η production may explain the experimental cross sections of $n + p \to n + p + \eta$ with respect to $p + p \to p + p + \eta$ [4].

4 $\bar{p}p \to \phi\pi$

The selection rules in C- and P-parity allow to write the following matrix element for this process: $\mathcal{M} = g_\phi \tilde{\chi}_2\ \sigma_2 \vec{\sigma} \cdot \vec{k} \times \vec{U}^* \chi_1$ which results in a single spin configuration of the interacting particles: $+1/2\ +1/2\ \to\ +1$, i.e. the ϕ-meson is produced with transversal polarization. If one assumes that the $\phi-$meson conserves the polarization of the $s\bar{s}-$ component inside the polarized proton, then necessarily, at small momentum transfer ($Q^2 \simeq 0.36\ GeV^2$) the spin of this component is **parallel** to the proton spin. Taking into account the results obtained in polarized deep inelastic scattering we have to conclude that the polarization properties of the $s\bar{s}-$ component must be Q^2-dependent, and the sign has to reverse at $Q^2 \simeq 1\ GeV^2$. A similar strong dependence in the region of $Q^2 \simeq 1\ GeV^2$ appears also in the Gerasimov-Drell-Hearn sum-rule, due to the contribution of the excitation of nucleonic resonances.

References

1. S. J. Brodsky, J. Ellis and M. Karliner, Phys. Lett. **B206** (1988) 309.
2. J. Reifenrother et al. Phys. Lett. **B267** (1991) 299.
3. S. Willocq et al. Z. Phys. C **53** (1992) 207,
   J. Ellis, D. Kharzeev and A. Kotzinian, CERN-TH/1995/135.
4. E. Chiavassa et al., Phys. Lett. **B337** (1994) 192.

Session 9
Strong Interaction Theory — QCD

LATTICE STUDY OF THE POMERON

D.S. HENTY[†], C. PARRINELLO[*], D.G. RICHARDS[†], J.I. SKULLERUD[†]
UKQCD COLLABORATION
[†]*Dept. of Physics & Astronomy, University of Edinburgh, Edinburgh EH9 3JZ, U.K.*
[*]*D.A.M.T.P., University of Liverpool, Liverpool L69 3BX, U.K.*

We investigate the phenomenology of the Landshoff-Nachtmann two-gluon-exchange model of the Pomeron using gluon propagators computed in the Landau gauge within quenched lattice QCD simulations. As the propagators have been evaluated entirely from QCD first principles, our results provide a consistency check of the model. Finally, we report a proposal to investigate the structure of the Pomeron-quark vertex, using a colour-singlet Pomeron source constructed from the gauge fields.

The Pomeron trajectory was introduced to describe certain features of hadronic scattering: total hadronic cross sections slowly rising with energy, their factorisation, the observation that the elastic amplitude is predominantly imaginary, and the quark-counting rule. This description has proved remarkably durable, and interest in pomeron physics has been renewed by recent results from HERA. In this talk we will investigate a particular QCD model of the pomeron, the non-perturbative Landshoff-Nachtmann (LN) two-gluon-exchange (2GE) model,[1] within a lattice gauge simulation. We will then suggest how the Pomeron may be investigated from a lattice gauge simulation at a more fundamental level.

The 2GE model has proved highly successful at describing Pomeron exchange. Crucial to the model is that the gluons be infra-red finite. Recently, the phenomenology has been explored using gluon propagators from the solution of the Schwinger-Dyson equation.[2,3] In this talk we will explore the phenomenology using lattice gluon propagators in the Landau gauge.

The momentum dependence of the gluon propagator we extract from the lattice simulation. The "effective" coupling of the gluon propagator to a quark we determine by requiring that the Pomeron-quark coupling, β_0, attain its phenomenological value. We perform the calculation at several volumes and lattice spacings to control systematic uncertainties due to the finite volume and non-zero lattice spacing.[5]

As an example of the calculation, we show in Figure 1 a comparison of the model with the *pp* differential cross section, using three lattices at different volumes and spacings; both the total cross section and the slope of the differential cross section are close to their phenomenological values. The calculation

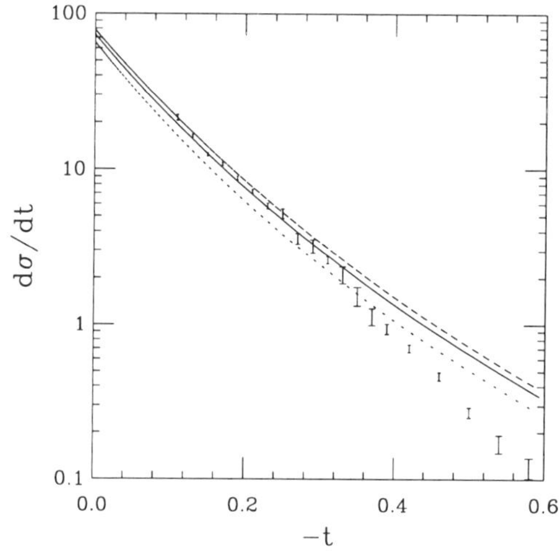

Figure 1: Data for the pp elastic cross section at $\sqrt{s} = 53$ GeV from Ref. 4, together with the lattice prediction corrected for energy dependence on 16^4 lattices at $\beta = 6.2$ (solid) and $\beta = 6.0$ (dots), and on a $24^3 \times 32$ lattice at $\beta = 6.0$ (dashes).

also exhibits two further features of diffractive scattering: the quark counting rule, and the weaker coupling of the Pomeron to mesons composed of heavy quarks.

Encouraged by this application of the LN model, we are tempted to go further, and explore the nature of the Pomeron-quark vertex.[6] Here we are using a colour-singlet source for the Pomeron constructed from the gauge fields, and the lattice quark propagator fixed to the Landau Gauge. This should provide a valuable clue to the mechanism for diffractive physics in QCD.

References

1. P.V. Landshoff and O. Nachtmann, Z. Phys. C35 (1987) 405.
2. F. Halzen, G. Krein and A.A. Natale, Phys. Rev. **D47** (1993) 295.
3. M.B. Gay Ducati, F. Halzen and A.A. Natale, Phys. Rev. **D48** (1993) 2324.
4. A. Breakstone *et al.*, Nucl. Phys. **B248** (1984) 253.
5. D.S. Henty, C. Parrinello and D.G. Richards, Phys. Lett. **B369** (1996) 130.
6. C. Parrinello, D.G. Richards and J.I. Skullerud (in preparation).

LIGHT HADRON MASSES ON COARSE LATTICES WITH IMPROVED ACTIONS

FRANK X. LEE,[a] DEREK B. LEINWEBER[b]
[a] *TRIUMF, 4004 Wesbrook Mall, Vancouver, British Columbia, Canada V6T 2A3*
[b] *Department of Physics, University of Washington, Seattle, WA 98195*

The advent of mean-field improved perturbation theory [1] has generated renewed interest in the on-shell improvement of lattice QCD actions. The remarkable success of the $\mathcal{O}(a^2)$-improved gauge-field action [2] has led us to reconsider the two-link $\mathcal{O}(a^2)$-improved fermion action of Hamber and Wu,[3] with mean-field improved operators.

Tree-level $\mathcal{O}(a^2)$-improved mean-field improved gauge configurations are generated in the quenched approximation via the Cabibbo-Marinari pseudo-heat-bath algorithm. Fermion propagators are obtained via the Stabilized Biconjugate Gradient algorithm.[4] We consider a local source and both local and gauge-invariant Gaussian smeared sinks.[5] We found smearing is necessary in order to extract observables prior to a loss of signal. Five quark masses ranging from the strange quark mass to one-third the strange quark mass are considered. The strange quark mass was taken to be approximately 180 MeV.

To evaluate scaling in the observables, two lattice spacings a are considered on equal physical volume lattices. The lattice parameters are summarized in table 1. A single-elimination jackknife estimate of the covariance matrix provides a correlated χ^2 which is used to select the time-slice analysis regime.

Dispersion is examined by calculating the speed of light squared from the relation $c^2 = (E_p^2 - M^2)/p^2$, to be compared with 1. E_p and M are extracted from lattice correlation functions and p is the smallest nontrivial momentum $(1,0,0)\,2\pi/L$. Table 2 compares ρ-meson results for $m_q \sim 125$ MeV with the $\mathcal{O}(a)$-improved SW action[6] and the $\mathcal{O}(a^2)$-improved D234 action[7]. The last column tests rotational symmetry for the Hamber Wu action (HW) by determining c^2 from the smallest two-dimensional diagonal momentum. The $\mathcal{O}(a^2)$-improved actions perform well at the spacing of 0.24 fm.

Hadron mass ratios within the baryon octet, decuplet and the vector meson nonet scale well and look very similar to experiment. However, mass ratios of hadrons with differing spins reveal less than satisfactory scaling for the

Table 1: Summary of lattice parameters.

Volume	β	$a_{\rm st}$	N_U	u_0	$\kappa_{\rm cr}$
$6^3 \times 12$	6.25	0.40 fm	90	0.820	0.1967(2)
$10^3 \times 16$	7.00	0.24 fm	37	0.866	0.1823(3)

Table 2: Evaluation of dispersion and rotational invariance via c^2.

a_{st}	SW (1,0,0)	D234 (1,0,0)	HW (1,0,0)	HW (1,1,0)
0.40	0.48(4)	0.93(3)	0.88(6)	0.81(9)
0.24	--	1.00(6)	0.96(9)	1.03(7)

larger lattice spacing. The N/ρ mass ratio summarized in figure 1 is the most revealing ratio. The Hamber and Wu action has reproduced the state of the art quenched QCD ratio on a tiny $10^3 \times 16$ lattice. This result bodes well for future explorations of hadron phenomenology beyond the quenched approximation.

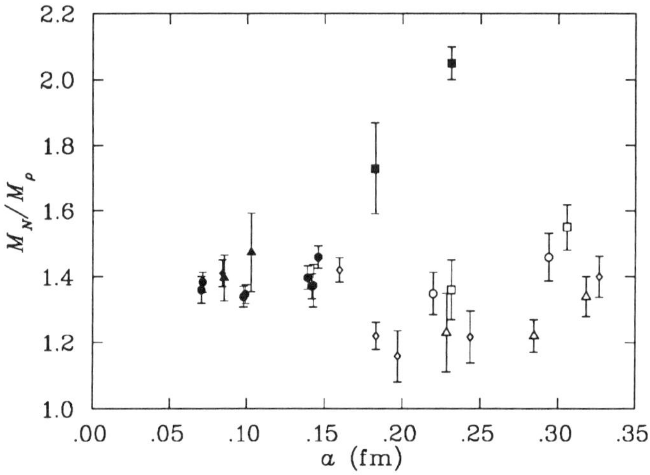

Figure 1: The N/ρ mass ratio as a function of the lattice spacing a_ρ for a variety of lattice QCD actions. Solid symbols denote the standard Wilson action. Open symbols denote improved actions including SW [6] (◇), D234 [7] (△), HW [8] (□), and HW (○) (this work).

This work is supported in part by the Natural Sciences and Engineering Research Council of Canada and U.S. DOE under grant DE-FG06-88ER40427.

1. G. P. Lepage and P. B. Mackenzie, Phys. Rev. D **48**, 2250 (1993).
2. M. Alford *et al.*, Phys. Lett. **B361**, 87 (1995).
3. W. Hamber and C. Wu, Phys. Lett. **133B**, 351 (1983).
4. A. Frommer *et al.*, J. Mod. Phys. **C5**, 1073 (1994).
5. S. Güsken, Nucl. Phys. B (Proc. Suppl.) **17**, 361 (1990).
6. S. Collins, *et al.*, Nucl. Phys. B (Proc. Suppl.) **47**, 366 (1995).
7. M. Alford, *et al.*, Nucl. Phys. B (Proc. Suppl.) **47**, 370 (1995).
8. H. R. Fiebig and R. M. Woloshyn, TRI-PP-96-1, hep-lat/9603001.

GLUEBALLS FROM IMPROVED LATTICE ACTIONS

COLIN MORNINGSTAR[a] and MIKE PEARDON[b]
[a] *University of California at San Diego, La Jolla, CA 92093-0319*
[b] *University of Kentucky, Lexington, KY 40506-0055*

The low-lying glueball masses and the hadronic scale r_0 are computed in lattice SU(3) gauge theory with the aim of establishing the effectiveness of the improved action approach in removing finite-spacing artifacts. The use of anisotropic lattices in which the temporal spacing is much smaller than that in the spatial directions allows much more efficient glueball mass measurements.

Glueballs and hybrid mesons are presently of great interest theoretically and experimentally. The lattice formulation of QCD provides an ideal setting in which to carry out theoretical studies of such systems from first principles using sophisticated numerical simulations. In order to extract the physical properties of glueballs and hybrid mesons from such simulations, systematic errors from the finite lattice spacing a must be removed or made acceptably small. There are two approaches to accomplishing this: (1) using finer grids or (2) using improved actions on coarse grids. The first approach is much simpler and has been used in almost all previous glueball and hybrid meson studies. However, this approach requires vast computational power. As the grid is made finer, many more lattice sites are needed to maintain the physical volume of the system. The simulation costs rise typically as $1/a^6$ as a is decreased. Because of this, lattice studies of glueballs have in the past been dominated by large collaborations using some of the world's fastest supercomputers.

Here, we show that the second approach, the use of improved actions, can be used to study glueballs much more efficiently. Improved actions have smaller lattice spacing errors, and hence, permit the use of much coarser lattices which can be simulated using modern computer workstations. The key to the success of the improvement program is the reliable determination of the couplings of the interactions terms in the action. Much effort over the past decade has been directed towards this problem. Recently, two competing methods have emerged, one which uses block-renormalization group transformations, another which advocates a judicious combination of mean field theory and perturbation theory. In this work, we use the latter approach.

A novel feature of our calculations is the use of anisotropic lattices in which the temporal spacing a_t is much smaller than the spacing a_s in the spatial directions. This allows much more efficient glueball mass measurements by exploiting the enhanced signal-to-noise of the glueball correlation functions at

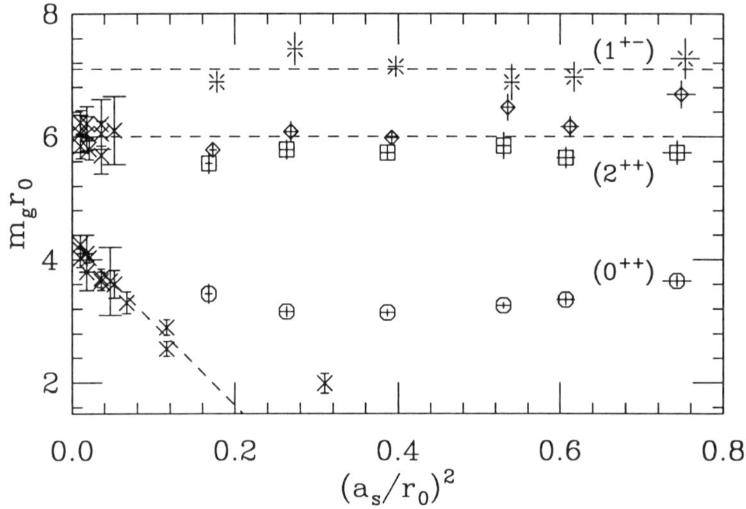

Figure 1: Low-lying glueball masses m_g against lattice spacing a_s in terms of the hadronic scale $r_0 \approx 1/2$ fm defined by $[r^2 dV(r)/dr]_{r=r_0} = 1.65$ where $V(r)$ is the static quark potential. The $\times$ denote results for the scalar and tensor glueballs using the simple Wilson action[1]. Results using the improved action for the A_1^{++}, E^{++}, T_2^{++}, and T_1^{+-} representations are indicated by $\circ$, $\square$, $\diamond$, and $*$, respectively.

smaller temporal separations. Mean-field link renormalization[2] is crucial for maintaining the proper renormalized anisotropy a_t/a_s.

Our results are shown in Fig. 1. The scalar glueball mass from the improved action exhibits dramatically reduced cutoff contamination compared to the Wilson action. Finite-a_s errors are seen to be small for the tensor and pseudo-vector glueballs, although differences between the E^{++} and T_2^{++} representations indicate small violations of rotational invariance, especially for large a_s. These results clearly show that glueballs can be studied without the use of supercomputers, provided that simulations are carried out using improved actions on anisotropic lattices.

References

1. J. Sexton, *et al.*, *Phys. Rev. Lett.* **75**, 4563 (1995); P. De Forcrand *et al.*, *Phys. Lett.* B **152**, 107 (1985); C. Michael and M. Teper, *Nucl. Phys.* B **314**, 347 (1989); UKQCD Collaboration, *Phys. Lett.* B **309**, 378 (1993).
2. G.P. Lepage and P.B. Mackenzie, *Phys. Rev.* D **48**, 2250 (1993).

CONFINEMENT IN LIGHT-FRONT QCD

M. BURKARDT
Department of Physics, New Mexico State University
Las Cruces, New Mexico 88003-0001 USA

Abstract

Numerical results for the calculation of the (rest-frame) $Q\bar{Q}$ potential, for light-front quantized QCD_{2+1} on a $\perp$ lattice are presented. Both in the longitudinal as well as the $\perp$ spatial directions one obtains linear confinement. The resulting potential is almost rotational symmetric.

Light-front (LF) quantization is the most physical approach to calculating parton distributions on the basis of QCD [1]. In the transverse lattice formulation of QCD [2], one keeps the time direction and one spatial direction continuous and discretizes the transverse spatial directions. The space time geometry is thus an array of $1+1$ dimensional sheets. One the one hand, this provide both UV and IV cutoffs in the transverse directions and on the other hand one can still perform LF quantization since the longitudinal directions are continuous. Furthermore, in the compact formulation, it is straightforward to implement Gauss' law as a constraint on the states, which helps avoid troublesome divergences for $k^+ \rightarrow 0$, which plague many other formulations of LF QCD.

Another advantage of the transverse lattice is that confinement is manifest in the limit of large $\perp$ lattice spacing $a_\perp$. The mechanism differs for the longitudinal and the $\perp$ directions: if one separates a $Q\bar{Q}$ pair longitudinally then, since $a_\perp$ is large, the fields in different sheets couple only weakly and the quarks interact only with fields in the same sheet, i.e. effectively the theory reduces to $1+1$ dimensional QCD, where confinement is known to be linear. In contrast, when one separates the $Q\bar{Q}$ pair transversely, gauge invariance demands that they are connected by a chain of (gluon) link fields. For large $a_\perp$, where there are only little fluctuations, this implies that the energy of such a configuration is given by the energy for creating one link quantum times the number of link quanta, i.e. linear confinement also in the $\perp$ direction. Since the confinement mechanisms are very different for these two cases, one might ask whether a rotationally invariant $Q\bar{Q}$ potential results in the continuum limit. In fact, in the limit of large $a_\perp$ one finds in $2+1$ dimensions[3]: $V(x_L, x_\perp) = \sigma\left(|x_L| + |x_\perp|\right)$, where σ is the string tension, which is clearly not rotationally invariant. In order to investigate this issue, I used DLCQ and a Lanczos algorithm to calculate the rest frame $Q\bar{Q}$ potential from the LF Hamiltonian for QCD_{2+1} on a $\perp$ lattice. The procedure for computing the rest frame potential in this formalism follows Ref.[5]. An approximation, where one allows at most one link field quantum per link, was used. Within this approximation, one obtains only a first order phase transition at the critical point, i.e. the lattice spacing always remains finite in physical units. The calculations were done at the critical point, where $a_\perp \approx 1.04\sigma^{-1/2}$. The resulting $Q\bar{Q}$ potential is shown in the Figure as a function of $r = \sqrt{x_\perp^2 + x_L^2}$.

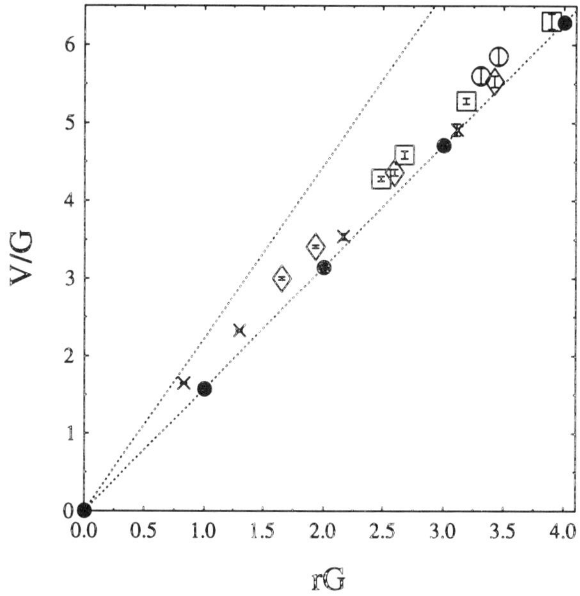

One free dimensionless parameter was adjusted in the calculation to get equal string tensions in the longitudinal and transverse directions. For large $a_\perp$, $V(x_\perp, x_L)$ fills the area between the dotted lines. The symbols are numerical results at the critical point for various directions. The small residual anisotropy is probably due to the truncation to one quantum per link.

References

1. M. Burkardt, Advances Nucl. Phys. **23**, 1 (1996).
2. W. A.Bardeen et al., Phys. Rev. D **21**, 1037 (1980).
3. M. Burkardt and B. Klindworth, hep-ph/9601289.
4. M. Burkardt, ELFE Meeting 1995, hep-ph/9510264.
5. M. Burkardt, LF-workshop, Zakopane, Aug 1994, hep-ph/9410219.

NEW QCD SUM RULES FOR NUCLEON AXIAL VECTOR COUPLING CONSTANTS

FRANK X. LEE,[a] DEREK B. LEINWEBER,[b] XUEMIN JIN[a]
[a] *TRIUMF, 4004 Wesbrook Mall, Vancouver, British Columbia, Canada V6T 2A3*
[b] *Department of Physics, University of Washington, Seattle, WA 98195*

We study the predicative ability of the QCD Sum Rule method for g_A, using a comprehensive Monte Carlo based uncertainty analysis [1]. This is the first application of such analysis to a three-point function.

In this approach, the complete QCD parameter space is explored, allowing a quantitative study of how the uncertainties in the QCD input parameters propagate to the phenomenological fit parameters. The Borel window over which the phenomenological and QCD sides of the sum rules are matched is determined by the following criteria: a) OPE convergence — the last term of the truncated OPE series is less than 10% of the total OPE side, b) ground-state dominance — all excited state contributions are no more than 50% of the phenomenological side. Those sum rules which do not meet these criteria are considered invalid and are discarded.

Two sets of new QCD sum rules for g_A are derived in the external field method, using generalized nucleon interpolating fields. Three sum rules are derived from the correlator $<\eta_{1/2}\bar{\eta}_{1/2}>$ of spin-1/2 interpolating fields: $\eta_{1/2} = \eta_1 + \beta\,\eta_2$ where β is a real parameter, $\eta_1 = \epsilon^{abc}\left(u^{aT}C\gamma_5 d^b\right)u^c$, and $\eta_2 = \epsilon^{abc}\left(u^{aT}Cd^b\right)\gamma_5 u^c$. The Ioffe current may be recovered by setting $\beta = -1$. However, Ioffe's choice is not optimal. Another set of 8 sum rules is derived from the mixed correlator $<\eta_{\mu,1/2}\bar{\eta}_{\nu,3/2}>$ of spin-1/2 and spin-3/2 fields: $\eta_{\mu,1/2} = \gamma_\mu\gamma_5\,\eta_{1/2}$ and $\eta_{\nu,3/2} = \epsilon^{abc}[(u^{aT}C\sigma_{\rho\lambda}d^b)\sigma^{\rho\lambda}\gamma_\nu u^c - (u^{aT}C\sigma_{\rho\lambda}u^b)\sigma^{\rho\lambda}\gamma_\nu d^c]$, the latter has both spin-1/2 and spin-3/2 components.

Our analysis reveals that the sum rule from the spin-1/2 interpolators which was chosen in previous works [2], does not have a valid Borel window. Therefore the previous predictions for g_A are unreliable. On the other hand, the sum rule at the structure $iZ_\mu\sigma^{\mu\nu}p_\nu\gamma_5$ has a valid Borel window of 0.91 GeV to 1.16 GeV with the optimal mixing [1] $\beta = -1.2$. A combined analysis of this sum rule together with the nucleon mass sum rules [1] yields $g_A = 1.48\pm^{1.06}_{0.65}$. The relative error of approximately 60% is large compared to a 10% error for the nucleon mass obtained from the same input parameters. Fig. 1 shows distributions for selected fit parameters drawn from 1000 sets of QCD input parameters. Fig. 2 shows two of the parameters that have significant correlations with g_A. The origin of the large error in g_A is mainly due to the poorly determined pole residue, as g_A is extracted from the form $\lambda^2\,g_A\,e^{-M_N^2/M^2}$. The vacuum susceptibility κ_v also contributes to this large uncertainty. Preliminary analysis of the sum rules from the mixed correlator shows no improvement over this conclusion. Analysis of other axial couplings is under way.

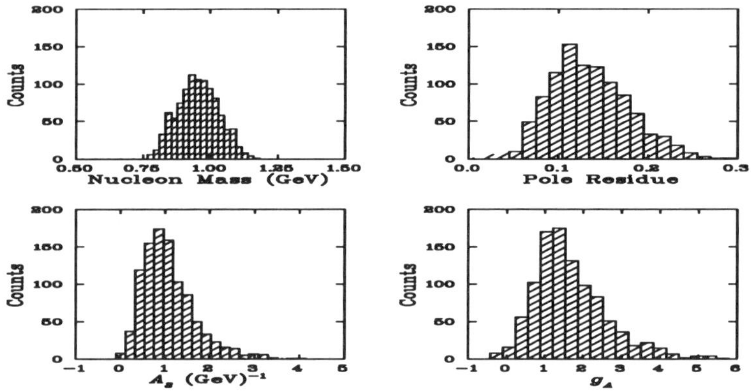

Figure 1: Histograms of selected fit parameters drawn from 1000 sets of QCD parameters. We find: nucleon mass $M_N = 0.94 \pm 0.10$ GeV, pole residue $\tilde{\lambda}^2_{1/2} = 0.12 \pm 0.6$ GeV^6, transition strength $A_3 = 1.08 \pm 0.61$ GeV^{-1}, and $g_A = 1.48 \pm^{1.06}_{0.65}$.

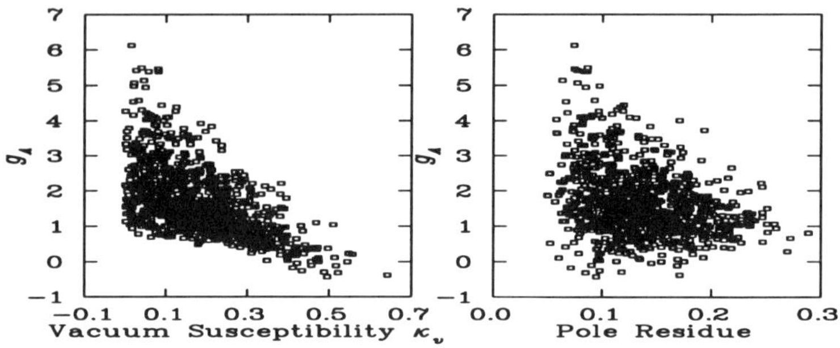

Figure 2: Scatter plots of correlations with g_A.

This work is supported in part by the Natural Sciences and Engineering Research Council of Canada and U.S. DOE under grant DE-FG06-88ER40427.

1. D. B. Leinweber, "QCD Sum Rules for Skeptics," UW-17-N95, nucl-th/9510051; Phys Rev. **D51**, 6383 (1995).
2. V.M. Belyaev and Ya.I. Kogan, Phys. Lett. **136B**, 273 (1984); C. Chiu, J. Pasupathy, and S. Wilson, Phys. Rev. D **32**, 1786 (1985); E.M. Henley *et al.*, Phys. Rev. D **46**, 431 (1992).

LOW-ENERGY QCD: CHIRAL COEFFICIENTS, $U_A(1)$ BREAKING AND THE QUARK-QUARK INTERACTION

T.MEISSNER
Department of Physics and Astronomy, University of South Carolina, Columbia, SC 29208, USA

M.R.FRANK
Institute for Nuclear Theory, University of Washington, Seattle, WA 98195, USA

1 A Path from QCD to Chiral Perturbation Theory

A global color symmetry model (GCM) that is based upon an effective quark-quark interaction arises from the QCD partition function by formally integrating over the gluon fields and truncating the expansion in gluon n point functions after $n = 2$.[1] This model truncation respects all global symmetries of QCD, in particular chiral symmetry. What is lost is invariance under local color gauge transformations. Fixing a model form for the gluon propagator $D(s)$, or, what is equivalent, the running coupling $\alpha(s)$, specifies a quark-quark interaction. Furthermore the model allows a $\frac{1}{N_c}$ expansion. The lowest order $(\mathcal{O}(N_c))$ consists in solving the Dyson-Schwinger equation for the quark self energy including rainbow gluon dressings and the ladder Bethe-Salpeter equation for mesonic bound states. One then is able to change the degrees of freedom from quarks to hadrons by performing an appropriate variable transformation in the generating functional path integral. For the moment only the Goldstone bosons are taken into account. We obtain a non-local hadronic interaction

Table 1: The chiral coefficients using a running coupling $\alpha(s)$. The parameter choices listed maintain $f_\pi = 86$MeV, which is the generic scale of dynamical chiral symmetry breaking. The numbers in parantheses are the "phenomenological" values.

$\alpha_1(s) = 3\pi s\chi^2 e^{-s/\Delta}/(4\Delta^2) + \pi d/\ln(s/\Lambda^2 + e)$				
Δ (GeV^2)	χ (GeV)	L_1(0.7±0.5)	L_3(-3.6±1.3)	L_5(1.4±0.5)
0.002	1.4	0.84	-4.4	1.0
0.02	1.5	0.82	-4.4	1.14
0.2	1.65	0.81	-4.0	1.66
0.4	1.84	0.80	-3.8	2.0

between the Goldstone fields, which can be systematically expanded in the momenta of the Golstone fields and the current quark mass leading exactly to the form given by Gasser and Leutwyler.[2] The chiral low energy coefficients L_i are now determined by the dynamics of the quark-quark interaction and can be numerically calculated from the quark self energy functions.[3] As one can see from Tab.1, L_1 and L_3 are practically independent on the quark-quark interaction, whereas L_5 depends noticeably on its form.

2 Triangle Daigrams, $U_A(1)$ Breaking and the η' Mass

Including higher mass mesonic states other than the Goldstone fields and integrating them out generates triangle diagrams such as

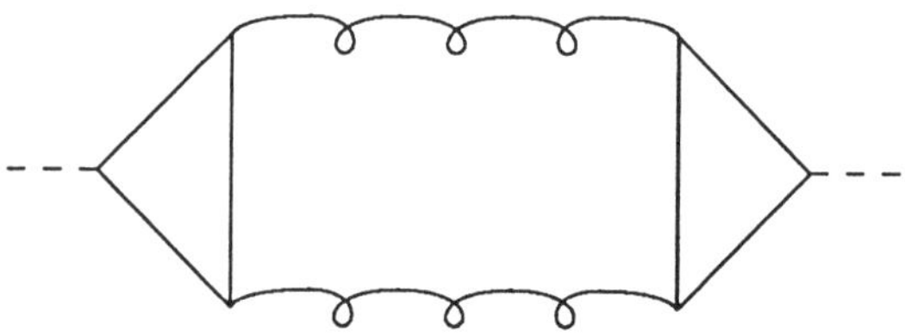

Those diagrams are supressed by one order of $\frac{1}{N_c}$ compared with ones mentioned in the last section. They break the $U_A(1)$ symmetry and depending on the behavior of the modelled running coupling $\alpha(s)$ in the infrared can create a mass for the iso-singlet pseudoscalar Goldstone boson η'. This provides a mechanism for an η' mass without employing explicitly topological gauge field configurations (instantons).[4]

References

1. C.Roberts, R.Cahill and J.Praschifka, *Ann.Phys.(N.Y.)* **188**, 20 (1988).
2. J.Gasser and H.Leutwyler, *Nucl. Phys.* B **250**, 465 (1985).
3. M.Frank and T.Meissner, *Phys. Rev.* C **53**, 2410 (1996).
4. M.Frank and T.Meissner, in preparation.

$< 0|T(A^3_\mu A^8_\nu)|0 >$: An Example of Non-Trivial Chiral Constraints on QCD Sum Rule Methods

Kim Maltman

York University, 4700 Keele St., North York, Ontario, Canada M3J 1P3

The axial current correlator $< 0|T(A^3_\mu A^8_\nu)|0 >$ is evaluated to one-loop in chiral perturbation theory and the results compared to those obtained from a QCD sum rule analysis which retains operators up to and including dimension 6 and keeps contributions to $\mathcal{O}(\alpha, \alpha_s)$. It is shown that the slope of the correlator obtained from the sum rule treatment is missing the leading terms in its chiral expansion, which accounts for the order of magnitude discrepancy between the value of the slope obtained via chiral perturbation theory and the sum rule method. The way in which the formal quark mass dependence of observables as obtained from chiral perturbation theory can be used to provide non-trivial constraints on other methods is stressed.

Owing to their intimate connections with QCD, QCD sum rules (QSR) and Chiral Perturbation Theory (ChPT) have become popular methods for treating hadronic systems. Although its realm of applicability is more restricted than that of the QSR approach, ChPT has the advantages of (1) a *rigorous* relation to QCD and (2) a systematic expansion procedure. The rigorous relation to QCD is a consequence of the fact that the effective Lagrangian of ChPT is *by construction* the most general such Lagrangian having all the symmetries of QCD and breaking the approximate chiral symmetries in the same manner as QCD breaks them[1]. The systematic ("chiral") expansion is an expansion in momenta and light quark masses. Because of its systematic nature, it can provide useful constraints even when, as is known to sometimes happen, the chiral series for a given observable is slowly converging since, the formal expansion in momenta and quark masses to a given order being a rigorous consequence of QCD, the absence of such terms in any other treatment necessarily signals the incompatibility of that treatment with QCD.

A specific example of how chiral methods can expose otherwise non-obvious problems is the case of the QSR treatment of the axial-current correlator

$$i\int d^4x\ e^{iq.x} < 0|T(A^3_\mu(x)A^8_\nu(0))|0 > \equiv \Pi_1(q^2)q_\mu q_\nu + \Pi_2(q^2)g_{\mu\nu}. \quad (1)$$

One has

$$\Pi_1 = \left[(g_\pi m_\eta^2 - g_\eta m_\pi^2) + q^2(g_\eta - g_\pi)\right] / [(q^2 - m_\eta^2)(q^2 - m_\pi^2)], \quad (2)$$

where g_η and g_π have been evaluated in Ref. [2] employing standard sum rule truncations (contributions to $\mathcal{O}(\alpha, \alpha_s)$ and operators up to and including dimension 6 are retained in the pQCD side of the sum rules). Because the expressions which result for

g_π and g_η involve the quark condensate and the meson masses, the chiral expansions for which are well-known [1], one can derive, from the QSR results, expressions for the leading term in the chiral expansion for the slope of the numerator with respect to q^2, as obtained via the QSR method [3]. (It can be shown that the leading term in the chiral expansion for the constant term in the numerator of the expression for Π_1 above is correctly reproduced by the QSR analysis, so we will not discuss it any further here.) One finds, from the QSR analysis,

$$(g_\eta - g_\pi)_{\ell o} = \theta_0 F^2 \left(8B_0^2(m_s - \hat{m})(m_s + 2\hat{m})/9M_B^4)\right) , \tag{3}$$

where $\theta_0 = \frac{\sqrt{3}(m_d - m_u)}{4(m_s - \hat{m})}$, is the leading chiral order π-η mixing angle, M_B is the Borel mass of Ref. [2] and all other notation is as in Ref. [1].

In contrast, treating the slope consistently to 1-loop in ChPT, one finds

$$(g_\eta - g_\pi)_{\ell o} = \theta_0 F^2 \left(\frac{(m_\pi^2 - \bar{m}_K^2)}{8\pi^2 F^2} \log(\bar{m}_K^2/\mu^2) - \frac{B_0(m_s - \hat{m})}{8\pi^2 F^2} \left[1 - \frac{256\pi^2}{3} L_5^r \right] \right) . \tag{4}$$

The sum rule treatment of the slope of the correlator is missing both the chiral log and those terms linear in the quark masses which occur in the last factor of Eqn. (2), and hence is incorrect. The numerical consequences are also serious, the slope factor being 53 MeV^2 in ChPT[3], versus only 3.9 MeV^2 in the sum rule treatment.

It is important to note that nothing out of the ordinary was done in the original QSR analysis. It would be a highly non-trivial matter to have discovered the problem with the sum rule analysis purely within the QSR context. Indeed, short of performing an improved treatment of the continuum on the phenomenological side, including operators of higher dimension and terms of higher order in α, α_s, one could not have identified the existence of a problem. Moreover, it is not, at present, clear how far one would have to proceed with these improvements before resolving the problem. In contrast, the problem was easily exposed using chiral methods. The key to using ChPT to obtain such useful constraints is in noting that the formal quark mass dependence of quantities such as the slope of the numerator for the correlator Π_1 is a rigorous consequence of QCD.

1. J. Gasser and H. Leutwyler, Nucl. Phys. **B250** (1985) 465
2. C. -T. Chan, E. M. Henley and T. Meissner, Phys. Lett. **B343** (1995) 7
3. K. Maltman, Phys. Lett. **B351** (1995) 56

$U_A(1)$ SYMMETRY RESTORATION IN QCD WITH N_f FLAVORS

Su Houng Lee
Department of physics, Yonsei University, Seoul 120-749, Korea

Tetsuo Hatsuda
Institute of Physics, University of Tsukuba, Tsukuba, Ibaraki 305, Japan

We discuss the question of whether correlation functions of mesonic currents are effectively symmetric under $U(N_f) \times U(N_f)$ at high temperature phase of QCD after chiral symmetry restoration. General statements are given concerning this question for QCD with N_f flavors

1 Main Derivation

Recently, there have been reports that after chiral symmetry restoration, the effect of $U_A(1)$ breaking does not appear in the correlation function of mesonic currents[1,2]. This implies that at high temperature QCD effectively recovers the original $U(N_f) \times U(N_f)$ symmetry that it had at the Lagrangian level.

However, taking into account the contributions from topologically non-trivial gauge field configurations, it can be shown that the answer depends on the number of quark flavors N_f and the instanton density at high temperature. Topologically non-trivial gauge field configurations in the partition function of QCD are always accompanied by the $n_+(n_-)$ number of zero modes satisfying $\not{D}\psi_0 = 0$. The number of zero modes are related to the topological winding number $\nu = \frac{g^2}{32\pi^2} \int d^4x F\tilde{F} \neq 0 = n_+ - n_-$. Writing out the QCD partition function with $\nu = 0$ and $\nu = 1$ part, we have

$$\begin{aligned} Z[J] &= \int D[A] e^{-S_{YM}} \mathrm{Det}[\not{D} + J] \\ &+ \int D[A]_{\nu=1} e^{-S_{YM}} \mathrm{Det}'[\not{D} + J] \times det_{st}\left(\int d^4x \bar{\psi}_0(x) J_{st} \psi_0(x)\right) \end{aligned} \quad (1)$$

where in the second line, we have explicitly separated out the zero mode contribution in the fermion determinant. Here, det_{sf} stands for the determinant in the flavor space which is denoted by the index $s, f = 1, \cdots N_f$. When $J = M$(mass matrix) $\rightarrow 0$, the topological configurations do not contribute. However, this is not true in the presence of external source such as looking at the correlation function. As an typeical example, let us look at the $U_A(1)$

effect in the the difference between the $\sigma-\sigma$ and $\delta-\delta$ two point functions. Starting from Eq.(1), we have

$$\begin{aligned}&\langle \bar{q}(x)q(x),\bar{q}(0)q(0)\rangle - \langle \bar{q}(x)\tau^3 q(x),\bar{q}(0)\tau^3 q(0)\rangle \\ &= \frac{1}{Z}\int D[A]_{\nu=0}e^{-S_{YM}}\mathrm{Det}'[\not{D}+m]\cdot \mathrm{tr}[S(x,x)]\mathrm{tr}[S(0,0)] \\ &+\frac{1}{Z}\int D[A]_{\nu=\pm 1}e^{-S_{YM}}\mathrm{Det}'[\not{D}+m]\cdot 4\cdot \bar{\psi}_0(x)\psi_0(x)\cdot \bar{\psi}_0(0)\psi_0(0)+O(m).\end{aligned}$$

The first term comes from the topologically trivial sector ($\nu=0$) and the second from the topologically non-trivial sector ($\nu=\pm 1$). In this case, the contribution from $\nu=\pm 1$ part remains even in the chiral limit at high temperature when chiral symmetry gets restored and $S(x,x)\propto \rho(\lambda=0)\sim O(m)$.

It should be noted, however, that for general N_f, effects of $\nu\neq 0$ sector would be of $O(m^{N_f-2})$ and vanish in the chiral limit for $N_f\geq 3$. For QCD in the real world, it is proportional to $m_s(\sim 150$ MeV) and the effect may not be negligible. Due to the chirality, this differnce remains if the we need a $U_A(1)$ transformation to go from the first current to the second current. From these kind of considerations, we arrive at the following result[3].

1) $\nu\neq 0$ sectors do not contribute to two point functions for $N_f\geq 3$ in the $SU(N_f)$ symmetric chiral limit. In this case, the physical spectrum will effectively be multiplets of $U(N_f)\times U(N_f)$, as was claimed in [2].

2) If $N_f=2$, $\nu\neq 0$ sectors do contribute differently to the two point functions of π,δ,σ,η even in the chiral limit and even when chiral symmetry is restored at high temperature. In this case, argument in [2] does not apply.

3) In the real world, $SU(3)$ is not exact and $m_s\sim$ 150MeV. In this case, $\nu\neq 0$ sectors do contribute with possible non-negligible effect. However, at high temperature, the topological configurations (Instanton density) are exponentially suppressed $exp(-\rho^2T^2)$ and the $U_A(1)$ breaking effect would also be suppressed.

References

1. E.V. Shuryak, Comments *Nucl. Part. Phys.* **21** (1994) 235.
2. T.D. Cohen, "The High Temperature Phase of QCD and $U(1)_A$ Symmetry", Univ. of Maryland preprint no.96-060, hep-ph/9601216. To be published in *Phys. Rev.* D.
3. Su H. Lee and T. Hatsuda, "$U_A(1)$ Symmetry restoration in QCD with N_f flavors", hep-ph/9601373 . To be published in *Phys. Rev.* D.

WHAT DOES A CHANGE IN THE QUARK CONDENSATE SAY ABOUT RESTORATION OF CHIRAL SYMMETRY IN MATTER?

Michael C. Birse
Theoretical Physics Group, Department of Physics and Astronomy, University of Manchester, Manchester, M13 9PL, UK

The contribution of nucleons to the quark condensate in nuclear matter includes a piece of first order in m_π, arising from the contribution of low-momentum virtual pions to the πN sigma commutator. Chiral symmetry requires that no term of this order appears in the NN interaction. The mass of a nucleon in matter thus cannot depend in any simple way on the quark condensate alone.

The quark condensate $\langle\bar{q}q\rangle$ decreases in nuclear matter, with a leading density dependence given by the model-independent form[1]

$$\frac{\langle\bar{q}q\rangle_\rho}{\langle\bar{q}q\rangle_0} = 1 - \frac{\sigma_{\pi N}}{f_\pi^2 m_\pi^2}\rho. \tag{1}$$

Using the value $\sigma_{\pi N} \simeq 45$ MeV from the determination of Gasser *et al.*[2], this corresponds to a $\sim 30\%$ reduction at the density of nuclear matter. Since $\langle\bar{q}q\rangle$ is an order parameter for the hidden chiral symmetry of QCD, this decrease is generally interpreted as a signal of partial restoration of the symmetry[3].

There has been much recent interest in this idea[4,5,3]. Corrections to (1) of higher order in the density have been estimated and found to be small, at least up to normal nuclear densities[6,7]. The range of the effect[8] and the role of correlations[9] have been examined. Many studies have been made of its possible consequences for the properties of hadrons in matter and it has been suggested that the scalar fields of Dirac phenomenology could be related to symmetry restoration, through either QCD sum rules[10] or scaling laws for hadron masses in matter[4,5]. However, chiral symmetry itself shows that the relation between hadron masses in matter and $\langle\bar{q}q\rangle$ cannot be so simple[11].

To see this, consider the mass of a nucleon in matter (at mean-field level):

$$M_N^* = M_N + V_s(0)\rho, \tag{2}$$

where $V_s(q^2)$ is the scalar-exchange part of the NN potential. Suppose now, for the sake of argument, that this mass contained a piece that depended on the quark condensate so that, to first order in the density, there was a term in M_N^* proportional to $\sigma_{\pi N}\rho/(f_\pi^2 m_\pi^2)$. This would imply the existence of a scalar piece of the NN interaction that was proportional to $\sigma_{\pi N}/m_\pi^2$ [8].

The chiral expansion of $\sigma_{\pi N}$ has the form[12]

$$\sigma_{\pi N} = A m_\pi^2 - \frac{9}{16\pi}\left(\frac{g_{\pi NN}}{2M_N}\right)^2 m_\pi^3 + \cdots, \tag{3}$$

where the constant A involves a counterterm corresponding to short-distance physics. In contrast, the $\mathcal{O}(m_\pi^3)$ term is nonanalytic in the current quark mass and so arises purely from the long-distance physics of the pion cloud surrounding a nucleon. If there were a piece of the NN interaction proportional to $\sigma_{\pi N}/m_\pi^2$ (with a coefficient independent of m_π), it would contain a nonanalytic term of order m_π. However Weinberg's dimensional counting rules[13] show that no such term can be present in the chiral expansion of the NN potential. Chiral symmetry thus rules out any interaction proportional to $\sigma_{\pi N}/m_\pi^2$ on its own. Hence there can be no direct dependence of the nucleon mass in matter on the quark condensate alone (unless its coefficient is of order m_π^2).

This presents a problem both for simple scaling laws[4,5] and for QCD sum rules in matter[10]. In the latter case, the operator product expansion involves local operators that cannot distinguish between short- and long-range effects. A better treatment of the πN continuum is required to get the correct leading nonanalytic behaviour of the nucleon mass from such sum rules[14].

In addition, this observation raises the question of whether the contributions of low-momentum pions to the quark condensate correspond to partial restoration of chiral symmetry in matter[11].

1. E. G. Drukarev and E. M. Levin, Nucl. Phys. **A511** (1990) 679; **A516** (1990) 715(E); T. D. Cohen, R. J. Furnstahl and D. K. Griegel, Phys. Rev. **C45** (1992) 1881.
2. J. Gasser, H. Leutwyler and M. E. Sainio, Phys. Lett. **B253** (1991) 252.
3. M. C. Birse, J. Phys. G: Nucl. Part. Phys. **20** (1994) 1537.
4. C. Adami and G. E. Brown, Phys. Reports **224** (1993) 1.
5. G. E. Brown and M. Rho, Phys. Reports **269** (1996) 233.
6. G. Q. Li and C. M. Ko, Phys. Lett. **338** (1994) 118.
7. R. Brockmann and W. Weise, Phys. Lett. **B367** (1996) 40.
8. M. C. Birse, Phys. Rev. **C49** (1994) 2212.
9. J. Delorme, G. Chanfray and M. Ericson, nucl-th/9603005.
10. T. D. Cohen *et al.*, Phys. Rev. Lett. **67** (1991) 961.
11. M. C. Birse, Phys. Rev. **C53** (1996) 2048.
12. J. Gasser and H. Leutwyler, Phys. Reports **87** (1982) 77.
13. S. Weinberg, Nucl. Phys. **B363** (1991) 1.
14. M. C. Birse and B. Krippa, hep-ph/9603387.

RECOVERING RELATIVISTIC NUCLEAR PHENOMENOLOGY FROM THE QUARK-MESON COUPLING MODEL

XUEMIN JIN and B.K. JENNINGS
TRIUMF 4004 Wesbrook Mall, Vancouver, B.C., Canada V6T 2A3

The quark-meson coupling (QMC) model for nuclear matter, which describes nuclear matter as non-overlapping MIT bags bound by the self-consistent exchange of scalar and vector mesons is modified by the introduction of a density dependent bag constant. It is found that when the bag constant is significantly reduced in nuclear medium with respect to its free space value, large canceling isoscalar Lorentz scalar and vector potentials for the nucleon in nuclear matter emerge naturally. Such potentials are comparable to those suggested by relativistic nuclear phenomenology. This suggests that the modification of the bag constant in the nuclear medium may play an important role in low- and medium-energy nuclear physics.

While the description of nuclear phenomena has been efficiently formulated using the hadronic degrees of freedom, new challenges arise from the observed small but interesting corrections to the standard hadronic picture such as the EMC effect which reveals the medium modification of the internal structure of nucleon. To address these new challenges, it is necessary to build theories that incorporate quark-gluon degrees of freedom, yet respect the established theories based on hadronic degrees of freedom.

A few years ago, Guichon[1] proposed a quark-meson coupling (QMC) model to investigate the direct "quark effects" in nuclei. This model describes nuclear matter as non-overlapping MIT bags interacting through the self-consistent exchange of mesons in the mean-field approximation[1,2]. The QMC model, however, predicts much smaller scalar and vector potentials for the nucleon than obtained in relativistic nuclear phenomenology. The latter is a general approach based on nucleons and mesons which has gained tremendous credibility during last twenty years[3]. It is known that the large and canceling scalar and vector potentials are central to the success of the relativistic nuclear phenomenology.

We observe that the bag constant is held to be at its free space value in the QMC model. This assumption can be questioned. The bag constant in the MIT bag model contributes $\sim 200 - 300$ MeV to the nucleon energy and provides the necessary pressure to confine the quarks. Thus, the bag constant is an inseparable ingredient of the bag picture of a nucleon. When a nucleon bag is put into the nuclear medium, the bag as a whole reacts to the environment.

As a result, the bag constant may be modified. There is little doubt that at sufficiently high densities, the bag constant is eventually melted away and quarks and gluons become the appropriate degrees of freedom. Therefore, It is reasonable to believe that the bag constant is modified and decreases as density increases. This physics is obviously bypassed in the QMC model by the assumption of $B = B_0$ [1,2].

We modify the QMC model by introducing a density dependent bag constant. We model the density dependence of the bag constant in two ways: one invokes a direct coupling of the bag constant to the scalar meson field, and the other relates the bag constant to the in-medium nucleon mass [4]. Both models feature a decreasing bag constant with increasing density. We find that the reduction of the bag constant in nuclear matter partially offsets the effect of the internal quark structure of the nucleon and effectively introduces a new source of attraction. This attraction needs to be compensated with additional vector field strength. The decrease of bag constant also implies the increase of bag radius in nuclear matter. This is consistent with the "swollen" nucleon picture discussed in the literature.

When the bag constant is reduced significantly in nuclear matter with respect to its free-space value, we find that our modified quark-meson coupling model predicts large and canceling scalar and vector potentials for the nucleon in nuclear matter, which is qualitatively different from the prediction of the simple QMC model. These potentials are consistent with those suggested by the relativistic nuclear phenomenology, implying that the essential physics of the relativistic nuclear phenomenology can be recovered. This suggests that the drop of the bag constant in nuclear medium may play an important role in low- and medium-energy nuclear physics.

Acknowledgments

This work was supported by the Natural Sciences and Engineering Research Council of Canada.

References

1. P. A. M. Guichon, *Phys. Lett.* B **200**, 235 (1988).
2. S. Fleck, W. Bentz, K. Shimizu, and K. Yazaki, *Nucl. Phys.* **A510**, 731 (1990); K. Saito and A. W. Thomas, *Phys. Lett.* B **327**, 9 (1994).
3. S. J. Wallace, *Ann. Rev. Nucl. Part. Sci.* **37**, 267 (1987).
4. X. Jin and B.K. Jennings, *Phys. Lett.* B **374**, 13 (1996); nucl-th/9604018.

HEAVY QUARK SYMMETRY WITHIN A LIGHT-FRONT CONSTITUENT QUARK MODEL

SILVANO SIMULA

Istituto Nazionale di Fisica Nucleare, Sezione Sanità
Viale Regina Elena 299, I-00161 Roma, Italy

Abstract

The elastic and transition electroweak form factors among heavy-light pseudoscalar mesons are investigated within a relativistic constituent quark model formulated on the light-front, showing that the constraints arising from Heavy Quark Symmetry are fulfilled at any value of the recoil. The Isgur-Wise form factor is calculated adopting meson wave functions derived from an effective $q\bar{q}$ interaction able to reproduce the meson mass spectra. Adopting the Heavy Quark Effective Theory, our result for the Isgur-Wise form factor is applied to a parameter-free extraction of the weak mixing matrix element $|V_{bc}|$ from the exclusive data available for the $B \to D^* \ell \nu_\ell$ decay channel. The branching ratios for the $B_s \to D_s \ell \nu_\ell$ and $B_s \to D_s^* \ell \nu_\ell$ decays are estimated.

The Isgur-Wise (IW) form factor is calculated within a relativistic constituent quark model formulated on the light-front [1], adopting meson wave functions derived from the effective $q\bar{q}$ interaction of Ref. [2], which nicely reproduces meson mass spectra. The limit of infinite heavy-quark masses is performed both for the space-like elastic form factor $H^{(el)}(q^2)$ and for the time-like transition form factors $h_+(q^2)$ between pseudoscalar mesons. It is shown that, according to the Heavy Quark Symmetry, the two limits coincide at any value of the recoil and for any given wave function describing the motion of the light spectator-quark. Moreover, in the heavy-quark limit it turns out that: i) the calculated form factor h_- vanishes at any value of the recoil; ii) the mass dependence of the form factor h_+ at the point of zero recoil is consistent with Luke's theorem [3]. For the $B \to D$ transition the value $\delta_{1/m^2} = -4\%$ is obtained, in agreement with the updated estimate $\delta_{1/m^2} = -5.5 \pm 2.5$ % from [4]. It is also shown that the results obtained for the leptonic decay constant f_{PS} of pseudoscalar mesons are in nice agreement with the predictions from lattice QCD simulations and QCD sum rules (see Ref. [5]), and, moreover, the mass dependence of f_{PS} is consistent with the prediction of the Heavy Quark Effective Theory ($HQET$), i.e. $\sqrt{M_{PS}} f_{PS} \to const.$ Our result for the IW form factor is presented and it compares very favourably with the one of recent lattice QCD calculations [6] as well as with the results obtained in [7] within a QCD sum rule analysis at next-to-leading order in renormalization-group improved perturbation theory. The slope of our IW form factor turns out to be 1.03 (1.14) for a spectator-quark mass of 0.220 (0.419) GeV (see [1]).

Using the $HQET$, the semileptonic $B \to D^* \ell \bar{\nu}_\ell$ decay is investigated and the Cabibbo-Kobayashi-Maskawa matrix element $|V_{bc}|$ is extracted from the experimental data, obtained by the $ARGUS$ [8], $CLEO-II$ [9] and $ALEPH$ [10] collaborations, with the recoil dependence fixed by our calculated IW form factor (see Fig. 1). It can clearly be seen that our prediction nicely reproduces with no free parameters the recoil dependence of the $CLEO$ data. Adopting $\eta_A = 0.96$ for the radiative corrections [4] and $\delta_{1/m^2} = -4\%$ for the power corrections (i.e., our result for $h_+^{B \to D}(1)$), the value $|V_{bc}| = 0.0419 \pm 0.0029$ is obtained from the $CLEO$ data (with $\chi^2/N_{D.F.} \simeq 2.1/6$). The sensitivity of the result to the choice of the light spectator-quark wave function is found to be quite small in comparison with the one related to present experimental uncertainties.

Within our quark model and adopting the $HQET$, the decay rate for the $B \to D \ell \nu_\ell$ and $B \to D^* \ell \nu_\ell$ transistions is calculated. Using the experimental branching ratios,

$Br(B^0 \to D^-\ell^+\nu_\ell) = 1.9 \pm 0.5$ % [12] and $Br(B^0 \to D^{*-}\ell^+\nu_\ell) = 5.47 \pm 0.16_{stat} \pm 0.67_{syst}$ % [11], and adopting for the B^0 lifetime the PDG value $\tau(B^0) = 1.50 \pm 0.11$ ps [12], the parameter $|V_{bc}|$ is estimated to be 0.0411 ± 0.057 and 0.0408 ± 0.030, respectively. The average of our predictions, $|V_{bc}| = 0.0413 \pm 0.024$, is in good agreement with the updated experimental determinations of $|V_{bc}|$ obtained from exclusive and inclusive semileptonic decays of the B-meson, viz. $|V_{bc}|_{excl} = 0.0373 \pm 0.0045_{exp} \pm 0.0065_{th}$ and $|V_{bc}|_{incl} = 0.0398 \pm 0.0008_{exp} \pm 0.0040_{th}$ [13]. Finally, adopting $\tau(B_s^0) = 1.59^{+0.17}_{-0.15}$ $(stat) \pm 0.03$ $(syst)$ ps [14], the branching ratios $Br(B_s \to D_s \ell \nu_\ell)$ and $Br(B_s \to D_s^* \ell \nu_\ell)$ are estimated to be 1.95 ± 0.30 % and 5.85 ± 0.90 %, respectively.

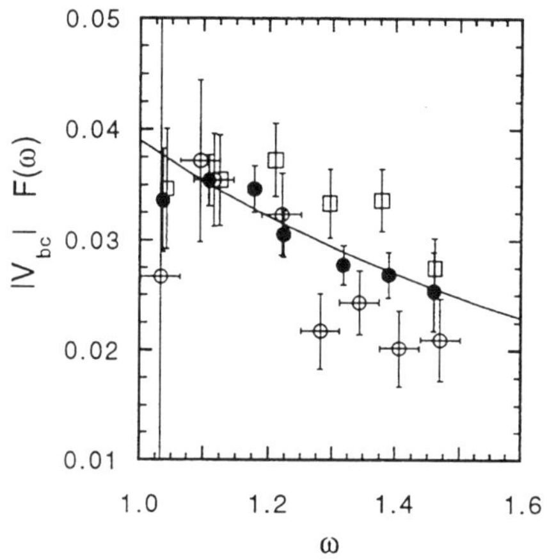

Fig. 1. The quantity $|V_{bc}|$ $F(\omega)$, extracted from the experimental data for the exclusive channel $B \to D^* \ell \nu_\ell$ versus the recoil ω. The open dots and squares are from Refs. [8] and [10], respectively. The full dots are the $CLEO-II$ data [9]. The solid line is our prediction, obtained through a least-χ^2 fit to the $CLEO$ data, with the recoil dependence fixed by the results of our relativistic calculation of the IW form factor.

References

[1] S. Simula: Phys. Lett. **B373** (1996) 193; and to be published.

[2] S. Godfrey and N. Isgur: Phys. Rev. **D32** (1985) 185.

[3] M.E. Luke: Phys. Lett. **B252** (1990) 447.

[4] M. Neubert: Phys. Lett. **B338** (1994) 84.

[5] F. Cardarelli, I.L. Grach, I.M. Narodetskii, E. Pace, G. Salmè and S. Simula: Phys. Lett. **B332** (1994) 1.

[6] UKQCD collaboration, K.C. Bowler et al.: Phys. Rev. **D52** (1995) 5067.

[7] M. Neubert: Phys. Rev. **D47** (1993) 4063.

[8] ARGUS Coll., H. Albrecht et al.: Z. Phys. **C57** (1993) 533.

[9] CLEO-II Coll., B. Barish et al.: Phys. Rev. **D51** (1995) 1014.

[10] ALEPH Coll., I. Scott et al.: Proc. of the 27th Int. Conf. on *High Energy Physics*, Glasgow (Scotland), 1994, eds. P.J. Bussey and J.G. Knowles (IOP, London, 1995).

[11] DELPHI Coll., P. Abreu et al.: preprint CERN-PPE/96-11.

[12] Review of Particle Properties, L. Montanet *et al.*: Phys. Rev. **D50** (1994) 1173.

[13] T. Skwarnicki, preprint HEPSY-95-307, to appear in the Proc. of the 17th Int. Conf. on *Lepton-Photon Interactions*, Beijing, China, August 1995.

[14] ALEPH Coll., D. Buskulic et al.: Phys. Lett. **B361** (1995) 221.

Faddeev approach to the baryons in the NJL model

N.ISHII, H.ASAMI, W.BENTZ and K.YAZAKI
Department of Physics, University of Tokyo, Hongo, Bunkyo-ku, Tokyo 113, Japan

The relativistic Faddeev equation in the NJL model is solved numerically in the ladder approximation. The main results of the above approach are reviewed. The effects of the "meson exchange" on the Faddeev equation in the ladder approximation are evaluated perturbatively.

1 The relativistic Faddeev equation in the ladder approximation

We solve the relativistic three quark Faddeev equation for the nucleon and the delta in the Nambu-Jona-Lasinio(NJL) model numerically in the ladder approximation[1]. We investigate the contributions of the two-quark interactions in the scalar ($J^\pi = 0^+$, iso-scalar, color $\bar{3}$) and the axial-vector ($J^\pi = 1^+$, iso-vector, color $\bar{3}$) diquark channels treating their coupling constants g_s and g_a as parameters. We study the dependence of the masses of the nucleon and the delta on these coupling constants and find that the following linear relations are approximately valid: $M_\mathrm{N} \simeq -1.1(g_\mathrm{s}/g_\pi) - 0.7(g_\mathrm{a}/g_\pi) + 1.7$ [GeV], $M_\Delta \simeq -0.76(g_\mathrm{a}/g_\pi) + 1.52$ [GeV]. We see that the interaction in the scalar diquark channel gives a main attraction to the nucleon state, and that the interaction in the axial-vector diquark channel gives a large attractive correction to it. We also see that the interaction in the axial-vector diquark channel contributes to the delta state attractively. These two relations give the nucleon-delta mass difference: $M_\Delta - M_\mathrm{N} \simeq 1.1(g_\mathrm{s}/g_\pi) - 0.06(g_\mathrm{a}/g_\pi) - 0.18$ [GeV]. It follows that the nucleon-delta mass difference is, in the Faddeev approach, due to the interaction in the scalar diquark channel.

2 The Faddeev approach and the mean-field approximation

The mean-field approximation is another approach to the baryons in the NJL model, which leads to the hedge-hog configuration. The relativistic Faddeev equation in the ladder approximation corresponds to the three-particle RPA in the nuclear structure theory, and it is only through the RPA-vacuum that the excitations of the negative energy states can be taken into account. We would like to include the effects of the mean-field by introducing a "meson exchange" interaction in addition to the bare one. It is so hard to solve the relativistic Faddeev equation with non-separable interactions that we use perturbation theory. The eigenvalue($\lambda(E)$) of the Faddeev kernel for the nucleon state including the first order deviation is plotted in Fig.1 (solid line) against the total energy E with the unperturbed one (dashed line) for $g_s = 0.675$ and $g_a = 0$. The nucleon mass is obtained by solving the equation $\lambda(E = M_N) = 1$. We see that the nucleon mass is much reduced if we include the perturbation. We have to note that the matrix elements of the perturbing interaction between the "excited" diquark states are neglected in this calculation. However, they are negligible in the quark-diquark limit, and the qualitative behaviors are not expected to change. Actually, most of the contributions come from the matrix element of the pion exchange interaction between the "ground" diquark state. In the future calculation, we are going to solve the Faddeev equation with the "renormalized interaction" which includes also the above matrix element. Since the interaction of this Faddeev equation is still separable, we can solve it without using the perturbation theory. Then, with this solution, we are going to evaluate perturbatively effects which are not yet included in this renormalized interaction.

Acknowledgments

One of the authors N.Ishii is supported by the JSPS fellowship.

Fig.1

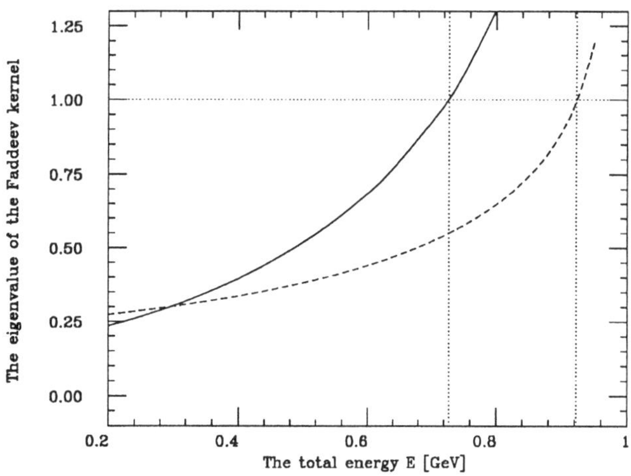

References

1. N.Ishii, W.Bentz and K.Yazaki, *Phys. Lett.* B **301**, 165 (1993), *Phys. Lett.* B **318**, 26 (1993).
   N.Ishii, W.Bentz and K.Yazaki, *Nucl. Phys.* A **587**, 617 (1995).
   N.Ishii, H.Asami, W.Bentz and K.Yazaki, *Phys. Rev.* C **51**, 3388 (1995).

Effective Chiral theory of mesons

Bing An Li
Department of Physics and Astronomy, University of Kentucky, Lexington, KY 40506, USA

Can we have a theory to unify all the studies done in the field of meson physics? This theory should be QCD inspired, self-consistent, and phenomenologically successful[1]. Following the idea of the bosonization of 1+1 field theory, we propose to simulate the meson fields by quark operators. For example, $\rho^i_\mu = -\frac{1}{g_\rho m^2_\rho}\bar\psi\tau_i\gamma_\mu\psi$. Using PCAC, current algebra, and this expression, we obtain $\frac{1}{2}f_{\rho\pi\pi}g_\rho = 1$ which is well known. The simulations are realized by a Lagrangian constructed by using $U(2)_L \times U(2)_R$ chiral symmetry and the minimum coupling principle,

$$\mathcal{L} = \bar\psi(x)(i\gamma\cdot\partial + \gamma\cdot v + \gamma\cdot a\gamma_5 - mu(x))\psi(x) - \bar{\psi(x)}M\psi(x) + \frac{1}{2}m_0^2(\rho^\mu_i\rho_{\mu i} + \omega^\mu\omega_\mu + a^\mu_i a_{\mu i} + f^\mu f_\mu) \tag{1}$$

where $a_\mu = \tau_i a^i_\mu + f_\mu$, $v_\mu = \tau_i\rho^i_\mu + \omega_\mu$, and $u = exp\{i\gamma_5(\tau_i\pi_i + \eta)\}$. Using the least action principle, we obtain

$$\begin{aligned}
\frac{\Pi_i}{\sigma} &= i(\bar\psi\tau_i\gamma_5\psi + ix\bar\psi\tau_i\psi)/(\bar\psi\psi + ix\bar\psi\gamma_5\psi),\\
x &= (i\bar\psi\gamma_5\psi - \frac{\Pi_i}{\sigma}\bar\psi\tau_i\psi)/(\bar\psi\psi + i\frac{\Pi_i}{\sigma}\bar\psi\tau_i\gamma_5\psi),\\
\rho^i_\mu &= -\frac{1}{m_0^2}\bar\psi\tau_i\gamma_\mu\psi, \quad a^i_\mu = -\frac{1}{m_0^2}\bar\psi\tau_i\gamma_\mu\gamma_5\psi,\\
\omega_\mu &= -\frac{1}{m_0^2}\bar\psi\gamma_\mu\psi, \quad f\mu = -\frac{1}{m_0^2}\bar\psi\gamma_\mu\gamma_5\psi,
\end{aligned} \tag{2}$$

where $u = e^{i\eta\gamma_5}(\sigma + i\gamma_5\tau\cdot\Pi)$, $\sigma = \sqrt{1-\Pi^2}$, and $x = tan\eta$.

Using the method of path integral to integrate out the quark fields, the effective Lagrangian of mesons is obtained. The Wess-Zumino-Witten Lagrangian with spin-1 fields is derived from the Lagrangian. All the vertices of the meson physical processes of normal parity and abnormal parity are derived. In this theory the parameters are: m_u, m_d, m_s, g, m. g is an universal coupling constant. Input f_π, we determine $m = 300MeV$ and choose $g = 0.35$.

All tree diagrams are of order $O(N_c)$, hence they are leading contributions. A diagram with loops is at higher order in large N_c expansion. A cut-off of momentum has to be introduced.

$$\frac{N_c}{(4\pi)^2}\{log(1+\frac{\Lambda^2}{m^2})+\frac{1}{1+\frac{\Lambda^2}{m^2}}-1\}=\frac{1}{16}\frac{F^2}{m^2}=\frac{3}{8}g^2,\ \Lambda=1.6GeV. \quad (3)$$

The masses of all mesons in this theory are below the cut-off.

The quark condensate is found to be

$$<0|\bar{\psi}(x)\psi(x)|0>=-m^3\frac{N_C}{8\pi^2}\{\frac{\Lambda^2}{m^2}-log(1+\frac{\Lambda^2}{m^2})\}. \quad (4)$$

To the leading order in quark mass expansion, the masses of the octet pseudoscalar mesons are derived(Gell-Mann, Oakes, Renner formulas), for example,

$$m_\pi^2=-\frac{2}{f_\pi^2}(m_u+m_d)<0|\bar{\psi}\psi|0>, \quad (5)$$

The KSFR sum rule derived in this theory leads to

$$m_\rho^2=6m^2=(0.751GeV)^2.$$

In the limit of $m_q=0$, the masses of the four low lying vector mesons originate from dynamical chiral symmetry breaking. From the spontaneous chiral symmetry breaking we obtain

$$(1-\frac{1}{2\pi^2g^2})m_a^2=6m^2+m_\rho^2,(1-\frac{1}{2\pi^2g^2})m_f^2=6m^2+m_\omega^2. \quad (6)$$

$$(1-\frac{1}{2\pi^2g^2})m_{K_1(1400)}^2=6m^2+m_{K^*(892)}^2,(1-\frac{1}{2\pi^2g^2})m_{f_1(1510)}^2=6m^2+m_\phi^2. \quad (7)$$

To summarize other results obtained by this theory. We obtain: Vector meson dominance(VMD); Weinberg's first sum rule is satisfied analytically; The scattering lengths and slopes of $\pi\pi$ scattering obtained by Weinberg are revealed from this theory. It is found that ρ meson exchange dominates $\pi\pi$ scattering; In the chiral limits, two coefficients of chiral perturbation theory are determined and in good agreement with CPT; The amplitude of $\pi^0\to\gamma\gamma$ obtained in this theory is the same as the triangle anomaly; The theory provides a unified description of meson physics at low energies. Theoretical results of the decay widths of pseudoscalar, vector, and axial-vector mesons agree with data well.

References

1. Bing An Li, *Phys. Rev.* D **52**, 5165 (1995), and *Phys. Rev.* D **52**, 5184 (1995).

BREAKDOWN OF CONVENTIONAL FACTORIZATION FOR ISOLATED PHOTON CROSS SECTIONS

EDMOND L. BERGER[a], XIAOFENG GUO[b] and JIANWEI QIU[b]
[a] *High Energy Physics Division, Argonne National Laboratory, Argonne, IL 60439, USA*
[b] *Department of Physics and Astronomy, Iowa State University, Ames, IA 50011, USA*

Using $e^+e^- \to \gamma + X$ as an example, we show that the conventional factorization theorem of perturbative QCD breaks down for isolated photon cross sections in a specific part of phase space. Implications are discussed.

Much of the predictive power of perturbative QCD derives from factorization theorems [1]. *Conventional* factorization expresses a physical quantity as the convolution of a partonic part with a nonperturbative matrix element, and it requires that the perturbatively calculated partonic part be infrared safe order by order in α_s. Predictions then follow when processes with different hard scattering but the same matrix elements are compared. Using $e^+e^- \to \gamma X$ as an example, we demonstrate that the perturbatively calculated partonic part for the isolated photon cross section is not infrared safe in a well-defined phase space.

The essence of isolation is that a cone of half-angle δ is drawn about the direction of the photon's momentum, and the isolated cross section is defined for photons accompanied by less than a specified amount of hadronic energy in the cone, *e.g.*, $E_h^{cone} \leq E_{\max}$.

At high energy, photons can result from long-distance fragmentation of quarks and gluons, themselves produced in short-distance hard collisions. In such fragmentation contributions, hadronic energy in the isolation cone has two sources: a) energy from parton fragmentation, E_{frag}, and b) energy from non-fragmenting final-state partons, $E_{partons}^{cone}$, that enter the cone. When the maximum hadronic energy allowed in the isolation cone is saturated by the fragmentation energy, $E_{\max} = E_{frag}$, there is no allowance for energy in the cone from other final-state partons. In particular, if there is a gluon in the final state, the phase space for this gluon becomes restricted. By contrast, isolation does not affect the virtual gluon exchange contribution. Therefore, for isolated photons, there is a possibility that the infrared singularity from the virtual contribution may not be cancelled completely by the restricted real contribution. We showed [2,3] that such incomplete cancellation of infrared singularities appears first at next-to-leading order (NLO) in the quark fragmentation con-

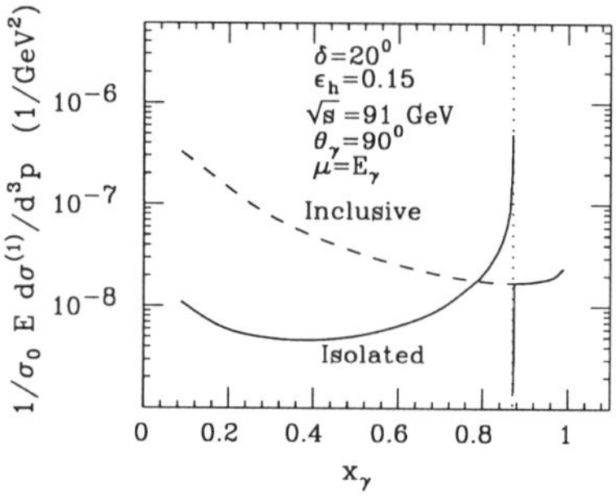

Figure 1: Comparison of the one-loop quark fragmentation contributions to the isolated cross section and the inclusive cross section in $e^+e^- \to \gamma X$ as a function of $x_\gamma = 2E_\gamma/\sqrt{s}$.

tributions.

If conventional factorization were true, the fragmentation contributions to the cross section for isolated photons would be expressed in the factorized form

$$E_\gamma \frac{d\sigma^{iso}_{e^+e^-\to\gamma X}}{d^3\ell} = \sum_c \int_{\max\left[x_\gamma, \frac{1}{1+\epsilon_h}\right]}^1 \frac{dz}{z}\, E_c \frac{d\hat{\sigma}^{iso}_{e^+e^-\to cX}}{d^3p_c}\left(x_c = \frac{x_\gamma}{z}\right) \frac{D_{c\to\gamma}(z,\delta)}{z}\,; \tag{1}$$

$x_\gamma = 2E_\gamma/\sqrt{s}$, $x_c = 2E_c/\sqrt{s}$, $\epsilon_h = E_{\max}/E_\gamma$, and the sum extends over $c = q, \bar{q}$ and g. $D_{c\to\gamma}(z,\delta)$ is the nonperturbative function that describes fragmentation of parton "c" into a photon. The lower limit of integration results from the isolation requirement with the assumption that all fragmentation energy is in the isolation cone [3]. Because of the isolation condition, the phase space constraints are different in three regions: a) $x_\gamma < 1/(1+\epsilon_h)$, b) $x_\gamma = 1/(1+\epsilon_h)$, c) $x_\gamma > 1/(1+\epsilon_h)$. We show below that the next-to-leading order partonic hard part for quark fragmentation, $E_q d\hat{\sigma}^{iso}_{e^+e^-\to qX}/d^3p_q$, is infrared sensitive.

When $x_\gamma < 1/(1+\epsilon_h)$, subprocesses with two-body final states do not contribute. Therefore, there is no leading-order quark (or antiquark) fragmentation contribution, and one-loop virtual diagrams do not contribute. The well-known $1/(1-x_q)$ infrared singularity of the real gluon emission diagrams, as $x_q = x_\gamma/z \to 1$, will remain in $\hat{\sigma}^{iso}_{e^+e^-\to qX}$. After convolution with $D_{q\to\gamma}(z)$, this inverse power infrared sensitivity yields a logarithmic divergence proportional to $\ell n(1/x_\gamma - (1+\epsilon_h))$. As shown in Fig. 1, as $x_\gamma \to 1/(1+\epsilon_h)$, the isolated cross section becomes larger than the inclusive cross section, which is certainly not physical. This infrared sensitivity in $\hat{\sigma}^{iso}_{e^+e^-\to qX}$ signals a breakdown of conventional perturbative factorization.

When $x_\gamma = 1/(1+\epsilon_h)$, $x_q = x_\gamma/z = 1$ is possible. Therefore, the one-loop virtual diagrams, which are proportional to $\delta(1-x_q)$ will contribute. However, isolation constraints limit the phase space of real gluon emission in the real subprocess, $e^+e^- \rightarrow q\bar{q}g$. Consequently, the infrared divergences in the real and virtual contributions do not cancel completely in the isolated case, unlike the inclusive case. In $n = 4 - 2\epsilon$ dimensions, we find [3]

$$E_q \frac{d\sigma^{(1)iso}_{e^+e^- \rightarrow qX}}{d^3p_q} \sim \left\{ \frac{1}{\epsilon^2} + \frac{1}{\epsilon}\left(\frac{3}{2} - \ell n \frac{\delta^2}{4}\right)\right\} \delta(1 - x_q) \; + \text{finite terms} . \quad (2)$$

At $x_q = 1$, corresponding to $x_\gamma = 1/(1 + \epsilon_h)$, the partonic part for quark fragmentation in Eq. (2) is infrared divergent, and the perturbative calculation is not well-defined. Conventional perturbative factorization again breaks down.

When $x_\gamma > 1/(1 + \epsilon_h)$, due to the finite cone size, there is a very small region of phase space where the isolated cross section is not exactly equal to the inclusive cross section. This region of phase space is proportional to $\epsilon_h \delta^2/4$. In this narrow region above $x_\gamma = 1/(1 + \epsilon_h)$, all one-loop contributions to the isolated cross section have a logarithmic divergence $\ell n(1 + \epsilon_h - 1/x_\gamma)$ as $x_\gamma \rightarrow 1/(1 + \epsilon_h)$ from above [3].

In summary, the NLO partonic part for the quark fragmentation contribution to the isolated cross section, calculated in perturbative QCD, is infrared sensitive when $x_\gamma \leq 1/(1 + \epsilon_h)$. Conventional perturbative factorization for the cross section of isolated photons in e^+e^- annihilation breaks down for x_γ in the neighborhood of $1/(1 + \epsilon_h)$. Breakdown of factorization means that the isolated cross section cannot be calculated reliably in perturbative QCD near $x_\gamma = 1/(1+\epsilon_h)$. This result challenges theorists to find a modified factorization scheme.

References

1. J. C. Collins, D. E. Soper and G. Sterman, in *Perturbative Quantum Chromodynamics*, ed. A. H. Mueller (World Scientific, Singapore, 1989).
2. E. L. Berger, X. Guo and J. Qiu, *Phys. Rev. Lett.* **76**, 2234 (1996).
3. E. L. Berger, X. Guo and J. Qiu, Argonne report, ANL-HEP-PR-95-48, hep-ph/9605324, *Phys. Rev.* D (in press).

THE NEUTRON STRUCTURE FUNCTION AT LARGE x

W.MELNITCHOUK
Department of Physics, University of Maryland, College Park, MD 20742

Reanalysis of the proton and deuteron structure function data, corrected for medium effects in the deuteron, gives a neutron to proton ratio consistent with the perturbative QCD expectation of 3/7 as $x \to 1$, but larger than the value of 1/4 obtained in earlier analyses. We suggest extracting the d/u ratio from W-boson asymmetries in $p\bar{p}$ collisions as an independent test.

The ratio of the neutron to proton structure functions at large x provides clues about the nature of the symmetry breaking among the nucleon's spin and flavor distributions. In nature SU(6) spin-flavor symmetry is of course broken — the d quark distribution is considerably softer than the u, and the F_2^n/F_2^p ratio falls well below the quark model prediction of 2/3. In fact, the "canonical" result quoted in the literature states that $d/u \sim (1-x)$ as $x \to 1$, or equivalently, that $F_2^n/F_2^p \to 1/4$.

This result poses somewhat of a dilemma, as it deviates significantly from the perturbative QCD expectation of $d/u \to 1/5$, or $F_2^n/F_2^p \to 3/7$, as $x \to 1$, which is a consequence of helicity conservation [1]. This apparent contradiction requires therefore a careful re-examination of the proton and deuteron data from which F_2^n was obtained. This is particularly necessary in view of the fact that typical analyses [2] usually only account for effects due to Fermi motion of nucleons in the deuteron, while assuming no nuclear EMC effect at $x \sim 0.6$. Recent calculations [3] based on more detailed treatments of Fermi motion and binding effects suggest, however, a 3–5% EMC effect in the deuteron at $x \sim$ 0.5–0.7. The differences between the F_2^D/F_2^N ratios in the binding plus Fermi motion, and Fermi motion only models, is illustrated in Fig.1(a).

While the theoretical curves in Fig.1(a) require some input for the neutron structure, in practice one needs to extract F_2^n from the data without incorporating any prejudice associated with modeling F_2^n itself. The best method to avoid such ambiguities is the so-called deconvolution method [2]. After subtracting from the deuteron data any non-convolution corrections (in the impulse approximation F_2^D is written as a sum of a convolution term plus relativistic, non-convolution, corrections [3]), the convolution component is unsmeared through an iterative procedure until convergence is achieved and the true F_2^n found. As expected, the resulting F_2^n/F_2^p ratio, shown in Fig.1(b), is somewhat larger for the binding plus Fermi motion model of the deuteron than the Fermi motion alone result. In particular, inclusion of the EMC effect in the deuteron leads to a result more consistent with a limiting value of 3/7 than with the older 1/4 result [4].

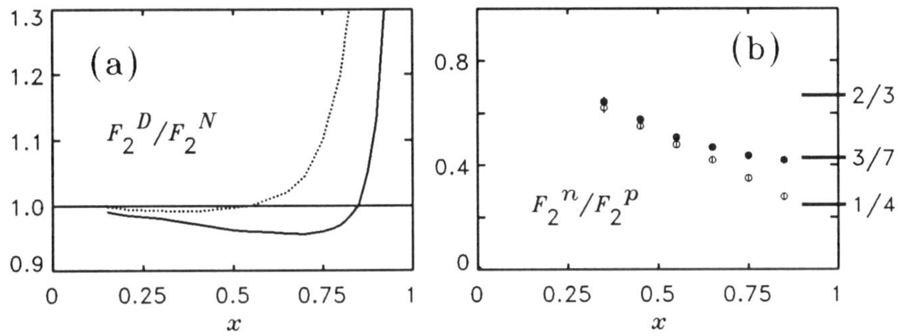

Figure 1: (a) Deuteron to nucleon ratio, with binding and Fermi motion effects (solid), and with Fermi motion only (dotted). (b) Neutron to proton ratio, with binding and Fermi motion corrections (full circles), and Fermi motion only (open circles). Data are from SLAC[2].

Given the sensitivity of F_2^n/F_2^p obtained from proton and deuteron data to theoretical inputs for the deuteron structure, to resolve the issue one would obviously like to appeal to data which are free of ambiguities associated with nuclear effects. An alternative to deep-inelastic scattering experiments is to measure the large-x d/u ratio through W-boson production in $p\overline{p}$ collisions. Because $W^+(W^-)$ bosons are created primarily through $u\overline{d}$ ($d\overline{u}$) annihilation, the ratio of W^+ to W^- cross sections at large W-boson rapidities will be quite sensitive to the details of the u and d quark distributions at large x. Estimates [5] suggest that measurement of the resulting lepton charge asymmetry at large lepton rapidities could enable one to distinguish between the two limiting cases of d/u.

References

1. G.R.Farrar and D.R.Jackson, Phys.Rev.Lett. 35 (1975) 1416.
2. J.J.Aubert et al., Nucl.Phys. B293 (1987) 740; L.W.Whitlow et al., Phys.Lett. B 282 (1992) 475.
3. W.Melnitchouk, A.W.Schreiber and A.W.Thomas, Phys.Lett. B 335 (1994) 11; Phys.Rev. D 49 (1994) 1183; A.Yu.Umnikov and F.C.Khanna, Mod.Phys.Lett. A 10 (1995) 91.
4. W.Melnitchouk and A.W.Thomas, Maryland preprint U.MD. 96-044, nucl-th/9602038, Phys.Lett. B (1996) in print.
5. W.Melnitchouk and J.C.Peng, work in progress.

NUCLEAR MODIFICATION OF THE FLAVOR ASYMMETRY $\bar{u} - \bar{d}$

S. KUMANO *
Department of Physics, Saga University
Honjo-1, Saga 840, Japan

In order to test the NMC finding of flavor asymmetry $\bar{u} - \bar{d}$ in the nucleon, existing Drell-Yan data for the tungsten target are often used. However, we have to be careful in comparing nuclear data with the nucleon ones. We investigate whether there exists significant nuclear modification of the $\bar{u} - \bar{d}$ distribution in a parton-recombination model. It should be noted that a finite $\bar{u} - \bar{d}$ distribution is theoretically possible in nuclei even if the sea is symmetric in the nucleon. In neutron-excess nuclei such as the tungsten, there exist more d-valence quarks than u-valence quarks, so that more $\bar{d}$-quarks are lost than $\bar{u}$-quarks due to parton recombinations. Our results suggest that the nuclear modification in the tungsten is a 2–10 % effect on the NMC $\bar{u} - \bar{d}$ distribution. The nuclear modification of the flavor asymmetry should be an interesting topic in connection with ongoing Drell-Yan experiments.

The New Muon Collaboration (NMC) suggested that the Gottfried sum rule should be violated in 1991. It indicates $\bar{d}$ excess over $\bar{u}$ in the nucleon. Since then, there have been efforts to investigate mechanisms of creating a flavor asymmetric distribution $\bar{u} - \bar{d}$ in the nucleon.[1] In order to test the NMC finding, Drell-Yan experiments are in progress at Fermilab. On the other hand, there exist Drell-Yan data for various nuclear targets, so that some people use, for example, tungsten data in investigating the flavor asymmetry. However, we have to be careful in comparing the NMC result with the tungsten data because of possible nuclear medium effects. In order to find whether such a comparison makes sense, we estimate a nuclear modification effect.[2] It in turn could be found experimentally by analyzing accurate Drell-Yan data in the near future.

We investigate the $\bar{u} - \bar{d}$ distribution in the tungsten nucleus. If isospin symmetry could be applied to parton distributions in the proton and the neutron, the distribution per nucleon becomes $x[\bar{u}(x) - \bar{d}(x)]_A = -\varepsilon x[\bar{u}(x) - \bar{d}(x)]_{proton}$ without considering nuclear modification. It is just the summation of proton and neutron contributions. The neutron-excess parameter ε is defined by $\varepsilon = (N - Z)/(N + Z)$, and it is 0.196 for the tungsten ${}^{184}_{74}W_{110}$. According to the above equation, the flavor distribution should be symmetric ($[\bar{u} - \bar{d}]_W = 0$) if it is symmetric in the nucleon. However, it is not the case in a parton-recombination model.

It is well known that shadowing phenomena occur in nuclear structure functions at small x. One of the ideas for explaining the shadowing is the recombination model.[3] In an infinite momentum frame, the longitudinal localization size of a parton with momentum xp_N exceeds the average longitudinal nucleon separation in a Lorentz contracted nucleus $[(2\text{ fm})M_A/P_A$=$(2\text{ fm})m_N/p_N]$ in the small x region ($x < 0.1$). Therefore, partons from different nucleons could interact, and the interaction is called parton recombination. We apply studies of this model in the structure function F_2 to the asymmetry $\bar{u} - \bar{d}$ in nuclei.

The flavor asymmetry could be created in the recombination model in the following way. In a neutron-excess nucleus ($\varepsilon > 0$) such as the tungsten, more $\bar{d}$ quarks are lost than

$\bar{u}$ quarks in the parton recombination process $\bar{q}q \to G$ because of the d quark excess over u in the nucleus. The $\bar{q}q \to G$ type recombination processes produce positive contributions at small x. In the $(\bar{u}-\bar{d})_N \neq 0$ case, the $\bar{q}(x)G \to \bar{q}$ process is the dominant one kinematically at small x. Its contribution to $\bar{u}(x) - \bar{d}(x)$ becomes negative due to the neutron excess. In the medium x region, the $\bar{q}G \to \bar{q}(x)$ process becomes kinematically favorable. Because it produces $\bar{q}$ with momentum fraction x, its contribution becomes opposite to the one at small x. These results are shown in Fig. 1.

We evaluate the recombination effects on the $\bar{u} - \bar{d}$ distribution at Q^2=4 GeV^2 in the tungsten nucleus $^{184}_{74}W_{110}$ (ε = 0.196).[2] Input parton distributions are those of the MRS-D0 (1993). In the $(\bar{u} - \bar{d})_N = 0$ case, $\Delta = 0$ is taken in the MRS-D0 distributions. Obtained results are shown in Fig. 1, where the solid curve shows recombination contributions to $\bar{u} - \bar{d}$ (per nucleon) in the tungsten nucleus, and the dashed curve shows the result in the $(\bar{u}-\bar{d})_N \neq 0$ case.

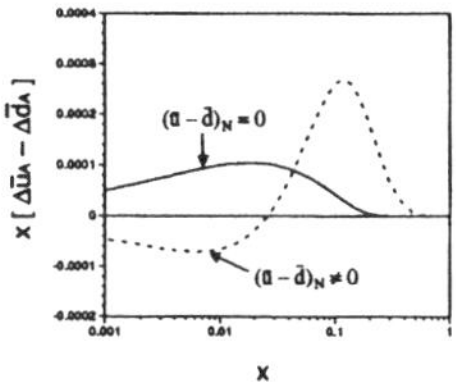

Figure 1: Recombination contributions to $\bar{u} - \bar{d}$ in the tungsten nucleus.

We briefly comment on Q^2 dependence of our calculation. Because the recombinations are higher-twist effects, there is a Q^2 dependent factor $\alpha_s(Q^2)/Q^2$ in the contributions. It may seem to be very large at small Q^2; however, parton distributions $p(x, Q^2)$ are also modified. As a consequence, the overall Q^2 dependence is not so significant. According to our estimate, there are merely factor-of-two differences between the asymmetric distribution at Q^2=4 GeV^2 and the one at $Q^2 \approx 1$ GeV^2. Considering this factor of two, we find that the nuclear modification is of the order of 2%–10% compared with the asymmetry suggested by the NMC.

Although our estimate is based on a special model, the results indicate roughly several % nuclear modification in the $\bar{u} - \bar{d}$ distribution of the tungsten nucleus. Therefore, the existing tungsten Drell-Yan data should not be compared with the NMC flavor asymmetry within several % magnitude. On the other hand, the modification itself is an interesting topic for future theoretical and experimental studies.

Acknowledgment

This research was partly supported by the Grant-in-Aid for Scientific Research from the Japanese Ministry of Education, Science, and Culture under the contract number 06640406.

* Email: kumanos@cc.saga-u.ac.jp. Information on his research is available at http://www.cc.saga-u.ac.jp/saga-u/riko/physics/quantum1/kumano.html.

References

1. S. Kumano, SAGA-HE-97-96, to be submitted for publication.
2. S. Kumano, Phys. Lett. B342 (1995) 339.
3. S. Kumano, Phys. Rev. C48 (1993) 2016; M. Miyama and S. Kumano, Comput. Phys. Commun. 94 (1996) 185.

INELASTIC J/Ψ PHOTOPRODUCTION OFF NUCLEI: GLUON ENHANCEMENT OR DOUBLE COLOR-EXCHANGE SCATTERING?

B.Z. KOPELIOVICH
Max-Planck-Institut für Kernphysik, Postfach 103980, 69029 Heidelberg, Germany

J. HÜFNER
Institut für Theor. Physik der Universität Philosophenweg 19, 69120 Heidelberg, Germany

AL.B. ZAMOLODCHIKOV
Laboratoire de Physique Mathématique, Université Montpellier II, Pl.E.Bataillon, 34095 Montpellier, France

The nuclear enhancement observed in inelastic photoproduction of J/Ψ should not be interpreted as evidence for an increased gluon density in nuclei. The nuclear suppression rate due to initial and final state interactions is calculated and a novel two-step color exchange process is proposed, which is able to explain the data.

Measuring inelastic photoproduction $\gamma A \to J/\Psi X$, the NMC collaboration [1] found a nuclear enhancement of the production rate per nucleon. The experimental ratio Sn/C of the cross sections per nucleon [1] shown in Fig. 1 exceeds the value one by a few percent. This enhancement was fully attributed in [1] to the gluon density, which was expected in some models to be enhanced in nuclei by a few percent. This conclusion, however, neglects initial and final state interactions, which lead to a nuclear suppression by $20 - 30\%$. To explain the unusually strong enhancement of inelastic photoproduction of J/Ψ we suggest a new mechanism.

The inelastic photoproduction cross section of the J/Ψ on a nucleus can be represented in the form

$$\left.\frac{d\sigma_{\Psi}^{\gamma A}}{dx_F}\right|_{GL} = S_{\Psi}^{\gamma A}\ A\ \frac{d\sigma_{\Psi}^{\gamma N}}{dx_F}\ . \tag{1}$$

The longitudinal momentum transfer in inelastic J/Ψ photoproduction is much larger than the Fermi-momentum, therefore contributions of different nucleons add up incoherently. However, coherent elastic rescatterings of J/Ψ are possible, so the longitudinal momentum transfer in elastic photoproduction, $q_{el} = (Q^2 + M_{\Psi}^2)/2\nu$ is very important. It governs the energy variation of the

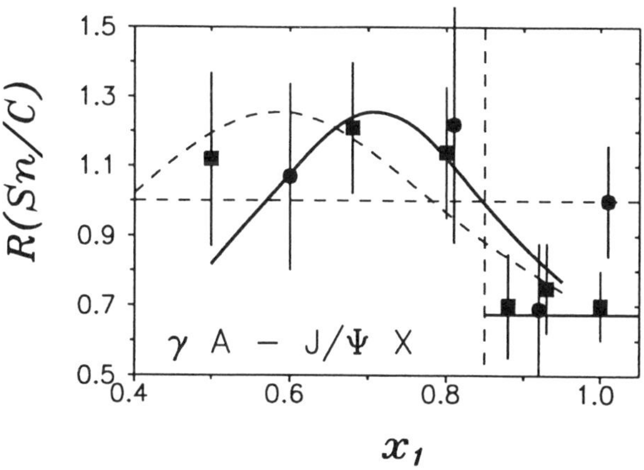

Figure 1:

nuclear suppression factor, which ranges dependent on energy within

$$S_{\Psi}^{\gamma A}|_{q_{el}\gg 1/R_A} = \frac{1}{\sigma_{in}^{\Psi N}}\int d^2b\,\left\{\frac{\sigma_{tot}^{\Psi N}}{\sigma_{in}^{\Psi N}}\left[1-e^{-\sigma_{in}^{\Psi N}T(b)}\right]\right.$$
$$\left. - \;\sigma_{el}^{\Psi N}\;T(b)\;e^{-\sigma_{in}^{\Psi N}T(b)}\right\} \approx 1-\frac{1}{2}\sigma_{tot}^{\Psi N}\;\langle T\rangle \qquad (2)$$

$$S_{\Psi}^{\gamma A}|_{q_{el}\ll 1/R_A} = \int d^2b\;T(b)e^{-\sigma_{in}^{\Psi N}T(b)} \approx 1-\sigma_{tot}^{\Psi N}\;\langle T\rangle\;, \qquad (3)$$

where the mean nuclear thickness $\langle T\rangle = A^{-1}\int d^2b\;T^2(b)$. We skip the full expression for $S_{\Psi}^{\gamma A}$, which is too lengthy and is derived for the first time in [2].

Results (2)-(3) are easily interpreted: at high energy the lifetime of the $c\bar{c}$ fluctuation of the photon, or the coherence time $t_c = 1/q_{el}$, is long and the mean path of the $c\bar{c}$ pair in the nucleus is twice as long as at low energy.

For the cross section of inelastic J/Ψ photoproduction on a nucleon in eq. (1) we use the triple-Pomeron parameterization

$$\frac{d\sigma(\gamma N\to J/\Psi X)}{dx_F} = \frac{B_{el}^{\Psi p}\,\sigma(\gamma^* p\to J/\Psi p)}{B_{el}^{pp}\;B_{in}^{\Psi p}\,\sigma_{el}^{pp}}\;\frac{G(pp\to pX)}{1-x_1} \approx \frac{8.9\;nb}{1-x_1} \qquad (4)$$

All the parameters here are known known, provided that the Pomeron factorization exists. Eq. (4) nicely describes the data from the NMC experiment [1] not only in the triple-Reggeon region, but down to mid x_F values.

The J/Ψ photoproduction on a nucleus can also result from the photoproduction on a bound nucleon of a $c\bar{c}$ pair in a color-octet state, which propagates then through the nuclear matter loosing energy for hadronization unless another color-exchange interaction with converts the pair back to a colorless state. This is a new mechanism of diffractive excitation of the nucleus. We assume a constant energy loss per unit of length $dE/dz = -\kappa$, leading to $x_1 = 1 - \kappa\Delta z/\nu$, where Δz is the longitudinal distance covered by the $c\bar{c}$ pair in color-octet state.

The contribution of this mechanism to the nuclear suppression factor can be estimated as follows

$$\Delta\left(\frac{d\sigma_{\Psi}^{\gamma A}}{dx_F}\right) = \frac{2\pi\ \nu}{\kappa} B_{\Psi N}\sigma(\gamma N \to J/\Psi N)\int d^2b \int\limits_{-\infty}^{\infty} dz_1\ \rho_A(b,z_1)\int\limits_{z_1}^{\infty} dz_2$$

$$\rho_A(b,z_2)\delta\left[z_2 - z_1 - (1-x_1)\frac{\nu}{\kappa}\right]\ \exp\left\{-\sigma_{in}^{\Psi N}\left[T(b) - T_{z_1}(b) + T_{z_2}(b)\right]\right\} \quad (5)$$

We assume that the first interaction at point (b, z_1), which converts the $c\bar{c}$ pair into the color octet state, corresponding to the cut Pomeron, has cross section $\sigma_{in}^{\Psi N}$. Important is that the $c\bar{c}$ pair accepts no longitudinal momentum transfer at this point, but the loss of longitudinal momentum, $(1 - x_1)\nu$, is completely due to hadronization. The second interaction at point (z_2, b) has the same cross section, except an extra factor 1/8, since one does not sum over final colors in this case.

The only unknown parameter in (5) is the density of energy loss κ. In string model it equals to the color-octet string tension, $dE/dz = -\kappa_8 \approx (2\pi\alpha'_P)^{-1} \approx 4-5\ GeV^{-2}$. Perturbative QCD provides a close value $\kappa_8 = 4.5\ GeV/fm$, because the mean color-octet charge is 9/4 times bigger than the color-triplet one, besides, and the radiation spectrum is doubled due to the second color-exchange interaction.

Combining (5) with standard Glauber contribution (1) we calculated the ratio of the cross sections Sn/C shown in Fig. 1 at $\nu = 100\ GeV$ (solid curve) and at $\nu = 70\ GeV$ (dashed curve), in nice agreement with data [1]. The data at $x_F > 0.9$ are dominated by quasielastic photoproduction, which was also calculated and is shown by the horizontal line.

We conclude that the results on inelastic J/Ψ photoproduction on nuclei cannot be explained by enhancement of the gluon density in nuclei. We suggest a new mechanism of double color-exchange diffractive interaction with a nucleus, which nicely explains the data in a parameter-free way.

References

1. P. Amaudruz et al., *Nucl. Phys.* **B371** (1992) 553
2. J. Hüfner, B.Z. Kopeliovich, Al.B. Zamolodchikov, paper in preparation

THE PARTONIC STRUCTURE OF THE GENERALIZED BFKL POMERON: AN OVERVIEW

M. Genovese[a], N.N. Nikolaev[c,d] and B.G. Zakharov[d]

[a] *Institut de Physique Nucléaire de Lyon, Université Claude Bernard 43 boulevard du 11 novembre 1918, F-69622 Villeurbanne Cedex, France*
[c] *IKP(Theorie), KFA Jülich, 5170 Jülich, Germany*
[d] *L. D. Landau Institute for Theoretical Physics, GSP-1, 117940, ul. Kosygina 2, Moscow 117334, Russia*

Diffractive Deep Inelastic Scattering can be understood (if one assumes single pomeron exchange) as DIS on Pomerons radiated by protons. Many authors assumed[1)] that the Pomeron can be treated as a particle with a well-defined partonic structure and flux $f(x_{\mathbf{P}})/x_{\mathbf{P}}$ in the proton. Under this strong assumption, one is lead to the operational definition of the Pomeron structure function $F_2^{(\mathbf{P})}(\beta, Q^2)$:

$$(M^2+Q^2)\frac{d\sigma_D(\gamma^*\to X)}{dt\,dM^2}\bigg|_{t=0} = \frac{\sigma_{tot}^{pp}}{16\pi}\frac{4\pi^2\alpha}{Q^2} f_{\mathbf{P}}(x_{\mathbf{P}})F_2^{(\mathbf{P})}(\beta,Q^2) \qquad (1)$$

(we use the Regge theory conventions). The factorization property (1) should be proved starting with a QCD treatment of diffractive DIS and it only makes sense if one can prove that the flux function $f_{\mathbf{P}}(x_{\mathbf{P}})$ can be defined so that the Q^2 dependence of the r.h.s. of (1) is concentrated in $F_2^{\mathbf{P}}(\beta, Q^2)$, which satisfies the conventional QCD evolution. We report here on the works[2)], based on the colour dipole approach to the BFKL Pomeron[3)], where we have found that the factorization (1) is broken.

The QCD process which corresponds to DIS on the valence $q\bar{q}$ pair is the diffraction excitation of the $q\bar{q}$ Fock state of the photon, which has the cross section

$$\frac{d\sigma_D}{dt}\bigg|_{t=0} = \frac{1}{16\pi}\int_0^1 dz \int d^2\vec{r}\,|\Psi_{\gamma^*}(Q^2,z,r)|^2\sigma(x,r)^2 = \frac{4\pi^2\alpha}{Q^2}\int\frac{dr^2}{r^2}\,W(Q^2,r)\left[\frac{\sigma(x,r)}{r^2}\right]^2, \qquad (2)$$

where Ψ_{γ^*} is the wave function of the $q\bar{q}$ pair of transverse size $\vec{r}$, z is the fraction of the photon light-cone momentum carried by the quark, and $\sigma(x,r)$ is the colour

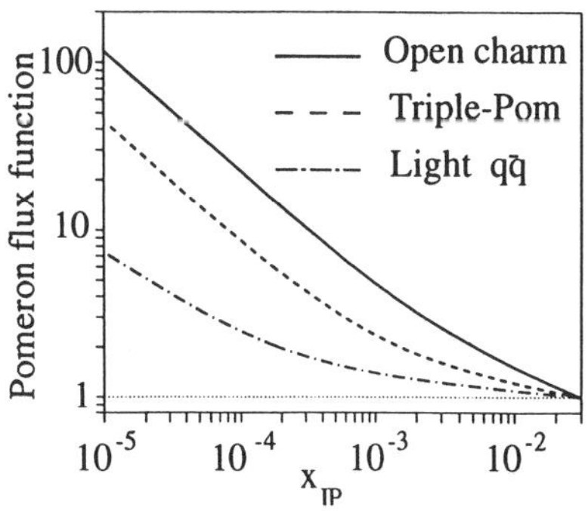

dipole cross section. The properties which allow reinterpretation of the cross section (2) as DIS on the valence $q\bar{q}$ state of the Pomeron are:
(i) The mass spectrum, which roughly follows[3)]

$$\left.\frac{d\sigma_D}{dM^2dt}\right|_{t=0} \propto \frac{M^2}{(Q^2+M^2)^3} = \frac{1}{Q^2}\cdot\frac{1}{Q^2+M^2}\cdot\beta(1-\beta)\,, \tag{3}$$

is x-independent to a good approximation.
(ii) At sufficiently high Q^2 the diffractive cross section satisfies the Bjorken scaling[3)].
(iii) $W(Q^2,r)$ is peaked at a large, Q^2-independent, hadronic size: this recalls the Q^2-independent large distance between the valence quark and antiquark in the pion, and we can analogously speak of DIS off the valence $q\bar{q}$ state of the Pomeron.

Using the colour dipole cross section[4)], which gives a good description of the HERA data on the proton structure, we find from (2) and (3) the form and absolute normalization (at $Q^2 = 10\mathrm{GeV}^2$) of the valence $q\bar{q}$ structure function of the Pomeron $F_{val}^{(\mathbf{P})}(\beta) = 0.27\beta(1-\beta)$. The flux function $\phi_{\mathbf{P}}(x_{\mathbf{P}})$, defined subject to the normalization $\phi_{\mathbf{P}}(x_{\mathbf{P}}^0 = 0.03) = 1$, equals

$$\phi_{\mathbf{P}}(x_{\mathbf{P}}) = \frac{\int_0^1 dz \int d^2\vec{r}\,|\Psi_{\gamma^*}(Q_{\mathbf{P}}^2,z,r)|^2\sigma(x_{\mathbf{P}},r)^2}{\int_0^1 dz \int d^2\vec{r}\,|\Psi_{\gamma^*}(Q_{\mathbf{P}}^2,z,r)|^2\sigma(x_{\mathbf{P}}^0,r)^2}\,. \tag{4}$$

The mass spectrum (3) for excitation of the $q\bar{q}$ state rapidly decreases at large $M^2 \gg Q^2$. The Born process for the $\propto 1/(M^2+Q^2)$ mass spectrum, typical of the so-called triple-Pomeron regime, is the diffraction excitation of the $q\bar{q}g$ Fock state of the photon. One obtains[3)] ($\vec{\rho}$ is the quark–gluon separation):

$$(Q^2+M^2)\left.\frac{d\sigma_D}{dtdM^2}\right|_{t=0} \simeq \int dz\,d^2\vec{r}\,|\Psi_{\gamma^*}(Q^2,z,r)|^2\cdot\frac{27}{32}\frac{\alpha_S(r)r^2}{2\pi^2}\int d\rho^2\left[\frac{\sigma(x_{\mathbf{P}},\rho)}{\rho^2}\right]^2\mathcal{F}(\mu_G\rho) \tag{5}$$

with an approximate factorization of Q^2- and $x_{\mathbf{P}}-$ dependence ($\mathcal{F}(z) = z^2[K_1(z)^2 + zK_1(z)K_0(z) + \frac{z^2K_0(z)^2}{2}]$). This factorization property allows us to define the corresponding structure function and the Pomeron flux function. Furthermore, one can introduce the two-gluon wave function of the Pomeron[3)]

$$|\Psi_{\mathbf{P}}(\beta,\vec{r})|^2 = \frac{1}{f_{\mathbf{P}}(x_{\mathbf{P}})}\cdot\frac{81}{8\pi^4}\cdot\frac{1-\beta}{\beta}\cdot\frac{1}{\sigma_{tot}^{pp}}\left[\frac{\sigma(x_{\mathbf{P}},r)}{r^2}\right]^2\mathcal{F}(\mu_G r)\,, \tag{6}$$

where $\vec{r}$ is the transverse separation of gluons in the Pomeron and β is a fraction of the Pomeron's momentum carried by a gluon. The $1/\beta$ dependence is the usual soft-gluon behaviour, the factor $(1-\beta)$ is an educated guess. In the wave function (6), the $x_{\mathbf{P}}$-dependence approximately cancels out, and it gives the $x_{\mathbf{P}}$-independent gluon structure function of the Pomeron $G_{\mathbf{P}}(\beta) = \beta\int d^2\vec{r}\,|\Psi_{\mathbf{P}}(\beta,\vec{r})|^2 = 0.28(1-\beta)$.

Then, Eq. (5) gives the input sea structure function of the Pomeron for $\beta \ll 1$:

$$F_{sea}^{\mathbf{P}}(Q_{\mathbf{P}}^2) = \frac{81}{32\sigma_{tot}^{pp}}\frac{1}{f_{\mathbf{P}}(x_{\mathbf{P}})}\frac{Q_{\mathbf{P}}^2}{\pi\alpha}\int d\rho^2\left[\frac{\sigma(x_{\mathbf{P}},\rho)}{\rho^2}\right]^2\mathcal{F}(\mu_G\rho)\int_0^1 dz d^2\vec{r}\,|\Psi_{\gamma^*}(Q_{\mathbf{P}}^2,r,z)|^2 r^2\alpha_S(r). \tag{7}$$

As a starting approximation, we take $F_{sea}^{\mathbf{P}}(\beta,Q_{\mathbf{P}}^2) = C_{sea}(1-\beta)^2 = 0.063(1-\beta)^2$, with the normalization following from Eq. (7).

From the previous discussion, it follows that the two fluxes $\phi_{\mathbf{P}}(x_{\mathbf{P}})$ and $f_{\mathbf{P}}(x_{\mathbf{P}})$

are not identical (see the figure). The different $x_{\mathbb{P}}$-dependence of these two fluxes is a dynamical prediction of this analysis, to be contrasted to conjectured forms of a universal flux of Pomerons[1]. Only at very large $\frac{1}{x_{\mathbb{P}}}$, well beyond the kinematical range of HERA, do the two fluxes have a similar $\frac{1}{x_{\mathbb{P}}}$-dependence.

Furthermore other factorization breaking effects are expected due to the heavy quarks[2a] and longitudinal[2b] valence contributions. In fact in both the cases the Ψ_{γ^*} of eq. (4) differs from the light quarks one for transverse polarized γ^*.

An explicit calculation shows that the relevant factorization scale is $q^2 = (k^2 + m_f^2)\frac{M^2+Q^2}{M^2}$ where k is the transverse momentum of quark with respect to the γ^*-pomeron collision axis and m_f the quark mass.

The transverse cross section is dominated by small k^2 and one finds that for the charm component the flux is $\phi_{\mathbb{P}}^{c\bar{c}} = \left[\frac{G(x_{\mathbb{P}},q^2)}{G(x^0_{\mathbb{P}},q^2)}\right]^2$, which acquires a β dependence through the factorization scale $q^2 = \frac{m_c^2}{(1-\beta)}$. The numerical evaluation shows that it is steeper than the light quarks one (see the figure).

For what concerns the longitudinal cross section, the calculation shows that it is dominated by large k^2, leading to the factorization scale $q_L^2 = \frac{Q^2}{4\beta}$. This requires to introduce a Q^2 dependent flux $\phi_{\mathbb{P}}^L = \left[\frac{G(x_{\mathbb{P}},q_L^2)}{G(x^0_{\mathbb{P}},q_L^2)}\right]^2$. Incidentally, this last flux, in contrast to the transverse photons case, is flavour-independent.

The longitudinal Pomeron structure function has the form $F^L_{2,\mathbb{P}} = \Sigma_f e_f^2 \frac{A_f}{Q^2}(1-2\beta)^2 \cdot \beta^3$ ($A_{ud} = 0.82\text{GeV}^2$, $A_s = 0.61\text{GeV}^2$ and $A_c = 0.05\text{GeV}^2$) and is dominant for large β, even at $Q^2 \approx 100GeV^2$.

Quite a different situation occurs for the sea flux defined by eq. (7), which is universal for all the flavours and γ^*-polarizations.

Finally, further factorization breaking effects are expected when one considers charge currents diffractive DIS[6].

Albeit these factorization breaking effects are still too elusive to be detected in the available experimental data, they should be observed when more precise data will appear.

References

[1] G.Ingelman and P.Schlein, *Phys. Lett.* **B152** (1985) 256; H.Fritzsch and K.H.Streng, *Phys.Lett.* **B164** (1985) 391; E.L.Berger *et al. Nucl.Phys.* **B286** (1987) 704; A.Donnachie and P.V.Landshoff, *Phys.Lett.* **B191** (1987) 309, *Nucl.Phys.* **B303** (1988) 634.

[2] a) M.Genovese, N.N.Nikolaev and B.G.Zakharov, J. Exp. Theor. Phys. **81** (1995) 625, b) DFTT 77/95 to be published in Phys. Lett. B, c) Lycen/9602 to be published in Phys. Lett. B

[3] N.N.Nikolaev and B.G.Zakharov, *Z. Phys.* **C49** (1991) 607, **C53** (1992) 331 and **C64** (1994) 631.

[4] ZEUS Collaboration: M.Derrick et al., *Phys. Lett.* **B315** (1993) 481; DESY 95-093.

[5] H1 Collaboration: T.Ahmed al., *Nucl. Phys.* **B429** (1994) 477; DESY 95-36.

[6] M. Bertini, M. Genovese, N.N. Nikolaev and B.G. Zakharov, work in progress.

Gluon Contribution to $g_1(x)$ and Quark Masses

F. M. Steffens and A. W. Thomas

Department of Physics and Mathematical Physics,
and Institute for Theoretical Physics,
University of Adelaide, Adelaide, SA 5005, Australia

We study the effect of the quark masses when calculating the gluonic contribution to $g_1(x)$ and find that charm has to be included in the present analysis of the data.

Since its release, the EMC data on the proton spin structure function has caused great excitement in the literature because the measured singlet axial charge was much smaller than expected. Two main mechanisms were suggested to explain the experimental value: a sizeable, negative, polarization of the proton sea quarks and/or a sizeable, *positive*, gluon polarization. Irrespective of which of the hypotheses is the correct one, the important conclusion from the theoretical activity which followed the EMC experiment is that, in fact, the quark spin content of the proton was never measured[1]. This can be easily seen from the fact that the measured, renormalized, gauge invariant, quark axial current has a divergence, even when the quark masses are set to zero because of the axial anomaly. This implies that the classical relation between chirality and helicity, as present in the Dirac equation, is lost. Hence, arguments based on conservation of chirality cannot be used to extract information on the quark helicities.

In any case, it was soon realized[2] that the measured axial charge could be related to the quark helicities if the anomalous gluon contribution was added to it. In general, it has been stated in the literature that the gluonic contribution is $\alpha_s \Delta g/2\pi$ for the u, d and s quarks and zero for all the other heavier quarks. Here we investigate the rôle that the quark masses play in calculating the gluonic contribution and conclude that charm should be present along with the three light flavours.

Basically, the gluon contribution to g_1 takes the form $C_g(x, Q^2/\mu^2) \otimes \Delta g(x, \mu^2)$, where C_g is the Wilson coefficient (or the hard part of the photon - gluon fusion cross section). A calculation of C_g, using a cut-off scheme, yields[3]:

$$C_g(x,Q^2) = \frac{\alpha_s}{2\pi}N_f\left\{(2x-1)\left[ln\left(\frac{Q^2}{\mu^2+m_q^2}\right)+ln\left(\frac{1-x}{x}\right)-1\right]\right.$$
$$\left.+ \quad (1-x)\frac{2m_q^2}{\mu^2+m_q^2}\right\}, \tag{1}$$

in the limit of very large momentum transfer, Q^2, and where μ^2 is the factorization scale. If we take for the charm mass $m_c = 1.5\ GeV$, we find that for $\mu^2 = 10\ GeV^2$ the gluonic contribution in the charm axial charge is around 80 % of the contribution of the light quarks. It is thus unjustified to disregard charm when extracting the quark helicities in the proton. Of course, the present experiments are performed at finite momentum transfer, so that these conclusions should be modified because complications from higher twists should be included. An extended, but by no means complete, treatment of the problem with finite momentum transfer has been given by Steffens and Thomas [3]. In that case it is found that for the momentum transfer corresponding to the present experiments it is important to include a fourth flavour. In particular, if we use chiral symmetry to extract Δg from the data, we obtain $\Delta g(Q^2 = 10\ GeV^2) = 3.04 \pm 1.4$ when the usual analysis with 3 active flavours is used and $\Delta g(Q^2 = 10\ GeV^2) = 2.32 \pm 1.06$ when charm is included – as it should be.

References

1. G. Veneziano, Mod. Phys. Lett. A4 (1989) 1615; G. M. Shore and G. Veneziano, Phys. Lett. 244B (1990) 75; Nucl. Phys. B381 (1992) 23; S. D. Bass and A. W. Thomas, J. Phys. G 19 (1993) 925.
2. G. Altarelli and G. G. Ross, Phys. Lett. B212 (1988) 391; R. D. Carlitz, J. C. Collins and A. H. Mueller, Phys. Lett. B214 (1988) 299; A. V. Efremov, J. Soffer and O. V. Teryaev, Nucl. Phys. B246 (1990) 97.
3. F. M. Steffens and A. W. Thomas, Phys. Rev. D 53 (1996) 1191.

The effective quark dynamics with instantons

B.O. Kerbikov, S.V. Bashinsky[1], D.S. Kuzmenko and Yu.A. Simonov

Institute of Theoretical and Experimental Physics
Moscow 117259, Russia

Abstract

The partition function is derived for quarks in the background of instantons for any number of flavours N_f. In addition to the known 't Hooft-type term containing $2N_f$ quark fields, there appear other terms, in particular the four-fermion interaction of the NJL type. For $N_f = 2,3$ gap equations are obtained from which the dynamical quark masses are estimated.

The effective quark Lagrangian is known to appear when the topological charges are present in the vacuum. This topic in particular was thoroughly studied in the framework of the instanton gas/liquid model [1–8]. However, it is the ansatz for the partition function (see Refs. [4, 5]) which lies at the core of most of the results obtained so far. Our purpose is to provide a rigorous derivation of the partition function based on a sole assumption that instantons (I's) and antiinstantons ($\bar{I}$'s) form a dilute gas of randomly orientated pseudoparticles. In case of $N_f = 3$ our partition function contains $4q$ NJL-like vertices in addition to the $6q$ ('t Hooft-type) ones.

Consider quarks in the instanton ensemble $A_\mu(x) = \sum_{i=1}^{N} A_\mu^{(i)}(x)$ with the fixed number of I's (N_+) and $\bar{I}$'s (N_-) in the Euclidean 4-volume V. The fermionic part of the partition function reads

$$Z = \int \mathcal{D}\psi^\dagger \mathcal{D}\psi \left\langle exp[- < \psi | S^{-1} | \psi >] \right\rangle_{\Omega,R}, \tag{1}$$

where

$$S^{-1} = (-i\partial\!\!\!/ - gA\!\!\!/ - im) \tag{2}$$

and $\langle \mathcal{O}(\Omega, R)\rangle_{\Omega,R}$ denotes the averaging of the operator $\mathcal{O}(\Omega, R)$ over the positions and color orientations of all pseudoparticles. We note in passing that confining configurations can be superimposed on instantons [7] leaving valid the arguments to follow. Next step is to write the inverse propagator as

$$S^{-1} = S_0^{-1} + \sum_{i,m,m'} S_0^{-1} | u_m^i > \varepsilon_{mm'} < u_{m'}^i | S_0^{-1}, \tag{3}$$

with S_0 being the free propagator, $|u_m^i >$ – the quark m-th eigenmode on the i-th instanton. The matrix elements $\varepsilon_{mm'}$ have to be determined from the gap equation which amounts to the Shwinger–Dyson equation in the mean-field approximation (see, e.g. [9]).

---

[1] Present address: MIT, Cambridge, MA, USA

Inserting (3) into (1) and performing the averaging using the cumulant (cluster) expansion one gets (only zero modes are kept)

$$Z = \int D\psi^+ D\psi exp\{- < \psi | S_0^{-1} | \psi > + \sum_{n=1}^{N_f} (Y_n^{(+)} + Y_n^{(-)})\} \qquad (4)$$

with the vertices

$$Y_n^{\pm} = \frac{(-1)^n}{n!} \left(\frac{V}{N_{\pm}}\right)^{n-1} \sum_{(f_1..f_n)\in(1..N_f)} (2\pi)^4 \delta(\sum_{i=1}^{n}(p_i - p_i')) \times$$

$$\times \int \prod_{i=1}^{n} \frac{d^4p_i}{(2\pi)^4} \frac{d^4p_i'}{(2\pi)^4} det_{i,j}^{(n)} [\psi_{f_i}^\dagger (p_i) M^{\pm}(p_i, p_j') \psi_{f_j}(p_j')] \qquad (5)$$

and the dynamical quark mass

$$M^{(\pm)}(p_k, p_l) = \frac{N_{\pm}}{VN_c} S_0^{-1} u_0^{(I/\bar{I})}(p_k) \epsilon_{00} u_0^{(I/\bar{I})\dagger}(p_l) S_0^{-1}. \qquad (6)$$

Expressions (4–6) are not complete without gap equations which we derived diagrammatically [9] and which yield the dynamical quark mass decreasing with the growth of N_f. As a final step we performed bosonization [10] (alternative recipe was proposed earlier in [6]) and revealed the chiral symmetry breaking.

The $4q$–vertices in the case $N_f = 3$ describe the OZI violating process $u + \bar{u} \to s + \bar{s}$ and the strange quarks admixture in hadrons.

The financial support under INTAS Grant CT93-0023 is gratefully aknowledged.

References

[1] C.G. Callan, R. Dashen, D.J. Gross, Phys. Rev. **D17**, 2717, 1978.

[2] E.V. Shuryak, Nucl. Phys. **B 203** 93, 116, 140, 1982.

[3] D.I. Diakonov and V.Yu. Petrov, Nucl. Phys. **B272** , 457, 1986.

[4] D.I. Diakonov and V.Yu. Petrov, LNPI Report 1153, 1986.

[5] M.A. Nowak, J.J.M. Verbaarschot and I. Zahed, Nucl.Phys. **B 324**,1, 1989.

[6] D.I. Diakonov and V.Yu. Petrov, In: Quark Cluster Dynamics, eds. K. Goeke, P.Kroll, H.-R.Petry (Springer Lecture Notes in Physics, 1992), p.p. 288-297.

[7] Yu.A. Simonov, Yad. Fiz. bf 57 1491, 1994.

[8] D.I. Diakonov, M.V. Polyakov and C. Weiss, preprints RUB TPH 27/95, PNPI-TH-2076

[9] S.P. Klevansky, Rev. Mod. Phys. **64**, 649, 1992.

[10] B.O. Kerbikov and Yu.A. Simonov, Preprint ITEP 16-95, HEP-TH 9503146.

FORM FACTORS OF PIONS AND QUARKS IN THE INSTANTON VACUUM OF QCD

Andree Blotz and Edward Shuryak
State University of New York at Stony Brook, Stony Brook, NY 11790, USA

Within an ensemble of instantons and anti-instantons in a liquid phase of the QCD vacuum we show that the interactions of pions and quarks with external currents can be successfully described in terms of Euclidean correlation functions. As results the electromagnetic formfactor of the pion with a monopole parametrization is reproduced and also the axial charge of the quarks is significantly reduced by the interacting instantons.

1 The instanton vacuum

In recent years a model of the QCD vacuum was developed which is based on the topological non trivial solutions of the classical Yang-Mills equations. These instantons and anti-instantons could be shown not only to explain the anomalous breaking of the axial $U(1)$ symmetry, which is related to the appearance of the η' meson as a non-Goldstone particle, but also explains qualitatively and quantitatively the spontaneous breaking of chiral symmetry [1]. Especially it describes a variety of mesonic and baryonic correlation functions[2]. The basic versions of the model, the Random Instanton Liquid Model (RILM) and the Interacting Instanton Liquid Model (IILM) [3] refer to the quenched and unquenched calculations of lattice QCD.

2 The electromagnetic formfactor of the pion

For the investigation of the pion electromagnetic formfactor we calculated a three point correlation function of two pion currents and an external electromagnetic field in coordinate space and for the RILM and the IILM. This correlator can be compared to experimental observables by making a simple model Ansatz for the spectral density, entering the phenomenological part of our correlator. The Ansatz corresponds to a pion pole clearly separated from a continuum contribution above some threshold value. For the pion two point function this leads to a nice description of the correlator basically reproducing the experimental pion mass. Using a similiar Ansatz for the three point function, we could show now, that not only the pion parameters of the correlation function are consistent with the two point function, but that the formfactor of the pion is well described (Fig. 1) by a monopole parametrization with a mass of $m_V = 680 \pm 50\text{MeV}$, which is within the experimental error bars (cf.[4] for future improvements). More importantly, this result is quite sensitive to the average size (Fig.1) of the instantons and the *experimental* instanton size corresponds to $\bar{\rho} \simeq 0.35 \pm 0.03$, which is the value favoured also by other correlation functions and basically found in lattice calculations.

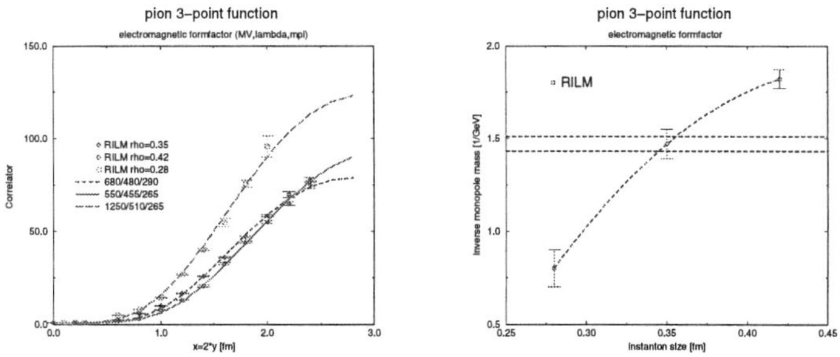

Figure 1: Left: The three-point correlation function of the pion and an external electromagnetic current, normalized to the free correlator and for three different instanton sizes in the random configuration (RILM). The instanton density is kept constant $\bar{n} = 1\text{fm}^{-4}$. The parameters of the pole contribution, m_V, λ_π and m_π in this order, are given in the legend. Right: The monopole mass as a function of the instanton size.

3 Axial formfactors of quarks

Similiarly to the pion three point function we calculated the three point function for the axial vector form factors of the quarks. As a result and a first step in understanding the proton spin, the axial charge of the quarks is significantly ($\simeq 50\%$) reduced in the presence of the interacting instanton medium.

Acknowledgments

The authors acknowledge the support of Alexander von Humboldt Foundation (AB) and Department of Energy grant DE-FG02-88ER40388.

References

1. E. Shuryak, Nucl. Phys. **B203**, 116 (1982)
2. E. Shuryak and J. Verbaarschot, Nucl. Phys. **B410**, 55 (1993).
3. T. Schaefer and E. Shuryak, Phys. Rev. **D53**, 6522 (1996)
4. D. Mack, The Charged Pion Form Factor, proposal CEBAF Experiment 93-021, 1993.

Elastic and Transition Form Factors of Mesons

Craig D. Roberts
Physics Division, Bldg. 203, Argonne National Laboratory
Argonne, IL 60439-4843, USA

The Dyson-Schwinger equations [DSEs] provide a Poincaré invariant, nonperturbative, renormalisable, continuum approach to solving gauge field theories, which also yields perturbation theory in the weak coupling limit.[1] The elementary quantities in this approach are the Schwinger functions, which are the moments of the measure of the field theory. Their analytic properties provide information about confinement and dynamical chiral symmetry breaking [DCSB]. For example, the absence of a Lehmann representation for the two-point quark and gluon Schwinger functions is a sufficient condition for confinement because it ensures the absence of quark and gluon production thresholds in colour-singlet to singlet $\mathcal{S}$-matrix amplitudes. These features make the approach well suited to studying hadronic observables and elucidating the observable effects of bound state substructure.

An essential feature of QCD is that gluonic contributions to the gluon vacuum polarisation provide a dramatic enhancement in the quark-quark interaction at $q^2 < 1\,\text{GeV}^2$; i.e., an integrable singularity in the form of a distribution.[2] This effect is transmitted to the two-point quark Schwinger function (propagator) and is responsible for confinement and DCSB. Such nonperturbative dressing of the Schwinger functions is fundamentally important in QCD.

In the weak-coupling limit the DSE for the quark propagator can be used to perturbatively calculate $B(p^2)$, the Dirac-scalar piece of the, necessarily momentum dependent, quark self energy. In perturbation theory each contribution is proportional to the current-quark mass, m_q. Hence, for $m_q = 0$, mass generation is impossible in perturbation theory. The DSE for the quark propagator is a nonlinear integral equation. When the quark-quark interaction is sufficiently strong at small q^2 then a direct solution yields $B(p^2) \neq 0$, even if $m_q = 0$. This is the case in QCD where this purely nonperturbative effect is caused by the infrared enhancement of the gluon propagator, which is due to gluon self interactions. The DSEs show that the gluon and quark condensates are correlated; i.e, they should not be treated as independent quantities.

A pion with total momentum P is described by a Bethe-Salpeter equation. In the chiral limit, $m_q = 0$, one finds that for $P^2 = 0$ this Bethe-Salpeter equation and the quark DSE are equivalent, which means that DCSB in the chiral limit necessarily entails, *without fine-tuning*, the existence of a massless, pseudoscalar bound state of a quark and antiquark, whose effective masses are

$\sim 300\,\text{MeV}$. The apparent dichotomy of the pion as both a low-mass bound state of heavy constituents and a Goldstone boson is thus easily understood.

Gluon DSE studies are quantitatively accurate for $q^2 > 1\,\text{GeV}^2$. For $q^2 < 1\,\text{GeV}^2$, however, they only provide qualitatively reliable information. Studies of hadron observables using DSEs therefore extrapolate the known UV behaviour into the infrared using a parametrisation of the quark-quark interaction at small q^2. This phenomenological application allows the calculation of observables at all momentum-scales, thereby providing an efficacious means of relating experiment to the behaviour of the quark-quark interaction in the infrared. It has been employed successfully in studies of a wide range of observables including: the electromagnetic form factors F_π, $F_{K^\pm}$ and F_{K^0};[3] and the anomalous $\gamma^*\pi \to \gamma$ and $\gamma\pi^* \to \pi\pi$ transition form factors.[4] The latter are of particular interest because their chiral-limit value is constrained by anomalous Ward identities, which are manifest *without fine-tuning* in this approach.[5]

The approach is particularly well suited to the study of exclusive processes such as these, which are the focus of experiments planned at CEBAF and Fermilab. In each case, irrespective of the q^2 of the probe, soft momenta are important in the transition from initial to final state. In all of the calculated form factors one finds signatures of bound state structure and final state reformation effects. For example, the elastic π and $K^{\pm,0}$ factors, F_M, exhibit a peak in $q^2\, F_M(q^2)$, whose position is characteristic of bound state size, confinement and the importance of soft momenta at the quark-photon vertex. These studies indicate that nonperturbative effects dominate the form factors at presently accessible momentum transfer. They illustrate and emphasise the importance of the nonperturbative dressing of QCD Schwinger functions in the infrared.

In this approach one has a single framework applicable to the calculation of observables at all momentum-scales. It involves a small, fixed number of parameters, which it is possible to relate to the fundamental parameters of QCD and which experiment can be used to probe and constrain.

This work was supported by the US Department of Energy, Nuclear Physics Division, under contract number W-31-109-ENG-38.

1. C. D. Roberts and A. G. Williams, *Prog. Part. Nucl. Phys.* **33**, 477 (1994).
2. N. Brown and M. R. Pennington, *Phys. Rev.* D **39**, 2723 (1989).
3. C. J. Burden, C. D. Roberts and M. J. Thomson, *Electromagnetic Form Factors of Charged and Neutral Kaons*, to appear in *Phys. Lett.* B.
4. M. R. Frank, *et al.*, *Phys. Lett.* B **359**, 17 (1995); R. Alkofer and C. D. Roberts, *Phys. Lett.* B **369**, 101 (1996).
5. C. D. Roberts, *Electromagnetic Pion Form Factor and Neutral Pion Decay Width*, to appear in *Nucl. Phys.* A.

THE INFRARED BEHAVIOUR OF α_S FROM MANDELSTAM'S APPROXIMATION TO THE GLUON DYSON-SCHWINGER EQUATION

A. HAUCK, R. ALKOFER
Institut für Theoretische Physik, Universität Tübingen,
Auf der Morgenstelle 14, 72076 Tübingen, Germany

L. VON SMEKAL
Physics Division, Argonne National Laboratory,
Argonne, Illinois 60439-4843, USA

Using Mandelstam's approximation to the gluon Dyson-Schwinger equation we calculate the gluon self-energy in a renormalisation group invariant fashion. We obtain a non-perturbative β function. The scaling behaviour near the ultraviolet stable fixed point is in good agreement with perturbative QCD. No further fixed point for positive values of the coupling is found, *i.e.*, α_S increases without bound in the infrared. The non-perturbative result for an infrared singular quark interaction can be used to relate the string tension σ to the QCD scale Λ.

The infrared behaviour of the running coupling in strong interactions, α_S, is of great interest, since it may provide an understanding of confinement. Its study is an intrinsically non-perturbative problem. One suitable framework to address this problem is provided by the Dyson-Schwinger equations of QCD. Studies of this infinite tower of equations rely on specific truncation schemes. In the present paper we focus on an approximation to the gluon Dyson-Schwinger equation originally proposed by Mandelstam.[1] This yields a simplified equation for the inverse gluon propagator in momentum space,[a]

$$D^{-1\,\mu\nu}(k) = D_0^{-1\,\mu\nu}(k) \quad + \quad \frac{g_0^2 N_C}{32\pi^4 k^2} \int d^4q \, \Gamma_0^{\mu\rho\alpha}(k, q-k, -q) \times D^{\alpha\beta}(q) D_0^{\rho\sigma}(k-q) \Gamma_0^{\beta\sigma\nu}(q, k-q, -k) \quad , \tag{1}$$

where D_0 and Γ_0 are the bare gluon propagator and the bare three-gluon vertex. The use of a second bare vertex in the gluon loop is combined with one of the gluon propagators being bare. This is to account for some of the dressing of the vertex as entailed by its Slavnov-Taylor identity.[1]

Solutions to this non-linear integral equation for the gluon self-energy have been studied and approximated with various numerical and analytic methods[2] showing that it results in an infrared singular gluon propagator diverging as

---

[a]In Euclidean space with a positive-definite metric; colour indices are suppressed.

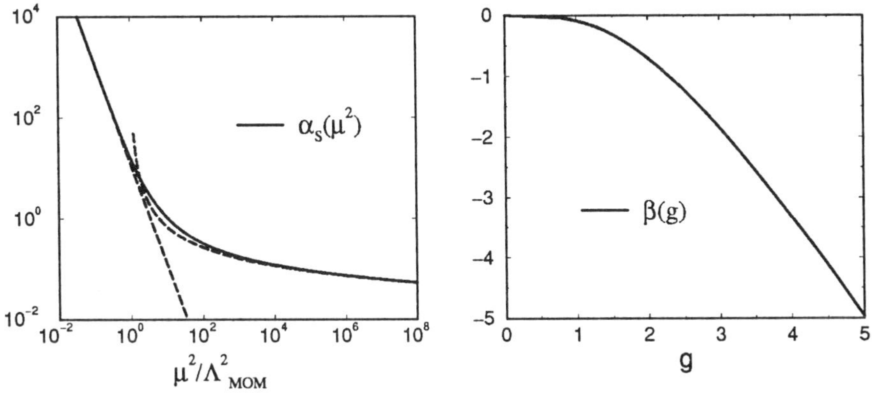

Figure 1: The strong coupling $\alpha_S(\mu^2)$ with its asymptotic limits and the non-perturbative β function of the Mandelstam approximation for a momentum subtraction scheme.[3]

$1/k^4$ for $k^2 \to 0$. It can be shown that in the Mandelstam approximation the product of the coupling and gluon propagator, $gD_{\mu\nu}(k)$, is a renormalisation group invariant, which is obtained as a selfconsistent solution to (1), and the infrared singular result allows to identify a physical scale, the string tension.[3]

The non-perturbative and renormalisation group invariant result relates the string tension σ to the QCD scale (the only parameter) by $\sigma = 2\Lambda^2$. Fixing σ to its phenomenological value (it may be extracted from heavy quarkonium spectra or the slope of Regge trajectories) gives a value for Λ of about 600 MeV. Alternatively, from $\alpha_S(m_Z^2) \simeq 0.108$ we find $\Lambda \simeq 640$MeV. For a momentum subtraction scheme with $N_f = 0$ this is the correct order of magnitude.

The scaling behaviour of the solution near the ultraviolet fixed point is determined by the coefficients $\beta_0 = 14$ and $\gamma_A^0 = 7$ which are reasonably close to the perturbative values for $N_f = 0$, *i.e.*, $\beta_0 = 11$ and $\gamma_A^0 = 13/2$. The running coupling, which is obtained for arbitrary scales from the renormalisation condition, resembles the two-loop perturbative form asymptotically (see fig. 1). It increases without bound in the infrared, no further fixed point exists.

References

1. S. Mandelstam, *Phys. Rev.* D **20**, 3223 (1979).
2. D. Atkinson, P. W. Johnson and K. Stam, *J. Math. Phys.* **23**, 1917 (1982); N. Brown and M. R. Pennington, *Phys. Rev.* D **39**, 2723 (1989).
3. A. Hauck, L. v. Smekal and R. Alkofer, *preprint*, hep-ph/9604430 (1996).

HEAVY QUARK MESONS FROM THE DYSON-SCHWINGER EQUATION TECHNIQUE

C.J. BURDEN
Department of Theoretical Physics,
Research School of Physical Sciences and Engineering,
Australian National University,
Canberra, A.C.T. 0200, AUSTRALIA

The Dyson-Schwinger equation (DSE) technique has proved to be highly successful for understanding the physics of light quarks[1]. Here the technique will be extended to the heavy quark sector in a way consistent with Heavy Quark Effective Theory.

1 The Heavy Quark Propagator

We present a formalism for calculating the heavy quark propagator directly in the region of the bare fermion mass pole $p^2 = m_Q^2$, where m_Q is the heavy current quark mass, and apply it to the heavy-quark–light-antiquark meson spectrum. Full details of the calculation will be presented elsewhere [2].

Consider the simplest viable truncation of the quark DSE,

$$S^{-1}(p) - i\gamma \cdot p - m_Q = \frac{4}{3} \int \frac{d^4 q}{(2\pi)^4} \gamma_\mu S(q) \gamma_\mu \Delta(p-q), \tag{1}$$

in which the quark–gluon vertex has been replaced by a bare vertex and the dressed gluon propagator is modelled by a representation in a Feynman-like gauge. Writing $S^{-1}(p) = i\gamma \cdot p A(p^2) + B(p^2)$, we set

$$A(p^2) = 1 + \Sigma_A(k)/m_Q, \qquad B(p^2) = m_Q \left(1 + \Sigma_B(k)/m_Q\right), \tag{2}$$

where we have introduced a new momentum variable via $p_\mu = i m_Q v_\mu + k_\mu$, with v_μ a constant unit 4-vector. To leading order in $1/m_Q$ the heavy quark propagator becomes

$$S(p) = \frac{1 + \not{v}}{2} \frac{1}{ik \cdot v + \Sigma(k \cdot v)}, \tag{3}$$

where $\Sigma(k \cdot v) = \Sigma_A - \Sigma_B$. The heavy quark DSE is then

$$\Sigma(k \cdot v) = \frac{4}{3} \int \frac{d^4 k'}{(2\pi)^4} \Delta(k - k') \frac{1}{ik' \cdot v + \Sigma(k' \cdot v)}. \tag{4}$$

This equation is numerically tractable once Δ is specified, and provides information regarding fermion self dressing in the vicinity of the bare fermion mass pole, $k_\mu \approx 0$.

Table 1: B and D meson mass differences.

calculation (GeV)	experiment (GeV)
0.267	0.099 ($D_s(1968) - D(1869)$)
0.267	0.096 ($B_s(5375) - B(5279)$)
0.640	0.551 ($D_1(2420) - D(1869)$)
0.640	0.410 ($D_1(2420) - D^*(2010)$)
0.318	0.568 ($D_{s1}(2536) - D_s(1968)$)

2 Heavy-light Mesons.

The homogeneous ladder approximation to the meson Bethe-Salpeter equation (BSE) in a Feynman-like gauge is,

$$\Gamma(p,P) = -\frac{4}{3}\int \frac{d^4q}{(2\pi)^4}\Delta(p-q)\gamma_\mu S_q(q-\xi P)\Gamma(q,P)S_Q(q+(1-\xi)P)\gamma_\mu. \quad (5)$$

Here $S_q(p)$ and $S_Q(p)$ are the full light and heavy quark propagators obtained from the DSEs. By choosing the meson momentum $P_\mu = i(m_Q+\delta)v_\mu$, and introducing a new momentum partitioning parameter $\eta = (m_Q+\delta)\xi/\delta$, the heavy quark current mass m_Q can be removed entirely from Eq. (5) which is then solved for the meson mass gap δ. We find that the degeneracies predicted by HQET (e.g. between the 0^- and 1^- states with light quark spin content $j=\frac{1}{2}$) are respected.

For an initial exploratory calculation we have solved the BSE with a simple gaussian gluon propagator. With the light current quark mass choices $m_{\mathrm{up/down}} = 6.5$ MeV and $m_{\mathrm{strange}} = 135$ MeV, we obtain the meson mass differences listed in Table 1. Our calculations overestimate somewhat the mass gaps between the pseudoscalar D_s and D or B_s and B mesons, and overestimate to a lesser extent the mass gaps between the D_1 and D mesons. In future we hope to carry out calculations with a more realistic model gluon propagator, and with input parameters specifically chosen to give agreement with the light quark spectrum.

References

1. C. D. Roberts and A. G. Williams, Prog. Part. and Nucl. Phys. **33**, 475 (1994).
2. C. J. Burden and D.-S. Liu, *A non-perturbative treatment of heavy quarks and mesons*, hep-ph 9605328.

FORM FACTORS OF KAON SEMILEPTONIC DECAYS

Andrei Afanasev[a,b] and W.W. Buck[b]
[a] *Thomas Jefferson National Accelerator Facility*
12000 Jefferson Ave., Newport News, VA 23606, USA

[b] *The NuHEP Research Center, Department of Physics*
Hampton University, Hampton, VA 23668, USA

A calculation of the semi-leptonic decays of the kaon (K_{l3}) is presented. The results are direct predictions of a covariant model of the pion and kaon introduced earlier by Ito, Buck, Gross. The weak form factors for K_{l3} are predicted with absolutely no parameter adjustments of the model. We obtained for the form factor parameters: $f_-(q^2 = m_l^2)/f_+(q^2 = m_l^2) = -0.28$ and $\lambda_+ = 0.028$, both within experimental error bars. Connections of this approach to heavy quark symmetry will also be discussed.

A covariant model based on coupled Bethe–Salpeter and Schwinger–Dyson equations with an NJL–like interaction was successfully used earlier to describe electromagnetic properties of pions and kaons.[1,2]

Using this model, we analysed form factors of K_{l3} decays (describing the processes $K \rightarrow \pi l \nu$). Assuming that only the vector weak current contributes to the decays, the matrix element may be written as

$$J_\mu = \frac{G_F \sin\theta_C}{\sqrt{2}}[f_+(t)(P_K + P_\pi)_\mu + f_-(t)(P_K - P_\pi)_\mu],$$

where P_K and P_π are the four–momenta of the K and π mesons, G_F is a Fermi coupling constant, θ_C is a Cabibbo angle, and f_+ and f_- are dimensionless form factors which can depend only on $t = (P_K - P_\pi)^2$. For the decay channel, the physical region is limited by the masses of the lepton, pion and kaon: $m_l^2 \leq t \leq (m_K - m_\pi)^2$.

In our model, the pion and the kaon both share the same light quark mass of 250 MeV. It is this mass, together with the corresponding mesonic vertex functions (wave functions) employed in the calculation of the elastic and transition form factors of the pion and kaon that is now employed in the K_{l3} decays addressed here. The pion calculation provides a non-perturbative solution to the axial anomaly for space–like Q^2 and the charged pion form factor[1] while the kaon wave function provides a good description of the charged and neutral kaon form factors.[2]

The main results of the predictive calculation of the K_{l3} decays are illustrated in the Table 1 and compared to selected theoretical models and experiment (see also Ref.[3]). Results are in quite good agreement with available empirical data for all observables studied and, when one quark is infinitely heavy, we reproduce the heavy quark symmetry limit for the ratio f_-/f_+.[4]

Table 1: Model predictions for the parameters of K_{l3} decay form factors. *From the corresponding values of λ_+ and λ_0; **From the corresponding values of λ_+ and $\xi_A(0)$.

	CPT	VMD[7]	ISGW2[8]	This work	Experiment[9]
λ_+	0.031[5] 0.0328[6]	0.0245	0.019	0.028	0.0286±0.0022 (K_{e3})
ξ_A	-0.164±0.047*[5] -0.235[6]	-0.28	-0.28	-0.28	-0.35±0.15 ($K_{\mu3}$)
λ_0	0.017±0.004[5] 0.0128[6]	0.0	-0.005**	0.0026	0.004±0.007 ($K^+_{\mu3}$) 0.025±0.006 ($K^0_{\mu3}$)

It should be noted that Jefferson Lab measurements of π charge form factor (E–93–021), kaon charge form factor (E–93–018), a $\gamma^* p \to \gamma p$ measurement at the π^0 pole[10] and K_{l3} transitions,[11] will further constrain theoretical models addressing all electroweak pion–kaon reactions. More to the point, all of these experiments are connected; in fact, we contend that they must be connected without new parameter fits or adjustments.

It is speculated that to employ these theoretical techniques in the study of the nucleon, our quark masses must be accompanied by a rather large inter–quark kinetic energy in order to bind the nucleon and extract form factors.

Acknowledgments

The work of A.A. was partially supported by the US Department of Energy under contract DE–AC05–84ER40150; the work of W.W.B. was partially supported by the National Science Foundation Grant Number HRD-9154080.

References

1. H. Ito, W.W. Buck, F. Gross, *Phys. Lett.* B **287**, 23 (1992); *Phys. Rev.* C **45**, 1918 (1992).
2. W.W. Buck, R.A. Williams, H.Ito, *Phys. Lett.* B **351**, 24 (1995).
3. A. Afanasev, W.W. Buck, Preprint CEBAF–TH–96–12; hep-ph/9606296.
4. N. Isgur, M.B. Wise, *Phys. Lett.* B **232**, 113 (1989); **237**, 527 (1990).
5. J. Bijnens *et al.*, Preprint hep-ph/9411311.
6. E.P. Shabalin, *Sov. J. Nucl. Phys.* **51**, 296 (1990).

7. L.–M. Chounet, J.–M. Gaillard, M.K. Gaillard, *Phys. Rep.* C **4**, 199 (1974).
8. D. Scora, N. Isgur, *Phys. Rev.* D **52**, 2783 (1992).
9. Review of Particle Properties, *Phys. Rev.* D **50**, 1173 (1994).
10. A. Afanasev in *Workshop on CEBAF at Higher Energies*, Eds. N. Isgur and P. Stoler (CEBAF, Newport News, 1988), p.185, and CEBAF Letter of Intent #LOI–94/005. Co–Spokesmen: A. Afanasev, J. Gomez, and S. Nanda.
11. J.M. Finn, *private communication.*

Probing Confined-Quark Dynamics via Nucleon Form Factors

Axel Bender
Physics Division, Bldg. 203, Argonne National Laboratory
Argonne, IL 60439-4843, USA

The QCD Dyson-Schwinger equations [DSEs] provide the basis for a phenomenology, in which the known large-q^2 behaviour of the gluon propagator is extrapolated to small-q^2 via a parametrisation that incorporates the qualitative features of many studies of the gluon DSE. This phenomenology has been successfully applied in the calculation of a wide range of hadron observables; for example: the meson ground-state spectrum; the elastic scattering and anomalous transition form factors of pseudoscalar mesons; and the electroproduction of vector mesons.[1] In this approach, applicable at *all* momentum scales, these observables probe the nonperturbative, confined-quark, bound state structure of the hadrons, which is their dominant, determining characteristic.[2]

The nonperturbative dressing of quarks and gluons, of which condensate formation is but one part, is of particular importance in QCD. For example, it ensures the absence of quark and gluon production thresholds in colour-singlet to singlet $\mathcal{S}$-matrix amplitudes, which is a sufficient condition for confinement. This in itself, however, neither precludes nor entails the existence of bound states, whether coloured or not. The existence of hadronic bound states can only be established via the Bethe-Salpeter equation (mesons) or the covariant Fadde'ev equation (baryons). These homogeneous bound state equations are derived from the inhomogeneous DSEs for the 4-point or 6-point quark Schwinger functions under the assumption that the associated $\mathcal{T}$-matrix has a pole in a given channel. The complex nature of the QCD vacuum and the intrinsically nonperturbative nature of bound-state problems suggest that one should employ dressed gluon and quark propagators in constructing the kernels of these equations. The DSEs facilitate this since they provide a natural framework for the unified treatment of the structure of the QCD vacuum and bound states, in which the global symmetries of QCD are manifestly preserved.

The simplest (lowest order) truncation of the two-body Bethe-Salpeter equation [BSE] is ladder approximation, in which the kernel is constructed from the model dressed gluon propagator and dressed-quark propagators obtained in rainbow approximation. With this kernel Goldstone's theorem is manifest and one can obtain a good description of the ground-state meson spectrum. However, it has the defect that it is purely attractive in the q-$\bar{q}$ colour-singlet and q-q colour-antitriplet channels. This entails the existence of colour-antitriplet quark-quark bound states, which are not observed, and can be said to underly the diquark-pole approximation to the 2-body $\mathcal{T}$-matrix in the solution of the covariant 3-body Fadde'ev equation for the baryon.

A systematic, Goldstone theorem preserving improvement of the ladder truncation demonstrates that repulsive terms appear in the kernel at every higher order.[3] In the meson channel these terms are cancelled by additional attractive terms that also appear. However, in the colour-antitriplet quark-quark channel they are not and act to eliminate the diquark bound states completely. It is this that provides for the absence of such coloured bound states in the strong interaction spectrum. Nevertheless, it is possible to simplify the study of the Fadde'ev equation by neglecting explicit 3-body interactions and approximating the 2-body $\mathcal{T}$-matrix as a sum of confined-diquark contributions.

In order to explore the efficacy of such a simplification, the electromagnetic form factors of the nucleon are calculated using a product Ansatz for the nucleon Fadde'ev amplitude: $\Psi = \sum_I \Gamma_I^{qq} D_I^{qq} \psi_I^{qq-q}$, where: Γ_I^{qq} is a quark-quark (confined-diquark) correlation amplitude; D_I^{qq} is the confined-diquark propagator; ψ_I^{qq-q} is the confined-diquark–quark correlation amplitude; and the index I ranges over all allowed colour-antitriplet, flavour and Dirac quark-quark correlations. The functions Γ_I^{qq}, D_I^{qq} and ψ_I^{qq-q} are parametrised such that no quark production thresholds occur in the kernel of the 6-point quark Schwinger function.

The preliminary results are encouraging.[4] Isoscalar-scalar and isovector-pseudovector confined-diquark correlations are found to provide the dominant contribution to the form factors. An isoscalar-scalar confined-diquark correlation alone is insufficient to describe both the electric and magnetic properties of the nucleon. The preliminary calculations indicate that the inclusion of the isovector-pseudovector correlation can remedy this problem. It is important to note that the length-scales that characterise Γ_I^{qq} and ψ_I^{qq-q} are approximately equal, which means that the diquark correlation and the nucleon itself have the same space-time extent.

This work was supported by the US Department of Energy, Nuclear Physics Division, under contract number W-31-109-ENG-38.

1. C. D. Roberts, *Elastic and Transition Form Factors of Mesons*, these proceedings.
2. R. Alkofer, A. Bender and C. D. Roberts, Int. J. Mod. Phys. **A 10** (1995) 3319.
3. A. Bender, C. D. Roberts and L. v. Smekal, *Goldstone Theorem and Diquark Confinement Beyond Rainbow–Ladder Approximation*, to appear in Phys. Lett. B.
4. A. Bender, M. R. Frank and C. D. Roberts, in progress.

Session 10
Strong Interaction Theory — Nuclear

RELATIVISTIC CALCULATION OF THE TRITON BINDING ENERGY

FRANZ GROSS[1,2] and ALFRED STADLER[1,3]
[1] *College of William and Mary, Williamsburg, VA 23187, USA*
[2] *Thomas Jefferson National Accelerator Facility, Newpoet News, VA 23606, USA*
[3] *Centro de Física Nuclear da Universidade de Lisboa, 1699 Lisboa Codex, Portugal*

Preliminary results for the triton binding energy obtained from the relativistic spectator or Gross equation are reported. The equations take the Dirac structure of the nucleons fully into account. Numerical results are presented for a family of realistic OBE models with off-shell scalar couplings. It is shown that these off-shell couplings improve both the fits to the two-body data and the predictions for the binding energy.

The three-body spectator equations were first introduced and applied to scalar particles in 1982,[1] and then extended to the case of three spin 1/2 particles in lectures given at the University of Hannover soon afterward. Recently, a more tractable form for the equations has been developed, and a full derivation of the equations will be published elsewhere.[2] Briefly, the three-particle spectator equations are covariant equations of the Faddeev type in which two of the three particles are always on-mass shell. For identical particles, the final form of the three-body spectator equation is

$$\langle J1(23)\rho|\Gamma^1\rangle = \sum_{j'm'} \sum_{\substack{\lambda_2''\lambda_3''\rho'' \\ \lambda_1'\lambda_2'\lambda_3'\rho'}} \int_0^{q_{\rm crit}} q'^2 dq' \frac{m}{E_{q'}} \int_0^{\pi} d\chi \sin\chi \, \langle j(23)\rho|M^1|j(2''3'')\rho''\rangle$$
$$\times \frac{m}{E_{\tilde p''}} g^{\rho''}(q,\tilde p'')\, \mathcal{P}_{12}^{\rho''\rho'}[1(2''3''),1'(2'3')] \frac{m}{E_{\tilde p'}} \langle J'1'(2'3')\rho'|\Gamma^1\rangle\,, \quad (1)$$

where $\rho = \pm$ is the ρ-spin of the off-shell particle, q and $\tilde p$ are the magnitudes of the three momentum of the on-shell spectator (particle 1) and on-shell particle 2 (defined in the rest frame of the 12 pair), M^1 is the two-body scattering amplitude for the 23 pair, $\mathcal{P}_{12}^{\rho''\rho'}[1(2''3''),1'(2'3')]$ is the matrix element of the permutation operator which interchanges particles 1 and 2, and $g^\rho(q,\tilde p)$ the propagator of the off-shell particle in different ρ-spin states. Note that Eq. (1) includes a sum over intermediate helicities and angular momentum quantum numbers, and an integration over the internal spectator momentum q' and the angle χ between the directions of $\mathbf{q}'$ and $\mathbf{q}$. The integration over q' has been limited to the finite interval $[0, q_{\rm crit}]$ (when $q' = q_{\rm crit}$ the interacting pair is recoiling at the speed of light, and contributions for $q' \geq q_{\rm crit}$ are neglected[2]).

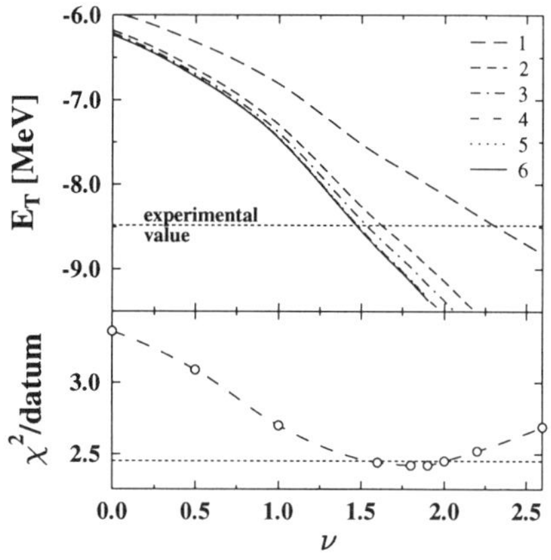

Figure 1: Triton binding energy E_t for specific values of the maximum two body momentum j_{max} (upper panel) and χ^2 for the fits to the two-body data (lower panel) versus the scalar meson off-shell parameter ν. The lower panel includes the line $\chi^2 = 2.45$ for reference.

We have solved Eq. (1) numerically for a variety of realistic NN models which include phenomenological scalar σ (with $I = 0$) and scalar δ ($I = 1$) exchanges with off-shell sNN couplings of the form

$$g_s \Lambda_s(p', p) = g_s \left[1 - \frac{\nu_s}{2m}(m - \not{p}' + m - \not{p}) + \frac{\kappa_s}{4m^2}(m - \not{p}')(m - \not{p})\right] . \quad (2)$$

The family of models discussed here has $\kappa_s = 0$, $\nu_\sigma = -0.75\,\nu$, and $\nu_\delta = 2.60\,\nu$.

Fig. 1 shows the rapid dependence of the binding energy on ν; a value in good agreement with experiment would be obtained for $\nu \simeq 1.5$, still in the range of ν's which give the best fit to the two-body data.[3]

Acknowledgments

This work could not have been done without the support of the US DOE under grant #DE-FG05-88ER40435 which is gratefully acknowledged.

References

1. F. Gross, Phys. Rev. C **26**, 2226 (1982).
2. A. Stadler, F. Gross, and M. Frank, in preparation.
3. A. Stadler and F. Gross, submitted for publication to Phys. Rev. Letters.

Relativistic Hamiltonians in many-body theories

P. Amore, M. B. Barbaro, A. De Pace
Dipartimento di Fisica Teorica dell'Università di Torino and Istituto Nazionale di Fisica Nucleare, Sezione di Torino, via P.Giuria 1, I-10125 Torino, Italy

The Foldy-Wouthuysen transformation is used to obtain relativistic potentials, which are then expanded in $1/m_N$. These potentials are utilized to recover, with high accuracy, the relativistic self-consistent equation for the effective mass of the nucleon in nuclear matter. A Coulomb sum rule, which saturates in the space-like region, can also be defined in a natural way.

Historically, two different methods have been used to obtain relativistic corrections in a potential description of nuclear matter, namely *Breit reduction* and *minimal relativity* [1]. In the first case one expands the relativistic scattering amplitude and interprets it as the matrix elements of the Fourier transform of the potential (Born term) between Pauli spin wave functions. In the case of *minimal relativity*, the potential is defined starting from the Blankenbecler-Sugar reduction of the Bethe-Salpeter equation, which, through a proper redefinition of the amplitudes, becomes identical to the nonrelativistic Lippmann-Schwinger equation. In both approaches one obtains velocity-dependent and non-local potentials.

The problem of defining a potential accounting for relativity may also be addressed in the Foldy-Wouthuysen (FW) formalism. Indeed, the FW transformation may be easily implemented in the language of Green's functions [2]. The FW unitary operator, $\mathcal{T}(\mathbf{k}) = \sqrt{(E_k + m_N)/2E_k}[1 + \boldsymbol{\gamma} \cdot \mathbf{k}/(E_k + m_N)]$, may be used to define the following FW noninteracting Green's function

$$G_0^{FW}(k, k_0) = P_+ \left[\frac{\vartheta(k - k_F)}{k_0 - E_k + i\varepsilon} + \frac{\vartheta(k_F - k)}{k_0 - E_k - i\varepsilon}\right] - P_- \frac{1}{k_0 + E_k - i\varepsilon}, \quad (1)$$

where $P_\pm \equiv (1 \pm \gamma^0)/2$ are operators projecting out the large/small components in the wave function and k_F is the Fermi momentum. Note how closely G_0^{FW} resembles the nonrelativistic noninteracting single particle Green's function.

The transition amplitude can be expressed in terms of the FW Green's function by a suitable redefinition of the interaction term $V = g^2\Gamma_1 D(q)\Gamma_2$ (g being the coupling constant, Γ_i the vertex operators and $D(q)$ the propagator of the interaction line). If one also projects on the large components of the wave functions, one can indeed define

$$V_{(+)}^{FW} = g^2 P_+ \mathcal{T}^\dagger(\vec{k}_1')\Gamma_1 \mathcal{T}^\dagger(\vec{k}_1) P_+ D(q) P_+ \mathcal{T}^\dagger(\vec{k}_2')\Gamma_2 \mathcal{T}^\dagger(\vec{k}_2) P_+ . \quad (2)$$

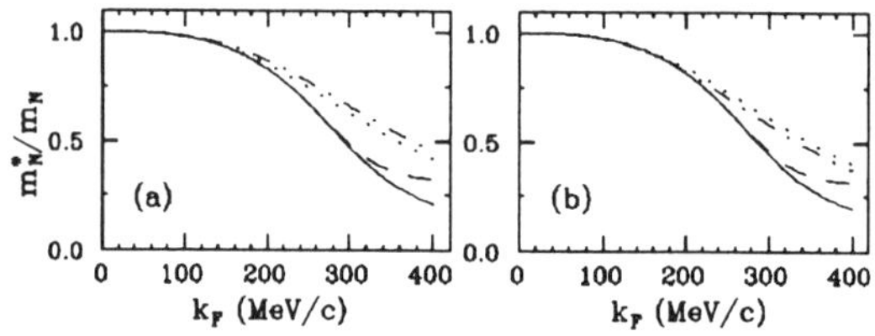

Figure 1: m^*_N as a function of k_F for the MFT (solid) and for the FW expansion (dash). The nonrelativistic m^*_N from the Breit reduced (dot) and the minimal relativity (dot-dashed) potentials is also displayed. (a) MFT parameters; (b) Bonn potential parameters.

By expanding $V^{FW}_{(+)}$ in $1/m_N$ the Breit reduced potential is obtained[2].

The FW procedure, however, provides a systematic way of dealing with $1/m_N$ expansions in a many-body context. As an example, it has been applied to the Hartree problem in nuclear matter with scalar and vector interactions, where it allows one to deal properly, — i. e., at the same order in the expansion, — with the issue of *self-consistency*. Indeed, at the leading order one gets the following equation for the effective nucleon mass

$$m^*_N = m_N - \frac{2k_F^3}{3\pi^2}\frac{g_s^2}{m_s^2}\left(1 - \frac{3k_F^2}{10{m^*_N}^2}\right) , \tag{3}$$

reproducing with high accuracy the relativistic mean field theory (MFT) results over a wide range of densities. This is not possible if one employs a relativistic potential together with the standard nonrelativistic expression for the effective mass (Fig. 1).

The FW transformation may also be used to define a relativistic Coulomb sum rule, which saturates as a result of the sum over particle-hole excitations only, as in the nonrelativistic case. In principle these excitations do not exhaust the sum rule and the Dirac sea should also be accounted for: Nonetheless, for the most direct contact with the experiment, a normalizing factor for the longitudinal response function, — which in the nonrelativistic limit reduces to the usual form factor $G^2_{E_p}$, — may be naturally introduced, in such a way to provide saturation already in the space-like region[4].

References

1. G.E. Brown, A.D. Jackson, and T.S. Kuo, *Nucl. Phys.* **A133**, 481 (1969).
2. P. Amore, M.B. Barbaro, and A. De Pace, in print in *Phys. Rev.* C.
3. B.D. Serot and J.D. Walecka, *Adv. Nucl. Phys.* **16**, 1 (1986).
4. P. Amore and A. Molinari, in preparation.

THE AXIAL CHARGE RENORMALIZATION IN A RELATIVISTIC FINITE NUCLEI DESCRIPTION

E. Oset[+], A. Gil[+], M. Kleinmann[++], H. Müther[++]

[+]*Departamento de Física Teórica and IFIC Centro Mixto Universidad de Valencia-CSIC; 46100 Burjassot (Valencia), Spain*
and
[++]*Institut für Theoretische Physik der Universität Tübingen, 72076 Germany*

Starting from a realistic One-Boson-Exchange model of the nucleon-nucleon interaction the relativistic mean field for nucleons is determined within the Dirac Brueckner Hartree Fock approach for finite nuclei. The matrix elements of the axial charge operator evaluated for the solutions of the Dirac equation with this selfenergy are investigated.

In our work we consider the relativistic features, the effects of correlations and the single-particle wavefunctions consistently. For that purpose we employ the results of the Dirac Brueckner Hartree Fock (DBHF) calculations of [1]. We have found renormalization effects due to the use of the relativistic wave functions, enhancing the axial charge in the direction found in earlier perturbative approaches for nuclear matter [2, 3, 4]. However, the quantitative results differ from the estimates derived for nuclear matter significantly. Using the G-matrix derived from a realistic meson exchange model of the NN interaction [5] a perturbative estimate of the heavy meson exchange current contribution to the axial charge at nuclear matter saturation density [6] would yield a renormalization factor of 1.4 and a non perturbative treatment would lead to enhancement as large as 1.7. For finite nuclei the enchancement factors found are considerably smaller, of the order of 1.2-1.3. We argue that this reduction of the renormalization effect is due to the smaller densities at the surface of finite nuclei, which are relevant for the evaluation of actual matrix elements.

The actual determination of the genuine relativistic effects is done in the following way: The relativistic potential determined in [1] is of the form

$$\Sigma(r) = \Sigma^{(s)}(r) + \gamma^0 \Sigma^{(v)}(r)\,. \tag{1}$$

With this potential we solve the Dirac equation

$$[i\vec{\gamma}\vec{\nabla} + M + \Sigma^{(s)}(r) + \gamma^0\Sigma^{(v)}(r)]\psi(r) = E\gamma^0\psi(r) \tag{2}$$

and determine the single particle relativistic wave functions and the matrix elements of the $\gamma^0\gamma^5$ operator between these wave functions. On the other hand we deduce the equivalent non relativistic potential related to the relativistic one of eq. (1) as described in ref. [7], solve with it the Schrödinger equation and determine the single particle wave functions and the matrix element of the non relativistic reduction of the $\gamma^0\gamma^5$ operator

$$\gamma^0\gamma^5 \rightarrow \frac{\vec{\sigma}(\vec{p}+\vec{p}')}{2M} \tag{3}$$

The ratio of the matrix elements for each transition evaluated in the relativistic and non relativistic form is what we call the renormalization factor. These factors are found to be 1.33 for the $1p_{1/2} \rightarrow 2s_{1/2}$ transition in $^{16}O(0^+ \rightarrow 0^-$ transition in all cases), 1.20 for the $1p_{3/2} \rightarrow 1d_{3/2}$ transition in and 1.24 for the $1d_{3/2} \rightarrow 2p_{3/2}$ transition of ^{40}Ca. As in the case of the perturbative calculation $\Sigma^{(s)}$ is the critical magnitude in the renormalization. But the results obtained for the renormalization with this relativistic non perturbative approach are smaller than found previously with other approaches.

References

[1] R. Fritz, H. Müther and R. Machleidt, Phys. Rev. Lett. 71 (1993) 46; R. Fritz and H. Müther, Phys. Rev. C 49 (1994) 633.

[2] J. Delorme, I.S. Towner, Nucl. Phys. A475 (1987) 720.

[3] M. Kirchbach, D.O. Riska and K. Tsushima, Nucl. Phys. A563 (1993) 525.

[4] I. S. Towner, Nucl. Phys. A542 (1992) 631.

[5] R. Machleidt, Adv. in Nucl. Phys. 19 (1989) 189.

[6] A. Gil, M. Kleinmann, H. Müther and E. Oset, Nucl. Phys. A 584 (1995) 621.

[7] M. Jaminon and C. Mahaux, Phys. Rev. C40 (1989) 454.

Relativistic passage of nucleon through nucleus

T. SUZUKI
Department of Physics, Shizuoka University, Shizuoka 422, JAPAN

Semiclassical approaches commonly serve to describe nuclear scattering with sufficiently small wavelength. However, time-dependent descriptions in terms of the wave packet reveal some novelties beyond the scope of the conventional semiclassical versions [1, 2]. The explicit energy-dependence of the Schrödinger equation equivalent potential in the Dirac approach leads to a singular motion of the wave packet, which behaves as if an attractive interaction acted upon the nucleon at higher intermediate energies [2]. In this paper we further examine this wave packet behavior.

The wave packet comprising positive energy spinors the size of which is taken to be greater than the Compton wavelength reads

$$\psi(\mathbf{r}, t)= \int d^3q/E(k')a(\mathbf{q})\exp\{-iE(k')t\}\,\psi(\mathbf{k}', \mathbf{r}), \tag{1}$$

where $\mathbf{k}'=\mathbf{k}+\mathbf{q}$ and $E(k)=(k^2+m^2)^{1/2}$. We discuss the behavior of the probability density $\rho=\psi^+\psi$ using the eikonal approach [3] to provide an explicit form. In order that the center of the relevant envelope function $G(\mathbf{r})$ with $\mathbf{r}=\mathbf{b}+z\mathbf{n}$ has nonvanishing impact parameter we replace $\mathbf{r}$ with $\mathbf{r}$-$\mathbf{s}$ in G, where $\mathbf{s}$ is parallel to $\mathbf{b}$. Assuming $a(\mathbf{q}) \propto \exp(-d^2\mathbf{q}^2)$ and disregarding the spin-orbit term the probability density becomes

$$\rho(\mathbf{r}, t)=A \exp[J-\{(\mathbf{b}-\mathbf{s})^2+L^2\}/(2d^2)], \tag{2}$$

$$L=z-ut+\partial(\mathrm{Re}\,\phi)/\partial k, \qquad \phi=-(m/k)\int_{-\infty}^{z} dz'\, U(r, k),$$

$$J=(2m/k)\int_{-\infty}^{z} dz'\,\mathrm{Im}U(r, k),\ \ U(r, k)=S(r)+(E/m)V(r)+V_2\,,\ \ V_2=(S^2-V^2)/2m,$$

where S and V are the Lorentz scalar and vector potentials, u being the incident velocity. The spreading rate is disregarded.

Now we investigate the motion of the peak of the density. Let it go along the line parallel to the z-axis with $b=s=|\mathbf{s}|$. The position of the peak is determined by $(\partial \rho / \partial z)_{b=s} = 0$:

$$z - ut + \int_{-\infty}^{z} dz' \, \gamma W = 2kd^2 (1 + \gamma W)^{-1} \gamma \, \mathrm{Im} U, \quad W = \mathrm{Re}[S + (m/E)V + V_2], \qquad (3)$$

where $\gamma = m/k^2$ and $r = (z^2 + s^2)^{1/2}$. The velocity of the peak is obtained from (3) by differentiation with respect to time as

$$v(z, s, k) = u / [1 + \gamma W - 2kd^2 \, \partial / \partial z \{ \gamma \, \mathrm{Im} U / (1 + \gamma W) \}] . \qquad (4)$$

At higher intermediate energies the term $(m/E)\mathrm{Re}V$ in W is not dominant compared to $\mathrm{Re}S<0$. This gives rise to $W<0$ and it, therefore, follows that $v/u = (1 + \gamma W)^{-1} > 1$ particularly near $z=0$ in spite of the repulsion $\mathrm{Re}U>0$. Such a result is due to the appreciable contribution of the lower component of the Dirac spinor. Unless the packet size d is small to some extent it is expected from (3) that two peaks appear with the passage of time. In order to have a single peak alone the size has to satisfy $2kd^2\{\partial / \partial z(\gamma \, \mathrm{Im} U)\} < 1$ for $z = (R^2 - s^2)^{1/2}$ in view of (4) with $|\gamma W| < 1$, R being the nuclear radius.

Let the peak pass the points $z_1 = -(R^2 - s^2)^{1/2}$ and $z_2 = (R^2 - s^2)^{1/2}$ at the times t_1 and t_2, respectively. Then from (3) the time delay for the peak motion is given as

$$\tau(s, k) = t_2 - t_1 - (z_2 - z_1)/u = 1/u \int_{-\infty}^{\infty} dz' \, \gamma W, \qquad (5)$$

where the spherical symmetry of U is taken into account. Hence we have $\tau < 0$ as if the nucleon were subject to some attractive potential.

The time delay given by the scattering amplitude is positive when $\mathrm{Im}U \neq 0$ [2]. But if $\mathrm{Im}S = \mathrm{Im}V = 0$ that for the partial wave amplitude with $b=s$ coincides with (5) and is negative. The positive time delay might be due to the imaginary part of U. Hence the amplitude as well would reveal its novel feature, which, however, differs from that of the wave packet.

[1] T. Suzuki, Ann. Phys. (N. Y.) **189**, 352 (1989).
[2] T. Suzuki and I. Maeda, Z. Phys. **A347**, 75 (1993).
[3] R. D. Amado et al., Phys. Rev. **C28**, 1663 (1983).

RELATIVISTIC BOUND-STATE EQUATIONS IN THREE DIMENSIONS

D.R. PHILLIPS AND S.J. WALLACE
Department of Physics, University of Maryland,
College Park, MD, 20742

A systematic procedure is derived for obtaining three-dimensional bound-state equations from four-dimensional ones. A simple four-dimensional equation with the correct one-body limit may be obtained by a reorganization of the generalized ladder Bethe-Salpeter kernel. Thus, our approach to deriving three-dimensional equations may be applied to that four-dimensional equation, thereby yielding a retarded interaction for use in the three-dimensional "ET" equation.

1 Reduction to three dimensions without the use of delta functions

For many years three-dimensional quasi-potential equations (QPEs) have been used as an approximation to Bethe-Salpeter equations (BSEs). The reduction to three dimensions is done by placing a delta-function constraint on the relative four-momentum. In some cases this results in unphysical singularities in both boosts of, and corrections to, the interaction [1]. We seek a reduction technique free of this flaw.

Consider the BSE for the bound-state vertex function, Γ,

$$\Gamma = KG_0\Gamma. \tag{1}$$

Splitting $K = K_1 + K_2$, with K_1 "three-dimensional", i.e. not depending on the zeroth component of the relative four-momenta, yields a set of equations:

$$\begin{aligned} \Gamma &= [1 - K_2G_0]^{-1}\,\Gamma_1; & (2)\\ \Gamma_1 &= K_1\langle\mathcal{G}\rangle\Gamma_1; & (3)\\ \mathcal{G} &= G_0 + G_0K_2\mathcal{G}; & (4)\\ \langle\mathcal{G}\rangle &\equiv \int \frac{dp_0'\,dp_0}{(2\pi)^2}\,\mathcal{G}(p_\mu', p_\mu; P). & (5) \end{aligned}$$

Eq. (3) is three-dimensional, so if K_1 is chosen to be a cm-frame instant interaction then (4) defines $\mathcal{G}$ order by order in K_2. There is, however, a general way to account for all the relative-energy integrations in $\langle\mathcal{G}\rangle$. Demanding that K_1 be chosen such that $\langle\mathcal{G}\rangle = \langle G_0\rangle$ produces

$$\begin{aligned} \Gamma_1 &= K_1\langle G_0\rangle\Gamma_1; & (6)\\ K_1 &\equiv \langle G_0\rangle^{-1}\langle G_0K\mathcal{G}\rangle\langle G_0\rangle^{-1}. & (7) \end{aligned}$$

In the absence of negative-energy states K_1 may be represented as a sum of time-ordered perturbation theory diagrams [2]. 3D equations like (6) have two major advantages over QPEs: they may easily be written in any frame, and they include meson-retardation effects present in the original 4D equation.

2 4D and 3D equations with the correct one-body limit

The BSE will only have the correct one-body limit if its kernel includes the full set of generalized ladder graphs [3]. Suppose that this kernel, K_X, is written

$$K_X = V + VG_C K_X. \tag{8}$$

A useful G_C is one which allows the crossed-box graph, $K_X^{(4)}$, to be written $K_X^{(4)} = K_X^{(2)} G_C K_X^{(2)}$, in the one-body limit, on-shell [2,4]. ($K_X^{(2)}$ is the ladder graph.) With such a choice of G_C, the fourth-order correction to V vanishes in that limit. So, a simple 4D equation with the correct one-body limit is

$$\Gamma = K_X^{(2)}(G_0 + G_C)\Gamma. \tag{9}$$

Our approach to deriving 3D equations may then be applied to Eq. (9) [2]. When done at "zeroth order" this yields the 3D "ET" equation [5]. At "first order" it produces a retarded interaction for use in that equation. This "retarded ET" 3D equation has the correct one-body limit, simple boost properties, and may be systematically improved upon. Its only deficiency is that certain meson-production thresholds in the bound-state vertex function are displaced. This can be corrected for in a well-defined way.

The quality of this 3D equation, and our general prescription for deriving such equations, is then tested by calculating bound-state properties in a scalar field theory using six 3D and 4D bound-state equations [2]. Equations obtained using the method espoused here approximate the bound-state properties calculated from their parent 4D equations significantly better than QPEs do.

References

1. N. K. Devine and S. J. Wallace, Phys. Rev. C **48**, 973 (1993).
2. D. R. Phillips and S. J. Wallace, Nuclear theory preprint archive no. nucl-th/9603008, accepted for publication in Phys. Rev. C.
3. F. Gross, Phys. Rev. C **26**, 2203 (1982).
4. S. J. Wallace in *Nuclear and Particle Physics on the Light Cone*, eds. M. B. Johnson and L. S. Kisslinger (World Scientific, Singapore, 1988).
5. S. J. Wallace and V. B. Mandelzweig, Nucl. Phys. **A503**, 673 (1989).

Vacuum Polarization and Meson Exchange Interactions in Finite Nuclei

T. Kohmura, T. Nagata and M. Nakano*)

Institute of Physics, University of Tsukuba, Ibaraki 305, Japan and

*)Univ. of Occupational and Environmental Health, Kitakyushu 807, Japan

We have shown using Bonn A parameters that Bruckner-Hartree-Fock (BHF) which takes into account nucleon-nucleon correlations improves the nuclear fits but not well enough; it repoduces the nuclear radius but not well the total energy of a nucleus. So far we have worked only in the positive energy nucleon sector. We may have to take into account negative energy nucleons in order to improve the nuclear theory.

Taking into account positive and negative energy nucleons on the equal footing, we have solved the equations in the relativistic Hartree approximation (RHA) for finite nuclei. We compare the result with that of the mean field theory (MFT) which uses only positive energy nucleons. In the both approximations we expand the nucleon and meson fields on a relativistic wave packet basis. This method needs not rely on the derivative expansion to sum the divergent densities and energies of negative energy nucleons.

In each of RHA and MFT we determine a best fit set of coupling constants (set a and set b) which reproduces well the nuclear total energy and the radius of ^{16}O and using these parameters we calculate various quantities of the nucleus. The MFT calculation yields also negative energy nucleon energies, when Dirac equation is solved. The results are shown in Table I. When we use a same set (set a, for example) of coupling constants for MFT and RHA calculations, the results show that the contribution of negative energy nucleons suppresses significantly the mean meson fields and increases the nuclear radius. Comparing the results of MFT and RHA using their best fit coupling constants, we see that they reproduce equally well the nuclear properties but that they predict quite differently negative energy nucleon levels. The RHA calculation predicts the

This expansion yields the corrected meson exchange interactions above to be expressed in terms of an effective meson mass and coupling constant. In Table II we show the calculated values of the vacuum polarization functions for σ and ω mesons in nuclear matter and the effective coupling constants and meson masses. The numerical results show that the vacuum polarization suppresses the meson exchange interactions by 30 per cent and increases their ranges also by 30 per cent. This result is consistent with the result of the above RHA calculation.

The density dependence of the renormalized polarization functions indicates that although the meson exchange interactions in a two-nucleon system are hardly affected by the vacuum polarization, they are significantly affected in finite nuclei. The nuclear interactions determined from two nucleon systems are not expected to reproduce nuclear properties, unless the vacuum polarization effect is taken into account.

We are analyzing the data of proton scattering from ^{58}Ni at 200-400 MeV obtained by H. Sakaguchi at al. at Osaka. When we apply the optical potential obtained from the relativistic Love-Franey nucleon-nucleon scattering t-matrix, the data require some suppression of the optical potential. This suppression is a possible evidence of the vacuum polarization correction to meson exchange interactions in nuclear medium.

The $N\bar{N}$ vacuum polarization correction to the BHF calculation may also solve the long standing problem in nuclear physics what is the nuclear formalism that reproduces fairly well both the nuclear total energy and radius.

This work is partly to be published in Nuclear Physics A.

Table II. Vacuum polarization functions, effective coupling constants and meson masses

$g_\sigma = 7.92, g_\omega = 8.92, m_\sigma = 550$ MeV

	σ	ω
$\Pi^{ren}{}_{vac}$	$0.44{m_\sigma}^2 + 1.04q^2$	$0.88q^2$

		σ	ω
$\Pi^{ren}{}_{vac}$	g^*	$0.70g_\sigma$	$0.73g_\omega$
	m^*	$0.84m_\sigma$	$0.73m_\omega$

binding energies of antinucleons to be less than 500 MeV, much smaller than the prediction of the MFT calculation.

Secondly we take the positive energy nucleon sector picture of a nucleus to elucidate the above result of the RHA calculation. In this picture the meson exchange interactions

$$V(q) = -\frac{g^2}{m^2 + q^2}$$

between positive energy nucleons should be corrected by the $N\bar{N}$ vacuum polarization excitations as

$$V(q) = -\frac{g^2}{m^2 + q^2 + \Pi(q)}$$

with the renormalized vacuum polarization function $\Pi(q)$. The renormalized vacuum polarization functions are dependent on nuclear density and the vacuum polarization correction is negligible in two-nucleon systems. Taking the static approximation $q_0 = 0$, we expand the polarization function $\Pi(q)$ up to second order in $\vec{q}$ as

$$\Pi(\vec{q}, q_0 = 0) = am^2 + b\vec{q}^2.$$

Table I. MFT and RHA calculations for ^{16}O

set a: $g_\sigma = 8.88,\ g_\omega = 9.11,\ m_\sigma = 550$ MeV

set b: $g_\sigma = 13.29, g_\omega = 16.87, m_\sigma = 550$ MeV

The binding energies are shown for single particle energies.

		MFT		RHA	experiment
		set a	set b	set a	
positive	$1s_{\frac{1}{2}}$(MeV)	147	43.0	47.0	45.7
energy	$1p_{\frac{3}{2}}$(MeV)	100	24.1	20.5	21.8
nucleons	$1p_{\frac{1}{2}}$(MeV)	65.2	14.2	15.3	15.7
negative	$1s_{\frac{1}{2}}$(MeV)	948	894	468	
energy	$1p_{\frac{1}{2}}$(MeV)	861	841	385	
nucleons	$1d_{\frac{3}{2}}$(MeV)	735	775	301	
$\sqrt{< r^2 >}$(fm)		1.71	2.72	2.61	2.72
E_{total}(MeV)		-792	-142	-93.0	-142

RECENT STUDIES ON FEW NUCLEON REACTIONS AT LOW ENERGIES

M. VIVIANI[(1)], A. KIEVSKY[(1)], S. ROSATI[(1,2)] and R.SCHIAVILLA[(3)]
[(1)] *Istituto Nazionale di Fisica Nucleare, Sezione di Pisa, I-56100 Pisa, Italy*
[(2)] *Physics Department, Pisa University, I-56100 Pisa, Italy*
[(3)] *CEBAF Theory Group, Newport News, VA 23606 and Old Dominion University, Norfolk, VA 23529*

1 Summary

In a series of recent papers, a variational technique for calculating the bound–state and scattering–state wave functions of systems with $A = 3$, 4 has been developed [1]. The method consists in the expansion of the wave functions on a basis of Correlated Hyperspherical Harmonic (CHH) functions and using the Ritz and Kohn variational principles to determine the bound–state and scattering–state wave functions, respectively. The Hamiltonian considered contains realistic two– and three–body interaction terms. The use of variational principles guarantees the best choice of the expansion coefficients of the wave function on a given basis functions.

The results obtained for the ground state of nuclei with $A = 3$ and the n–d elastic scattering states agree with those given by other accurate techniques. Also, accurate calculations of the various p–d scattering observables are feasible. For this reaction, highly quality measurements are today available and a detailed comparison can be performed in order to make evident the inadequacies of the various interaction models. The results obtained show that the theoretical underestimation of A_y and iT_{11} polarization observables is directly related with the potential in the ^3P channel. To this respect, it is important to understand better the effects of the Three Nucleon Interaction (TNI) on the above mentioned observables.

The CHH wave functions (including also Δ-isobar admixtures) have been used to study the thermal neutron radiative capture on deuterium, and the ^{2}H($\vec{p}$,γ)^{3}He and p($\vec{d}$,γ)^{3}He reactions in the center of mass energy range 0–100 keV. The nuclear electromagnetic current includes one- and two-body components [2]. Large effects due to two-body currents, in particular the long-range ones associated with the tensor component of the nucleon-nucleon interaction, have been observed. For example, including the one–body electromagnetic current components, the calculated value of the zero–energy astrophysical factor $S(0)$ for the p–d capture reaction is $S(0) = 0.138$ eV b ($S(0) = 0.191$ eV b when the two–body currents are also included), to be compared with the experimental value $S(0)_{exp} = 0.18 \pm 0.01$ eV b. Good, quantitative agreement after the inclusion of two–body components in the current between theory and experiment has been found for all observables, with the exception of the vector analyzing power for which the calculated values underestimate the data by about 30 %.

The CHH expansion has been also used to calculate four–nucleon properties. The estimate of the alpha–particle binding energy has been improved by enlarging the number of channels included in the expansion. So far, we have considered all the channels with $\sum_{i=1,3} \ell_i \leq L = 4$ (80 channels) where ℓ_i is the orbital angular momentum quantum number associated with the i–th jacobi vector, producing an increase of approximately 0.2 MeV in the binding energy with respect to the numbers reported in our previous article [1]. Calculations are now in progress to consider all the channels with $L \leq 6$. It is hoped that the binding energy estimations could reach a good convergence to settle the problem of the discrepancy between the GFMC [3] and Faddeev–Yakubovsky [4] results.

The calculation of the $T = 1$ scattering states of a $A = 4$ system (n–^{3}H and p–^{3}He elastic scattering) are almost completed. The contributions from S, P and D–waves have been considered so that quantitative comparison between the experimental and theoretical cross section and polarization observables are possible. In particular, a discrepancy between the theoretical and empirical estimation of A_y, similar to that one reported for the $A = 3$ system, has been obtained.

To conclude, an overall, quantitative agreement between theory and experiment suggests that the nuclear dynamics in these complicated low-energy processes is fairly well understood. The remaining discrepancies, such as the 30 % underprediction for the vector analyzing powers, indicate that further refinements are necessary in the present theoretical framework. Examples of these are: 1) the analysis of new TNI terms and 2) the inclusion of relativistic corrections.

1. A. Kievsky, S. Rosati, M. Viviani, Nucl. Phys. **A577**, 511 (1994); M. Viviani, A. Kievsky, S. Rosati, Few-Body Systems **18**, 25 (1995).
2. M. Viviani, R. Schiavilla A. Kievsky, Phys. Rev. C, in press.
3. J. Carlson, Phys. Rev. **C38** (1988) 1879.
4. W. Glockle and H. Kamada, Nucl. Phys. **A560** (1993) 541.

THE SPIN STRUCTURE OF ^{3}He

J. Sowinski
for the CE-25 Collaboration
Indiana University Cyclotron Facility and Department of Physics, Bloomington, Indiana 47405

The $^3\overrightarrow{He}(\vec{p},pN)$ reaction has been used to investigate the spin dependent momentum distributions of protons and neutrons in ^{3}He. We have found that with large momentum transfer the PWIA is valid, thus allowing their extraction directly from the data. The experimental results are compared to distributions determined from spectral functions generated with Faddeev techniques. It is concluded that there is good theoretical understanding of the spin of ^{3}He at the nucleon level.

The spin structure of ^{3}He, how the nucleons and their spin produce the spin of the ^{3}He nucleus, is of current interest for a number of reasons. For nuclear physicists it is an important testing ground as ^{3}He is a tightly bound nucleus where all the nucleons are essentially valence nucleons and hence important to the spin of the nucleus. Moreover the structure of ^{3}He is exactly solvable, under the assumptions of non-relativistic equations and a potential for the N-N interaction, via the Faddeev formalism. The structure of ^{3}He is calculated to be dominated (~90%) by an S-state with proton spins paired off, indicating that the spin of ^{3}He is primarily due to the neutron. The protons gain some small negative polarization from mixed symmetry S′-states and the neutrons at high Fermi momentum become unpolarized (and even somewhat negatively polarized) due to D-states in the wave function. The interest to particle physicists is that these properties make polarized ^{3}He a good substitute for a polarized neutron target and it has, in fact, been used this way in a number of the nucleon spin structure measurements discussed at this conference.

We have used the $^3\overrightarrow{He}(\vec{p},pN)$ reaction to investigate the spins of the nucleons in ^{3}He as a function of their missing momentum. Experimental details can be found in Ref. 1. In PWIA the polarization of the struck nucleons is determined by comparing the quasi-free spin observables, target analyzing power A_t and spin correlation C_{NN}, to free N-N observables under the same center of mass kinematics. By detecting the energy and angles of the two nucleons the missing momentum and missing energy can be reconstructed using momentum and energy conservation. In PWIA these are interpreted as the momentum of the struck nucleon and the excitation energy of the recoil, respectively. We have found [2] that measurements at high momentum transfer are necessary for the PWIA to be valid.

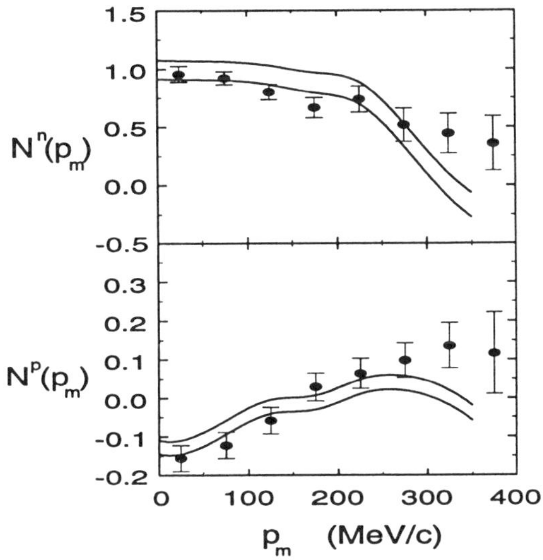

Figure 1: The spin asymmetry in the momentum distribution of neutrons (upper panel) and protons (lower panel) in ^{3}He as determined from N-N quasi-free scattering. The theoretical curves are based on Faddeev calculations as described in the text.

The spin dependent momentum distributions extracted from C_{NN} at 200 MeV are shown in Fig. 1. These are compared to curves generated from spin dependent spectral functions,[3] the probability of finding a nucleon with a given spin and momentum in a nucleus whose removal results in a given missing energy, by integrating over missing energy and using monte carlo techniques to account for experimental resolutions.[4] The error band on the curves is due to the uncertainty in the free N-N observables from different phase shift analyses. Note that the presence of S′-states is confirmed by the negative values of the spin distribution for protons at low missing momentum. We conclude there is reasonable agreement out to 300 MeV/c and that there is good theoretical understanding of the spin of ^{3}He at the nucleon level.

I gratefully acknowledge my collaborators, the authors of Refs. 1,2 and 4.

References

1. C. Bloch *et al.*, *Nucl. Instrum. Methods* A **354**, 437 (1995).
2. M.A. Miller *et al.*, *Phys. Rev. Lett.* **74**, 654 (1995).
3. R.-W. Schultze and P.U. Sauer, *Phys. Rev.* C **48**, 38 (1993).
4. R.G. Milner *et al.*, *Phys. Lett.* B **379**, 67 (1996).

Some results in the 4 nucleon low energy scattering problem

F. Ciesielski, J. Carbonell, C. Gignoux
Institut des Sciences Nucléaires, 53. Av. des Martyrs, 38026 Grenoble, France

Results of solving the Faddeev-Yakubowsky (FY) equations in configuration space for the low energy 4 nucleon problem in the S-wave approximation are presented.

1 Introduction

The continuum spectrum of the 4N system provides a rich variety of resonant structures [3] and makes a link between the relative simplicity of the A=2,3 problems and the nuclear complexity. The powerful tools developped in the recent years in the few-body community makes possible an accurate solution of the 4-body scattering problem. The understanding of the A=4 chart in terms of the fundamental NN forces constitutes an exciting theoretical challenge. We have started a program with such a purpose by solving the FY equations in configuration space. The first results concern the scattering reactions below the first three particle break-up threshold. They have been obtained with the MT I-III S-wave interaction without Coulomb forces. Within this approximation the ^{3}H and ^{3}He are degenerate with E=-8.53 MeV.

2 Results

The scattering length (a) and effective range (r_0) for the different spin-isospin (S,T) elastic 1+3 channels are shown in Table 1 together with the comparable calculations and the existing experimental values.

Table 1: Low energy parameters for the N-3N scattering.

S	T	a [1]	a [2]	a	r_0	a_{exp} [4]
0	0	-25.6	-	14.75	6.5	-
1	0	2.65	-	3.25	1.8	-
0	1	4.09	4.0	4.13	2.0	4.45
1	1	3.61	3.6	3.73	1.8	3.32

2.1 n-t

The n-t reaction is a pure T=1. The corresponding scattering lengths in Table 1 are to be compared to the last published experimental values. The measured zero energy cross section is 170.18 ± 9.7 fm^2 whereas we found 184.6. The S-wave alone is unable to reproduce the observed structure at $E_{cm} \approx 3$ MeV, attributed to P-wave resonances [3]. We have included L=1 states in the n-t relative motion. The results, shown in Fig. 1, indicate that although a substantial part of the structure is reproduced, there is a clear need for P-wave interaction.

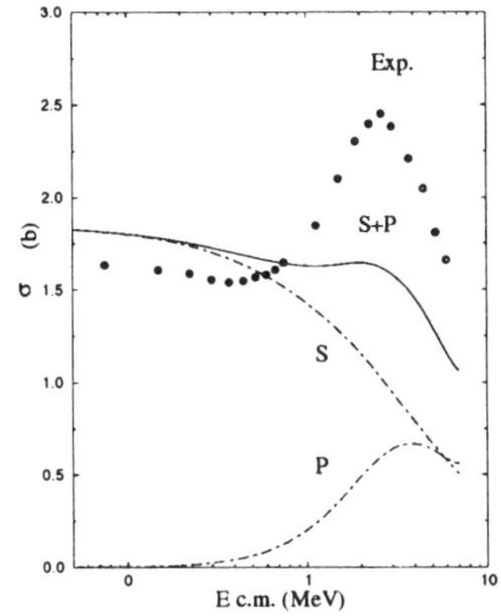

Figure 1: n-t total cross section.

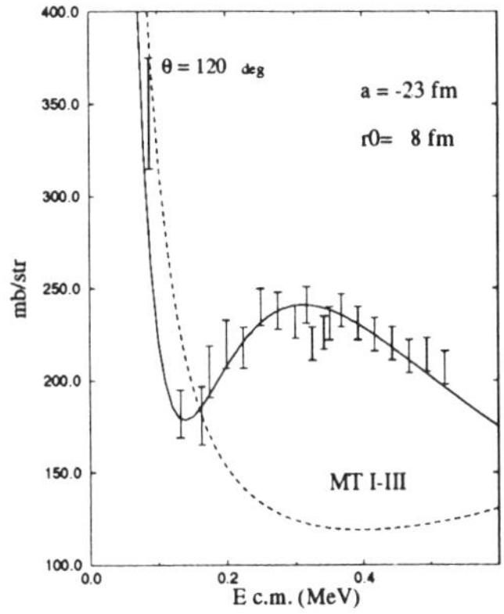

Figure 2: p-t differential cross section.

2.2 p-t

Few data were found concerning the p-t reaction below the n-^{3}He threshold. In Fig. 2 is shown the differential cross section at fixed c.m. angle $\theta = 120°$. Neglecting Coulomb, the p-t strong scattering lengths are simply obtained by averaging the isospin values, i.e. $a_0^{pt} = 9.44$ and $a_1^{pt} = 3.49$. These quantitites, inserted in the Coulomb corrected Bethe formula, provide the low energy strong amplitude f_{sc}. The total amplitude f is obtained by adding the Coulomb one, so that $f = f_c + f_{sc}$. The results dramatically fail to reproduce the data (dashed line). To account for the observed strong+Coulomb interference figure, a large negative a_{pt} value is required. A good fit is obtained with $a_{pt} = -23$ fm and $r_0 = 8$ fm. The physical origin of that is discussed below.

2.3 ^{4}He

The binding energy for the ^{4}He is 30.2 MeV, in agreement with several authors. A first excited state was found with $E = -0.27$ MeV below the 3N threshold. The proximity of this state to the threshold is responsible for the big value of a in the S=T=0 channel. This state appears as a $J^\pi = 0^+$ resonance, 0.4 MeV above p-t in [3]. Its position is compatible with the negative value of the scattering length required by the p-t differential cross section. It's worth to notice that the right position of this resonance is a crucial point for any attempt to describe the low energy p-t and n-^{3}He data. The Coulomb repulsion could be enough for that. By slightly modifying the parameters of the potential we failed in pulling the state out of the bound region, the 3N threshold moving in the same direction. An ad-hoc 4-body term was successful in giving a negative scattering length without susbtantially modifying the 3 and 4N ground states.

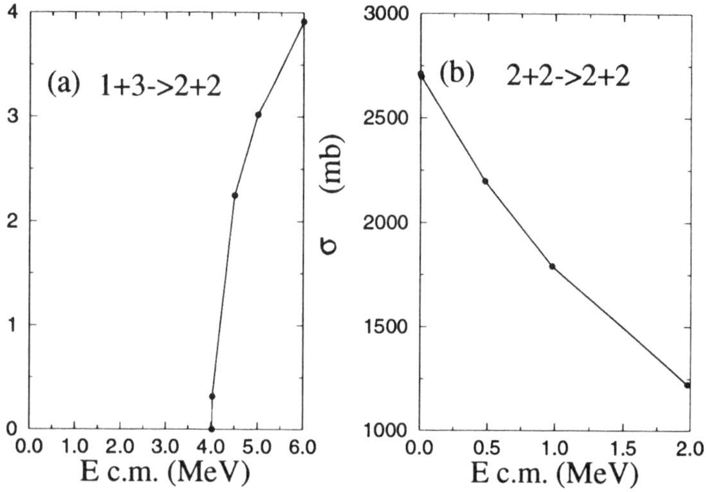

Figure 3: First inelastic channel: (a) 1+3→ 2+2 and (b) 2+2→2+2 cross section.

2.4 d-d

The 2×2 S matrix for the first inelastic reaction $N + 3N \to NN + NN$ was obtained and cross sections are shown in Fig. 3. The comparison with the experimental data is still premature due to the simplicity of the interaction.

References

1. J.A. Tjon, Phys. Lett **63B** (1976) 391
2. S.L. Yakovlev, I.N. Filikin, *Phys. At. Nucl.* **58**, 754 (1995)
3. D.R. Tilley, H.R. Weller and G.M. Hale, *Nucl. Phys.* B **541**, 1 (1992)
4. G.M. Hale et al. *Phys. Rev.* C **42**, 438 (1990)
5. N. Jarmie et al., *Phys. Rev.* **130**, 1987 (1963)

Boson Exchange NN Potential with Nonlinear Boson Fields

L. Jäde and H. V. von Geramb
Theoretische Kernphysik, Universität Hamburg
Luruper Chaussee 149, D-22761 Hamburg

Motivated by chiral invariant meson–baryon Lagrangians and spontaneous symmetry breaking, we use a Lagrangian

$$\mathcal{L}_\Phi = \frac{1}{2}\left(\partial_\mu\partial^\mu\Phi^2 - m^2\Phi^2\right) - \frac{\lambda_1}{2p+2}\Phi^{2p+2} - \frac{\lambda_2}{4p+2}\Phi^{4p+2} + \mathcal{L}_{int}, \tag{1}$$

in which Φ stands for the meson fields and $\mathcal{L}_{int}$ accounts for the interaction with nucleons and other fields. The nonlinearity is of utmost importance. The decoupled single meson field satisfies a nonlinear Klein–Gordon equation (KGE)

$$\partial_\mu\partial^\mu\Phi + m^2\Phi + \lambda_1\Phi^{2p+1}\lambda_2\Phi^{4p+1} = 0. \tag{2}$$

As physical relevant solutions Φ we take an expansion in terms of wave solutions φ of the free KGE. This is known as method of base equations [1]. This substitution reduces (2) to an ordinary nonlinear differential equation, which permits direct integration. The solutions are expressed as a power series in terms of φ and Gegenbauer polynomials

$$\Phi(\varphi) = \sum_{n=0}^{\infty} C_n^{1/2p}(w)\, b^n\, \varphi^{2pn+1}, \tag{3}$$

with

$$b = \left\{\left[\frac{\lambda_1}{4(p+1)m^2}\right]^2 - \frac{\lambda_2}{4(2p+1)m^2}\right\}^{\frac{1}{2}} \quad \text{and} \quad w = \frac{1}{b}\frac{\lambda_1}{4(p+1)m^2}.$$

These particular solutions $\Phi(\varphi)$ are *self–interacting solitary meson fields*. Quantization is achieved by using the quantized solutions of the free KGE

$$\varphi(x,k) = \frac{1}{\sqrt{2\omega_k V D_k}}\, a(k)\, e^{-ikx} \quad \text{and} \quad \varphi^\dagger(x,k) = \frac{1}{\sqrt{2\omega_k V D_k}}\, a^\dagger(k)\, e^{ikx}, \tag{4}$$

where V is the volume of the system and D_k is a momentum dependent Lorentz–invariant function which is later used to guarantee finite self–scattering amplitudes. Setting

$$\Phi(x,k) = \Phi(\varphi(x,k)) = \sum_{n=0}^{\infty} C_n^{1/2p}(w)\, b^n\, \frac{1}{(2\omega_k V D_k)^{pn+1/2}}\, a^{2pn+1}(k)\, e^{-ikx(2pn+1)} \tag{5}$$

one obtains the annihilation and creation operators for solitary mesons. Using

$$\Phi_N^\dagger(x,k) = \frac{1}{\sqrt{N!}}|N,k\rangle\langle N,k|\Phi^\dagger(x,k) \quad \text{and} \quad \tilde{\Phi}(x) = \sum_{\vec{k},N}\left(\Phi_N(x,k) + \Phi_N^\dagger(x,k)\right), \tag{6}$$

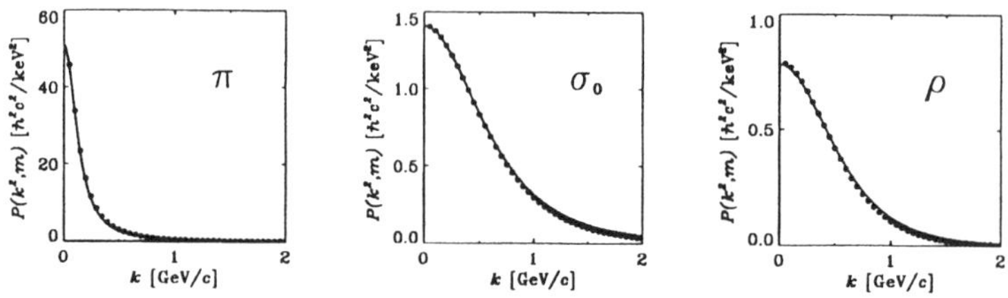

Figure 1: Solitary meson propagators (full) compared to Feynman propagators multiplied with Bonn–B form factors (circles).

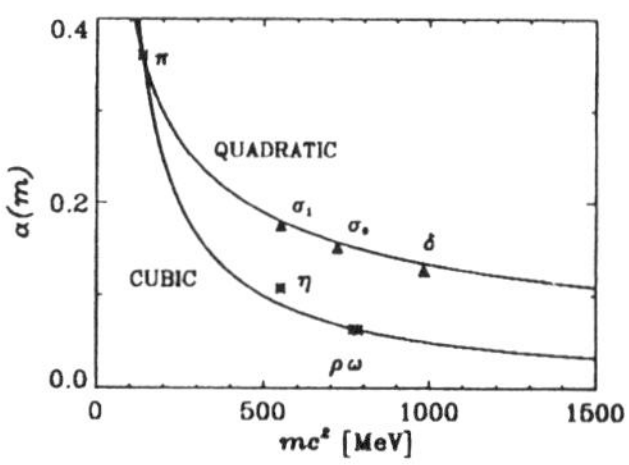

Figure 2: Self-interaction coupling constants predicted by the scaling law versus empirical values corresponding to the Bonn–B amplitudes.

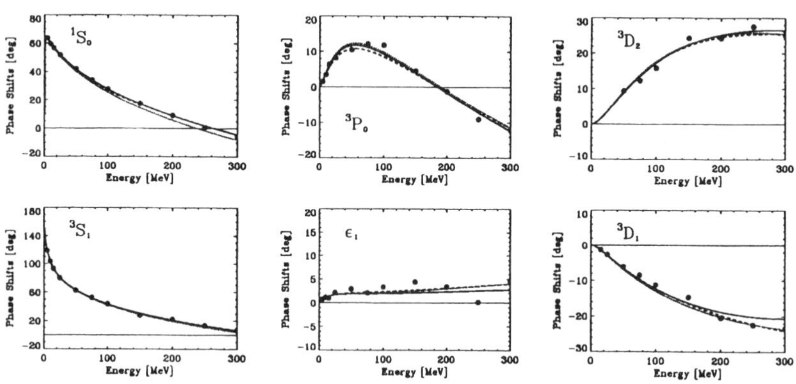

Figure 3: *np* phase shifts: Arndt SM95 (circles), Nijmegen PWA (dashed), Bonn–B (dotted) and OSBEP (full).

Triton Binding Energy and Minimal Relativity

F. SAMMARRUCA AND R. MACHLEIDT

Department of Physics, University of Idaho,
Moscow, Idaho 83843, U.S.A.

Abstract

For relativistic three-body calculations, essentially two different approaches are in use: field theory and relativistic 'direct interactions'. Results for relativistic corrections of the triton binding energy obtained from the two approaches differ even in their sign, which is most puzzling. One fundamental conceptual difference between the two approaches is the use of an invariant two-body amplitude in the field-theoretic approach. We show that this increases the triton binding energy by about 0.30 MeV, which may explain a large part of the discrepancy.

In recent years, many careful calculations of the triton binding energy with realistic nucleon-nucleon (NN) interactions have been performed. Employing local two-nucleon potentials of high quality, the triton binding energy is predicted to be 7.62(1) MeV [1]—in 34-channel charge-dependent nonrelativistic Faddeev calculations. Performing the same type of three-body calculation, the result of 8.00 MeV is obtained when a high-precision non-local NN potential (CD-Bonn) is used [2]. The experimental value for the triton binding energy is 8.48 MeV.

However, most calculations employing realistic two-body forces have been performed in a non-relativistic framework. Due to the presence of high momentum components in the triton wave function, it is not clear that the non-relativistic approximation is justified. Moreover, even if relativistic effects are small, they may be large as compared to the small discrepancy between theory and experiment. In any case, it is important to have an accurate accounting of relativistic corrections to the triton binding energy.

For calculations of the relativistic three-body bound-state, essentially two different theoretical frameworks have been applied:

1. field theory,

2. relativistic direct interactions.

Within the first scheme, Rupp and Tjon [3] used Bethe-Salpeter equations for two and three particles. They constructed covariant multirank interactions from phenomenological as well as meson-exchange potentials (Paris and Bonn). They find that relativity increases the triton binding energy by 0.20 to 0.38 MeV (depending on the two-body interaction applied).

The field-theoretic approach has also been pursued in our own work, in which we applied the relativistic three-dimensional version of the Bethe-Salpeter equation

Table 1: Deuteron Properties

	OSBEP	Exp.		OSBEP	Ex
E_B (MeV)	2.224589	2.22458900(22)	A_s ($\mathrm{fm}^{-1/2}$)	0.8803	0.880
μ_D	0.8529^b	0.857406(1)	D/S	0.0257	0.025
Q_D (fm^2)	0.2686^b	0.2859(3)	r_{rms} (fm)	1.9566	1.962

[b]Meson exchange current contributions not included

the *solitary meson propagator* is defined by

$$iP(y-x) = \langle 0|T\tilde{\Phi}(y)\tilde{\Phi}(x)|0\rangle,$$

in momentum space this propagator has the form

$$iP(k^2,m) = \sum_n \left[C_n^{\frac{1}{2p}}(w)\right]^2 (m^p\alpha)^{2n} \frac{(2pn+1)^{2pn-2}}{D_k^{2pn+1}\left(\vec{k}^{\,2}+M_n^2\right)^{pn}} i\Delta_F(k,M_n)$$

where $\alpha = b/(2mV)^p$, $M_n = (2pn+1)m$ and $i\Delta_F(k^2,m)$ denotes the
propagator.

In application to NN interactions this framework of nonlinear boson fie
the following essential results [2]:
i.) The phenomenological form factors of the Bonn–B potentials [3] are re
the dynamics of solitary mesons. Fig. 1 shows a representative comparison
the solitary meson propagators for π, σ_0 and ρ and a corresponding Feynn
agator multiplied with the Bonn–B form factor. *ii.*) We deduce an empiric
law which relates any meson nonlinearity to the pion self–interaction. Fig
this scaling relation. *iii.*) Computed NN phase shifts are in excellent a
with experimental data, see Fig. 3 for some selected partial waves. *iv.*)
properties are extremely well reproduced, see Table 1. The D–state prob
this NN interaction model amounts to be $P_D = 4.7\,\%$.

References

[1] P. B. Burt, *Quantum Mechanics and Nonlinear Waves* (Harwood
New York, 1981).

[2] L. Jäde and H. V. von Geramb, LANL Preprint Server nucl–th/960
mitted to Phys. Rev. C. (1996).

[3] R. Machleidt, Adv. in Nuclear Physics **19**, 189 (1989).

proposed by Blankenbecler and Sugar, together with relativistic one-boson exchange potentials [4]. We find an increase of the triton binding energy due to relativistic effects of 0.19 MeV using the Bonn B potential. The same result is obtained for a more recent, high-precision version of the Bonn potential, resulting in a total triton binding energy of 8.19 MeV for the CD-Bonn potential [2].

In the approach that uses relativistic direct interactions, a relativistic Hamiltonian is defined as the sum of relativistic one-body kinetic energies and two-body interactions and their boost corrections [5,6,7,8]. The latter are derived from commutation relations of the Poincaré group. Using this relativistic framework, it is found that relativistic effects *reduce* the triton binding energy by approximately 0.30 MeV [9].

In summary, when comparing results from the two different relativistic approaches listed above, even the sign of the relativistic corrections is controversal. This fundamental difference in the results calls for a serious investigation of the issue.

It is the purpose of this paper to start such an investigation.

If for the relativistic approach, the framework of relativistic quantum field theory is chosen, then one may distinguish between the following three sources for relativistic effects in the three-body system:

- Relativistic (invariant) two-nucleon amplitude (t-matrix),
- relativistic kinematics,
- relativistic Faddeev propagator.

To ensure maximal transparency of a systematic investigation, one may focus on one of these effects at a time.

Recently, we have started to investigate the first point of this list, namely the effects caused by using a relativistic (invariant) two-body t-matrix, t_{rel}, as input to the Faddeev equations, as compared to the usual nonrelativistic t-matrix, t_{NR}, used in the conventional approach. The two t-matrices can be related by

$$t_{rel}(q', q) = \sqrt{\frac{E'}{m}} t_{NR}(\mathbf{q}', \mathbf{q}) \sqrt{\frac{E}{m}} \tag{1}$$

where $E = \sqrt{m^2 + \mathbf{q}^2}$, $E' = \sqrt{m^2 + \mathbf{q}'^2}$, and m is the nucleon mass. This relationship has become known as *minimal relativity* [10].

Our result is that the effect on the triton binding energy due to Eq. (1) is attractive and, using a wide range of potentials, *increases the binding energy by about 0.30 MeV.*

Of course, the other aspects of relativity mentioned above also play a role, but they introduce uncertainties of a different nature. For instance, various three-dimensional relativistic three-body equations are available, which all have in common the elimination of the relative time component in the original four-dimensional integration over internal momenta. This procedure is carried out in such a way as to preserve relativistic invariance and three-particle unitarity, but it is not unique. Also, the effect of relativistic kinematics is dependent on the way relative momenta of the interacting pair are defined [11]. Thus, *even within the same field-theoretic approach*, some ambiguities are inherent to any three-dimensional reduction of the relativistic equation. These require a separate investigation.

On the other hand, the effect of minimal relativity, which we have chosen to single out in this work, is clear-cut and its origin is easy to pin down. It is a consequence of manifest covariance, and it is absent from approaches based on relativistic direct interactions. Therefore, it may shed some light on the discrepancy between the two major theoretical frameworks.

References

[1] J.L. Friar *et al.*, Phys. Lett. B **311**, 4 (1993).

[2] R. Machleidt, F. Sammarruca, and Y. Song, Phys. Rev. C **53**, 1483 (1996).

[3] G. Rupp and J.A. Tjon, Phys. Rev. C **45**, 2133 (1992).

[4] F. Sammarruca, D.P. Xu, and R. Machleidt, Phys. Rev. C **46**, 1636 (1992).

[5] B. Bakamjian and L. H. Thomas, Phys. Rev. **92**, 1300 (1952).

[6] L. L. Foldy, Phys. Rev. **122**, 275 (1961); R. A. Krajcik and L. L. Foldy, Phys. Rev. D **10**, 1777 (1974).

[7] B. D. Keister and W. N. Polyzou, Adv. Nucl. Phys. **20**, 225 (1991).

[8] B. D. Keister, AIP Conf. Proc. **334**, 164 (1995).

[9] J. Carlson, V.R. Pandharipande, and R. Schiavilla, Phys. Rev. C **47**, 484 (1993); J.L. Forest, V.R. Pandharipande, and J.L. Friar, *ibid.* **52**, 568 (1995); J.L. Forest, V.R. Pandharipande, J. Carlson, and R. Schiavilla, *ibid.* **52**, 576 (1995).

[10] G.E. Brown, A.D. Jackson, and T.T.S. Kuo, Nucl. Phys. **A133**, 481 (1969).

[11] A. Stadler and F. Gross, AIP Conf. Proc. **334**, 867 (1995).

STUDY OF $pp \to pp\pi^o$ REACTION IN CHIRAL PERTURBATION APPROACH

K. KUBODERA, B.-Y. PARK, F. MYHRER, T. MEISSNER, J.R. MORONES
Department of Physics and Astronomy, University of South Carolina, Columbia, South Carolina 29208, USA

The recent high-precision experiments[1] on the $pp \to pp\pi^0$ reaction near threshold indicate that the observed total cross sections are much larger (by a factor of ~5) than the prediction of the conventional calculations [2]. This surprising feature has invited many theoretical investigations.[3-5] Motivated by these developments, we have attempted[6] at a systematic treatment of this reaction with the use of chiral perturbation theory (χPT) or, to be more specific, with the use of the heavy-fermion formalism of χPT. It is widely believed that χPT[7] provides a general guiding principle for describing low-energy hadron dynamics.

Our calculation indicates that a systematic treatment of all the diagrams (up to the lowest meaningful order of chiral counting) gives significantly different results from those of the conventional treatments, both for the one-body Born term and for the two-body rescattering term. In particular, the off-shell $\pi\pi NN$ vertex obtained in χPT gives rise to a large enhancement of the pion rescattering term. Hernández and Oset[4] demonstrated earlier an off-shell enhancement of the rescattering diagram using the phenomenological parametrization of the $\pi\pi NN$ vertex. We have found,[6] however, that the sign of the enhanced rescattering vertex obtained in χPT is opposite to that reported in Ref. 4. This flip of the sign leads to a destructive interference between the one-body and rescattering terms, resulting in theoretical cross sections that are much too small (typically by a factor of ~100) compared with the experimental values. This feature, which was also noted by Cohen, Friar, Miller and van Kolck[8], seems to indicate that the threshold $pp \to pp\pi^0$ reaction is highly sensitive to processes of much shorter ranges than the conventional Born and rescattering terms.

The σ-meson exchange diagram considered by Lee and Riska[3] is a prominent, if not the unique, candidate of these short-range processes. However, in order to determine whether the σ-meson exchange process is the true agent to remove the large discrepancy between the observed and theoretical cross sections, we need to examine the following issues. First, the above-mentioned extremely small transition amplitude resulting from cancellation between the Born and rescattering terms implies that, even after the heavy-meson-exchange contributions are added, the calculated cross sections remain much smaller than the experimental values.[8] This is indicative of the extremely delicate nature of the "good" agreement achieved in Ref. 3 by invoking the σ-meson exchange mechanism. Having said that, we must quickly add that both of the existing χPT calculations[6,8] of the $pp \to pp\pi^0$ reaction share the same approximation which in Ref.6 was referred to as the "fixed kinematics approximation". Namely, to facilitate the Fourier transformation of the two-body transition operator from $\mathbf{p}$- to $\mathbf{r}$-representation, some of the kinematical variables in the rescattering Feynman diagrams are fixed at the values that correspond to the strict threshold kinematics with initial or final nucleon-nucleon interactions turned off. Hanhart *et al.*'s detailed calculation in momentum space[5] indicates that the "fixed kinematics approximation" needs to be carefully examined.

We are currently carrying out an improved χPT calculation[9] in which the nuclear transition matrix elements are evaluated with the use of momentum-space nuclear wave functions. A direct $\mathbf{p}$-space calculation allows us to avoid the problematic "fixed kinematics approximation". It is of great interest to see to what extent this improved calculation modifies the results of the existing calculations. We are particularly interested in the role of the σ-exchange diagram in this kinematically improved treatment

The work reported here is supported in part by the National Science Foundation, USA, Grant No. PHYS-9310124.

References

1. H.O. Meyer *et al.*, Phys. Rev. Lett. **65**, 2846 (1990); Nucl. Phys. **A539**, 633 (1992); A. Bondar *et al.*, Phys. Lett. **B356**, 8 (1995).
2. D.S. Koltun and A. Reitan, Phys. Rev. **141**, 1413 (1966).
3. T.-S.H. Lee and D.O. Riska, Phys. Rev. Lett. **70**, 2237 (1993).
4. E. Hernández and E. Oset, Phys. Lett. **B350**, 158 (1995).
5. C. Hanhart et al, Phys. Lett. **B358**, 21 (1995).
6. B.-Y. Park, F. Myhrer, R. Morones, T. Meissner and K. Kubodera, Phys. Rev. C **53**, 1519 (1996).
7. S. Weinberg, Physica **96A**, 327 (1979); J. Gasser and H. Leutwyler, Ann. Phys. **158**, 142 (1984); E. Jenkins and A.V. Manohar, Phys. Lett. **B255**, 558 (1991); for a recent review, see e.g., V. Bernard, N. Kaiser and U.-G. Meissner, Int. J. Mod. Phys. **E4**, 193 (1995).
8. T.D. Cohen, J.L. Friar, G.A. Miller and U. van Kolck, preprint: nucl-th/9512036
9. T. Sato (Osaka University), T.-S.H. Lee (Argonne National Laboratory) and the University of South Carolina Group, in progress

CHIRAL LAGRANGIANS AND QUARK CONDENSATE IN NUCLEI

M. ERICSON , G. CHANFRAY and J. DELORME
Institut de Physique Nucléaire et IN2P3, CNRS, Université Claude Bernard Lyon I, 43 Bd. du 11 Novembre, F-69622 Villeurbanne Cedex, France

The density evolution of the quark condensate is governed by the nuclear sigma commutator Σ_A (per nucleon) with the exact relation [1]:

$$\bar{q}q(\rho)/\bar{q}q(0) = 1 - \rho\Sigma_A(\rho)/f_\pi^2 m_\pi^2 \qquad (1)$$

When the PCAC relation holds, Σ_A is given by the scattering amplitude of soft pion on the nucleus A (T_A^0 per nucleon), according to $\Sigma_A = -f_\pi^2\, T_A^0$. The condensate question can then be treated as a scattering problem.

One example is the linear sigma model. In the impulse approximation $T_A^0 = T_N^0 = -\Sigma_N/f_\pi^2$ which leads to the linear drop with density of the condensate. Various corrections apply:

i) The distortion of the soft pion wave (or coherent rescattering) with [2] $T_A^0 = T_{Born}^0/(1-\rho T_{Born}^0/m_\pi^2)$. Taking the low density limit $T_{Born}^0 = T_N^0 = -\Sigma_N/f_\pi^2$ replaces the linear drop by an hyperbolic one, preventing the full restoration to occur.

ii) However Birse and Mc Govern [3] pointed out that the model imposes certain σ meson exchange corrections to the amplitude (graphs 1 a-c). To first order in ρ, $T_{Born}^0 = -(\Sigma_N/f_\pi^2)(1+\frac{5}{2}\rho\Sigma_N/f_\pi^2 m_\pi^2)$, helping the restoration.

iii) Besides the coherent rescattering, the incoherent one (correlations) can also play an important role [4].

iv) Correlations should be introduced also in the σ meson exchange terms [5].

The effects *iii*) and *iv*) have to be treated together due to partial cancellation between them [5], which are necessary to fulfill the requirement of chiral symmetry that the amplitude should behave as m_π^2. The total amplitude (including distortion) is, to first order in ρ: $T_A^0 = -(\Sigma_N/f_\pi^2)(1+\frac{3}{2}\Sigma_N\rho/f_\pi^2 m_\pi^2)$. The bracket is larger than unity, thus helping the restoration process. However the σ exchange graphs which are sea-gull ones also apply to physical pion. Their contribution to the s-wave optical potential parameter is $= -0.023 m_\pi^{-1}$, too large in view of the total measured value $(b_0)_{eff} = -0.029 m_\pi^{-1}$ which arises from a large part from other known sources. This was our motivation to go to the non-linear sigma model [5].

We start from the linear model, modifying the chiral symmetry breaking

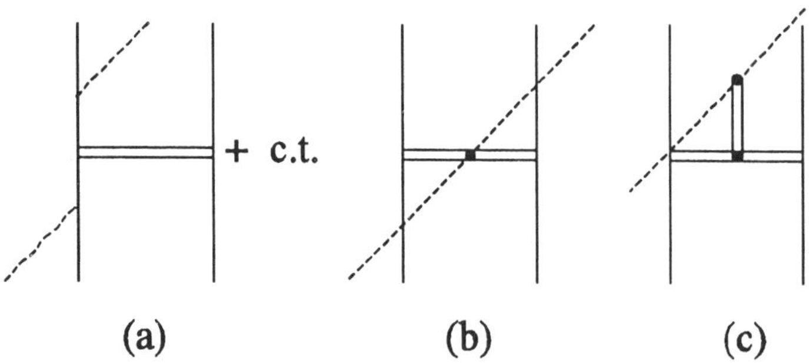

Figure 1: σ meson exchange graphs for scattering of soft and physical pions.

Lagrangian, but keeping the same chiral transformation properties:

$$\mathcal{L}^{\sigma}_{CSB} = f_\pi m_\pi^2 \sigma - \frac{\Sigma_N}{f_\pi^2}\sigma\, \overline{\psi}(\sigma + i\tau.\pi\gamma_5)\psi \ . \tag{2}$$

We eliminate the σ field, writing $\sigma = f_\pi \cos(F)$ and redefine the pion field $\pi = f_\pi \hat{\phi} \sin(F)$. We also transform the nucleon field in order to recover the free nucleon sigma commutator in the tree approximation. We have adopted two forms for the function F. The first one, $\tan(F/2) = \phi/2f_\pi$, leads to a Weinberg type version (see *e.g.* ref. [6]) which does not verify PCAC. The second one $\sin(F) = \phi/f_\pi$ together with a suitable transformation of the pion field leads to the PCAC version. In both cases there is one common correction due to the interaction between the nucleons to the relation $\Sigma_A = \Sigma_N$. It represents the contribution to the sigma commutator of the exchanged pions and is given by the correlated part of the expectation value of the squared pion field:

$$\Sigma_A = \Sigma_N + \frac{1}{2} m_\pi^2 \left[\langle 0|\phi^2|0\rangle\right]_{correl.} \ . \tag{3}$$

In addition, other corrections appear in the PCAC version. In particular Σ_A, proportional to the soft amplitude, is consequently affected by the coherent and incoherent rescattering. Moreover a contact term $\phi^2(\bar{N}N)^2$ replaces the σ meson exchange of the linear sigma model and affects T_A^0 and hence Σ_A . However there are total cancellations between the corrections: one occurs (to all orders) between the distortion and the contact terms, another one between the incoherent rescattering and the effect of correlations on the contact term, irrespective of the correlation profile. Altogether there is no further correction than the one of eq.(3) and the condensate is independent of the version of the Lagrangian, a result also reached by Kirchbach and Wirzba [7].

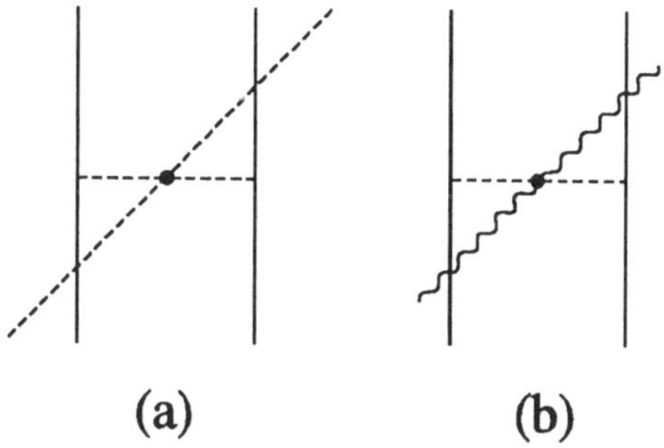

Figure 2: Sea-gull graphs for pion (a) and photon (b) scattering on exchanged pions.

In terms of PCAC the correction term in eq. (3) represents the scattering of soft pions off the exchanged pions (sea-gull graph of fig. 2a). It occurs as well as a pion exchange correction in the case of physical pions, *i.e.* in the s-wave optical potential, but also in the forward Compton amplitude [8] (fig. 2b) after appropriate scaling. Thus the question of the quark condensate is linked to pion and photon physics. As an example, in the case of the deuteron, from the calculated exchange contribution in the Compton amplitude $F_d^{\pi\, exch.} \approx (e^2/8M_N)\times 0.25$ [8], one can deduce the exchange contribution to the sigma commutator $\Sigma_d^{\pi\, exch.} = 1.6$ MeV and to the π-d scattering length: $\delta_{\pi-d}^{\pi\, exch.} = -0.002 m_\pi^{-1}$, which is significant at the level of the present experimental accuracy.

Another exploitation of this link is the extraction of the correction in ϕ^2 of eq. (3) in nuclei from the photoabsorption data [9].

References

1. G. Chanfray and M. Ericson, *Nucl. Phys.* A **556**, 427 (1993).
2. M. Ericson, *Phys. Lett.* B **301**, 11 (1993).
3. M.C. Birse and J.A. Mc Govern, *Phys. Lett.* B **309**, 231 (1993).
4. T. Ericson, *Phys. Lett.* B **321**, 312 (1994).
5. J. Delorme, G. Chanfray and M. Ericson, *preprint LYCEN 9604*, to appear in *Nucl. Phys. A*.
6. B.W. Lynn, *Nucl. Phys.* B **402**, 281 (1993).
7. M. Kirchbach and A. Wirzba, *preprint IKDA 96/7, nucl-th/9603017.*
8. M. Ericson and M. Rosa-Clot, *Phys. Lett.* B **188**, 11 (1987).
9. M. Ericson and M. Rosa-Clot, *in preparation.*

Effective Meson Masses in Nuclear Matter Based on A New Evaluation of Vacuum Polarization

Masahiro Nakano

University of Occupational and Environmental Health, Kitakyushu 807, Japan

Nobuo Noda, Tomohiro Mitsumori, Kazuharu Koide, Hiroaki Kouno and Akira Hasegawa

Saga University, Saga 840, Japan

Toshitake Kohmura

Tukuba University of Tukuba, Ibaraki 305, Japan

(May 24, 1996 Talk at PANIC'96)

Effective meson masses of two renormalizable mesons σ and ω, and two nonrenormalizable meson π and ρ are calculated in the random phase approximation.

We show a possible method to evaluate the meson self-energy in the time-like region. Two distinct deference can be found when we compare to the previous method.

First is the representation of the nucleon propagator. Different from the traditional density-Feynman representation of the nucleon propagator, we use the particle, hole and antiparticle propagator in order that the meson self-energy naturally represents the physical processes such as particle-hole or particle-antiparticle excitations, and that it includes no unphysical components. It is important that only particle-antiparticle components contribute to the meson mass in this representation, thus we calculate only the particle-antiparticle part of the meson self-energy. This is one of the merits of this representation.

Second is the problem of the evaluation of the divergent integrals. Commonly, infinities are subtracted according to the usual prescription of renormalization. However, there exist arguments that the description of the interaction of mesons and nucleons should naturally lead to an introduction of a momentum cutoff because of the finite size of hadrons.

Furthermore, renormalization prescription has another important problem when we treat nonrenormalizable mesons such as ρ or π mesons, the problem is how to estimate the vacuum contribution of nonrenormalizable mesons.

In our study, we approached this problem by introducing form factors. But the simple use of the form factor leads us unphysical results, strong cutoff dependence. In order to reduce the cutoff dependence, we subtract the divergent terms so as to obtain the convergent value at the large cutoff.

Our prescription is the introduction of form factor following the idea of renormalization. In this prescription, the form factor works as a regulator to make the divergent integrals finite. We subtract the following 4 terms to cancel the divergent terms in meson self-energy.

$$f_{total} \equiv f_0 - f_a - f_b - f_c - f_d \quad (1)$$

where the original integral f_0 and the subtracted terms include the same form factor and are given by

$$\begin{aligned}
f_0 &= \Pi_\Lambda(R)\\
f_a &= \Pi_\Lambda(R)|_{\rho_B=0,x=0}\\
f_b &= (M^* - M)\frac{\partial}{\partial M^*}\Pi_\Lambda(R)|_{\rho_B=0,x=0}\\
f_c &= \frac{1}{2}(M^* - M)^2\frac{\partial^2}{\partial M^{*2}}\Pi_\Lambda(R)|_{\rho_B=0,x=0}\\
f_d &= x\frac{\partial}{\partial x}\Pi_\Lambda(R)|_{\rho_B=0,x=0}\\
x &\equiv R_\mu^2 + m^2
\end{aligned} \tag{2}$$

These subtracted terms are the same as the conventional renormalization in the σ meson case. It is important, however, all the terms are finite and tractable in this prescription because they include the form factors. The renormalization point here is taken at $\rho_B = 0$ (zero baryon density) $x \equiv R_\mu^2 + m^2 = 0$ (on-shell momentum with the rest mass).
This choice ensures that the meson mass at zero-baryon density coincides with the free meson mass, and this condition is important in order that this subtraction using form factors is the same as the conventional renormalization in the σ meson case.
What we calculate here is the meson mass shift due to the finite baryon-density.
Since the above prescription is applicable to any meson, we now present numerical results for the effective meson masses based on the relativistic Hartree approximation. In this calculation, coupling constants are given by $g^\sigma = 7.93$, $g^\omega = 8.98$, $f^\pi = 0.98$,$g^\rho = 2.72$ and $\kappa^\rho = 6.0$.
First, the cutoff parameter Λ dependence of the σ and ω meson mass is shown in Fig.1. From this figure, we can see the behavior at the large cutoff parameter Λ. We found that the meson mass using the original integral f_0 depends strongly on Λ. This means that the calculation of simple use of form factor gives an unphysical result, namely, very large mass at a small cutoff. The calculation with fewer subtraction of $f_0 - f_a - f_d$ also depends strongly on the cutoff parameter. It does not give a convergent value. Only the fully subtracted meson self-energy gives convergence, and it is important the Λ dependence is very weak in this full subtraction. Almost the same meson mass is obtained at any cutoff over around $\Lambda = 2000$MeV.
In the case of the σ meson, we can show that the large limit of the Λ coincides with the renormalized meson self-energy. For this case, we obtain the analytic expression of the finite value after subtraction. We can compare our results with this renormalized meson self-energy.
In Fig.2, the full line shows the conventional renormalized meson self-energy, and the dotted lines are the fully subtracted meson self-energies using form factors with various cutoff. This figure shows that the fully subtracted meson self-energy using form factor approaches to the renormalized value when the cutoff parameter Λ becomes large. This means that this prescription is a kind of renormalization using form factors as a regulator in the renormalizable meson case. Thus we apply this prescription to the nonrenormalizable mesons.

The Λ dependence of the effective masses of π and ρ mesons are shown in Fig.3. As in the case of σ meson, the original integral f_0 and the insufficient subtraction do not give convergence. In this case, the mesons are nonrenormalizable, but the convergence is obtained to subtract the divergent 4 terms as in σ meson. It is shown that these nonrenormalizable meson self-energy has the second order and the logarithmic divergence, and that they do not have higher divergence. Thus a similar subtraction gives the convergent value.

Fig.4 shows that the density dependence of the meson mass (at the $\Lambda = 5000$ MeV). From this figure, we notice that the σ meson mass reduces slightly but the other mesons have very weak density dependence. Especially ω and ρ meson masses do not reduce greatly, which does not agree with previous studies. The difference comes from the method of subtraction. Many calculations use dimensional regularization and subtract only two terms, while we subtracted four terms to cancel the divergent Λ dependence.

In summary, this prescription is the introduction of form factor following the idea of renormalization. Actually this is the same as the conventional renormalization in the case of σ meson. The form factor works as a regulator to make the divergent integrals finite and tractable. Since this prescription is not the renormalization but a kind of subtraction to cancel the divergent terms, we can apply this prescription to the nonrenormalizable mesons. It is an important conclusion that the predicted mass does not change so much even if the cutoff parameter is changed in this prescription.

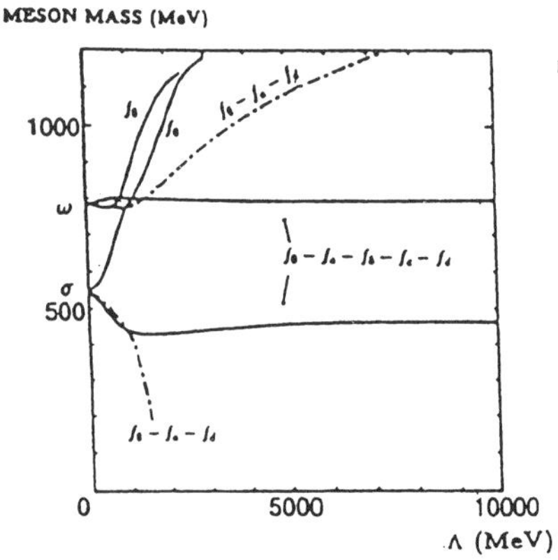

Fig.1 The cutoff dependence of the effective sigma and omega meson masses.

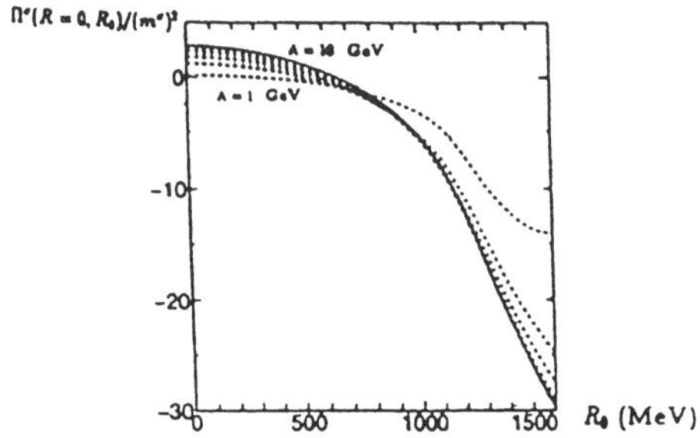

Fig.2 The R_0 dependence of the sigma meson self-energy given by renormalization result (full line) and the cutoff field theory (dadahed lines).

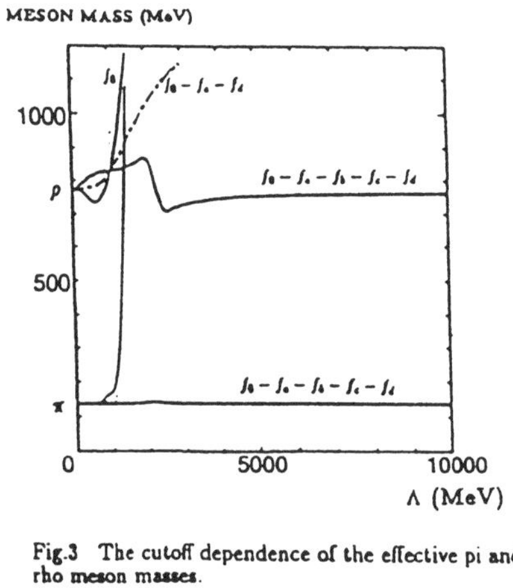

Fig.3 The cutoff dependence of the effective pi and rho meson masses.

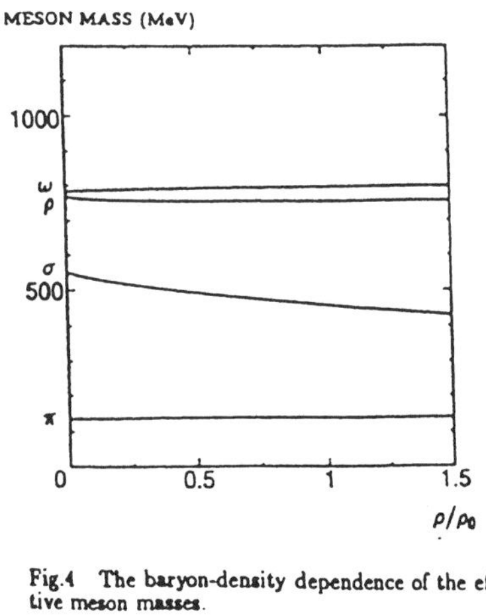

Fig.4 The baryon-density dependence of the effective meson masses.

MICROSCOPIC OPTICAL POTENTIALS FOR ELASTIC NUCLEON-NUCLEUS SCATTERING

Ch. Elster, S.P. Weppner
Department of Physics and Astronomy, Ohio University, Athens, OH 45701, USA

C.R. Chinn, R.M. Thaler
Department of Physics and Astronomy, Vanderbilt University, Nashville, TN 37235, USA

The theoretical basis of this work is the spectator expansion of multiple scattering, which orders the scattering process according to the number of participating target nucleons. Thus the first order term involves interactions between the projectile and one target nucleon. The optical potential describing elastic scattering of nucleons from nuclei is then in first order given by

$$\langle \mathbf{k}'|U|\mathbf{k}\rangle = \langle \mathbf{k}'\Phi_A \mid \sum_{i=1}^{A} \tau_i \mid \mathbf{k}\Phi_A\rangle \tag{1}$$

Here Φ_A stands for the target ground state. Introducing a projector on the ground state, $P = |\Phi_A\rangle\langle\Phi_A|/\langle\Phi_A|\Phi_A\rangle$, and the nucleon-nucleon (NN) interaction represented by v_{oi}, the operator τ_i is given as

$$\begin{aligned}\tau_i &= v_{oi} + v_{oi}G_0(E)(1-P)\tau_i \\ &= \hat{\tau}_i - \hat{\tau}_i G_0(E) P \tau_i\end{aligned} \tag{2}$$

with
$$\hat{\tau}_i = v_{oi} + v_{oi}G_0(E)\hat{\tau}_i = v_{oi} + v_{oi}\frac{1}{E + i\varepsilon - h_o - H_A}\hat{\tau}_i. \tag{3}$$

Here h_o represents the projectile kinetic energy and H_A the target Hamiltonian.

The many-body nature of the propagator for the projectile+target system is treated in first order by coupling the struck target nucleon via a mean field force to the residual nucleus, i.e. the channel Green's function

$$G_0(E)^{-1} = (E + i\varepsilon - E_i) - h_o - h_i - W_i - (H^i - E_i) \tag{4}$$

becomes in first order

$$G_i(E)^{-1} = (E + i\varepsilon - E_i) - h_o - h_i - W_i \tag{5}$$

where W_i represents the mean field force on the target particle i.
With this Eq. (3) can be rewritten as[1]

$$\hat{\tau}_i = t_{oi} + t_{oi} g_i W_i G_i(E)\hat{\tau}_i \tag{6}$$

where $t_{oi} = v_{oi} + v_{oi} g_i t_{oi}$ can be identified with the NN t-matrix. The two-nucleon propagator is given as $g_i^{-1} = [(E + i\varepsilon - E_i) - h_o - h_i]$. Introducing the operator $\mathcal{T}_i = W_i + W_i g_i \mathcal{T}_i$, Eq. (6) becomes

$$\hat{\tau}_i = t_{oi} + t_{oi} g_i \mathcal{T}_i g_i \hat{\tau}_i. \tag{7}$$

The second term in Eq. (7) describes the coupling of the struck target nucleon to the residual nucleus and though relatively small compared to the leading term t_{oi} (the conventional $t\rho$ approximation), this term is important for a correct prediction of the elastic scattering observables at projectile energies lower than 200 MeV.

We performed for the first time a completely consistent calculation of the first order term of the resulting optical potential for nucleon-nucleus elastic scattering, which involves the convolution of a realistic fully off-shell NN t-matrix with a realistic nuclear density matrix generated by this mean field. The spin structure of the mean field force is fully incorporated, whereas three-body kinematics is treated approximately[2]. One of the most striking results is the correct prediction of the spin observables for elastic proton scattering from light (^{16}O) as well as heavy (^{90}Zr, ^{208}Pb) nuclei as low as 65 MeV. Another important result is the correct prediction of total cross section measurements for neutron scattering from light (^{12}C, ^{16}O) as well as heavy (^{90}Zr, ^{208}Pb) nuclei in the energy regime between ~75 to 600 MeV[a].

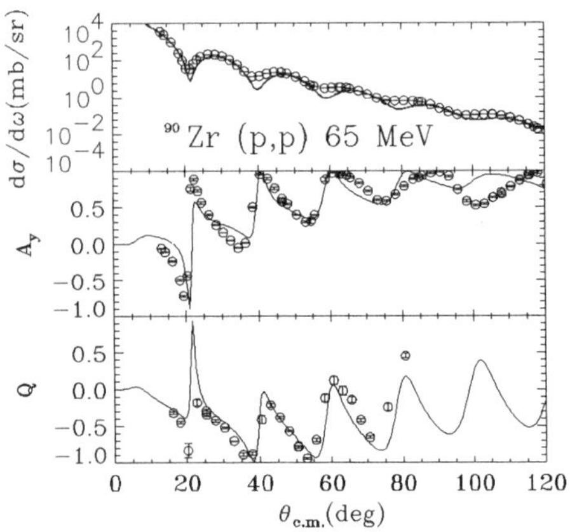

$d\sigma/d\omega$, A_y, and Q for ^{90}Zr

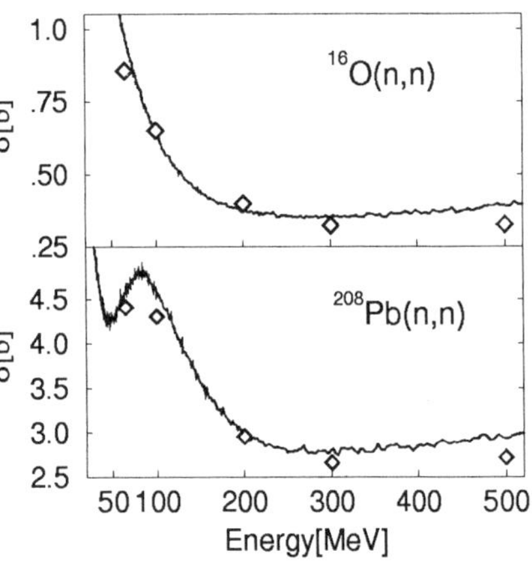

Theory(diamonds) and data(−).

References

1. C.R. Chinn, Ch. Elster, R.M. Thaler, and S.P. Weppner, *Phys. Rev.* C **52**, 1992 (1995).
2. C.R. Chinn, Ch. Elster, R.M Thaler, *Phys. Rev.* C **48**, 2956 (1993).

---

[a]This work was in part supported by the U.S. Department of Energy. The computational support of the Supercomputer Centers OSC, NERSC and PSC are acknowledged.

Session 11
New Experimental Techniques

MEASURING THE ENERGY OF COSMIC RAY MUONS IN A SAMPLING CALORIMETER

L. G. BUGEL
Stratton Mountain School, Stratton Mountain, VT 05155, USA

A. P. CHIKKATUR, A. BODEK, P. AUCHINCLOSS, P. deBARBARO, H. BUDD, D. A. HARRIS, W. K. SAKUMOTO, U. K.YANG
University of Rochester, Rochester, NY 14627, USA

C. G. AROYO, A. O. BAZARKO, J. CONRAD, J. H. KIM, B. J. KING, S. KOUTSOLIOTAS, C. McNULTY, S. R. MISHRA, P. Z. QUINTAS, A. ROMOSAN, F. J. SCIULLI, W. G. SELIGMAN, M, H. SHAEVITZ, P. SPENTZOURIS, E. G. STERN
Columbia University, New York, NY 10027, USA

R. H. BERNSTEIN, G. KOIZUMI, M. J. LAMM, W. MARSH, K. S. McFARLAND
Fermi National Accelerator Laboratory, Batavia, IL 60510, USA

T. BOLTON, D. NAPLES
Kansas State University, Manhattan, KS 66506, USA

R. B. DRUCKER
University of Oregon, Eugene, OR 97403, USA

R. A. JOHNSON, M. NUSSBAUM, L. PERERA, M. VAKILI
University of Cincinnati, Cincinnati, OH 45221, USA

T. KINNEL, W. H. SMITH
University of Wisconsin, Madison, WI 53706, USA

W. C. LEFMANN, R. STEINER
Adelphi University, Garden City, NY 11530, USA

P. NIENABER
Xavier University, Cincinnati, OH 45207, USA

M. J. OREGLIA, B. A. SCHUMM
University of Chicago, Chicago, IL 60637, USA

A technique for measuring the energy of muons in the TeV range using the energy deposited in a coarse-grained iron-scintillator calorimeter is presented. The new technique uses a likelihood fit comparing energies deposited in the calorimeter to a GEANT-based detector Monte Carlo to determine the muon energy. Results are presented from application of this technique to the 1987 CCFR cosmic ray data.

1 Analysis

The CCFR detector consists of a 690 ton iron target-calorimeter instrumented with 84 scintillators and 42 drift chambers, followed by a 420 ton magnetized iron muon spectrometer. The detector is completely described in ref. 1.

The muon energy loss measurements used in this study are from a one month cosmic ray exposure. Only muons that traversed the entire detector are used. Details of this data sample are discussed in ref. 2.

The analysis is a variation on the pair meter technique presented in ref. 3. A GEANT 3.21 detector simulation was run at 16 muon energies, and the predicted energy deposits are converted to MIPS using the truncated mean method described in ref. 4. This allows direct comparison between GEANT and data. A maximum likelihood fit of measured energy loss over 80 scintillation counters is then used to find the most probable muon energy.

2 Results

This calorimetric method has been applied to the CCFR cosmic ray data, and the energy measured is compared to the energy as determined by the toroid spectrometer. For each muon, the fractional error is computed:

$$\text{Frac.Error} = \frac{E_\mu - E_{cal}}{E_\mu}, \tag{1}$$

where E_{cal} is the muon energy calculated using the calorimetric method, and E_μ is the muon energy as measured by the CCFR spectrometer.

The RMS fractional error for $E_\mu > 400$ GeV is ~52%. This is comparable to the estimated theoretical error of 53%. For muons of low energy the technique fails.

References

1. W.K. Sakumoto et al., CCFR Collaboration: Nucl. Instr. and Meth. A 234 (1990) 179; B.J. King et al.: Nucl. Instr. and Meth. A 302 (1991) 254.
2. W.K. Sakumoto et al., Phys. Rev. D 45 (1992) 3042.
3. R.P. Kokoulin, A.A. Petrukhin: Sov. J. Part. Nucl. 21 (1990) 332.
4. P.S. Auchincloss et al.: Nucl. Instr. and Meth. in Phys. Res. A 343 (1994) 463.

THE QUARTZ FIBER ZERO-DEGREE CALORIMETER FOR THE NA50 EXPERIMENT

E. CHIAVASSA, G. DELLACASA, N. DEMARCO, M. GALLIO, P. GUAITA, A. MUSSO, M. PEZZETTA, A. PICCOTTI, E.SCOMPARIN and E. VERCELLIN
[a] *Dipartimento di Fisica Sperimentale dell'Università di Torino and Istituto Nazionale di Fisica Nucleare, Sezione di Torino, via P.Giuria 1, I-10125 Torino, Italy*

Within the frame of the NA50 experiment[1] at the CERN SPS we have built and operated a zero degree hadron calorimeter[2] (ZDC) to measure the energy carried by spectator nucleons and nuclear fragments emitted at very small angles in Pb-Pb collisions at 160 GeV/nucleon. Since the zero-degree energy is directly related to the collision geometry, its measurement allows to determine the impact parameter of the Pb-Pb collision.

The NA50 ZDC is a "spaghetti" calorimeter with quartz optical fibers as active material instead of the conventional scintillating fibers. In this calorimeter the Cherenkov light produced in the fibers by the particles of the shower is exploited. This technique[3] offers the advantage of being highly radiation resistant (up to several Grads), very fast (the signal width is of the order of 10 ns) and intrinsically insensitive to induced radioactivation.

The ZDC has a front face surface of 5x5 cm^2 and 65 cm length; the fibers are embedded in Tantalum slabs with a fibre to Tantalum volume ratio of about 1/17. The 900 fibers, uniformly distributed in the Tantalum volume, are parallel to the beam direction and have a diameter of 0.365 mm. The front face of the calorimeter is located 165 cm downstream the target and it covers the pseudorapidity interval $\eta \geq 6.14$, delimited by a copper collimator. The overall dimensions of the detector were imposed by the geometrical constraints of the experiment. The fibers are divided in four groups, corresponding to the four quadrants of the ZDC, and each group is read-out by a XP-2242B photomultiplier.

The ZDC operated successfully in the 1994 and '95 NA50 runs with lead beam. As it can be seen in fig. 1, the resolution (σ_E/E) on the lead peak is 5% (7%) with 18 ns (12 ns) ADC gate width and a beam intensity of about $3 \cdot 10^6$ (10^7) Pb-ions/s. A pile-up detection system, based on multiple delayed ADC gates, allowed to identify (and recovery) a large fraction of superimposed events even in such severe conditions. The ZDC was also tested with 160 GeV/c

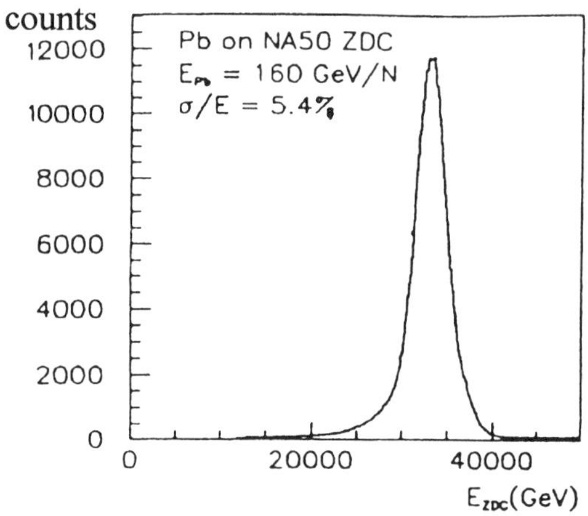

Figure 1: Response of the ZDC to 160 GeV/A Pb beam with 18ns ADC gate

proton beams; the observed resolution was $\sigma_E/E \simeq 30\%$. These results turn out to be in agreement with the predictions of a monte-carlo simulation based on the GEANT code.

Determining the centre of gravity of the energies deposed in the four quadrants we can measure the coordinates of the incident particle on the front face of the ZDC with a resolution of about 2mm r.m.s. for 160 GeV protons and better that 0.5 mm r.m.s. for 160 GeV/A lead ions.

During the run of November '95 the ZDC absorbed a dose of about 2 Grad with a signal attenuation of 30%, due to both fibre and PM ageing. No effect of induced radioactivation on the pedestal value was observed during and after the run.

The construction of a set of 6 quartz fiber calorimeters is foreseen for the ALICE[4] detector at LHC, were they will be used in the trigger system to select central Pb-Pb collision.

References

1. E.Vercellin, for the NA50 collaboration, *proceedings of this conference, par. sess. 3*
2. E.Chiavassa et al., *Nucl. Instrum. Methods* A **367**, 267 (1995).
3. P. Gorodetzky et al., *Nucl. Instrum. Methods* A **361**, 161 (1995).
4. ALICE Technical Proposal, *CERN/LHCC 95-71, LHCC/P3*

OPTIMIZATION OF A LASER-DRIVEN POLARIZED DEUTERIUM INTERNAL TARGET FOR NUCLEAR PHYSICS EXPERIMENTS

W. J. Cummings, J. A. Fedchak, C. E. Jones, R. S. Kowalczyk
Argonne National Laboratory, Argonne, IL 60439

Storarge ring experiments which measure the spin dependent properties of the deuterium nucleus require an intense source of nuclear polarized deuterium atoms. For this purpose, we have developed a laser driven deuterium source (LDS). There does not currently exist a laser which can directly pump the ground state of hydrogen and its isotopes, but it has been demonstrated that H and D can be electronically polarized by the technique of spin-exchange optical pumping [1]. This technology is being developed as an alternative to the traditionally used atomic beam source as it offers the possibility of greater target thickness.

Of course, for use as an internal target, the LDS must provide polarized nuclei. Previously, it was generally considered necessary to induce RF transitions in order to obtain nuclear polarization from the LDS because it operates in a high magnetic field. However, it was realized that even in the presence of high magnetic fields, collisions in a sufficiently dense sample of polarized hydrogen or deuterium atoms drive the ensemble to a spin-temperature distribution, where the net polarization is shared between the valence electrons and the nuclei[2, 3]. In this paper, we report the first measurements of nuclear tensor polarization, p_{zz}, thus demonstrating that the technology is capable of producing nuclear polarized D targets.

The LDS apparatus has been described in detail elsewhere, therefore only the nuclear polarimeter will be described [1]. Polarized D atoms flow into a storage cell of about 44 cm in length and 23 mm in diameter. The atoms are ionized by a 2 keV electron beam directed along the length of the cell. A 300 gauss magnetic field prevents electrons from striking the cell wall. D^+ ions are then extracted from the cell via electric fields and accelerated to about 42 keV into a tritiated foil (Ti^3H). Neutrons resulting from $^3H(D,n)\alpha$ are detected in two NE110 scintillators placed at 0 and 90 degrees with respect to the polarization axis. The tensor analyzing power of $^3H(D,n)\alpha$ produces an asymmetry in the signals detected in the two scintillators and may then be used to measure p_{zz}. Further details of the polarimeter can be found in Price and Haeberli [4].

The measured values of the p_{zz} versus flow of deuterium nuclei are shown in Figure (1). Target thickness is directly proportional to D number density and hence flow. As can be seen from the figure, the tensor polarization is essentially flat with flow for flows greater than about 3×10^{17} s^{-1}. This behavior is expected if one has obtained a high enough target thickness to achieve spin-temperature. In these conditions, increasing the number density should not

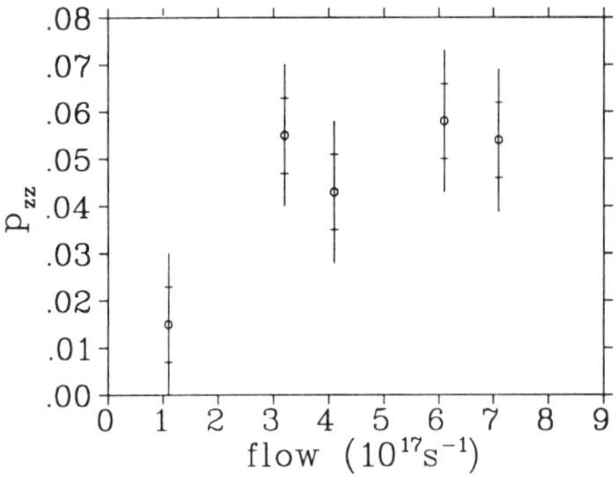

Figure 1: Nuclear Tensor Polarization versus flow of D nuclei. The error bar on each point is dominated by systematic effects. The statistical error on each point is shown as the inner error bar.

appreciably change the polarization. However, increasing the deuterium density will eventually lead to a decrease in D polarization. D polarization is limited by the spin-exchange rate with K which is proportional to K number density and K polarization. These factors are limited by spin-relaxation in wall collisions [5] and radiation trapping [6].

Our initial p_{zz} measurements verify that the technique of spin-exchange optical pumping is capable of producing nuclearly polarized D. Efforts to optimize the LDS, in terms of target thickness and polarization, for use in internal targets are currently under way. These efforts include improvements in surface coatings and glassware geometry to minimize the detrimental effects of wall collisions.

This work was supported by the U.S. Department of Energy, Nuclear Physics Division, under contract No. W-31-109-ENG-38.

References

[1] M. Poelker et al, *Nucl. Instr. and Methods* **A 364**, 58 (1995), and references cited therein.

[2] L. W. Anderson, F. M. Pipkin, and J. C. Baird, *Phys. Rev.***120**, 1279 (1960).

[3] T. Walker and L. W. Anderson, *Nucl. Instr. and Methods* **A 334**, 313 (1993).

[4] J. S. Price and W. Haeberli, *Nucl. Instr. and Methods* **A326**, 416 (1993).

[5] D. R. Swenson and L. W. Anderson, *Nucl. Instr. and Methods* **B 29**, 627 (1988).

[6] D. Tupa et al, *Phys. Rev.* **A 33**, 1045 (1986).

NEW POSSIBILITY OF A SEARCH FOR NEUTRON EDM BY POLARIZATION METHOD IN NEUTRON DIFFRACTION

V.V.Fedorov, E.G.Lapin, O.I.Sumbaev, V.V.Voronin
St.Petersburg Nuclear Physics Institute, 188350 Gatchina, Russia

A new method of measuring a neutron EDM, using neutrons diffracted by a non-centrosymmetric quartz-crystal, is proposed. The method is based on the effect of depolarization of the diffracted neutron beams (direct and reflected) passed through the crystal. Description of the effect and a set-up is given.

Recently we have shown an existence of a strong crystal field E_g (up to 10^9 V/cm) acting on neutrons diffracted by a non-centrosymmetric crystal and have measured it [1]. The measured value has coincided with the calculated one. It was shown also [2], that for Bragg angles θ_B close to $\pi/2$ (for Laue scheme of diffraction) the time of a neutron passage through crystal can essentially increase, because it is determined as $t = L/v_{\|} \sim (\pi/2 - \theta_B)^{-1}$, where L is the crystal thickness and $v_{\|} = v \cos\theta_B$ is the velocity of diffracting neutron that moves along the crystallographic planes for Laue diffraction. As a result the sensitivity of diffraction method may be comparable with that of the most sensitive now magnetic resonance method using ultracold neutrons (UCN-method) [3, 4]. The comparison is given in table.

In the present work, we propose a polarization method for searching the neutron EDM that is more simple than the method discussed in [2] and which can be also usable for Bragg angles θ_B close to $\pi/2$. The possible set-up is shown in fig.

Table. Data presented here are given for cold neutrons of the vertical channel of the PNPI reactor WWR-M and for UCN from the universal liquid-hydrogen source of cold and ultracold neutrons. The crystal dimensions are 3.6 x 14 x 14 cm^3. The crystallographic planes $(11\bar{2}0)$ are used (2d=4,9 Å) and $\pi/2 - \theta_B = 1/30$.

	UCN-method PNPI[3]	Diffraction of cold neutrons (Polarization method)
E (kV/cm)	10 - 15	$2{,}1 \times 10^5$
t (s)	70 (v =5-6 m/s)	1.3×10^{-3} ($v_{\|} = 27$ m/s)
Et (kV s/cm)	700 - 1050	280
N (neutrons/s)	60 - 80	2500

The essence of the method is the following (see fig). For some crystal thickness $L = L_0$ and for the polarization of initial neutron beam along its direction (i.e. normal to the Schwinger magnetic field $\vec{H}_g = [\vec{E}_g \times \vec{v}_{||}]/c$) the spins for two kinds of the neutron Bloch waves inside the crystal will rotate in opposite directions by the angles $\pm\pi/2$, so the first half of neutrons will have the spins antiparallel to the other half. Therefore if the EDM is absent the polarization of the diffracted beams (direct and reflected) will be zero.

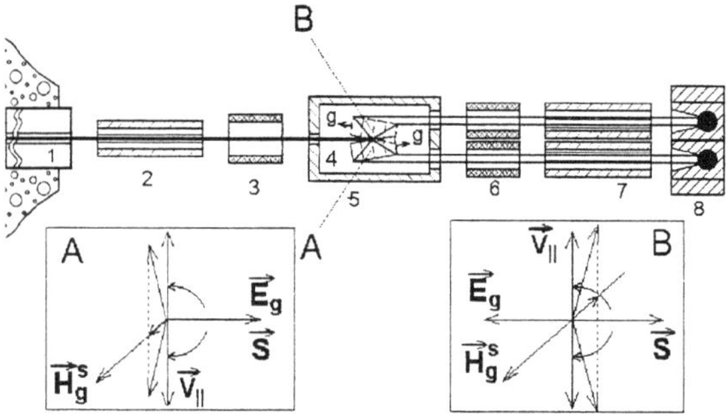

Fig. The possible set-up for EDM searching. 1 is the reactor interchannel neutron-guide, 2 is the multislit neutron-guide polarizer, 3, 6 are the flippers, 4 is the α-quartz monocrystal, 5 is the magnetic shield, 7 is the duplex multislit neutron-guide analyzer, 8 is the duplex detector. A and B are the crystal positions different by the rotation angle $2\theta_B \simeq 180°$, $\vec{g}$ is the crystal reciprocal-lattice vector, $\vec{E}_g \parallel \vec{g}$.

The existence of the EDM in this scheme leads to a slight beam polarization along the Schwinger magnetic field $\vec{H}_g$:

$$P_h \approx \frac{2DE_g L_0}{\hbar v_{||}}. \tag{1}$$

For crystallographic planes $(11\bar{2}0)$, EDM $D = 10^{-25}$ e·cm and $\pi/2 - \theta_B = 1/30$ ($L_0 = 3.6$ cm, $E_g = 2.1 \times 10^8$ V/cm) we shall have $P_h \approx 0.8 \times 10^{-4}$.

The polarization will have the opposite sign for the different crystal positions A and B (we should note that the residual polarization, if it exists without EDM, will have the same sign for both positions of the crystal).

We note that the maximum corresponding value of the spin rotation effect for the Bragg diffraction scheme [5] is $\sim 2.5 \times 10^{-6}$ (for the same crystal and thickness) that is 30 times less than in our case. Such a value for the Laue diffraction in hypothetical absorbing WC-crystal [6] is 7×10^{-6}. There is no possibility to increase these both effects using the Bragg angles close to $\pi/2$ (as we can do in our case) because of the Bragg diffraction scheme in the first case [5] and the absorption in the second one [6].

This work was supported by the Russian Foundation for Fundamental Research (No 94-02-04792) and by Grant No J4X100 from the International Science Foundation and Russian Govenment.

References

[1] Alexeev V.L. et all. Nucl. Instr. and Meth., 1989, V. A284. P181.

[2] V.V.Fedorov et all. Preprint LNPI-1644, Leningrad, 1990. J.Phys., 1992, V. G18. P.1133.

[3] I.S.Altarev et all. Phys. Lett., 1992. V. B276. P.242. Preprint PNPI-2055, Gatchina, 1995.

[4] K.F.Smith et all. Phys. Lett., 1990. V. B234. P.191.

[5] M.Forte. J.Phys., 1983. V. G9. P.745.

[6] V.G.Baryshevskii et all. Izvestiya Vuzov SSSR, ser. fiz., 1985. V.8. P.110.

A POLARIZED DEUTERIUM INTERNAL TARGET

M. FERRO-LUZZI[*,†], Z.-L. ZHOU[‡], J.F.J. VAN DEN BRAND[†,‡,§], H.J. BULTEN[‡] and J. LANG[*], for the 91-12 collaboration

[*] *Inst. für Teilchenphysik, ETH, CH-8093 Zürich, Switzerland*
[†] *NIKHEF, 1009 DB Amsterdam, The Netherlands.*
[‡] *Dept. of Physics, University of Wisconsin, Madison, WI 53706, USA*
[§] *Vrije Universiteit, 1081 HV Amsterdam, The Netherlands*

We have developed a polarized deuterium internal target with a thickness of about 2×10^{13} cm^{-2} (luminosity $\simeq 10^{31}$ cm^{-2}s^{-1} at beam currents $I_e \simeq$ 100 mA). Target tensor polarizations up to $P_{zz}^{\pm} = +0.5/-0.9$ have been achieved and measured in-situ with an accuracy better than 5 %. Fig. 1 shows a schematic outline of the internal target setup[1]. A polarized deuterium beam ($\sim 2 \times 10^{16}$ at/s) produced in an atomic beam source (ABS) is fed into a cooled storage cell located inside the AmPS electron storage ring at NIKHEF. The ABS consists of an RF dissociator (D), a cooled nozzle (CH), collimators, Stern-Gerlach sextupole magnets (S1, S2) and two RF transition units (MFT, SFT)[2] that are used to switch every 10 s the tensor polarization of the atomic beam from +1 to −2 (at zero vector polarization). The target spin axis can be oriented at will in the electron scattering plane by setting the appropriate holding field. A Breit-Rabi polarimeter allows for tuning and monitoring of the injected atomic beam intensity and polarization.

The absolute target polarization $P_{zz} = n_{\rm D} \cdot P_{\rm zz}({\rm D}) + 2n_{{\rm D}_2^{\rm rec}} \cdot P_{\rm zz}({\rm D}_2^{\rm rec})$ (with $n_{\rm D} + 2n_{{\rm D}_2^{\rm rec}} + 2n_{{\rm D}_2^{\rm unp}} = 1$) is determined in-situ with a novel polarimetry technique[3]. The relative contributions to the target thickness of the atoms ($n_{\rm D}$), of unpolarized molecules from background gas or undissociated molecular beam ($n_{{\rm D}_2^{\rm unp}}$), and of potentially polarized molecules due to recombination in the storage cell ($n_{{\rm D}_2^{\rm rec}}$) are measured by ion extraction: ions produced in the storage cell by the circulating electron beam are extracted with electrostatic devices (RL, EL, SD, AL) and mass-analyzed with an $\mathbf{E} \times \mathbf{B}$ Wien filter (WF). $P_{\rm zz}({\rm D})$ is measured directly by accelerating the $\rm D^+$ ions (60 keV) onto a tritiated titanium foil and by using the neutron anisotropy of the ^{3}H(d,n)^{4}He reaction. Neutron (background) singles rates of about 5 kHz (0.5 kHz) at $I_e = 100$ mA were obtained, which allowed fast and accurate measurements of $P_{\rm zz}({\rm D})$. Finally, $P_{\rm zz}({\rm D}_2^{\rm rec})$ was determined by comparing the tensor asymmetries $\rm A^{Cu,Al}$ in the ^{2}H(e, e′d) reaction obtained with two different storage cells, a copper cell and an aluminum cell coated with Teflon. Fig. 2 shows the mass spectra obtained by measuring the ion current as a function of the WF magnetic field. It is seen that the polarized atoms largely recombine to molecules on a Cu surface whereas on a Teflon-coated Al cell recombination is strongly suppressed. Combining these data with the asymmetries A^{Cu}/A^{Al} and the results obtained with the $^3H(d,n)^4He$ polarimeter, we extracted $P_{zz}(D_2^{rec})/P_{zz}(D) = 0.85 \pm 0.33$.

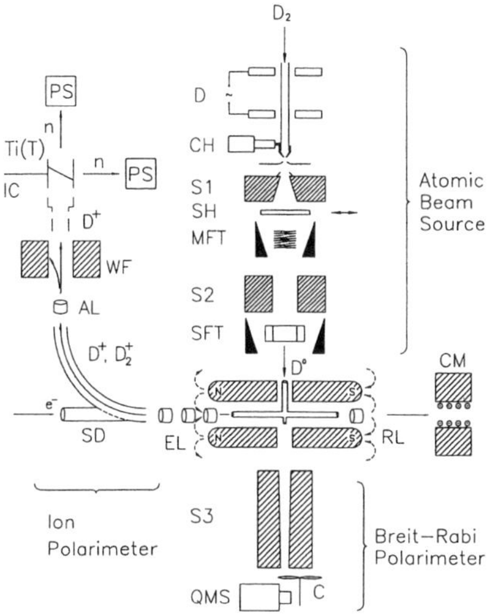

Figure 1: Polarized deuterium internal target (description in the text).

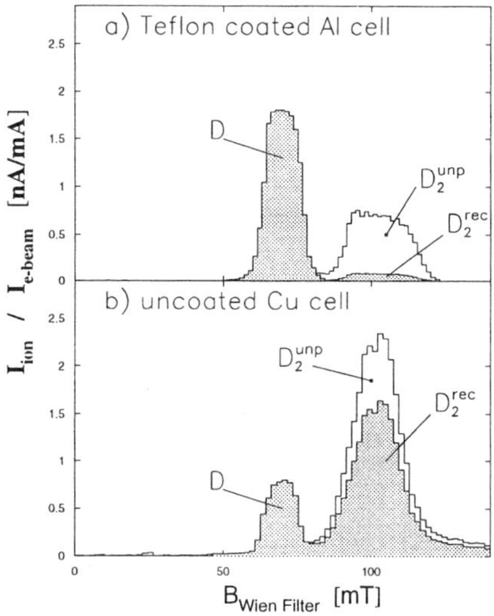

Figure 2: Target mass-spectra measured with the ion polarimeter.

Recently, the internal target has been used in two experiments to measure tensor analyzing powers in elastic and quasi-free (e,e′p) scattering from tensor polarized deuterons [4]. For future experiments, newly designed ABS focusing magnets and an optimized storage cell will permit to increase the figure of merit of the setup by at least an order of magnitude, allowing for high precision measurements of electromagnetic spin observables for hydrogen and deuterium.

References

1. Z.-L. Zhou, M. Ferro-Luzzi, J.F.J. van den Brand, H.J. Bulten *et al.*, accepted (April 1996) for publication in *Nucl. Instrum. Methods* A.
2. M. Ferro-Luzzi, Z.-L. Zhou, H.J. Bulten and J.F.J. van den Brand, *Nucl. Instrum. Methods* A **364**, 44 (1995).
3. Z.-L. Zhou, M. Ferro-Luzzi, J.F.J. van den Brand, H.J. Bulten *et al.*, accepted (May 1996) for publication in *Nucl. Instrum. Methods* A.
4. Three contributions to this conference by M. Bouwhuis, J.F.J. van den Brand and Z.-L. Zhou.

BEAM ENERGY MEASURMENT AT TJNAF USING THE HALL C ARC BEAMLINE AS A SPECTROMETER

P. Guèye[1], P. Adderley[2], D. Barker[3], K. Capek[2], R. Carlini[2], J. Dahlberg[2]
S. Danagoulian[4], W. Hinton[1], J. Karn[2], B. Kross[2], R. Lauze[2], R. Li[2],
E. Martin[2], D. Neuffer[2], W. Oren[2], S. Wood[2], C. Yan[2]

[1] *Hampton University, Nuclear/High Energy Physics Center of Excellence (NuHEP), Hampton, VA 23668*
[2] *Thomas Jefferson National Accelerator Facility (TJNAF), 12000 Jefferson avenue, Newport News, VA 23606*
[3] *California Institute of Technology, LIGO Project, MS 51-33, Pasadena, CA 91125*
[4] *North Carolina A&T State University, Department of Physics, 1601E, Market Street, Greensboro, NC 27411*

Abstract: *The Hall C arc transport beamline has been used as a spectrometer to measure the beam energy and energy spread of the electron beam at TJNAF for one year. It has a dispersion of* 12.5 $cm/\%$ *with all the quadrupoles, sextupoles, and beam correctors switched off. The transverse position and the angle of the beam at the entrance, the mid-point and the exit of the arc is precisely measured by three pairs of wire scanners (superharps) with absolute beam position readout* $\Delta x \simeq 10\ \mu m$. *A complete error analysis shows that the absolute beam energy measurment accuracy is* $\Delta E/E = 6.23\ 10^{-4}$ *and the relative energy measurment has an accuracy of* 10^{-4} *using the pair of superharps at the mid-point.*

1 The Hall C arc transfer line

The arc achromat section of Hall C beamline consists of: 8 dipoles, 12 quadrupoles, 8 sextupoles and 8 pairs of beam correctors (vertical and horizontal), which transports the beam with second order achromaticity into the end station C of the Continuous Electron Beam Accelerator Facility (TJNAF). The incoming electron beam is bent to a 34.3^o angle in the horizontal direction which corresponds to a total path length of 41.6 m downstream from the point of tangency.

2 Method

The beam energy measurment by using an arc spectrometer was proposed and developed by [1,2,3,4]. The method consists of:

1. Alternate the arc optics from the normal achromatic mode to the dispersive mode by setting all non dipole elements (quadrupoles, sextupoles) to a zero $\int Bdl$ value. The beam correctors are used to compensate the terrestrial field.

2. Measure the absolute transverse beam position and orientation at the entrance and at the end of the Hall C arc line. This information is determined by using a set of two pairs of wire scanner systems (called superharps) equiped with absolute position encoders.

3. Center the beam at the end of the arc by adjusting the arc magnet current. For this purpose, the current in the calibrated bending magnets is varied to set the position of the beam to be along the central ray of the dipoles in the arc. Note that all dipole magnets are connected in serie.

4. Read the current through pre-calibrated $\int Bdl - I$ data. Transfer this information into a total $\int Bdl_T$ for all 8 dipoles: $\int Bdl_T = \Sigma_{i=1}^{8} \int Bdl$.

5. Determine the beam momentum P_0.

When all the quadrupoles, sextupoles and beam correctors are switched off, the total dispersion at the end of the arc is 12.5 $cm/\%$. The effective length of an arc dipole is 300 cm, $B = 4.9922$ kG at 6 GeV and the total $\int Bdl$ is about 1.5 10^6 $Gauss.cm$.

This mehtod requires accurate position and angle measurments at the superharps and accurate determination of the magnetic field integral ($\int Bdl$) as a function of current (I) in the arc.

3 Error analysis

As described above, this method requires an accurate position/angle determination. The position of the beam can be known at a 10 μm accuracy from each superharp device. The distance between 2 superharps is around 2.5 m. The corresponding error on the energy measurment is 0.5 10^{-3}.

The other requirement for this method is a precise determination of the field integral $\int Bdl$ of the arc bending magnets. The $\int Bdl$ is measured by a moving wire device with comparison to a reference magnet. The relative accuracy of this moving wire device is about a few 10^{-5}. In order to obtain precision $\int Bdl$ data of the arc magnets, the longitudinal field profile mapping for at least two production BA magnets must be done. From the relative $\int Bdl$ data, the absolute $\int Bdl$ data for each of the arc magnets can then be extracted. The corresponding error on the energy measurment is 0.13 10^{-3}.

Including other misalignment of optical elements (magnet positioning, orientation ...) and miscelaneous error sources (girder sink ...), the total systematic errors of this method contribute to a 10^{-3} level.

4 Results

The five different beam energies allowed at TJNAF were measured with the method described above. After optimization of the RF cavity phases in the north and south linac, the energy spread was found to be around $\Delta P_0/P_0 \sim 10^{-4}$.

5 Further development

An online absolute energy monitoring can also be available. For that purpose, the beam energy momentum P_0 (dispersive optics) is transferred into the normal achromatic transportation mode on the target by energizing the quadrupoles and sextupoles. Under the asumption that the beam energy doesn't change during this transition period, the absolute momentum calibration to the achromatic mode will be sensitive to relative shifts in the beam momentum. These variations will transfer to correponding shifts in the beam position at the mid-point of the arc which will be measured by the use of another pair of superharps located at the middle of the arc.

References

[1] C. Yan, R. Carlini, J. Napolitano, D. Neuffer, *CEBAF-R-92-003*, **March**, (1992)
[2] D. Neuffer, C. Yan, R. Carlini, *CEBAF-TN-92-054*, **November**, (1992)
[3] R. V. Servranckx, *CEBAF-TN-92-044*, **September**, (1992)
[4] D. Neuffer, C. Yan, R. Carlini, *CEBAF-PR-93-004*, **April**, (1993)

LARGE ACCEPTANCE MAGNETIC SPECTROMETERS FOR POLARIZED DEEP INELASTIC SCATTERING

YURY G. KOLOMENSKY
Department of Physics and Astronomy, University of Massachusetts
Amherst, MA 01003

REPRESENTING SLAC E154 COLLABORATION

We describe the design and performance of two magnetic spectrometers used in the polarized deep inelastic electron scattering experiment E154 at the Stanford Linear Accelerator Center.

1 The Experiment

The experiment E154 [1] at SLAC precisely measured the spin dependent structure functions $g_1^n(x,Q^2)$ and $g_2^n(x,Q^2)$ of the neutron in the kinematic range $0.014 \leq x \leq 0.8$ and $1\ \text{GeV}^2 \leq Q^2 \leq 17\ \text{GeV}^2$ using a polarized ^{3}He target and a polarized 48.3 GeV electron beam. Computing the integral $\int_0^1 g_1^n(x)dx$, we test the Ellis-Jaffe sum rule [2] and using data from neutron and hyperon decays we extract the total quark contribution to the nucleon helicity. This measurement, combined with the previous determination of the proton spin structure function $g_1^p(x)$, tests the Bjorken sum rule [3] and provides valuable Q^2-dependent information to check QCD predictions. The high statistical and systematic precision of the experiment will provide a strong basis for understanding the spin structure of the nucleon.

2 The Spectrometer System

Scattered electrons are detected in two independent magnetic spectrometers centered at 2.75° and 5.5° relative to the beam line. Each spectrometer is based on a pair of dipoles bending in the opposite direction. Such an "S-bend" configuration [4] increases the acceptance and significantly extends the kinematic range compared to conventional systems that use bends in the same direction. The design does not allow photons from the target to reach the detectors without bouncing at least twice off the magnet and collimator apertures, thus significantly reducing neutral background. The schematic view of the spectrometers is shown in Fig. 1.

The "S-bend" design results in a moderate momentum resolution which is adequate for asymmetry measurements. The momentum resolution of both

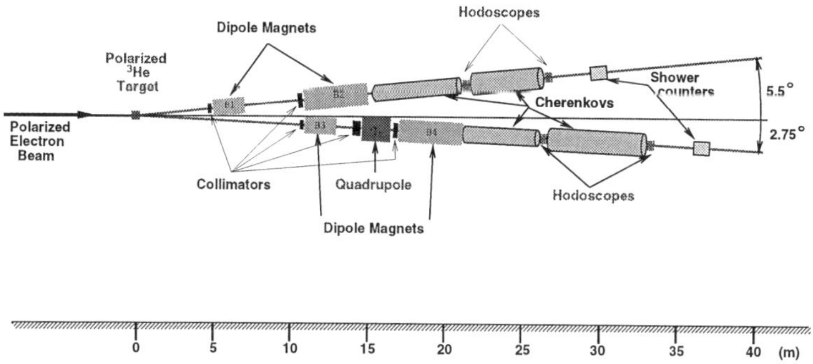

Figure 1: The plan view of the spectrometer systems.

spectrometers varies with momentum from 2% to 4.5%. The angular resolution is 0.5 msr in the non-bend plane and 1 msr in the bend plane. For comparison, the lead glass energy resolution $\sigma(E)/E = 0.025 + 0.065/\sqrt{E}$ is of comparable size [5].

Both detectors consisted of a pair of gas Cherenkov detectors, a pair of scintillation hodoscopes, and a lead-glass calorimeter in fly's eye configuration to provide efficient electron and pion identification and sufficient angular and momentum resolution. The detector segmentation is dictated by the high rate environment of a short (250 ns) beam pulse of SLAC. The total of 535 (2.75°) and 233 (5.5°) hodoscope channels were read out via multihit TDCs, each of 200 lead glass blocks was equipped with an ADC and 1-3 TDC channels, and Cherenkov phototubes were read out via the Flash ADCs by the VME-based DAQ system.

All spectrometer systems performed well throughout the course of the experiment. The overall dead time was determined by the intrinsic electronic dead time of every channel, thus being negligible. The contamination of electron sample with charged hadrons did not exceed 5% while electron efficiency was above of 90%.

References

1. The E154 Proposal, E. W. Hughes *et al.* (1993) (unpublished).
2. J. Ellis and R. Jaffe, *Phys. Rev.* **D9** (1974) 1444.
3. J. D. Bjorken, *Phys. Rev.* **148** (1966) 1467.
4. G. G. Petratos *et al.*, Report No. SLAC-PUB-5678 (1991) (unpublished).
5. H. Borel *et al.*, IEEE trans. of Nucl. Sci. **42**, No. 4 (1995).

ECONOMICAL DETECTORS BASED ON EXTRUDED GROOVED SCINTILLATORS AND SAFE LIQUID SCINTILLATORS

S. MAJEWSKI, R. WOJCIK, B. KROSS, H. FENKER, and C. ZORN.
TJNAF, 12000 Jefferson Ave., Newport News, VA 23606

S. STEPANYAN, G. ASRYAN, M. TARVERDYAN, E. ARARATYAN, K. EGIYAN, and S. MAILYAN
Yerevan Physics Institute, Yerevan, Armenia

Results are presented of a 1 cm thick extruded grooved scintillator module. An average signal of 35-40 photoelectrons per cosmic particle was measured. Results are also presented of new highly efficient and safe liquid scintillators.

1 Extruded Grooved Scintillators

Embedded waveshifting (WVS) fiber readout is a well known technique where plastic WVS fibers are placed in grooves machined in a scintillator to capture and transmit light to photomultiplier tubes (PMT). Machining the scintillator, however, is labor intensive and exposes scintillators to the possibility of damage. When scintillators with straight grooves are needed, this problem can be solved by using pre-grooved extruded scintillators. With a special attachment, produced at TJNAF, the Yerevan Physics Institute produced such extruded grooved scintillators for tests at TJNAF[1].

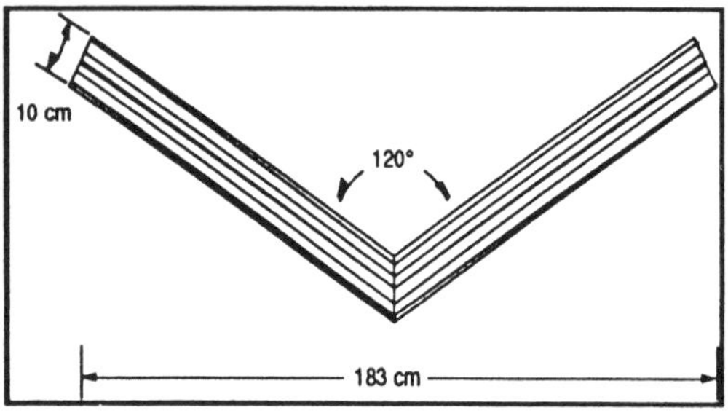

Figure 1: Extruded groove scintillator module.

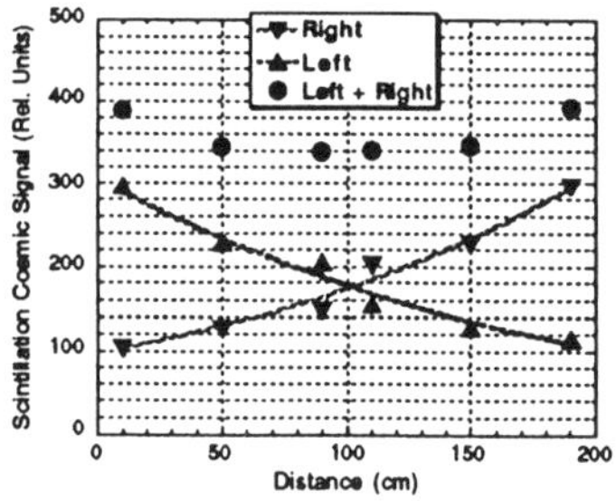

Figure 2: Uniformity of response.

Extruded polystyrene scintillators of a 1 x 10 cm cross-section were produced with five equally spaced 2 mm half-circular grooves. The surface quality of the grooves is as good as the flat surfaces. A scintillator module was produced by gluing two 1 m long pieces at a 60° angle (fig. 1). Two 1 mm Kuraray double-clad Y11 WVS fibers were glued into each groove using UV curable adhesive[2]. The module was wrapped with Teflon tape and the fibers connected to R580-17 Hamamatsu PMTs. An average summed signal (left + right PMT) of 35-40 photoelectrons per cosmic particle was measured (fig. 2).

2 Economical Safe Liquid Scintillators

When one needs complicated scintillator shapes, using a naturally flexible medium such as liquid scintillators can reduce the expense, production time and material waste involved. The questions are how to produce efficient and economical light transporting containers and handle the many hazards usually associated with liquid scintillators. Using the liquid light guide technology developed by TJNAF[3] one can easily produce economical large diameter flexible scintillator light guide modules (typically up to 0.5" diameter, but up to 2" possible with less flexibility).

ScintiVerse BD, ScintiSafe Econo F, and ScintiSafe 30% formulas[4] are liquid scintillators based on either phenyl-o-xylyl ethane (PXE) or Linear Alkylbenzenes. They do not contain aromatic solvents or release harmful vapors thus eliminating the toxicity, flammability, and disposal problems associated with standard liquid scintillators. Bicron[5] prepared for us a scintillator based on PXE named BC599-16 using the same fluors as in their BC408 plastic scintillator. PXE's high index of refraction (1.566) results in a numerical aperture (N.A.) of 0.81 when used with Teflon. Plastic scintillating fibers have a N.A. of only 0.58. We also prepared our own scintillator based on PXE with 1.3% by weight of a large Stokes shift fluor, 1-phenyl-3-mesityl-2-pyrazoline (PMP)[6]. BC599-16 was found to have very high light output with a decay time of only 2.7 ns (table 2, figure 3). Our non-optimized homemade PXE/PMP liquid scintillator also showed rather high light output.

Detailed articles on the above work to appear in Nuclear Instruments and Methods A are in preparation.

Table 2: Liquid scintillator light output.

Scintillator	Rel. Light Yield (%)
NE224	100
Solvent-Free	25
ScintiVerse BD	48
ScintiSafe Econo F	77
PXE/PMP	86
Bicron 599-16	109

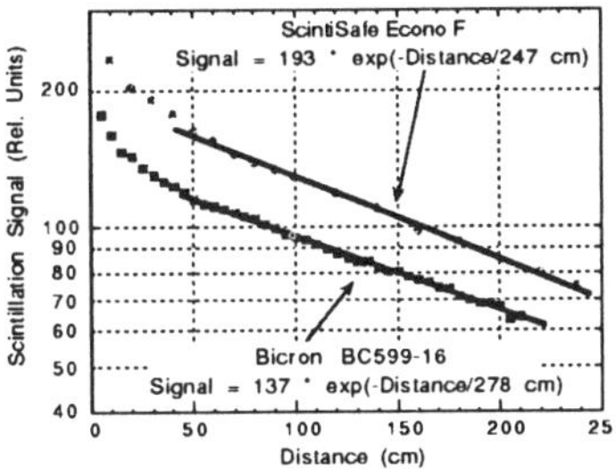

Figure 3: Attenuation of selected scintillators

References

1. R. Wojcik et al., Nucl. Instr. and Meth. A342 (1994) 416.
2. Type 3-20262, Dymax Corporation, Torrington, CT 06790.
3. L. Majewski et al., Nucl. Instr. and Meth. A337 (1993) 628.
4. Fisher Scientific, Fair Lawn, NJ.
5. Bicron Corporation, Solon, OH 44139.
6. P. Destruel et al., Nucl. Instr. and Meth. A276(1989)69.

The 1 m^3 CF_4 Time Projection Chamber: A Detector for Neutrino Physics

A. Paić
(for the MUNU collaboration: Grenoble, Münster, Neuchâtel, Padua, Zürich)
Institut de Physique, Université de Neuchâtel, Rue A.L. Breguet - 1
CH 2000 Neuchâtel, Switzerland

A cylindrical 1 m^3 Time Projection Chamber with CF_4 drift gas is being tested. For the first time tracks were drifted in CF_4 over a 1.60 m distance at 3 bar, and the drift length was measured to be 22^{+14}_{-6} m. The chamber is to be used for a $\overline{\nu}_e e^-$ scattering experiment at the Bugey reactor, but a similar detector can also be used for $\beta\beta$ decay, solar neutrino and dark matter experiments.

1 The MUNU experiment

The MUNU experiment is a $\overline{\nu}_e e^-$ scattering experiment with low recoil electron energy threshold ($T_{min} \sim 300 - 500$ keV). The cross section of the process has a weak component and, in the event of a nonvanishing neutrino magnetic moment, a magnetic one. The relative importance of the latter increases at small recoil energies, hence the desire for low threshold. MUNU expects to achieve a sensitivity to $\mu_\nu \sim 3 \cdot 10^{-11} \mu_B$ after 1 year of running.

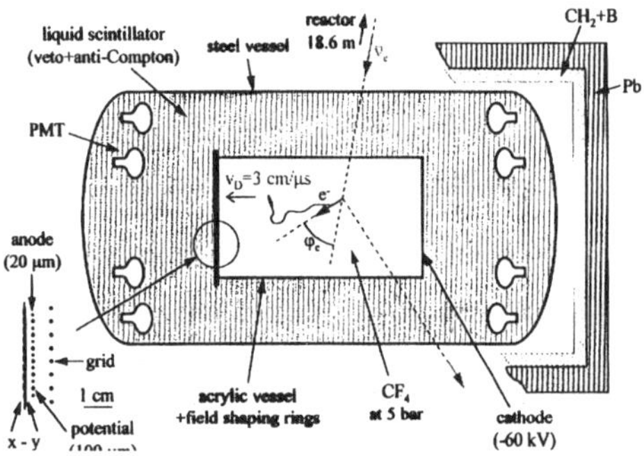

Figure 1: The MUNU experiment. The TPC inside performs tracking and calorimetry of the recoil electrons. The outside vessel contains the anti-Compton liquid scintillator.

2 The detector

The Time Projection Chamber (TPC) is the central component of the MUNU experiment, serving as an electron target, tracker and calorimeter. The (anti)neutrinos (from a reactor) scatter off an electron in the CF_4 (tetrafluoromethane) gas, the electron ionizes the gas along its path forming a track, which then drifts in the uniform electric field towards the multiwire proportional anode, where multiplication of the charge occurs. Next to the anode wire plane, at a 3 mm distance, the xy−plane is a thin mylar foil with copper strips etched on either side - horizontal on one side and vertical on the other. This plane is used to read out the induced signals for the reconstruction of the track topology in x and y, while the z coordinate is deduced from the time of arrival. These signals are amplified and read out by 25 MHz FADC's.

The 1 m^3 acrylic TPC is to be immersed in a 20 m^3 liquid scintillator anti-Compton veto detector, equipped with 48 ø20 cm low radioactivity photomultipliers, triggering at 10 photoelectrons. The experiment is to be located at the Bugey nuclear power plant which will provide the neutrino source, while the electron target is the drift gas itself, 1 m^3 of CF_4 at 5 bar. It is expected to yield about 10 $\overline{\nu}e^-$ scattering events/day, and about 5 events/day due to background.

The advantage of CF_4 lies with its high density as an electron target, low Z which minimizes multiple Coulomb scattering at low energy (and thus permits track discrimination through topology), radiopurity, low cost, and above all the absence of free protons that could lead to unwanted inverse β decays.

The recoil energy and scattering angle can be measured in the TPC.

3 Detector performance

The TPC has been tested at 1 and 3 bar. The CF_4 gas was purified using an Oxysob filter and a cold trap at -90°C, 10 degrees above the CF_4 condensation temperature. Only 16 out of a total of 256 x and as many y strips were read out. A drift field of 50-80 V/cm·bar was applied. At the latter value, a drift velocity of 3.2 cm·μs was measured. A coincidence measurement with two plastic scintil-

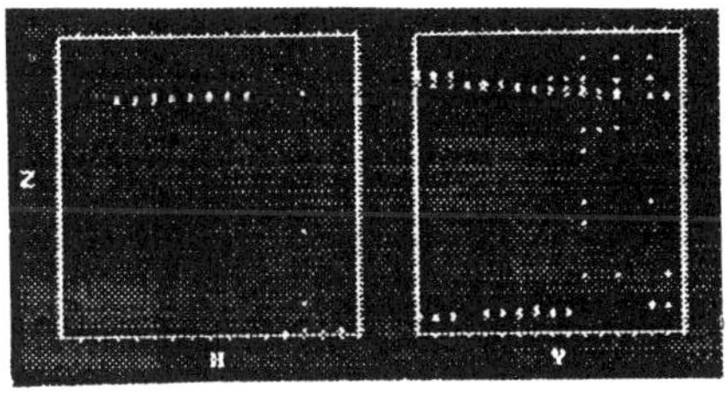

Figure 2: Tracking performance of the detector. A muon track is shown in two views: x-z and y-z.

lators enabled us to measure signals from near-vertical cosmic muons, depositing 0.48 and 1.44 MeV at 1 and 3 bars respectively. The anode signal was recorded at two positions 1 m apart. From the shift of the peak position in these two spectra, we deduce a mean drift length of:

$$\ell_{drift} = 22^{+14}_{-6}\text{m at } 50\text{V/cm} \cdot \text{bar}.$$

The tracking is also found to work satisfactorily - muon tracks can clearly be seen.

These results are encouraging, especially the high drift length which means that we can envisage a scaled-up version (4×50 m^3) of the same detector to perform solar neutrino spectroscopy at low energy, thus permitting possibly to resolve the missing ^{7}Be solar neutrino puzzle.

The same TPC filled with ^{136}Xe could be used for $0\nu\beta\beta$ decay search with a sensitivity to $m_\nu \sim 0.3 - 0.5$ eV. With CF_4, a dark matter search on the spin 1/2+ ^{19}F nucleus might be possible.

HERA Polarimeter Operation in 1995

M. Spengos[a]
University of Pennsylvania, Philadelphia, USA

In 1995 HERA routinely delivered polarized positron beam for the HERMES experiment. The transverse beam polarization was measured with a Compton polarimeter. A pair of spin rotators provided longitudinal spin polarization at the experiment. The value of the polarization during HERMES data taking varied between 45% and 70%, with typical values near 53%. The beam polarization was measured every minute with a statistical error of 2% to 4% and a systematic error of approximately 5.5%.

1 Introduction

A laser backscattering Compton polarimeter is used at the HERA ring to measure and monitor the transverse electron/positron polarization which developes naturally through the Sokolov-Ternov effect [1]. An Argon-ion laser provides a 10 Watt continuous beam of $2.41eV$ photons. Lenses and mirrors transport the beam over more than $200m$ to the $e\gamma$ interaction point. The laser light helicity is switched with a frequency of 83.8 Hz using a Pockels cell. Energy and vertical position of the backscattered Compton photons are measured in a tungsten-scintillator calorimeter about $65m$ from the interaction point. When the stored electrons or positrons are vertically polarized, the Compton photons will show an up-down asymmetry proportional to the spin polarization, which changes sign when switching between left and right circular polarized laser photons. This asymmetry shifts the center of gravity of the vertical position of the Compton photons on the calorimeter by $300\mu m$ for 100% electron polarization.

2 Measurements

In electron/positron storage rings the electron/positron spin can become polarized antiparallel to the magnetic bending field as a result of synchrotron radiation emission (Sokolov-Ternov effect). The polarization increases in exponentially according to $P(t) = P_{max}(1-e^{-(t-t_0)/\tau})$, where P_{max} is the asymptotic polarization and τ is the polarization build-up time. For HERA P_{max} is 92.4%. The value of the polarization during HERMES data taking was typically 55%, with values ranging between 45% and 70%. The overall polarization performance of HERA throughout the data taking of the HERMES experiment is displayed in figure (1).

---

[a]Representing HERMES and the Polarimeter group

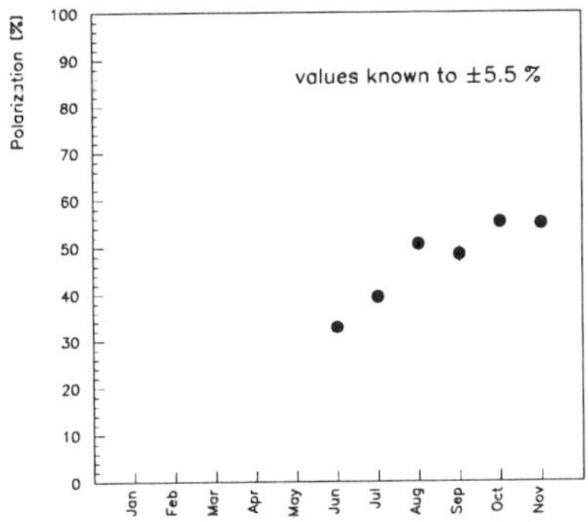

Figure 1: Average polarization per month for HERA for 1995. HERMES took data after July .

Typical statistical errors were $\Delta P/P = 2\%$ to 4% for a measurement time of 1 minute. This error should be significantly reduced in further running since the laser beam transport to the interaction point has been improved. The systematic error was approximately 5.5%. The main contribution to this 5.5% error comes from the calibration procedure of the absolute normaliztion of the beam polarization which is about 3.2% . That procedure, also called rise time calibration, is based on the fact that the slope of the exponential polarization build-up gives an independent determination of the asymptotic polarization [1].

3 Future

To improve the accuracy of the rise time calibration method, it is planned to selectively depolarize a fraction of the bunches, while the other bunches remain polarized. This will allow to check the machine stability by examining the still polarized bunches and also obtain rise-time curves for calibrations by the means of the depolarized bunches, simultaneously. The hardware and software of this new partial depolarizer has been implemented [3] and will be tried out soon. A further effort to reduce the systematic error of the beam polarization is the installation of a second Compton polarimeter close to the HERMES detector which will measure the longitudinal electron/positron spin polarization between the spin rotators [2]. It will be commissioned in 1996.

References

[1] D.P. Barber et al., NIM **A338** (1994) 166-184.

[2] W. Lorenzon, "Proposal to DESY for a Longitudinal electron Polarimeter", May 1995.

[3] F. Zetsche, Amsterdam SPIN96 conference, in preparation.

The HERMES Polarized ^{3}He Target

J. Stewart For the HERMES Collaboration
University of Liverpool, Liverpool L69 3BX, England

The HERMES experiment at DESY has recently completed the 1995 data taking period where a polarized gaseous ^{3}He target was used in conjunction with the HERA polarized positron beam to study the spin structure of the neutron. The parameters and performance of this target under operating conditions are presented.

The HERMES internal polarized ^{3}He target combines the polarizing method of laser optical pumping and metastability exchange[1] with the storage cell technique to produce a target whose characteristics match the requirements for an internal target in a high energy storage ring. Figure 1 shows the major components of the target. The ^{3}He atoms are fed into the optical pumping cell through a precision capillary C1 and flow into the target storage cell through capillary C2. The input pressure and conductances are selected to give a pumping cell pressure of 0.5 Torr and a flow rate of 10^{17} atoms/sec. A weak RF discharge is established in the pumping cell which excites about 1:10^6 atoms from the 1^1S_0 ground state to the 2^3S_1 metastable hyperfine multiplet. Circularly polarized light with λ =1083 nm from a 10 W CW Nd:LNA laser excites one metastable hyperfine state to the $2^3P_{0,1,2}$ levels,[2] thereby transferring angular momentum to the electronic state. These atoms then decay, depleting the population of the initial hyperfine state and polarizing the metastable atoms. The polarization is then transferred to the ground state atoms through metastablilty exchange collisions whereby the metastable atom and a ground state atom exchange electron quantum numbers during a collision. The storage cell is a 29 mm by 9.8 mm elliptical tube 400 mm long made of 125 μm thick high purity aluminum. The cell is mounted in the HERA beam and cooled to 25 K. By confining the atoms near the beam and cooling the atoms a target density of 10^{15} nucleons/cm^2 is obtained.

The ^{3}He polarization is measured by exciting atoms to a long lived state through electron impact. In this situation the nuclear polarization is transferred to the electronic system through the hyperfine interaction and is determined by measuring the circular polarization of the light emitted when the state decays. This technique is used in the pumping cell where the atoms are excited by the RF discharge. The degree of polarization of the 667 nm light from the $3^1D_2 \rightarrow 2^1P_1$ transition is a function of the nuclear polarization. This has been calibrated to 2% using NMR techniques.[3] The final precision is 5%

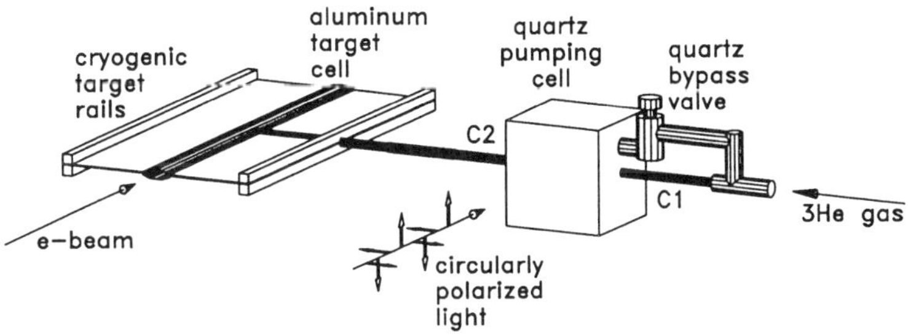

Figure 1: The HERMES ^{3}He target.

when measurement uncertainties are included.

A Target Optical Monitor (TOM) uses essentially the same technique to obtain a relative measure of the polarization inside the target cell by looking at atoms directly excited by the HERA beam.[4] The TOM can be used to measure the polarization inside the target cell to 7% after cross calibration with the pumping cell polarimeter. It also served as a backup polarimeter and was used to investigate beam-target effects. The target polarization changed by less than 5%, at the 1σ confidence level, for beam currents in the range 14 to 28 mA. Also, no statistically significant change in target polarization with temperature between 19 K and 64 K was observed.

The target operated nearly continuously throughout the 1995 data taking period. The average polarization was 0.47 ± 0.03 and a nominal target density of 10^{15} nucleons/cm^2. The polarization rise time was about 12 sec and the polarization was reversed every 10 minutes. The maximum operating target density was limited by the acceptable reduction in HERA beam lifetime.

References

[1] T.E. Chupp, R.J. Holt, R.G. Milner, Ann. Rev. Nuc. Part. Sci. **44**, 373 (1994), and references therein.

[2] S.F. Pate *et al.* in Int. Workshop on Polarized Beams and Polarized Gas Targets, ed. Hans Paetz Gen. Schieck and Lutz Sydow (World Scientific London 1996).

[3] W. Lorenzon et al., Phys. Rev. **A47**, (1993) 468.

[4] M.L. Pitt *et al.* in Int. Workshop on Polarized Beams. Op. Cit.

POMME-HYPOM: A VECTOR AND TENSOR DEUTERON POLARIMETER UP TO 2 GEV.

E. TOMASI-GUSTAFSSON, J. BALL, M. BOIVIN, T. HENNINO, R. KUNNE, V. P. LADYGIN, J. YONNET
LNS, CE Saclay, 91191 Gif-sur-Yvette Cedex, France

L. BIMBOT, Y. BISSON, J. L. BOYARD, PH. COURTAT, R.GACOUGNOLLE, M. MORLET, R. SKOWRON
IPN, 91400 Orsay, France

N.E. CHEUNG, M. K. JONES, C. F. PERDRISAT
The College of William and Mary, Williamsburg, Virginia 23185, USA

V. PUNJABI
Norfolk State University, Norfolk, Virginia 23504, USA

YU. BORZOUNOV, L. GOLOVANOV, N. M. PISKUNOV, I. M. SITNIK, E. A. STROKOVSKY, A.P. TSVINEV
JINR-LHE, 141980 Dubna, Moscow region, Russian Federation

We present the last results obtained at Saturne on the measurement of vector and tensor analyzing powers of deuterons at high energy. The calibration of the effective analyzing powers of POMME has been completed up to 1.8 GeV. Moreover a tensor deuteron polarimeter HYPOM based on d,p elastic scattering, has been tested under beam at 1.6 GeV.

1 Introduction

Many experimental programs at intermediate energy require the measurement of the polarization of the outgoing proton or deuteron. We show here the last results obtained with the polarimeter POMME, and its modification into a new vector and tensor deuteron polarimeter, HYPOM.

2 High energy calibration of POMME

POMME is a proton and vector deuteron polarimeter, working at Saturne II since 1989 [1]. It is based on the measurement of the azimuthal asymmetry from the inclusive deuteron scattering on a carbon target. It has been recently calibrated up to 1.8 GeV deuteron [2] incident energy and 3 GeV proton [3]. The polarimeter consists of three front wire chambers (which measure the incident trajectory and are set in the focal plane of a spectrometer), a carbon analyzer (up to 30 cm thick) and three large size rear chambers which detect the trajectory of the scattered particle. A set of scintillators, in front of the carbon block constitutes the trigger. The inclusive vector analyzing powers remain large up to 1.8 GeV (Fig. 1) and the figure of merit is of the order of 0.07, since all inclusive reactions contribute. However to obtain large tensor analyzing powers an exclusive reaction has to be selected. Low energy (up to 500 MeV) deuteron polarimeters (AHEAD [3], Polder [4]) are based on the $(d, 2p)$ charge exchange reaction or d, p elastic scattering. In particular the last reaction has large cross section as well as vector and tensor analyzing powers at high energies.

3 Results of the first tests of HYPOM

HYPOM is an extended focal plane polarimeter which works in the range of deuteron energies between 1-2 GeV . It is based on a (reversible) modification of POMME which allows the measurement of the azymuthal asymmetry in the d, p elastic scattering reaction. The target has to cover a large part of the acceptance of the spectrometer detection and to have a large thickness to increase the efficiency. Also, it must have as little material surrounding the liquid hydrogen as possible to minimize absorption of the low energy recoil protons. This ensemble has been designed as three modules, each of them composed of a cylindrical liquid hydrogen target of 50 *cm* length and 7 *cm* diameter [4] surrounded by layers of straw detectors which give the position of the recoil protons in one plane and a scintillator for timing.

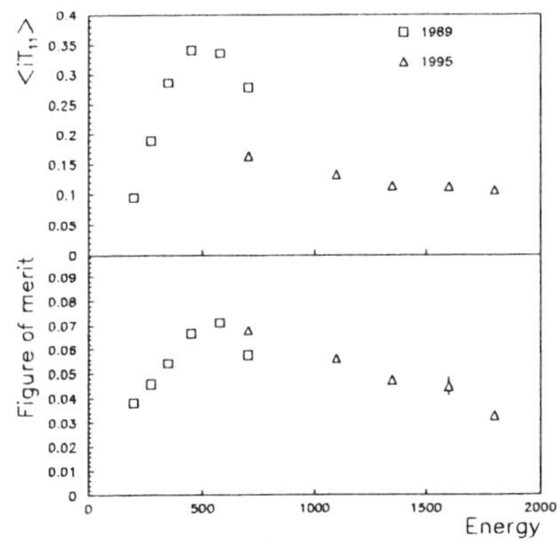

Fig. 1 Vector analyzing powers and figure of merit of Pomme.

One module has been built and was tested with a 1.6 GeV vector and tensor polarized deuteron beam at Saturne. The polarimeter was placed at the focal plane of the spectrometer SPES4. The measurements were performed at 0°. The polarized deuteron beam was incident directly on the polarimeter. The lowest possible intensity from the machine $\simeq 10^8$ particles per beam burst was further reduced to approximately $10^4 - 2 \cdot 10^4$ with the help of beam slits and by defocalizing the quadrupoles of the spectrometer.

Typical values of the vector and tensor polarizations were $\rho_{10} = -0.40 \pm 0.02$ and $\rho_{20} = 0.65 \pm 0.02$.

The sign of the polarization changed at each beam burst allowing to cancel eventual instrumental asymmetries. The off-line analysis consists in the reconstruction of the incoming and outgoing deuteron trajectories, using the informations on the wire chambers. The straw detectors gave information on the projected angle of the recoil proton. Software requirements on the vertex and on the coplanarity of (d, p) elastic scattering were used to select the *good events* which filled a two-dimensional spectrum for each beam state (i), as a function of the polar- θ and azimuthal- ϕ scattering angles. A fitting procedure allow to extract the vector and tensor analyzing powers from the dependence on $cos\phi$ and $cos2\phi$ and a constant coefficient. A large fraction of the protons emitted at 90^0 are absorbed in the target, reducing the efficiency to $1^0/_{00}$. However the corresponding deuterons are scattered at very small angles, where the analyzing powers are negligible and finally the figure of merit is not affected. The analyzing powers have been compared to the result of a MonteCarlo simulation based on previously measured cross sections and analyzing powers [7] and the agreement with the expected values is very good. The next step of the analysis will be the reconstruction of the beam polarization, from the spectra corresponding to each beam state. This implies the experimental determination of a function that takes into account the unpolarized cross section and the geometrical and kinematical cuts. Improvements of the trigger and of the surrounding material of the target are under way.

The first experiment which plans to use this polarimeter is the study of the ^{3}He structure through the measurement of the induced tensor deuteron polarization in the ^{1}H(^{3}He,$\vec{d}$)X reaction.

References

1. B. Bonin et al., Nucl. Instr. and Meth. A288 (1991) 379 and A288 (1991) 389.
2. E. Tomasi-Gustafsson et al., Nucl. Instr. and Meth. A366 (1995) 96.
3. N.E. Cheung et al., Nucl. Instr. and Meth. A363 (1995) 561.
4. S. Kox et al., Nucl. Instr. and Meth. A346 (1994), 527.
5. M. Garçon et al., Phys. Rev. **C49** (1994) 2516.
6. L.B.Golovanov et al., LNS/Ph/96-04, to be published in Nucl. Instr. and Meth. A
7. E. Tomasi-Gustafsson, J. Yonnet and V. Ladygine, LNS/Ph/94-07, LNS/Ph/95-13,

DIAMOND MICROSTRIP NUCLEAR DETECTOR

B. Vlahovic, A. Soldi, J.M. Dutta, O. Petifer North Carolina Central University, Durham, NC
H. Fenker, S. Majewski, Continuous Electron Beam Accelerator Facility, Newport News, VA,
H. A. Wynands, Kobe Steel USA Ltd., Durham, NC
L. McNeal University of North Carolina at Chapel Hill, NC
Z. Sitar, S. Han, North Carolina State University, Raleigh, NC
V. Borjanovic, ETF, Zagreb University, Zagreb, Croatia

I. Introduction

Precise tracking of charged particles within a high particle rate environment is difficult. The silicon microstrip detectors used at Fermilab, CERN, KEK and other laboratories can track charged particles precisely. However, they are destroyed by the high luminosity that makes them necessary. Conventional wire chambers are not able to handle the high rates. Recently, diamond microstrip detectors have attracted significant attention and they are being studied as a possible solution to the above problem[1,2].

In semiconductor detectors, energy loss from particles and radiation creates electron-hole pairs which drift apart under the influence of an applied electric field. It is the movement of these charges that constitute the signal current. Silicon detectors employ a reverse biased junction diode structure which must exhibit a leakage current that is small in comparison to the size of the signals. However, exposure to radiation results in progressive dopant type inversion thus destroying the very diode structure on which the detector depends.

Diamond has a high intrinsic resistivity, greater than 10^{12} Ω-cm, hence low noise and low leakage current. Because of this advantage one can build a diamond detector using a simpler ohmic structure, as shown in Fig. 1, thus eliminating the radiation damage inherent in the diode structure.

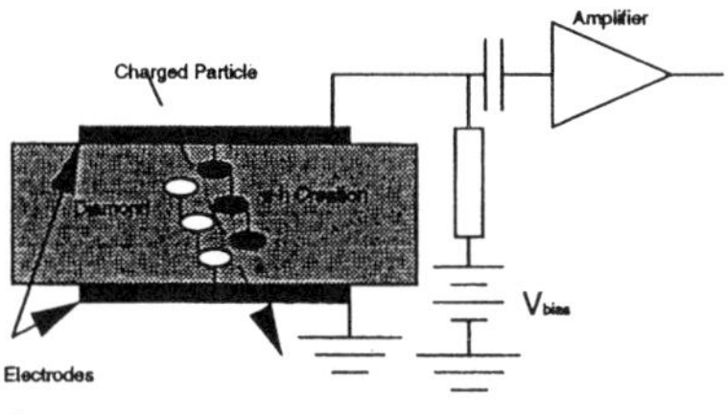

Fig. 1.

Particles and radiation will produce dislocations, point and p-type defects in any semiconductor crystal. However, dosage rates high enough to destroy the junction diode in silicon do not generate sufficient number of these defects to significantly increase the leakage current or destroy diamond detectors. No increase in leakage current was observed in a Schottky type detector for exposure of up to $10^{16}/\text{cm}^2$ of 1.5 MeV electrons[3,4]. Diamond has the advantage of a neutron transmutation cross section better than 20 times lower than silicon, while the cohesive energy of diamond is 7.37 eV/atom as compared to 4.63 eV/atom in silicon. Leakage current in diamond can be reduced further by incorporating a Schottky barrier structure at the metal contact.

Diamond has also thermal conductivity two to four times larger than Cu.

Thus diamond can be exposed directly to high intensity beams without increasing its temperature greatly. Other significant advantages of diamond are the high carrier mobility, and a saturated drift velocity $\sim$ 220 μm/ns or 2.7 times higher than silicon. These properties should make diamond detectors intrinsically faster and able to operate at extremely high rates.

II. Results and discussion

We prepare diamond by a microwave enhanced chemical vapor deposition process (CVD). Hydrocarbon methane, hydrogen and oxygen are mixed and deposited at 60 Torr on a 800°C hot silicon substrate.

After removing the deposited diamond film from the silicon substrate the metal electrode strips are deposited with a structure that provides a proper ohmic contact (as for instance an annealed titanium/gold bilayer). The metal strips are wire-bonded with Au wires to micro pins and connected to a preamplifier.

The test station at CEBAF consists of amplifiers, data collection equipment (currently configured as a multi-channel analyzer), and various radioactive sources. We have taken measurements of pulse height using alpha particles and are currently working on improved electronics so that we can observe the signals of minimum-ionizing particles. An x-ray generator is available for testing diamond detectors configured as beam profile-monitors.

We are working with high quality polycrystalline, with highly oriented (100 textured) and with homoepitaxial diamond films. Both ohmic and Schottky diodes type detectors are being produced.

We have begun testing the quality of these films by electron scanning microscopy, Raman spectroscopy, and electrical and optoelectrical measurements. Some preliminary results for polycrystalline films are shown on fig. 2. and 3.

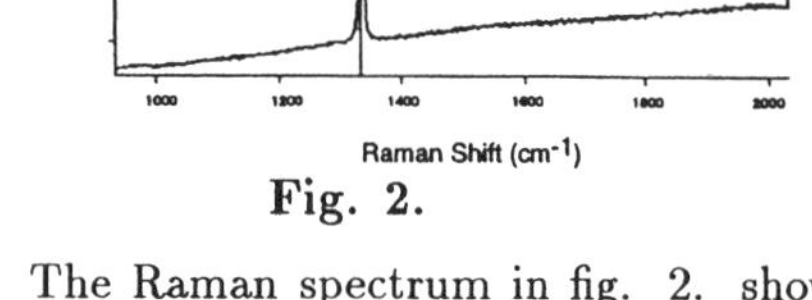

Fig. 2.

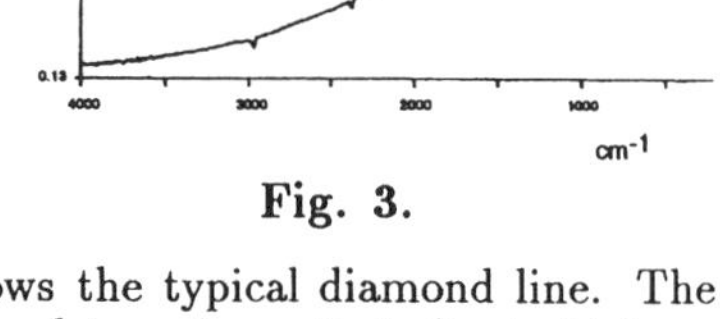

Fig. 3.

The Raman spectrum in fig. 2. shows the typical diamond line. The line FWHM of about $\sim$4 cm^{-1} and the high signal to noise ratio indicate high quality diamond film. The sharp Raman line at 1333 cm^{-1}, which is characteristic of diamond, is followed by extremely weak and broad bands between 1400 and 1600 cm^{-1} which are characteristic of amorphous carbon left at the polycrystalline diamond grain boundaries. The IR transparency spectrum in fig. 3 shows the very small C-H stretch infrared absorption band at 3.5 μm, which is a good indication that there is little hydrogen present. Thus the diamond sample is very pure.

Fig. 4. shows results obtained with an Electron Scanning Microscope (ESM) on randomly oriented polycrystalline, highly oriented (100) textured, and homoepitaxial diamond films respectively.

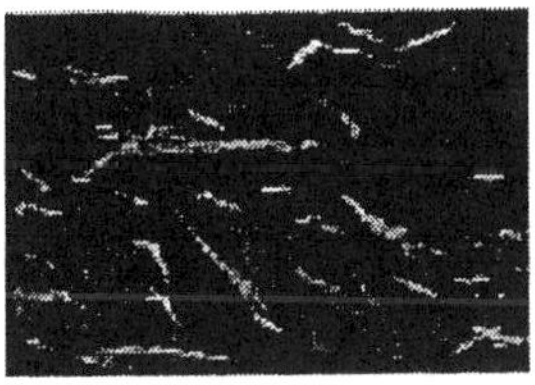
Fig. 4. a)

Fig. 4. b)

Fig. 4. c)

We have also begun preliminary conductivity, photoconductivity, and luminosity measurements and radiation hardness tests for both diamond films and diamond Schottky diodes. Up to now no increase in leakage current is observed after Schottky diode exposure to radiation of up to 10^{14}n/cm^2.

Fig. 5.a shows the pedestal and an alpha energy spectrum from Am^{241} α source as detected by single crystal diamond Schottky diode detector. The specimen thickness is 250 μm while the bias voltage is 30 V. On the vertical axis are counts in logarithmic scale, and on the horizontal axis the channel numbers. The pulse shape of an α particle on the same diode detector is shown in fig. 5.b. The long decay time is related to the decay time of the charge sensitive AmpTek A250 preamplifier, which is $\approx$ 300 μs.

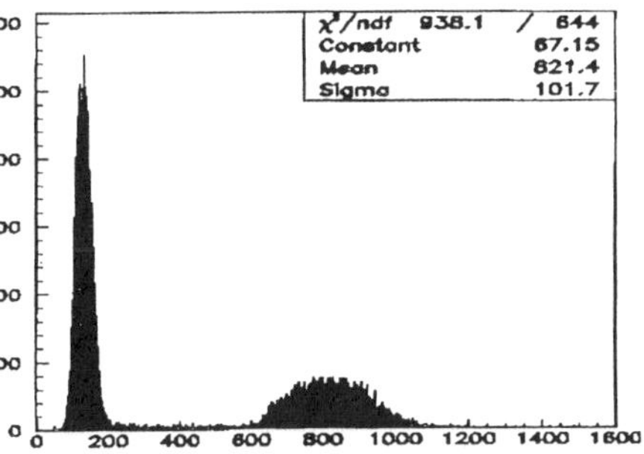

Fig. 5. a)

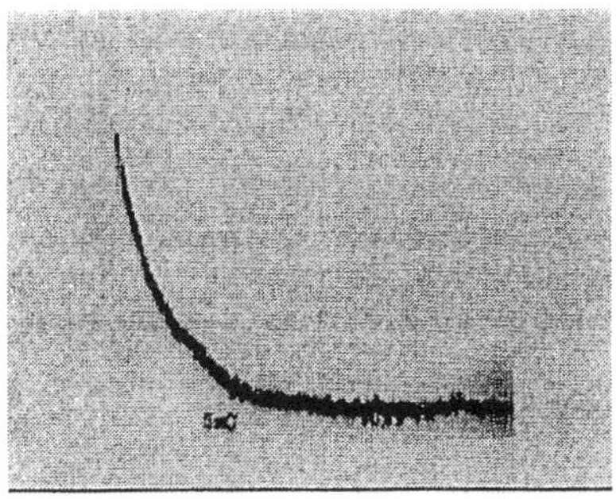
5. b)

In addition to a 250 μm thick ohmic diamond detector, and several Schottky diode detectors, four thin, 10 to 25 μm, beam position and profile detectors have been produced. They will be inserted in the beam at CEBAF during this summer to test their performance and design.

Highly textured (100) oriented film detectors under construction will be tested and characterized during this summer and we hope to have the first detector built from homoepitaxial film by the end of this summer.

III. References

1. F.Borchelt et al., Nucl. Instr. and Meth. in Phys. Res., **A354**, 318 (1995).
2. W. Dulinski et al., CERN PPE/94-222, (1994).
3. M. Franklin et al., Nucl. Instr. and Meth. in Phys. Res., **A315**, 39 (1992).
4. S. F. Kozlov et al., IEEE Trans. Nucl. Sci., **NS-22**, 160 (1975).

Session 12
New Facilities

NEW RESULTS ON B TO CHARM DECAYS FROM CLEO

RAINER WANKE
Physics Department, Ohio State University, 174 W 18th Avenue, Columbus, Ohio 43210, USA

(Representing the CLEO collaboration)

Using data taken with the CLEO II detector on the $\Upsilon(4S)$ resonance, we present new (preliminary) results on charm production in B decays We find evidence for the upper vertex D^0/D^+ production process $B \to \bar{D}X$ and place an upper limit on the inclusive rate of penguin decays. We also studied the branching fraction and kinematics of the semileptonic decay $\bar{B}^0 \to D^+ l^- \bar{\nu}$ and measured the product $|\mathcal{F}(1)\, V_{cb}|$.

1 Upper Vertex D^0/D^+ Production and $\mathcal{B}(b \to sg)$

One of the puzzles in B decays is the low observed semileptonic decay rate.[1] Two possible explanations are a significant production of D^0/D^+ mesons from the W decay ("upper vertex production") or a surprisingly high rate of penguin decays ($b \to sg$). Both quantities can be measured with D–lepton correlations in $B\bar{B}$ decays.

Using a dataset of $\approx 2.4\,\text{fb}^{-1}$ collected on the $\Upsilon(4S)$ resonance, we select D^0 and D^+ mesons by their decays to $K^-\pi^+$ and $K^-\pi^+\pi^+$ respectively. The leptons are required to have a momentum $p > 1.5$ GeV, to ensure that they are from a semileptonic B decay. Examining the angle between D and lepton enables us to separate between D–lepton combinations from same and from different B mesons, the charge correlation is used to determine the D origin.

The ratio between the observed numbers of $D\,l^-$ and $D\,l^+$ combinations from different B mesons leads (after corrections for $B\bar{B}$ mixing and mistags) to the ratio of upper vertex to lower vertex D^0/D^+ production:

$$\Gamma(B \to DX)/\Gamma(B \to \bar{D}X) = 0.107 \pm 0.029 \pm 0.018.$$

The measured yield of $D\,l^+$ combinations from different B's divided by the yield of $D\,l^-$ combinations from the same B gives us the ratio P_g/P_s of the probabilities of $\bar{D}$ production in generic (P_g) and semileptonic (P_s) B decays. On the other hand this ratio is given by adding up all possible B decay modes: $P_g/P_s = 1 + |V_{ub}/V_{cb}|^2 - \frac{1}{2}Prob(D_s;\text{low.vtx.}) - Prob(c\bar{c}) - Prob(baryons) - Prob(b \to sg)$. Using mainly CLEO measured numbers for the various contributions[1–5] and measuring $P_g/P_s = 0.886 \pm 0.037 \pm 0.019$, we can place an upper limit on the inclusive rate of penguin decays:

$$\mathcal{B}(b \to sg) < 8.5\% \text{ at the 90\% confidence level.}$$

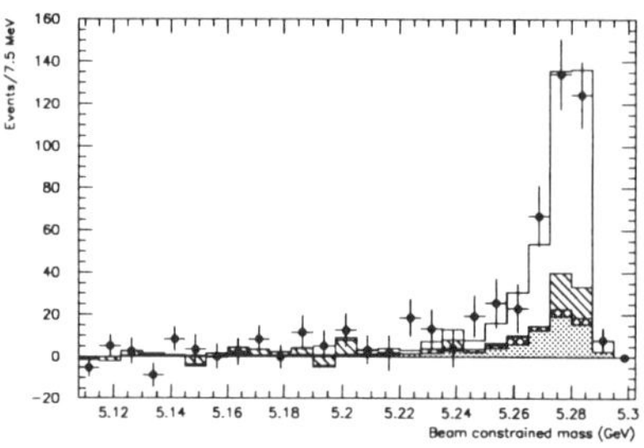

Figure 1: Beam-constrained mass of $\bar{B^0} \to D^+ l^- \bar{\nu}$. Continuum and D^+ backgrounds are subtracted. Shown are $\bar{B^0} \to D^+ l^- \bar{\nu}$ (white), other $B \to X l \nu$ (hatched) and other backgrounds (shaded).

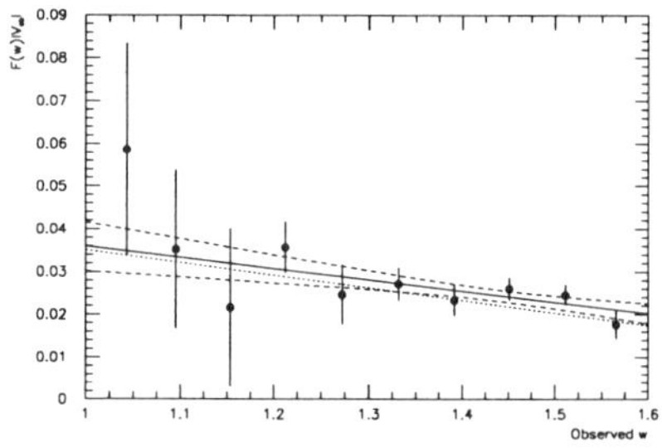

Figure 2: Measured values of $|\mathcal{F}(w)V_{cb}|$. with the result of the fit to $d\Gamma/dw$ (solid), 1 σ statistical errors (dashed) and result from $B \to D^* l \nu$ [6] (dotted).

2 Study of $\bar{B^0} \to D^+ l^- \bar{\nu}$

Semileptonic decays offer a unique possibility for understanding heavy quark decay dynamics and measuring elements of the CKM matrix.

Using the hermeticity of the CLEO detector we reconstruct the ν energy and momentum by looking for the missing energy and momentum of the event. We require the missing invariant mass to be close to zero, we allow only net event charges of $0, \pm 1$, and we reject events with a second lepton. The D^+ mesons are selected via their decay $D^+ \to K^- \pi^+ \pi^+$. Backgrounds from continuum events are subtracted using data taken below the $\Upsilon(4S)$ and those from random $K\pi\pi$ combinations using D^+ sidebands. The beam–constrained mass distribution is shown in fig. 1. We find 238 ± 27 signal events, resulting in $\mathcal{B}(\bar{B^0} \to D^+ l^- \bar{\nu}) = (1.89 \pm 0.22 \pm 0.35)\%$, where the systematic error is dominated by uncertainties in the ν detection.

In the heavy quark limit the Isgur-Wise function $\xi(w)$ (with $w \equiv (m_B^2 + m_D^2 - q^2)/2m_B m_D$) is closely related to the form factor $\mathcal{F}(w)$. From a fit to $d\Gamma/dw$ we obtain $|\mathcal{F}(1)V_{cb}| = (3.60 \pm 0.57) \times 10^{-2}$ (see fig. 2).

References

1. G. Buchalla, I. Dunietz and H. Yamamoto, Fermilab-PUB-95/167-T.
2. R. Ammar *et al.* (CLEO), CLEO–CONF95-9, EPS0165 (1995).
3. X. Wu *et al.* (CLEO), CLEO–CONF95-11, EPS0169 (1995).
4. R. Balest *et al.* (CLEO), *Phys. Rev.* D **52**, 2661 (1995).
5. G. Crawford *et al.* (CLEO), *Phys. Rev.* D **45**, 725 (1992).
6. B. Barish *et al.* (CLEO), *Phys. Rev.* D **51**, 1014 (1995).

THE BNL AGS ACCELERATOR COMPLEX--STATUS AND FUTURE PLANS*

MITSUYOSHI TANAKA
AGS Department, Brookhaven National Laboratory
Upton NY11973, USA

This paper describes the present performance and capability of the BNL AGS accelerator complex and possible future intensity upgrade plans. In 1995, the AGS reached its design upgrade goal of $6.0 \cdot 10^{13}$ ppp with the Booster. The AGS with a new fast extracted beam (FEB) system is able to perform single bunch multiple extraction at 30 Hz per AGS cycle for the g-2 experiment and for RHIC injection.

1. Introduction

In July 1960, the BNL Alternating Gradient Synchrotron (AGS) accelerated protons to 30 GeV at an intensity of $\sim 10^{10}$ ppp--becoming the world's most powerful atom smasher. Since then, the AGS has been periodically upgraded, improved, and has evolved to satisfy its continuously changing user's wishes; e.g. the 50 MeV Linac was replaced with the 200 MeV Linac in 1970, the AGS was connected to the existing Tandem Van de Graaff Accelerator in 1983, the 1.5 GeV Booster was added in 1991, and the associated AGS subsystem upgrades were completed.

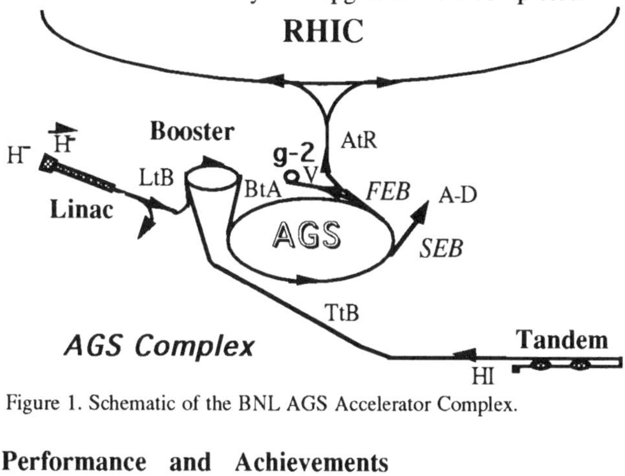

Figure 1. Schematic of the BNL AGS Accelerator Complex.

2. Recent Performance and Achievements

The AGS now accelerates heavy ions up to gold (Au^{77+}) at various energies for the heavy ion physics program and has recently increased the slow-extracted proton beam (SEB) intensity above $6.0 \cdot 10^{13}$ ppp at 25 GeV/c with a 1.6 sec uniform spill for various intensity hungry particle physics experiments such as rare K-decay experiments.

---

*Work performed under the auspices of the US Department of Energy.

The accelerator also recently became equipped with a new FEB extraction system that allows the AGS to perform single bunch multiple extraction of a heavy ion or a (polarized) proton beam at a time interval of 33 ms up to 12 times per AGS cycle for filling of the RHIC collider or for 3.0 GeV/c π/μ production for the 14 m diameter g-2 muon storage ring. In the fall of 1995, a single bunch of $0.25 \cdot 10^9$ gold ions was successfully extracted at 11.2 GeV/c/N from the AGS and transported through the 600 m long AGS-to-RHIC (AtR) transfer line to an internal beam dump. The RHIC injection test, which includes the full AtR line and 1/6 of the superconducting ring, is scheduled for the end of 1996. The commissioning of V/V1 lines with protons is currently taking place for the g-2 ring.

3. Possible Future Upgrade Plans

So far we have focused mainly on increasing extracted beam intensity without much attention to the emittance blowup or beam loss. A better understanding of the critical high intensity effects can lead to better machine tuning, hence it could increase the intensity further. However, beyond 10^{14} ppp, various upgrades of the AGS are needed: 1) the AGS proton beam intensity could be further increased to 1.5×10^{14} ppp with the implementation of a barrier cavity rf system by avoiding bunched beam stacking during injection; 2) by adding a 1.5 GeV accumulator ring in the AGS tunnel to accept beam continuously from the Booster during 1.2 - 2.5 sec of the AGS cycle time, two to four times more beam pulses can be accumulated and be transferred into the AGS ring; 3) by increasing the AGS repetition rate to 2.5 Hz, the average beam current could be pushed to 40 μA.

4. Summary

There were/are several proposals (LAMPF II, KAON, JHP etc.) for a high intensity hadron facility around the world, but only one that is already *in operation* and *under construction*. It is the BNL AGS complex. It is being done a step-wise, physics driven approach in the most cost-effective away to minimize the downtime to the ongoing physics program and the cost without a major up front commitment in funding. The AGS average beam current was 1 μA a just few years ago, is now 3 μA, becomes 5-7 μA soon with a barrier bucket rf system and tuning. With an accumulator, which is under design, it will reach 12 μA.

After 1999, the AGS becomes an injector for RHIC to fill its two rings with gold or polarized proton beams in a few minutes every 8-10 hours, the AGS could be available between fillings for forefront particle and nuclear physics program at an incremental cost.

Acknowledgments

The work described in this paper has been performed collectively by the accelerator staff in the AGS Department and the RHIC Project.

The COMPASS project

The COMPASS collaboration
presented by S. Paul
Max Planck Institut für Kernphysik, Heidelberg, Germany

We describe the main physics issues and experimental challenges for the new COMPASS project, which has recently been proposed at CERN as the only high energy fixed target facility to run beyond the year 2000, starting up in 1999.

1.1 Introduction

A new experiment has been proposed at the CERN SPS aiming at a continuation of the study of hadronic structure using high energy beams of muons and hadrons [1]. This experiment will allow a first determination of a possible gluon polarization in the nucleon, the study of semi leptonic and leptonic decays of charmed hadrons and the search for doubly charmed baryons as well as an extension of the search for glueballs into the mass range above 2 GeV/c^2.

The study of the polarization of the gluons in a nucleon requires the use of a spin dependent process sensitive to the gluon degrees of freedom. One possibility is the production of open charm hadrons by polarized photons interacting with a polarized nucleon. The analyzing power of the photon-gluon production process is large enough to promise good sensitivity for ΔG. The polarized photons are obtained from the scattering of polarized muons on the target. Both, beam and target polarizations will be longitudinal and the asymmetry in the D-meson yield with respect to a flip of the target polarization is directly connected to ΔG. Within 2 years of measuring time a sensitivity of about 0.1 should be obtained for the kinematical region of $.08 \leq \eta \leq .3$ (check !).

The study of semi leptonic decays offers a clean way to obtain insights into the decay dynamics of hadrons and thus into the hadronic structure. A measurement of decay rates and asymmetries as function of the four momentum transfer allows an extraction of form factors for which definite predictions exist within the framework of HQET. Leptonic decays of mesons are a key to the heavy mesons decay constants, one of the basic hadronic parameters calculable theoretically (e.g. via lattice QCD). Double charmed baryons in turn constitute a novel type of baryon, similar in structure to a heavy meson with the two heavy quarks playing the role of a heavy diquark surrounded by a light quark. They may offer the link between simple mesons and the much more complexe effective 3-body systems.

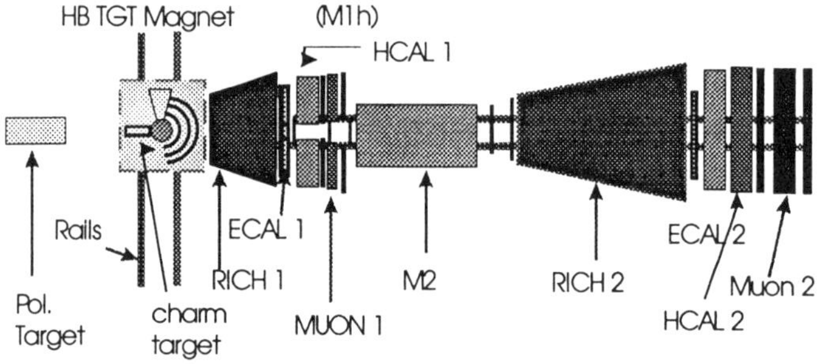

Figure 1: The COMPASS spectrometer layout (only shown is the hadron layout).

The existence of glueballs, predicted by QCD, may recently have been demonstrated using a large variaty of data ranging from $\bar{p}$p-annihilation to J/Ψ decays. Present calculations have many open parameters which may well be adjusted by the observation of the glueball ground state. The key information, besides their proof of existence, is the determination of the relevant mass scales which can only be achieved if the excitation spectrum of these objects is known. The spectroscopy of such states will thus provide us with stringent tests of our conception for this type of bound states.

1.2 Experimental Setup

The experimental program outlined above requires a very modern apparatus capable to withstand very high particle fluxes, interaction and trigger rates the realization of which constitues a very challenging task. Although the experimental program is very diverse it can be achieved within a single apparatus based on a two stage magnetic spectrometer outlined in fig. 1. Both spectrometer parts provide full particle identification based on RICH-technology, electromagentic and hadronic calorimetry as well as muon identification. The need of very different target configurations, however, requires the use of two different large angle spectrometer magnets for the muon and the hadron beam program.

References

1. The COMPASS collaboration, Proposal P297, CERN/SPSLC/96-14
   The COMPASS collaboration, P297 Addendum I, CERN/SPSLC/96-30

Relativistic Polarized Neutrons at the Laboratory of High Energy Physics, JINR

A.Kirillov, L.Komolov, A.Kovalenko, E.Matyushevsky, A.Nomofilov, P.Rukoyatkin, V.Sharov, A.Starikov, L.Strunov, A.Svetov

Lab. of High Energy Physics, Joint Institute For Nuclear Research, Dubna, Russia

presented by **P.Rukoyatkin** [1]

Using slowly extracted polarized deuterons, available at the accelerator facility of the Laboratory of High Energy Physics[1], polarized quasi-monochromatic neutrons with momenta from 1.1 to 4.5 GeV/c have been generated. Depending on momentum, from 10^4 to 10^6 polarized neutrons per accelerator cycle were produced. At present, the polarized neutrons are mainly intended for measuring the $(\vec{n},\vec{p})$ total cross section differences (the Delta-Sigma experiment[2]) with the Polarized Proton Target (PPT)[3].

Deuterons provide the best way to generate neutron beams with well defined characteristics. The breakup cross section constitutes a significant ($\simeq 30\%$) part of the total (dA) cross section, which does not depend on energy. Also, the higher the energy of the produced beams, the better their properties: at high neutron velocities $v_n/c \gg k_{int}^{rms}/m_n$, the angular spread and yield at zero degree in the laboratory frame behave as $\sigma_{\theta_x} \sim 1/p_n$, $Y \sim {p_n}^2$, which means that the stripping neutron beam self-collimates as the energy rises. Starting from the chrestomathic experiment of Helmholz et al., neutron beams based on deuteron breakup were realized in different laboratories as soon as accelerated deuterons became available at their facilities[4]. Polarized neutron beams are obtained from vector polarized deuterons. The question of polarization of such beams is not so obvious *a priori.* In[5], the polarization transfer from deuterons to protons $\kappa = P_p/P_d$ at 0° and $p_p = p_d/2$ at breakup were studied. It was shown that the predominant part of the secondary protons inherits the polarization of primary deuterons entirely, i.e. $\kappa \simeq 1$. As neutron beams are formed under the same conditions, one may assume that $P_n \simeq P_d$.

The neutron beam lines of the Laboratory of High Energy Physics operate on the basis of a slowly extracted deuteron beam from the Sychrophasotron. The maximum momentum of extracted beams is $9GeV/c$, the typical spill length - $0.5sec$. Acceleration cycles repeat each 8-10sec, depending on the energy. The nominal intensity of polarized deuterons at a maximum momentum is $10^9 ppc$ (up to $5 \cdot 10^{11} ppc$ can be extracted in the unpolarized mode). The deuteron polarization is oriented along the vertical axis and can be flipped each cycle, which is important from a methodical point of view. Its absolute value is measured by a specially dedicated fast polarimeter[6]. It was shown that there are no depolarizing resonances in the entire energy range of the Synchrophasotron. During recent operations, the average polarization of the deuteron beams was 0.535 ± 0.009[2]. We relate this value to our polarized neutron beams, also.

There are two neutron beam lines, both situated in the main experimental hall. The first was prepared at an early stage of the Delta-Sigma experiment working up and was used for methodical purposes. The channel was organized at the end of the VP-1[1] line where the slowly extracted beam dumps. The beam dump, whose central solid part was replaced by an insertion with a hole, served as the $3.5m$ long first section of the neutron collimator. It was followed by a $2.5m$ long iron block having a collimation hole $3cm$ in diameter. The distance from the neutron producing target to the collimator exit was $12m$, determining the solid angle $\simeq 5\mu sr$. At the certain bend angle of the sweeping magnet, it was possible

[1] e-mail: rukoyat@sunhe.jinr.dubna.su

to extract stripping protons from the same target into the 5V beam line where part of the flux was measured by a scintillator telescope. Using the measured value and the known relation between the acceptances for neutrons and protons, the flux in the neutron beam was calculated. The neutron intensity was also measured by activation of a CH−sample placed in the neutron beam. This is summary of the measurements (errors $\simeq 15-20\%$) carried out with a $20cm$ CH target-stripper are:

$P_n, GeV/c$	1.13	1.50	1.77	2.25	4.5
$I_n/I_d(10^9), ppc$	$7.6 \cdot 10^3/0.1$	$3.5 \cdot 10^4/0.3$	$7.7 \cdot 10^4/0.5$	$1.6 \cdot 10^5/0.8$	$1.0 \cdot 10^6/1.0$

The first four points are calculations from the proton fluxes and the last was taken from the activation measurements.

The spatial distribution in the neutron beam was checked by means of an assemblage consisting of three scintillators, S_A, S_1, S_2, a convertor ($10-30cm$ of polyethylene) and a multiwire proportional chamber, set up in the order: $S_A - Conv. - MWPC - S_1 - S_2$. "Neutral" logic $\bar{S_A} \cdot S_1 \cdot S_2$ triggered the proportional chamber, whereas "charged" $S_A \cdot S_1 \cdot S_2$ gave a charge contamination to the neutron beam. The neutrons interacted with the convertor matter and events with small angles, restricted by the S_2 counter ($\pm 0.03r$), projected the neutron beam spot on the chamber planes. Event selection solely by this trigger was insufficient at neutron momenta $P_n \gtrsim 2GeV/c$ due to multiplicity growth, and an additional criteria, "only single events in the chamber planes", was then applied. The variances of measured and "true" distributions are connected via: $(\sigma_x^{(meas.)})^2 = (\sigma_x^{(beam)})^2 + L^2[(\sigma_{\theta_x}^{(beam)})^2 + (\sigma_{\theta_x}^{(sec.)})^2]$, where L is the distance from the convertor center to the chamber. In this way, it was obtained that $\sigma_x^{(meas.)} \cdot \sigma_y^{(meas.)} \simeq 11 \cdot 11mm^2$, where additions to the "pure" beam sizes does not exceed 10%. An example of the distribution, measured with a $22cm$ polyethylene convertor is shown in Fig.1a (no selection for multiplicity).

As the Delta-Sigma experiment required a large floor space in a beam zone, the existing 1V line was adopted for this purpose (see Figure). By means of the B1-B3 bending magnets, primary deuterons were bent from the host direction of the extracted beam (VP-1) at the f4 point and directed to a neutron producing target of $17cm$ $Be + 6cm$ C (T). The Q1,Q2 quadrupole dublet performed the required beam focusing. As it was important to retain the incident beam at strictly 0°, five multiwire ionization chambers were installed along the beam line, including at positions just before the target and at the beam dump (BD) exit. Also, the directions after each bending magnet were preliminarily traced by means of a thin

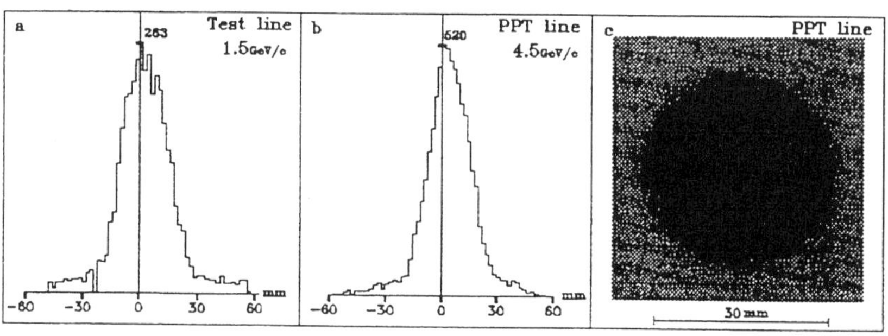

Figure 1: Neutron space distributions: a,b - from MWPC, c - exposed emulsion

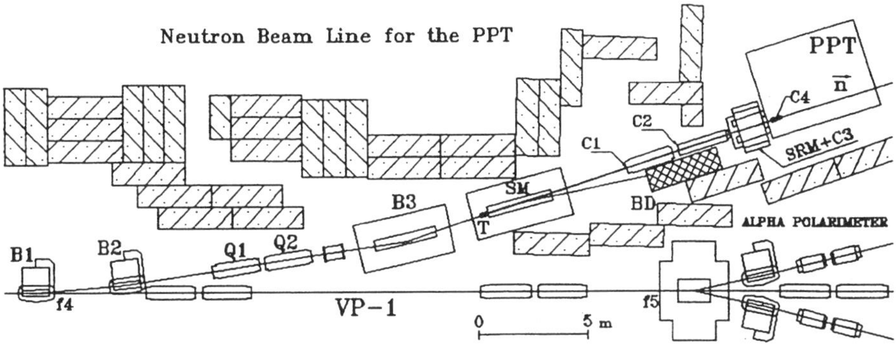

current-carrying wire. To turn neutron spins from the vertical direction to a longitudinal one, the beam line was equipped with a magnet (SRM) of 2.7 *Teslameters* maximum field integral. The neutron beam was formed by a collimator composed of four stages: C1,C2 ($\oslash 4cm, \oslash 3cm$, iron) and C3,C4 ($\oslash 2.8cm$, brass) $6m$ in total length.The collimator defined $\Delta\Omega \simeq 3\mu sr$, a $1.2mrad$ angular divergence and a beam spot in the PPT, that fitted into a $\oslash 3cm$ circle. The neutron space distribution was continuously monitored during data taking runs using a MWPC placed $50cm$ downbeam from the PPT. A capsule with the target material, situated in a $2.6T$ longitudinal magnetic field, served as a convertor. The vertical distribution at $P_n = 4.5GeV/c$ is shown in Fig.1b (selection for multiplicity applied). A very visual test of the neutron spot was produced by an emulsion irradiated at the PPT entrance (Fig.1c). Neutron flux was measured by the setup detectors, calibrated using the activation method during a preliminary run. Date taking runs were carried out at 1.92, 3.31 and $4.50GeV/c$ neutron beam momenta. The intensities of the polarized neutrons, averaged over each run, were $2.7 \cdot 10^4$, $2.0 \cdot 10^5$ and $4.7 \cdot 10^5 \vec{n}/cycle \pm 10 - 15\%$ at deuteron intensities on the target $3.0 - 6.5 \cdot 10^8 \vec{d}/cycle$. Charge contamination was negligible. Momentum spread of the neutrons was 5% (FWHM).

This work was supported in part by the ISF and the Russian Government through Grant JHW 100, and by the Russian Foundation for Basic Research through Grants RFBR-93-02-16715, RFBR-96-02-18736a.

[1] I.Issinsky et al., Acta Physica Polonica, **B25**, 673 (1994)

[2] V.Sharov et al., In: Proc. VI Workshop on Spin Phenomena in High Ener. Phys., 1995, Protvino, to be publ.; Prep., DAPHIA/SPP 96-03, to be publ. in Z. Physik C v.71, N1, 1996;

[3] F.Lehar et al., Nucl. Instr.&Meth. **356**, 58 (1995)

[4] A.Helmholz, Phys. Rev. **72**, 1003 (1947); Yu.Kazarinov, et al., Zh. Eksp. Teor. Fiz. **43**, 35 (1962); G.Bižard et al., Nucl. Instr.&Meth. **111**, 451 (1973); Bevatron Exper. Handbook (1973); A.Gasparyan, JINR Comm., 1-9111 (1975); G.Korolev et al., Phys. Lett. **B165**, 262 (1985); H.Noumi et al., In: Proc. 9th Symp. on Acceler. Science, Tsukuba, p.205 (1993)

[5] E.Cheung et al., Phys. Lett. **B284**, 210 (1992); A.Nomofilov et al., Phys. Lett. **B325**, 327 (1994)

[6] V.Ableev et al., Nucl. Instr.&Meth. **A306**, 73 (1991)

GRAAL[a]: A POLARIZED GAMMA RAY BEAM (0.4-1.5GeV) AT ESRF

J. Ajaka[1], M. Anghinolfi[2], V. Bellini[3], G. Berrier[1], N. Bianchi[4], J.P.Bocquet[5], P. Calvat[5], M.Capogni[6], L.Casano[6], M. Castoldi[2], P. Corvisiero[2], A.D'Angelo[6], J.P. Didelez[1], R. Di Salvo[7], Ch. Djalali[1], M.A. Duval[1], F. Faure[5], R. Frascaria[1], G. Gervino[7], F.Ghio[8], P. Girard[1], B.Girolami[8], E. Guinault[1], E. Hourani[1], V. Kuznetsov[9], A. Lapik[9], P. Levi Sandri[4], S. Lo Nigro[3], D.Moricciani[6], M. Morlet[1], V. Muccifora[4], V. Nedorezov[9], C.Perrin[5], E. Polli[4], D.Rebreyend[5], A.R. Reolon[4], G. Ricco[2], M. Rigney[1], M. Ripani[2], L. Rosier[1], P. Rossi[4], F. Roudot[1], Th. Russew[5], M. Sanzone[2], C.Schaerf[6], A. Scire[3], M. L. Sperduto[3], A. Sudov[9], M. Taiuti[2], A. Turinge[10], J. Van de Wiele[1], D. Vinciguerra[3], A. Zucchiatti[2].

1) Institut de Physique Nucléaire d'Orsay, France
2) INFN sezione di Genova and Università di Genova, Italy
3) INFN Laboratori Nazionali del Sud and Università di Catania, Italy
4) Laboratori Nazionali di Frascati, Italy
5) Institut des Sciences Nucléaires de Grenoble, France
6) INFN sezione di Roma II and Università di Tor Vergata, Italy
7) INFN sezione di Torino and Università di Torino, Italy
8) INFN sezione Sanità and Instituto Superiore di Sanità, Roma, Italy
9) Institute for Nuclear Research, Moscow, Russia
10) Kurchatov Institute of Atomic Energy, Moscow, Russia

The GRAAL facility now produces a highly polarized gamma-ray beam by Compton scattering of laser photons on the electrons of the European Synchrotron Radiation Facility (ESRF) at Grenoble. Preliminary results have been obtained with the LAGRANGE detector showing its excellent performances. The first experiment is scheduled for September 96.

The GRAAL facility has recently become operational. It now produces a tagged and polarized gamma-ray beam (400-1500 MeV) by Compton scattering of laser photons on the 6 GeV electrons stored in the ring of the ESRF in Grenoble, France. As compared to the bremsstrhalung method, Compton scattering provides two main advantages. First, the gamma-ray energy distribution is rather flat and furthermore, low energy photons can be significantly reduced by collimation, resulting in a maximal flux at the Compton edge. Secondly, thanks to the small spin-flip amplitude Compton scattering allows to

[a]GRenoble Anneau Accélérateur Laser

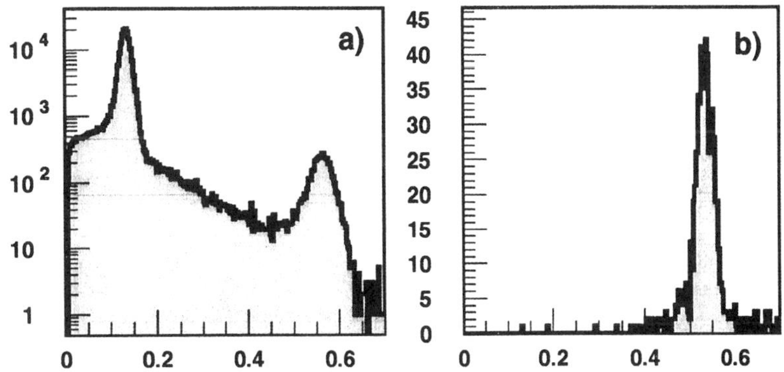

Figure 1: Invariant mass for a) two pkotons b) six photons in the BGO.

produce highly polarized gamma-rays. The degree of polarization is $\approx$ 100% at the Compton edge for both linear and circular polarization and then decreases with energy ($\approx$ 50% at $E_{max}/2$ for linear). The flux is typically 2-3 $10^6\gamma/s$ and is mainly limited by the ESRF.

The 4π detector LAGRANGE with its associated electronics and data acquisition system is nearly complete. It consists of an electromagnetic BGO calorimeter (480 cristals covering $\Delta\phi=2\pi$, $\Delta\theta$=25-155°) completed by a forward shower detector (with good angular and poor energy resolution). Tracking of charged particles is performed by multi-wire proportional chambers (2 cylindrical and 2 planes) while particle identification is obtained by time-of-flight and dE/dx measurements.

A first test experiment has been carried out between April and May. Figure 1 shows an invariant mass spectrum calculated from the BGO information for 2 and 6 gamma-rays. One can clearly identify π^0 and η in the first case while the simultaneous observation of 6 gamma-rays selects unambiguously the decay channel $\eta \rightarrow 3\pi^0$, thus demonstrating the high performances of the detector. Beam asymmetries for the reaction γ+p$\rightarrow$ η+p could therefore be measured using the high degree of polarization of the beam. The first experiment scheduled in September will be dedicated to the η (reaction mechanism with the polarization observable Σ and study of its decay channels).

CELSIUS AS AN η FACTORY

R. BILGER[14], M. BLOM[6], D. BOGOSLAWSKY[8], A. BONDAR[4], W. BRODOWSKI[14], H. CALÉN[6], B. CHERNYSHEV[11], I. CHUVILO[7], H. CLEMENT[14], S. DAHLGREN[6], V. DUNIN[8], C. EKSTRÖM[13], C.-J. FRIDÉN[13], K. FRANSSON[6], M. GORNOV[11], Y. GUROV[11], L. GUSTAFSSON[6], H. HIRABAYASHI[9], S. HÄGGSTRÖM[6], B. HÖISTAD[6], H. IKEGAMI[12], A. JANSSON[6], A.JOHANSSON[6], T. JOHANSSON[6], K. KILIAN[10], G. KOLACHEV[4], M. KOMOGOROV[8], L. KOMOGOROVA[8], S. KULLANDER[6], A. KUPSC[2], A. KUZMIN[4], A. KUZNETSOV[8], P. MARCINIEWSKI[2], A. MARTEMYANOV[7], Y. MIZUNO[12], B. MOROSOV[8], J. MOEHN[6], A. NAWROT[2], W. OELERT[10], Z. PAWLOWSKI[5], A. POVTOREJKO[8], T. PURLATZ[4], D. REISTAD[13], R. RUBER[6], S. SANDUKOVSKY[8], R. SHAFIGULLIN[11], B. SHWARTZ[4], V. SIDOROV[4], V. SOPOV[7], J. STEPANIAK[2], A. SUKHANOV[4], A. SUKHANOV[8(6)], V. TCHERNYSHEV[7], V. TIKHOMIROV[8], A. TUROWIECKI[1], G. WAGNER[14], Z. WILHELMI[1], A. YAMAMOTO[9], A. YAMAOKA[9], Y. YUASA[12], J. ZABIEROWSKI[3], A. ZERNOV[8], J. ZLOMANCZUK[1(6)]

[1] *IEP, Warsaw University, Poland;* [2] *INS, Warsaw, Poland;* [3] *INS, Lodz, Poland;* [4] *BINP, Novosibirsk, Russia;* [5] *IRE, Warsaw Technical University, Poland;* [6] *ISV, Uppsala University, Sweden;* [7] *ITEP, Moscow, Russia;* [8] *JINR, Dubna, Russia;* [9] *KEK, Tsukuba, Japan;* [10] *KFA, Jülich, Germany;* [11] *MEPI, Moscow, Russia;* [12] *RCNP, Osaka, Japan;* [13] *TSL, Uppsala, Sweden;* [14] *Tübingen University, Germany*

The rare decays of the η meson are of great interest for investigations of P, C and CP symmetries and for tests of chiral pertubation theory in strong and electromagnetic interactions. The η meson has unique properties and is well suited for such studies: high mass, small mass width and the quantum numbers as those of the vacuum. High-precision experiments on rare decays of light mesons are currently being prepared within the WASA project - a high-luminosity experiment using a close to 4π detector configuration[1, 2] at the CELSIUS ion storage ring at the The Svedberg Laboratory, Uppsala.

The CELSIUS ring[2] is a combined accelerator and storage ring with a circumference of 81.8 m. The ring consists of four 90^0 magnet quadrants and four straight sections. The latter are used for injection of the ions from the Gustaf Werner cyclotron, electron cooling, and experiments at two target stations with a cluster-jet[3] and a pellet target[4], respectively. The beams in CELSIUS range from protons to argon ions with maximum energies of 1360 MeV for protons and 470 MeV/u for ions with charge-to-mass ratio of 1/2. An application for funds has been submitted to increase the magnetic field by 30% in order to reach 9 Tm.

The target system to be used in the WASA experiments has to fulfil a number of strong requirements; a target thickness about 20 times that of the cluster-jet target[3], a target beam travelling abot 2 m in a narrow tube through the central detector, a good vertex definition and minimum pertubation to the stored beam and to the ring vacuum. To fulfil these requirements, a pellet target generator providing a stream of frozen hydrogen microspheres has been proposed[5] and developed[6].

The production of the pellets by breaking up a liquid jet into droplets at triple point conditions and the injection into vacuum is described in detail in Ref. 6. The pellets of 40 μm diameter, later reduced to 30 μm by using a more narrow injection nozzle, have a speed of 60 m/s and a total flow of about 7×10^4 per second. The pellet stream moves with an angular divergence of ±0.04° about 4 meters before being trapped in a cryogenic beam dump. The pellets have been exposed to 200, 310 and 900 MeV protons and 570 MeV deu-terons circulating in the CELSIUS ring. The results from the test experiments[4] show that an optimal target thickness of 5×10^{15} atoms/cm^2 will give acceptable lifetimes of the circulating ion beam, a few minutes at energies around 1 GeV, as well as acceptable vacuum conditions, about 10^{-6} mbar, in the scattering chamber. All major requirements on the target system to be used in the WASA experiments have been shown to be fulfilled by the hydrogen pellet target. Future developments include the diagnostics for the pellet tracking, and the production of deuterium pellets.

The WASA experimental setup includes the pellet-target system described above, a central detector for particles scattered isotropically, a forward detector for particles scattered at angles below 18°, and a tagging spectrometer located in the magnet quadrant following the WASA setup for He recoils emitted at 0°. The main emphasis of the WASA experimental programme is put on the decay channels of the η meson. The CELSIUS ring with the pellet target will serve as an efficient η factory producing about 3×10^7 η mesons per day at a luminosity of 10^{32} cm^{-2}s^{-1} and a duty factor of 75%. A positive outcome of the developments of deuterium pellets will increase the production rate a factor of 5.

1. S. Kullander (spokesman), *WASA: Wide Angle Shower Apparatus for Particle Physics at CELSIUS*, Proposal 1987.
2. *TSL Progress Report 1994-1995*, ed. A. Ingemarsson, The Svedberg Laboratory, Box 533, S-751 21 Uppsala, Sweden.
3. C. Ekström, *Internal Targets at the CELSIUS Storage Ring*, Proc. 19th INS Int. Symp. on Cooler Rings and their Application, Tokyo, Japan, 1990, p. 228.
4. C. Ekström et al., *Nucl. Instr. and Meth. in Phys. Res.* A371(1996)572.
5. S. Kullander, *CELSIUS information day*, 1984-10-01.
6. B. Trostell, *Nucl. Instr. and Meth. in Phys. Res.* A362(1995)41.